Using Autodesk Inventor® 6

Using Autodesk Inventor® 6

RON K. C. CHENG

autodesk Press

THOMSON

DELMAR LEARNING

Australia • Canada • Mexico • Singapore • Spain • United Kingdom • United States

THOMSON

DELMAR LEARNING

autodesk Press

Using Autodesk Inventor® 6
Ron K. C. Cheng

Autodesk Press Staff

VP, Technology and Trades SBU:
Alar Elken

Executive Director:
Sandy Clark

Senior Acquisitions Editor:
James DeVoe

Senior Development Editor:
John Fisher

Executive Marketing Director:
Maura Theriault

Channel Manager:
Fair Huntoon

Marketing Coordinator:
Sarena Douglass

Executive Production Director:
Mary Ellen Black

Production Manager:
Andrew Crouth

Production Editor:
Stacy Masucci

Senior Art & Design Coordinator:
Mary Beth Vought

Editorial Assistant:
Mary Ellen Martino

Library of Congress Cataloging-in-Publication Data
ISBN 1-4018-2713-6

CONTENTS

PREFACE

Because computer and computer-aided design technologies are rapid evolving, you can now use advanced 3D solid modeling systems on your personal computer throughout your design. Autodesk Inventor® is mechanical engineering design software, an assembly-centric 3D solid modeling system for constructing 3D parametric solid parts, assemblies of solid parts, presentations of assemblies, and engineering drawings of solid parts, assemblies, and presentations. It uses four kinds of data files: you construct 3D solid parts in solid part files, assemblies of solid parts in assembly files, exploded views and animations of exploded views in presentation files, and engineering drawings of solid parts, assemblies, and presentations in drawing files.

Normally there are two stages in constructing a 3D solid part: analysis and synthesis. First you think about how you will break the 3D object into simple 3D solid features and combine the features to form the 3D object. Then you think about how to construct the features and recombine them accordingly. Basically, there are two kinds of solid features: sketched and placed. Sketched solid features are derived from sketches; you extrude, revolve, sweep, or loft the sketches to form sketched features. Placed solid features are preconstructed; you select a placed solid feature from the menu and specify the location and parameters. To maintain proper relative positions between features, you need work features: work planes, work axes, and work points.

In general, constructing sheet metal components is similar to making other kinds of solid parts. However, there are some major differences. A sheet metal component is uniform in thickness throughout, because a sheet metal component is constructed by the cutting and folding of sheet metal. When you fold sheet metal, you have to consider the bends and the seams. In thinking about the design of a 3D sheet metal component, you need to think about how to unfold the 3D object to a 2D flat pattern. To manage the design of sheet metal, you use the special set of sheet metal application tools that take care of the bends, seams, and flat pattern.

An assembly is a collection of parts put together properly to serve a purpose. In the computer application, an assembly file is linked to a set of solid part files. There are three approaches to constructing an assembly. In the first approach, you design and construct all the solid parts; then you start an assembly file and place the solid parts in the assembly. In the second approach, you start an assembly file; then you design and create the solid parts while you are working in the assembly environment, where you can perceive, visualize, and use the dimensions and features of the existing solid parts while designing new solid parts. The third approach is a hybrid approach; you design and construct some solid parts before constructing the assembly file. Then you put the solid parts together in the assembly and design other solid parts in the assembly environment. In the assembly, you can apply assembly constraints to align and mate the parts together. When an assembly is complete, you can generate a bill of materials that can be linked to other databases. In order to explain how various components of an assembly are related to each other, you can construct presentation files that show the exploded view and animation of an assembly.

In a modern digital factory, you transmit digital design data about the solid parts and assemblies of components to the downstream operating departments. However, there are occasions when 2D engineering drawings are necessary for the purpose of communication among the operators and designers. A drawing has two kinds of objects: drawing sheets and drawing views. You prepare a drawing sheet, and the computer generates the orthographic drawing views on the drawing sheets. Then you add annotations and sketched geometry to complete the drawing.

In addition to the general 3D solid modeling technology, Autodesk Inventor offers a number of advanced modeling technologies. These techniques are explained logically in the book in various chapters.

Spreadsheet Control	To control a set of dimensions in a solid part and across a number of solid parts, you can use a spreadsheet.
iFeatures	To reuse a solid feature in other designs, you can export a feature of a solid part to become an element of the design catalog, and you can import the element to other designs.
iParts	To generate a family of parts of similar shapes but different sizes, you construct an iPart. When you place in iPart in an assembly, you specify the parameters.
iMate	To speed up the process of assembling components in an assembly, you define an assembly interface in a solid part, called the iMate.
Welded Structure	To illustrate various stages of manufacturing a welded structure, you construct solid parts to depict the individual components of the structure and construct a welded structure from the parts. In the structure, you add pre-welding machining processes, weldments, and post-welding processes.
Deriving from an Assembly	To add flexibility to the linear approach in feature-based solid modeling, you can construct a number of solid parts, put them in an assembly, and derive a solid part from the assembly.
Adaptive Technology	By using adaptive technology, you can ensure that dimensions of corresponding parts in an assembly fit properly with each other by setting the dimensions of a feature of a solid part to adapt to the dimensions of the solid features of another solid part.

2D Layout Drawings	While modeling in 3D is a global trend in designing, proper deployment of 2D layout drawings is very useful in the initial design stage. You can use 2D layout drawings to validate and evaluate a mechanism. You construct 3D solid parts based on the 2D layout drawings.
Workgroups	To take full advantage of the Internet and intranet in design, a team of designers can work together by using a common workgroup search path and library search path.

ORGANIZATION OF THIS BOOK

This book comprises ten chapters, with each chapter covering a major topic. Each chapter includes a summary and review questions. Tutorials are included in Chapters 3 through 10, so that you can practice the concepts learned in the chapter. There are also two appendices that discuss importing and editing solids from other applications.

Chapter 1 outlines the concepts of computer modeling, delineating the three kinds of computer models that we use in the computer to represent 3D objects. It also introduces the concepts of parametric solid modeling, assembly modeling, and associative engineering drafting. Finally, it stresses the importance of managing upstream and downstream manufacturing processes.

Chapter 2 provides a brief introduction to the functions of Inventor, explaining the four kinds of data files you will use. It also familiarizes you with the design support system and design management system that help you while you are designing.

Chapter 3 explains the key concepts of parametric feature-based solid modeling. You will learn how to construct parametric sketches and build two kinds of solid features from the sketches. You will also learn how to combine features made from sketches and incorporate pre-constructed features in your model. Besides learning how to construct solid parts, you will apply lighting to the environment and set material and color for the solid part.

Chapter 4 is the second part of solid part modeling. You will learn how to construct artificial planes, axes, and points to help construct sketched solid features and pre-constructed solid features. In addition, you will learn how to construct features of complicated shapes from sketches, derive a solid part from another solid part, manipulate design parameters, and incorporate textual and graphical design information. You will also learn how to construct iParts from which you can generate a family of parts, and to construct time-saving user-defined custom features for incorporating with other solid parts.

Chapter 5 depicts the concepts of sheet metal modeling. You will construct 3D sheet metal parts and 2D flat patterns of 3D sheet metal parts, and you will convert 3D solid parts to 3D sheet metal parts.

Chapter 6 introduces the concepts of assembly modeling and explains the three design approaches in making a design. You will learn how to put into an assembly a set of solid parts that you already constructed and solid parts from a content library. You will also learn how to construct solid parts in the context of the assembly. Finally, you will construct a bill of materials (BOM), or parts list.

Chapter 7 is the second part of assembly modeling. It introduces more advanced techniques in assembly modeling, including the snap and go technique, constructing the assembly interface, animating an assembled mechanism by manipulating the constraints, and setting relative motions between the components in the assembly. Finally, it explains how to construct features in the context of the assembly and how to construct a welded structure.

Chapter 8 deals with advanced modeling techniques: constructing solid features by using adaptive technology, deriving a solid part from an assembly, promoting features among solid parts in an assembly, using 2D layout in conceptual design, and working in a collaborative environment.

Chapter 9 explains the concepts of assembly presentation and delineates the way to construct an exploded assembly and animate the explosion.

Chapter 10 delineates the concepts of engineering drafting. You will generate orthographic views from solid parts and orthographic and presentation views from assemblies of components.

Accompanying this book is a CD that contains all the data files for each of the tutorials presented in the book. The CD also includes PDF files that contain additional Exercises which correspond to the material presented in Chapters 3 through 10.

ONLINE COMPANION

If any updates to Autodesk Inventor 6 occur, we will create additional resources related to the content and post them online at: http://www.autodeskpress.com/resources/olcs/index.asp

ACKNOWLEDGMENTS

This book would never have been realized without the contribution of many individuals. Special thanks go to the professionals who reviewed the previous edition.

- Steve Brown, College of the Redwoods, Eureka, California
- John Clauson, Oakton Community College, Des Plaines, Illinois
- Scott Ertel, West Irondequoit High School, Rochester, New York
- David Rousch, Ohio Northern University, Ada, Ohio
- Mark Kurdi, Sheridan College, Brampton, Ontario, Canada
- Tom Singer, Sinclair Community College, Dayton, Ohio

Several people at Delmar Publishing Company also deserve special mention, particularly James Devoe, acquisitions editor; John Fisher, the developmental editor who worked closely with me on this book; Alar Elken, the publisher; Mary Ellen Martino, the engineering editorial assistant; Stacy Masucci, the production editor; Mary Beth Vought, the art and design coordinator; John Shanley of Phoenix Creative Graphics, the compositor; David Cohn, the technical editor who reviewed the current manuscript in detail; and Gail Taylor, the copy editor.

Ron K. C. Cheng

CHAPTER 1

Introduction to Computer Modeling

OBJECTIVES

The aims of this chapter are to give an overview of computer modeling and to introduce the concepts of parametric solid modeling, assembly modeling, and associative engineering drafting. After studying this chapter, you should be able to

- Explain the principles of solid modeling, assembly modeling, and associative engineering drafting

OVERVIEW

You use computer models to represent your design ideas, to facilitate product or system design, and to manage upstream and downstream computerized design and manufacturing systems.. By using computer design systems, you can construct three kinds of 3D models to represent a product or system: 3D wireframe model, 3D surface model, and 3D solid model. Among them, the 3D solid model is superior because it contains all the information regarding the vertices, edges, faces, and volume of the 3D object that it describes.

To make easy changes and modifications to a computer model, you use the parametric feature-based modeling approach; a computer model is composed of a number of features, which are modifiable by changes to their parameters. Based on a set of parametric solid models of the individual components of a product or system, you construct a virtual assembly. Using the virtual assembly, you evaluate and arrange the product or system as a whole. From the set of 3D solid parts and assemblies that you constructed, you can produce electronic data files for computerized manufacturing processes and 2D engineering drawings for conventional production methods.

COMPUTER MODELING CONCEPTS

With traditional manual methods, you use sketches, engineering drawings, or models to express your design ideas in the various design stages of product or system development. Using models or drawings, you explore, visualize, and discover more about the products or systems. However, physical models are expensive to make and they are inflexible because once they are made, modification is difficult or sometimes impossible. On the other hand, 2D engineering drawings can be ambiguous and inadequate when used to illustrate complex geometric shapes.

With the advent of computer and computer-aided design applications, you can construct 3D models in the computer to represent the product or system. You evaluate your design, generate conventional 2D engineering drawings, and produce electronic data for making rapid prototypes and for making the product or system by using downstream computerized manufacturing system.

Nowadays, the computer-aided design tool is a prime means of communication and a medium for developing design ideas, products, and systems. Using the computer, you design the component parts and construct 3D models, assemble the 3D component parts as virtual assemblies, generate 2D projection drawings for making the parts and assemblies by traditional methods, and output the 3D models and assemblies in electronic format for downstream computerized processes.

REPRESENTATION

Because a model, whether physical or digital, is used to represent a product or a system, the first thing to consider before you start constructing a model in the computer is what kind of attributes of the product or system you intend to represent. You can categorize these attributes in three main groups: geometry (shape, profile, and silhouette), appearance (color and texture), and properties (mass, center of gravity, and other physical properties). Here our concern is to represent the product or system's geometric shape as well as its appearance and properties.

3D MODELS

To design and develop 3D objects in the computer, you use 3D models. Many years ago, when 3D was first introduced in computer applications, objects were represented as simple wireframe models. Over the years, computer-aided design applications have developed from simple electronic drawing boards to sophisticated 3D design and development tools. Now surfaces and solids can both be represented in the computer.

3D WIREFRAMES

The 3D wireframe model is the most primitive type of 3D object. It is a set of unassociated line and arc segments that are put together in the 3D space. The line and arc segments serve only to give the pattern of a 3D object; they are not related one to another. As such, the model lacks any surface or volume information; it describes only the edges of the 3D object. Because of the scant information provided by the model, the use for a 3D wireframe model is limited. Figure 1.1 shows a 3D wireframe model.

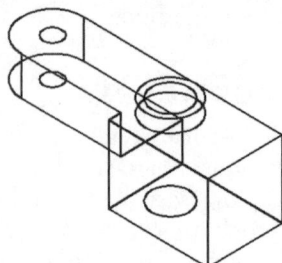

Figure 1.1 *Simple wireframe model*

3D SURFACES

The second type of 3D model, the surface model, is a set of surfaces that are put together in a 3D space to give the figure of a 3D object. When compared to a 3D wireframe model, a surface model has, in addition to edge data, information on the contour and silhouette of the surfaces. You use surface models to represent complicated free-form objects, which can be used in a computerized manufacturing system, or to generate photorealistic rendering or animation. Figure 1.2 shows a surface model.

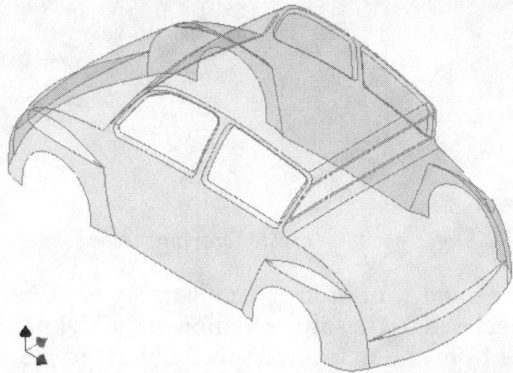

Figure 1.2 *Surface model*

3D SOLIDS

A 3D solid model is superior to the other two kinds of models with regard to information, because it contains integrated mathematical data not only about the surfaces and edges but also about the volume of the object that the model describes. In addition to visualization and manufacturing, you use a solid model in more sophisticated design evaluation tasks. There are many ways to construct a 3D solid in a computer. Among them, the most popular method is to construct wireframe profiles and use the profiles to construct 3D solid objects in four basic ways: extruding, revolving, lofting, and sweeping.

To extrude, you construct a 2D profile on a plane and translate (extrude) the profile in a direction perpendicular to the plane. The volume enclosed by the extrusion motion defines an extruded solid. (See Figure 1.3.)

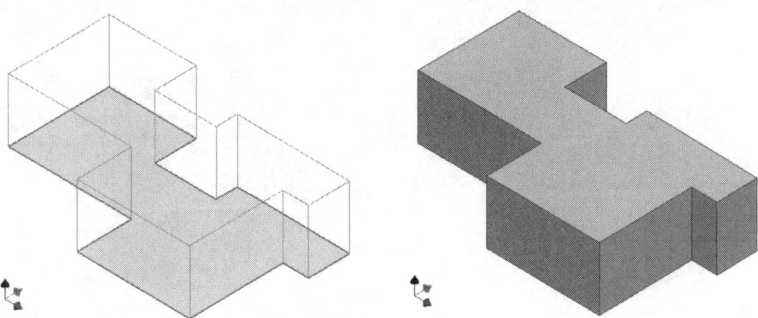

Figure 1.3 *Extruding a 2D profile to form a 3D extruded solid*

To revolve, you construct a 2D profile and an axis on a plane and revolve the profile about the axis. The volume enclosed by the revolving motion defines a revolved solid. (See Figure 1.4.)

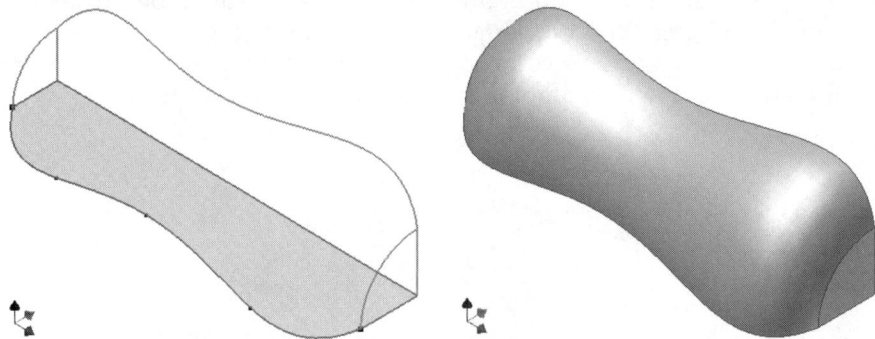

Figure 1.4 *Revolving a 2D profile about an axis to form a revolved solid*

To loft, you construct a number of profiles on a number of planes and loft through the profiles. The volume enclosed by the lofting motion defines a loft solid. The cross section of a loft solid changes from the first profile to the next profile and then to the next. Optionally, you can add guide rails while lofting. (See Figure 1.5.)

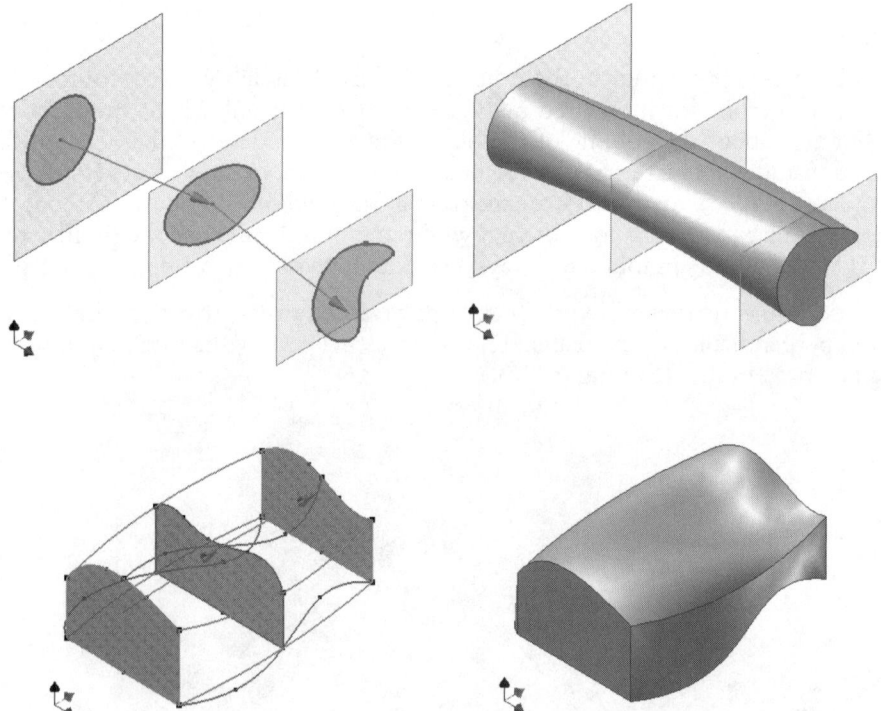

Figure 1.5 *Lofting along a series of 2D profiles to form a loft solid*

To sweep, you construct a profile and a path and transit (sweep) the profile along the path. The volume enclosed by the sweeping motion defines a sweep solid. The cross section of a sweep solid is constant along the sweeping path. (See Figure 1.6.)

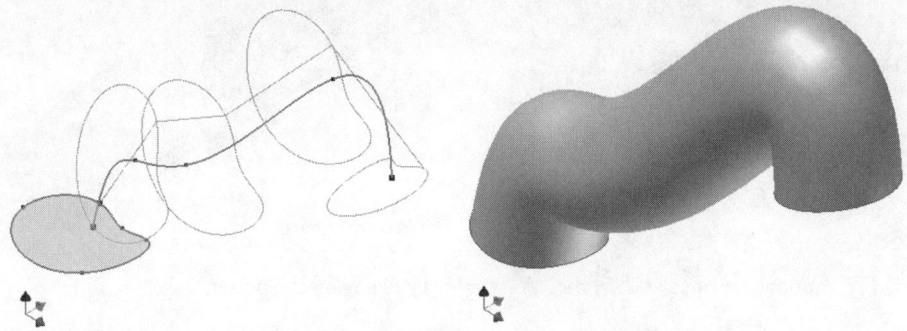

Figure 1.6 *Sweeping a profile along a path to form a sweep solid*

In addition to these four basic kinds of solids, there are other kinds of pre-constructed solids that you can use in a 3D model. You will learn about them in Chapter 3.

PARAMETRIC AND NON-PARAMETRIC APPROACHES

To construct the profiles for making the solids, there are two major approaches: parametric and non-parametric. The conventional way of constructing wireframes in the computer is to specify the location, orientation, and dimension of the lines accurately and explicitly. Once the lines are produced and the 3D solids are constructed from the wireframes, it is difficult to redefine the lines and modify the solids. This is commonly known as the non-parametric approach.

In the parametric approach, you start constructing the profiles by making free-hand sketches. During the initial sketching stage, the geometry and size of the lines need not be accurate and precise. You concentrate on form and shape. To develop a proper geometric shape in accordance with your design intent, you specify the geometric relationships between the line segments of the sketch by assigning geometric constraints such as horizontal, vertical, and parallel.

After developing the geometric shape, you specify the size by assigning parametric dimensions to the sketch. While geometric constraints and parametric dimensions are added to the sketch, the shape and size of the sketch change accordingly. With the parametric wireframes, you construct parametric 3D solids. Not only can you change a sketch before making a solid from it, but you can also alter the parametric dimensions and geometric shape of the cross section of the solid feature at a later stage. In fact, you change the parameters of the wireframes and the solids flexibly whenever you want. Figure 1.7 shows a modification made to a solid by changing its dimension parameters.

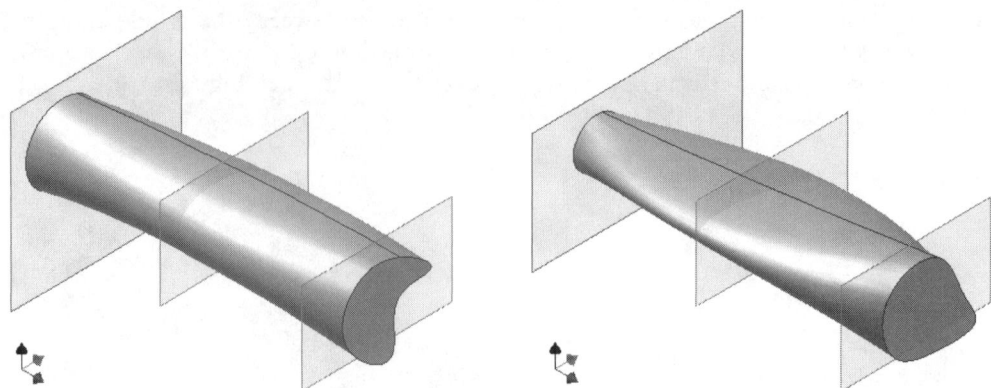

Figure 1.7 *Parameters of solid (left) modified (right) by changing the parameters*

ASSEMBLY MODELING

A product or a system usually has more than one component part. After you construct the individual component parts in 3D solids, you put them together in an assembly to depict the entire product or system. Figure 1.8 shows the assembly of a number of component parts.

Simply stated, an assembly is a collection of component parts put together to form a useful whole. However, making an assembly in a computer application goes far beyond translating solid parts together in 3D space. You have to specify a positional relationship between component parts by assigning assembly constraints to selected features of a solid part in relation to selected features of another solid part. Features that you can select for placing assembly constraints are the vertices, edges/axes, and faces of a solid part.

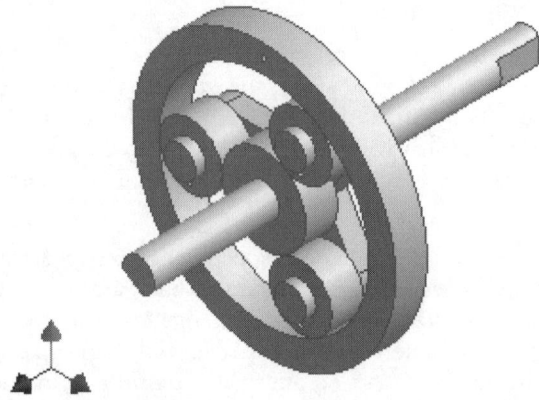

Figure 1.8 *Assembly of component parts*

It is normal engineering practice to translate (explode) the components of the assembly to illustrate how the parts are assembled together. Through the computer application, you can illustrate and animate the assembly explosion. Figure 1.9 shows the exploded presentation of an assembly.

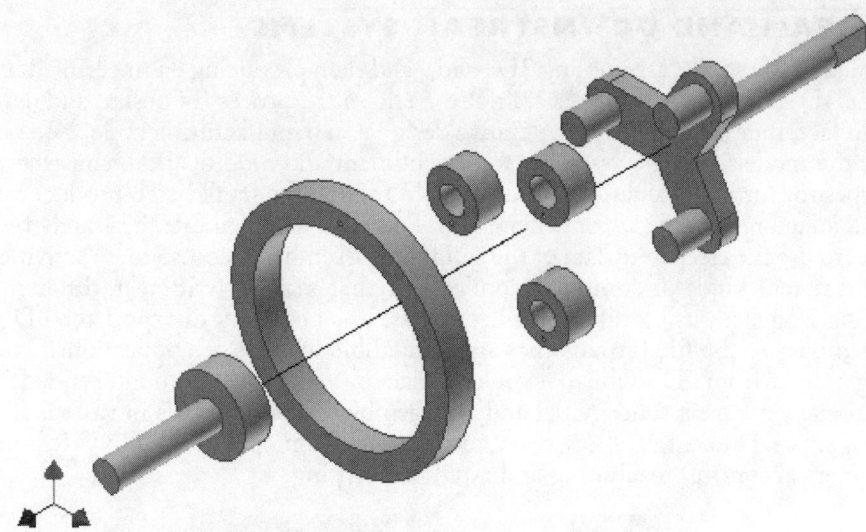

Figure 1.9 *Exploded presentation of an assembly*

ASSOCIATIVE ENGINEERING DRAFTING

Conventionally, 3D objects are represented in a 2D drawing through projection views. The most common projection method is the orthographic projection method, in which a 3D object is projected onto a 2D plane to obtain a drawing view. In the traditional approach to constructing an engineering drawing, you think about how an object would look when viewed in a certain direction. Then you construct the 2D drawings in accordance with your perception. We all know that this process is very time consuming and can be inaccurate and incomplete. With 3D solid models, the creation of 2D projection views is automatic. You select an object and specify a projection direction. The computer generates 2D views from the 3D objects. Figure 1.10 shows an engineering drawing derived from a parametric solid part.

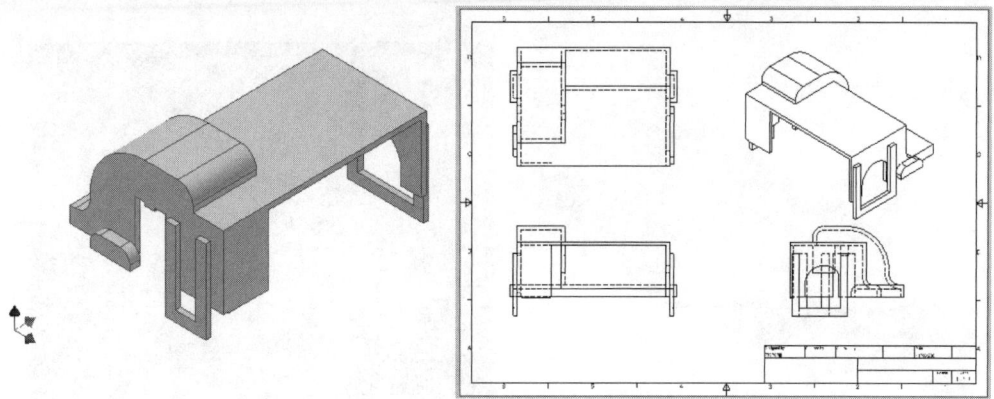

Figure 1.10 *A parametric solid and an engineering drawing derived from it*

UPSTREAM AND DOWNSTREAM SYSTEMS

Constructing 3D solids, assembling 3D solids, and then producing engineering drawings from the 3D solids and assemblies are not the end in the process of design and development. In fact, they are a part in the iterative design and manufacturing cycle. You need to analyze the models and the assembly for amendment and modification, construct rapid prototypes for further evaluation, produce rapid tooling for small batch production, and mass manufacture the objects for general sale. To enable and facilitate these activities, it is essential to use the electronic data of the solids and assembly. Because these activities will require different kinds of computer applications that may be written in different programming languages using different file formats, it is necessary to export the 3D solids and assemblies to the file formats that are compatible with those applications. To reuse existing designs from 3D solids that you construct by using other computer-aided design applications, you save a solid model and an assembly of solid models in various file formats. Figures 1.11 and 1.12 show two typical downstream operations: CNC (computerized numerical control) machining and rapid prototyping.

Figure 1.11 *Downstream computerized operations: CNC machining*

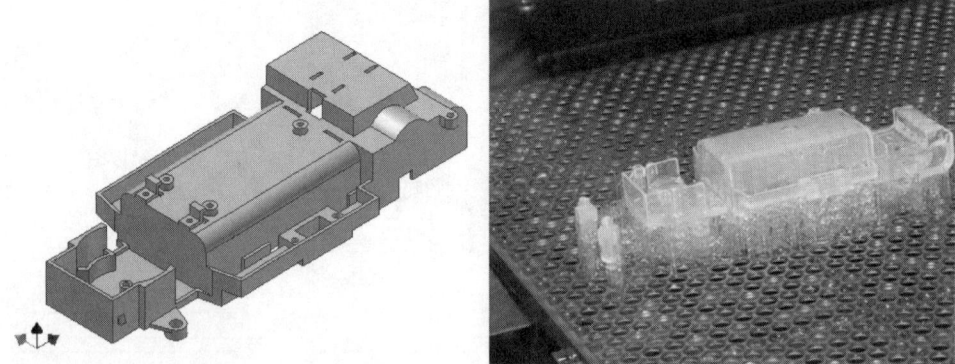

Figure 1.12 *Computer model and rapid prototype constructed from a STL file derived from the model*

SUMMARY

There are three kinds of 3D models in a computer: 3D wireframe, 3D surface, and 3D solid. Among them, the 3D solid model is superior because it provides integrated information about the vertices, edges, faces, and volume of the 3D object that it represents.

The most common method for constructing a 3D solid is to make sketches and extrude, revolve, loft, or sweep the sketches to form an extruded solid, revolved solid, loft solid, or sweep solid.

There are two major approaches to constructing 3D sketches and 3D solids: parametric and non-parametric. With the non-parametric approach, the parameters of lines and the solids constructed from the lines are static and cannot be changed. With the parametric approach, you make free-hand sketches, assign geometric constraints, and specify parametric dimensions. Unlike a non-parametric solid, a parametric solid is flexible, and you can modify it simply by specifying new parameters to the sketches and the solid.

With a set of 3D parametric solid models, you put them together in a virtual assembly by constraining the relative positions of their vertices, edges, and faces. To illustrate how the parts are assembled in an assembly, you construct an exploded view of the assembly.

To work with computerized manufacturing systems, you produce the 3D parts and 3D assembly in an electronic format. To manufacture the parts in a more conventional way, you output 2D projection drawings from the 3D solids and assemblies.

REVIEW QUESTIONS

1. Briefly explain the three kinds of 3D models in a computer application.

2. State the difference between the parametric and non-parametric approaches.

3. Give an outline of the design process, from making 3D models to producing the parts and assembly.

CHAPTER 2

Introduction to Autodesk Inventor

OBJECTIVES

The aims of this chapter are to outline the key functions of Autodesk Inventor and to introduce the four kinds of Inventor data files, the Inventor user interface, the design support system, and the design management tools. After studying this chapter, you should be able to

- Describe the key functions of Autodesk Inventor
- List the four kinds of Inventor files
- Use the Inventor user interface
- List the kind of file formats supported by Autodesk Inventor
- Use Autodesk Inventor design support system
- Use Autodesk Inventor design management tools

OVERVIEW

Autodesk Inventor is a 3D parametric computer-aided design (CAD) application. It enables you to design and construct 3D solid parts with free-form surface features, assemblies of solid parts, exploded presentations of assemblies, and engineering drawings. In addition, it enables you to save 3D solids, assemblies, and engineering drawings in various file formats for use in downstream computerized manufacturing operations, and to open various kinds of file formats so that you can re-use computer models produced by other computerized engineering applications. To help you design and then manage your design, Inventor provides a set of design support systems and file management tools.

AUTODESK INVENTOR FUNCTIONS

A product, except for a very simple object such as a ruler or a piece of eraser, consists of a number of component parts. Using the computer as a tool to design and manufacture products, you construct computer models to represent each individual component of the product and put the computer models together to form a virtual assembly, to explore and evaluate the integrity of the design. To illustrate how the components of an assembly are related to each other, you construct an exploded view from the assembly. Although it is very common to use electronic data of computer models directly in downstream computerized manufacturing operations, there are times when conventional 2D engineering drawings are required. To meet this requirement, you output engineering drawings from the computer models of the individual parts, assemblies, and exploded assemblies. Serving these design requirements, Autodesk Inventor has four basic functions:

- Constructing 3D parametric solid parts
- Constructing assemblies of solid parts
- Constructing exploded presentations of assemblies
- Constructing engineering drawings

CONSTRUCTING SOLID PARTS AND SHEET METAL PARTS

The prime function of Autodesk Inventor is to represent a component part in the computer in the form of 3D parametric solid parts. To design and construct a solid part, you start by making rough sketches that reflect your design intent. While making the sketch, you concentrate on form and shape. Then you refine the sketch by specifying geometric constraints and parametric dimensions. Using the sketches, you extrude, revolve, loft, or sweep to make a 3D object. Figure 2.1 shows a solid part consisting of a number of solid features that are constructed one by one and combined together.

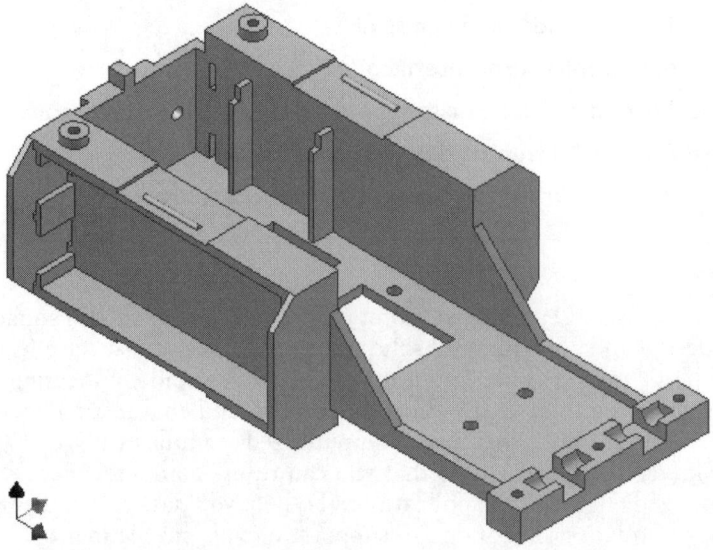

Figure 2.1 *Solid part of a component*

You can also construct or import 3D free-form surfaces and incorporate them in a solid part, thus expanding the repertoire of form and shape. Figure 2.2 shows the solid part for the model of the body of a toy car.

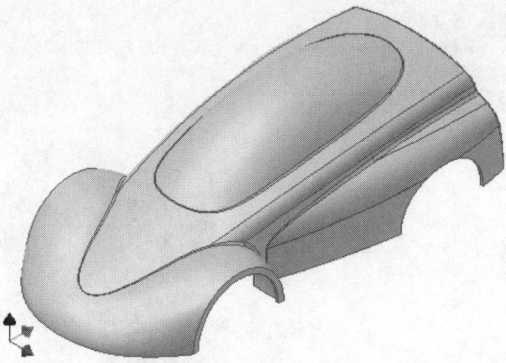

Figure 2.2 *Solid part with free-form surface features*

Sheet metal parts are a special kind of solid part. In reality, you make a sheet metal component by cutting and folding a sheet of metal of uniform thickness. To meet the manufacturing requirements of providing rounded bends at the joints of faces, relieves at the bends, hems at the edges, and seams at the joints, you need a special kind of solid modeling tool. Figure 2.3 shows the model of a sheet component.

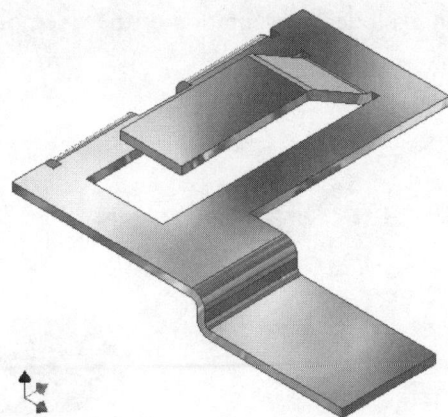

Figure 2.3 *Sheet metal component*

CONSTRUCTING ASSEMBLIES OF SOLID PARTS

The next function of Autodesk Inventor is to construct a virtual assembly of 3D solid parts. An assembly is a device consisting of a number of component parts. There are three approaches to making an assembly: the bottom-up, top-down, and hybrid approaches. In the bottom-up approach, you construct all the solid parts of an assembly in individual files and then link them together in an assembly file. In the top-down approach, you construct the parts while working in the assembly file. The hybrid approach is a combination of the bottom-up and top-down approaches. No matter which approach you use, you place the component parts in a proper location and orientation relative to each other and maintain a proper alignment among the parts in an assembly by adding assembly constraints. Figure 2.4 shows the assembly of the axle of a toy car.

Figure 2.4 *Assembly of a set of components*

CONSTRUCTING EXPLODED PRESENTATION OF ASSEMBLIES

The third function of Autodesk Inventor is to construct an exploded presentation of assemblies. To illustrate how various parts of an assembly are put together, you construct a presentation of an assembly. In the presentation, you explode or tweak the components apart. Figure 2.5 shows an exploded presentation of the assembly shown in Figure 2.4.

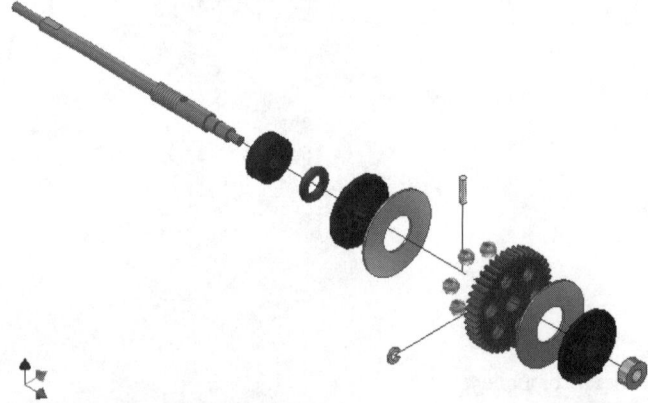

Figure 2.5 *Components of an assembly tweaked apart*

CONSTRUCTING ENGINEERING DRAWINGS

The fourth function of Autodesk Inventor is to construct engineering drawings. An engineering drawing is an communication tool that depicts a 3D design in 2D engineering drawing views. You can construct an engineering drawing in two ways: You can select a solid part or an assembly, and the computer application automatically generates 2D orthographic views of 3D solid parts, sheet metal parts, and assemblies, flat patterns of sheet metal, and exploded views of assemblies. You can also construct a parametric 2D engineering drawing by using the drafting tools provided. Figure 2.6 shows the engineering drawing derived from the solid part of a toy car body, and Figure 2.7 shows a 2D draft drawing constructed with the drafting tools.

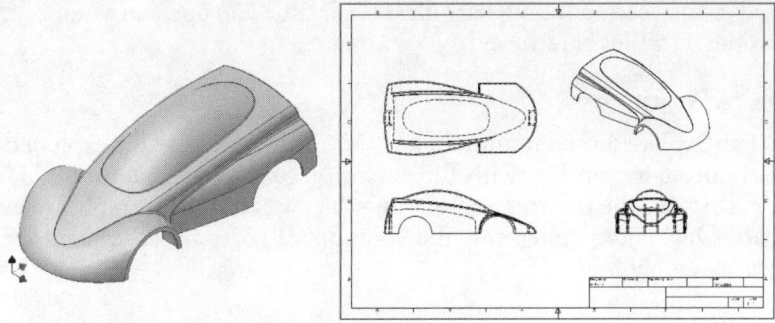

Figure 2.6 *Engineering drawing derived from a solid part*

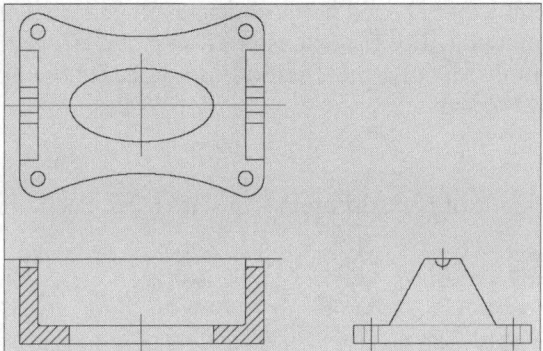

Figure 2.7 *2D draft drawing without any reference to 3D computer models*

AUTODESK INVENTOR FILE TYPES

To manage the four design functions, Autodesk Inventor uses four kinds of files:

Part files	For constructing 3D solid parts and sheet metal parts
Assembly files	For assembling solid parts and/or sub-assemblies
Presentation files	For exploding an assembly
Drawing files	For deriving engineering drawings from 3D solid parts, sheet metal parts, assemblies, and presentation and for constructing 2D draft drawings

PART FILES

You construct a 3D solid part or a sheet metal part in a part file, with the file extension *.ipt*. A part file stores the definition of the parametric 3D solid part.

ASSEMBLY FILES

To construct an assembly or a sub-assembly, you use an assembly file, with file extension *.iam*. An assembly file links to a set of parametric 3D solid parts and/or sub-assembly of parametric 3D solid parts. It stores only the information about how the component parts are assembled together—information regarding the details of the parametric 3D solid parts is stored in the corresponding part files. Each time you open an assembly file, information from the part files is retrieved.

PRESENTATION FILES

To construct an exploded presentation of an assembly or animate the exploded presentation, you use a presentation file, with file extension *.ipn*. A presentation file links to an assembly file. It stores the information on how the parts of the assembly are exploded or tweaked apart. Details regarding how the component parts are assembled are stored in the respective assembly file.

DRAWING FILES

A drawing file serves two purposes: to derive a 2D engineering drawing of a parametric 3D solid part, an assembly of 3D solid parts, and an exploded view of an assembly, and to construct a 2D draft drawing. The file extension is *.idw*. An associative drawing file links to a part file, an assembly file, or a presentation file. It stores the information about the 2D presentation of 3D objects.

FILE ICONS

To depict these four kinds of files, different icons are used. See Figure 2.8.

Standard.ipt   Standard.iam   Standard.idw   Standard.ipn

Figure 2.8 *Icons (from let to right) for part file, assembly file, presentation file, and drawing file*

STARTING AUTODESK INVENTOR

Now start Autodesk Inventor by selecting Autodesk Inventor icon on your desktop. In the What To Do panel of the Open dialog box (see Figure 2.9), the four icons (Getting Started, New, Open, and Projects) enable you to perform the four different kinds of tasks.

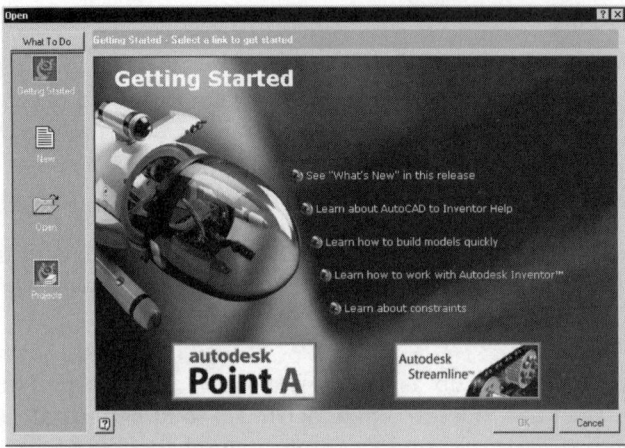

Figure 2.9 *Open dialog box with Getting Started from the What To Do panel selected*

GETTING STARTED PANEL

After selecting Getting Started from the What To Do panel, you can select a link to one of the following topics:

- See "What's New" in this release
- Learn about AutoCAD to Inventor Help
- Learn how to build models quickly
- Learn how to work with Autodesk Inventor
- Learn about constraints

See "What's New" in this release

This section provides information on new features of the current release. In the What's New in Autodesk Inventor dialog box shown in Figure 2.10, select a topic.

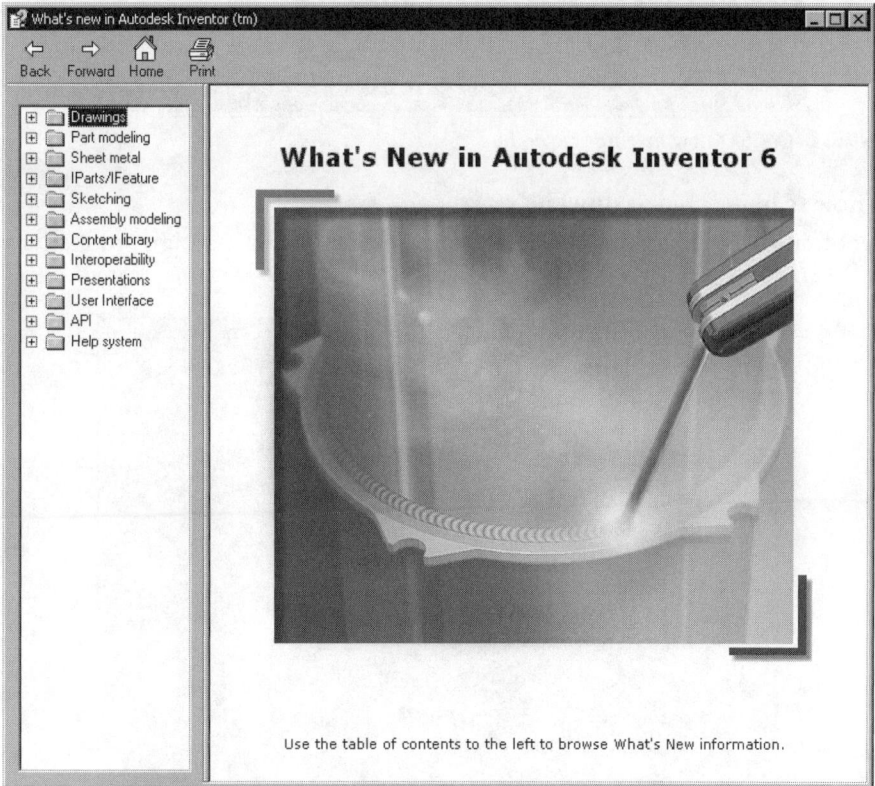

Figure 2.10 *What's New in Autodesk Inventor dialog box*

Learn about AutoCAD to Inventor Help

This is a set of help topics to help AutoCAD users migrate to Autodesk Inventor. Figure 2.11 shows the Get to Know Inventor dialog box.

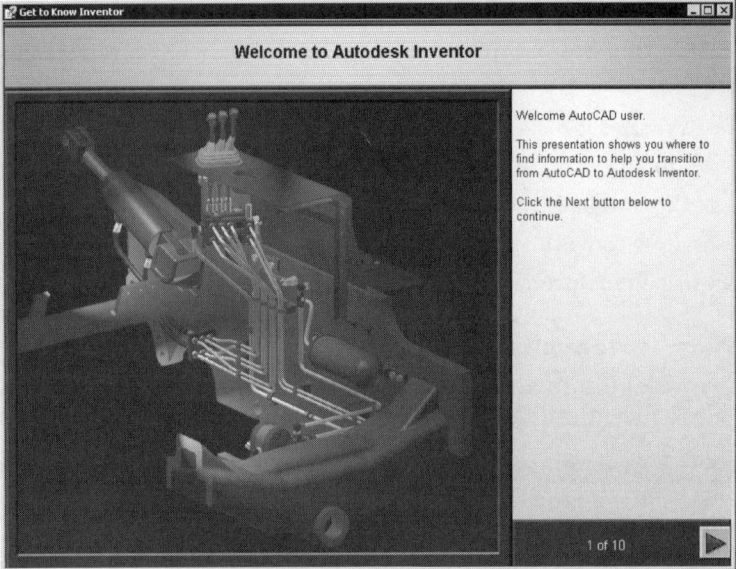

Figure 2.11 *Get to Know Inventor dialog box*

Learn how to build models quickly

This set of tutorials helps you use Autodesk Inventor to increase design productivity. (See Figure 2.12.)

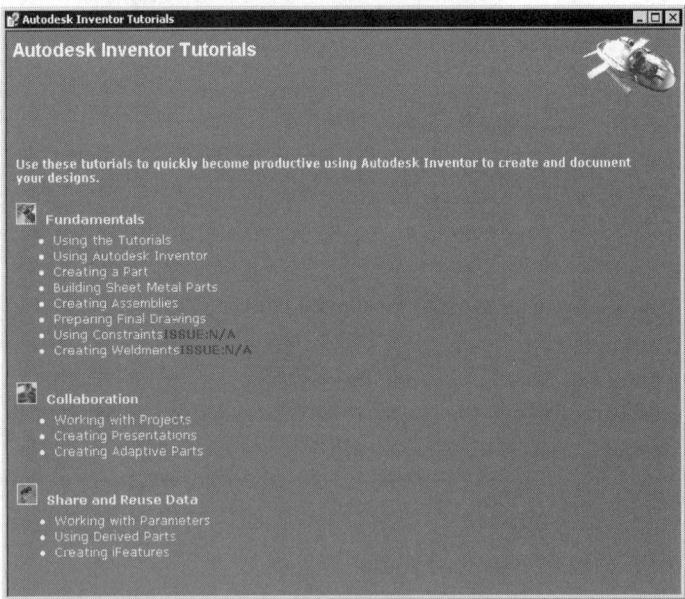

Figure 2.12 *Autodesk Inventor Tutorials on fundamentals, collaboration, and share and reuse data*

Learn how to work with Autodesk Inventor

This set of tutorials introduces you to using Autodesk Inventor to design. (See Figure 2.13).

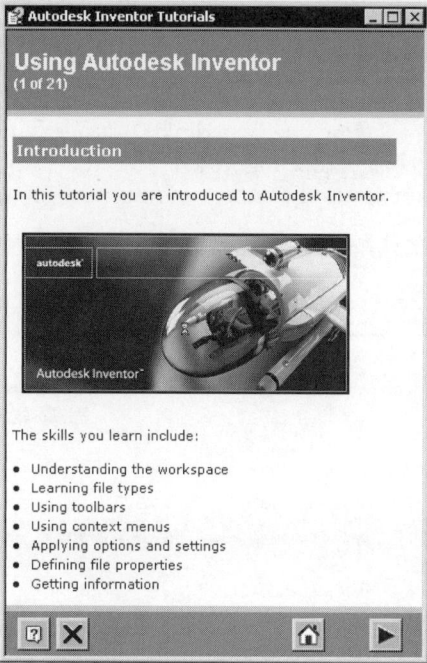

Figure 2.13 *Autodesk Inventor Tutorials on introduction to Autodesk Inventor*

Learn about constraints

This dialog box explains the concepts of constraints in part and assembly modeling (See Figure 2.14.)

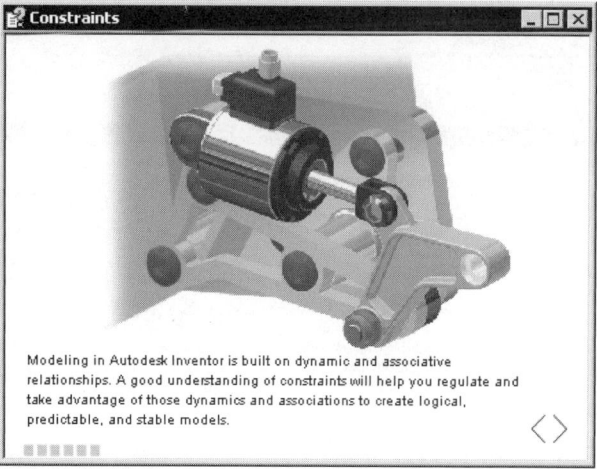

Figure 2.14 *Constraints dialog box*

Hyperlinks

In addition to the four topics, there are two hyperlinks in the Getting Started panel:

Autodesk Point A (*http://pointa.autodesk.com*)

Autodesk Streamline (*http://www.streamline.autodesk.com*)

If your computer is already connected to the Internet, you can visit Autodesk Point A™ and Autodesk Streamline™ by selecting the hyperlinks. Autodesk Point A is an online design resource and community Web site. (See Figure 2.15). Autodesk Streamline is a hosted service for sharing digital design data across the entire manufacturing team. (See Figure 2.16).

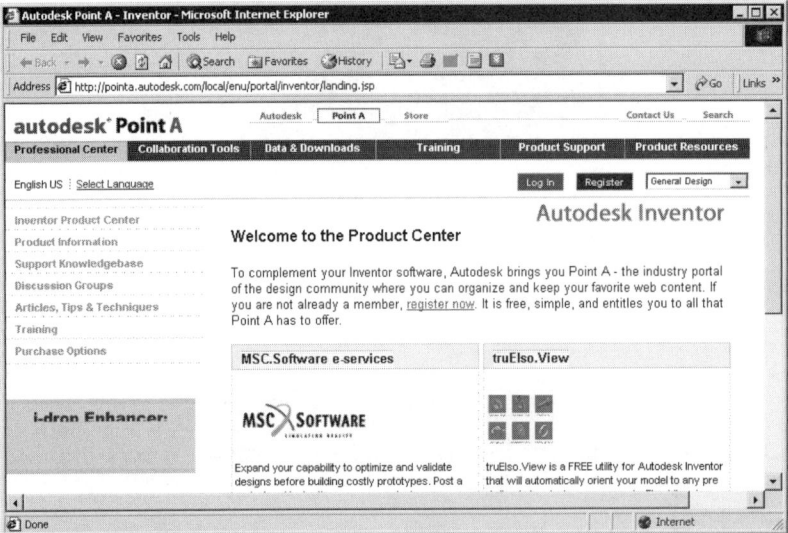

Figure 2.15 *Autodesk Point A Web site*

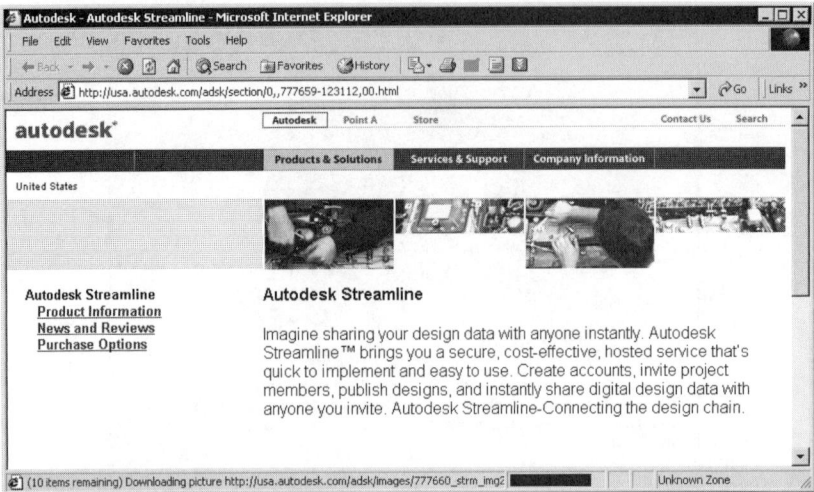

Figure 2.16 *Autodesk Streamline Web site*

NEW PANEL

The second panel of the Open dialog box is the New panel. To start a new file, you select a template here and click the OK button. Depending on what kind of tasks you are going to perform, you select a solid part file template, a sheet metal file template, a presentation file template, an assembly file template, or a presentation file template.

In this panel, there are three tabs: Default, English, and Metric. On the Default tab, the templates conform to a drafting system (English or Metric) that you configure when you install Inventor. The English tab contains templates that conform to English standard, and the Metric tab contains templates that conform to Metric standard. In each of the tabs, you will find four kinds of file template icons (*.ipt*, *.iam*, *.ipn*, and *.idw*). Each of these icons represents a template file saved in the Autodesk Inventor's template directory.

Template files are computer files having all the necessary settings that might be required in repetitive work. When you need to construct a series of files that have similar settings, you can make use of a template file. You can make any Autodesk Inventor file available as a template file by saving it in the template directory. If you add a sub-folder in the template directory, the folder will appear as an additional tab in the New panel of the Open dialog box. Figure 2.17 shows the New panel.

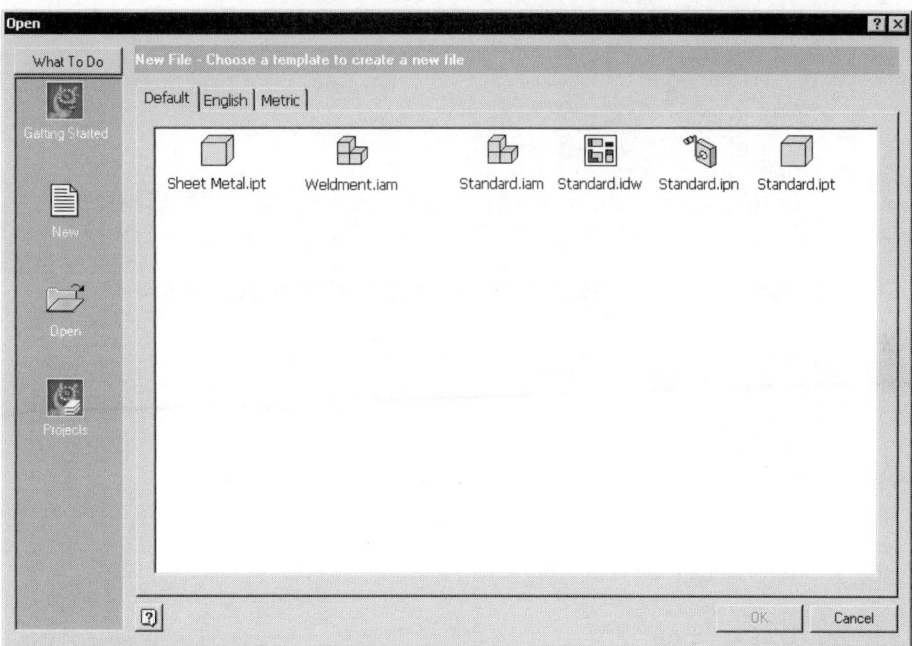

Figure 2.17 *Choose a template to create a new file*

OPEN PANEL

If you want to open an existing file, select Open in the What To Do panel. (See Figure 2.18.) Later in this chapter, you will learn more about the kinds of files that you can open, in addition to Autodesk Inventor files.

PROJECTS PANEL

Autodesk Inventor makes extensive use of "project files" to manage various kinds of files used in a project; a project file is a text file that specifies the locations of the files that make up a project. To select a project file, select Projects in the What To Do panel and select a project. (See Figure 2.19.) You will learn how to construct a project file later in this chapter.

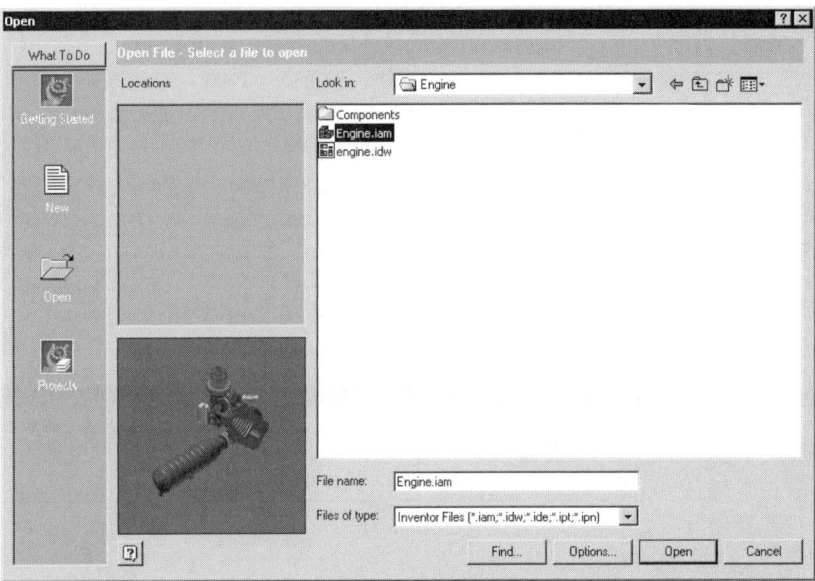

Figure 2.18 *Open File dialog box*

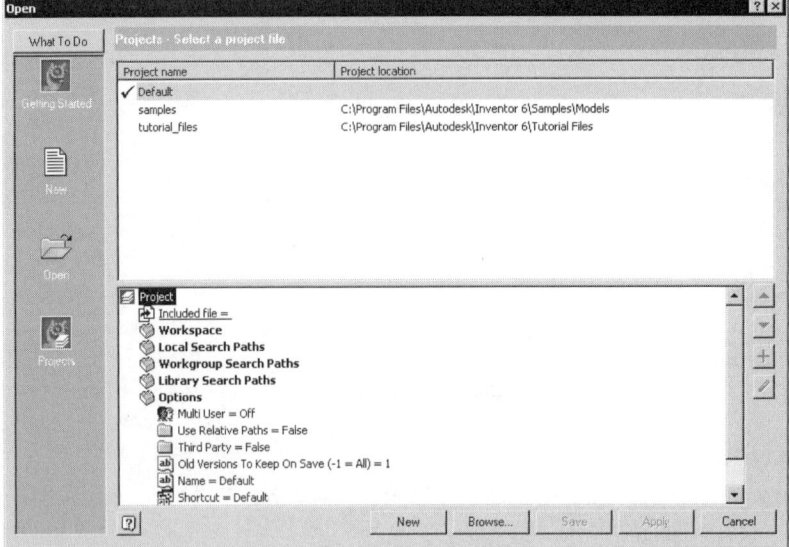

Figure 2.19 *Projects panel*

APPLICATION WINDOW

Now click the Cancel button in the Open dialog box. This brings you to the Inventor application window. (See Figure 2.20.) In the application window, there are six major areas. At the top of the window there is a set of pull-down menus. Below the menu bar is a toolbar. Further down the screen, you will find a Browser Bar, a command panel, and the graphics area. At the bottom of the window, you will find a status bar.

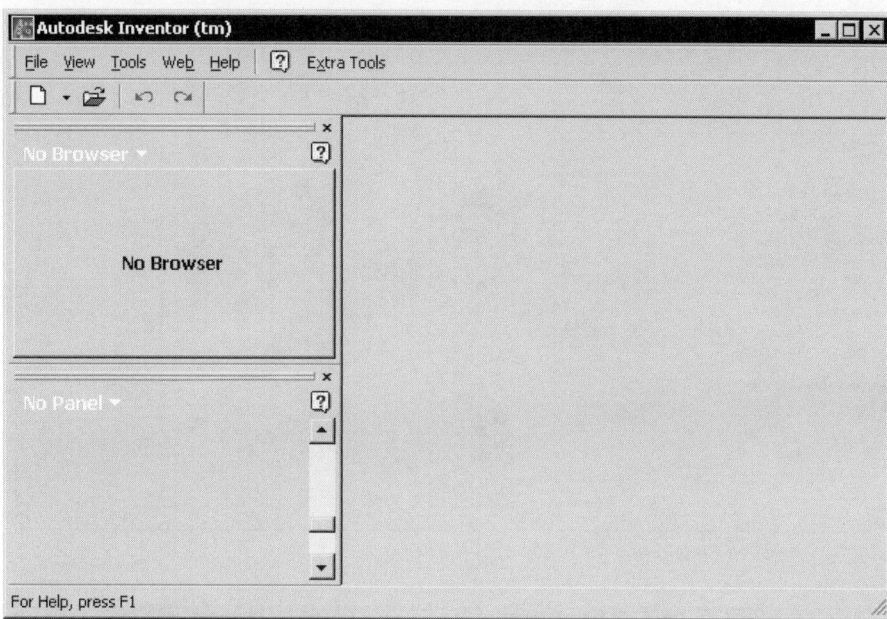

Figure 2.20 *Application window*

STARTING A NEW FILE

Having decided what to do (construct a part, assembly, presentation, or drawing file), you start a new file by selecting New from the File menu. In the Open dialog box, select New from the What To Do panel. As we have said, you will find three tabs: Default, English, and Metric.

New Part File

In each tab of Open dialog box, there are two kinds of part file templates: *Standard.ipt* and *Sheet Metal.ipt*. You can select the *Standard.ipt* template to construct a solid part (you will use this template in the Chapters 3 and 4) and you can select the *Sheet Metal.ipt* template to construct a sheet metal part (you will learn about sheet metal parts in Chapter 5). Figure 2.21 shows the application window for a solid part.

New Assembly File

You can select the *Standard.iam* template to construct an assembly of parts. (You will learn about assembly in Chapters 6 and 7.) Figure 2.22 shows the application window for an assembly.

New Presentation File

You can select the *Standard.ipn* template to construct a presentation of assemblies. (You will learn about exploded presentation in Chapter 9.) Figure 2.23 shows the application window for a presentation of an assembly.

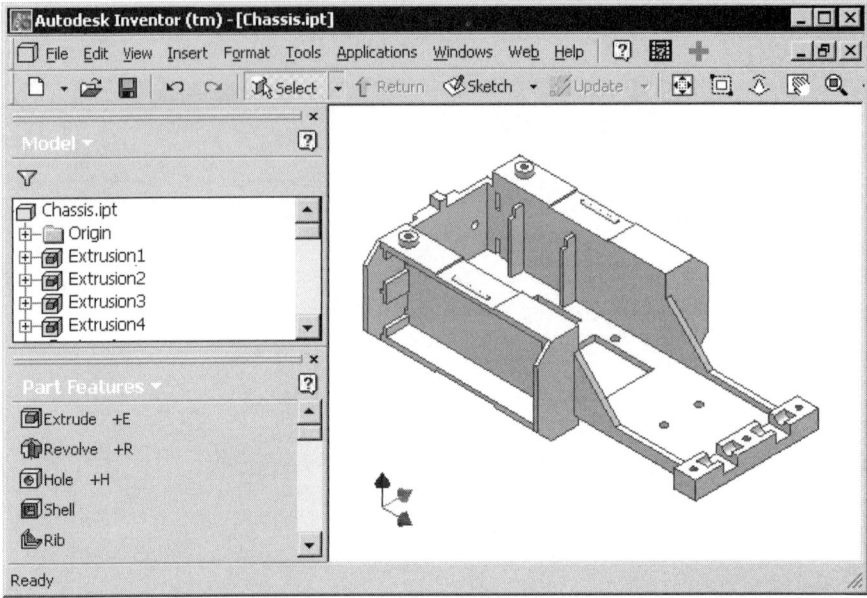

Figure 2.21 *Application window for a 3D solid part file*

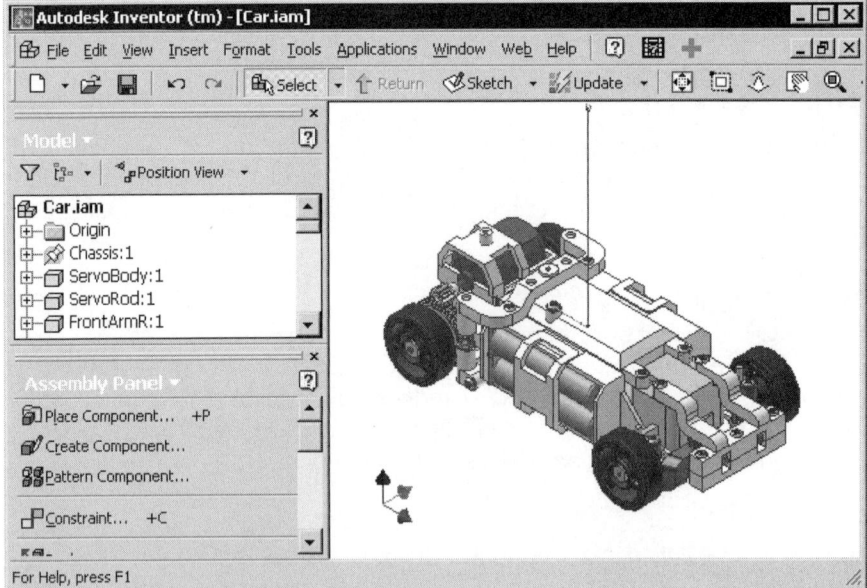

Figure 2.22 *Application window for an assembly file*

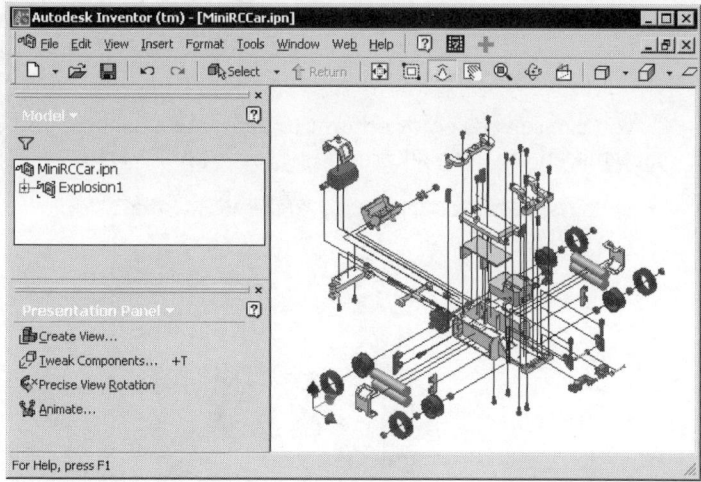

Figure 2.23 *Application window for a presentation file*

New Engineering Drawing File

You can select the *Standard.idw* template to construct an engineering drawing. (You will learn about engineering drawing in Chapter 10.) Figure 2.24 shows the application window for constructing an engineering drawing.

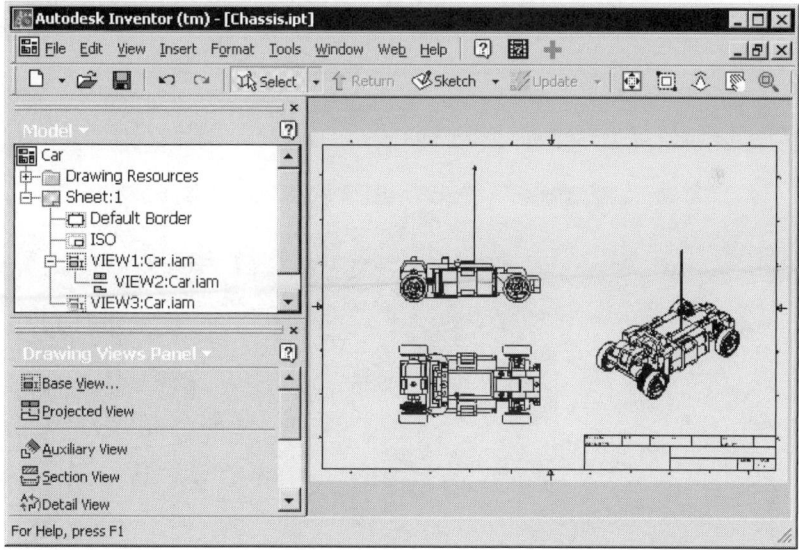

Figure 2.24 *Application window for a drawing file*

GRAPHICS AREA

The graphics area is the working area where you construct sketches and solid features in a solid part file, construct assemblies of solid parts and sub-assemblies in an assembly file, construct presentations of an assembly in a presentation file, and construct engineering drawings in an engineering drawing file. You can configure the color style and the background of the graphics area by selecting Tools > Application Options. In the Options dialog box, shown in Figure 2.25, select the Colors tab.

On the Colors tab, you can select a color scheme and choose a background image. Figure 2.26 shows an image placed on the background of the graphics area.

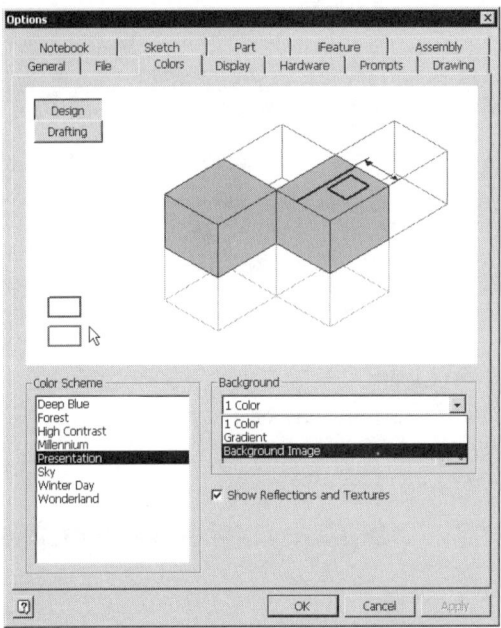

Figure 2.25 *Colors tab of the Options dialog box*

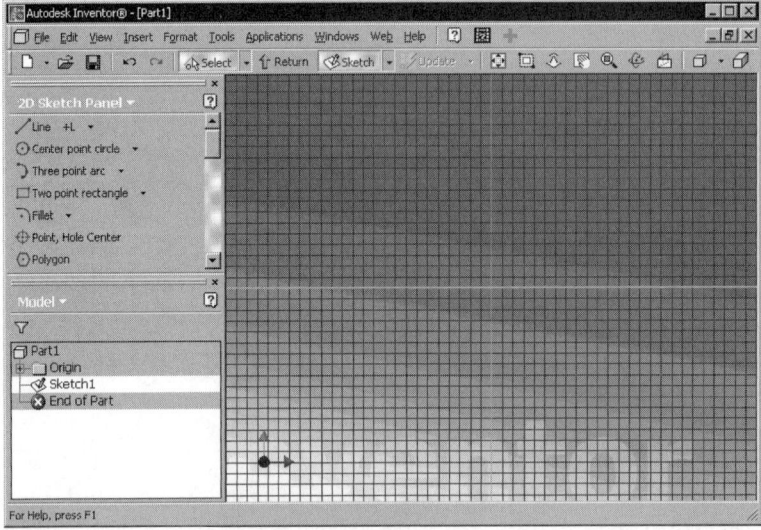

Figure 2.26 *Graphics area with an image background*

PANEL BAR

The Panel Bar has a number of palettes that enable you to access various design tools. It is context sensitive, so the tools available will vary according to the current design context. In each palette of the Panel Bar, there are two modes available: general mode and expert mode. In general mode, text accompanies each icon to depict the command. In expert mode, text is not displayed until you place the cursor on the icon. Figure 2.27 shows the Sketch Panel Bar in expert mode.

Figure 2.27 *2D Sketch Panel Bar in expert mode*

BROWSER BAR

The Browser Bar shows a hierarchy of objects in the file. Figure 2.28 shows the Browser Bar for a part file.

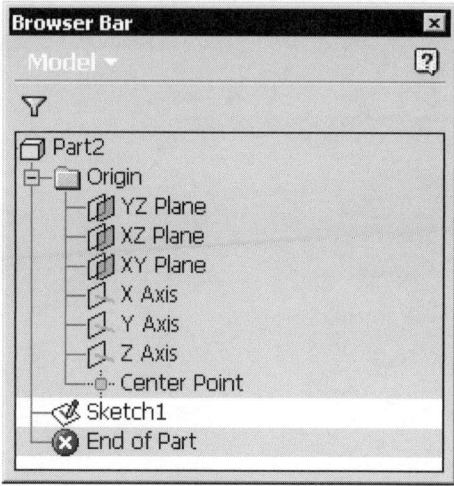

Figure 2.28 *Browser Bar showing hierarchy of objects in a file*

CUSTOMIZING TOOLBARS

Toolbars are configurable in several ways. To configure a toolbar, select Tools > Customize. In the Customize dialog box, shown in Figure 2.29, select the Environments tab. Here you can configure which toolbars to display in various design environments.

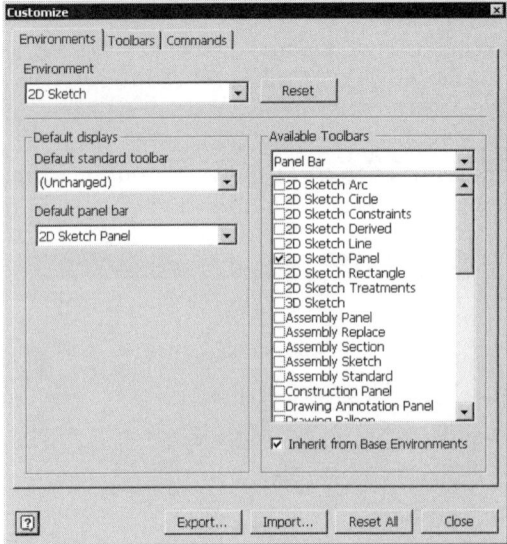

Figure 2.29 *Environments tab of the Customize dialog box*

Altogether, there are quite a lot of toolbars. Displaying all of them will take up considerable space. Therefore, by default they are not displayed. To display a toolbar, select the Toolbars tab (Figure 2.30) from the Customize dialog box, select a toolbar from the list, and click the Show button. In addition, you can select to display other toolbars, or even combine often-used tools on a customized toolbar.

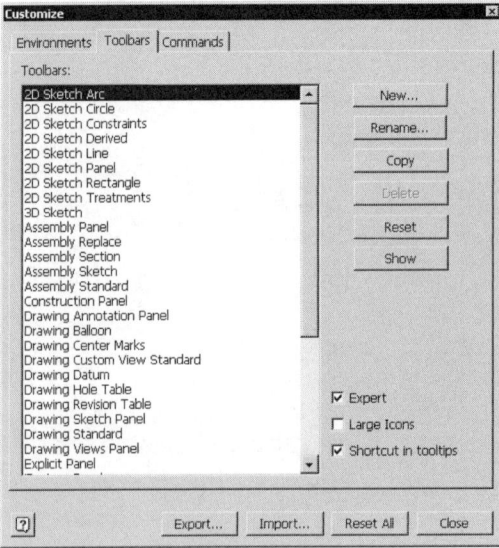

Figure 2.30 *Toolbars tab of the Customize dialog box*

To include a particular command icon in a toolbar, select the Commands tab (Figure 2.31) in the Customize dialog box, select a command, and drag it to a toolbar.

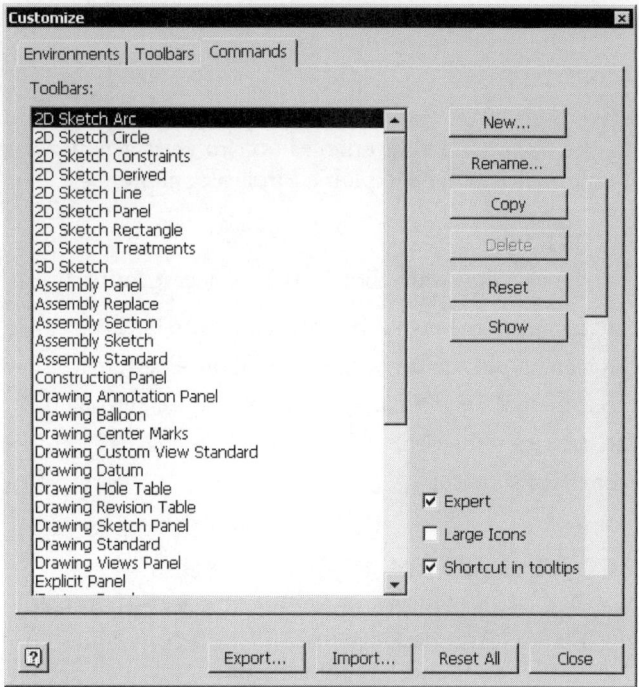

Figure 2.31 *Commands tab of the Customize dialog box*

RIGHT-CLICK MENUS

Normally, your mouse has two buttons. You use the left button to select an object and use the right mouse button to activate a context-sensitive menu. Depending on the location of the cursor and the kind of file you are working on, right-clicking the mouse will bring up different kinds of shortcut menus. You will use the right button very often in the following chapters while you work through the tutorials. Figure 2.32 shows two right-click menus when you place your cursor in the graphics area of a part file.

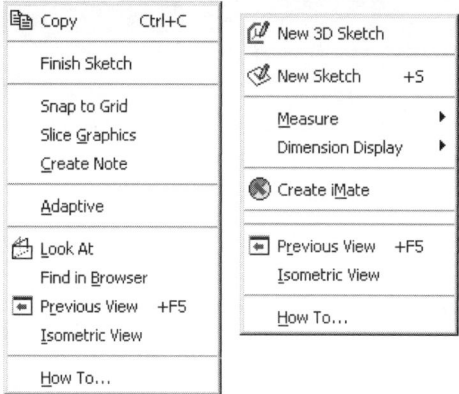

Figure 2.32 *Context-sensitive right-click menus*

If you have a roller-wheel mouse, you can use the roller to zoom in and zoom out the display.

COMMAND SHORTCUT

To speed up command selection, you can use command shortcuts. For example, pressing the S key executes the command that enables you to start a new sketch. Shortcuts for different kinds of commands will be explained in later chapters.

COMMAND SELECTION

To summarize, there are several ways that you can select a command.

- Select an item from the menu with the left mouse button
- Select an icon on the toolbar with the left mouse button
- Select an icon on the Panel Bar with the left mouse button
- Select an item from the right-click menu
- Use a command shortcut

DESIGN SUPPORT SYSTEM

To help you design, Autodesk Inventor provides a design support system with six major components: Help Topics, Autodesk Online, What's New about Autodesk Inventor, Visual Syllabus, Design Doctor, and Sketch Doctor.

HELP TOPICS

The Autodesk Inventor Help dialog box, as shown in Figure 2.33, provides a comprehensive set of help topics on various aspects of Inventor. To access help topics, you can select the Help Topics from the Help pull-down menu or the Help Topics button on the Inventor Standard toolbar.

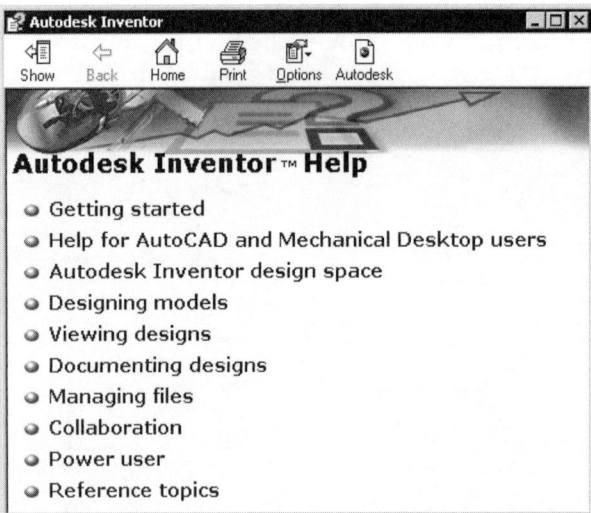

Figure 2.33 *Autodesk Inventor Help dialog box*

Help for AutoCAD Users

To help AutoCAD users to become familiar with the use of Autodesk Inventor, the Help for AutoCAD Users is available in the Getting Started dialog box and the Help Topics dialog box. To learn how to migrate from AutoCAD to Inventor by using the Getting Started dialog box, select Help for AutoCAD Users from the Help menu. (See Figure 2.34.)

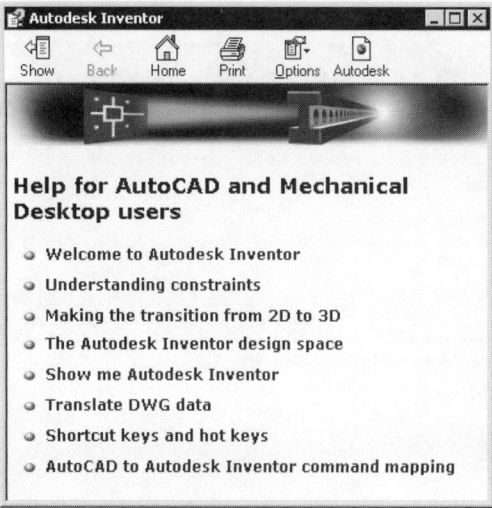

Figure 2.34 *Help for AutoCAD and Mechanical Desktop users*

Support Assistance Help

The support assistance provides technical support information about Autodesk Inventor and other resources. To find out the new enhancements in the Support Assistance, select Support Assistance Help from the Help menu. (See Figure 2.35.)

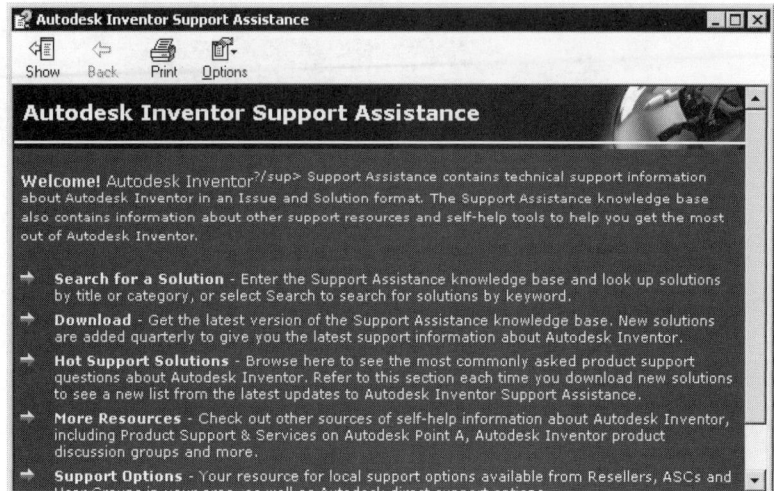

Figure 2.35 *Autodesk Inventor Support Assistance*

Programming Help

To access the Autodesk Inventor Application Programming Interface reference guide, select Programming Help from the Help menu. The programming help includes objects, methods, properties, events, and enums. (See Figure 2.36.)

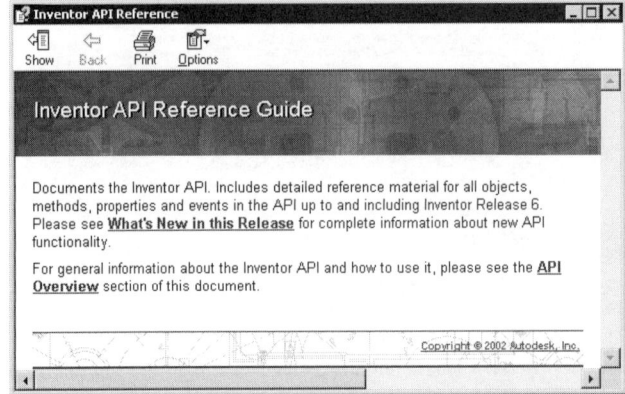

Figure 2.36 *Inventor API Reference Guide*

AUTODESK ONLINE

The Autodesk Online dialog box has four buttons: Autodesk Inventor Home Page, Autodesk Point A for Autodesk Inventor, Autodesk Streamline, and Autodesk Inventor Support and BigFix service. Because these buttons bring you to the appropriate Web sites, you need to have your computer connected to the Internet before you use Autodesk Online. To access Autodesk Online, select Autodesk Online from the Help menu. (See Figure 2.37.)

Figure 2.37 *Autodesk Online dialog box*

WHAT'S NEW ABOUT AUTODESK INVENTOR

To discover all the new features of the current release of Inventor in comparison to the previous release, you may access What's New about Autodesk Inventor from the Getting Started menu in the QuickStart dialog box shown in Figure 2.10 or by selecting What's New from the Help menu.

VISUAL SYLLABUS

The Visual Syllabus is a collection of lessons, organized in five categories: Part Modeling, Sheet Metal, Assembly Modeling, Presentations, and Drawings. To access these lessons, select the Visual Syllabus icon on the Inventor Standard toolbar. Figure 2.38 shows the Visual Syllabus dialog box.

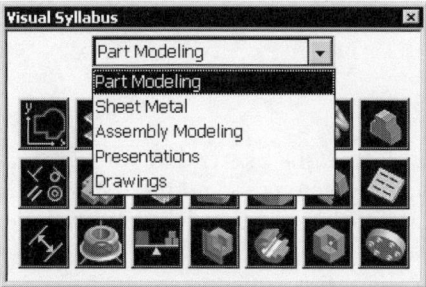

Figure 2.38 *Visual Syllabus dialog box*

DESIGN DOCTOR

The Design Doctor is a context-sensitive diagnostic tool. It is available only when you encounter some error or problem. Figure 2.39 shows a problem encountered in updating a shell feature of a solid part. To access Design Doctor, you select the feature with a problem from the browser (marked with the symbol "!"), right-click, and select Recover.

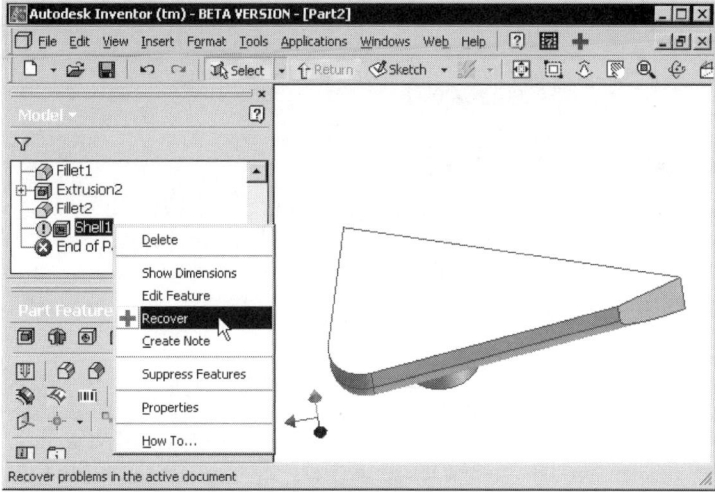

Figure 2.39 *Problem encountered and being recovered*

Following the instructions outlined in the Design Doctor dialog box as shown in Figure 2.40, you fix the error.

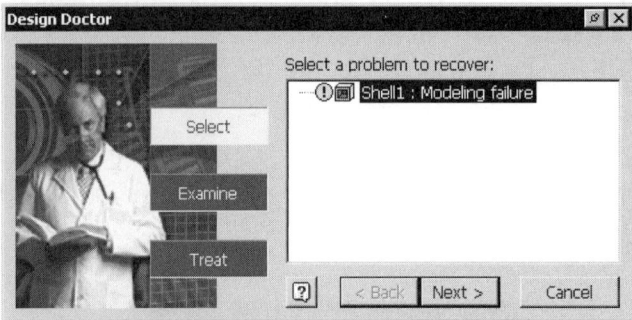

Figure 2.40 *Design Doctor dialog box*

SKETCH DOCTOR

Very similar to the Design Doctor, the Sketch Doctor is also a context-sensitive diagnostic tool. It is available when a problem is encountered in sketching. Figure 2.41 shows the Sketch Doctor dialog box after an attempt was made to extrude a self-intersecting closed-loop curve.

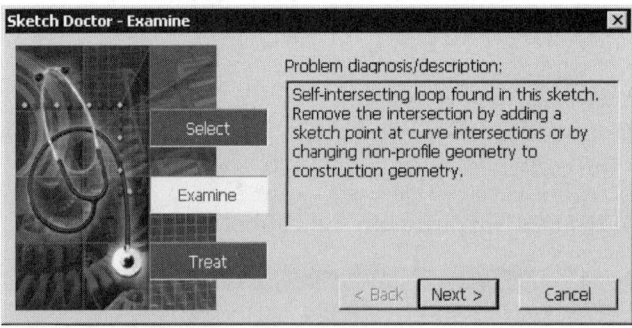

Figure 2.41 *Sketch Doctor dialog box*

OPEN, INSERT, AND SAVE AS

To re-use computer models or to incorporate objects constructed by using other applications, you open these files in Autodesk Inventor and continue with your design work or insert them in your Inventor file. To facilitate downstream computerized manufacturing systems using a different kind of file format, you can save Autodesk Inventor files to various file formats.

Apart from opening and saving the four basic kinds of Autodesk Inventor files (part file, assembly file, presentation file, and drawing file), you can open an iFeature file (You will learn more about iFeature in Chapter 4) and various kinds of file formats in Autodesk Inventor, and you can save Inventor files to various file formats. The file formats and their description are listed in Table 2.1.

Table 2.1 *Open and Save As file formats*

Open File Format	Save As File Format	Description
DWG	—	Standard AutoCAD drawing format.
DXF	—	Drawing exchange format.
PRT	—	Pro/Engineer part file format.
ASM	—	Pro/Engineer assembly file format.
—	BMP	Windows Bitmap (BMP) is a Windows image file format. The current screen display is saved to a 2D image file.
IGES	IGES	Initial Graphics Exchange Specification (IGES) format is an American standard established by the American National Standards Institute in 1979. Because it is a format for translating 3D wireframe lines and surfaces, an Inventor solid that you save in IGES format will be displayed as a surface model. Naturally, volume data will be lost.
SAT	SAT	Save As Text (SAT) is a file format of the ACIS object-oriented 3D geometric modeling engine by Spatial Technology Inc. It is a format for translating 3D lines, surfaces, and solids. ACIS supports two kinds of file formats: SAT and SAB. (SAB stands for Save As Binary.)
STEP	STEP	STandard for the Exchange of Product Model Data (STEP) is a product model data exchange standard that was initially developed by the International Organization for Standardization. Like SAT files, it is used for translating lines, surfaces, and solids.
—	STL	Stereolithography (STL) is a standard file format for use in most rapid prototyping machines. An STL file is a list of triangular surfaces that depict the 3D model. A 3D solid model saved in STL format is downgraded to a 3D model approximated by a set of triangular flat surfaces. The STL file translates the current 3D solid model to a set of triangular flat surfaces.
—	PTP	Streamline Part Packages is a standard file format for Autodesk Streamline.
—	XGL	X Windows Graphics Library (XGL) is a standard file format that captures all the 3D information that can be rendered by the OpenGL rendering library. The XGL file translates the 3D surface, texture mapping, and transparency of the current 3D model.
—	ZGL	Compressed XG (ZGL) is a compressed XGL format that is about 10 times smaller than XGL files.

FILE OPEN

In Inventor, you can open the following kinds of file formats:

- Inventor iFeature file (*.ide)
- AutoCAD Drawing (*.dwg)
- DXF (*.dxf)
- IGES (*.igs, *.ige, *.iges)
- Pro/Engineer File (*.prt*, *.asm*)

- SAT (*.sat)
- STEP (*.stp, *.ste, *.step)

To open these files, select Open from the File menu, select the file types in the Files of types box, and select a file. You can construct additional features as if you are working on an Inventor file.

Open Pro/Engineer Files

To work on a Pro/Engineer part or assembly file, you open a Pro/Engineer part or assembly file.

Open ACIS or STEP Files

You can open a solid part that is saved in SAT or STEP format. The SAT or STEP solid part becomes a base solid part in Inventor. Although there is no parametric information in the imported SAT or STEP solid, you can edit imported ACIS and STEP solids by manipulating their faces. (See Appendix A.)

Open IGES Files

You can open an IGES file as a solid or surfaces. You can use imported data as a base solid or as surfaces.

Open Mechanical Desktop Files

If you open a Mechanical Desktop file, all the parametric information in the DWG file will be imported and displayed as objects on the Browser Bar. The Mechanical Desktop parametric solid part becomes an Inventor parametric solid part file. (See Appendix B.)

FILE INSERTION

Insert AutoCAD File

To use 2D AutoCAD drawing entities in a sketch, you activate a sketch or start a new sketch and select Insert AutoCAD file on the 2D Sketch Panel Bar or toolbar.

Insert Objects

Along with geometric data about the 3D solids, you can incorporate additional information to your part file; you can insert various kinds of OLE objects in your Inventor part files. You can also insert BMP images, Excel spreadsheets, and Word documents.

Insert IGES Files

You can insert IGES data into an existing part file as surface objects.

Insert SAT Files

You can insert SAT files as construction objects.

SAVE AS

It is very important that you can use an Inventor solid part file in other computerized activities without having to reconstruct the parts again. To use the integrated data of a 3D solid model and an assembly of 3D solids in design analysis, rendering and visualization, or computerized manufacturing systems, you save the Autodesk Inventor files to various file formats.

You can save an Inventor part file, in addition to an Inventor part file (*.*ipt*), to

- BMP (*.*bmp*)
- IGES (*.*igs*, *.*ige*, *.*iges*)
- SAT (*.*sat*)
- STEP (*.*stp*, *.*ste*, *.*step*)
- STL (*.*stl*)
- Streamline Part Packages (*.*ptp*)
- XGL Files (*.*xgl*)
- ZGL Files (*.*zgl*)

You can save an Inventor drawing file, in addition to an Inventor drawing file (*.*idw*), to

- Drawing Files (*.*dwg*)
- DXF (*.*dxf*)
- BMP (*.*bmp*)
- Drawing Web Format (*.*dwf*)
- Streamline Drawing Packages (*.*dwp*)

You can save an Inventor Presentation file, in addition to an Inventor presentation file (*.*ipn*), to

- BMP (*.*bmp*)
- Streamline Presentation Package (*.*pnp*)

You can save an Inventor assembly file, in addition to an Inventor assembly file (*.*iam*), to

- BMP (*.*bmp*)
- IGES (*.*igs*, *.*ige*, *.*iges*)
- SAT (*.*sat*)
- STEP (*.*stp*, *.*ste*, *.*step*)
- Streamline Assembly Packages (*.*amp*)
- XGL Files (*.*xgl*)
- ZGL Files (*.*zgl*)

DESIGN MANAGEMENT

There are many ways to manage Autodesk Inventor files. You use project files, Design Assistance, Volo View, and Pack and Go wizard.

PROJECT FILE

To help manage Autodesk Inventor files used in a project, you set up a number of folders and construct a project file. In each project, you should have a home folder and a number of local folders, network folders, and library folders. The home directory is the working directory. The local and network folders are locations where Autodesk Inventor will search to open. The library folders are locations where standard components are saved. Search paths are useful in managing an assembly, which is a collection of component parts and linked part files. When you open an assembly file, Inventor will search first the library path, then the workspace path, then the local paths, and finally the workgroup paths.

Workspace	Your working directory, the default location where you save your files
Local Search Paths	Locations in your computer where Autodesk Inventor searches for files to open
Workgroup Search Paths	Locations in your computer or in a computer on the network that you and your teammates in a design group will access when opening a file
Library Search Paths	Locations to store the library parts

To store the information about these paths, you use a project file saved in the home directory. Besides specifying the folder location, a project file also enables you to specify other project files to be included in the search path and a number of options. In the options area, you specify multiple users, name of the project, shortcut to the project, location of the home folder of the project file, and the number of versions to keep.

There are three ways to construct and modify a project file. The first method is to select Projects from the File menu or select the Projects icon from the Open dialog box. (See Figure 2.19.) Another way to manage a project file without starting Autodesk Inventor is to select Start from the Windows main menu. From the Programs folder, select Inventor 6, then Tools, and then Project Editor. (See Figure 2.42.)

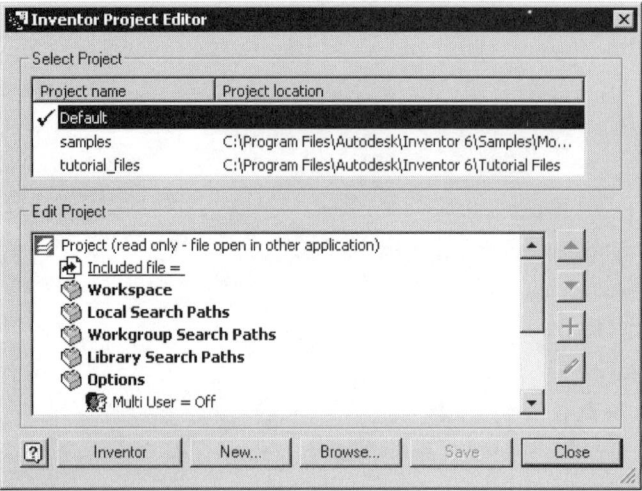

Figure 2.42 *Inventor Project Editor*

You can also use a text editor such as Notepad to open and edit a project text file. (See Figure 2.43.)

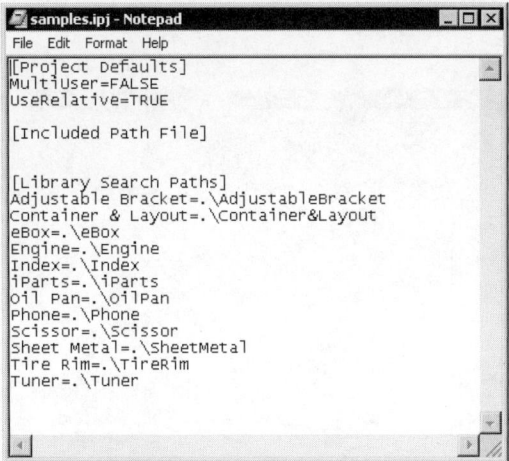

Figure 2.43 *Notepad editing a project file*

DESIGN ASSISTANT

Design Assistant is a tool that helps you find, track, and maintain Autodesk Inventor files together with related word processing, spreadsheet, and text files. There are three ways to use the Design Assistant:

> Select Design Assistant from the File menu.
>
> Select Design Assistant from the Windows start menu.
>
> Select a file or folder in the Windows Explorer, right-click, and select Design Assistant.

In the Design Assistant dialog box, there are three panes. The left pane of the Design Assistant dialog box has three buttons: Properties, Preview, and Manage; the center pane shows the selected folder or file; and the right pane shows the properties of the files in the selected folder if the Properties button is selected in the left pane. Figure 2.44 shows the properties of the files in the selected folder.

Figure 2.44 *Properties of the files in the selected folder*

To see a preview image of files, select the Preview button from the left pane. (See Figure 2.45.)

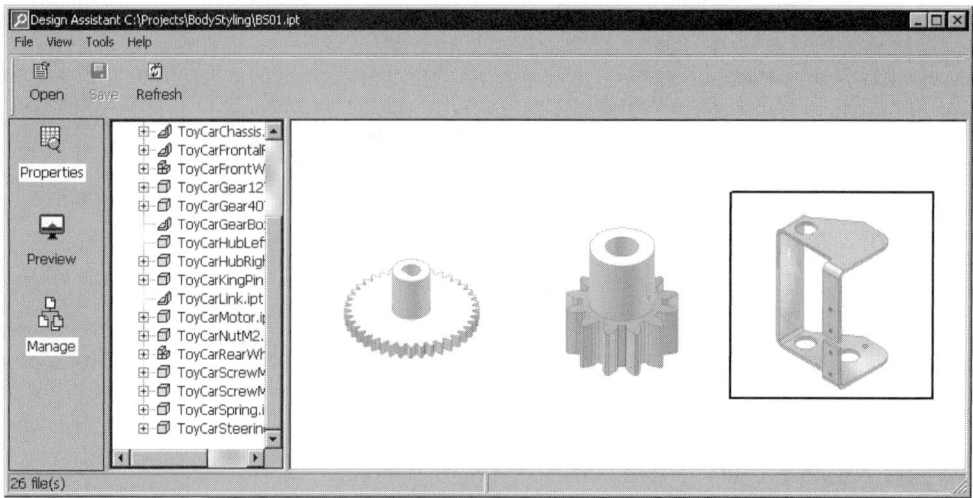

Figure 2.45 *Preview of files in the selected folder*

To manage files, select the Preview button from the left pane. (See Figure 2.46.)

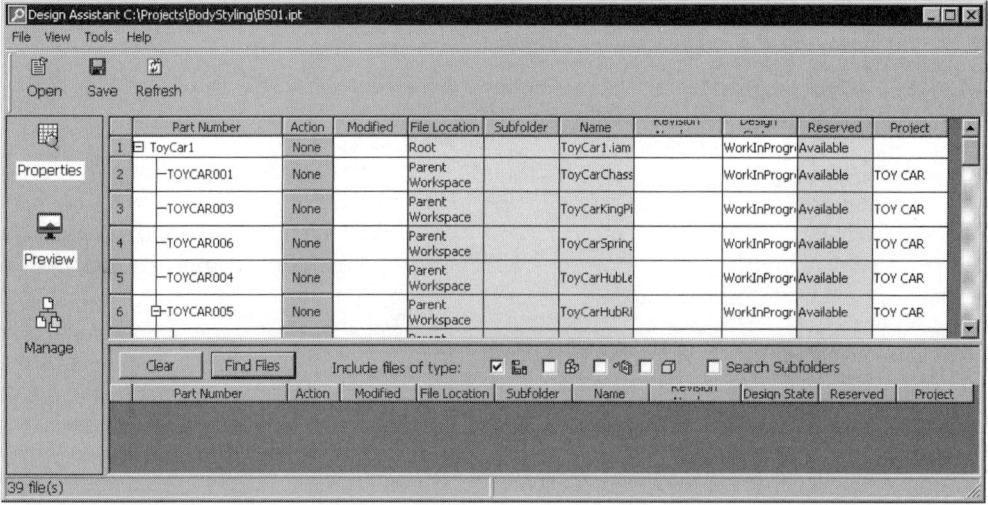

Figure 2.46 *Managing files in the selected folder*

Select a component, right-click, and select iProperties to discover the properties of the selected component. (See Figure 2.47.)

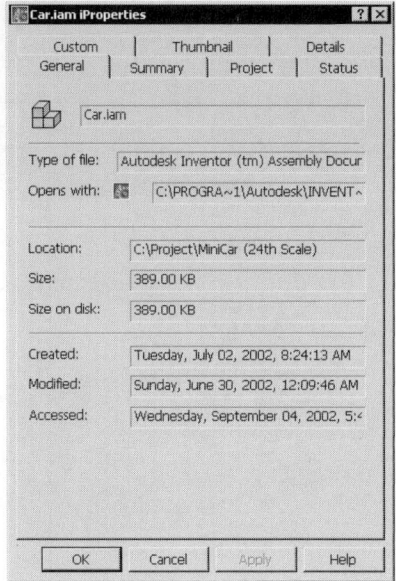

Figure 2.47 *iProperties dialog box*

PACK AND GO

To facilitate transportation of a set of files related to a project, you archive them into a single compressed file by using the Pack and Go wizard. Note that the compressed file has no reference to the original files. Instead, it is a set of individual files. You can access the Pack and Go wizard from the Design Assistant (Figure 2.48) or Windows Explorer (Figure 2.49). In the Design Assistant or Windows Explorer, you select the file, right-click, and select Pack and Go.

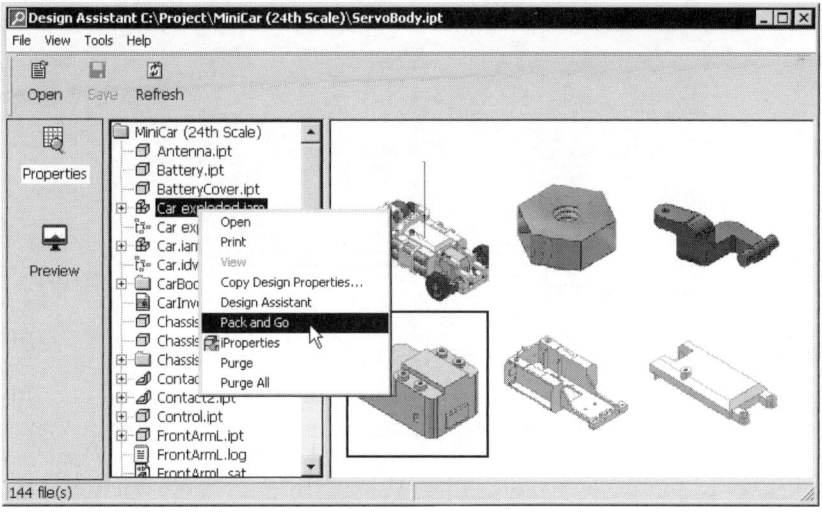

Figure 2.48 *Selecting Pack and Go from the Design Assistant*

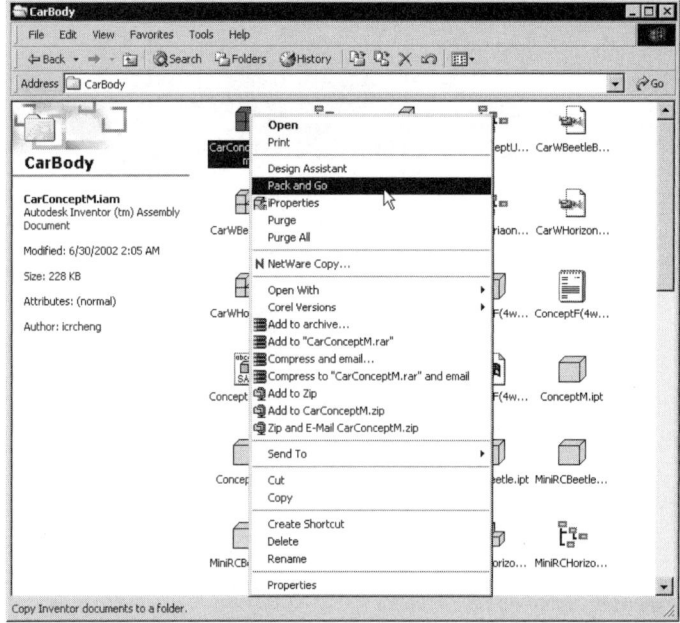

Figure 2.49 *Selecting Pack and Go from Windows Explorer*

If you select a part file, only the selected part file is archived. If you select an assembly file, all the components of the assembly file are included in the archived file. Figure 2.50 shows the Pack and Go dialog box.

Figure 2.50 *Pack and Go dialog box*

VOLO VIEW AND VOLO VIEW EXPRESS

Volo View is a design-viewing application that enables you to view files constructed by using Autodesk Inventor 5 (or later releases) and AutoCAD 2002. Figure 2.51 shows an Inventor file viewed in the Volo View application. By using the Volo View application, you view, mark up, and print Inventor files. You can also use Volo View Express, which is freeware.

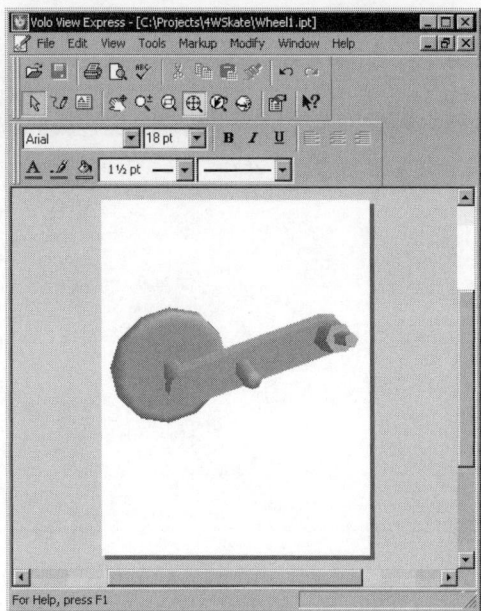

Figure 2.51 *Volo View Express application window*

SYSTEM SETTINGS

The way you construct solid parts, assemblies, presentations, and engineering drawings is affected by the system settings. You can modify system settings in the Options dialog box. To access the dialog box, select Application Options from the Tools menu. You can alter the settings through twelve tabs: General, File, Colors, Display, Hardware, Prompts, Drawing, Notebook, Sketch, Part, iFeature, and Assembly. The General, File, and Hardware tabs are discussed here. Other tabs in the Options dialog box are discussed in later chapters.

In the General tab shown in Figure 2.52, there are eight settings:

Undo file size (MB)	Determines the maximum size of the undo temporary file. Set the value to 64.
Locate tolerance	Sets the distance tolerance from which you select an object in terms of screen pixels. Set the tolerance to 6.
"Select Other" delay (sec)	Set the delay time before the Select Other dialog box is displayed after you move the cursor over a selected object. Set it to 1 second.
Annotation scale	Sets the scale of annotation.
Show startup dialog	Controls the Startup dialog box displayed when you start Autodesk Inventor. Accept the default.
Show 3D indicator	Enables the display of the 3D indicator in the bottom left corner of the screen. Accept the default.

Username	Specifies the user's name.
Text appearance	Specifies text font style.

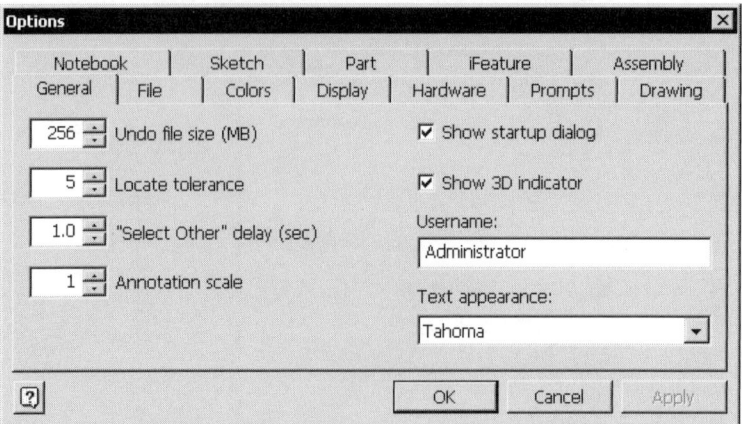

Figure 2.52 *Options dialog box: General tab*

The File tab shown in Figure 2.53 concerns location of undo file, template files, projects folder, workgroup design data, and default VBA project. The Hardware tab shown in Figure 2.54 concerns diagnostic of the graphics driver. Modify the settings if necessary.

Figure 2.53 *Options dialog box: File tab*

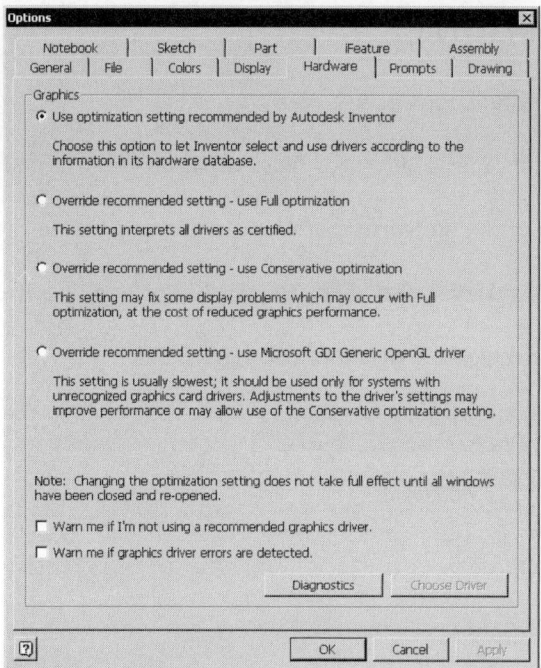

Figure 2.54 *Options dialog box: Hardware tab*

SUMMARY

Autodesk Inventor is a 3D parametric solid modeling tool. It enables you to construct parametric solid models, to compose an assembly from a set of solid parts or a set of assemblies of parts, to construct a presentation of an assembly in which component parts are exploded, and to construct 2D engineering drawings by deriving from the solid parts and assemblies or by making draft views.

Autodesk Inventor uses four kinds of files: construct a 3D solid model in a part file (.ipt), put components together to form an assembly in an assembly file (.iam), use a presentation file to manage exploded views of an assembly (.ipt), and construct engineering drawings in a drawing file (.idw).

To assist you in your design work, Autodesk Inventor's design support system has six components: Help Topics, Autodesk Online, What's New about Autodesk Inventor, Visual Syllabus, Design Doctor, and Sketch Doctor

To facilitate downstream computerized operations and the re-use of computer models constructed by using other applications, Inventor lets you open files of various formats and save your work to standard file exchange formats.

To help manage Autodesk Inventor files, you can use project files, Design Assistant, Pack and Go, and Volo View.

REVIEW QUESTIONS

1. What are the key functions of Autodesk Inventor?

2. What are the four kinds of Inventor files? What do you use them for?

3. Explore the use of the design support system and summarize the system.

4. Give a summary of file formats that you can save to and open from.

5. Outline the four design management tools provided.

CHAPTER 3

Part Modeling I

OBJECTIVES

The aims of this chapter are to delineate the key concepts of feature-based parametric solid modeling and the ways to construct sketches, to build simple solid features from sketches, to combine sketched solid features, to incorporate pre-constructed solid features, and to construct pattern features. Also explained are ways to set various display modes, to slice a shaded graphic display to show internal details of parts, and to edit a parametric feature-based solid. Finally, ways to measure a solid part, to set lighting in the 3D environment, and to assign material and color to a solid part are outlined. After studying this chapter, you should be able to

- Explain the concept of parametric feature-based solid modeling
- Construct parametric sketches and build simple solid features from sketches
- Combine sketched solid features
- Incorporate pre-constructed solids features in a 3D solid
- Construct pattern features
- Set various display modes
- Slice a shaded graphic
- Modify parametric feature-based solids
- Measure distance, angle, loop, and area
- Set lighting in the 3D environment
- Assign color and material properties to a solid part

OVERVIEW

The solid model of a 3D object is an integrated mathematical representation depicting the vertices, edges, faces, and volume of the object. Because any individual 3D object is unique in shape, it is impossible to derive a general mathematical expression that can represent all kinds of 3D objects. Over the years, various mathematical methods of representing 3D objects in the computer have been developed. Quite recently, the feature-based approach has become a commonly used 3D solid modeling method. Using this method, you decompose a complex 3D solid part into simple features that can be represented in the computer, construct the features accordingly, and combine the features.

Using Autodesk Inventor, you construct solids by making various kinds of features. Sketched solid features are features derived from sketches—you make sketches and let the computer construct the solid features from the sketches. To combine the sketched solid features, you use a join, cut, or intersect operation. The second kind of solid features, placed solid features, are pre-constructed solid features that you select from a menu, specifying a location and parameters. In addition, you can construct patterns of features. A major advantage of a parametric solid modeling system is that you can modify the parameters of the solid parts any time after you construct the solid. In addition to form, shape, and volume, you add to the solid part color, texture, and material properties to represent the appearance and facilitate engineering evaluation.

FEATURE-BASED PARAMETRIC MODELING CONCEPTS

Autodesk Inventor is a feature-based parametric solid modeling application. To construct a 3D solid model, you think about how to decompose an object into features, construct the features one by one, and combine them. Because Autodesk Inventor is a parametric system, parameters of all the features in a solid are modifiable. You can make a solid its approximate shape and size and modify the parameters at a later stage.

FEATURE-BASED MODELING APPROACH

In a feature-based modeling system, any 3D object, no matter how complex in form and shape, is constructed by composing a set of simple features. Using Autodesk Inventor, you construct simple features of a complex solid object in two basic ways: In the first way, you construct a sketch or a number of sketches and use the sketch or sketches to construct a solid feature or a surface feature. In the second way, you select a pre-constructed solid feature from the menu and place it in your solid part. If you want to repeat any feature in a solid, you construct patterns.

Using a feature-based solid modeling system, constructing a 3D solid model involves two major tasks: analyzing and synthesizing. With a 3D object in mind, you analyze it to determine what kinds of features make up the object. You think about how to construct the sketched features and how to combine them together. You also have to think about when to include the placed solid features in the solid model and when to construct a pattern. Furthermore, you have to decide the sequence of operation. After this careful analysis, you start making the features and synthesizing them to make the 3D solid.

To help establish a reference plane for making sketched features, placing pre-constructed solid features, and making patterns, you construct reference geometric objects. These reference objects are called work features.

PARAMETRIC SYSTEM

The word *parametric* is derived from the word parameter. Parameter here refers to the parameters of the geometric objects. In a parametric modeling system, parameters of all the objects are modifiable at any time. Naturally, you can assign an arbitrary value to a parameter initially and change it later. Consequently, you can use the computer as an electronic sketching pad to record your design idea. You can concentrate on forms and shapes rather than on dimensions. In the early stage of your design, you usually have only a shape of the object in your mind, but you might not have decided on any precise dimensions. In the absence of exact dimensions, you construct a rough parametric free-hand sketch. As the name implies, rough sketches are not precise at all—lengths are approximate. After constructing the sketch, you refine the geometry by specifying geometric constraints (such as horizontal, vertical, and parallel) and modify the size by specifying dimensions. After sketching, you make a solid or surface feature from the sketch or sketches.

SKETCHED SOLID AND SURFACE FEATURES

Sketched features derive from sketches. You construct a sketch or a number or sketches and let the computer system derive a feature from the sketch or sketches. You can construct two types of sketched features: solid and surface. You construct a solid or a surface from a closed-loop sketch or sketches, and you construct a surface from an open-loop sketch or sketches. In a parametric modeling system, both the sketches and the features derived from the sketches are parametric. You can edit the parameters of the sketches and the features any time to modify the size and shape of the feature.

There are four basic kinds of sketched solid and surface features: extruded, revolved, loft, and sweep. In addition, there are other kinds of sketched solid features: coil, rib, emboss, and split. In this chapter, you will learn about extruded and revolved solid features. In the next chapter, you will learn about the other kinds of sketched features.

Extruded Feature

To make an extruded feature, you make a sketch on a plane and extrude the sketch in a direction perpendicular to the plane. If you extrude a closed-loop sketch, you can construct a solid or a surface feature. If you extrude an open-loop sketch, you construct a surface feature. (See Figure 3.1.) If the opening of an open-loop sketch is closed by the body of the solid, you can construct a solid feature, which is equivalent to making a web. (You will learn how to construct a web and rib in the next chapter.) While extruding a sketch, you can extrude it in either direction or from mid-plane. (See Figure 3.2.)

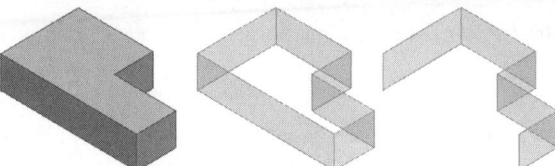

Figure 3.1 *Sketch extruded to form a solid (left) and extruded to form surface (middle and right)*

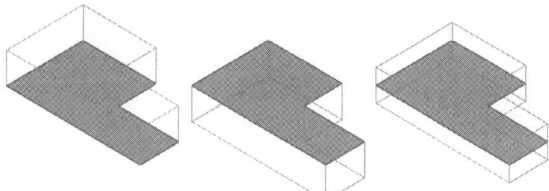

Figure 3.2 *Sketch extruded in either direction (left and middle) and from mid-plane (right)*

Besides specifying an extrusion distance, you can extrude a sketch to terminate at a face or from a face to another face. In Figure 3.3, the terminating faces are extruded surfaces. Basically, solid features form the main ingredients of a solid part, and surfaces serve several purposes. You will learn more about the use of surfaces in the next chapter. If a terminating surface to which the sketch is extruded provides two possible solutions, you can choose a minimum termination method or a maximum termination method. (See Figure 3.4.)

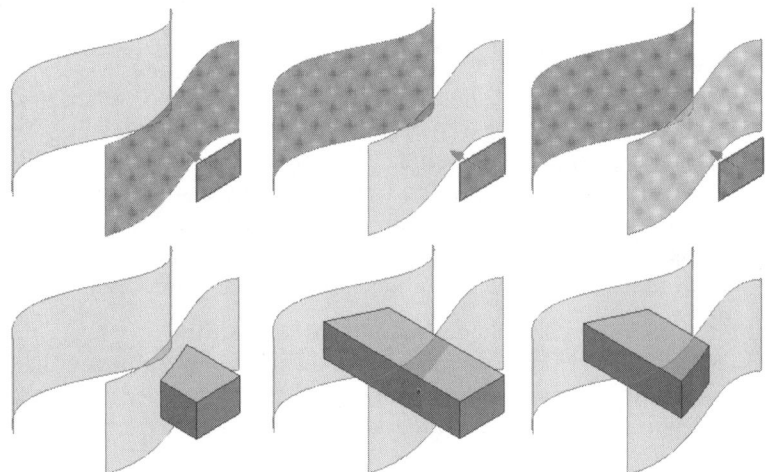

Figure 3.3 *Sketch extruded to a face (left and middle) and from a face to another face (right)*

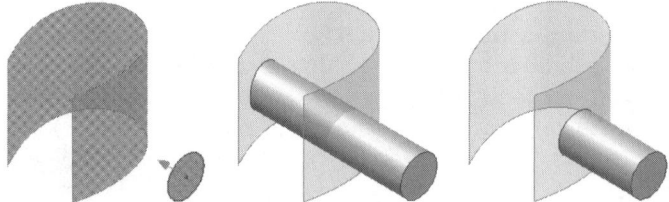

Figure 3.4 *Sketch extruded to a face (left), normal solution (middle), and minimum solution (right)*

Revolved Feature

To make a revolved feature, you make a sketch on a plane and revolve it about an axis in either direction or from mid-plane. (See Figure 3.5.)

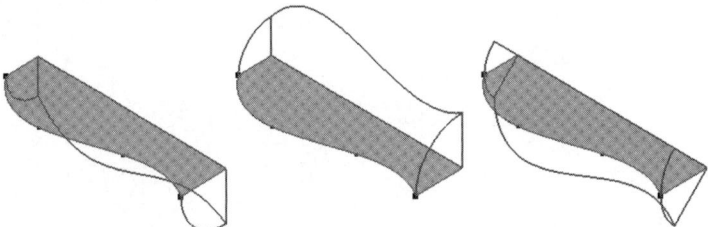

Figure 3.5 *Sketch revolved in either direction (left and middle) and from mid-plane (right)*

You can revolve a closed-loop sketch to form either a solid feature or a surface feature, and you can revolve an open-loop sketch to form a surface feature. (See Figure 3.6.)

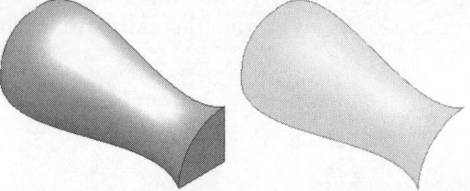

Figure 3.6 *Revolved solid feature (left) and revolved surface feature (right)*

SKETCH PLANES

As we have said, sketched features are derived from sketch(es). To construct a sketch, you need a sketch plane. In a part file, there are three default planes where you can construct a sketch plane: the XY, XZ, and YZ planes. In addition, you can use any existing planes of a solid feature or use artificial planes (work planes) established by using work features. (You will about learn work features in the next chapter.) Figure 3.7 shows a sketch constructed on a face of a solid part.

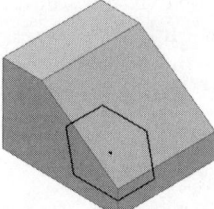

Figure 3.7 *Sketch constructed on a sketch plane established on a face of a solid*

BOOLEAN OPERATIONS

The first sketched solid feature you construct in a 3D solid part is called the base solid feature. It serves as the foundation on which you construct additional features by joining, cutting, and intersecting.

Join

The join operation enables you to join the new sketched solid feature with the existing solid. The resulting solid has the volume enclosed by the new solid feature and the existing solid. Figure 3.8 shows an extruded solid feature joined to the existing solid.

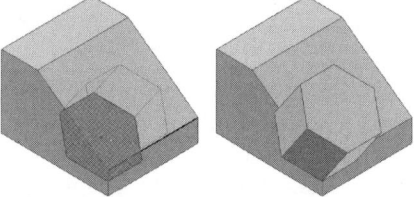

Figure 3.8 *Sketched solid feature joined to the existing solid*

Cut

The cut operation enables you to cut the new sketched solid feature from the existing solid. The resulting solid has the volume of the new sketched solid feature removed from the existing solid. Figure 3.9 shows an extruded solid feature cut from the existing solid.

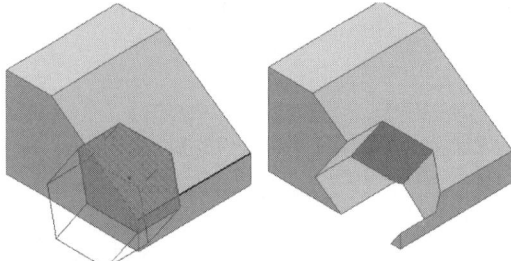

Figure 3.9 *Sketched solid feature cut from the existing solid*

Intersect

The intersect operation enables you to intersect the new sketched solid feature with the existing solid. The resulting solid has a volume that contains the portion common to both the new sketched solid feature and the existing solid. Figure 3.10 shows an extruded solid feature intersected with an existing solid.

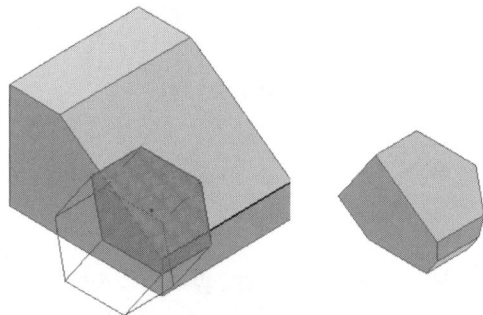

Figure 3.10 *Sketched solid feature intersected with the existing solid*

PRE-CONSTRUCTED (PLACED) SOLID FEATURES

In addition to constructing features from sketches, you can construct a feature by selecting a shape from the menu and specifying the parameters. We call these pre-constructed solid shapes placed solid features. Autodesk Inventor has six kinds of placed solid features: hole, thread, shell, fillet, chamfer, and face draft.

Hole Feature

A hole feature is a cylindrical feature cut on a solid. While cutting a hole, you can incorporate an internal thread and a countersink or counterbore. To construct a hole, you make a sketch to depict the center point of the hole and place a hole feature by specifying type, size, and location. Figure 3.11 shows (from left to right) a blind hole, a through hole, a counterbore hole, a countersink hole, and a through hole with thread.

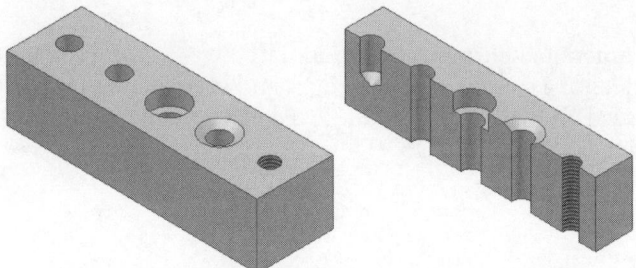

Figure 3.11 *An outside view and a cutaway view of a rectangular solid with a drilled blind hole, drilled through hole, counterbored hole, countersunk hole, and a threaded through hole*

Thread Feature

There are two ways to construct a thread. One way is to construct a helical coil feature and unite it to or subtract it from a cylindrical feature. (You will learn about the coil feature in the next chapter.) Because it takes time to construct the coil and it takes considerable memory space to store the data, you use a cosmetic thread in most circumstances. We call this feature the thread feature. To construct a thread feature, you select a cylindrical feature and specify the pitch and length of the thread. Figure 3.12 shows thread features placed on an external and an internal cylindrical feature of a solid.

Figure 3.12 *Thread features placed on cylindrical faces of a solid part*

Shell Feature

To make a solid hollow, you place a shell feature and state the thickness of the shell. Making a solid part hollow involves constructing a feature or a set of features, with the faces of the new features offsetting from the existing solid, and subtracting the feature or features from the original solid part. For a complicated solid part with many features, the shelling process might fail if one of the faces of the features cannot be offset. Figure 3.13 shows a shell feature placed on a solid part, making the solid hollow.

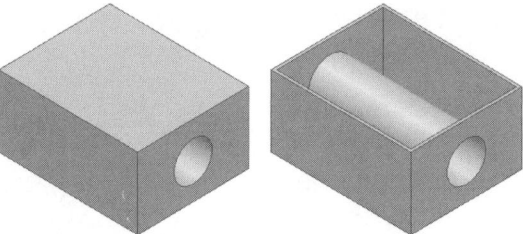

Figure 3.13 *Solid part (left) made hollow (right)*

Fillet Feature

For cosmetic or functional purposes, we round off edges of a component. A fillet feature rounds off the edges of a solid; you select edges and specify fillet radii. Figure 3.14 shows a fillet feature placed on the edges of a solid. At a vertex where three fillet edges meet, you can apply a setback. (See Figure 3.15.)

Figure 3.14 *Fillet feature placed on edges of a solid*

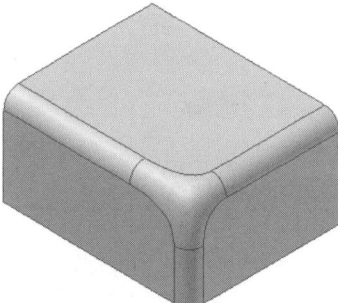

Figure 3.15 *Setback applied at a vertex where three fillet edges meet*

Chamfer Feature

Similar to filleting, we sometimes bevel the edges of a component. To bevel the edges, you place a chamfer feature. While chamfering, you select edges and specify bevel distances or the bevel distance and bevel angle. Figure 3.16 shows two kinds of chamfering treatments at a vertex where three chamfer edges meet.

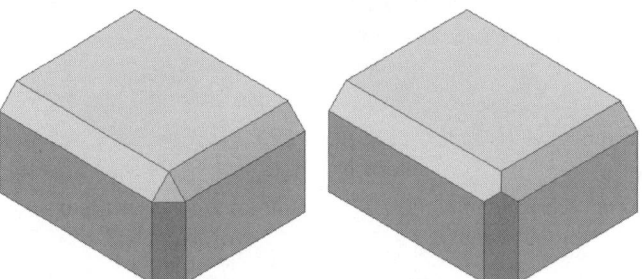

Figure 3.16 *Chamfer feature placed on edges of a solid*

Face Draft Feature

To facilitate the removal of the component from a mold, you taper the side faces of a solid part by placing a face draft feature. You select an edge or a split line and specify a draft angle. Figure 3.17 shows a face draft feature placed on the vertical faces of a solid part.

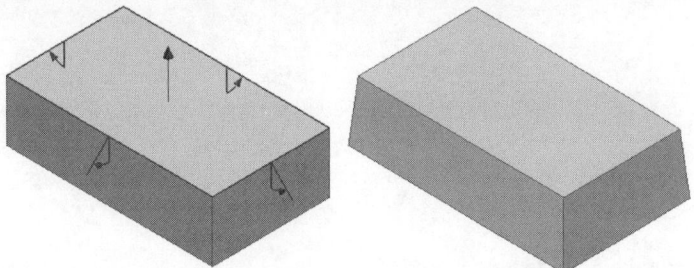

Figure 3.17 *Solid part (left) and face draft feature placed on its side faces*

PATTERN FEATURES

Quite often, we repeat features in a component. For example, we repeat the buttons of a device in either a rectangular or circular pattern. If a component is symmetric about a plane or an axis, you repeat features in the form of mirror copies. In all, there are three kinds of patterns: rectangular pattern, circular pattern, and mirror.

Rectangular Pattern

A rectangular pattern repeats a feature or features in two ways. It repeats a feature or a set of features in two linear directions that are not necessarily orthogonal to each other. (See Figure 3.18.) It also repeats features along two curves. (See Figure 3.19.)

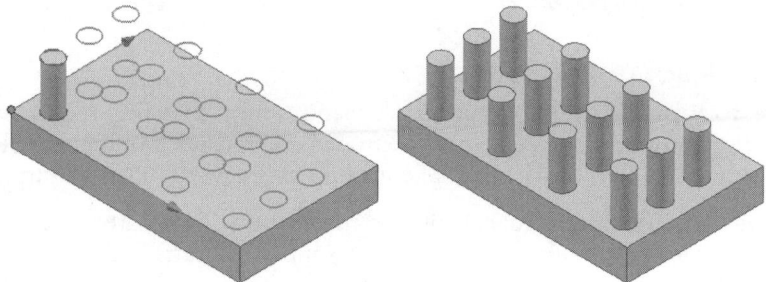

Figure 3.18 *Feature repeated in a pattern defined in two directions*

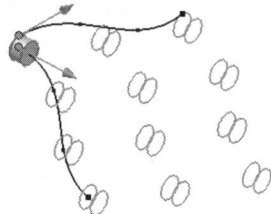

Figure 3.19 *Feature being repeated in a pattern defined by two paths*

Circular Pattern

To repeat a solid feature in a circular array, you select the feature and specify an axis, angular distance, and number of repetitions. Figure 3.20 shows a feature repeated in a circular array.

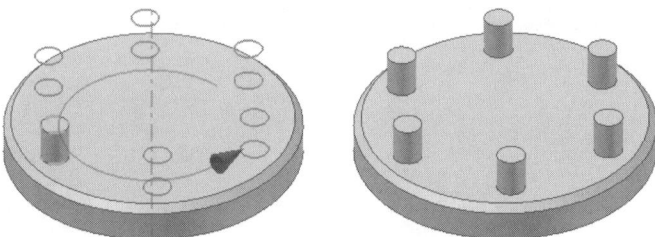

Figure 3.20 *Circular pattern feature placed*

Mirror Feature

To construct a mirror copy of a solid feature, you select a feature and specify a mirror plane. Figure 3.21 shows a shaped slot mirrored about an artificial work plane.

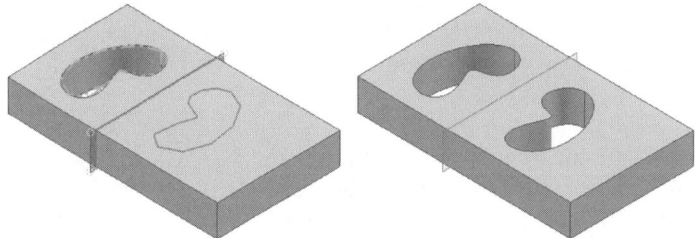

Figure 3.21 *Mirror feature placed*

SUMMARY OF SOLID FEATURES

To reiterate, you can construct solid or surface features from sketches. In this chapter, you will explore two kinds of sketched features, extruded and revolved. In the next chapter, you will learn about other kinds of sketched solid features and ways to construct surface features.

In addition to extruded and revolved solid features, you will explore six kinds of placed solid features (hole, thread, shell, fillet, chamfer, and face draft) and three kinds of pattern features (rectangular, circular, and mirror).

SYSTEM SETTINGS FOR PART MODELING

The Sketch, Part, and Display tabs in the Options dialog box affect the way a solid part is constructed. You should take some time to study the related system and make changes if necessary. The Options dialog box (select Tools > Application Options) lets you manipulate system settings. Among the twelve tabs of the dialog box, three of them (Sketch, Part, and Display) are discussed in the steps that follow.

1. Select Application Options from the Tools menu and select the Sketch tab.

SKETCH TAB

The sketch tab shown in Figure 3.22 sets the preferences for sketching.

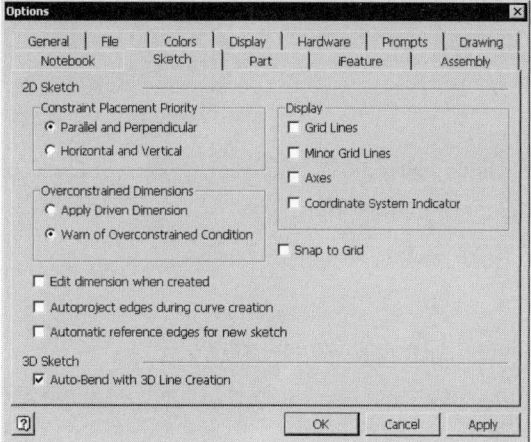

Figure 3.22 Options dialog box: Sketch tab

This tab has two major areas: 2D Sketch and 3D Sketch. In the 2D Sketch area, there are three areas (Constraint Placement Priority, Overconstrained Dimensions, and Display) and four check boxes.

Constraint Placement Priority	When you construct lines in a sketch, select the kind of constraint (Parallel and Perpendicular or Horizontal and Vertical) to have higher priority over the other kinds of constraint.
Overconstrained Dimensions	When you over-dimension a sketch, you can select Apply Driven Dimension to have it become a driven dimension or Warn of Overconstrained Condition to receive a warning message.
Display	Four check boxes in this area enables you to decide whether grid lines, minor grid lines, axes, and coordinate system indicators are displayed.
Edit dimension when created	If you select this check box, when you add dimensions to a sketch, you will be prompted to edit the dimension instead of having default dimensions applied.
Autoproject edges during curve creation	If you select this check box, when you construct curves, you can rub an existing geometry to project reference geometry.
Automatic reference edges for new sketch	If you select this check box, when you start a new sketch on a face, all the edges of the selected face will be projected onto the new sketch as reference geometry.

Snap to Grid	If you select this check box, the cursor will snap to the grid points while you sketch.

The 3D area has a single check box:

Auto-Bend with 3D Line Creation	If you select this check box, tangent corner bends will be placed on the 3D lines as you sketch them.

2. Select Parallel and Perpendicular and select Warn of Overconstrained Condition
3. Check all the boxes in the Display box.
4. Clear other check boxes.

PART TAB

The part tab sets the defaults for part construction.

5. Now select the Part tab of the Options dialog box to modify the settings related to the construction of solid parts.

The Part tab shown in Figure 3.23 has two check boxes (Parallel view on sketch creation and Auto-hide in-line work features) and two areas: Sketch on New Part Creation, with four options, and Construction (surface opacity). They concern the sketch plane and the display of a new sketch.

No new sketch	If selected, there will be no default sketch on new part creation. You have to set up a sketch plane.
Sketch On X-Y Plane	If selected, a sketch on the XY Plane is set up on new part creation.
Sketch On Y-Z Plane	If selected, a sketch on the YZ Plane is set up on new part creation.
Sketch On X-Z Plane	If selected, a sketch on the XZ Plane is set up on new part creation.
Opaque Surfaces	If this check box is selected, whenever a surface is constructed, it is set opaque.
Parallel view on sketch creation	If this check box is selected, whenever a new sketch plane is set up, the display will be changed to the top view of the new sketch plane.
Auto-hide in-line work features	If this check box is selected, whenever a new in-line work feature is constructed, it is set invisible.

6. Select the Sketch On X-Y Plane option, and clear the Parallel View On Sketch Creation and the Auto-hide in-line work feature check boxes. This way, a sketch is automatically constructed on the XY plane of the new part file, the display will not change when a new sketch plane is set up, and in-line work features are not hidden.

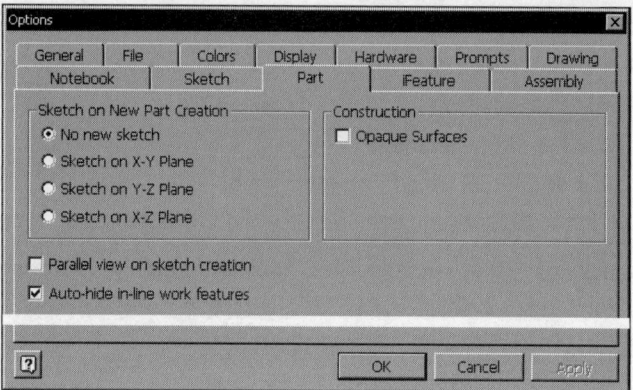

Figure 3.23 *Options dialog box: Part tab*

DISPLAY TAB

The display tab shown in Figure 3.24 lets you customize the wireframe and shaded display of the model.

7. Now select the Display tab.

The Display tab has three areas that control wireframe display, shaded display, and display quality. It also has two slider bars for adjustment of view transition time and minimum frame rate, and a box that controls dimming of hidden lines.

8. Click the OK button to close the Options dialog box.

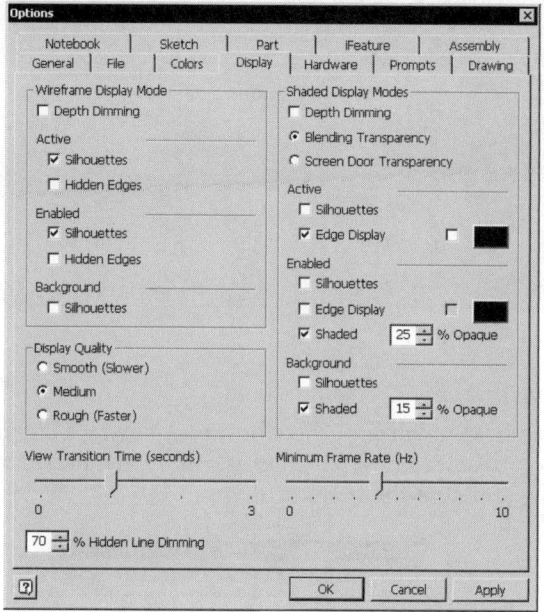

Figure 3.24 *Options dialog box: Display tab*

USER INTERFACE FOR PART MODELING

Now you will start a new part file and familiarize yourself with the Autodesk Inventor part file user interface.

1. Select New from the File menu.

2. In the New dialog box, select the Metric tab.

3. Select the *Standard.ipt* template and click the OK button.

After you start a new part file, the Inventor Standard toolbar is expanded to include a number of additional icons. Unless you selected the No New Sketch button from the Sketch panel of the Options dialog box, you will have a sketch object activated on the Browser Bar and a set of 2D sketching command icons on the command Panel Bar. (See Figure 3.25.)

INVENTOR STANDARD TOOLBAR

Below the menu bar, you will find the Inventor Standard toolbar. The Inventor Standard toolbar of a part file has a number of buttons and list boxes. If it is not displayed, select View > Toolbar and then check the Standard Bar check box.

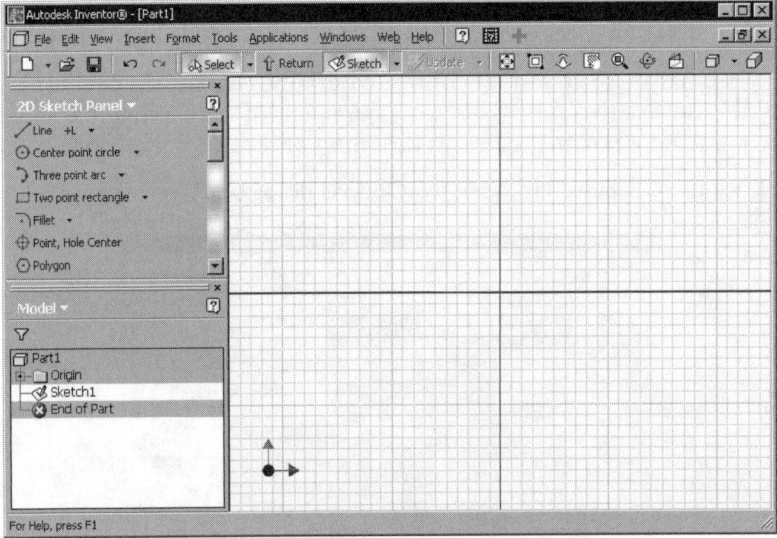

Figure 3.25 *New part file started*

New, Open, and Save Buttons

At the left of the toolbar, you will find three buttons: New, Open, and Save. (See Figure 3.26.) They enable you to start a new part file, assembly file, presentation file, or drawing file, open existing file, and save the current file.

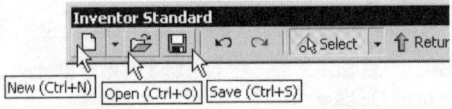

Figure 3.26 *New, Open, and Save buttons*

Undo and Redo Buttons

These buttons enable you to undo the last command and redo the undone command. (See Figure 3.27.)

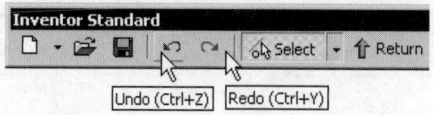

Figure 3.27 *Undo and Redo buttons*

Select List Box

The selection priority list box shown in Figure 3.28 determines the priority of selection. When you select an object from your graphics area and there is more than one object at the location of your cursor, selection priority settings will determine the kind of object that you will select. For example, when you place your cursor on a solid part and you set the priority to faces, you will select a face. On the other hand, if you set priority to features, you will select the feature.

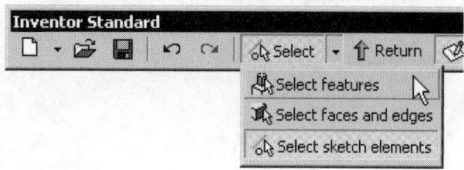

Figure 3.28 *Selection priority list box*

Selection priorities available depend on the kinds of file you open. (See Table 3.1.)

Table 3.1 *Selection priority settings available for each file type*

File	Selection priority
Part File	Feature, Face, Sketch
Assembly File	Component, Leaf Part, Feature, Face, Sketch
Presentation File	Component, Leaf Part
Drawing File	Edge, Feature, Part

When the cursor is in the graphics area, you can set selection priority by holding down the SHIFT key and right-clicking to bring up the selection priority shortcut menu. Figure 3.29 shows the priority shortcut menu for part files.

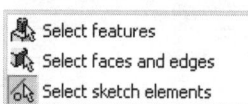

Figure 3.29 *Selection priority shortcut menu*

Return Button

After you finish a sketch, select the Return button to return to feature modeling mode. (See Figure 3.30.)

Figure 3.30 *Return button*

Sketch List Box

The Sketch list box has two buttons, for starting a new 2D sketch or a new 3D sketch. You will use one of these buttons to start a new sketch for making sketched solid features. You will use the 2D Sketch button in this chapter for making 2D sketches. In the next chapter, you will use the 3D Sketch button to construct 3D sketches. (See Figure 3.31.)

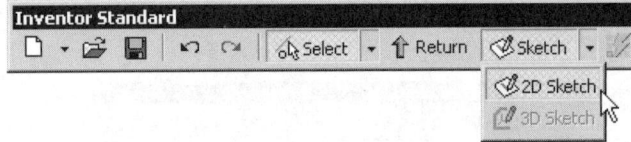

Figure 3.31 *Sketch list box*

Update Button

You will use the Update button to update changes that you make to the solid part. (See Figure 3.32.)

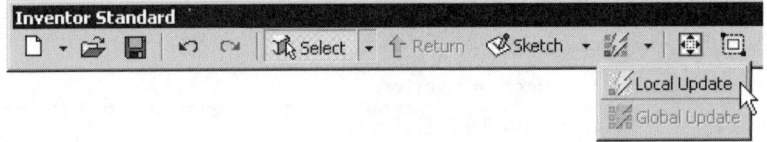

Figure 3.32 *Update button*

Display Control Buttons

The buttons shown in Figure 3.33 (from left to right) enable you to zoom all, zoom window, zoom in and out dynamically, pan, zoom to selected object, rotate the view, and zoom to the top view of a selected face (Look At).

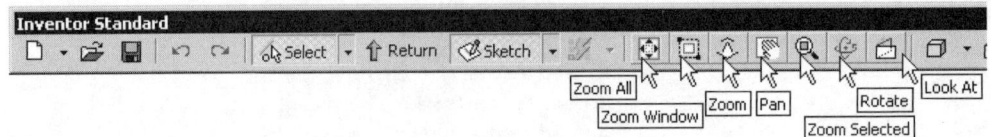

Figure 3.33 *Zoom all, Zoom Window, Zoom, Pan, Zoom Selected, Rotate, and Look At buttons*

If you are using a wheel mouse, you can perform the zoom operation by turning the wheel.

Model Appearance Buttons

Five sets of buttons shown in Figure 3.34 concern model appearance. The Shaded Display/Hidden Edge Display/wireframe display buttons enable you to select a display mode. The orthographic camera/perspective camera buttons enable you to set the display to orthographic or perspective. The no ground shadow/ground shadow/x-ray ground shadow buttons enable you to display shadow on an imaginary plane. The analyze face button turns on and off zebra display.

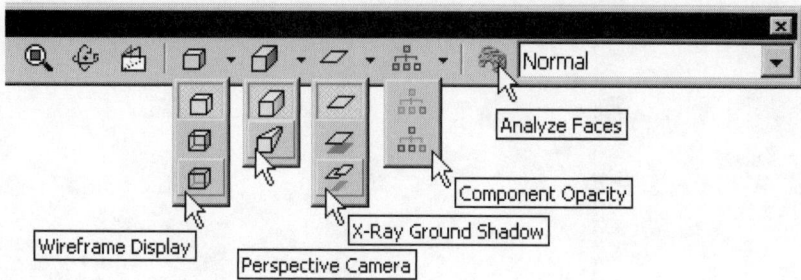

Figure 3.34 *Model appearance control buttons*

Style/Color List Box

The style/color list box shown in Figure 3.35 lets you set the style of sketch lines or the color of a solid. Sketch line styles choices are Normal, Construction, or Centerline. Construction lines assist in making sketches; they are not used in defining the cross sections or paths for making sketched solid features. Centerlines help establish an axis for making revolved solid features.

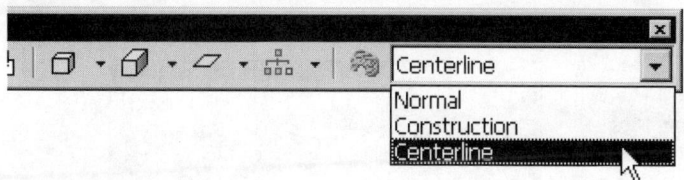

Figure 3.35 *Style/Color list box*

COMMAND PANEL BAR

The command Panel Bar is context sensitive. It has a number of palettes that enable you to access various design tools in accordance with the current design context. If it is not shown in your screen, select View > Toolbar and then check the Panel Bar button to display it. In a new part file with a new sketch started, the 2D Sketch Panel Bar is displayed.

Initially, the Panel Bar is docked to one side of the application window (either left side or right side). To change a docked Panel Bar to a floating Panel Bar and vice versa, double-click the header bar of the Panel Bar. Figure 3.36 shows the docked Sketch Panel Bar and Figure 3.37 shows the floating Panel Bar, both in expert mode.

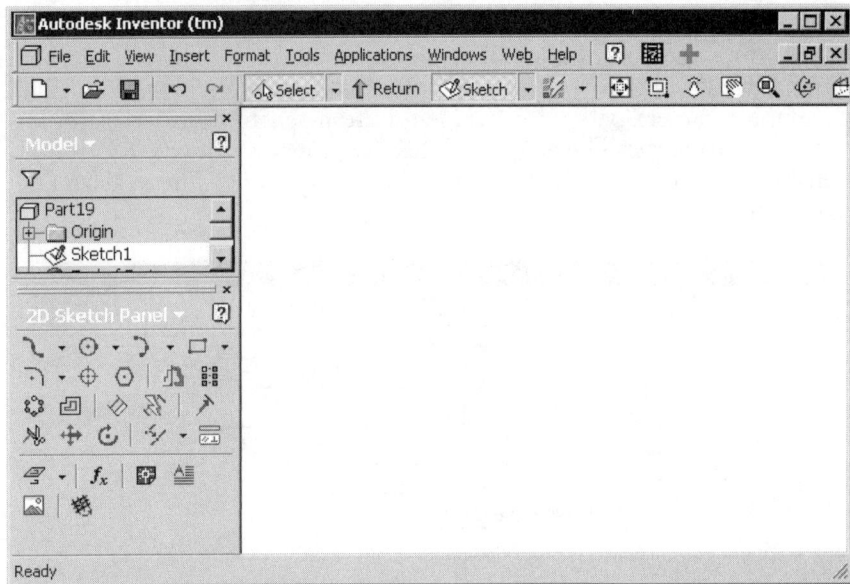

Figure 3.36 *Docked Panel Bar in expert mode*

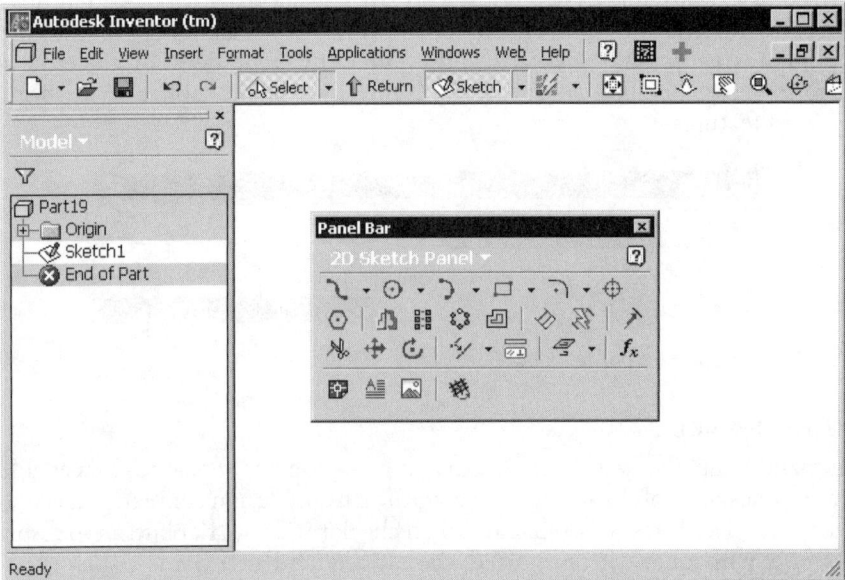

Figure 3.37 *Floating Panel Bar in expert mode*

A Panel Bar has two display modes, general mode and expert mode. In the general mode shown in Figure 3.38, command names are given along side each command icon. To hide the command names, right-click in the command panel and select the Expert box. To display the command names after you hide them, right-click in the command panel and clear the Expert box.

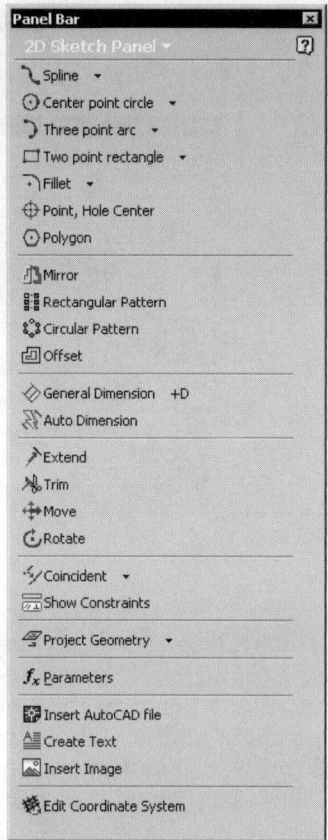

Figure 3.38 *Floating Panel Bar in general mode*

BROWSER BAR

Along with the Panel Bar is the Browser Bar, which shows the hierarchy of objects in a file. (See Figure 3.39.) For example, sketches and features that you construct will be displayed as objects in the hierarchy. If the Browser Bar is not shown in your screen, select View > Toolbar and then check the Browser Bar button to display it.

As with the Panel Bar, you can double-click its header bar to change it from a docked Browser Bar to a floating Browser Bar, or vice versa. Figure 3.39 shows a floating Browser Bar.

Initially, the Browser Bar has an origin that consists of the YZ, XZ, and XY planes, X, Y, and Z axes, and the center point. It also has a default sketch plane if one of the boxes (Sketch on X-Y Plane, Sketch on Y-Z Plane, or Sketch on X-Z Plane) of the Part tab of the Options dialog box shown in Figure 3.23 is checked.

When browsing the items on the Browser Bar, you can use the HOME, END, PGUP, and PGDN keys on the keyboard. To expand the hierarchy, right-click and select Expand All Children. To collapse the hierarchy, right-click and select Collapse All Children.

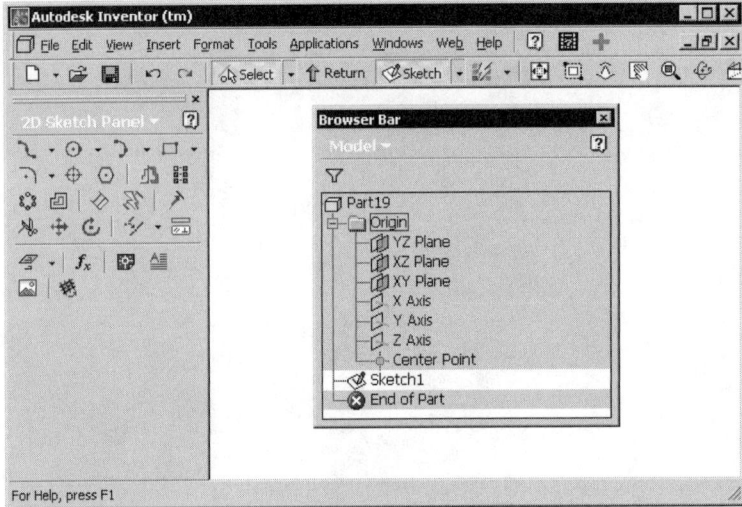

Figure 3.39 *Browser Bar*

GRAPHICS WINDOW

The graphics window is your main working area. You construct your solid part here. The appearance of the graphics area is configurable, depending on the settings you made in the Options dialog box. Select a color scheme on the Colors tab of the Options dialog box. In the graphics window, you might find a number of grid lines, two grid axes depicting the X and Y axes, a coordinate system indicator at the intersection of the X and Y axes, and a 3D indicator at the lower left corner. (See Figure 3.40.)

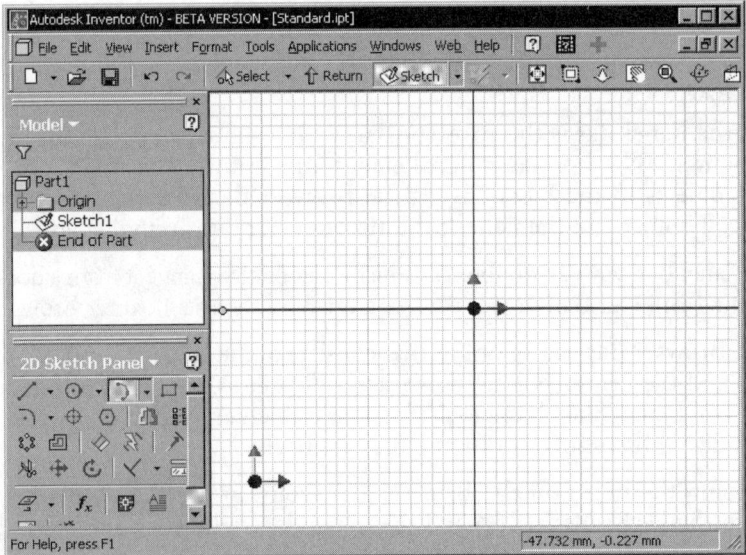

Figure 3.40 *Grid lines, grid axes, coordinate system indicator, and 3D indicator displayed in the graphics area*

These major and minor grid lines, grid axes, and coordinate system indicators are visual aids that help you perceive the approximate size of the current graphics display. You can turn them off by clearing the boxes (Grid Lines, Minor Grid Lines, Axes, and Coordinate System Indicator) on the Sketch tab of the Options dialog box shown in Figure 3.22.

At the lower left corner of the graphics window, there is a 3D indicator. If the indicator is not displayed, select Tools > Application Options and then select the General tab of the Options dialog box. On the General tab of the Options dialog box, select the Show 3D Indicator check box, and then click the OK button to exit.

STATUS BAR

At the bottom of the application window, there is a status bar providing text prompts.

SHORTCUT KEYS

Besides using the appropriate command panels and toolbars, you can use the shortcut keys shown in Tables 3.2 through 3.5.

Table 3.2 *Part modeling shortcut keys*

Shortcut Key	Function
S	Starts a new sketch.
L	Constructs line segments.
D	Constructs dimensions.
E	Constructs an extruded feature.
R	Constructs a revolved feature.
H	Places a hole feature.

Table 3.3 *Utilities shortcut keys*

Shortcut Key	Function
SHIFT + Right-click	Activates selection tool menu.
CTRL + ENTER	Disables inferring when using precise input.

Table 3.4 *Display control shortcut keys*

Shortcut Key	Function
F2	Pans the graphics display.
F3	Zooms the graphics display in or out.
F4	Rotates the graphics display.
SHIFT + F4	Rotates the display automatically.
F5	Returns the display to the previous view.
SPACE	Toggles common views on or off in view rotation.

Table 3.5 *Windows shortcut keys*

Shortcut Key	Function
CTRL+ Z	Undoes the last command.
CTRL+ Y	Redoes the last undone command.
CTRL+ C	Copies selected objects.
CTRL+ V	Pastes copied objects.
CTRL + S	Saves the current file.
CTRL + O	Opens a file.
CTRL+ N	Starts a new file.
CTRL + P	Prints the current file.
F1	Brings up Help for the current command or the Help dialog box.
ESC	Aborts the current command.
BACKSPACE	Clears the last selection.
DELETE	Deletes selected objects.
TAB	Alternates between input fields.

DOCUMENT SETTINGS FOR PART MODELING

Before you start designing a solid part, you should take some time examining the document's settings. Document settings concern default unit system and dimension precision for part modeling.

1. Select Tools > Document Settings to display the Document Settings dialog box.

The Documents Settings dialog box has four tabs, enabling you to set the default unit system and dimension precision, snap and grid settings for 2D sketches, adaptivity and snap setting in 3D sketches, and default tolerance values.

UNITS TAB

Use the Units tab to specify length, time, angle, and mass units and determine linear, angular dimension display precision, and dimension display mode. (See Figure 3.41.)

2. In the Document Settings dialog box, select the Units tab, if it is not already selected.

3. Now set the Length units to millimeter, Angle units to degree, Linear Dim Display precision to 3 decimal places, Time units to second, Mass units to kilogram, Angular Dim Display precision to 2 decimal places, and Display as value.

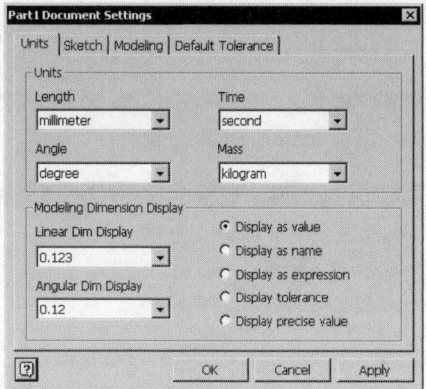

Figure 3.41 *Document Settings dialog box: Units tab*

SKETCH TAB

Use the Sketch tab to specify snap and grid spacing in 2D sketch and bend radius in 3D sketch.

4. Select the Sketch tab. (See Figure 3.42.)

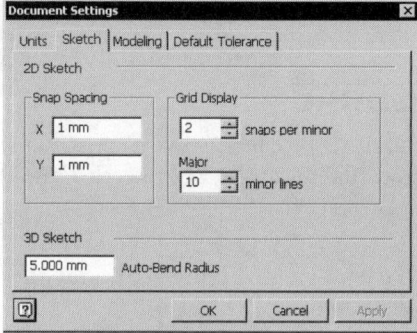

Figure 3.42 *Document Settings dialog box: Sketch tab*

MODELING TAB

Use the Modeling tab to specify snap spacing in 3D sketch.

5. Select the Modeling tab. (See Figure 3.43.)

Figure 3.43 *Document Settings dialog box: Modeling tab*

DEFAULT TOLERANCE TAB

Use the Default Tolerance tab to specify model tolerance.

6. Select the Default Tolerance tab. (See Figure 3.44.)

7. Click the OK button to close the dialog box.

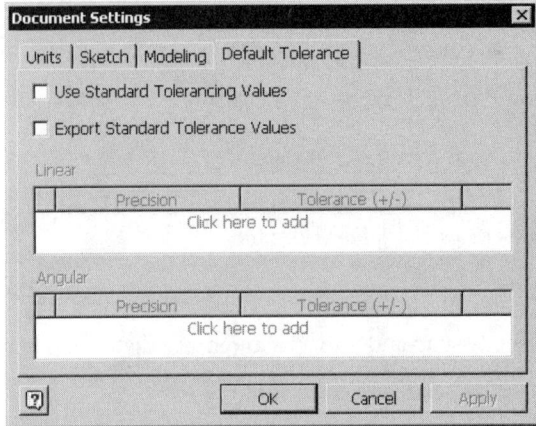

Figure 3.44 *Document Settings dialog box: Default Tolerance tab*

SKETCHING PROCEDURE

To construct a sketched solid or surface feature, you construct sketches. To make a sketch, you use the sketching tools available on the 2D Sketch Panel Bar or toolbar.

SKETCHING TOOLS

In the 2D Sketch Panel Bar shown in Figure 3.38 and the 2D Sketch Panel toolbar shown in Figure 3.45, there are twenty-five button areas. Table 3.6 describes the choices.

Figure 3.45 *2D Sketch Panel toolbar*

Table 3.6 *2D Sketch Panel Bar and toolbar options*

Option	Description
Line/ Spline	Line constructs a line. Spline constructs a spline.
Center Point Circle/ Tangent Circle/Ellipse	Center Point Circle constructs a circle with the center and a point on the circumference of the circle specified. Tangent Circle constructs a circle with three coplanar lines or edges selected. Ellipse constructs an ellipse.
Three Point Arc/ Tangent Arc/ Center Point Arc	Three Point Arc constructs an arc with the endpoints and a point on the arc specified. Tangent Arc constructs an arc that is tangential to a selected line or arc. Center Point Arc constructs an arc with first the center and then the endpoints specified.

Table 3.6 *2D Sketch Panel Bar and toolbar options (continued)*

Option	Description
Two Point Rectangle/ Three Point Rectangle	Two Point Rectangle constructs a rectangle with horizontal and vertical sides with the diagonal points specified. Three Point Rectangle constructs a rectangle with two points defined: an edge and a diagonal point.
Fillet/Chamfer	Fillet constructs a rounded corner at the intersection of two non-parallel lines. Chamfer constructs a bevel at the intersection of two non-parallel lines.
Point, Hole Center	Constructs a point or a hole center.
Polygon	Constructs a regular polygon with the number of sides and the size of the inscribed or circumscribed circle specified.
Mirror	Constructs a set of mirrored sketch geometry about a mirror line.
Rectangular Pattern	Constructs a rectangular array of sketch geometry in two directions.
Circular Pattern	Constructs a circular array of sketch geometry about a point.
Offset	Constructs an offset geometry from a selected geometry.
General Dimension	Constructs a parametric dimension or a driven dimension.
Auto Dimension	Automatically adds dimensions to fully constrain a sketch in a single step.
Extend	Extends a line or an arc to meet another line or arc.
Trim	Trims away a portion of a line or arc.
Move	Moves or copies sketch geometry from one specified point to another specified point
Rotate	Rotates sketch geometry or a copy of the selected sketch geometry about a specified point.
Perpendicular/Parallel/ Tangent/Coincident/ Concentric/Collinear/ Horizontal/Vertical/ Equal/Fix/Symmetric	See Table 3.7 describing geometric constraints.
Show Constraints	Displays the constraint applied to a sketch element. To delete unwanted constraints, select them, right-click, and select Delete.
Project Geometry/ Project Cut Edges/ Project Flat Pattern	Project Geometry constructs sketch objects by projecting selected geometry to the current sketch plane. Project Cut Edges constructs sketch objects by projecting the intersecting edges between the solid part and the current sketch plane. Project Flat Pattern works on sheet metal components. It unfolds selected disjointed face(s) onto the sketch plane so that sketch constraints can be referenced to the edges of the disjointed face(s).
Parameters	Constructs user parameters.
Insert AutoCAD file	Inserts AutoCAD 2D drawings onto the current sketch plane.
Create Text	Constructs text objects.
Insert Image	Inserts a bitmap image onto a sketch.
Edit Coordinates System	Moves the coordinate system to selected geometry of the solid part. You will find it useful when using the Precise Input tool.

MAKING A SKETCH

With the parametric modeling system, sketches need not be accurate and precise from the outset. You should concentrate only on the form and shape of the sketch in accordance with your design intent. Use lines, splines, circles, ellipses, arcs, rectangles, fillet, chamfer, points, polygon, and offset to construct sketch objects. You can repeat sketch elements by using mirror, rectangular pattern, and circular pattern. To modify a sketch, you trim, extend, move, and rotate. If you wish to enter precise coordinates while making a sketch, you can use the Inventor Precise Input toolbar. However, it must be emphasized that no matter how precise your sketch is, you still have to add parametric dimensions to fully constrain the sketch, because a fully constrained sketch is more predictable when it is used in making a solid feature.

CONSTRAINING THE SKETCH

To properly define the shape and size of a sketch, you apply constraints to the sketches. There are two kinds of constraints: geometric constraints and dimension constraints. Geometric constraint controls the shape and dimension constraint controls the size. In general, you should first apply geometric constraints and then add dimension constraints.

Adding Geometric Constraints

After you are satisfied with the general form and shape of your sketch, you refine it by specifying the geometric relationships between the elements of the sketch. For example, you might want to set a line to be perpendicular to another line and set an arc to be concentric with a circle. To specify these geometric relationships, you assign geometric constraints to the sketch. There are eleven kinds of geometric constraints, as described in Table 3.7.

Table 3.7 *Eleven kinds of geometric constraints*

Geometric Constraint	Description
Perpendicular	Sets two lines or axes of ellipses to be perpendicular to each other.
Parallel	Sets two lines or axes of ellipses to be parallel to each other.
Tangent	Sets a line, an arc, or a circle to be tangential to an arc or circle.
Coincident	Sets a point, the endpoint of a line, or the center of a circle or an arc to be coincident with another point, the endpoint of another line, or the center of another circle or arc.
Concentric	Sets the center of an arc, a circle, or an ellipse to be coincident with the center of another arc, circle, or ellipse.
Collinear	Sets two lines or a point and a line to be collinear.
Horizontal	Sets a line or axis of an ellipse to be horizontal or a point, the endpoint of a line, the center of a circle or an arc to lie horizontally with another point, the endpoint of another line, or the center of another circle or arc. Horizontal refers to parallel to the X axis.
Vertical	Sets a line or axis of an ellipse to be vertical or a point, the endpoint of a line, the center of a circle or an arc to lie vertically with another point, the endpoint of another line, or the center of another circle or arc. Vertical refers to parallel to the Y axis.

Table 3.7 *Eleven kinds of geometric constraints (continued)*

Geometric Constraint	Description
Equal	Sets two lines to have equal length or two arcs or circles to have equal radius.
Fix	Fixes the length of a line, the radius of a circle or an arc, or the location of a point, the endpoint of a line, and the center of a circle or an arc.
Symmetric	Causes lines and arcs to become aligned symmetrically about a selected mirror line.

The commands for adding geometric constraints and displaying geometric constraint symbols are indicated in Figure 3.46.

Figure 3.46 *Geometric constraint commands found on the 2D Sketch Panel Bar and toolbar*

Deleting Unwanted Geometric Constraints

While you make your sketch, Inventor automatically assigns some geometric constraints to your sketch in accordance with the way you construct your sketch. This is called inferred constraint. For example, if you construct a line that is nearly horizontal, Inventor will assign a horizontal constraint to it.

If you do not want a geometric relationship that Inventor assigned to your sketch or you want to delete a relationship that you assigned earlier, select Show Constraints on the 2D Sketch Panel Bar or toolbar and select a sketch element to display its geometric constraint symbols. To delete a constraint, select the symbol, right-click, and select Delete. Figure 3.47 shows geometric constraints of a sketch element displayed.

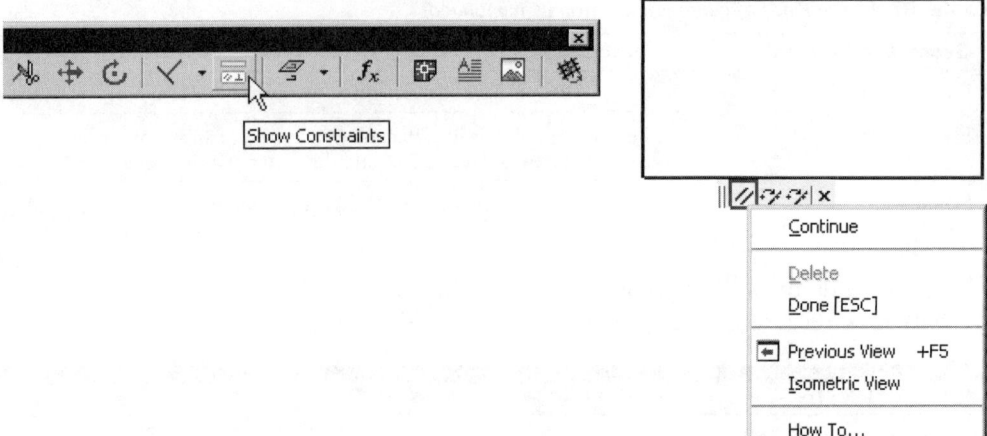

Figure 3.47 *Constraint symbols displayed*

Adding Dimension Constraints

General Dimension and Auto Dimension on the 2D Sketch Panel toolbar (Figure 3.48) and 2D Sketch Panel Bar enable you to construct parametric dimensions. General Dimension lets you add individual dimensions, and Auto Dimension lets you add dimensions to a sketch in one single step.

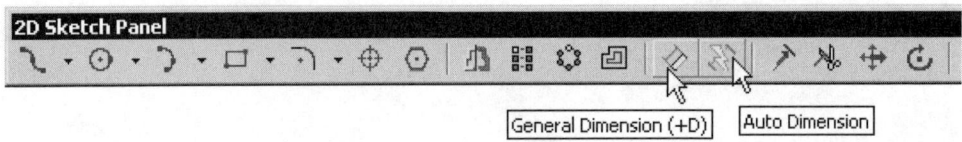

Figure 3.48 *Dimension commands*

A parametric dimension serves two purposes: It reports the size and it controls the size of the geometry. For example, if you construct a line that is 12.5 units long, when you add a parametric dimension to the line, the exact length is displayed. You can accept the length or you can enter a new dimension. If you enter a length of 14 units, the length of the line will change to 14 units.

Dimension Tolerance

As we all know, it is not feasible to manufacture a component to its exact size. Instead, we have to allow for deviation in size. In your design, the allowable deviation in size can be expressed and incorporated into the solid part in terms of dimension tolerances. To add tolerances to a dimension, select the dimension, right-click, and select Dimension Properties. (See Figure 3.49.)

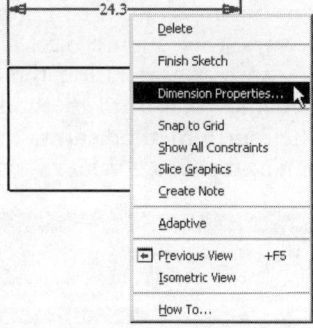

Figure 3.49 *Right-click menu after selecting a dimension .*

The Dimension Properties dialog box has two tabs, Dimension Settings and Document Settings. Figure 3.50 shows the Dimension Settings tab. This tab has three areas: Settings, Tolerance, and Evaluated Size. The Settings area displays the dimension's name, display precision, and the dimension's exact value. Each dimension of a sketch has a name. By default, the name begins with a letter "d" followed by a number starting from zero. The name of the first dimension that you add to a solid part is called d0, the second one d1, and so on. You can change the dimension's name by entering a name here. Display precision concerns the number of decimal places used when a dimension is displayed. Because the dimension here has an exact value of 24.25 mm and the display precision is 0.1, the dimension is shown as 24.3. The Tolerance area enables you to specify tolerances in various modes. If you already specified tolerances to a dimension, you can instruct the computer to evaluate the part to its nominal size, maximum value, or the minimum value in the Evaluated Size area.

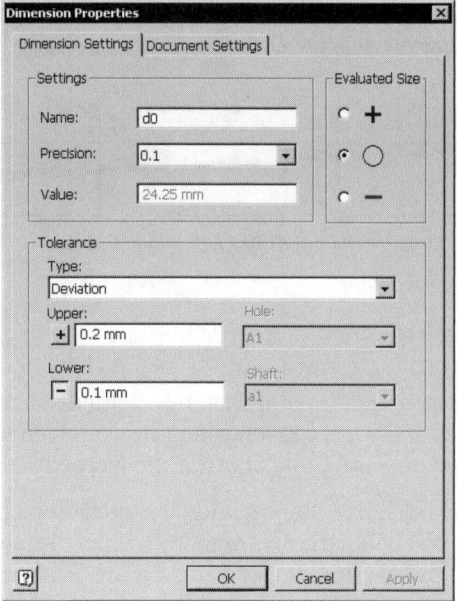

Figure 3.50 *Dimension Settings tab of the Dimension Properties dialog box*

The second tab is the Document Settings tab. (See Figure 3.51.) Here you can set the display of a dimension in five ways. (See Figure 3.52.) The first method, the default method, shows the value, taking into consideration the display precision. The second method shows the dimension's name. The third method shows the dimension's name together with its exact value. The fourth method shows the tolerances of the dimension. The fifth method shows the dimension's exact value.

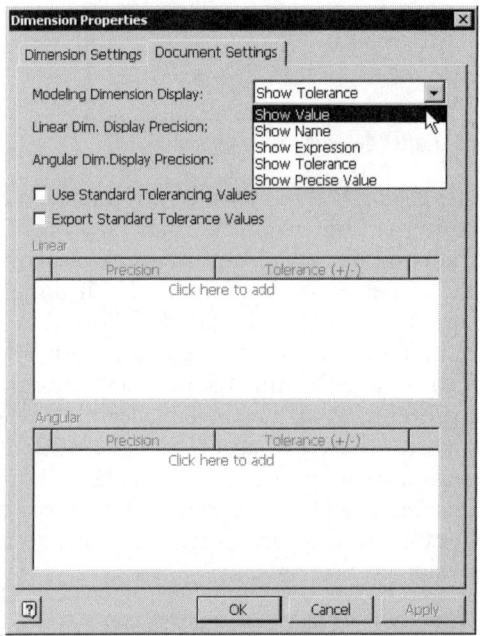

Figure 3.51 *Document Settings tab of the Dimension Properties dialog box*

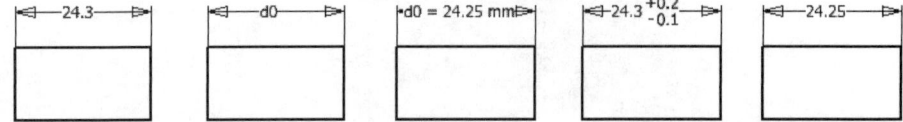

Figure 3.52 *Dimension display methods (left to right, value, name, expression, tolerances, and exact value)*

Notes on Dimensioning Sequence

Because parametric dimensions control the size of a sketch, the sketch's size will change as you assign dimensions. For example, if a sketch element's actual length is 20 mm and you specify a dimension of 200 mm, you lengthen the sketch element. On the other hand, if you specify a dimension of 5 mm, you shorten the element.

As a result of lengthening or shortening individual sketch elements, you might distort the overall shape of the sketch. To illustrate how an individual dimension can distort a sketch, Figure 3.53 shows a sketch of unknown size and Figure 3.54 shows the distorted sketch after you assign a dimension of extremely small value for the distance between A between C (Figure 3.53). Distortion occurs because distance AB (Figure 3.53) does not change.

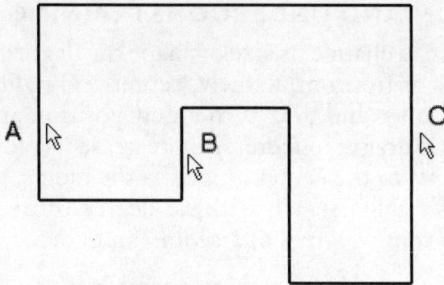

Figure 3.53 *Sketch of unknown size*

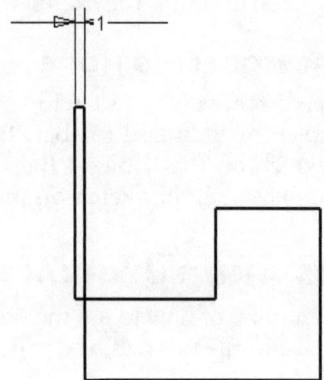

Figure 3.54 *Sketch distorted*

To avoid distorting the sketch, you add a dimension of any value (very small or very large) to the distance between A and B (Figure 3.53), and the relationship between the sketch elements will not be changed. Figure 3.55 shows two extreme cases of dimension value. As a rule of thumb, always place dimension to the innermost (or the smallest dimension) first and work from inside to outside.

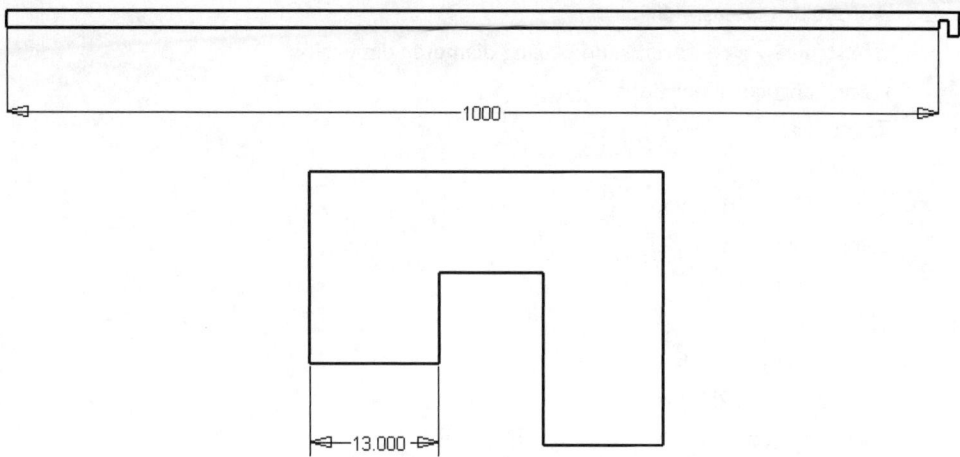

Figure 3.55 *Two extreme cases of dimension values*

DEGREES OF FREEDOM AND UNDERCONSTRAINING

We call the ways a sketch can change its size or shape the degrees of freedom. A circle, for example, has two degrees of freedom, namely, center and radius. After you construct a circle, you can select its center and drag it to a new position and you can select a point along its circumference and drag it to increase or decrease its radius. If you apply geometric or dimension constraints to the center as well as the radius, the circle is fixed and has no degree of freedom. We call a sketch with no degree of freedom a fully constrained sketch, and a sketch with some degrees of freedom an underconstrained sketch.

To use a sketch for subsequent feature-making operations, you do not necessarily have to fully constrain it. However, a fully constrained sketch is more predictable than an underconstrained one. Consider constraining the sketch to the origin point.

SKETCHING AND FEATURE MODELING MODE

After you are satisfied with the sketch, you exit sketching mode by selecting the Return button (Figure 3.30) on the Inventor Standard toolbar. If, for any reason, you want to modify the sketch, select the sketch on the Browser Bar (Figure 3.39), right-click, and select Edit Sketch or simply double-click the sketch on the Browser Bar.

SKETCHING METHODS AND TUTORIALS

To help you gain a better understanding of how to use the sketching tools in design, you will work on a set of sketching tutorials. In the tutorials, you will explore the following topics:

Constructing a polygon

Use of origin objects

Projection of geometry

Ways to save preview picture

Applying inferred geometric constraints while sketching

Constructing a 2D fillet

Automatic dimensioning

Constructing a centerline and placing diameter dimension

Placing angular dimensions

Trimming

Extending

Using construction lines

Ellipse construction

Spline construction

Offsetting

Constraining to midpoint

Constructing a 2D chamfer

Mirroring geometry

Rectangular pattern

Circular pattern

Precise input

Moving and rotating

Importing an AutoCAD file

Sketches that you construct in the following sketching tutorials will be used for making the solid parts of an assembly shown in Figure 3.56. You will make most of the components in this chapter and the remaining components in the next chapter. In Chapter 6, you will put them together to form an assembly.

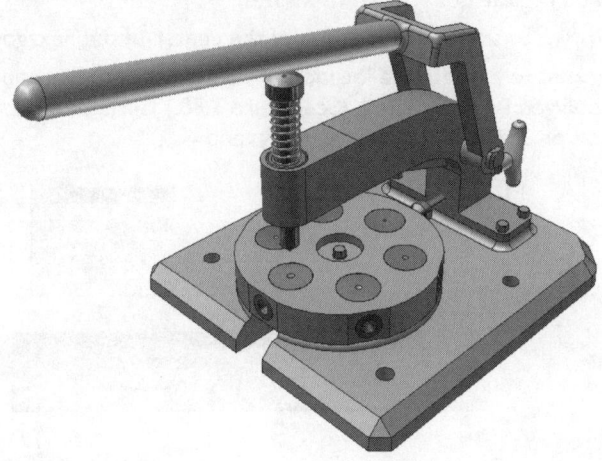

Figure 3.56 *Punch set*

CONSTRUCTING A POLYGON

A polygon is a set of contiguous line segments with geometric constraints governing the length of the segments (equal length) and the angle between adjacent segments (equal angle). You can construct a polygon by specifying the center point and a point on the inscribed or circumscribed circle.

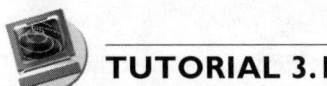

TUTORIAL 3.1

In Tutorials 3.1 through 3.3, you will construct a regular polygon for making the hexagonal head of a bolt shown in Figure 3.57.

Figure 3.57 *Completed model of a bolt*

1. Start a new part file. Use the metric template.

Tips: By default, a sketch plane is already established on the XY plane. You can choose not to have any default sketch plane or have the sketch plane established on XZ plane or YZ plane by selecting the appropriate boxes in the Part tab of the Options dialog box.

If there is no default sketch plane, you can establish a sketch plane by selecting the 2D Sketch button on the Inventor Standard toolbar and selecting one of the planes on the Browser Bar.

2. Select Polygon on the 2D Sketch Panel Bar or toolbar.

3. In the Polygon dialog box, select Inscribed and set the number of sides to 6 to construct a regular polygon with six sides.

4. Select a point to specify the location of the center of the hexagon.

5. Select a point to indicate the location of the radius of the inscribed circle. A regular polygon is constructed. (See Figure 3.58.) Click Done in the Polygon dialog box or press ESC to end the command.

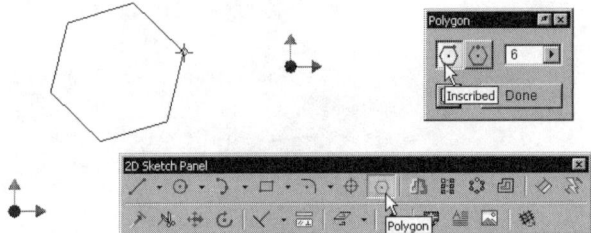

Figure 3.58 *Polygon constructed*

Now examine the constraints inherited in the sketch and add a geometric constraint.

6. Select Show Constraints on the 2D Sketch Panel Bar or toolbar.

7. Select edge A in Figure 3.59. (The constraint symbols are displayed.)

Tip: To remove an unwanted geometric constraint, select a symbol representing the constraint you want to delete, right-click, and select Delete.

8. Select Horizontal on the 2D Sketch Panel Bar or toolbar.

9. Select edge A in Figure 3.59. The selected line becomes horizontal. (Select the "x" sign from the constraint symbol list to turn off the display of constraint symbols.)

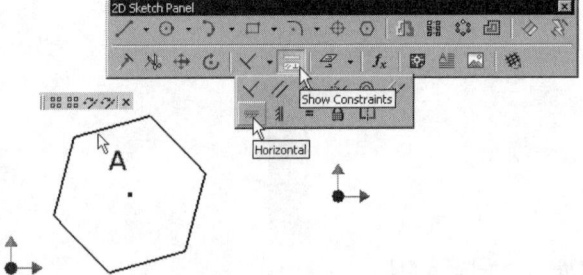

Figure 3.59 *Constraint symbol displayed and a horizontal constraint being added*

Now add a dimension.

10. Select General Dimension on the 2D Sketch Panel Bar or toolbar.

11. Select center point A and then edge B in Figure 3.60.

12. Select a point at C in Figure 3.60 to indicate the location of the dimension.

13. Double-click the dimension value to display the Edit Dimensions dialog box, if it is not displayed automatically.

 Tips: You can set the Edit Dimensions dialog box to be displayed automatically each time you add a dimension by selecting the Edit dimension when created box on the Sketch tab of the Options dialog box.

If you want to assign a tolerance to a dimension, you can select the dimension in the Edit Dimensions dialog box, right-click, and select Tolerance.

14. In the Edit Dimensions dialog box, set the dimension value to 6 mm and select the Checkmark to close the dialog box.

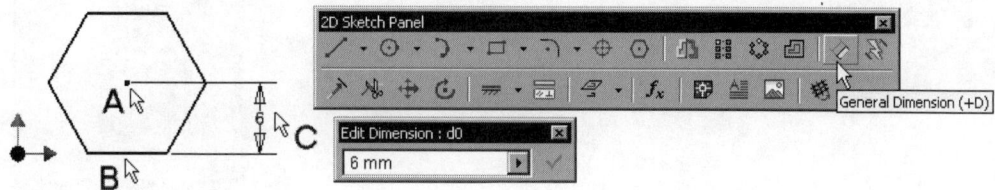

Figure 3.60 *Dimension being added*

A hexagon is constructed. You will construct the center point of the hexagon in the next tutorial.

USE OF ORIGIN OBJECTS AND PROJECTION OF GEOMETRY

A solid part file has an origin that consists of three default planes (XY, XZ, and YZ), three default axes (X, Y, and Z), and a center point. To use the origin or other geometry as references in sketching, you project them onto the current sketch plane. If you already constructed one or more solid features and you are constructing a new sketch, you can project edges and loops of edges from any existing features. Projected geometry is fully associated with the edges and loops of edges. If they are changed, the project geometry also changes.

 ## TUTORIAL 3.2

In this tutorial, you will project the origin's center point and use the projected geometry as reference to position the polygon. First you will expand the objects on the Browser Bar.

1. Move the cursor over the Browser Bar, right-click, and select Expand All.

2. Select Project Geometry on the 2D Sketch Panel Bar or toolbar.

3. With reference to Figure 3.61, select Center Point A on the Browser Bar to project a reference point B onto the sketch plane.

 Tip: You can project the origin's planes and axes as well.

4. Select Coincident on the 2D Sketch Panel Bar or toolbar.

5. Select center point C of the polygon and the projected reference point B in Figure 3.61 to apply a coincident constraint.

6. Select Return on the Inventor Standard toolbar to exit sketch mode.

 Tip: If you want to modify the sketch after you exit sketch mode, select the sketch on the Browser Bar and double-click or right-click and select Edit Sketch. Do not select the 2D Sketch button on the Inventor Standard toolbar because this button constructs a new sketch.

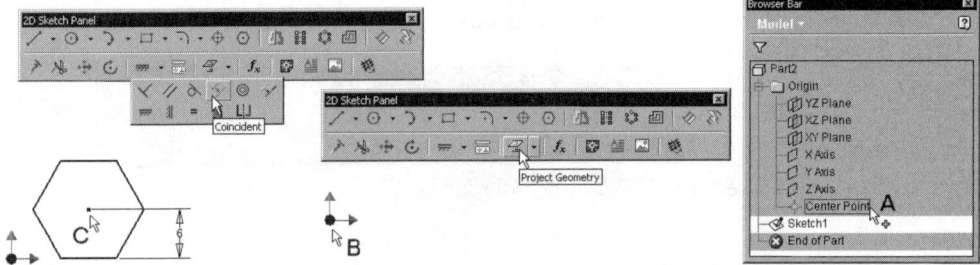

Figure 3.61 *Reference center point projected and polygon being relocated*

WAYS TO SAVE PREVIEW PICTURE

The sketch is complete. Save and close your file (file name: *PunchBolt.ipt*). To help manage the files in a project, set up a folder (folder name: *Punch*) in your workspace and save this file and the other files related to the punch set in the folder.

 TUTORIAL 3.3

In this tutorial, you will save your file and select a preview picture option while saving.

1. Select File > Save.

A preview picture helps identify a file before you open it. There are three ways to save a preview picture: Active Window On Save, Active Window, and Import From File.

2. In the Save As dialog box, select the Options button.

3. In the File Save Options dialog box, select one of the boxes: Active Window On Save, Active Window, or Import From File. (See Figure 3.62.)

4. Click the OK button to close the File Save Options dialog box.

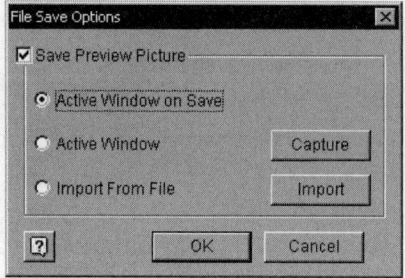

Figure 3.62 *File Save Options*

Active Window on Save	The preview picture will be the most updated graphics window.
Active Window	Use this option in conjunction with the Capture button to specify a preview picture to be saved in the file—the graphics window that you capture.
Import From File	Use this option in conjunction with the Import button to specify a 120 × 120 pixel image as the preview picture.

APPLYING INFERRED GEOMETRIC CONSTRAINTS WHILE SKETCHING

Geometric constraints can be applied to a sketch in two ways. You explicitly assign an appropriate geometric constraint after the sketch is made, or you implicitly express your design intent while you construct the sketch. For example, to implicitly assign a horizontal constraint to a line segment, select a point to indicate the first endpoint, move the cursor to either left or right in a nearly horizontal direction so that the horizontal constraint symbol is displayed at the cursor, and select a point to confirm the endpoint location.

 **TUTORIAL 3.4**

In Tutorials 3.4 through 3.6, you will construct a sketch to make a feature of the handle bar shown in Figure 3.63. In this tutorial, you will learn how to implicitly apply inferred geometric constraints while constructing a sketch.

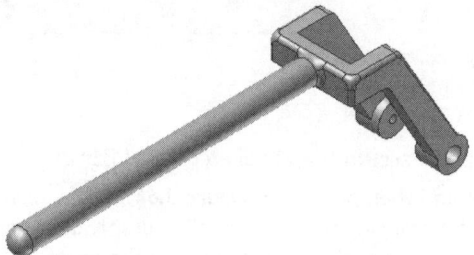

Figure 3.63 *Handle bar*

First you will construct a set of line segments.

1. Start a new part file. Use the metric template.

2. Select Line on the 2D Sketch Panel Bar or toolbar.

3. Select location A in Figure 3.64 to specify the start point.

4. Select location B in Figure 3.64 to specify the endpoint of the first line segment. This line is not constrained.

5. Move the cursor slowly to the left in a nearly horizontal direction. You will see a horizontal constraint symbol (looks like a "—" sign) at the cursor.

6. While the symbol is still displayed, select location C in Figure 3.64 to specify the endpoint of the second line segment. Now you have a horizontal line segment.

7. Move the cursor slowly upward and select a point at location D in Figure 3.64 in a nearly vertical direction. You will see a perpendicular constraint symbol (looks like the inverted "T"). It denotes that the line segment is perpendicular to the previous line segment.

8. Move the cursor to touch line segment BC and then move to a location E in Figure 3.64. You will see a parallel constraint symbol (looks like a "//" sign). By selecting a point while the symbol is displayed, you construct a parallel line segment.

9. Move the cursor slowly to location F until you find a parallel constraint symbol displayed at the cursor and a dotted line displayed between F and A. Selecting a point while the symbol and the dotted line are displayed, you construct a parallel line segment EF with its endpoint F perpendicular to line AB.

10. Right-click and select Done to exit line sketching.

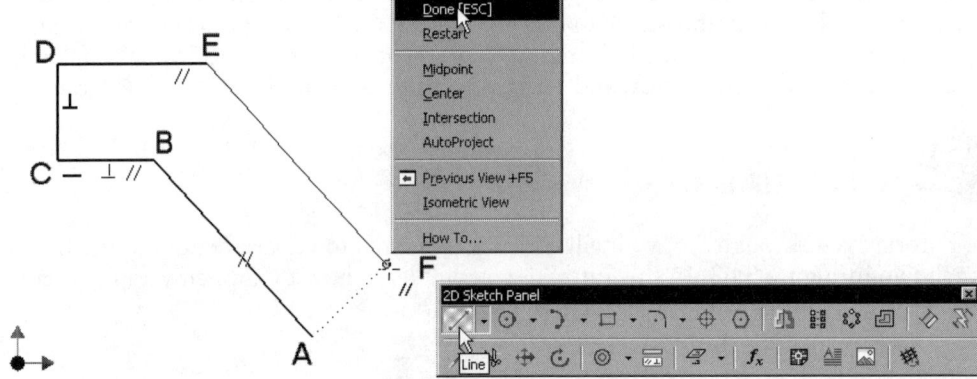

Figure 3.64 *Line segments constructed*

Now construct an arc.

11. Select Three point arc on the 2D Sketch Panel Bar or toolbar.

12. Select point A and then point B in Figure 3.65. While selecting points A and B, you should see a coincident symbol at the cursor, denoting that the endpoints of the arc are coincident with the endpoints of the selected line segments.

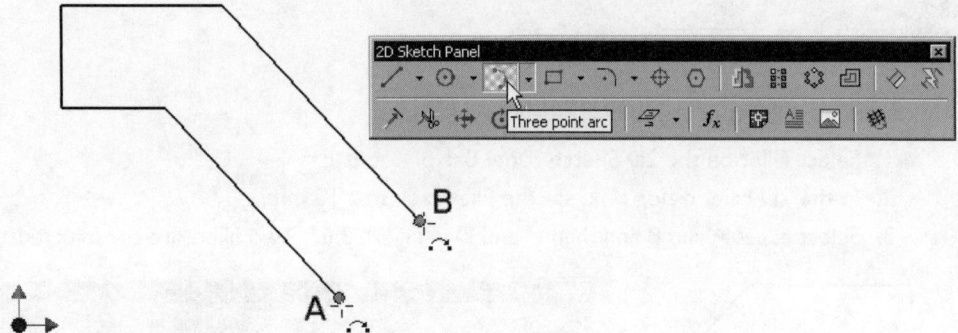

Figure 3.65 *Coincident constraints incorporated while selecting the endpoints of the arc*

13. Move the cursor slowly to location A in Figure 3.66 until you see two tangent constraint symbols displayed at the endpoints of the arc. If you select a point while the tangent constraint symbols are still displayed, you construct a tangent arc. If there is only one tangent constraint symbol or no tangent constraint symbol at all, you can assign the tangent constraint manually by selecting Tangent on the 2D Sketch Panel Bar or toolbar and then selecting the arc and line.

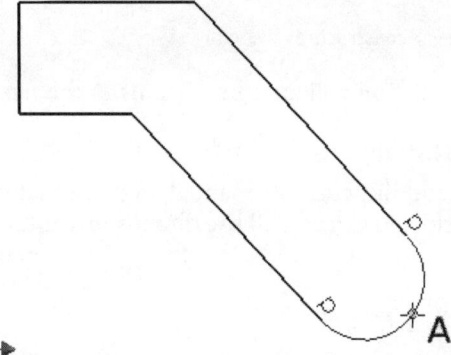

Figure 3.66 *Tangential arc being constructed*

Tip: Use the Show Constraint command to find out the kind of geometric constraints applied to a sketch. Delete those geometric constraints that you do not want and add those geometric constraints that are missing.

Do not close the file. You will construct 2D fillets in the next tutorial.

CONSTRUCTING A 2D FILLET

A 2D fillet is an arc constructed at the intersection of two sketch elements. After you construct a fillet, two tangent constraints governing the endpoints of the arc in relation to the selected sketch elements and a dimension governing the radius of the arc are placed automatically.

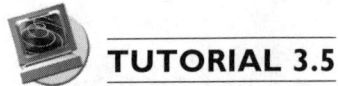

TUTORIAL 3.5

In this tutorial, you will construct 2D fillets.

1. Select Fillet on the 2D Sketch Panel Bar or toolbar.
2. In the 2D Fillet dialog box, set the fillet radius to 10 mm.
3. Select edges A and B and then C and D in Figure 3.67. Two fillets are constructed.

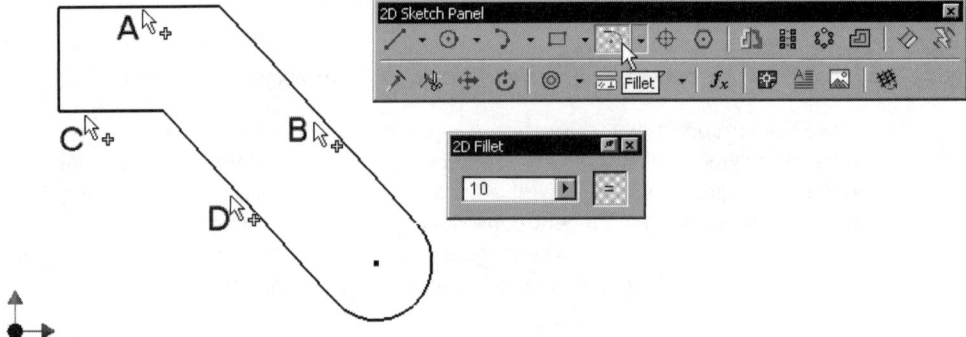

Figure 3.67 *2D fillets being constructed*

2D fillets are constructed. You will add dimensions in the next tutorial.

AUTOMATIC DIMENSIONING

There are two ways to add dimensions: manual and automatic. You can add dimensions one by one manually, and you can add all the dimensions automatically in one single step.

TUTORIAL 3.6

In this tutorial, you will add dimensions to the sketch in one single step.

1. Select Auto Dimension on the 2D Sketch Panel Bar or toolbar.
2. In the Auto Dimension dialog box, click the Apply button and then the Done button. (See Figure 3.68.)

 Tip: The advantage of using the Auto Dimension command is that you can save time by adding dimensions to a sketch in one step. However, you need to realize that the dimensions placed this way may not be fully congruent with your design intent.

Now delete two dimensions.

3. Select dimensions A and B in Figure 3.68 one by one and press DELETE. Two dimensions are deleted.

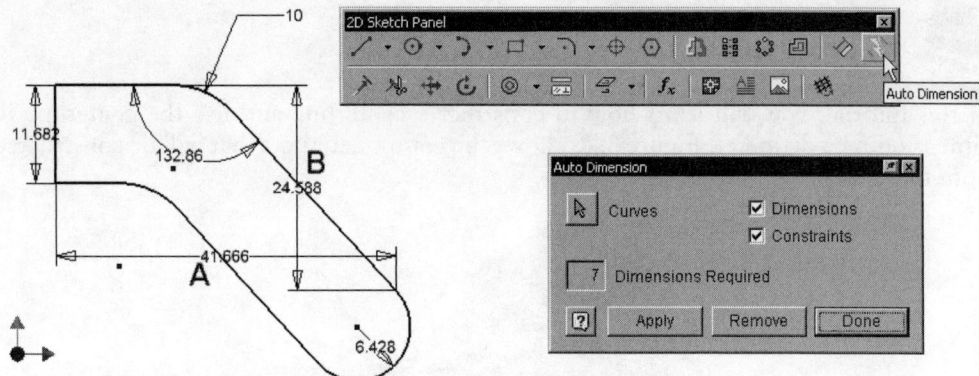

Figure 3.68 *Auto Dimension dialog box*

Now add two dimensions and modify three dimensions.

4. With reference to Figure 3.69, add two dimensions D (60 mm) and E (150 mm) and modify three dimensions A (30 mm), B (120 deg), and C (15 mm).

5. Select Return on the Inventor Standard toolbar to exit sketch mode.

The sketch is complete. Save and close your file (file name: *PunchHandle.ipt*). Remember to put this file in the folder (*Punch*).

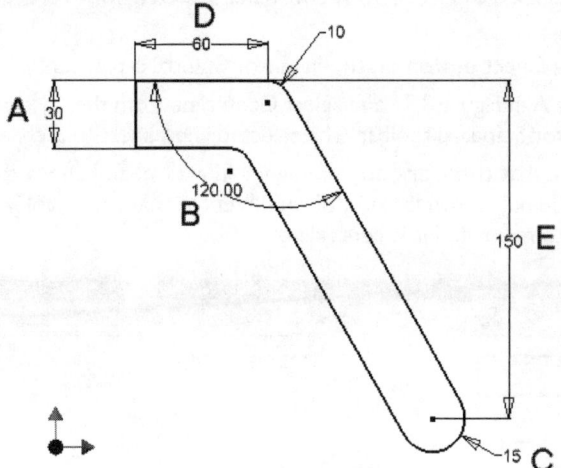

Figure 3.69 *Two dimensions deleted, three dimensions modified, and two dimensions added*

CONSTRUCTING A CENTERLINE AND PLACING DIAMETER DIMENSION

There are three kinds of linestyle: normal, construction, and centerline. Normal lines are sketch elements that will be used when you derive a solid feature from the sketch. Construction lines are used in establishing relationships between sketch elements. Centerlines specify axes and help establish diameter dimensions.

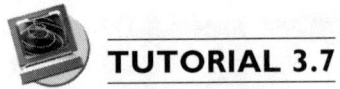

TUTORIAL 3.7

In this tutorial, you will learn how to construct a centerline and use the centerline in dimensioning a diameter. Figure 3.70 shows the component part that will be constructed from the sketch.

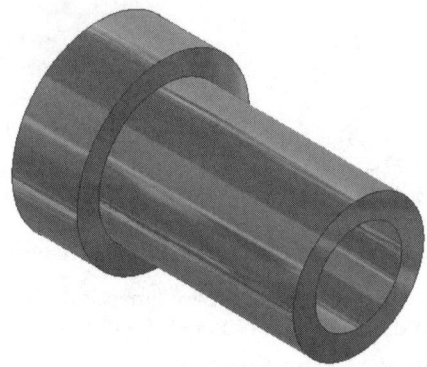

Figure 3.70 *Bush*

1. Start a new part file. Use the metric template.
2. With reference to Figure 3.71, construct a sketch with vertical and horizontal line segments.
3. Select the Select button on the Inventor Standard toolbar.
4. Select line A in Figure 3.71 and select Centerline from the Style pull-down list box on the Inventor Standard toolbar. The selected line's linestyle is changed to centerline.
5. Make sure that there are no commands in action, and check the linestyle specified in the Style pull-down list box of the Inventor Standard toolbar. Change the linestyle to normal, if it is centerline.

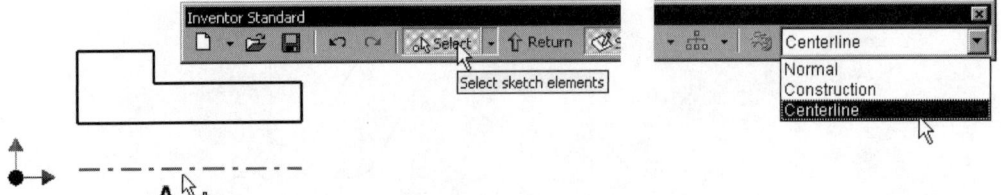

Figure 3.71 *Lines constructed and a line's linestyle changed*

Now add dimensions to the sketch.

6. Select General Dimension on the 2D Sketch Panel Bar or toolbar.
7. Select lines A and B in Figure 3.72.
8. Select location C in Figure 3.72.
9. In the Edit Dimension dialog box, change the dimension value to 40. A diameter dimension is constructed.

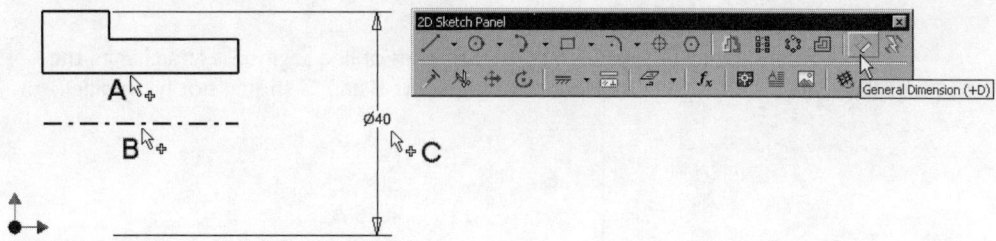

Figure 3.72 *Diameter dimension placed*

10. With reference to Figure 3.73, add dimensions to complete the sketch.

11. Select Return on the Inventor Standard toolbar.

The sketch is complete. Save and close your file (file name: *PunchBush.ipt*, folder name: *Punch*).

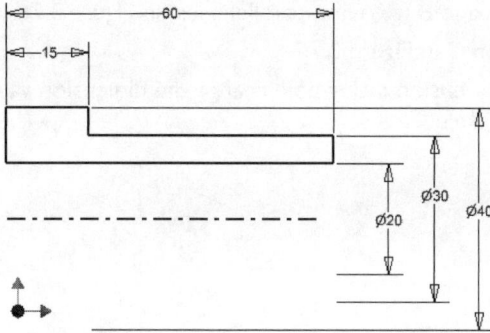

Figure 3.73 *Dimensions completed*

PLACING ANGULAR DIMENSIONS

The same command used for adding linear or radial dimensions can be used to construct angular dimensions. To place an angular dimension, you select two non-parallel lines and specify a location.

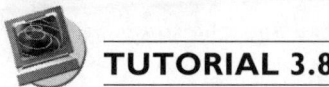

TUTORIAL 3.8

In this tutorial, you will learn how to place an angular dimension by working on the sketch of a feature of a component shown in Figure 3.74.

Figure 3.74 *Index bush*

1. Start a new part file. Use the metric template.
2. With reference to Figure 3.75, construct a set of line segments and change the linestyle of line A to centerline. Line segments B and C should not be parallel.

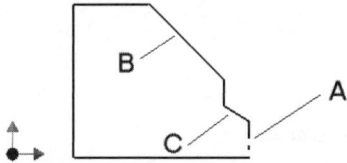

Figure 3.75 *Lines constructed and a line's linestyle changed*

Now construct an angular dimension.

3. Select General Dimension on the 2D Sketch Panel Bar or toolbar.
4. Select lines A and B (two non-parallel lines) in Figure 3.76.
5. Select location C in Figure 3.76.
6. In the Edit Dimension dialog box, change the dimension value to 45. An angular dimension is placed.

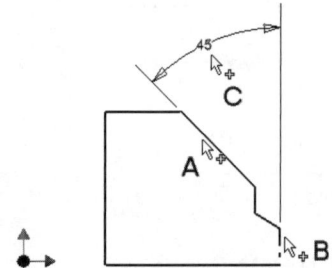

Figure 3.76 *Angular dimension being constructed*

7. With reference to Figure 3.77, add the remaining dimensions.
8. Select Return on the Inventor Standard toolbar.

The sketch is complete. Save and close your file (file name: *PunchIndexBush.ipt*, folder name: *Punch*).

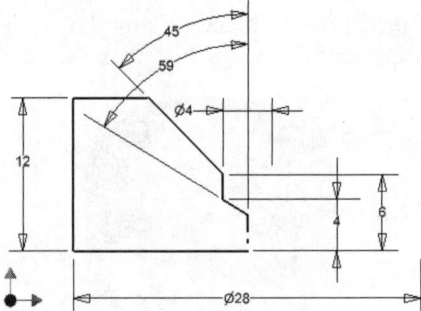

Figure 3.77 *Dimensions completed*

TRIMMING

The unwanted portion of a sketch element can be trimmed or removed.

TUTORIAL 3.9

In this tutorial, you will learn how to trim sketch elements by working on the sketch for a feature of the component shown in Figure 3.78.

Figure 3.78 *Hinge*

1. Start a new part file. Use the metric template.
2. With reference to Figure 3.79, construct a circle and two horizontal lines.

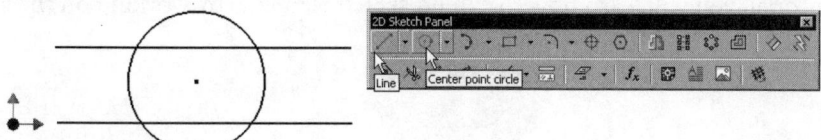

Figure 3.79 *Circle and lines constructed*

Now trim the circle and the lines.

3. Select Trim on the 2D Sketch Panel Bar or toolbar.
4. Select A, B, C, D, E, and F (portions of the line and circle to be trimmed) in Figure 3.80.

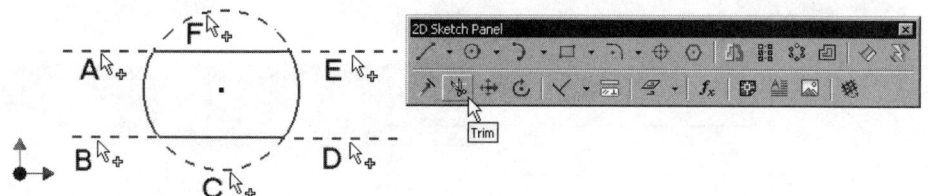

Figure 3.80 *Circle and lines being trimmed*

Now add geometric constraints and dimensions to the sketch.

5. Select Equal on the 2D Sketch Panel Bar or toolbar.

6. Select A and B (Figure 3.81) and then C and D (Figure 3.81) to apply equal constraints to the length of the lines and the radius of the arcs.

7. With reference to Figure 3.81, add two dimensions.

8. Select Return on the Inventor Standard toolbar.

The sketch is complete. Save and close your file (file name: *PunchHinge.ipt*, folder name: *Punch*).

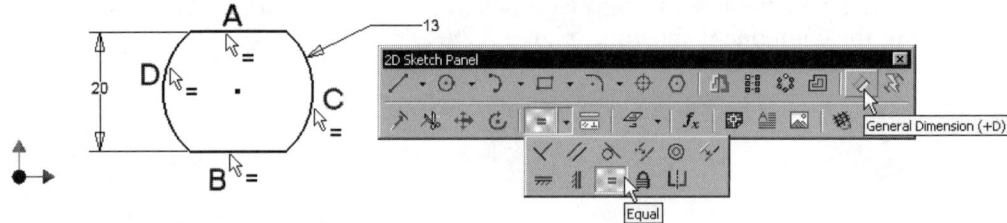

Figure 3.81 *Dimensions added*

EXTENDING

In contrast to trimming a sketch element, you can extend a sketch element.

TUTORIAL 3.10

In this tutorial, you will learn how to extend sketch elements by working on the sketch of a feature of a component shown in Figure 3.82.

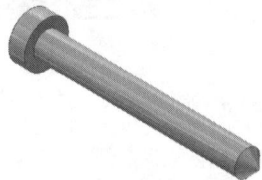

Figure 3.82 *Punch pin*

1. Start a new part file. Use the metric template.

2. With reference to Figure 3.83, construct five line segments and an arc and change line A's linestyle to centerline.

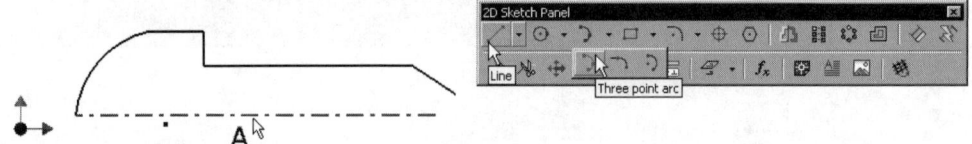

Figure 3.83 *Lines constructed and a line's linestyle changed*

Now extend two line segments and add a coincident constraint.

3. Select Extend on the 2D Sketch Panel Bar or toolbar.

4. Select A and B in Figure 3.84. The lines are extended.

5. Select Coincident on the 2D Sketch Panel Bar or toolbar.

6. Select center point of the arc C and line D in Figure 3.84. The arc's center is coincident with the centerline.

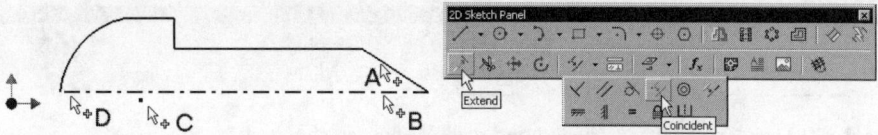

Figure 3.84 *Lines being extended and center point being repositioned*

7. With reference to Figure 3.85, add dimensions to constrain the sketch.

8. Select Return on the Inventor Standard toolbar.

The sketch is complete. Save and close your file (file name: *PunchPin.ipt*, folder name: *Punch*).

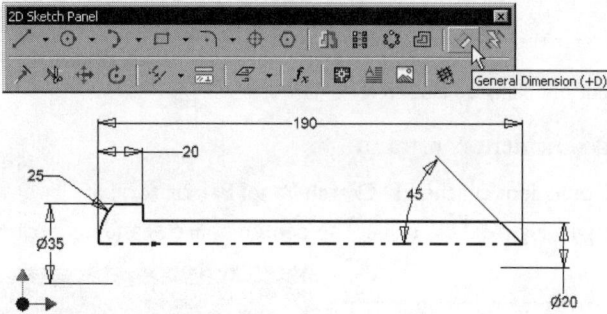

Figure 3.85 *Dimensions added*

USING CONSTRUCTION LINES

As we have mentioned, there are three kinds of linestyle: normal, construction, and centerline. Construction lines help you establish relationship among sketch elements and can be used as mirror lines.

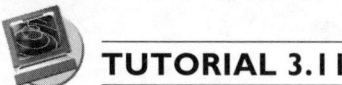

TUTORIAL 3.11

In this tutorial, you will learn how to use a construction line to aid dimensioning of a sketch by working on a sketch for the feature of a component shown in Figure 3.86.

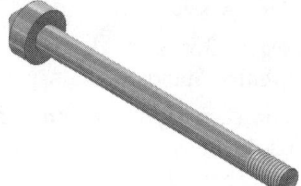

Figure 3.86 *Index pin*

1. Start a new part file. Use the metric template.
2. With reference to Figure 3.87, construct six line segments and change a line's linestyle to centerline.

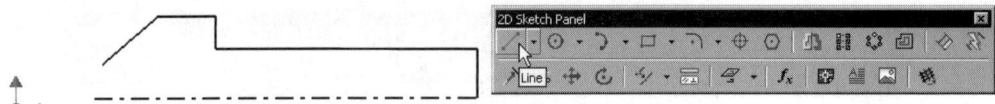

Figure 3.87 *Line segments constructed and a line's linestyle changed*

Now construct a tangent arc.

3. Select Tangent Arc on the 2D Sketch Panel Bar or toolbar.
4. Select endpoint A and then endpoint B in Figure 3.88.

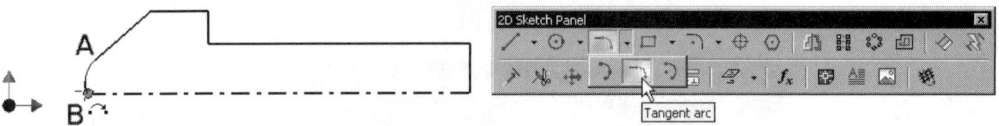

Figure 3.88 *Tangent arc being constructed*

Now add a coincident constraint.

5. Select Coincident on the 2D Sketch Panel Bar or toolbar.
6. Referring to Figure 3.89, select the center of arc A and centerline B.

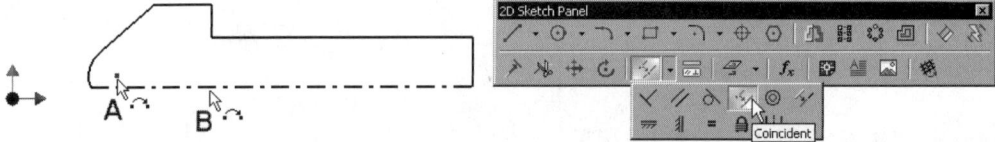

Figure 3.89 *Arc's center being constrained*

Now construct two construction lines and add collinear constraints to the sketch.

7. Referring to Figure 3.90, construct two line segments A and B.
8. Select Collinear on the 2D Sketch Panel Bar or toolbar.
9. Select lines A and C in Figure 3.90.

 Note: Depending on how you constructed these two lines, they may already be collinear, in which case you will see a message telling you that adding the constraint will overconstrain the sketch.)

10. Select lines B and D in Figure 3.90.
11. Select lines A and B in Figure 3.90 and select Construction from the style pull-down list box on the Inventor Standard toolbar.
12. Make sure that there is no command in action and check the current linestyle specified in the Style pull-down list box of the Inventor Standard toolbar. Change linestyle to normal, if it is Construction.

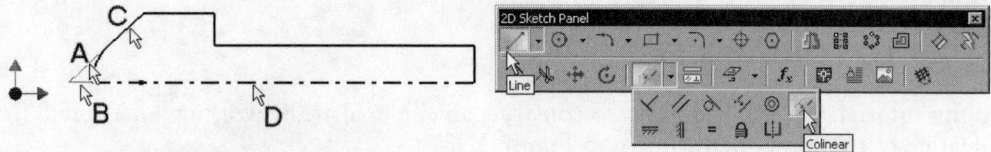

Figure 3.90 *Line segments constructed, collinear constraints applied, and linestyle changed*

Now dimension the sketch.

13. Select General Dimension on the 2D Sketch Panel Bar or toolbar.

14. Select endpoints A and B in Figure 3.91 and then location C to construct a dimension.

Tip: Construction lines help establish relationships between sketch elements.

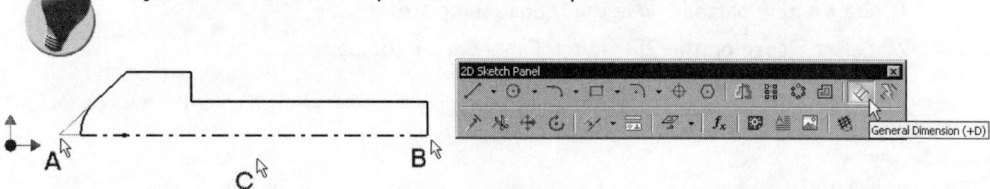

Figure 3.91 *Dimensions referenced to the construction lines*

15. With reference to Figure 3.92, add the other dimensions.

16. Select Return on the Inventor Standard toolbar.

The sketch is complete. Save and close your file (file name: *PunchIndex.ipt*, folder name: *Punch*).

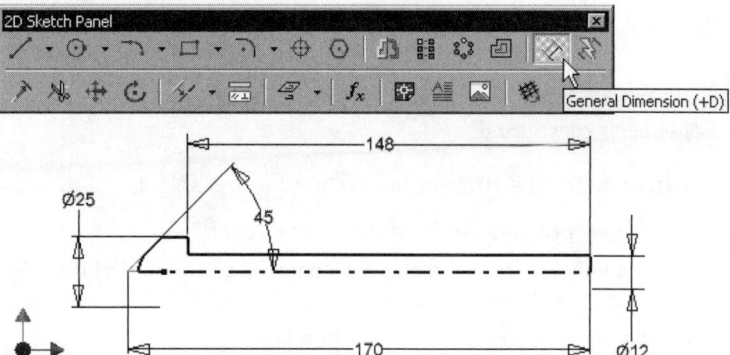

Figure 3.92 *Dimensions completed*

ELLIPSE CONSTRUCTION

You can construct an ellipse by specifying the center point and the locations of the minor and major axes. To construct an elliptical arc, you trim an ellipse. To dimension an ellipse, you specify its major and minor axes.

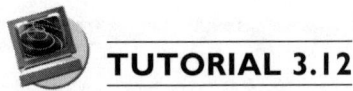

TUTORIAL 3.12

In this tutorial, you will learn how to construct an elliptical arc by working on a sketch for a feature of the component shown in Figure 3.93.

Figure 3.93 *Hand knob*

1. Start a new part file. Use the metric template.
2. Select Ellipse on the 2D Sketch Panel Bar or toolbar.

 Note: The Ellipse tool is on the same flyout on the toolbar as the Center Point Circle tool.

3. With reference to Figure 3.94, select point A to specify the ellipse's center, select point B to specify the direction and location of the first axis, and select point C to specify the location of the second axis. To learn how to orient an ellipse's axes direction, make sure that the axis along A and B is not horizontal.

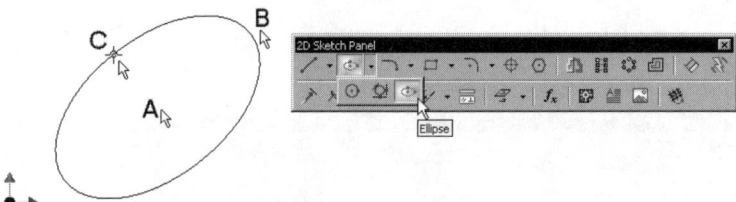

Figure 3.94 *Ellipse being constructed*

Now construct two construction lines. (See Figure 3.95.)

4. Construct a vertical line AB from the ellipse's center A.
5. Construct a line AC from the ellipse's center A to the one of the ellipse's axis C, collinear with the ellipse's major axis.
6. Change the lines' linestyle to construction line.

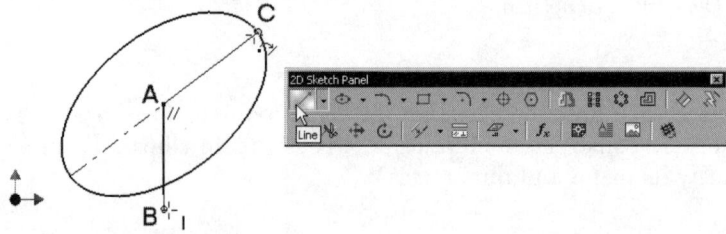

Figure 3.95 *Construction lines being constructed*

Now add dimensions to the sketch.

7. With reference to Figure 3.96, add a dimension.

8. Set the dimension value to 90 deg.

 Tip: The construction lines, one vertical and one collinear with an axis of the ellipse, help control the orientation of the ellipse.

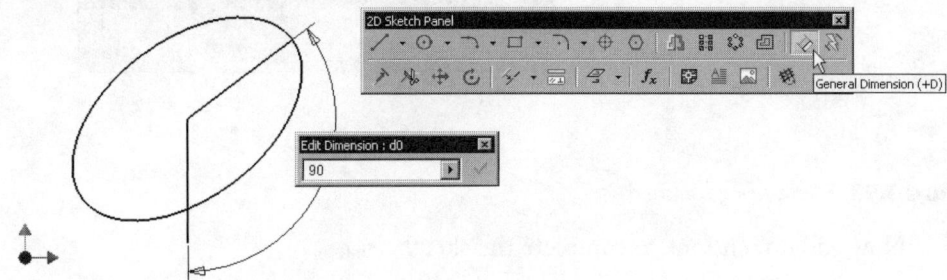

Figure 3.96 *Ellipse being placed*

Now re-position the center of the ellipse.

9. Select Project Geometry on the 2D Sketch Panel Bar or toolbar.

10. Select the center point A of the origin in Figure 3.97 on the Browser Bar to project the center point onto the sketch plane to obtain a reference point B.

11. Select Coincident on the 2D Sketch Panel Bar or toolbar.

12. Select the ellipse's center C and the projected center point B in Figure 3.97. (The location of C in your screen might be different.)

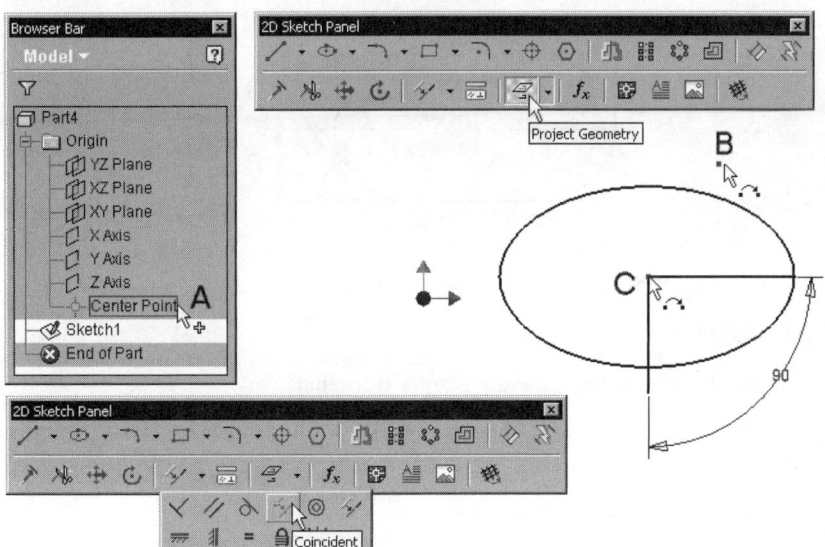

Figure 3.97 *Center point of the origin projected and ellipse being re-oriented*

Now trim the ellipse.

13. With reference to Figure 3.98, construct a horizontal line with its endpoints coincident with the circumference of the ellipse.

14. Trim portions A and B of the ellipse.

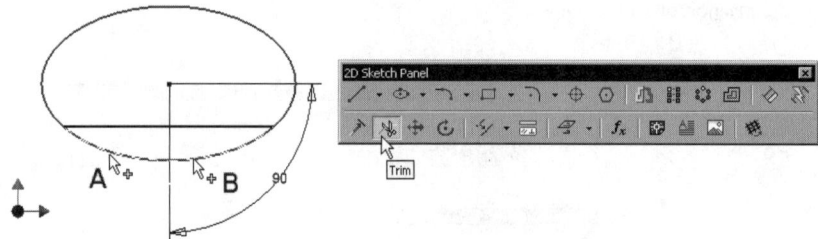

Figure 3.98 *Ellipse being trimmed*

Now add dimensions to complete the sketch.

15. Select General Dimension on the 2D Sketch Panel Bar or toolbar.

16. Select ellipse A and then location B in Figure 3.99. The ellipse's half axis distance is dimensioned.

17. Select ellipse A and then location C in Figure 3.99 to dimension the other half axis.

18. Select ellipse A, line D, and then location E. The distance from the line to the center of the ellipse is dimensioned.

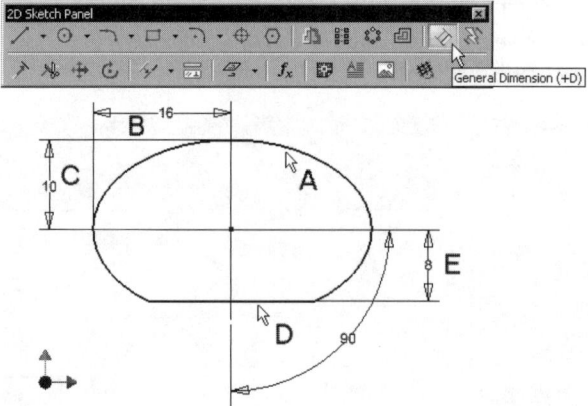

Figure 3.99 *Ellipse dimensioned*

19. Select Return on the Inventor Standard toolbar.

The sketch is complete. Save and close your file (file name: *PunchIndexKnob.ipt*, folder name: *Punch*).

SPLINE CONSTRUCTION

A spline is a free-form curve defined by specifying two endpoints and one or more intermediate points. The endpoints of the spline are called fit points; they are positionally constrained. The intermediate points along the spline are called shape points; they are unconstrained. In addition to manipulating the spline's fit and shape points, you can modify a spline by changing the spline's shape and curvature, changing the fit method, setting the spline's tension, and closing the spline.

Fit Points and Shape Points

You can locate the positions of a spline's fit points and shape points by specifying horizontal and vertical dimensions between them or selecting and dragging them. To increase or decrease the complexity of the spline, you insert or delete shape points.

Handle Bar and Curvature Bar

Without changing the positions of the fit and shape points, you can change a spline's shape by manipulating the handle bar and the curvature bar at the fit and shape points.

Fit Method

You fit the spline in one of three ways: smooth, sweet, and AutoCAD. A smooth fit constructs a spline with smooth continuity between fit and shape points. A sweet fit constructs a spline with smooth continuity and better curvature distribution. An AutoCAD fit uses AutoCAD's fit method.

Spline Tension

You can set the tightness of a spline against the fit and shape points. A low tension allows the spline to balloon out and a high tension tightens the spline between the points.

Closing a Spline

Closing a spline means inserting a spline segment between the start and end fit points to form a closed loop.

Curvature Display

To find out how the curve is fitted along the fit and shape points, you can display the curvature graph.

TUTORIAL 3.13

In Tutorials 3.13 and 3.14, you will construct a sketch for making the column shown in Figure 3.100. In this tutorial, you will learn how to construct and modify a spline.

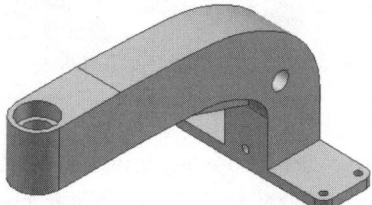

Figure 3.100 *Column*

1. Start a new part file. Use the metric template.
2. Select Spline on the 2D Sketch Panel Bar or toolbar.

Note: The Spline tool is on the same flyout on the toolbar as the Line tool.

3. Select three points A, B, and C in Figure 3.101 to specify two fit points and one shape point.
4. Right-click and select Continue. A spline is constructed.

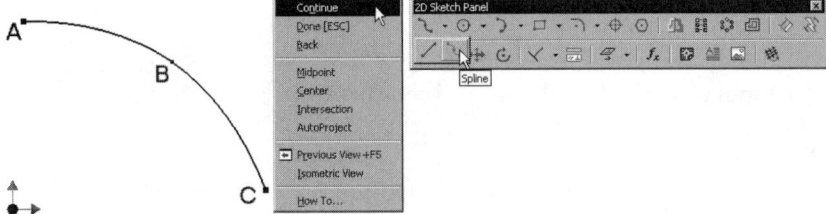

Figure 3.101 *A spline with two segments (two fit points and one shape point) constructed*

A spline can be closed. Now close the spline.

5. Select the spline, right-click, and select Close Loop. The spline is closed. (See Figure 3.102.)

Because we need an open-loop spline here, undo the last command.

6. Select Edit > Undo. The last command is undone.

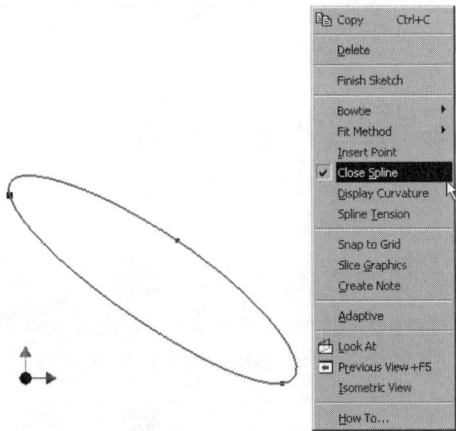

Figure 3.102 *Open-loop spline converted to a closed-loop spline*

Now add two shape points.

7. Select the spline, right-click, and select Insert Point. (See Figure 3.103.)
8. Select a point A along the spline. A shape point is inserted.
9. Repeat steps 7 and 8 to insert another shape at point B on the spline.

In contrast to fit points, you can select a shape point, right click, and select Delete to remove it.

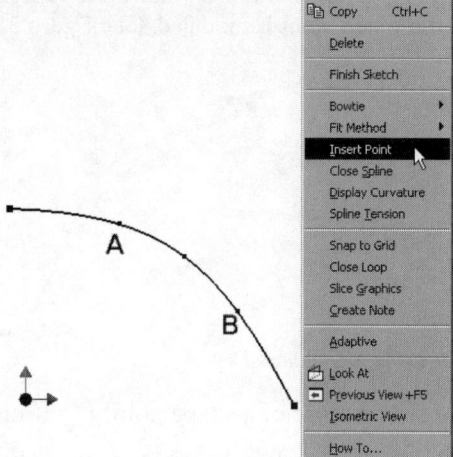

Figure 3.103 *Closing undone and two shape points inserted*

Now learn how to modify the shape of a spline by manipulating the handle bar at a shape point.

10. Select the spline, right-click, select Bowtie, and select Handle. You will find a handle bar displayed at a shape point near to the location where you select the spline.

11. Select and drag a grip point of the handle bar to discover how the location of the endpoint of the handle bar affects the shape of the spline. (See Figure 3.104.)

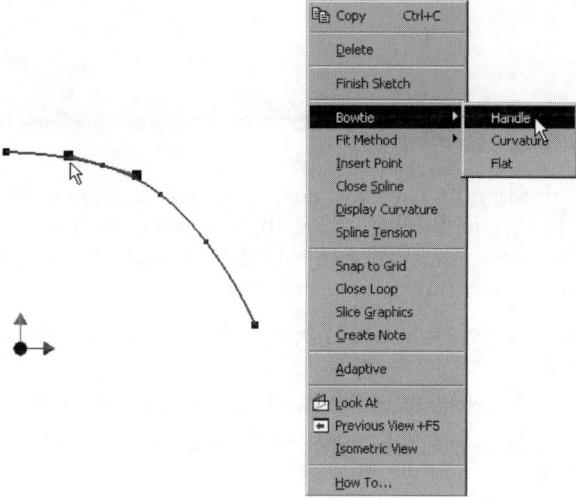

Figure 3.104 *Handle bar being manipulated*

Now display and manipulate the curvature bar at a shape point.

12. Select the spline, right-click, select Bowtie, and select Curvature. You will find a curvature bar displayed at the selected shape point.

13. Select and drag the grip point of the curvature bar to discover how the curvature of the spline at the shape point is modified. (See Figure 3.105.)

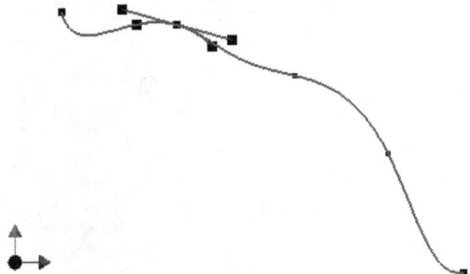

Figure 3.105 *Curvature bar being manipulated*

Now set the radius of curvature of a shape point to infinity.

14. Select the spline, right-click, select Bowtie, and select Flat. A flat curvature is a curvature with an infinite radius.

15. Select and drag the grip point of the handle bar to discover how a flat curvature affects the shape of the spline. (See Figure 3.106.)

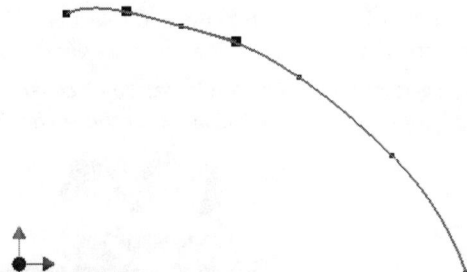

Figure 3.106 *Flat curvature specified*

Now disable the handle bars and curvature bars and then discover the radius of curvature of the spline by displaying the spline's curvature comb, and appreciate the three kinds of spline fitting methods. Because the spline here differs from yours, the curvature comb will be different.

16. Select the shape points one by one, right-click, select Bowtie, and clear the Handle box.

17. Select the spline, right-click, and check Display Curvature. (See Figure 3.107.)

18. Select the spline, right-click, select Fit Method, and then select AutoCAD. (See Figure 3.108.)

19. Select the spline, right-click, select Fit Method, and then select Sweet. (See Figure 3.109.)

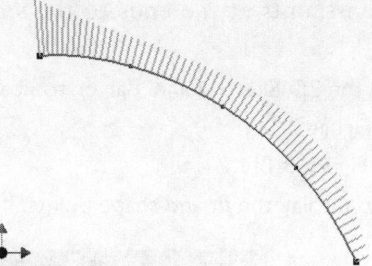

Figure 3.107 *Curvature comb displayed*

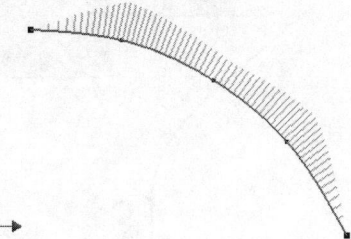

Figure 3.108 *AutoCAD fit method*

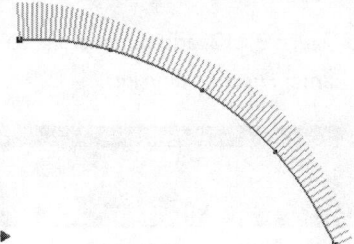

Figure 3.109 *Sweet fit method*

Now turn off the curvature comb and construct two line segments.

20. Select the spline, right-click, and clear Display Curvature.

21. With reference to Figure 3.110, construct a horizontal line with one of its endpoints coincident with the spline's endpoint A and a vertical line with one of its endpoints coincident with the spline's endpoint B.

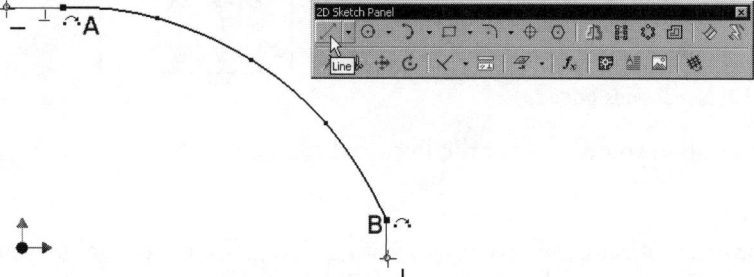

Figure 3.110 *Vertical and horizontal lines at the endpoints of the spline constructed*

Now add tangent constraints at the ends of the spline and add dimensions to complete the sketch.

22. Select Tangent on the 2D Sketch Panel Bar or toolbar.
23. Select A and B in Figure 3.111.
24. Select A and C in Figure 3.111.
25. Select the spline to display the fit and shape points, if they are not displayed.

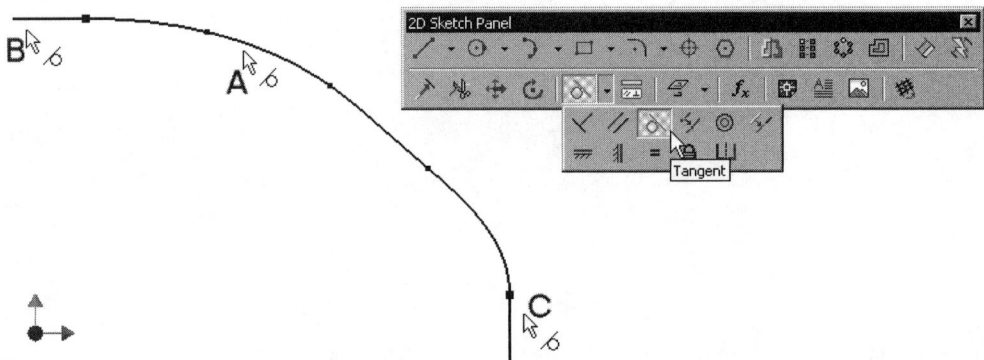

Figure 3.111 *Tangent constraint specified*

26. With reference to Figure 3.112, add dimensions to complete the sketch.
27. Save your file (file name: *PunchColumn.ipt*).

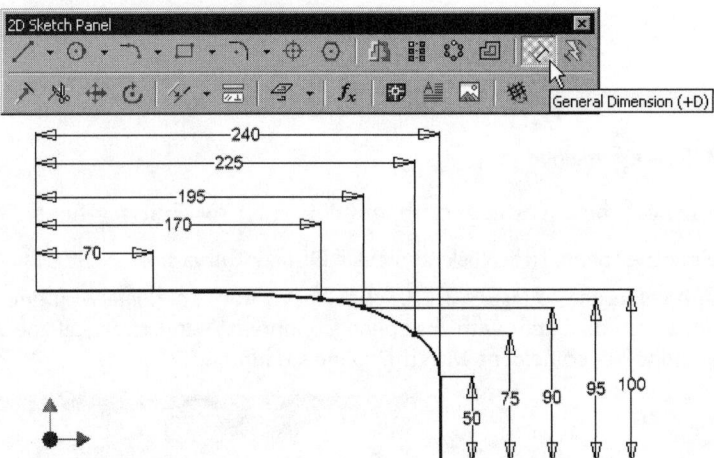

Figure 3.112 *Dimensions added*

In the next tutorial, you will offset the lines and spline.

OFFSETTING

You can construct offset geometry from existing lines, arcs, circles, splines, and ellipses.

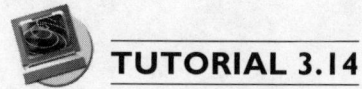

TUTORIAL 3.14

In this tutorial, you will continue to work on the file (*PunchColumn.ipt*).

1. Double-click the sketch on the Browser Bar or select the sketch on the Browser Bar, right-click, and select Edit Sketch to re-activate sketching mode.
2. Select Offset on the 2D Sketch Panel Bar or toolbar.
3. Select the spline at A in Figure 3.113 and then select a point near B in Figure 3.113. An offset curve is constructed.

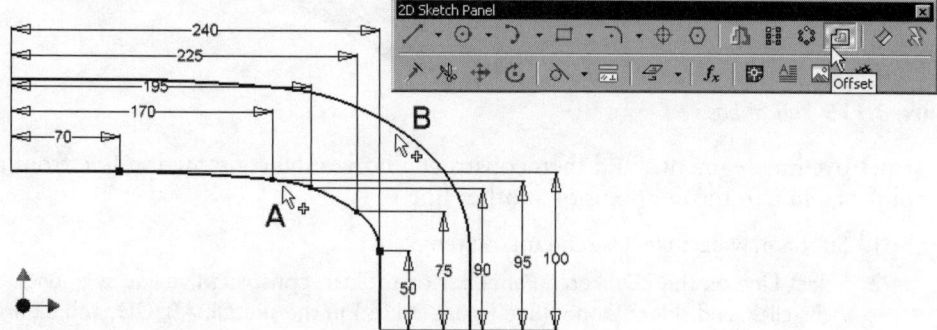

Figure 3.113 *Offset curve being constructed from the contiguous spline and lines*

4. Construct two line segments AB and CD and add a dimension (60 mm) at E. (See Figure 3.114.)

The sketch is complete. Save and close your file.

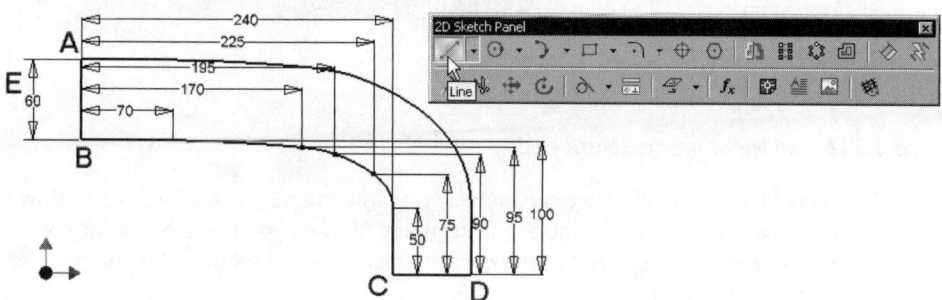

Figure 3.114 *Lines constructed and a dimension added*

CONSTRAINING TO MIDPOINT

Lines and arcs have selectable midpoint. In addition to endpoints, you can apply constraints to their midpoint.

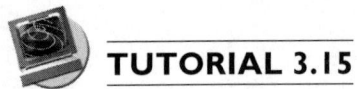

TUTORIAL 3.15

In Tutorials 3.15 through 3.18, you will construct a sketch for making a feature of the component shown in Figure 3.115. In this tutorial, you will learn how to constrain to the midpoint of a line.

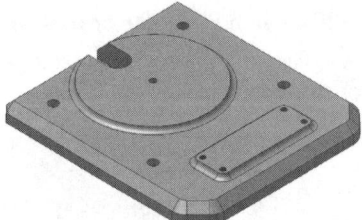

Figure 3.115 *Punch base*

Construct five line segments, and then construct a horizontal construction line from the midpoint of a line to the endpoint of another line.

1. Start a new part file. Use the metric template.

2. Select Line on the 2D Sketch Panel Bar or toolbar, construct five line segments, right-click, and select Done. (See Figure 3.116.) In the sketch, AB, CD, and EF are vertical lines, and BC and DE are horizontal lines.

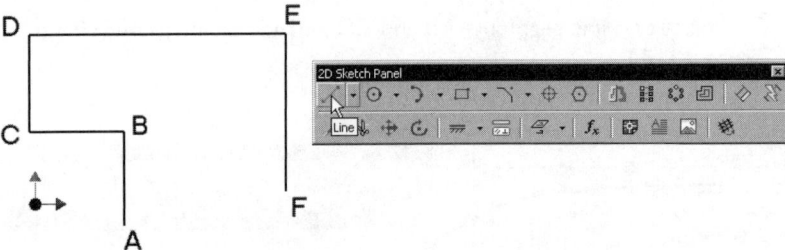

Figure 3.116 *Five line segments constructed*

3. Select Line on the 2D Sketch Panel Bar or toolbar again, move the cursor slowly to a point near the midpoint of line segment AB until you find a midpoint constraint symbol C displayed at the cursor, select the midpoint, and then endpoint D. (See Figure 3.117.)

Figure 3.117 *Lines constraining to the midpoints of AB and CD constructed*

4. Apply a horizontal constraint to line A in Figure 3.118.

5. Change the linestyle of line A in Figure 3.118 to construction line.

Figure 3.118 *Line changed to a construction line and horizontal constraint applied*

Do not close the file. You will construct a 2D chamfer in the next tutorial.

CONSTRUCTING A 2D CHAMFER

A chamfer is a bevel-shaped line segment at the intersection of two sketch elements. There are three ways to specify a 2D chamfer: equal distance, unequal distance, and distance and angle. In an equal distance chamfer, you specify a distance from the edge. In an unequal distance chamfer, you specify the two distances from the edge. In a distance and angle chamfer, you specify a chamfer distance and a chamfer angle.

TUTORIAL 3.16

In this tutorial, you will construct a 2D chamfer.

1. Select Chamfer on the 2D Sketch Panel Bar or toolbar.

Note: The Chamfer tool is located in the same flyout on the toolbar as the Fillet tool.

2. In the 2D Chamfer dialog box, set the equal distance to 10 mm.

3. Select edges A and B in Figure 3.119.

4. Click the Done button in the 2D Chamfer dialog box.

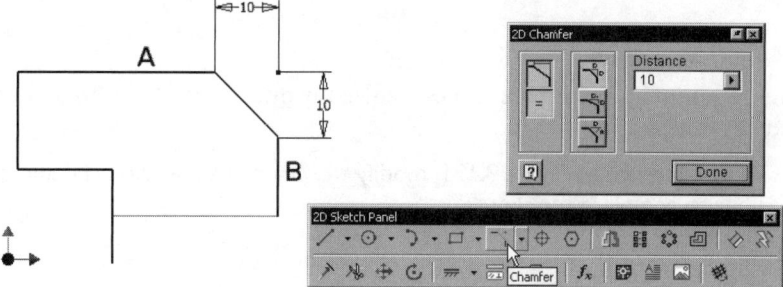

Figure 3.119 *2D chamfer constructed*

Do not close the file. You will construct a set of mirrored sketch elements in the next tutorial.

MIRRORING GEOMETRY

To construct a sketch with elements that are symmetrical about a mirror line, you select the elements and select a construction line or a centerline as the mirror line. Symmetric constraints will be applied automatically to relevant sketch elements in the process of mirroring.

 ## TUTORIAL 3.17

In this tutorial, you will construct a set of mirrored sketch elements

1. Select Mirror on the 2D Sketch Panel Bar or toolbar.
2. In the Mirror dialog box, click the Select button, if it is not already selected.
3. Select A, B, C, D, and E in Figure 3.120.
4. In the Mirror dialog box, click the Mirror Line button.
5. Select F (a construction line) in Figure 3.120.
6. In the Mirror dialog box, click the Apply button and then the Done button.

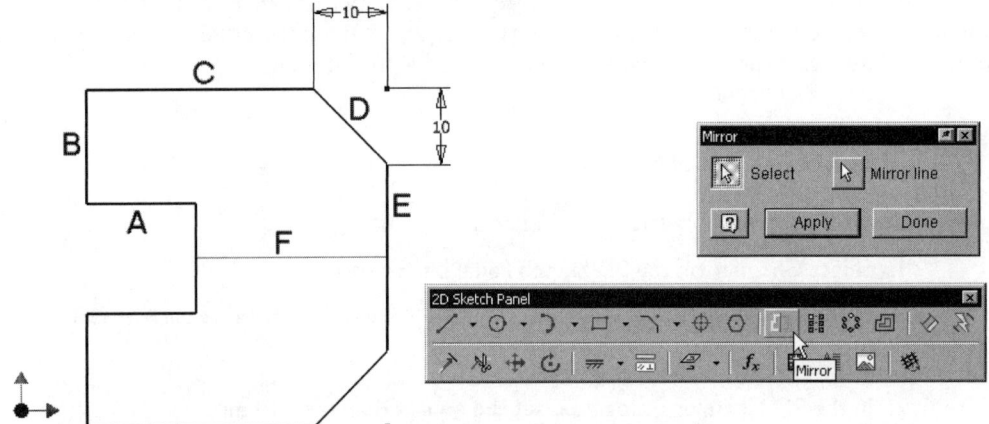

Figure 3.120 *Sketch elements being mirrored*

 Tip: Both construction lines and centerlines can be used as mirror line.

Line segments are mirrored. You will continue to work on this sketch in the next tutorial.

7. With reference to Figure 3.121, modify the chamfer dimension and add dimensions to complete the sketch.

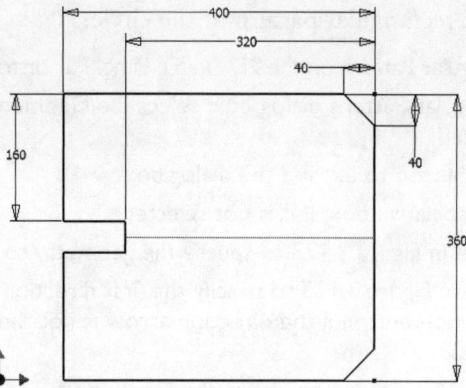

Figure 3.121 *Sketch dimensioned*

In the next tutorial, you will construct a rectangular pattern.

RECTANGULAR PATTERN

A rectangular pattern repeats selected sketch elements in multiple columns and multiple rows at specified distances. Each occurrence of the objects in the pattern is called an instance. You can set the instances to be associative to the original geometry and you can suppress individual instances. In an associative pattern, if the source geometry changes, the instances will change accordingly. A suppressed instance's linestyle is changed to construction line and the suppressed instance is disregarded when you construct a sketched solid feature from the sketch.

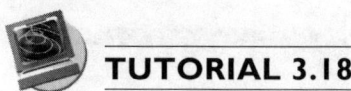

TUTORIAL 3.18

In this tutorial, you will construct a rectangular pattern.

1. With reference to Figure 3.122, construct a circle and place three dimensions (18 mm, 50 mm, and 60 mm).

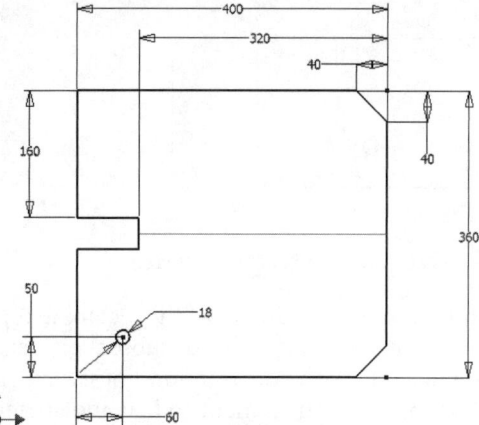

Figure 3.122 *A circle constructed and dimensioned*

Now construct a rectangular pattern of the circle.

2. Select Rectangular Pattern on the 2D Sketch Panel Bar or toolbar. (See Figure 3.123.)

3. In the Rectangular Pattern dialog box, select the Geometry button, if it is not already selected.

4. Click the >> button to expand the dialog box.

5. Select the Associative box, if it is not selected.

6. Select circle A in Figure 3.123 to specify the geometry to be repeated.

7. Select edge B in Figure 3.123 to specify the first direction of the pattern. Select the Flip Direction button if the direction arrow is not the same as that indicated in Figure 3.123.

8. In the Rectangular dialog box, select the Direction 1 direction button, set Count 1 to 2 and set Spacing to 200 mm.

9. Select edge C in Figure 3.107 to specify the second direction of the pattern. Select the Flip Direction button if the direction arrow is not the same as that indicated in Figure 3.123.

10. In the Rectangular dialog box, select the Direction 2 direction button and set Count 2 to 2 and set Spacing to 260 mm.

Now suppress an instance of the pattern.

11. Click the Suppress button in the Rectangular Pattern dialog box and select instance D in Figure 3.123 to suppress it.

12. Click the OK button.

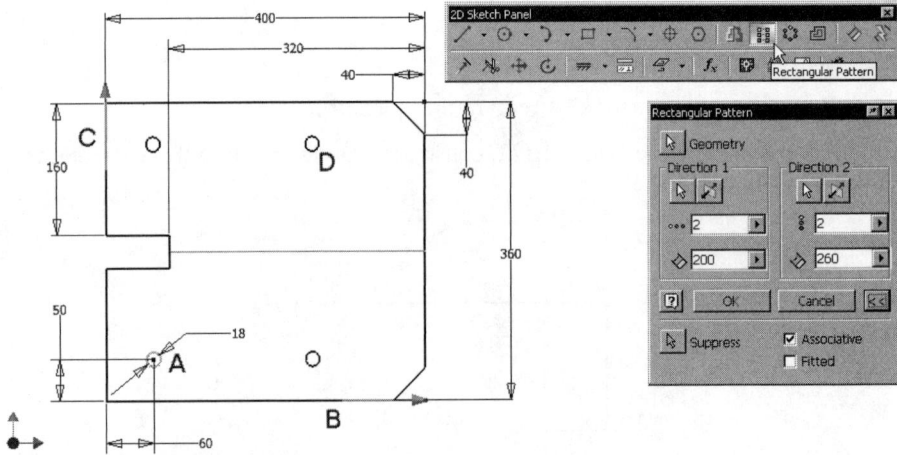

Figure 3.123 *Rectangular sketch pattern being constructed*

The linestyle of the suppressed instance of the rectangular pattern sketch is changed to construction, and thus it is not active in subsequent feature construction operation. For example, a circle with construction linestyle is not extruded. To revert a suppressed instance of a sketch element to become an active sketch element, you will change its linestyle back to normal.

13. Select the suppressed instance D (Figure 3.123) and select Normal from the Style pull-down box on the Inventor Standard toolbar.

14. Select Return on the Inventor Standard toolbar to exit sketch mode.

A rectangular sketch pattern is constructed. (See Figure 3.124.) Save and close your file (file name: *PunchBase.ipt*, folder name: *Punch*).

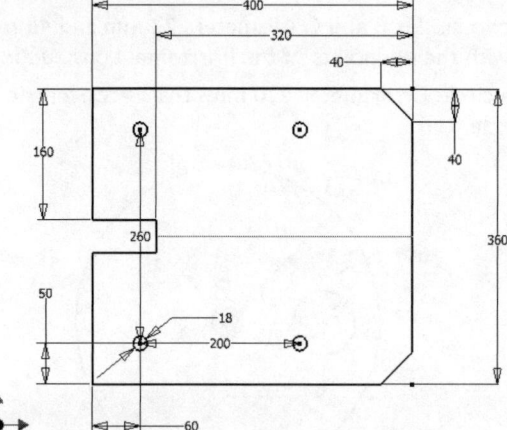

Figure 3.124 *Rectangular sketch pattern constructed*

CIRCULAR PATTERN

A circular pattern repeats selected sketch elements in a polar array about a selected center point or axis. Like a rectangular pattern, instances of a circular pattern can be associative, and individual instances can be suppressed.

 **TUTORIAL 3.19**

In this tutorial, you will construct a sketch for making a feature of the component shown in Figure 3.125. In the sketch, you will construct a circular pattern.

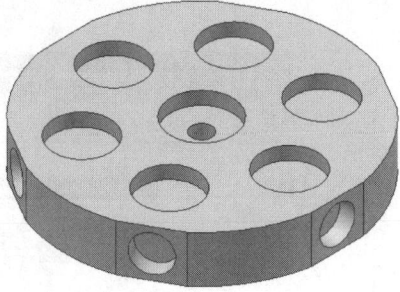

Figure 3.125 *Punch table*

1. Start a new part file. Use the metric template.

2. With reference to Figure 3.126, construct a horizontal line A and add a dimension (70 mm).

3. Change the linestyle of the horizontal line to construction line.

Note: You need to change the current linestyle back to Normal after changing the horizontal line to a construction line.

4. Construct two circles B and C (diameters 22 mm and 45 mm) with their centers coincident with the endpoints of the horizontal construction line.

5. Construct a circle D (diameter 220 mm) that is concentric with the 22 mm diameter circle.

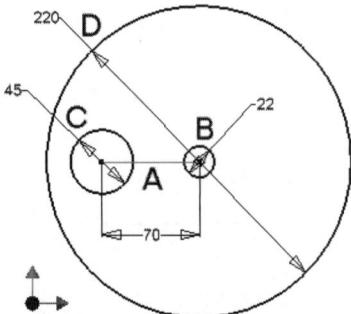

Figure 3.126 *Circles constructed and dimensions placed*

Now construct a circular pattern.

6. Select Circular Pattern on the 2D Sketch Panel Bar or toolbar.

7. Select center B in Figure 3.127 as the pattern axis.

8. Select circle A in Figure 3.127 as the geometry to repeat.

9. In the Circular Pattern dialog box, accept the default count (6) and default angle (360 deg) and click the OK button.

10. Select Return on the Inventor Standard toolbar to exit sketch mode.

The sketch is complete. Save and close your file (file name: *PunchTable.ipt*, folder name: *Punch*).

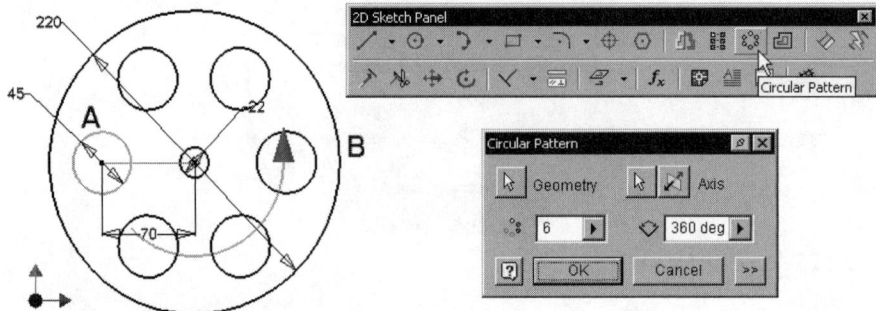

Figure 3.127 *Circular pattern being constructed*

PRECISE INPUT

If you prefer to enter precise lengths and coordinates of points while making a sketch, you can use precise input tools. Although a sketch constructed this way is precise, you should add dimensions to it. Otherwise, the solid feature derived from it will be unpredictable when edited.

TUTORIAL 3.20

In this tutorial, you will learn how to use precise input tools to construct a sketch for a feature of the component shown in Figure 3.128.

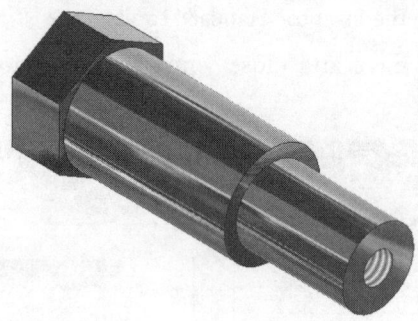

Figure 3.128 *Post*

1. Start a new part file. Use the metric template.
2. Select View > Toolbar > Inventor Precise Input to display the Inventor Precise Input toolbar.
3. Select Line on the 2D Sketch Panel Bar or toolbar.
4. Select a point on the screen. (The exact location is unimportant.)
5. In the Inventor Precise Input dialog box, check the Precise Relative and Precise Relative buttons.
6. Set delta X to 0 and delta Y to 10 and press ENTER. (See Figure 3.129.)

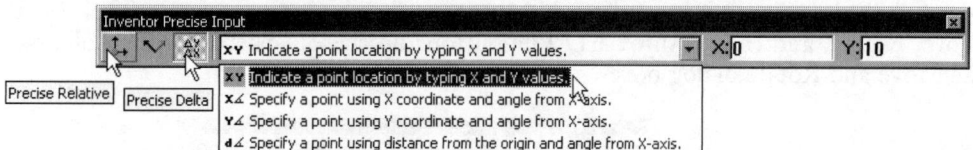

Figure 3.129 *A line segment constructed precisely*

7. Set delta X to 8 and delta Y to 0 and press ENTER.
8. Set delta X to 0 and delta Y to −2 and press ENTER.
9. Set delta X to 27 and delta Y to 0 and press ENTER.
10. Set delta X to 0 and delta Y to −2 and press ENTER.
11. Set delta X to 18 and delta Y to 0 and press ENTER.

12. Set delta X to 0 and delta Y to −6 and press ENTER.
13. Select the start point A in Figure 3.130.

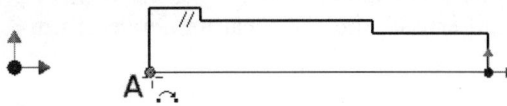

Figure 3.130 *Line segments constructed*

14. Change the linestyle of line A (in Figure 3.131) to centerline.
15. Select Auto Dimension on the 2D Sketch Panel Bar or toolbar.
16. Click the Apply and Done buttons.
17. Select Return on the Inventor Standard toolbar.

The sketch is complete. Save and close your file (file name: *PunchPost.ipt*, folder name: *Punch*).

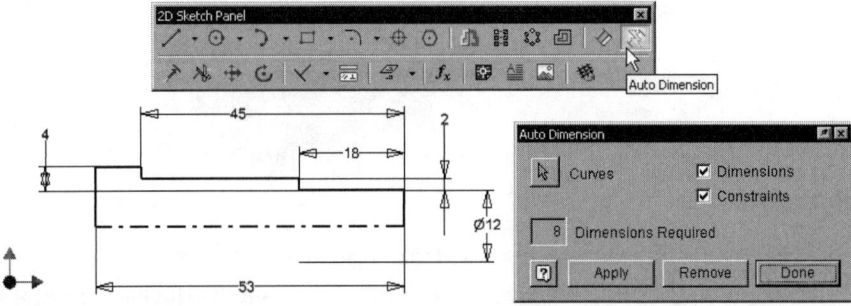

Figure 3.131 *Dimensions being placed*

MOVING, ROTATING, AND IMPORTING AN AUTOCAD FILE

You can move sketch elements by selecting the elements, specifying a base point and the destination point. You can also rotate sketch entities by selecting the entities, specifying the center point of rotation and the angle of rotation. Both Move and Rotate can be useful in conjunction with precise input. If you already constructed sketches in an AutoCAD file, you can import them to Inventor's sketch plane. Figure 3.132 shows the Move, Rotate, and Insert AutoCAD File buttons on the 2D Sketch Panel toolbar and the Move and Rotate dialog boxes.

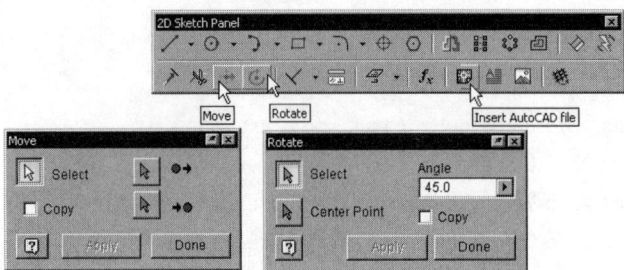

Figure 3.132 *Move, Rotate, and Insert buttons and the Move and Rotate dialog boxes*

SUMMARY ON SKETCHING

You make a sketch on a plane to depict your design intent. Initially, you should concentrate on the shape of the sketch. To properly confine the shape, you use geometric constraints that can be applied implicitly through inference while sketching or explicitly by adding geometric constraints after sketching. To alter the shape of the sketch, you can add or delete some of the constraints. After you are satisfied with the shape, you add dimensions. To position a sketch in relation to existing objects, you can use reference geometry projected from the origin objects or from edges of other features already constructed. After sketching, you select the Return button to exit sketching mode. If you want to edit the sketch again, double-click the sketch on the Browser Bar or select the sketch, right-click, and select Edit Sketch. Remember that the 2D Sketch button on the Inventor Standard toolbar is not for editing purposes.

FEATURES CONSTRUCTION TOOLS

After you select Return on the Inventor Standard toolbar, the 2D Sketch Panel Bar automatically changes to the Part Features Panel Bar. To construct sketched and other kinds of features, you use this Panel Bar or the Part Features toolbar. (See Figure 3.133.) To display the Part Features toolbar, select Customize from the Tools menu. On the Toolbars tab of the Customize dialog box, select Part Features from the toolbar list, click the Show button and then the Close button. The Part Features Panel Bar or toolbar has 31 button areas. Table 3.8 describes the choices. You will learn about some of the feature construction tools in this chapter and some in the next chapter.

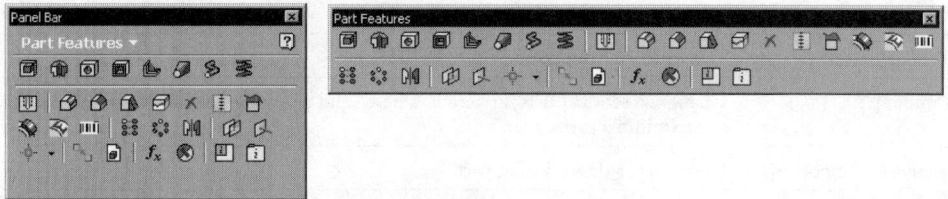

Figure 3.133 *Part Features Panel Bar and toolbar*

Table 3.8 *Part Features Panel Bar and toolbar options*

Option	Description
Extrude	Constructs an extruded feature from a sketch.
Revolve	Constructs a revolved feature from a sketch.
Hole	Places a hole feature.
Shell	Places a shell feature.
Rib	Constructs a rib or a web by extruding an open-loop sketch.
Loft	Constructs a loft feature from sketches.
Sweep	Constructs a sweep feature from sketches.
Coil	Constructs a coil feature from sketches.

Table 3.8 *Part Features Panel Bar and toolbar options (continued)*

Option	Description
Thread	Places a cosmetic thread feature.
Fillet	Places a fillet feature.
Chamfer	Places a chamfer feature.
Face Draft	Places a face draft feature.
Split	Splits a face into two faces or splits a solid into two solids and removes one of them.
Delete Face	Deletes faces from a set of surfaces or a solid.
Knit Surface	Knits a set of faces into a single quilt.
Replace Face	Replaces a face of a solid by a surface.
Thicken/Offset	Thicken thickens a surface to become a solid, and Offset constructs an offset surface.
Emboss	Constructs an embossment.
Decal	Maps a decal to a solid.
Rectangular Pattern	Places a rectangular pattern of features.
Circular Pattern	Places a circular pattern of features.
Mirror Feature	Places a mirror feature.
Work Plane	Constructs a work plane.
Work Axis	Constructs a work axis.
Work Point	Constructs a work point.
Promote	Changes a set of IGES surfaces to a base solid or stitches a set of disconnected IGES surfaces into a surface quilt.
Derived Component	Constructs a derived solid part.
Parameters	Manipulates model parameters.
Create iMate	Constructs an assembly interface to selected features of a component.
Insert iFeature	Inserts an iFeature.
View Catalog	Displays the iFeature catalog.

CONSTRUCTING A BASE SOLID FEATURE

The first feature of a solid part is the foundation upon which you construct additional features to compile a complex solid part. It is called the base solid feature, and it is a sketched solid feature. As we have mentioned, there are several kinds of sketched solid features. In the following tutorials, you will explore the following topics:

> Constructing an extruded solid feature
>
> Display control
>
> Constructing a revolved solid feature

EXTRUDED SOLID FEATURE AND DISPLAY CONTROL

An extruded solid is a sketched solid feature constructed by extruding a sketch in a direction perpendicular to the plane of the sketch. You can extrude a set of 2D sketches in several ways: extrude the profile a distance in one direction, in the other direction, and in both directions, extrude the profile to a selected plane, and extrude the profile from a plane to another plane.

TUTORIAL 3.21

In this tutorial, you will construct an extruded solid feature for the base of the punch set.

1. Open the file *PunchBase.ipt*, if you already closed it.
2. To facilitate working in 3D, the display is changed to isometric view. (See Figure 3.134.)

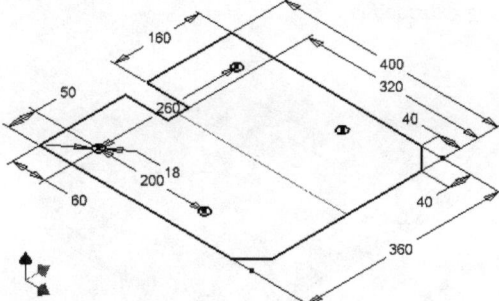

Figure 3.134 *Display changed to isometric view*

Tip: If you continue to work from sketching mode and the display is still showing the top view of the sketch, you can change the display to isometric view by right-clicking and selecting Isometric.

If you want to modify the sketch before you extrude it, select the sketch on the Browser Bar, and double-click or right-click and select Edit Sketch. Do not select the 2D Sketch button because it will create a new sketch.

3. Select Extrude on the Part Features Panel Bar or toolbar.

Note: If this is the first time you use this command, Inventor displays a What's New In Autodesk Inventor dialog box offering to describe information available for this command. You can view this information, or simply close the dialog box.

4. In the Extrude dialog box, click the Profile button, if it is not already selected.
5. Select the area of the sketch highlighted in Figure 3.135.

Tip: Do not select the areas inside the circles. If selected, the areas inside the circle will be extruded as well.

6. Select Distance, set the extrusion distance to 30 mm, and click the OK button.

The base solid feature is complete. (See Figure 3.136.) Save your file.

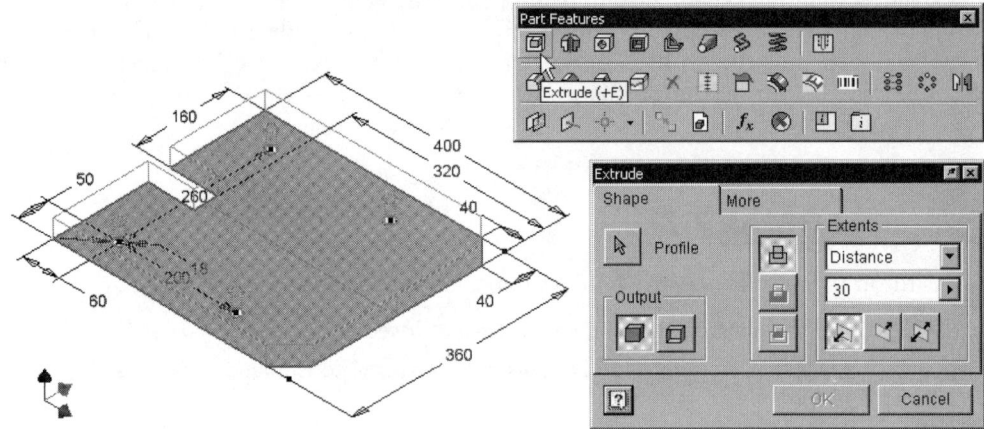

Figure 3.135 *Sketch being extruded*

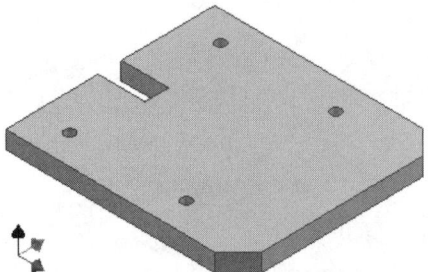

Figure 3.136 *Sketch extruded and display mode changed*

There are three kinds of display mode: shaded display, hidden edge display, and wireframe display. In addition, you can project a shadow onto the XY plane.

In shaded display mode, the 3D object is shaded.

In hidden edge display mode, the object is shaded and the edges are also shown.

In wireframe display mode, the object is not shaded and all the edges are shown.

By default, the display changes to shaded mode after the sketch is used to construct a sketched feature. To change the display mode, select Shaded Display, Hidden Edge Display, or Wireframe Display on the Inventor Standard toolbar.

Normally, the display projects the object orthogonally onto the screen. If you want to change to perspective view, select Perspective on the Inventor Standard toolbar.

To examine a 3D object in various viewing directions, you can rotate the display by selecting Rotate on the Inventor Standard toolbar.

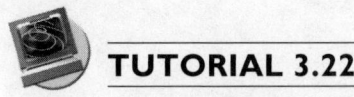

TUTORIAL 3.22

In this tutorial, you will set the display to perspective mode, view the solid part in different directions, and specify a shadow.

1. Continuing to work with the file *PunchBase.ipt*, select Perspective Camera on the Inventor Standard toolbar. (See Figure 3.137.)

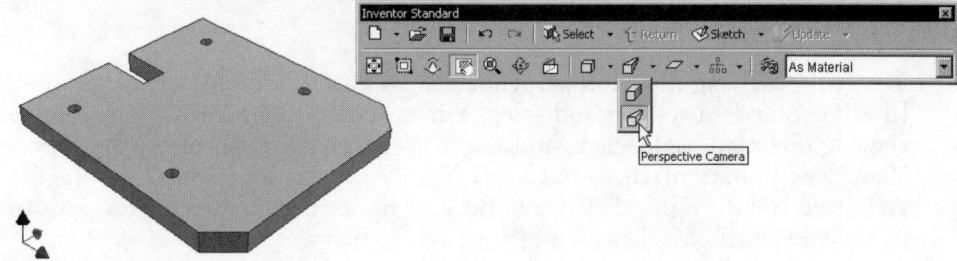

Figure 3.137 *Perspective display*

2. Select Orthographic Camera on the Inventor Standard toolbar to return to orthographic mode.

3. Select Rotate on the Inventor Standard toolbar. (See Figure 3.138.)

To rotate the view around the center mark, you hold down the left mouse button and drag. To rotate the display about the horizontal or vertical axis, you select the horizontal or vertical axis mark, hold down the left mouse button, and drag. One of the horizontal axes is indicated at A in Figure 3.138.

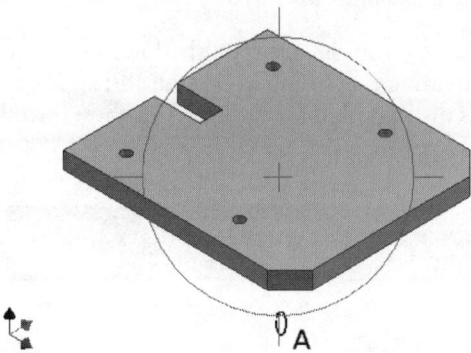

Figure 3.138 *View being rotated*

There are fourteen standard viewing directions: front, back, right side, left side, top, bottom, and eight isometric views. If you want to select these standard views, you press SPACE while the view is being rotated to display the 3D View Direction Selector and select one of the arrows. To close the 3D View Direction Selector, press SPACE again or right-click and select Done.

4. Press SPACE. (See Figure 3.139.)

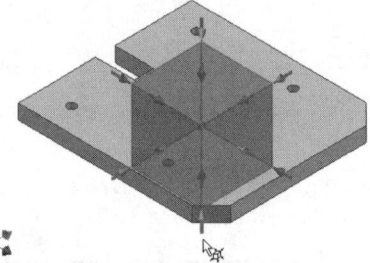

Figure 3.139 *3D View Direction Selector*

To set the viewing direction perpendicular to a face, you select Look At on the Inventor Standard toolbar and select a face. To set the display to the previous viewing direction, right-click, and select Previous View. To revert the Previous View operation, right-click, and select Next View. Figure 3.140 shows the Look At button and the right-click menu. Because right-click menu is context sensitive, menu items displayed depend on the current activity.

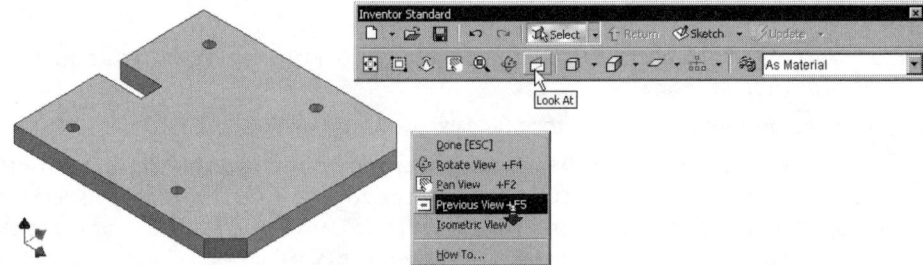

Figure 3.140 *Look At button and right-click menu*

To enhance the display, you can specify a shadow. There are two kinds of shadow: an ordinary shadow treating everything as opaque and an X-ray kind of shadow, making individual features distinguishable. Figure 3.141 shows two kinds of shadows.

5. Now save and close your file.

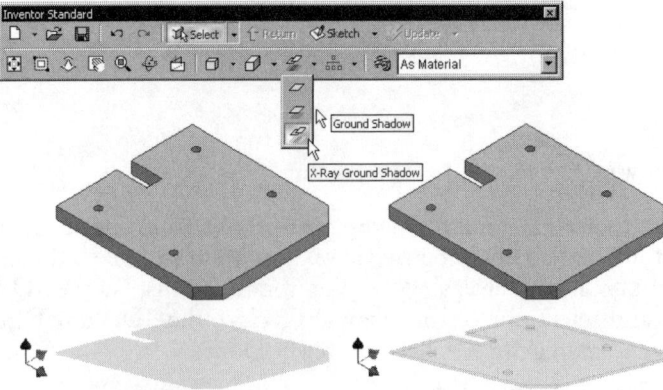

Figure 3.141 *Two kinds of shadows*

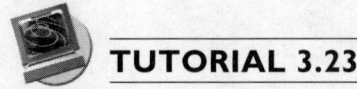

TUTORIAL 3.23

In this tutorial, you will construct an extruded solid feature for the bolt of the punch set.

1. Open the file *PunchBolt.ipt*.
2. Select Extrude on the Part Features Panel Bar or toolbar.
3. Set extrusion distance to 6 mm and click the OK button. (See Figure 3.142.)

Tip: The profile is automatically selected because there is only one closed loop in the sketch.

4. Save and close your file.

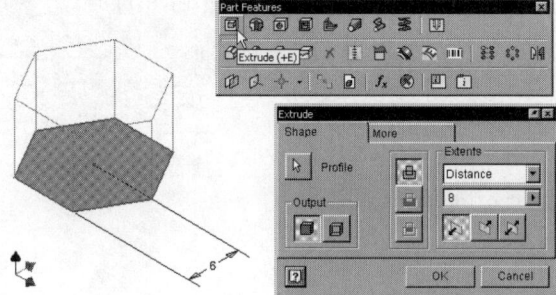

Figure 3.142 *Sketch being extruded*

TUTORIAL 3.24

In this tutorial, you will construct an extruded solid feature for the handle of the punch set.

1. Open the file *PunchHandle.ipt*.
2. Select Extrude on the Part Features Panel Bar or toolbar, set the extrusion distance to 20 mm, and click the OK button. (See Figure 3.143.)
3. Save and close your file.

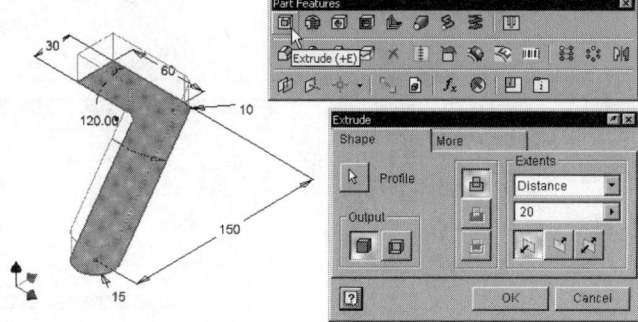

Figure 3.143 *Profile being extruded*

TUTORIAL 3.25

In this tutorial, you will construct an extruded solid feature for the hinge of the punch set.

1. Open the file *PunchHinge.ipt*.
2. Extrude the profile a distance of 8 mm. (See Figure 3.144.)
3. Save and close the file.

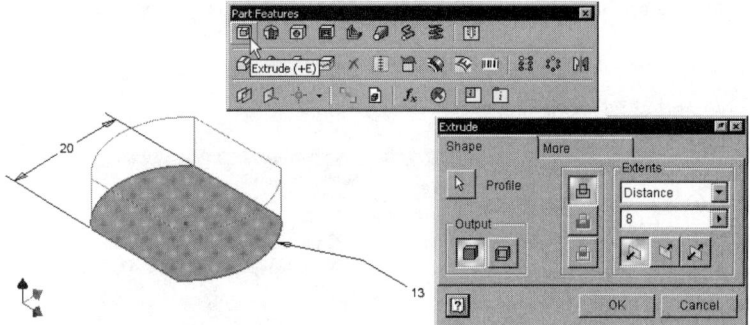

Figure 3.144 *Profile being extruded*

TUTORIAL 3.26

In this tutorial, you will construct an extruded solid feature for the table of the punch set.

1. Open the file *PunchTable.ipt*.
2. Extrude the profile a distance of 10 mm. (See Figure 3.145.)

 Note: In this case you need to select the area shown in Figure 3.145, but be careful not to select the individual circles that will become holes. If you mistakenly selected the holes, hold down the CTRL key and select the hole again to deselect it.

3. Save and close the file.

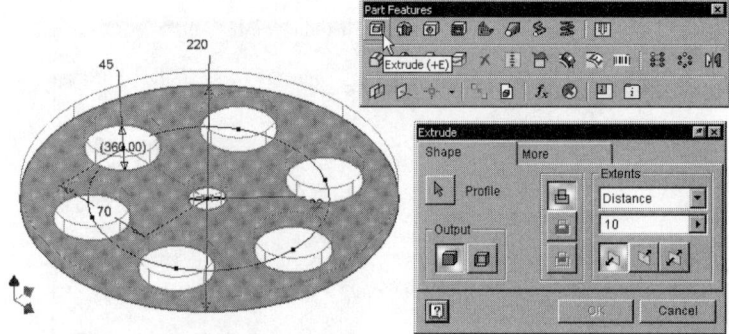

Figure 3.145 *Profile being extruded*

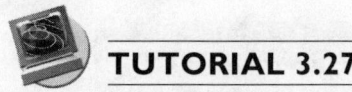

TUTORIAL 3.27

In this tutorial, you will construct an extruded solid feature for the column of the punch set.

1. Open the file *PunchColumn.ipt*.
2. Extrude the profile a distance of 50 mm from mid-plane. (See Figure 3.146.)
3. Save and close the file.

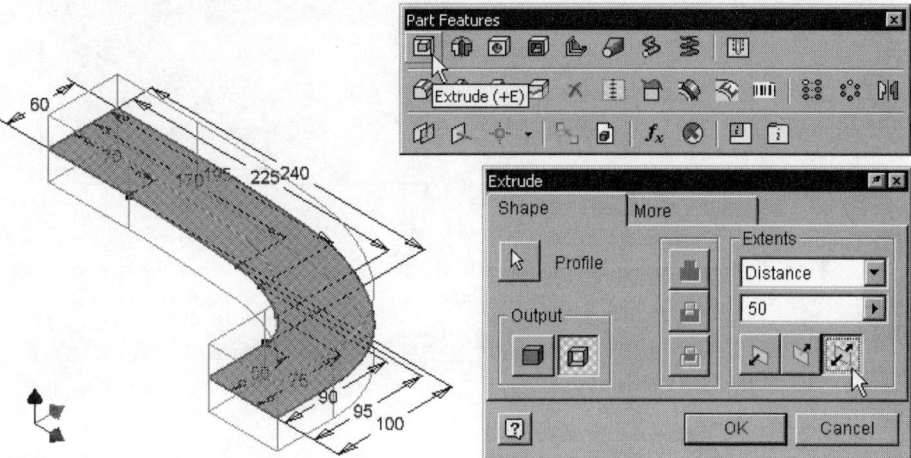

Figure 3.146 *Column being extruded from mid-plane*

REVOLVED SOLID FEATURE

A revolved solid is a sketched solid feature. You construct a sketch and revolve the sketch about an axis. The axis can be a line of the sketch, an existing edge, or a work axis. If you already constructed a centerline on the sketch, the centerline will be used as the default revolving axis.

Now work on the following tutorials to learn how to construct revolved solid features from sketches that you constructed earlier in this chapter.

TUTORIAL 3.28

In this tutorial, you will construct a revolved solid feature for making the bush of the punch set.

1. Open the file *PunchBush.ipt*.
2. Select Revolve on the Part Features Panel Bar or toolbar. (See Figure 3.147.)
3. In the Revolve dialog box, accept the default full revolving and click the OK button.

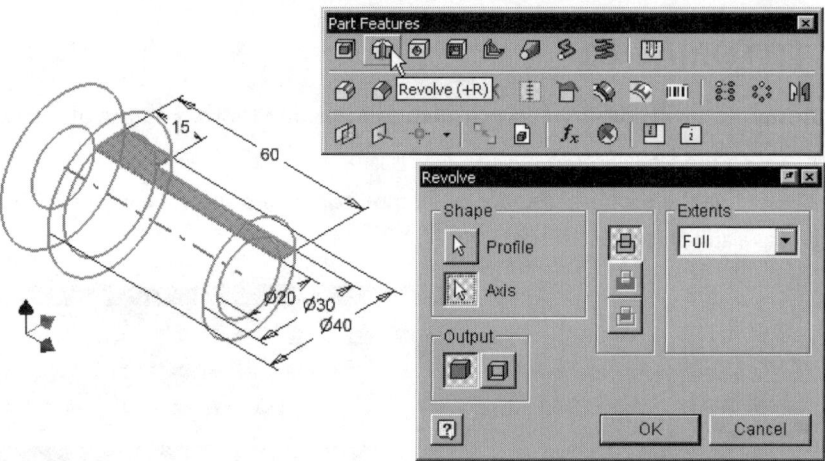

Figure 3.147 *Profile being revolved*

 Tip: The centerline is automatically selected as the axis of revolution. If there is no centerline in the sketch, you have to select the Axis button and select a line, an edge, or an axis.

4. Save and close your file.

 ## TUTORIAL 3.29

In this tutorial, you will construct a revolved solid feature for the index of the punch set.

1. Open the file *PunchIndex.ipt*.
2. Revolve the profile about the centerline 360 deg. (See Figure 3.148.)
3. Save and close your file.

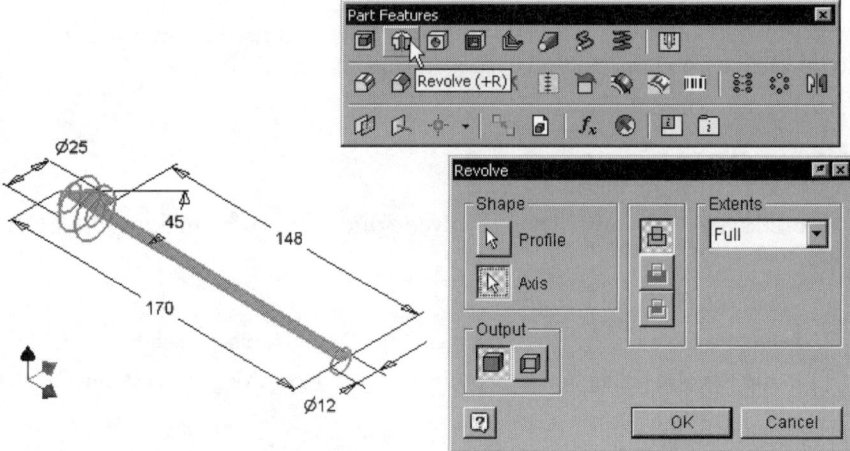

Figure 3.148 *Profile being revolved*

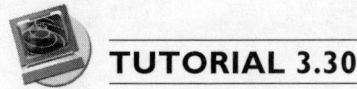

TUTORIAL 3.30

In this tutorial, you will construct a revolved solid feature for the index bush of the punch set.

1. Open the file *PunchIndexBush.ipt*.
2. Revolve the profile about the centerline 360 deg. (See Figure 3.149.)
3. Save and close your file.

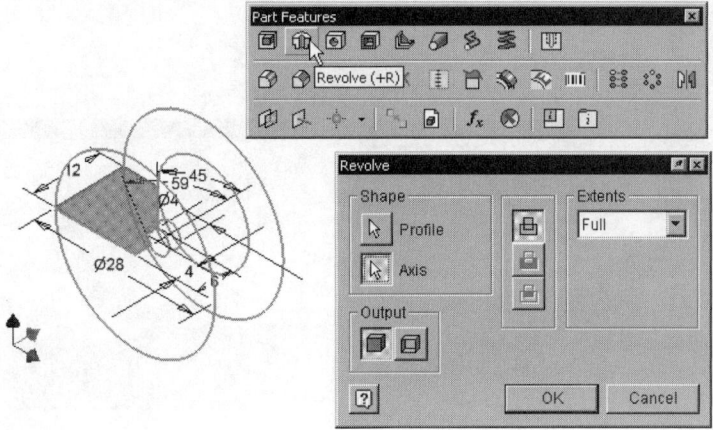

Figure 3.149 *Profile being revolved*

TUTORIAL 3.31

In this tutorial, you will construct a revolved solid feature for the pin of the punch set.

1. Open the file *PunchPin.ipt*.
2. Revolve the profile about the centerline 360 deg. (See Figure 3.150.)
3. Save and close your file.

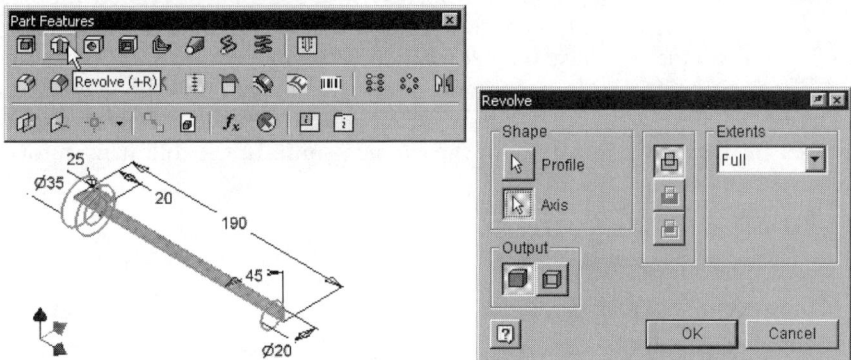

Figure 3.150 *Profile being revolved*

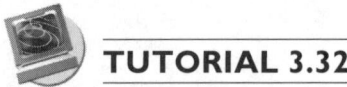

TUTORIAL 3.32

In this tutorial, you will construct a revolved solid feature for the post of the punch set.

1. Open the file *PunchPost.ipt*.
2. Revolve the profile about the centerline 360 deg. (See Figure 3.151.)
3. Save and close your file.

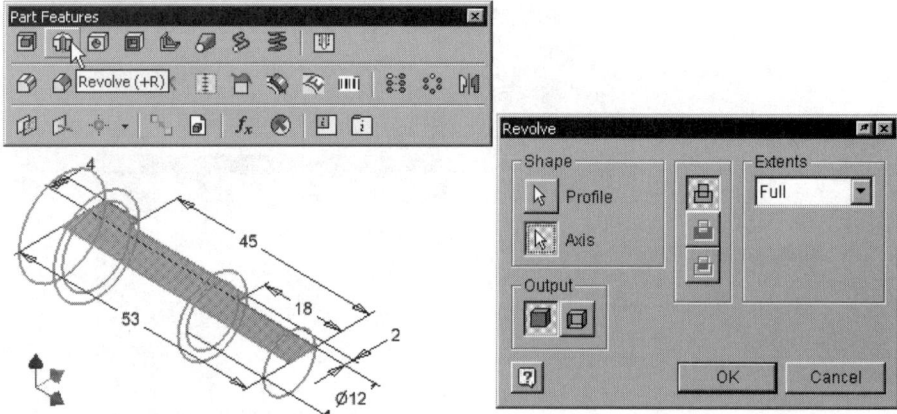

Figure 3.151 *Profile being revolved*

CONSTRUCTING ADDITIONAL SKETCHED SOLID FEATURES

To make a complex solid part that has a number of sketched solid features, you construct them one by one and combine them with the existing solid part sequentially. To make these additional sketched solid features, you construct sketches on a sketch plane that you establish on the three basic planes (XY, YZ, and XZ), the faces of a solid part, or artificial faces. (Artificial faces refer to work planes. You will learn how to construct work planes in the next chapter.)

As we have said, the first sketched solid feature in a solid part is called the base solid feature. While constructing the second or subsequent sketched solid feature of a solid part, you have to decide how it will be combined with the existing solid by using one of the three Boolean operations: join, cut, and intersect. You can join the new solid feature to the existing solid. You can cut the new solid feature from the existing solid. You can also intersect the new solid feature with the existing solid. In the following tutorials, you will explore the following topics.

Extruding an open-loop sketch

Slicing graphics

Constructing with join

Constructing with cut

Constructing with intersect

GRAPHICS SLICING, EXTRUDING AN OPEN-LOOP SKETCH, AND JOINING

Joining a sketched solid feature to a solid part produces a solid that has the volume enclosing the new solid feature and the existing solid part. Volumes in the original solid and the new solid feature are included in the resulting solid part.

Graphics slicing is a technique in which the shaded solid part is sliced along the current sketch plane and the Z direction of the sketch plane is clipped. To slice graphics, move the cursor over the graphics area, right-click, and select Slice Graphics. Now you will slice the graphics and set the display to a section view across the sketch plane.

 TUTORIAL 3.33

In this tutorial, you will construct three extruded solids one by one and join them to the base solid. One of the extruded solids will be made from an open-loop sketch. You will also learn how to slice graphics.

1. Open the file *PunchColumn.ipt*, if it is already closed.

2. Select 2D Sketch on the Inventor Standard toolbar and select the XY plane on the Browser Bar. A sketch is constructed on the XY plane.

 Tip: Do not select 2D Sketch and select a face more than once. Otherwise, you will end up with too many redundant sketches.

3. Right-click and select Slice Graphics. The display is sliced along the sketch plane.

4. Construct a line. (See Figure 3.152.)

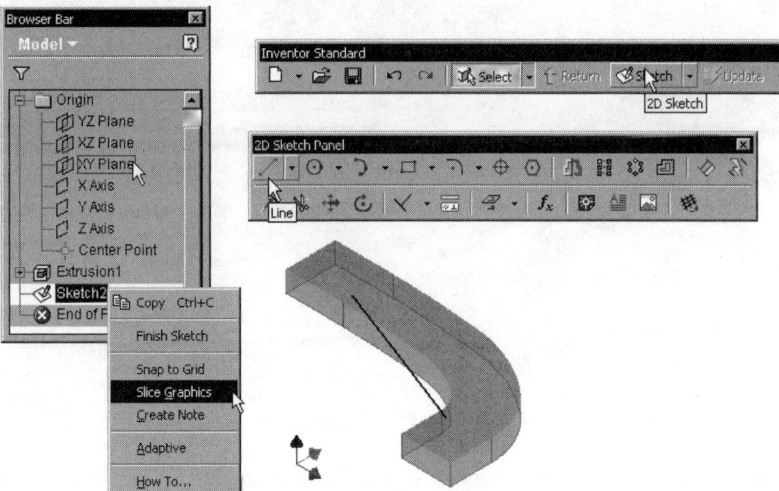

Figure 3.152 *Sketch constructed on XY plane and graphics sliced*

5. With reference to Figure 3.153, add four dimensions (10 mm, 70 mm, 10 mm, and 50 mm).

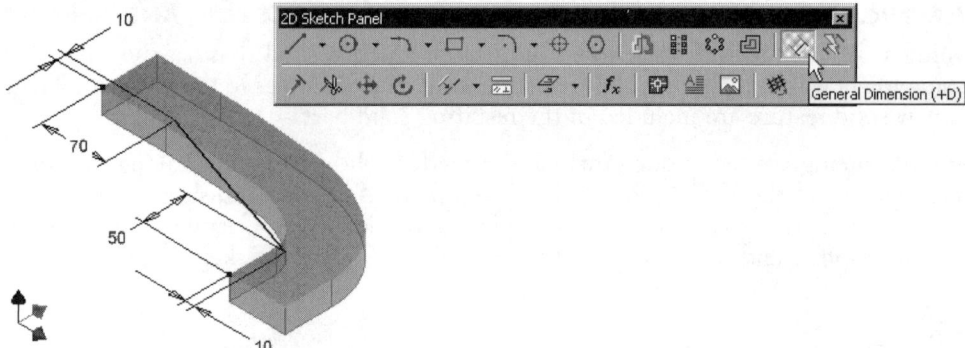

Figure 3.153 *Dimensions added to the sketch*

6. Select Return on the Inventor Standard toolbar to exit sketch mode.

7. Select Extrude on the Part Features Panel Bar or toolbar.

8. In the Extrude dialog box, click the Profile button, if it is not already selected.

9. Select line A in Figure 3.154.

10. Select Distance, select the extrude from mid-plane button, and set the extrusion distance to 8 mm.

11. Move the cursor to position B (Figure 3.154) to specify the direction of the extruded solid feature.

12. Click the OK button. An extruded solid feature is constructed from an open-loop sketch. (Figure 3.154.)

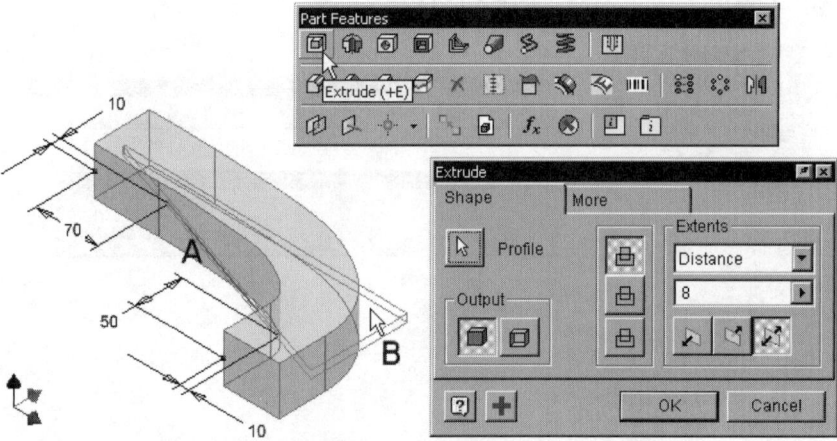

Figure 3.154 *Open-loop sketch being extruded and joined to the solid part*

13. Select 2D Sketch on the Inventor Standard toolbar and select face A in Figure 3.155 to construct a sketch.

14. Construct a rectangle with its vertical edges collinear with the edge of the solid. Add two dimensions (65 mm and 180 mm). Select Return to end sketching.

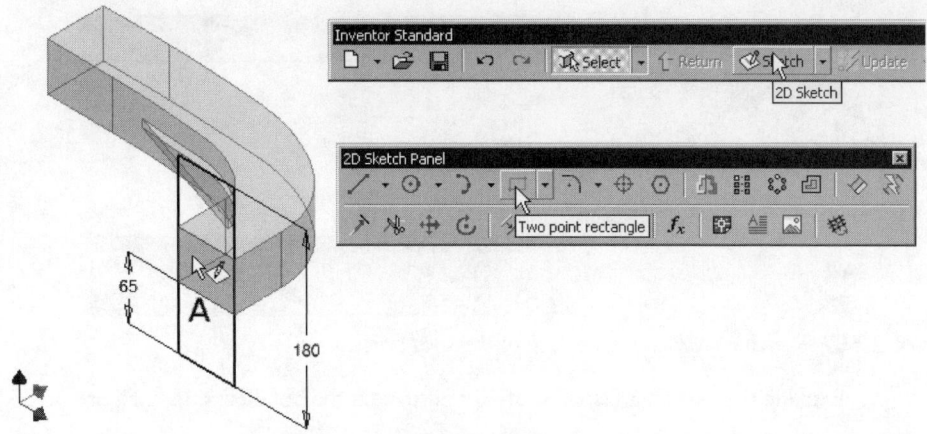

Figure 3.155 *Sketch constructed on a face of the solid part*

15. Extrude the sketch a distance of 10 mm to join the solid part. (See Figure 3.156.)

 Note: The extrusion direction is away from the existing solid, as shown in the figure.

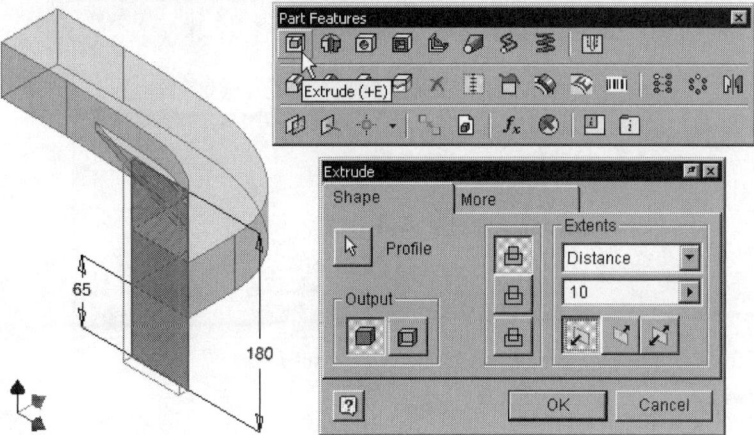

Figure 3.156 *Sketch being extruded and joined to the solid part*

16. Select 2D Sketch on the Inventor Standard toolbar and select face A in Figure 3.157.

17. Construct a circle with its center constrained to the midpoint of an edge. Add a dimension of 50 mm to the sketch.

 Note: You can also dimension the center of the circle as 25 mm from the edge of the existing solid or project the edge before constructing the circle.

18. Select Return to exit sketch mode.

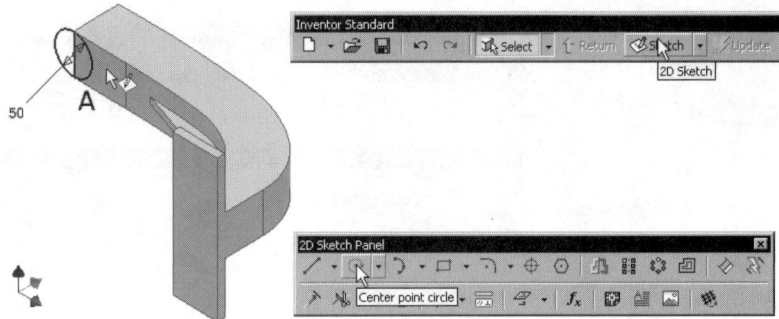

Figure 3.157 *Sketch constructed on a face of the solid part*

19. Extrude the sketch a distance of 60 mm to join the solid part. (See Figure 3.158.)

 Note: Change the direction of extrusion, if necessary, to match the direction indicated in the figure.

Two extruded solid features are constructed and joined to the base solid. Save and close your file.

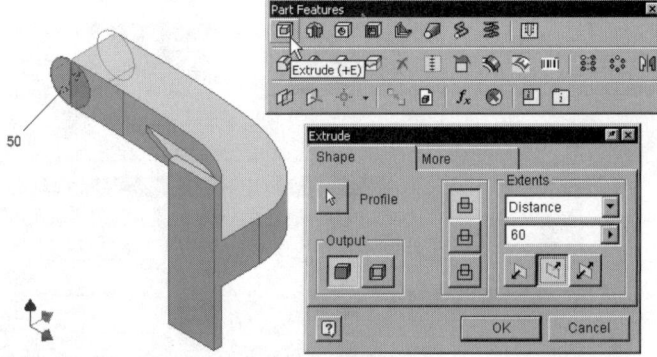

Figure 3.158 *Sketch being extruded and joined to the solid part*

TUTORIAL 3.34

In this tutorial, you will construct additional sketched solid features in a solid part and join the features to the base solid part.

1. Open the file *PunchHandle.ipt*, if you already closed the file.

2. Select 2D Sketch on the Inventor Standard toolbar and select the face A in Figure 3.159. A sketch plane is established on the selected face.

3. Select Project Geometry on the 2D Sketch Panel Bar or toolbar.

4. Select edges B, C, and D in Figure 3.159 to project the selected edges onto the sketch plane.

Tip: If you select the Automatic reference edges for new sketch box of the Sketch tab of the Options dialog box before you establish a new sketch, all the edges will automatically be projected onto the sketch plane. However, you might end up with too much redundant projected geometry, causing confusion.

If you select the Autoproject edges during curve creation box on the Sketch tab of the Options dialog box, you can project geometry onto the sketch by selecting and "rubbing" the geometry with the cursor.

Figure 3.159 *Sketch established and geometry from edges of existing solid feature being projected*

5. With reference to Figure 3.160, construct a line and add a dimension.

Figure 3.160 *Line constructed and dimensioned*

Now extrude the sketch to join the base solid.

6. Select Return on the Inventor Standard toolbar to exit sketch mode.
7. Select Extrude on the Part Features Panel Bar or toolbar.
8. In the Extrude dialog box, select Join operation.
9. Select the area indicated in Figure 3.161 and extrude it a distance of 55 mm.

 Note: Change the direction of extrusion, if necessary, to match the direction indicated in the figure.

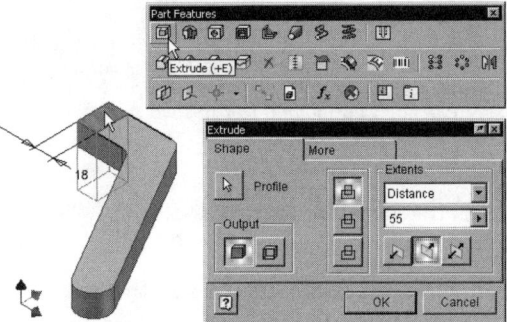

Figure 3.161 *Profile being extruded and joined to the base solid*

Now construct another extruded solid feature.

10. Select 2D Sketch on the Inventor Standard toolbar.

11. Select the face indicated in Figure 3.162 to establish a sketch plane.

12. Select Project Geometry and select circular edge A in Figure 3.162. The circular edge and its center point are projected.

13. Construct a circle with the center coincident with the projected center point. (A concentric constraint will be inferred as you construct the circle.)

14. Add a dimension of 40 mm.

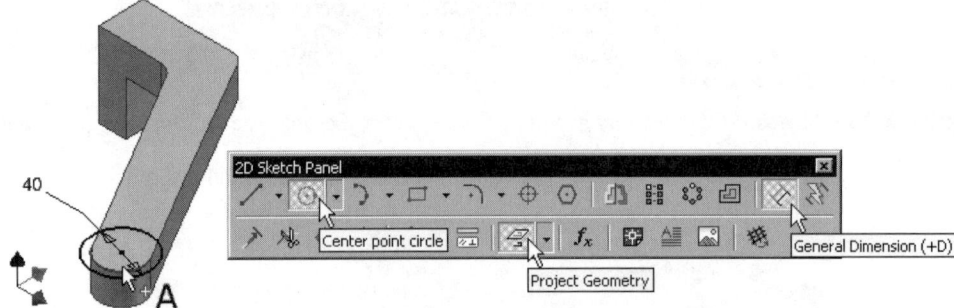

Figure 3.162 *Sketch constructed on a face of the solid part*

15. Select Return on the Inventor Standard toolbar to exit sketch mode.

16. Extrude the sketch a distance of 30 mm. (See Figure 3.163.)

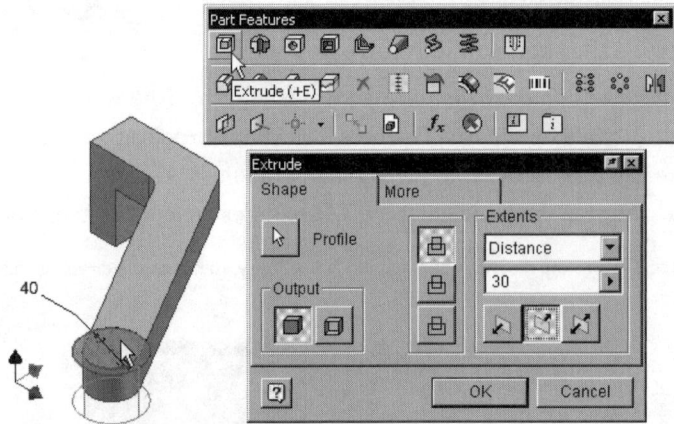

Figure 3.163 *Sketch being extruded and joined to the solid part*

Now construct another sketched feature.

17. Select Rotate on the Inventor Standard toolbar.

18. Press SPACE.

19. With reference to Figure 3.164, select a standard view.

20. Right-click and select Done.

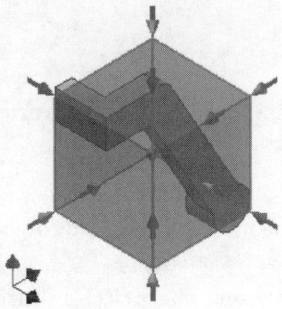

Figure 3.164 *Standard view being selected*

21. With reference to Figure 3.165, establish a sketch plane and construct a circle with a diameter of 30 mm. The center of the circle should be constrained to the midpoint of an edge.

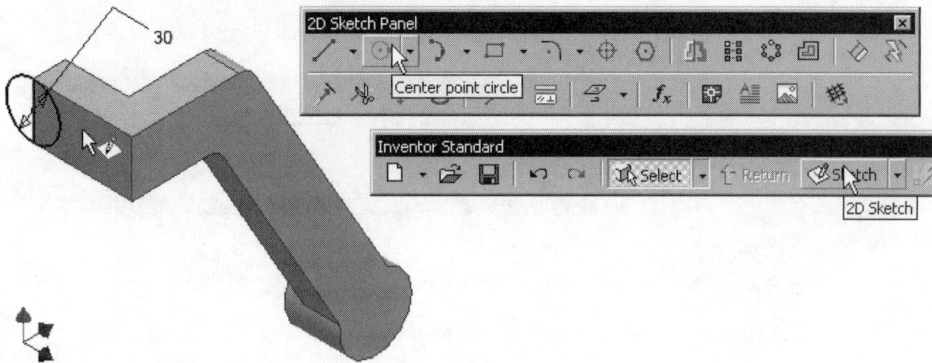

Figure 3.165 *Sketch constructed on a face of the solid part*

22. Extrude the sketch a distance of 400 mm to join the solid part. (See Figure 3.166.)

The sketched solid features are complete. Save your file.

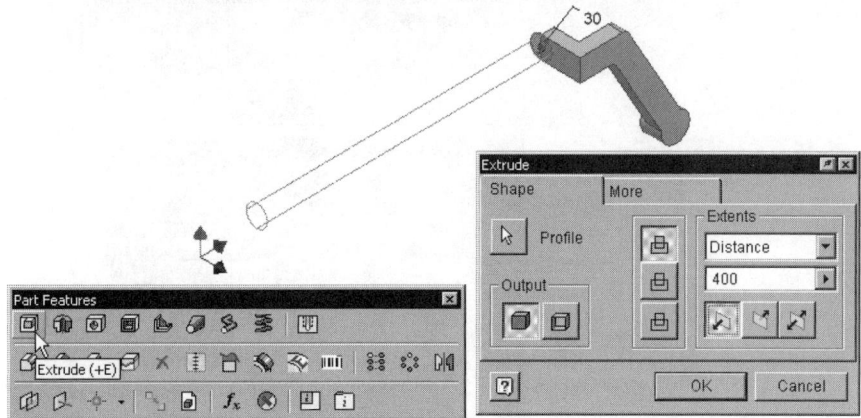

Figure 3.166 *Sketch being extruded and joined to the solid part*

TUTORIAL 3.35

In this tutorial, you will construct extruded solid features that join to the base solid feature of the hinge of the punch set.

1. Open the file *PunchHinge.ipt*.
2. Select 2D Sketch on the Inventor Standard toolbar and select the face A in Figure 3.167 to establish a sketch plane.
3. Select Project Geometry, and select circular edge B to project the edge and its center point.
4. Select Center Point Circle on the 2D Sketch Panel Bar or toolbar and select the projected center point C to use it as the center. (This way the circle will become concentric with the projected geometry.)
5. Select a point to indicate a point on the circumference.
6. Select General Dimension and place a diameter dimension of 20 mm.

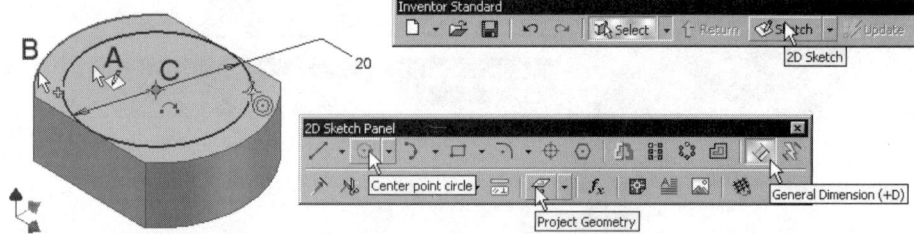

Figure 3.167 *Sketch constructed on a face of the solid part*

7. Select Return to exit sketch mode.
8. With reference to Figure 3.168, extrude the sketch a distance of 80 mm to join the solid part.

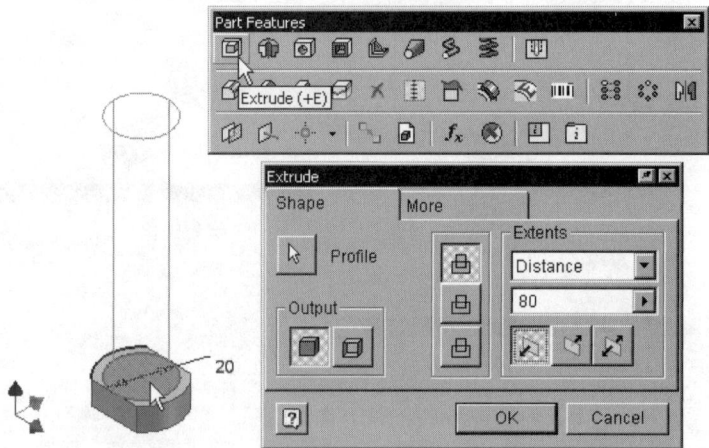

Figure 3.168 *Sketch being extruded and joined to the solid part*

Now construct another extruded solid feature to join the solid part.

9. Set up a sketch plane on face A in Figure 3.169.

10. Select Project Geometry on the 2D Sketch Panel Bar or toolbar and select the circular edge.

11. Construct a circle 12 mm that is concentric to the projected circular edge.

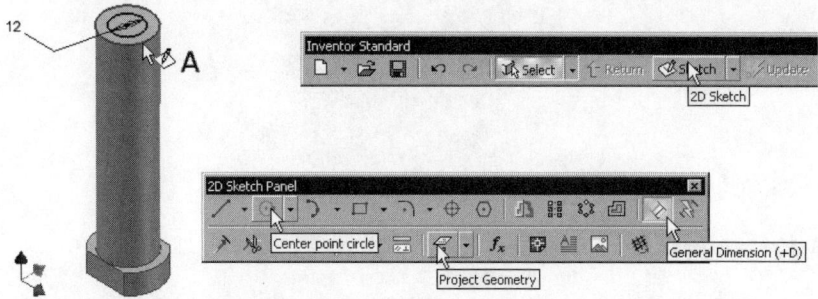

Figure 3.169 *Sketch constructed on a face of the solid part*

12. Select Return to exit sketch mode.

13. With reference to Figure 3.170, extrude the sketch a distance of 30 mm to join with the solid part.

14. Save your file.

Figure 3.170 *Sketch being extruded and joined to the solid part*

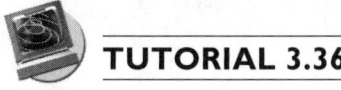

TUTORIAL 3.36

In this tutorial, you will construct an extruded solid feature to join with the base solid part of the table of the punch set.

1. Open the file *PunchTable.ipt*.

2. Select Rotate on the Inventor Standard toolbar.

3. Press SPACE to display the 3D View Direction Selector, if it is not displayed.

4. Select arrow A in Figure 3.171.

5. Right-click and select Done.

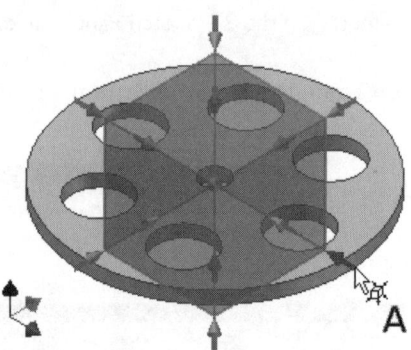

Figure 3.171 *3D view direction selector*

Now establish a sketch plane and construct a sketch.

6. Set up a sketch plane on face A in Figure 3.172.

7. Select Project Geometry on the 2D Sketch Panel Bar or toolbar, and select edge B in Figure 3.172.

8. Select Center Point Circle on the 2D Sketch Panel Bar or toolbar, select the projected center, and select a point inside the projected circular edge to construct a concentric circle.

9. Add a dimension of 16 mm to the circle.

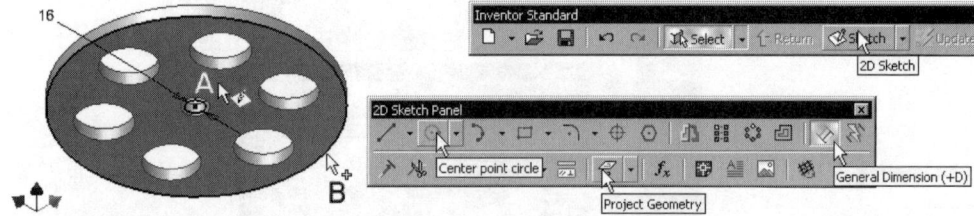

Figure 3.172 *Sketch constructed on a face of the solid part*

Now extrude the sketch to join with the solid part.

10. Exit sketch mode.

11. Extrude the sketch a distance of 26 mm to join the solid part. (See Figure 3.173.)

12. Save your file.

Figure 3.173 *Sketch being extruded and joined to the solid part*

CUTTING

Cutting a sketched solid feature from a solid part produces a solid that has the volume of the original solid part without that of the new solid feature. To remove a volume or construct voids or cavities in a solid part, you construct sketched solid features and cut them from the solid.

 TUTORIAL 3.37

In this tutorial, you will continue to work on the table of the punch set. You will construct two more sketched solid features that are to be cut from the solid part.

1. Open the file *PunchTable.ipt*, if you already closed it.

2. Right-click and select Isometric.

3. Establish a sketch plane on face A in Figure 3.174.

4. Construct a rectangle, add tangent constraints to edges B and C and edges D and E, and place two dimensions (108 mm and 30 mm) in accordance with Figure 3.174.

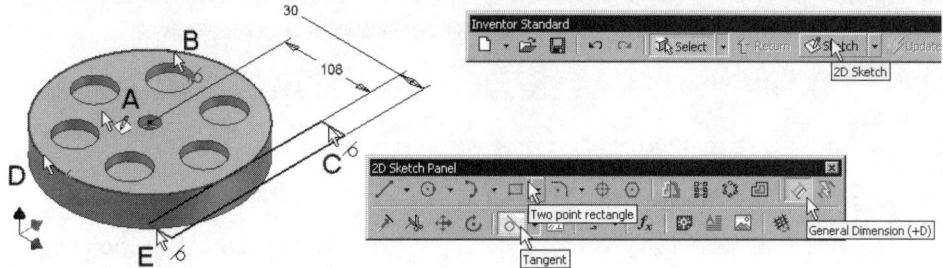

Figure 3.174 *Sketch constructed on a face of the solid part*

Now extrude the sketch to cut the solid part.

5. Select Return on the Inventor Standard toolbar.

6. Select Extrude on the Part Features Panel Bar or toolbar and select the rectangle as the profile to extrude.

7. In the Extrude dialog box, select All in the Extents pull-down list box, select Cut button, and click the OK button. (See Figure 3.175.)

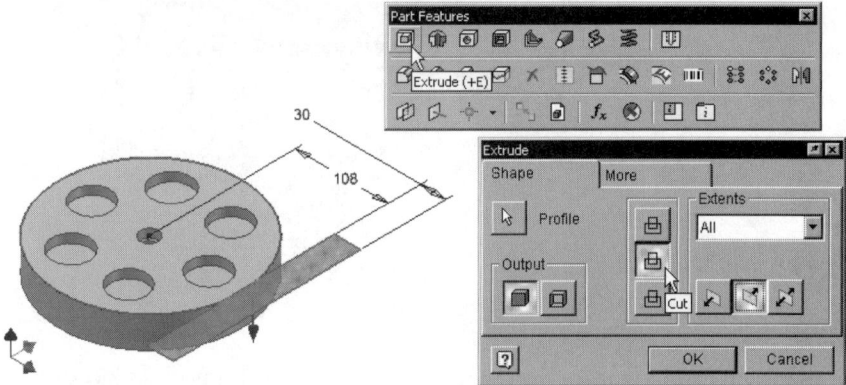

Figure 3.175 *Rectangular being extruded to cut the solid part*

Now construct another extruded cut feature.

8. Establish a sketch plane on face A in Figure 3.158.

9. Select Project Geometry on the 2D Sketch Panel Bar or toolbar and select edge B in Figure 3.176.

10. Construct a circle and add three dimensions in accordance with Figure 3.176. Dimension C is the diameter of the circle (28 mm), dimension D is the distance of the circle's center from one of the circular faces (18 mm), and dimension E (accept the default dimension value) is the distance of the circle's center from the end of the projected edge. (Do not exit sketch mode at this time.)

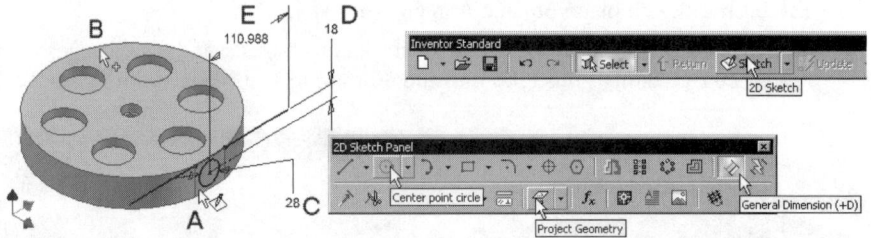

Figure 3.176 *A sketch constructed with a circle and three dimensions*

Now modify a dimension's value.

11. While still working in sketch mode, select Extrusion1 (the first extrusion object) on the Browser Bar, right-click, and select Show Dimension.

12. Double-click dimension A in Figure 3.177.

13. Select dimension B in Figure 3.177 to put the dimension's name in the Edit Dimension dialog box.

14. In the Edit Dimension dialog box, append the dimension B's name with the string /2 (meaning divided by 2) and select the Checkmark.

Tip: You can establish a mathematical relationship between dimensions by incorporating their name in an expression.

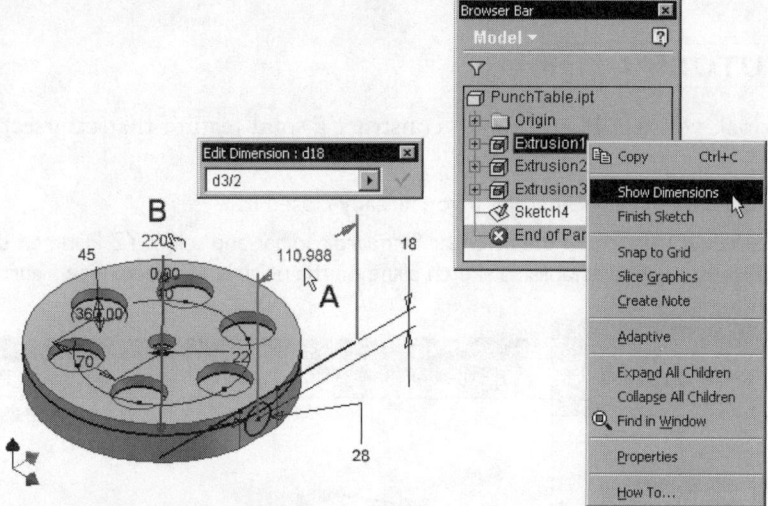

Figure 3.177 *Dimension being modified*

15. Exit sketch mode.
16. Extrude the sketch a distance of 12 mm to cut the solid part. (See Figure 3.178.)
17. Save your file.

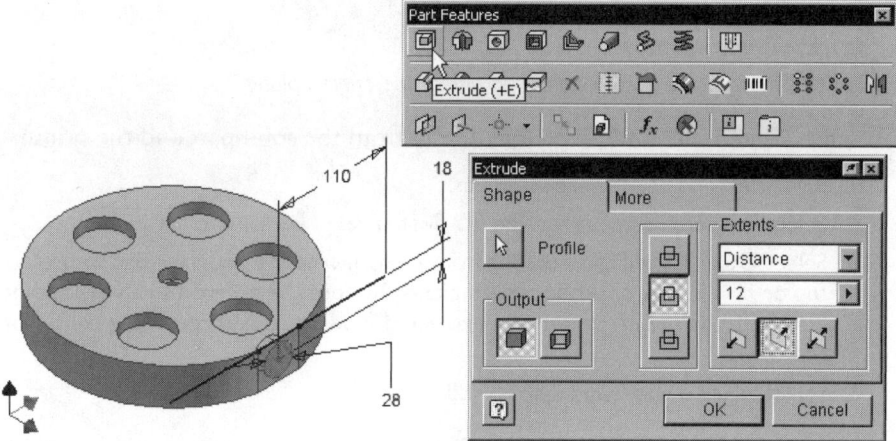

Figure 3.178 *Sketch being extruded to cut the solid*

INTERSECTING

Intersecting a sketched solid feature with a solid part produces a solid that has the volume common to the existing solid part and the new sketched solid feature.

 ## TUTORIAL 3.38

In this tutorial, you will learn how to construct a solid feature that intersects with the solid part.

1. Open the file *PunchBolt.ipt*, if you already closed it.
2. Select 2D Sketch on the Inventor Standard toolbar and select YZ Plane on the Browser Bar to establish a sketch plane on the origin's YZ plane. (See Figure 3.179.)

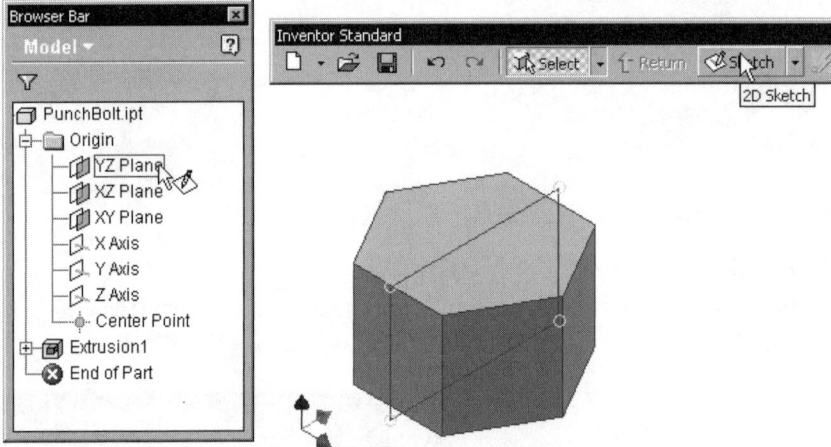

Figure 3.179 *Sketch plane being established one of the origin's plane*

Now slice graphics and project geometry from the solid part and the origin.

3. Right-click and select Slice Graphics.
4. Select Project Geometry on the 2D Sketch Panel Bar or toolbar.
5. Select Z axis (A of Figure 3.180) from the origin of the Browser Bar to project the origin's Z axis onto the sketch plane. Because the polygon's center is coincident with the origin' center, the projected Z axis passes through the center of the solid.
6. Select edges B and C in Figure 3.180.

 Note: You might need to change the display to Hidden Edge to make it easier to select edge C.

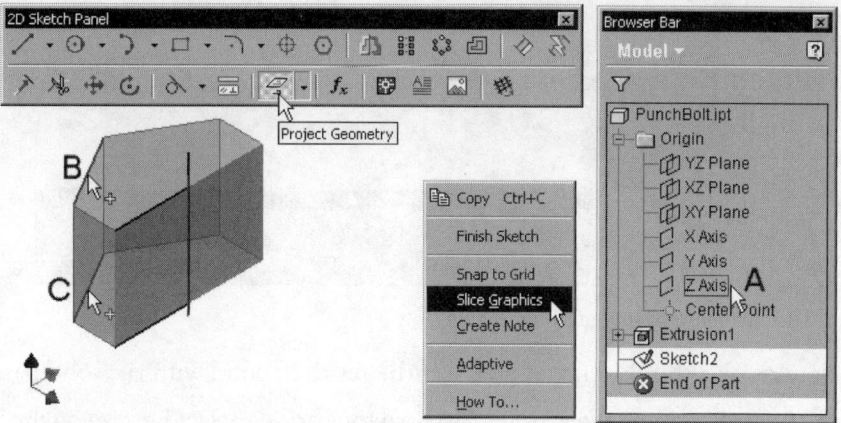

Figure 3.180 *Graphics sliced and edges and Z axis of the origin projected*

Now construct a sketch.

7. With reference to Figure 3.181, construct five line segments.

8. Select Collinear on the 2D Sketch Panel Bar or toolbar.

9. Select A and B (Figure 3.181).

10. Select C and D (Figure 3.181).

11. Select E and F (Figure 3.181).

12. Select Coincident on the 2D Sketch Panel Bar or toolbar.

13. Select G and H (Figure 3.181).

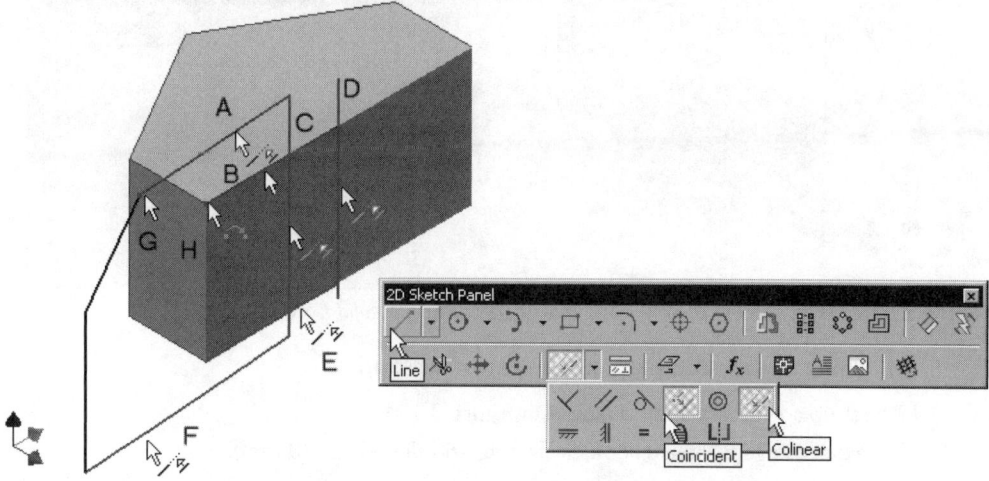

Figure 3.181 *Line segments constructed*

14. Add two dimensions (5 and 30 deg) in accordance with Figure 3.182.

Figure 3.182 *Dimensions added to the sketch*

Now revolve the sketch and intersect the revolved solid with the solid part.

15. Select Return on the Inventor Standard toolbar, and select Revolve on the Part Features Panel Bar or toolbar.

16. Select area A highlighted in Figure 3.183 as the profile to revolve.

17. Select the projected axis B as the axis of revolving.

18. In the Revolve dialog box, select Intersect operation, select Full extents, and click the OK button.

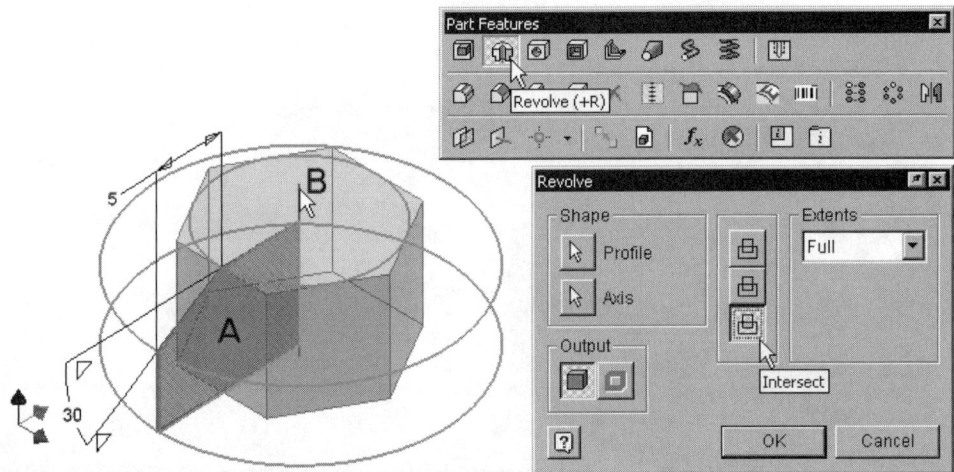

Figure 3.183 *Sketch being revolved and intersected with the solid part*

To complete the solid part, add an extruded solid.

19. Set up a sketch plane on face A in Figure 3.184.

20. Project edge B, construct a circle, and add dimension (8 mm).

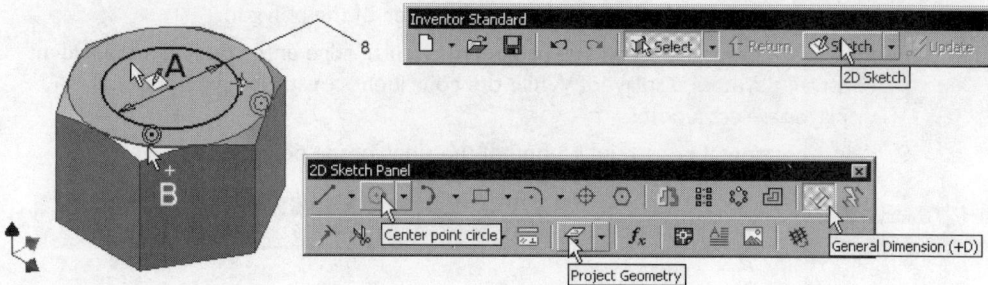

Figure 3.184 *Sketch constructed on a face of the solid part*

21. Exit sketch mode.
22. Extrude the circle a distance of 36 mm to join the solid. (See Figure 3.185.)
23. Save your file.

Figure 3.185 *Sketch being extruded and joined to the solid part*

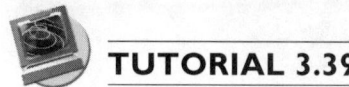

TUTORIAL 3.39

In this tutorial, you will construct an extruded solid feature to intersect with the solid part.

1. Open the file *PunchPost.ipt*, if you already closed it.
2. Use the 3D View Direction Selector to set the display in accordance with Figure 3.186.
3. Select 2D Sketch on the Inventor Standard toolbar and select face A in Figure 3.186.
4. Select Project Geometry on the 2D Sketch Panel Bar or toolbar and select edge B in Figure 3.186.
5. Select Polygon on the 2D Sketch Panel Bar or toolbar. (Figure 3.186.)
6. In the Polygon dialog box, check the Inscribed button, and set the number of sides to 6.

7. Select the projected center point as the center of the polygon.

8. Move the cursor slowly to the projected circular edge until you find a coincident constraint symbol displayed. While the coincident constraint symbol is still displayed, select a point.

9. Add a horizontal constraint to one of the sides of the polygon.

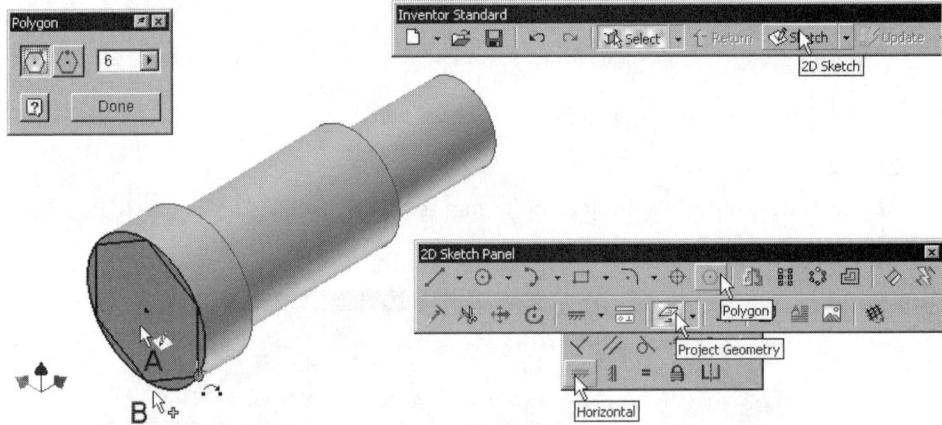

Figure 3.186 *Polygon constructed*

Now extrude the sketch and intersect the extruded feature with the solid.

10. Select Return on the Inventor Standard toolbar.

11. Select Extrude on the Part Features Panel Bar or toolbar.

12. Select the polygon, if it is not already selected.

13. In the Extrude dialog box, select Intersect operation, set Extents to All, and click the OK button. (Figure 3.187.)

Save your file.

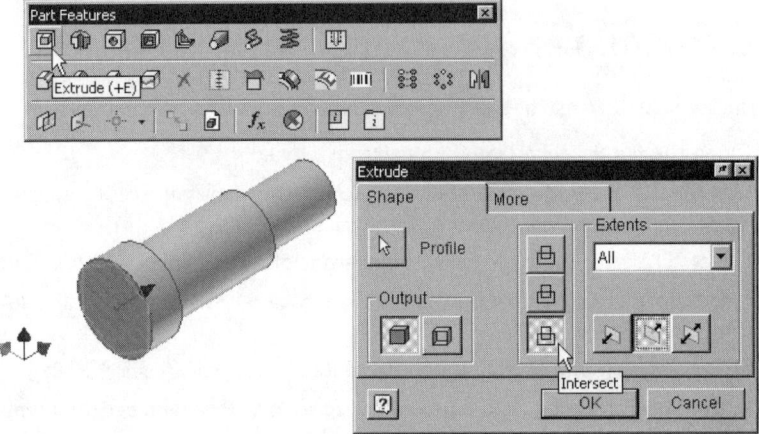

Figure 3.187 *Polygon being extruded and intersected with the solid part*

PLACED SOLID FEATURES AND PATTERNS

Placed solid features are pre-constructed features. You select them from the menu and specify their parameters. There are six kinds of placed solid features and three kinds of patterns. In the following tutorials, you will explore the following topics:

> Constructing a mirror feature
>
> Constructing a fillet feature
>
> Constructing a chamfer feature
>
> Constructing a shell feature
>
> Constructing a hole feature
>
> Constructing a thread feature
>
> Constructing a rectangular pattern
>
> Constructing a circular pattern
>
> Constructing a face draft feature

MIRROR FEATURE

Quite often, we have components that are symmetrical about a plane along its centerline. To construct a mirror image of selected features, you place a mirror feature by selecting features to be mirrored and specifying a mirror plane, which can be a face, sketch plane, or work plane. In the More panel of the Mirror dialog box, you select a creation method (identical or adjust to model).

 ## TUTORIAL 3.40

In this tutorial, you will construct a mirror feature.

1. Open the file *PunchHandle.ipt*, if you already closed it.
2. Select Rotate on the Inventor Standard toolbar.
3. Press SPACE, if the 3D View Direction Selector is not displayed.
4. Set the display in accordance with Figure 3.188
5. Select Mirror Feature on the Part Features Panel Bar or toolbar.
6. Select the extruded solid features A, B, and C highlighted in Figure 3.188.
7. Click the Mirror Plane button in the Mirror dialog box.
8. Select plane D in Figure 3.188.
9. Click the OK button.

A mirror feature is placed. (See Figure 3.189.) Save your file.

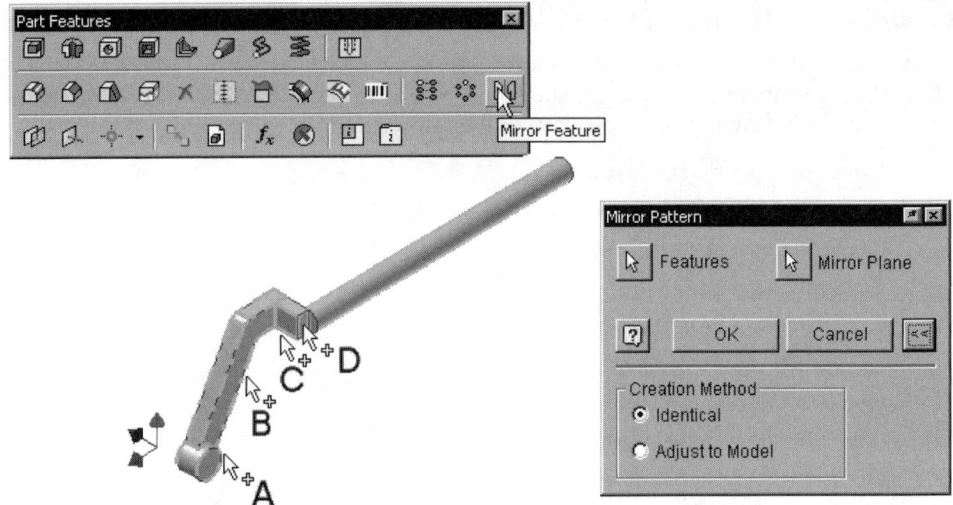

Figure 3.188 *Mirror feature being constructed*

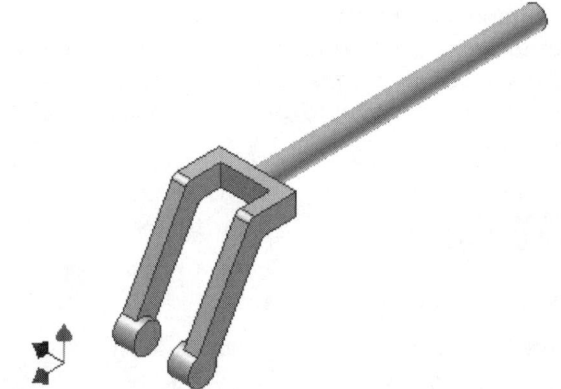

Figure 3.189 *Mirror feature constructed*

FILLET FEATURE

Sharp edges on a component can be hazardous and cause stress concentration. Therefore, we round them off. One way to remove the sharp edges is to construct a fillet feature.

A fillet feature rounds off the edges of a solid part as if an imaginary ball is rolled along the edges. You can construct constant-radius fillets, variable-radius fillets, and fillets of different sizes, in one operation. To define tangent continuous transition between the fillets at the intersecting edges, you add setback to the fillets.

In the More panel of the Fillet dialog box, you can set options to determine how the imaginary ball is rolled along the edges (Roll along sharp edges and Rolling ball where possible), to decide whether all tangent edges are selected when an edge is selected (Automatic edge chain), and to determine whether all edges intersecting with the fillet are calculated when a fillet is being constructed (Preserve all features).

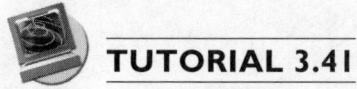

TUTORIAL 3.41

In this tutorial, you will construct a fillet feature on the handle of the punch set.

1. Continuing with the file *PunchHandle.ipt*, select Fillet on the Part Features Panel Bar or toolbar.

2. Select edge A in Figure 3.190.

3. In the Fillet dialog box, set the fillet radius to 15 mm.

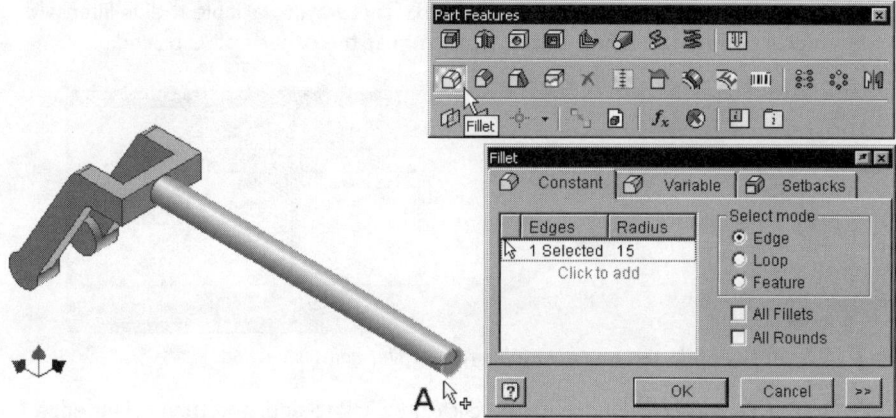

Figure 3.190 *An edge is selected to place a fillet*

Now add six more edges and specify their fillet radius.

4. On the Constant tab, click on Click to add.

5. Select edges A, B, C, D, E, and F in Figure 3.191 and set the fillet radius for these edges to be 5 mm.

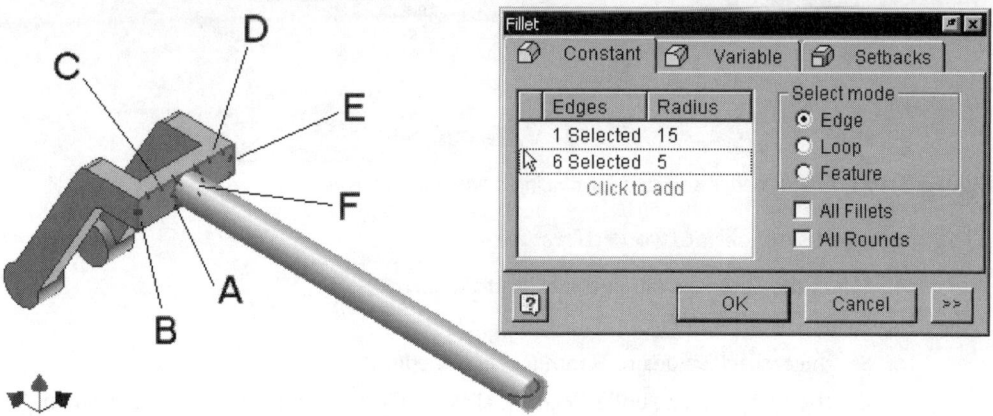

Figure 3.191 *Second set of fillet edges selected and fillet radius specified*

Now construct a set of variable radius fillets.

6. Select the Variable tab.

7. Select edge A in Figure 3.192.

 Tip: To specify the start point radius and the endpoint radius of a variable radius fillet, select Start and End one by one, note which of the endpoints of the selected edge is highlighted, and type an appropriate value in the Radius box of the dialog box.

8. In Figure 3.192, endpoint B is highlighted when Start is selected.

9. Type **5** in the Radius box.

10. Select End and type **1** in the Radius box. This way, a variable radius fillet with a radius of 5 mm at B and a radius of 1 mm at the other end is placed.

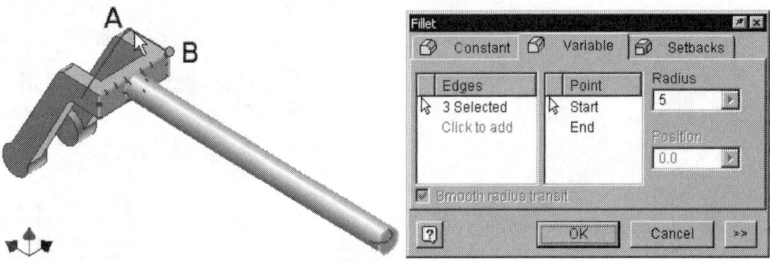

Figure 3.192 *An edge selected and a variable-radius fillet being specified*

11. In the Variable tab, under Edges, click on Click to add, and then select edge A in Figure 3.193.

12. Here endpoint B is highlighted; type **5** in the Radius box.

13. Select End and type **1** in the Radius box.

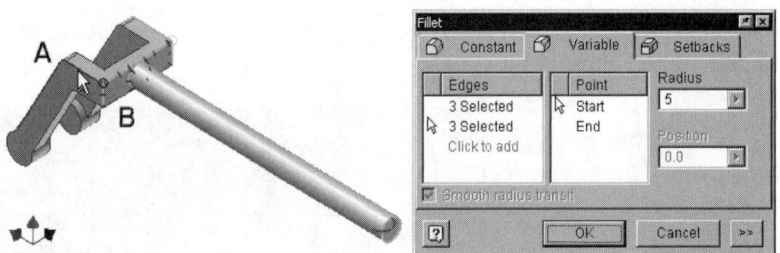

Figure 3.193 *Another edge selected for placing a variable-radius fill*

Now add setbacks at two vertices.

14. Select the Setbacks tab in the Fillet dialog box.

15. Select vertex A in Figure 3.194.

16. Set the setback values to 8 mm for all the edges.

17. On the Setbacks tab, under Vertex, click on Click to add and then select vertex B in Figure 3.194.

18. Click the OK button. The resulting part should look like Figure 3.195.

Save your file.

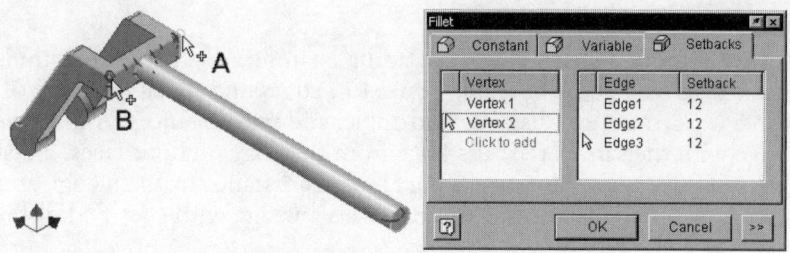

Figure 3.194 *Setbacks at two vertices being specified*

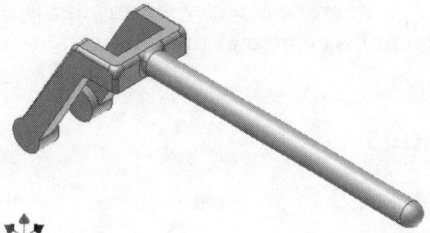

Figure 3.195 *Fillet feature with constant-radius fillets, variable-radius fillets, and setbacks placed.*

 TUTORIAL 3.42

In this tutorial, you will place a fillet feature on the base of the punch set.

1. Open the file *PunchBase.ipt*, if you already closed it.
2. With reference to Figure 3.196, rotate your display.
3. Select Fillet on the Part Features Panel Bar or toolbar.
4. Select edges A and B in Figure 3.196 and specify a radius of 20 mm.
5. In the Fillet dialog box, select Click to add.
6. Select edges C, D, E, and F Figure 3.196 and specify a radius of 30 mm.
7. Click the OK button.

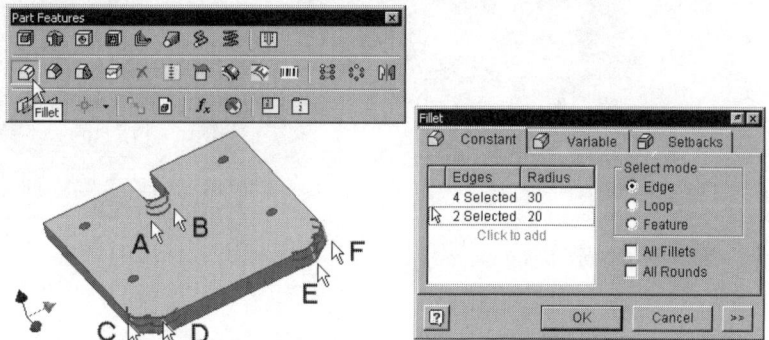

Figure 3.196 *Fillet feature being placed*

A fillet feature with two sets of fillet edges is placed. Save your file.

CHAMFER FEATURE

Another way to remove a sharp edge is to add a chamfer. A chamfer feature is a placed feature; it causes an edge to be beveled. You select edges and specify the size of the bevel in one of three ways: distance, distance and angle, and two distances. A distance chamfer bevels an edge with the same offset distance from the edge on both faces, a distance and angle chamfer defines a bevel by specifying an offset distance from an edge and an angle from one face, and a two distances chamfer bevels an edge with specified distances from each face.

In the More panel of the Chamfer dialog box, you can set options to determine whether all edges sharing a tangent point are selected when an edge is selected (Edge Chain), and whether a flat or a corner point is created at the intersection of chamfer edges (Setback).

TUTORIAL 3.43

In this tutorial, you will place a chamfer feature on the base of the punch set.

1. Continuing with the file *PunchBase.ipt*, select Chamfer in the Part Features panel bar or toolbar.
2. In the Chamfer dialog box, select the Distance button and set the offset distance to 15 mm.
3. Select edges A, B, C, D, and E in Figure 3.197.
4. Click the OK button.

A chamfer feature is placed. Save your file.

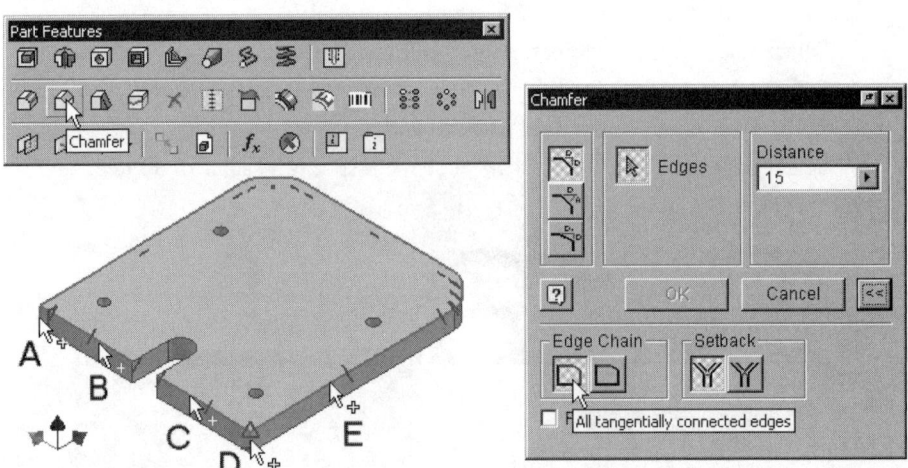

Figure 3.197 *Chamfer feature being placed*

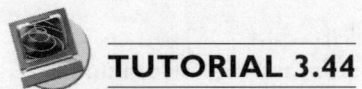

TUTORIAL 3.44

In this tutorial, you will place a chamfer feature on the hinge of the punch set.

1. Open the file *PunchHinge.ipt*, if you already closed it.
2. Select Chamfer on the Part Features Panel Bar or toolbar.
3. In the Chamfer dialog box, select the Distance button, and set the offset distance to 3 mm.
4. Select edges A and B in Figure 3.198.
5. Click the OK button.

A chamfer feature is placed. Save your file.

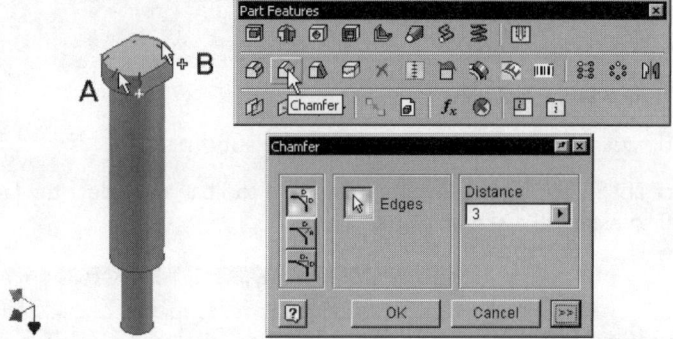

Figure 3.198 *Chamfer feature being placed*

SHELL FEATURE

Hollow objects are lighter in weight than solid objects without a cavity. To make a solid part hollow, you place a shell feature. A shell feature cuts a cavity in a solid by offsetting faces of the solid part. You can offset the faces in three ways: inside, outside, and both. Inside offsets the shell walls toward the solid part's interior, outside offsets the shell walls toward the part's exterior, and both offsets the shell walls equal distances to the inside and outside of the solid part. You can apply an uniform offset distance to the entire shelled solid part and you can also specify different offset distances to different faces. You can make openings on a shelled solid part by removing some of the faces while shelling. If you do not remove any faces, the shell cavity will be entirely enclosed within the solid part.

TUTORIAL 3.45

In this tutorial, you will construct a shell feature on the base of the punch set and then add a few more extruded features.

1. Open the file *PunchBase.ipt*, if you already closed it.
2. Set the display with reference to Figure 3.199.

3. Select Shell on the Part Features Panel Bar or toolbar.

4. In the Shell dialog box, select the Remove Faces button and the Inside button (if they are not already selected) and set the thickness to 8 mm.

5. Select face A in Figure 3.199.

6. Click the OK button. A shell feature is placed.

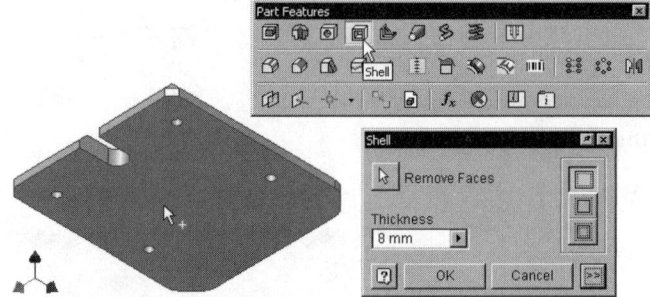

Figure 3.199 *Shell feature being placed*

Now add two more extruded features to the solid part.

7. Select 2D Sketch on the Inventor Standard toolbar and select the face in Figure 3.200 to establish a sketch plane.

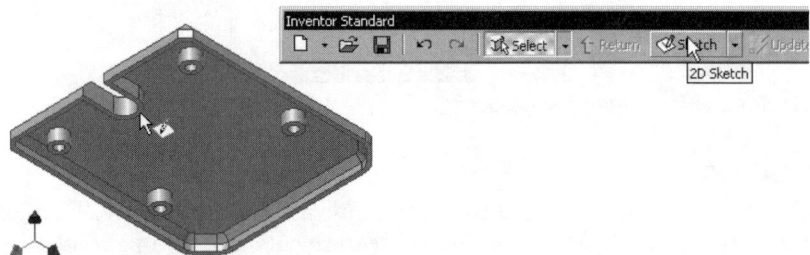

Figure 3.200 *Sketch plane being established*

8. With reference to Figure 3.201, construct a circle (diameter of 60 mm, 130 from edge A, and 180 mm from edge B).

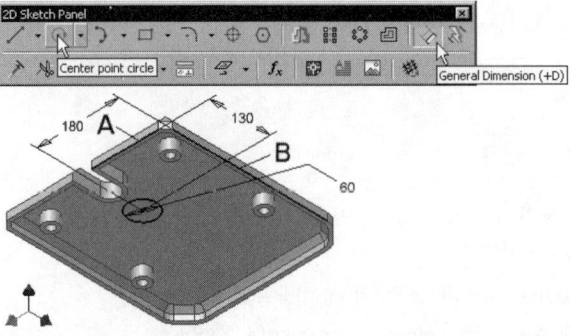

Figure 3.201 *Sketch constructed*

9. Select Return on the Inventor Standard toolbar.

10. Select Extrude on the Part Features Panel Bar or toolbar and select the circle that was just sketched.

11. In the Extrude dialog box, select To Next from the Extents pull-down list box and click the OK button. (See Figure 3.202.)

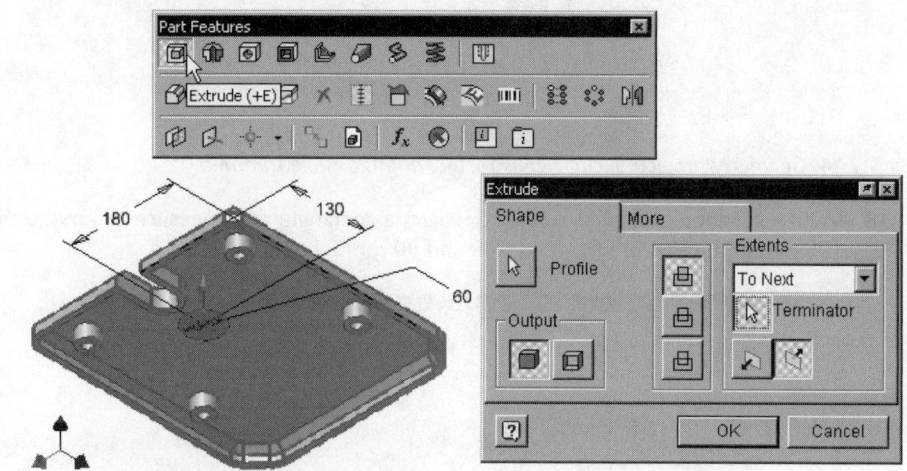

Figure 3.202 *Sketch being extruded*

12. Set the display to isometric view by right-clicking and selecting Isometric.

13. Select Wireframe Display from the Standard toolbar to set the display to wireframe mode.

14. Select 2D Sketch on the Inventor Standard toolbar and select face A in Figure 3.203.

15. Select Project Geometry on the 2D Sketch Panel Bar or toolbar and select edges B, C, D, and E in Figure 3.203.

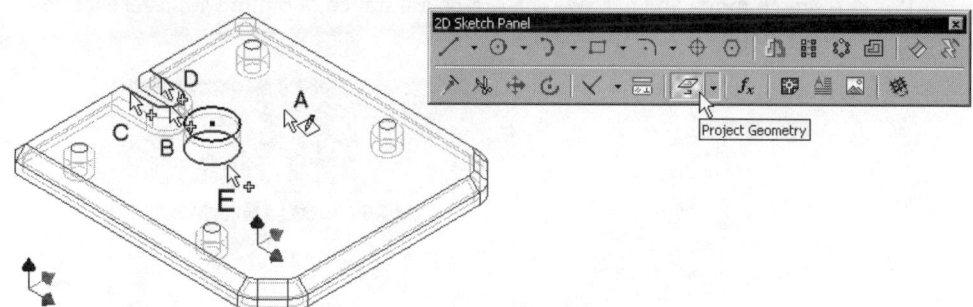

Figure 3.203 *Display set and sketch plane being established*

16. Construct a concentric circle with a diameter of 22 mm.

17. Trim the circle. (See Figure 3.204.)

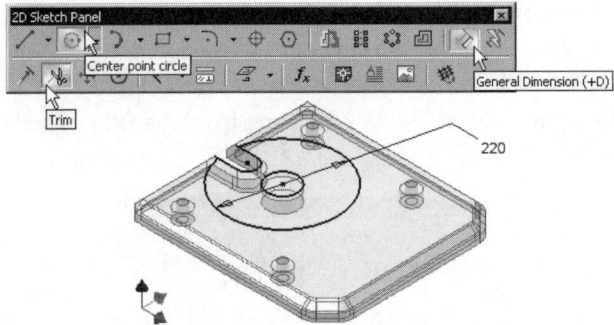

Figure 3.204 *Geometry projected, circle constructed, and the circle trimmed*

18. With reference to Figure 3.205, construct a rectangle that measures 60 mm times 180 mm and is 40 mm from edge A and 90 mm from edge B.

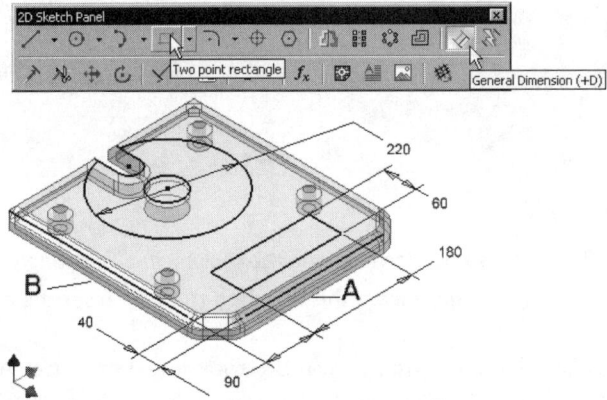

Figure 3.205 *Rectangle constructed*

19. Exit sketch mode and extrude the sketch a distance of 8 mm with reference to Figure 3.206.

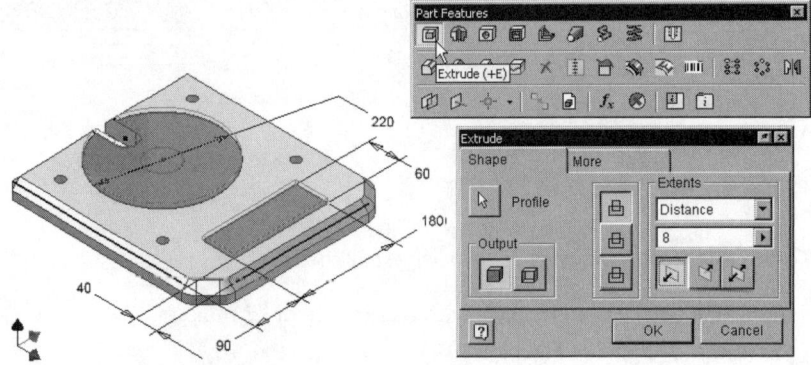

Figure 3.206 *Sketch being extruded*

Now add two fillet features.

20. With reference to Figure 3.207, add a fillet feature of 10 mm radius to four edges.

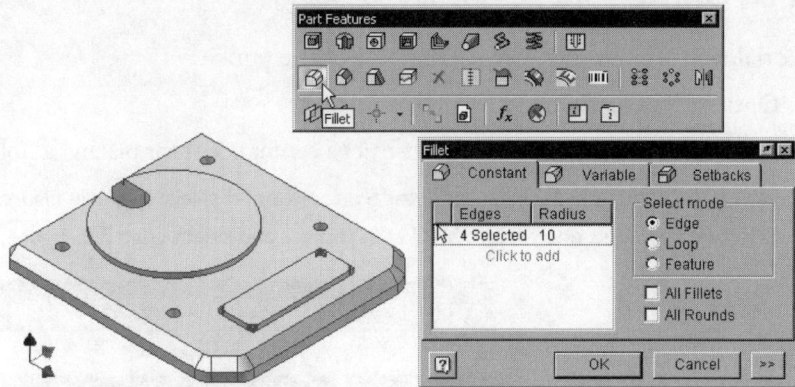

Figure 3.207 *Four edges being filleted*

21. With reference to Figure 3.208, add a fillet feature of 8 mm radius to two edges.

 Note: The bottom edges of the rectangle are being filleted.

Save your file.

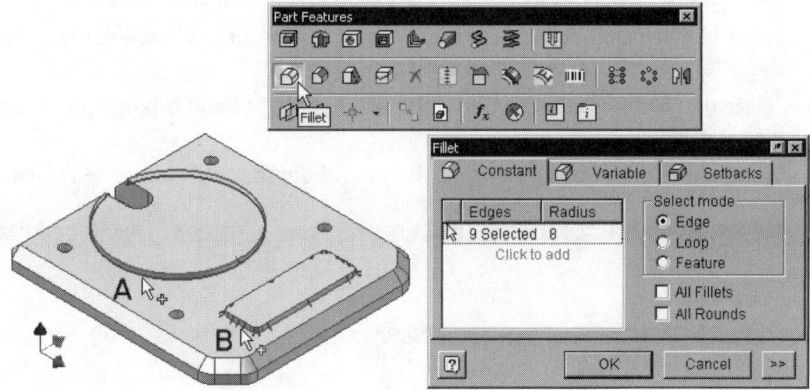

Figure 3.208 *Two edges being filleted*

HOLE FEATURE

Holes are used for many purposes in engineering and aesthetic design. A hole feature is a placed solid feature. There are three kinds of holes: drilled hole, countersink hole, and counterbore hole. To place a hole feature, you set up a sketch plane, construct center points to indicate the location of the holes, and specify the hole parameters (termination method, type of the hole, and whether the hole is threaded). Termination refers to the depth of a hole. You can drill a hole a specified distance, to a selected plane, and all the way through the solid part. If you place a blind hole, you can specify the drill tip angle. If you place a threaded hole, you need to specify the thread size.

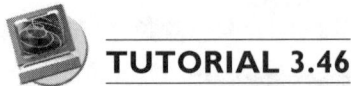

TUTORIAL 3.46

In this tutorial, you will place a hole on the post of the punch set.

1. Open the file *PunchPost.ipt.*

Now construct a sketch plane and construct a center point for placing a hole feature.

2. Select 2D Sketch on the Inventor Standard toolbar and select face A in Figure 3.209.

3. Select Project Geometry from the Sketch toolbar and select edge B in Figure 3.209.

Figure 3.209 *Sketch plane established and circular edge being projected*

4. Select Return on the Inventor Standard toolbar.

5. Select Hole on the Part Features Panel Bar or toolbar.

6. Select the center point A in Figure 3.210, if it is not already selected.

7. In the Holes dialog box, select the Centers button (if it is not already selected) and select the projected center point.

8. Select the Threads tab, check the Tapped and Right hand boxes, and select ANSI Metric M Profile.

9. Set the hole depth to 20 mm and the thread depth to 15 mm. (See Figure 3.210.)

 Note: Set the hole depth directly in the preview image of the hole in the Holes dialog box.

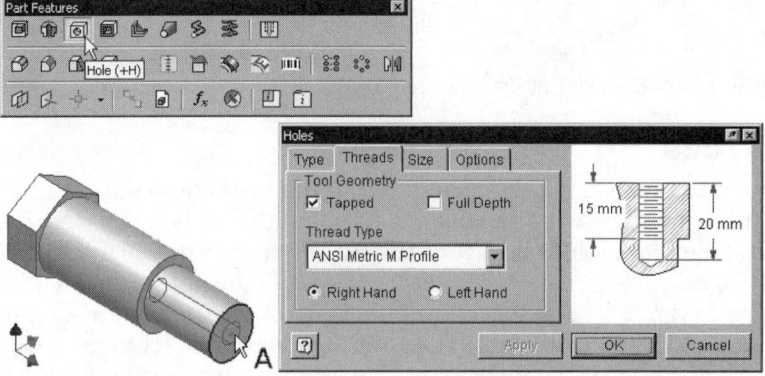

Figure 3.210 *Threads tab*

10. Select the Size tab, and set the nominal size to 6. (See Figure 3.211.)
11. Click the OK button.

A hole feature with thread is placed. The post is complete. Save and close your file.

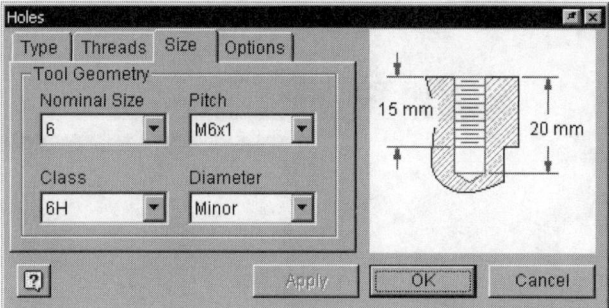

Figure 3.211 *Size tab*

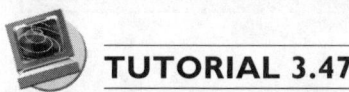

TUTORIAL 3.47

In this tutorial, you will place two hole features on the base of the punch set.

1. Open the file *PunchBase.ipt*, if you already closed it.
2. Select 2D Sketch on the Inventor Standard toolbar and select face A in Figure 3.212.
3. Select Point, Hole Center on the 2D Sketch Panel Bar or toolbar and select location B in Figure 3.212.
4. With reference to Figure 3.212, add two dimensions (both 10 mm) to locate the hole center.
5. Select Return on the Inventor Standard toolbar.

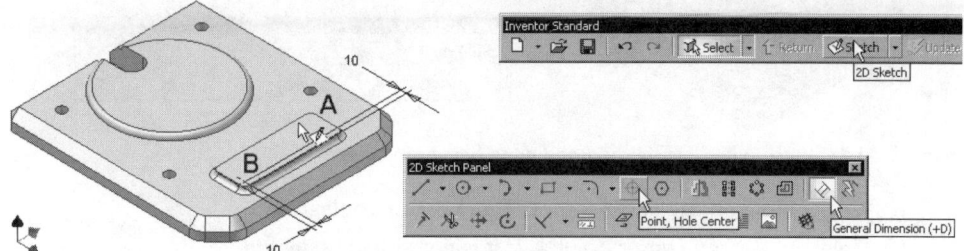

Figure 3.212 *Sketch plane established and hole center constructed*

6. Select Hole on the Part Features Panel Bar or toolbar.
7. Select hole center A, if it is not already selected.
8. In the Holes dialog box, select the Type tab, if it is not already selected.
9. Set operation to Drilled and termination to Through All.
10. Select the Threads tab.

11. Check the Tapped and Full Depth boxes, and select ANSI Metric M Profile.

12. Select the Size tab and set nominal size to 12 mm.

13. Click the OK button. (See Figure 3.213.)

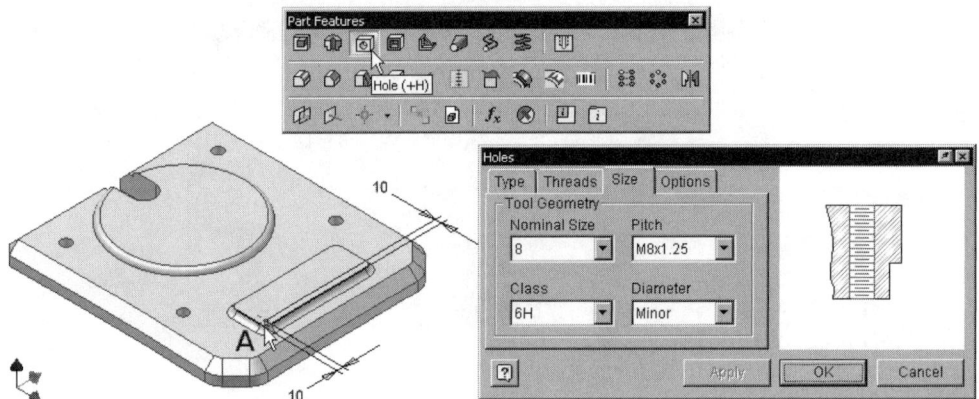

Figure 3.213 *Threaded hole feature being placed*

Now place another hole.

14. Rotate the display in accordance with Figure 3.214.

15. Select 2D Sketch on the Inventor Standard toolbar and select face A in Figure 3.214.

16. Select Project Geometry and select edge B in Figure 3.214.

17. Select Return on the Inventor Standard toolbar.

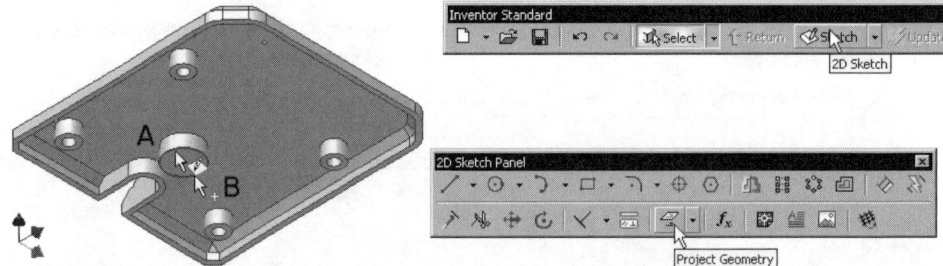

Figure 3.214 *Sketch plane established and edge being projected*

18. Select Hole on the Part Features Panel Bar or toolbar.

19. Select projected center point A, if it is not already selected.

20. In the Holes dialog box, select the Type tab, if it is not already selected.

21. Select Counterbore operation, and set Termination to Through All.

22. Set the drill diameter to 12 mm, counterbore diameter to 40 mm, and counterbore depth to 10 mm. (See Figure 3.215.)

Note: You may need to go to the Threads tab and clear the Tapped check box.

23. Click the OK button.

Two hole features are placed. Save your file.

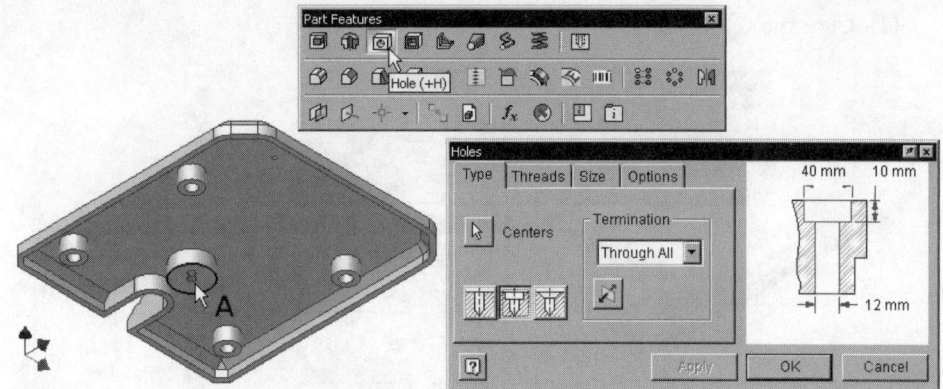

Figure 3.215 *Counterbore hole being placed*

TUTORIAL 3.48

In this tutorial, you will place a hole feature on the handle of the punch set.

1. Open the file *PunchHandle.ipt*, if you already closed it.

2. Select 2D Sketch on the Inventor Standard toolbar and select face A in Figure 3.216 to establish a sketch plane.

3. Select Project Geometry on the 2D Sketch Panel Bar or toolbar and select edge B in Figure 3.216 to project a geometry.

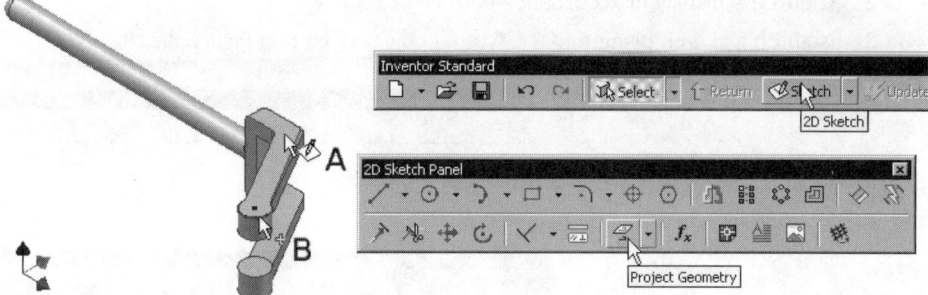

Figure 3.216 *Sketch plane established and geometry being projected*

4. Exit sketch mode and select Hole on the Part Features Panel Bar or toolbar.

5. Select the projected center point A in Figure 3.217.

6. Select the Type tab in the Holes dialog box, if it is not already selected.

7. Select the Counterbore button and set Termination to Through All.

8. Select the counterbore diameter to 20 mm and counterbore depth to 50 mm.

9. Select the Threads tab, and select the Tapped and Full Depth check boxes, and select the ANSI Metric M Profile thread type.

10. Select the Size tab, and set nominal size to 12 mm.

11. Click the OK button.

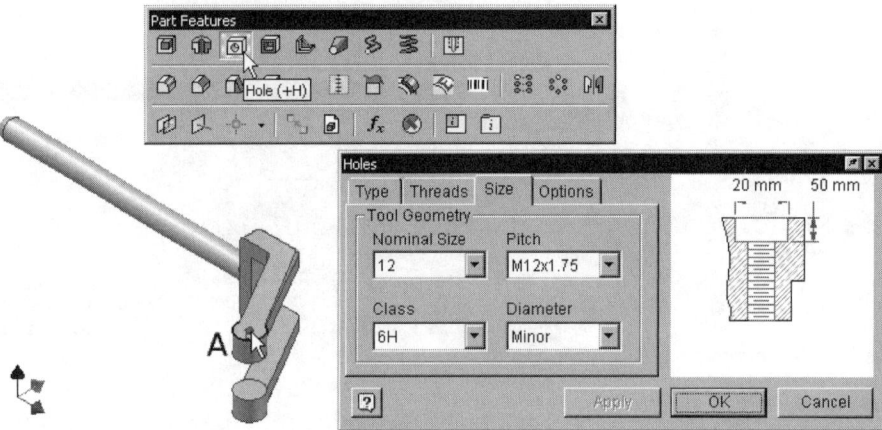

Figure 3.217 *Hole feature being placed*

The solid part is complete. Save and close your file.

 **TUTORIAL 3.49**

In this tutorial, you will place a number of holes on the column of the punch set.

1. Open the file *PunchColumn.ipt*, if you already closed it.

2. Rotate the display in accordance with Figure 3.218.

3. Establish a sketch plane on face A and project edge B in Figure 3.218.

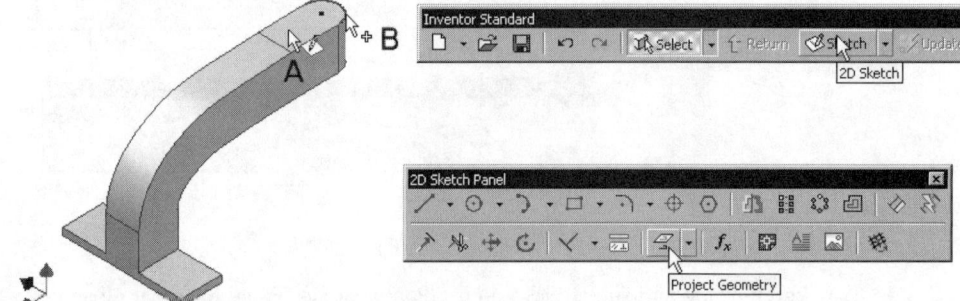

Figure 3.218 *Sketch established and edge projected*

4. Select Return on the Inventor Standard toolbar.

5. Select Hole on the Part Features Panel Bar or toolbar.

6. Select the projected center hole A in Figure 3.219, if it is not already selected.

7. In the Holes dialog box, select Counterbore operation, set termination to Through All, and set counterbore diameter to 40 mm, counterbore depth to 15 mm, and drill size to 30 mm. (See Figure 3.219.)

8. Click the OK button.

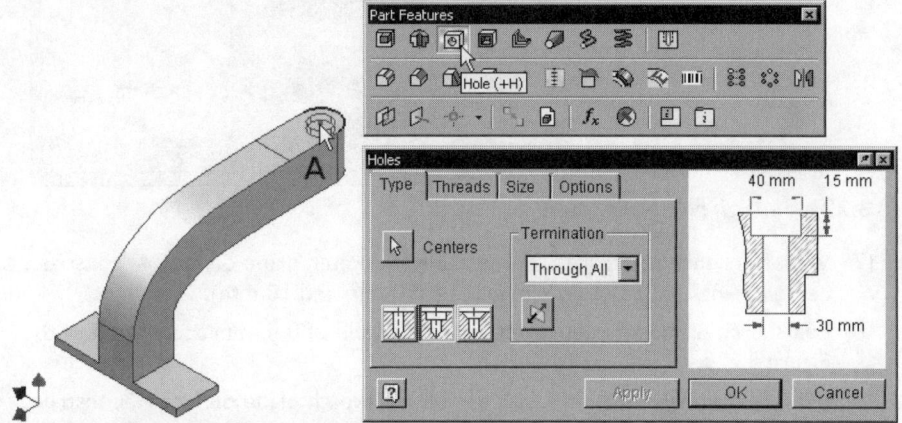

Figure 3.219 *Counterbore hole being placed*

9. Establish a sketch on face A in Figure 3.220.

10. Select Point, Center Hole on the 2D Sketch Panel Bar or toolbar and select location B in Figure 3.220.

11. Add two dimensions (46 mm and 80 mm). (See Figure 3.220.)

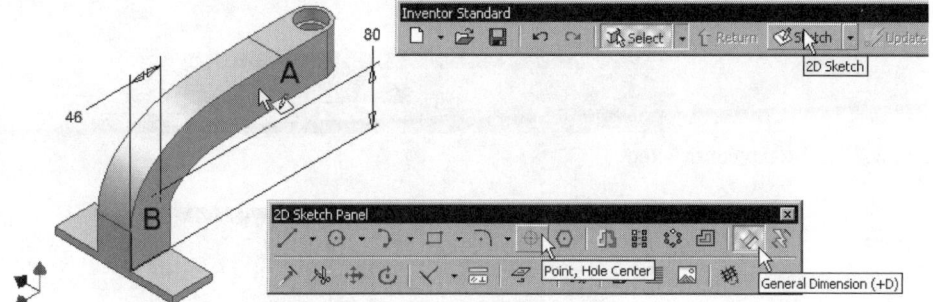

Figure 3.220 *Center point constructed*

12. Select Return on the Inventor Standard toolbar.

13. Select Hole on the Part Features Panel Bar or toolbar.

14. Select the projected center hole A in Figure 3.221, if it is not already selected.

15. In the Holes dialog box, select drilled operation, set termination to Through All, and set the diameter of the hole to 20 mm. (See Figure 3.221.)

16. Click the OK button.

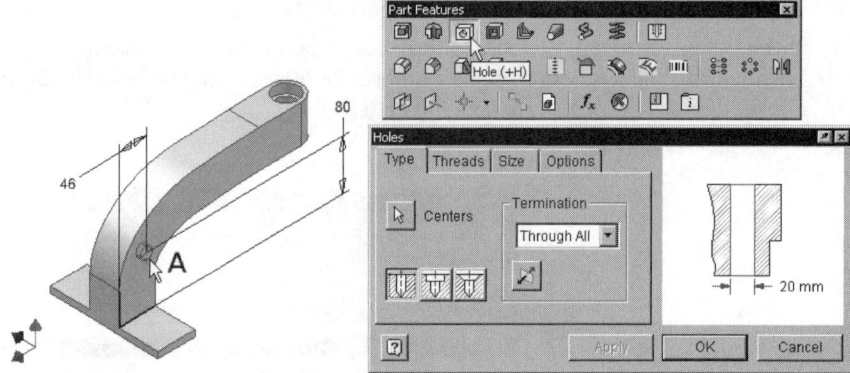

Figure 3.221 *Through hole being placed*

17. With reference to Figure 3.222, establish a sketch plane on face A, construct a center point, and add two dimensions (90 mm and 18 mm).

18. Exit sketch mode and construct a through hole of 8 mm in accordance with Figure 3.223.

19. With reference to Figure 3.224, establish a sketch plane on face A, construct a center point, add two dimensions (both 10 mm), and construct a through hole of 8 mm diameter.

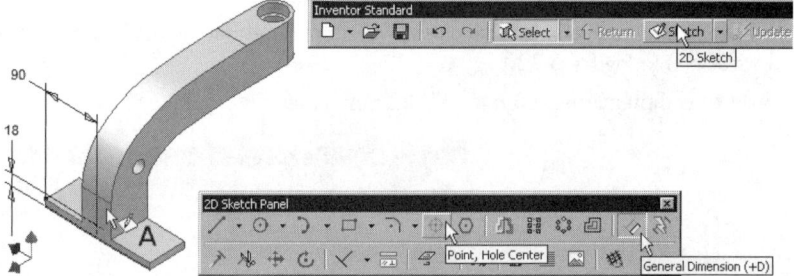

Figure 3.222 *Sketch constructed*

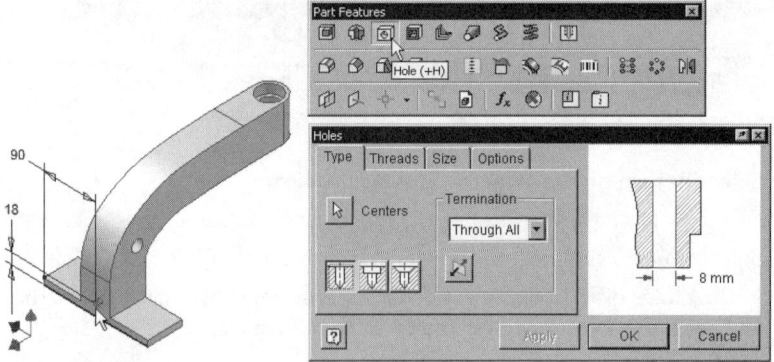

Figure 3.223 *Through hole being placed*

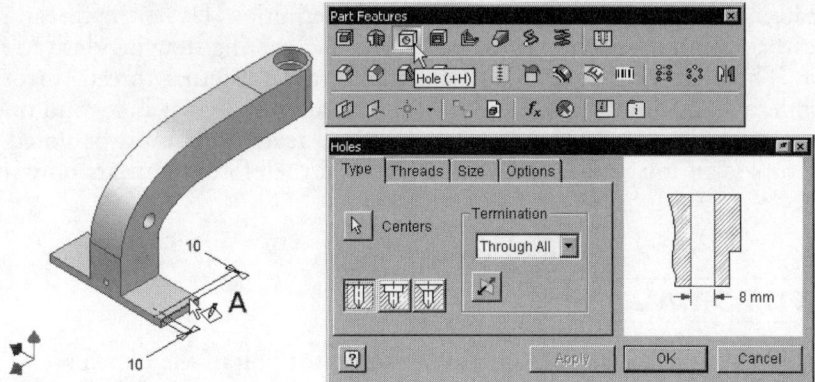

Figure 3.224 *Through hole being placed*

20. With reference to Figure 3.225, construct a fillet feature of 10 mm radius on the selected edges.

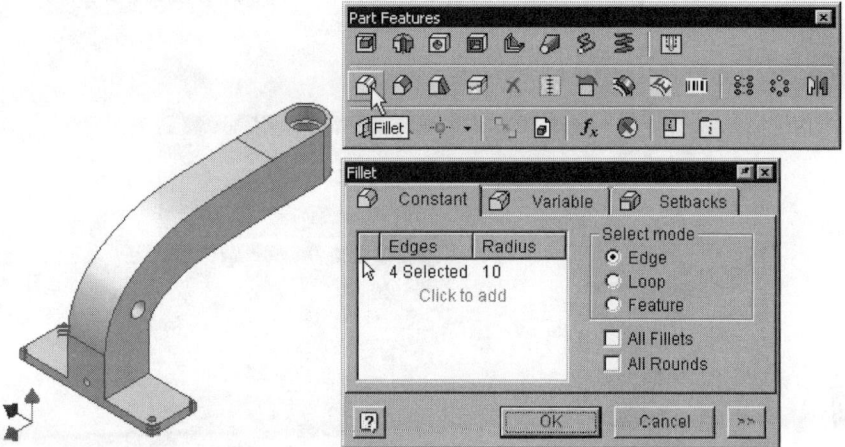

Figure 3.225 *Fillet being constructed*

Holes are constructed. Save your file.

THREAD FEATURE

A thread is a helical groove cut on a shaft or a hole. It is an essential engineering element and is commonly found in fasteners such as bolts and nuts. Because screw thread forms have to conform to national or international standards and the work involved in constructing a helical groove is tedious, requiring a lot of memory space to store in the computer, we use symbols in conjunction with appropriate notes in our drawings to depict a thread instead of constructing the helical curves. The same is true for making a thread feature in a solid part.

To represent a thread in a solid part, you do not have to actually cut a groove on the solid part. Instead, you simply place a cosmetic, shaded image on a selected circular feature with thread information stored in the solid part's definition. Using the thread information, thread convention will be generated on the engineering drawing derived from the solid part. There are two ways to construct a cosmetic internal thread (incorporate a thread while making a hole and place a thread feature to circular hole) and one way to construct a cosmetic external thread (place a thread feature). It must be noted that the thread feature is an image mapped on the face and therefore it appears only in shaded display mode.

 ## TUTORIAL 3.50

In this tutorial, you will place a cosmetic thread on the bolt of the punch set.

1. Open the file *PunchBolt.ipt*, if you already closed it.
2. Select Thread on the Part Features Panel Bar or toolbar.
3. Select cylindrical face A in Figure 3.226.
4. In the Thread Feature dialog box, clear the Full Length box, set length to 15 mm and offset to 0 mm, and click the OK button.

The solid part is complete. Save and close your file.

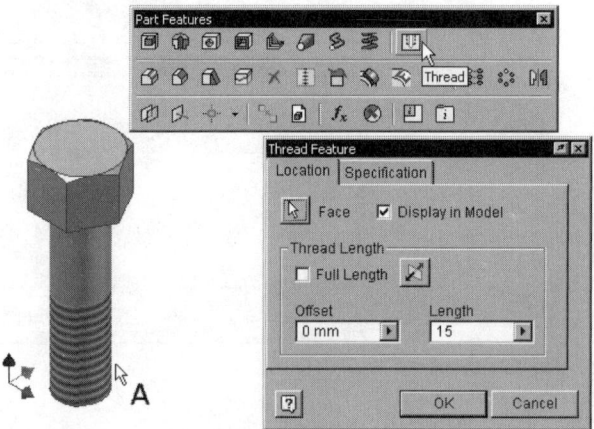

Figure 3.226 *Thread feature being placed on the bolt*

 ## TUTORIAL 3.51

In this tutorial, you will place a thread feature on the hinge of the punch set.

1. Open the file *PunchHinge.ipt*, if you already closed it.
2. With reference to Figure 3.227, construct a thread with full length on cylindrical face A.

The solid part is complete. Save and close your file.

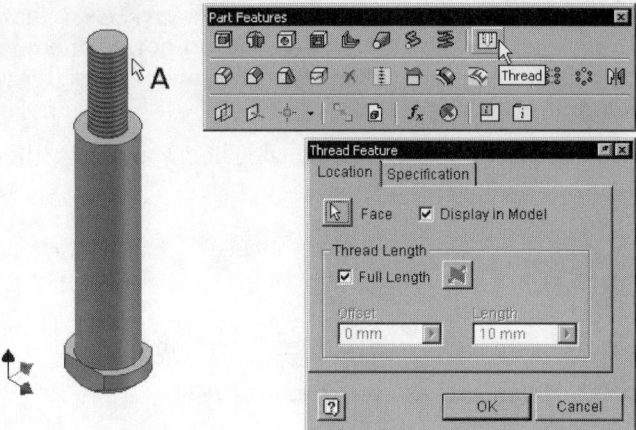

Figure 3.227 *Thread feature being placed on the hinge*

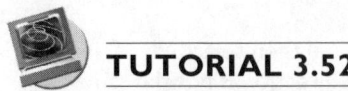

TUTORIAL 3.52

In this tutorial, you will place a thread feature on the index pin of the punch set.

1. Open the file *PunchIndex.ipt*, if you already closed it.
2. Construct a thread with a thread length of 20 mm on cylindrical face A. (See Figure 3.228.)

The solid part is complete. Save and close your file.

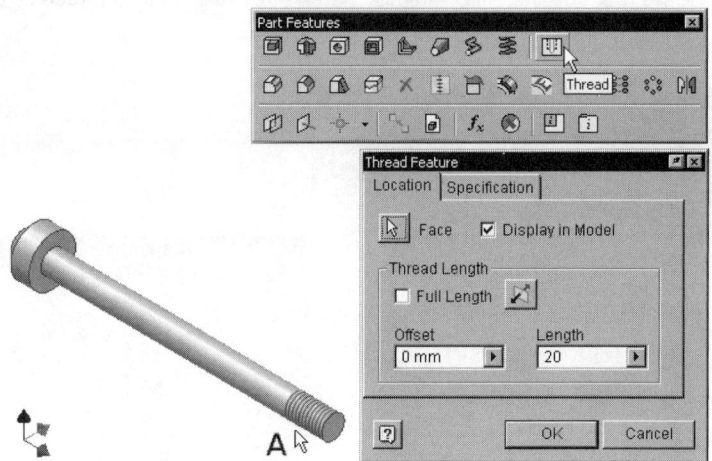

Figure 3.228 *Thread feature being placed on the hinge*

RECTANGULAR PATTERN FEATURE

To repeat selected solid features in a rectangular pattern, you place a rectangular pattern feature. You select features to be repeated, specify one or two directions, and state the spacing between repeated occurrences of the selected features. As we have said, directions of the pattern can be linear or along curves. If you do not want some of the repeated occurrences of the pattern, you select them on the Browser Bar after you construct the pattern, right-click, and select Suppress.

In the More panel of the Rectangular Pattern dialog box, you specify the creation method of the pattern: identical or adjust to the model.

TUTORIAL 3.53

In this tutorial, you will construct a rectangular pattern on the base of the punch set.

1. Open the file *PunchBase.ipt*, if you already closed it.
2. Set the display to an isometric view.
3. Select Rectangular Pattern on the Part Features Panel Bar or toolbar.
4. Select hole A in Figure 3.229.
5. Select the Direction button in the Direction 1 box and then select edge B in Figure 3.229. (Select the Flip direction button if the direction is not the same as that indicated in Figure 3.229.)
6. Select the Direction button in the Direction 2 box and then select edge C in Figure 3.229 (Select the Flip direction button if necessary.)
7. In the Rectangular Pattern dialog box, set the number count in Direction 1 to 2, spacing in Direction1 to 160 mm, the number count in Direction 2 to 2, and spacing in Direction 2 to 40 mm.
8. Click the OK button. A rectangular pattern feature is placed. (See Figure 3.229.)

Save your file.

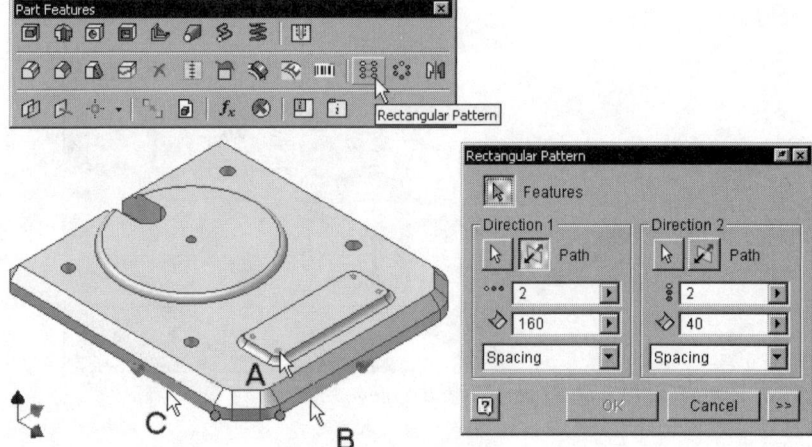

Figure 3.229 *Rectangular pattern feature being placed*

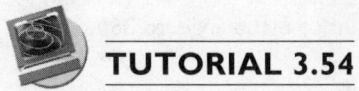

TUTORIAL 3.54

In this tutorial, you will construct a rectangular pattern feature on the column of the punch set.

1. Open the file *PunchColumn.ipt*, if you already closed it.
2. With reference to Figure 3.230, construct a rectangular pattern of hole A in direction B (spacing 160 mm) and direction C (40 mm).

The column is complete. Save and close your file.

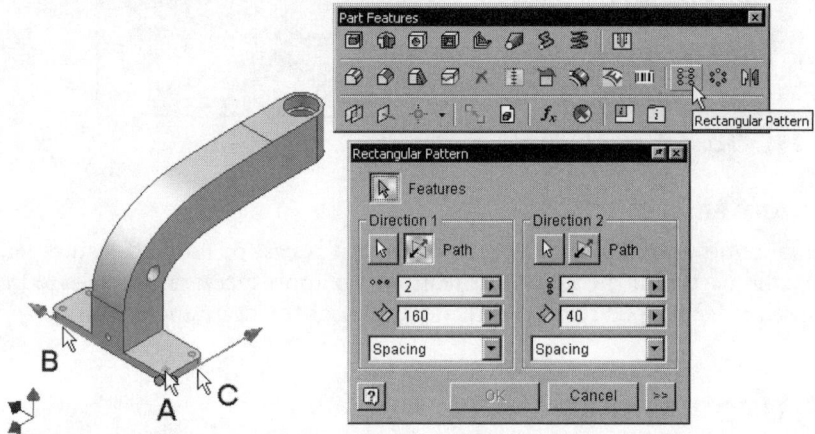

Figure 3.230 *Rectangular pattern being placed*

CIRCULAR PATTERN FEATURE

To repeat selected solid features in a circular pattern, you place a circular pattern feature. You select features to be repeated, specify an axis of the circular pattern, state the included angle or angle between occurrences, and indicate the direction of the pattern. Similar to placing a rectangular pattern, you can suppress individual occurrences by selecting them on the Browser Bar after they are constructed, right-click, and select Suppress.

In the More panel of the Circular Pattern dialog box, you specify the creation method (identical or adjust to model) and select position method (incremental or fitted).

TUTORIAL 3.55

In this tutorial, you will construct a circular pattern on the table of the punch set.

1. Open the file *PunchTable.ipt*, if you already closed it.
2. Select Circular Pattern on the Part Features Panel Bar or toolbar.
3. Select the Features button and select features A and B in Figure 3.231.
4. Select the Rotation Axis button and select feature C in Figure 3.231.

5. In the Circular Pattern dialog box, set number count to 6 and angle to 360 deg.

6. Click the OK button.

The solid part is complete. Save and close your file.

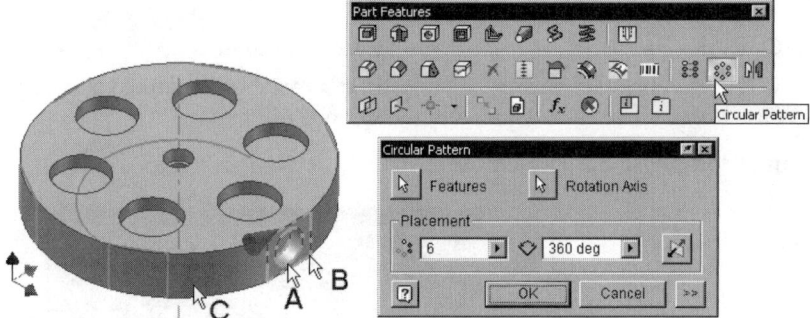

Figure 3.231 *Circular feature being placed*

FACE DRAFT FEATURE

To facilitate removal of a part from a mold after it is cast or molded, you taper selected faces of a solid part with the face draft feature. To apply face draft to selected faces of a solid, you select faces, specify draft direction, and set the draft angle.

 ## TUTORIAL 3.56

In this tutorial, you will place face draft features on the base of the punch set.

1. Open the file *PunchBase.ipt*, if you already closed it.

2. Select Face Draft on the Part Features Panel Bar or toolbar.

3. Select face A in Figure 3.232 to specify the direction.

4. Select face B in Figure 3.232 to specify the faces to apply face draft.

5. In the Face Draft dialog box, set the face draft angle to 3 deg.

6. Click the OK button. A face draft feature is placed.

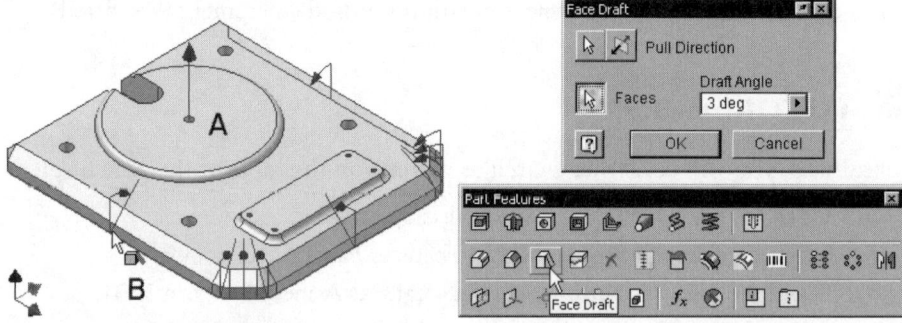

Figure 3.232 *Face draft feature being placed*

7. Set the display in accordance with Figure 3.233.

8. Add face draft feature to faces A, B, C, and D highlighted in Figure 3.233.

The solid part is complete. Save your file.

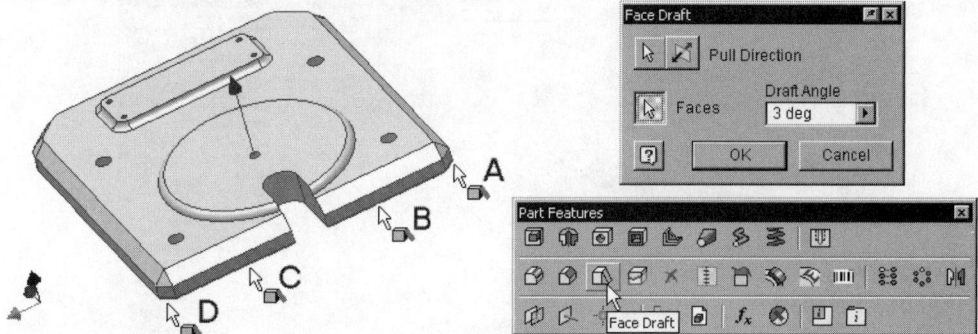

Figure 3.233 *Display rotated and face draft being placed*

EDITING METHODS

You can edit the features of a solid part in several ways. You can display the dimensions of the feature, change them, and perform an update. You can select the feature on the Browser Bar, right-click, select Edit Feature, and change the parameters in the feature creation dialog box. If you are dealing with a sketched solid feature, you can select the sketch on the Browser Bar, right-click, select Edit Sketch, modify the sketch, and update the solid part.

EDIT DIMENSIONS

The simplest way to modify a solid feature is to display, select, and edit its dimensions.

TUTORIAL 3.57

In this tutorial, you will display the dimensions of the sketch of a sketched solid feature and modify one of the dimensions.

1. Open the file *PunchTable.ipt*.

2. Select extruded solid feature A in Figure 3.234 (Extrusion1) on the Browser Bar, right-click, and select Show Dimensions.

Tip: On the Browser Bar, the name of the each feature is unique. The first extruded solid feature is Extrusion1, the second is Extrusion2, and so on. If a feature is deleted, its name will not be used again in the solid part.

3. Select and double-click dimension B in Figure 3.234 and change it to 50 mm.

4. Select Update on the Inventor Standard toolbar.

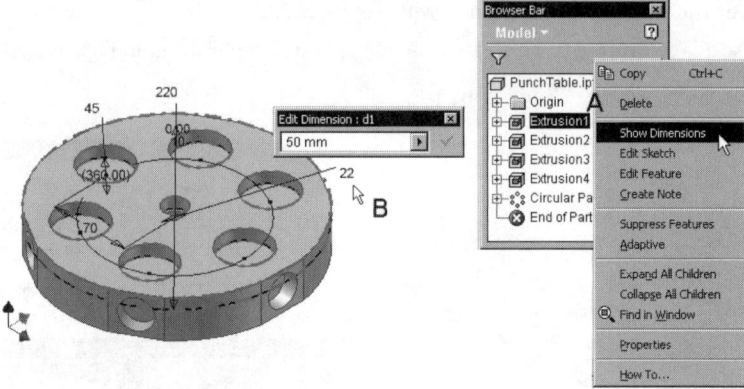

Figure 3.234 *Dimension being changed*

The solid part is modified. Save and close your file.

 TUTORIAL 3.58

In this tutorial, you will display a dimension of a placed solid feature and modify its value.

1. Open the file *PunchBase.ipt*.
2. Select the shell feature on the Browser Bar, right-click, and select Show Dimensions.
3. Double-click dimension A in Figure 3.235 (this dimension depicts the shell thickness) and change it to 10 mm.
4. Select Update on the Inventor Standard toolbar.

The thickness of the shell feature is modified. Save and close your file.

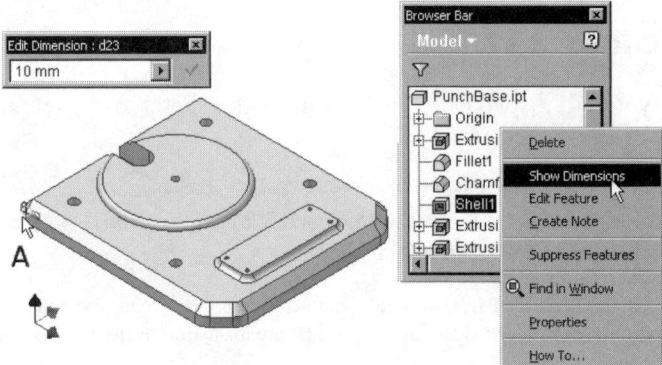

Figure 3.235 *Shell thickness being modified*

EDIT FEATURE

To change the way a feature is constructed or the way the feature is combined with the solid part, you edit the feature.

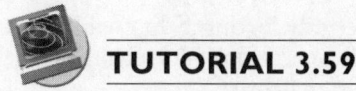

TUTORIAL 3.59

In this tutorial, you will change a join operation of an extruded solid feature to a cut operation, use the Design Doctor to help rectify the error, suppress a feature, and delete a feature.

1. Open the file *PunchBolt.ipt*.
2. Select File > Save Copy As and specify a new file name (*PunchNut.ipt*).
3. Close the file.
4. Open the file *PunchNut.ipt*.
5. Select the extruded feature in Figure 3.236 on the Browser Bar, right-click, and select Edit Feature.
6. In the Extrusion dialog box, select the Cut button and click the OK button. (This changes the join operation to a cut operation.)

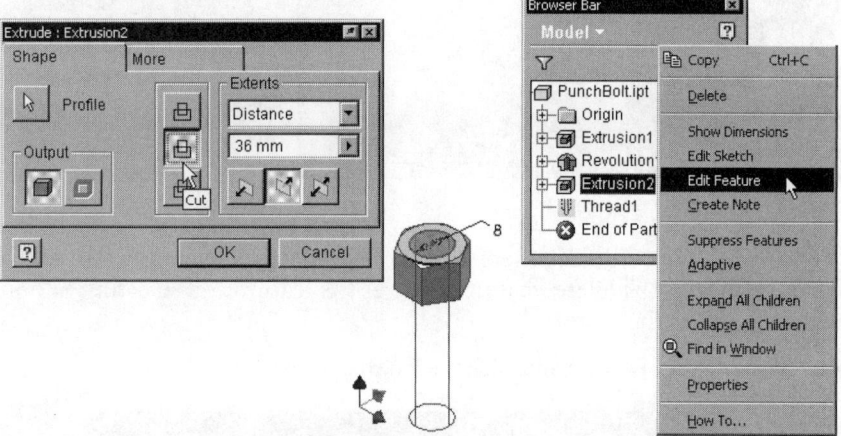

Figure 3.236 *Join operation being changed to cut operation*

DESIGN DOCTOR

After you change the join operation to a cut operation, the thread feature that is built on the extruded solid becomes invalid because the extruded solid has changed from an external cylindrical object to an internal cylindrical object. As a result, an error message is displayed. (See Figure 3.237.)

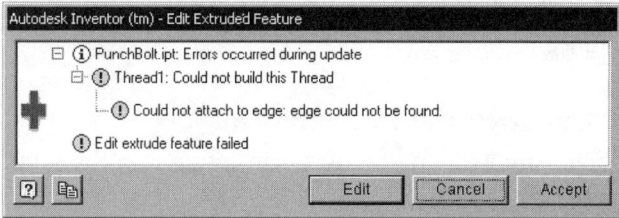

Figure 3.237 *Error message*

Now you have three choices: to edit, cancel, or accept. Selecting Edit enables you to do the editing again, selecting Cancel aborts the operation, and selecting Accept accepts the change and any invalid features that may arise.

7. Select the Accept button.

Now use Design Doctor to help correct the error.

8. Select Recover on the Inventor Standard toolbar. (See Figure 3.238.)

9. In the Design Doctor dialog box, select the Next button twice.

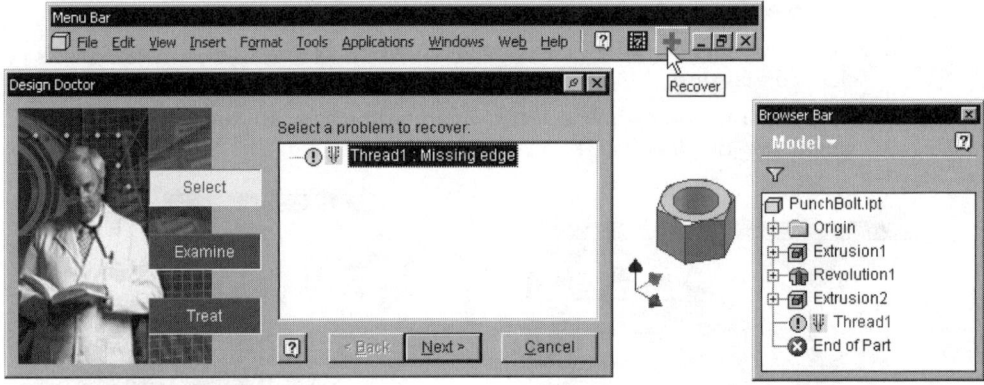

Figure 3.238 *Recovering an error*

Referring to Figure 3.239, there are two suggestions: Delete Feature and Suppress Feature. Selecting Delete Feature removes the feature and selecting Suppress Feature suppresses the feature.

10. Select Delete Feature and click the Finish button.

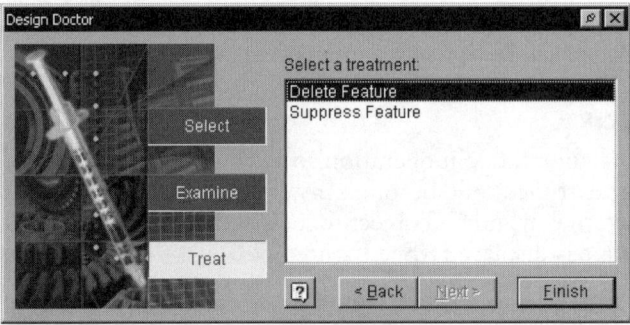

Figure 3.239 *Design Doctor dialog box*

SUPPRESS FEATURE AND DELETE FEATURE

To remove a feature with an option of reinstating it in the solid, you suppress the feature. A suppressed feature is disregarded but can be retrieved by unsuppressing. Any time you want to recover a suppressed feature, you can select it on the Browser Bar, right-click, and deselect Suppress.

DELETE FEATURE

If you delete a feature from the solid part, and the part file is saved and re-opened, you can no longer retrieve the feature.

11. Select the extruded feature in Figure 3.240, right-click, and select Delete.

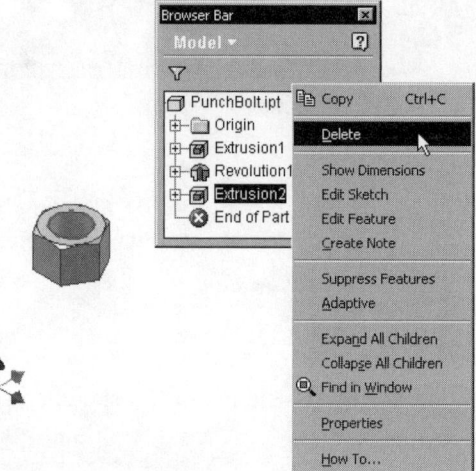

Figure 3.240 *Feature being deleted*

Because the feature that you are going to delete is a sketched solid feature, you have to decide whether to delete the sketch or not.

12. Clear the sketches of selected features box and click the OK button in the Delete Features dialog box. The feature is deleted but its sketch is not deleted. (See Figure 3.241.)

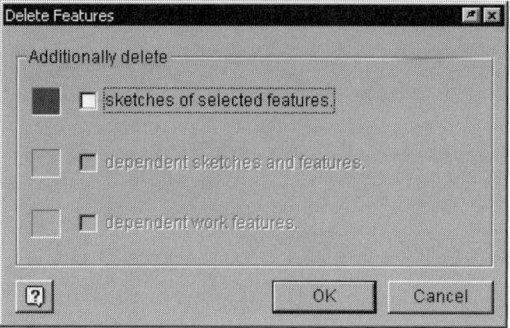

Figure 3.241 *Delete Features dialog box*

Now complete the solid part by adding a threaded hole.

13. With reference to Figure 3.242, activate the sketch, delete the circle, and place a threaded hole of M8 nominal size.

A hole with internal thread is constructed. The nut is complete. Save and close your file.

Figure 3.242 *Threaded hole placed*

EDIT SKETCH

To change the basic shape of the sketch, you modify the sketch. You can delete or add objects, and you can change the geometric constraints and parametric dimensions.

TUTORIAL 3.60

In this tutorial, you will edit the sketch elements of a sketched solid feature.

1. Open the file *PunchBush.ipt*.
2. Select the Revolved feature on the Browser Bar, right-click, and select Edit Sketch.
3. Select Chamfer on the 2D Sketch Panel Bar or toolbar.
4. In the 2D Chamfer dialog box, set the chamfer distance to 2 mm.
5. Select edges A and B in Figure 3.243 and then click Done.

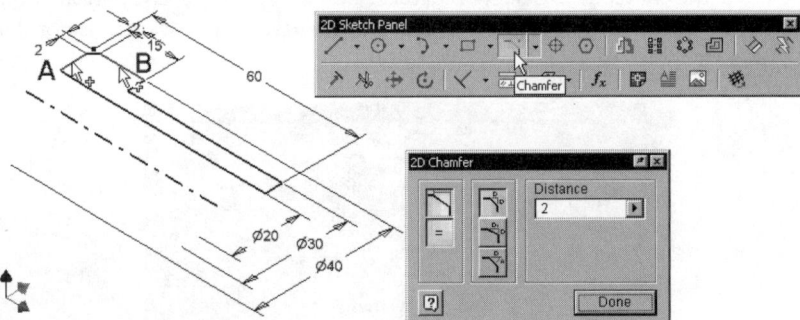

Figure 3.243 *Sketch being edited*

6. Select Return on the Inventor Standard toolbar

The part is modified and updated. Save and close your file.

FEATURE RE-ORDER

Features that you construct in a solid part form a hierarchy in the solid part's database. The hierarchy is exhibited on the Browser Bar. In many cases, the final outcome of the solid part depends largely on the sequence of feature construction. The following tutorial will help you appreciate how the sequence of operation affects the outcome.

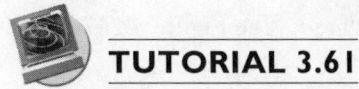

TUTORIAL 3.61

In this tutorial, you will construct a solid part with three features. You will first construct an extruded solid feature. Then you will place a hole feature. After that, you will place a shell feature. Upon finishing the solid part, you will re-order the sequence to place the shell feature before the hole feature.

1. Start a new part file. Use the metric template.
2. Construct a rectangle that measures 40 mm by 30 mm.
3. Extrude the rectangle a distance of 20 mm to form an extruded solid feature. (See Figure 3.244.)

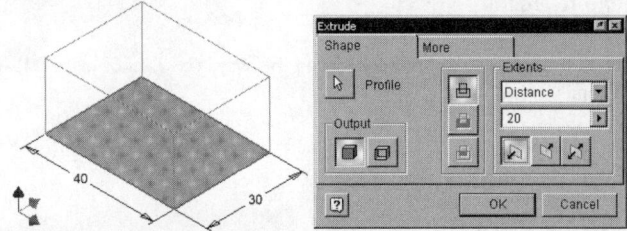

Figure 3.244 *Rectangle being extruded to form an extruded solid*

4. With reference to Figure 3.245, place a hole (10 mm diameter) at the middle of the top face of the extruded solid.

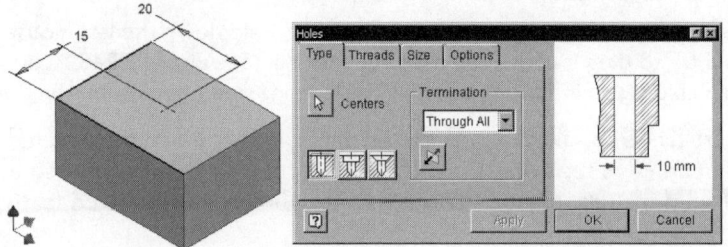

Figure 3.245 *Hole feature being placed*

5. Add a shell feature with a shell thickness of 1 mm and remove a face. (See Figure 3.246.)

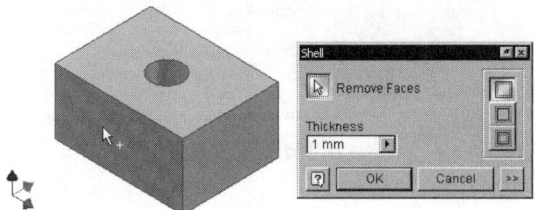

Figure 3.246 *Shell feature placed*

6. Select the Shell feature on the Browser Bar, hold down the left mouse button, and drag it above the hole feature on the Browser Bar.

With the release of the mouse button, the feature is re-ordered. (See Figure 3.247.) Save and close your file (file name: *Reorder.ipt*).

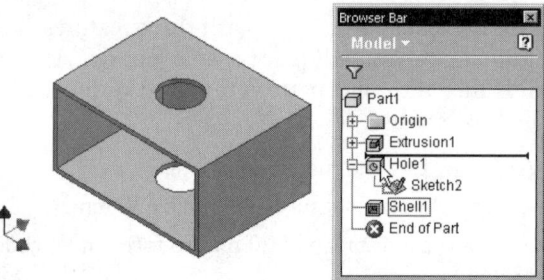

Figure 3.247 *Feature re-ordered*

 Tip: Not all the features can be re-ordered this way. You cannot place a feature before a feature it depends on.

 # TUTORIAL 3.62

To appreciate the meaning of dependency, you will try to re-order the features of the nut that you construct.

1. Open the file *PunchNut.ipt*.
2. Select the hole feature on the Browser Bar, hold down the left mouse button, and try to drag it above the revolved feature. (See Figure 3.248.) You will find a prohibited sign indicating that this re-ordering operation cannot be done.

This solid part has three features. Because the hole feature is constructed on a sketch projected from an edge resulting from the intersection of the revolved solid feature and the extruded solid feature, placing the hole feature before the revolved feature causes the hole's sketch to be invalid.

Figure 3.248 *Selected feature cannot be re-ordered*

MEASUREMENT OF OBJECTS

It might be necessary to perform measurement while you are constructing a solid part. Through the Tools menu, you can measure distance, angle, loop, and area.

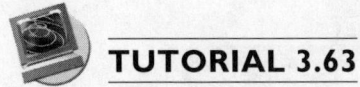

TUTORIAL 3.63

In this tutorial, you will measure the nut. First you will measure the radius.

1. Open the file *PunchNut.ipt*, if you already closed it.
2. Select Tools > Measure Distance.
3. With reference to Figure 3.249, select the circular edge to measure the radius.

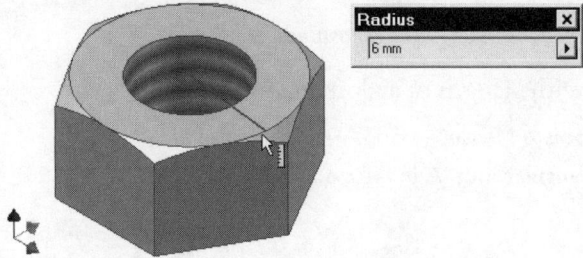

Figure 3.249 *Radius measured*

Now measure the distance between the top face's center and the bottom face.

4. With reference to Figure 3.250, select edge A, right-click, and select Select Other.
5. Cycle through the Select Other dialog box until the center of the top face is highlighted.
6. Select face B, right-click, and select Select Other.
7. Cycle through the Select Other dialog box until the bottom face is highlighted.

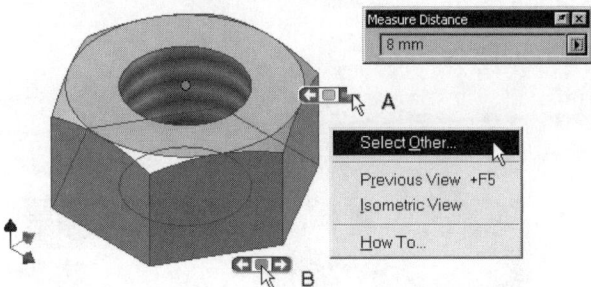

Figure 3.250 *Distance between the center of the top face and the bottom face being measured*

Now measure an angle.

8. Select Tools > Measure Angle.
9. Select edge A in Figure 3.251.
10. Select edge B in Figure 3.251.
11. To reset the measurement reading, select the Arrow in the Angle dialog box and select Reset.

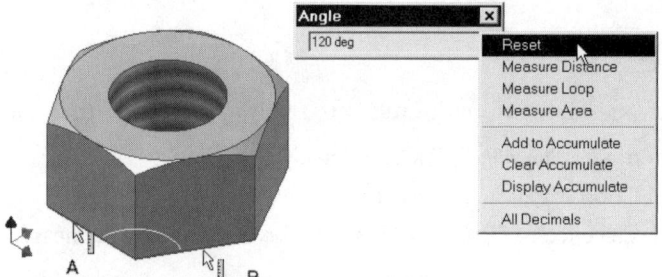

Figure 3.251 *Angle between two edges measured*

Now measure the length of a closed loop.

12. Select Tools > Measure Loop.

13. Select triangular loop A in Figure 3.252.

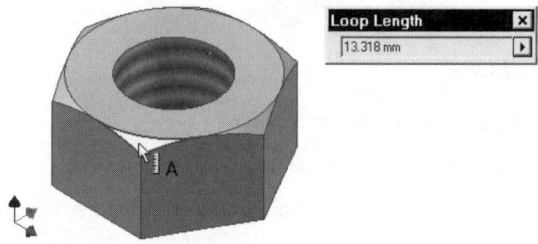

Figure 3.252 *Loop length being measured*

Now measure the area of a face.

14. Select Tools > Measure Area.

15. Select face A in Figure 3.253.

The measuring is finished. Close your file.

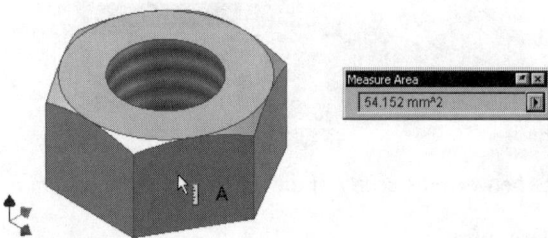

Figure 3.253 *Area being measured*

APPEARANCE AND PROPERTIES

In addition to representing a 3D object's vertices, edges, faces, and volume, you can incorporate a solid part's aesthetic appearance and physical properties. Now you will learn how to set the lighting style to manage the general lighting environment of the display, assign material and color to the solid part, and set properties for the solid part.

LIGHTING STYLE

When you set the display to hidden edge display or shaded mode, the solid part is rendered and is illuminated by a default light. To modify the way an object is illuminated, you set up lighting styles. In each lighting style, you set up a number of lights, position the lights, and assign color to the light.

TUTORIAL 3.64

In this tutorial, you will modify the lighting of a solid part. First you will change the location of the light.

1. Open the file *PunchNut.ipt*, if you already closed it.
2. Select Format > Lighting. (See Figure 3.254.)
3. In the Style Name area of the Lighting dialog box, select Default (a one-light setting). For each light style, you can have a maximum of four lights (numbers 1 through 4). The On/Off box enables you to turn on or off these lights. To modify the setting of a light, select the light in the Settings area to display the effect of the light.
4. In the Settings area, select the horizontal scroll bar and move it horizontally from left to right to appreciate the change in lighting effect when the light is moved in a horizontal direction.
5. In the Settings area, select the vertical scroll bar and move it from top to bottom to see the change in lighting effect when the light's position is changed vertically.

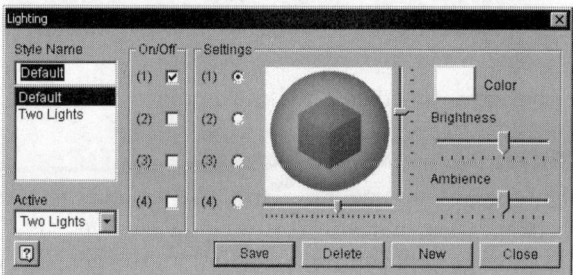

Figure 3.254 *Lighting dialog box*

Now change the light's color.

6. Select the Color button to change the color.
7. Select the Define Custom Color >> button to expand the Color dialog box. (See Figure 3.255.)

Tip: There are three ways to define a color: specify the color's hue, saturation, and luminous values, specify the color's red, green, and blue values, and select a color from the color swatch.

8. Select and drag the marker in the color pad horizontally to select a hue value. Hue value ranges from 0 to 256, representing colors ranging from red through purple.

9. Select and drag the marker in the color pad vertically to select a saturation value. You can regard zero value saturation as colorless.

10. Select and drag the vertical marker to select a luminous value. Zero luminous is complete darkness—color, regardless of the hue and saturation value, will become black. On the other hand, maximum luminous value causes all color to become white.

11. Click the Add to Custom Colors button and then click the OK button.

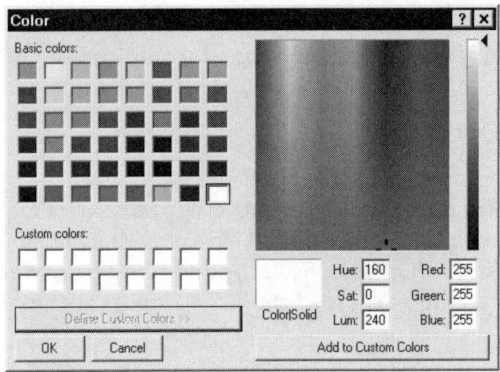

Figure 3.255 *Color dialog box*

Now set Brightness and Ambience.

12. In the Lighting dialog box, select and drag the Brightness scroll bar to select a brightness value (affecting how bright the light will be.)

13. Select and drag the Ambience scroll bar to select an ambient value (the contrast between the lighted and unlighted areas of a face).

14. Click the Save button and then the Close button.

MATERIAL PROPERTIES

To incorporate mechanical and physical properties in the database of the 3D solid model, you assign material. Properties that you can include are Density, Young's Modulus, Poisson's Ratio, Yield Strength, Ultimate Tensile Strength, Thermal Conductivity, Linear Expansion, Specify Heat, and Rendering Style.

 ## TUTORIAL 3.65

In this tutorial, you will assign material to a solid part.

1. Select Format > Materials. (See Figure 3.256.)

2. In the Material List of the Materials dialog box, select Steel, Mild.

3. Click the Save button and then the Close button.

 Tip: You can add new materials by clicking the New button, specifying a new material name, and entering properties.

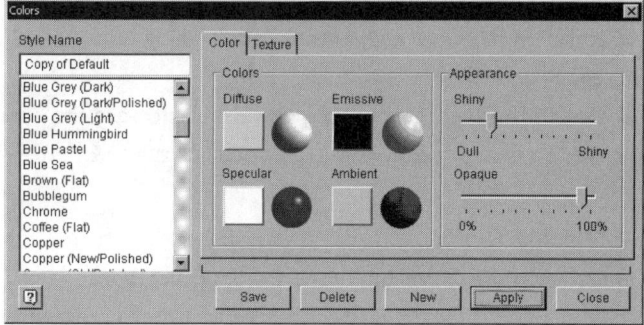

Figure 3.256 *Materials dialog box*

COLOR STYLE

To control the appearance of the 3D solid part in your graphic screen, you set color and assign texture. You can add color and texture to the entire solid part or selected faces of the solid part.

The faces of a 3D object have three color zones. The first zone (diffuse) is the portion of the 3D face under directed light. The second zone (ambient) is the portion of the 3D face not under directed light. The third zone (specular) is the portion of the 3D face that reflects the light. In addition to assigning colors to these color zones, you can assign emissive color, such that the object appears as if a light source is contained within the object. To set the level of reflectiveness for the specular color, you adjust the shininess level. To set the level of opacity, you adjust the opacity level.

To enhance reality, you can apply a texture map on component parts by using predefined textures or a texture that you construct.

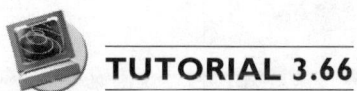 **TUTORIAL 3.66**

In this tutorial, you will assign colors and add texture maps on a solid part.

1. Select Format > Colors. (See Figure 3.257.)

Figure 3.257 *Color dialog box*

Now set the three color zones of the solid part.

2. In the Colors dialog box, select the New button.

3. In the Style Name box, specify a color name.

4. Select the Color tab, if it is not already selected.

5. On the Color tab, select Diffuse.

6. In the Color dialog box (the same as that shown in Figure 3.255), select the Define Custom Colors >> button.

7. Select a color, click the Add to Custom Colors button, and click the OK button. (The diffuse color is set.)

8. Select Ambient on the Color tab of the Colors dialog box.

9. Select the Define Custom Colors >> button.

10. Select the custom color that you just defined for the diffuse color.

11. Select and move the vertical scroll bar downward to decrease the luminous value. (See Figure 3.258.)

12. Click the Add to Custom Colors button, and click the OK button. (The ambient color is set.)

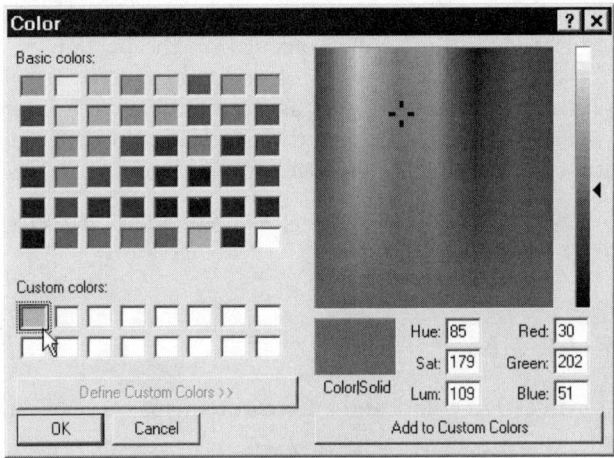

Figure 3.258 *Diffuse color being modified to become the ambient color*

13. Select Specular on the Color tab of the Colors dialog box.

14. Repeat Steps 9 and 10.

15. Select and move the vertical scroll bar upward to increase the luminous value.

16. Click the Add to Custom Colors button, and click the OK button. (The specular color is set.)

Now set the emissive color.

17. Select Emissive on the Color tab of the Colors dialog box.

18. Select the black color. (Selecting black means that there is no emissive color.)

Now set the shininess and opacity of the solid part.

19. Select and move the Shiny scroll bar horizontal to specify a shininess value.

20. Select and move the Opaque scroll bar horizontal to specify an opacity value.

21. Click the Apply button.

Now select a texture map.

22. Select the Texture tab. (See Figure 3.259.)

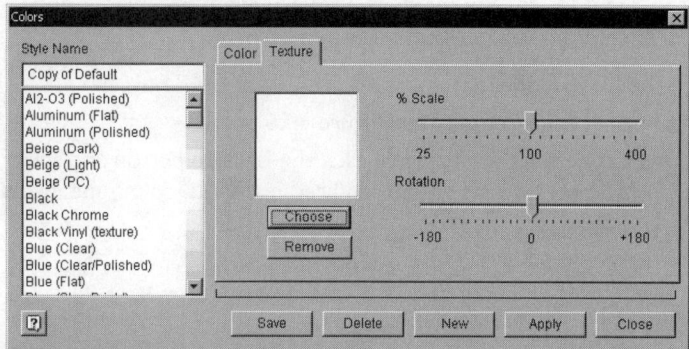

Figure 3.259 *Texture tab of the Color dialog box*

23. On the Texture tab, click the Choose button. (See Figure 3.260.)

Figure 3.260 *Texture Chooser dialog box*

24. In the Texture Chooser dialog box, select a texture, move the horizontal scroll bar to adjust the scale of the texture map, and click the OK button.

25. On returning to the Colors dialog box, move the % Scale scroll bar to further adjust the scale of the map, move the Rotation scroll bar to rotate the texture map in relation to the 3D object, and click the Apply button.

26. If you are satisfied with the appearance of the solid part, click the Save button and then the Close button.

The color and texture of the solid part are set. Save your file.

ORGANIZER

If you want to repeat the same kind of lighting styles, material properties, and color styles in several part files, you can copy them from a source file containing the required settings to another file.

TUTORIAL 3.67

In this tutorial, you will learn how to use the organizer to copy the lighting styles, material properties, and color styles from one file to another.

1. Open the file *PunchBolt.ipt*.
2. Select Format > Organizer. (See Figure 3.261.)
3. To open a source file for copying, click the Browse button and select a solid part file (*PunchNut.ipt*) that has all the required materials, color, and light settings.
4. Select the materials, color styles, or lighting styles
5. Select the Copy>> button to copy to the current solid part file.

Save and close your file.

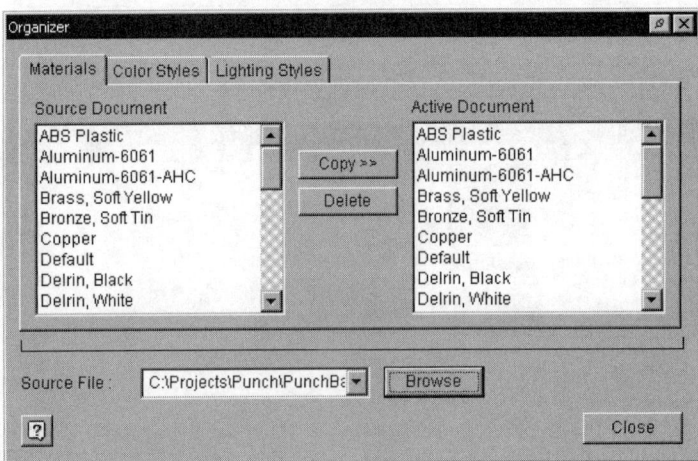

Figure 3.261 *Organizer dialog box*

iPROPERTIES

You can include information in the file about the physical properties of the solid part by using the Properties dialog box. The Properties dialog box is separate from the traditional Microsoft Window's properties dialog box. Content within the dialog box will remain intact when transiting from one country to another.

There are two ways to access the Properties dialog box: you can select iProperties from the File menu, and you can select an Inventor file in Window's explorer, right-click, and select iProperties.

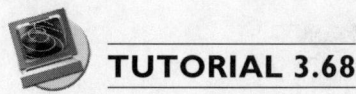

TUTORIAL 3.68

In this tutorial, you will assign properties to the nut by using the Properties dialog box.

1. Open the file *PunchNut.ipt*, if you already closed it.
2. Select File > iProperties. (See Figure 3.262.)

The Properties dialog box has seven tabs: Summary, Project, Status, Custom, Save, and Physical.

General	Provides general information about the file.
Summary	Saves the general information about a 3D solid. It concerns title, subject, author, manager, company, category, keywords, and comments of the solid.
Project	Saves the information about the project. It concerns part number, description, revision number, project, cost center, estimated cost, creation date, vendor, and web link.
Status	Saves the status of the 3D solid. It concerns status, checked by, checked date, engineer approved by, engineer approved date, manufacture approved by, manufacture approved date, reserved by, reserved, last reserved by, and reserve removed.
Custom	Saves any customized data. You specify data name, type, and value.
Save	Saves save information. It concerns the save preview picture, active window on save, active window, and import from file.
Physical	Saves physical properties assigned to the solid part. It concerns material, density, accuracy, mass, area, volume, center of gravity, and inertia properties.

3. Select the Physical tab and change the material to Stainless Steel, Austenitic.

As can be seen from the Properties dialog box, details regarding mass, surface area, volume, center of gravity, and principal moments are evaluated automatically. Among them, you can override the evaluated mass and volume data. Now save and close your file.

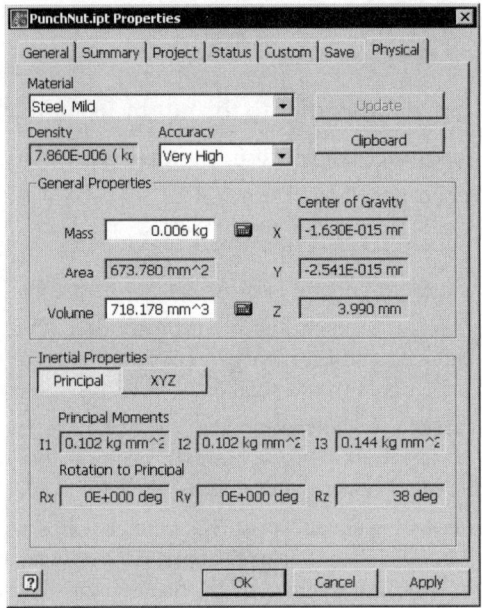

Figure 3.262 *Properties dialog box*

ENGINEERING DRAWING OUTPUT

Now you have completed several solid parts. If you wish to learn how to produce engineering drawings from them, you can proceed to Chapter 10. Figure 3.263 shows an engineering drawing constructed from the solid part of the hexagonal nut.

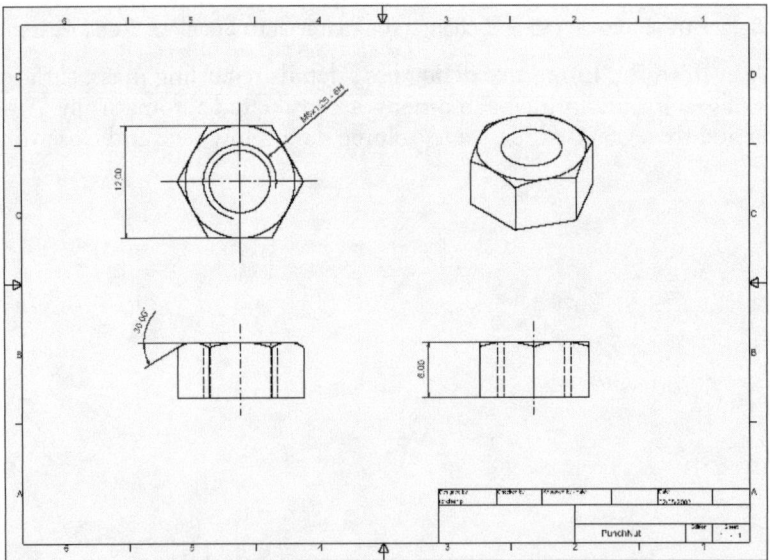

Figure 3.263 *Engineering drawing of the hexagonal nut*

SUMMARY

Using a feature-based approach to construct a 3D solid, you deduce a complex 3D object into solid features of simple shape, construct the features one by one, and combine them together as you construct them. There are two major kinds of solid features: sketched and placed.

Sketched solid features are derived from sketches. Using a parametric approach, sketches need not be precise. During the initial design stage, you should concentrate on the form and shape of the profiles. Then you apply geometric constraints to modify the geometric relationships between the sketch elements and set the size of the sketch by applying parametric dimensions. After sketching, you construct solids from the sketches.

In this chapter, you explored various sketching methods and two basic ways to construct solids from sketches: extruding a sketch in a direction perpendicular to the sketch to form an extruded solid, and revolving a sketch about an axis to form a revolved solid. The first sketched solid feature you construct in a solid part is called the base solid feature. For each additional sketched solid feature, you use a Boolean operation to combine them in one of three ways: join, cut, and intersect.

While constructing a 3D solid part, you can set the display to either shaded or wireframe mode. In wireframe mode, all the edges are displayed. In shaded mode, the interiors are hidden. When a sketch is constructed on a plane inside an existing solid part, you can set the display to shaded mode and slice the graphics along the sketch plane to clip away the portion of the solid part in front of the sketch plane. This way, you see the sketch plane more clearly.

Besides making sketches to construct solid features, you can incorporate pre-constructed solid features in your solid model. Unlike working with sketched solid features, you do not need to construct any sketch—you simply select the solid feature from the menu and place it on the solid part. These pre-constructed solid features are called placed solid features.

Because Autodesk Inventor is a parametric system, you can modify the parameters of the features any time during and after you construct the 3D solid. You can also change the way sketched solid features are combined. For example, you can change a join operation to a cut or intersect operation.

In addition to representing an object's form and shape, you can incorporate additional information on appearance and properties.

REVIEW QUESTIONS

1. Explain the meaning of the feature-based parametric modeling approach.

2. List the kinds of geometric constraints that you use in sketching.

3. With the aid of sketches, explain how extruded and revolved solid features are constructed.

4. Use simple examples to illustrate the three ways to combine sketched solid features. What kinds of features can you construct by simply specifying location and parameters of the feature? Use sketches to illustrate.

5. Outline the ways that you can modify a parametric feature-based solid part.

6. Compare wireframe display with shaded display. How can a shaded display be sliced along the current sketch plane?

7. State the parameters of a solid part that you can measure.

8. Outline how material and color properties can be incorporated into a solid part.

9. What information other than geometric shape and size can you incorporate in a solid part?

10. Explain the three ways to save a preview image in a file.

CHAPTER 4

Part Modeling II

OBJECTIVES

The aims of this chapter are to delineate the ways to construct work features, map decals on solid parts, and construct rib, emboss, loft, sweep, and coil features. This chapter also illustrates the use of surface features and imported surface data in solid construction and the use of planes and surfaces in splitting a solid part. In addition, this chapter explains how to derive a solid part from an existing solid, set up design parameters, use a design notebook, construct a family of parts, and construct user-defined customized features. After studying this chapter, you should be able to

- Use work features in solid part modeling
- Map decals to solid parts
- Construct rib, emboss, coil, loft, and sweep features
- Construct surface features and use surfaces in solid construction
- Split a solid part using planes and surfaces
- Construct derived solid parts
- Set up design parameters
- Use a design notebook
- Construct a family of parts (iPart)
- Construct and use iFeatures (design elements)

OVERVIEW

A feature-based parametric solid part depicting a 3D object consists of a base solid feature that is a sketched solid feature and, depending on the complexity of the solid part, a number of additional features. The additional features can be sketched features, placed solid features, or feature patterns. In Chapter 3, you learned how to construct 2D sketches on the default planes (XY, XZ, and YZ planes) and on faces of the solid part, construct two kinds of sketched solid features (extruded and revolved), combine sketched solid features, and construct various kinds of placed features and feature patterns. In this chapter, you will explore other kinds of sketched solid features (rib, emboss, loft, sweep, and coil) and surface features (extrude, revolve, loft, and sweep).

In making various kinds of features, you might need a plane, an edge, an axis, or a vertex. For example, you need a plane for establishing a sketch plane, a plane for constructing a mirror pattern, and an axis or edge for making a circular pattern. If these objects are not readily available from the solid part's origin or the body of the existing solid part, you need to construct artificial features—work planes, work axes, and work points. You will learn how to construct work features in this chapter.

In making a sweep feature, you need a profile depicting the cross section and a sketch depicting the path. To make a loft feature, you need two or more profiles depicting the cross sections of the feature and, optionally, rails to guide the transition of the profiles. To make sweep and loft features of complex shapes, you need 3D sketches. In this chapter, you will learn about 3D sketches.

Using sketches, you can construct surfaces as well as solids. You can construct extruded, revolved, loft, and sweep surfaces. Apart from these four kinds of surfaces, you can import surface data. Using a surface as splitting tool, you can split the faces of a solid into two sets of faces or cut away a portion of a solid. You can also use a surface as a termination surface while making sketched solid features such as extruded and revolved features.

This chapter will also address the following topics: derived solid parts, design parameters, design notebook, family of parts (iPart), and customized features (iFeature).

To construct a solid part that references to an existing solid part, you derive a new solid part and construct additional features. To control the dimensions of a solid part or a set of solid parts globally, you construct a spreadsheet and use the spreadsheet to guide the parameters of the solid part or the set of solid parts. To keep a design record in the solid part for future reference, you incorporate a design notebook to record design histories, intent, and information along with a part's graphical and geometric data. To construct a family of parts with the parameters and properties maintained in a spreadsheet embedded in the part, you construct an iPart. One of the ways to speed up the design process is to re-use solid features that are constructed in existing solid parts. You can export those reusable sketches and solid features as design elements to form as iFeatures in a design catalog. Then you can insert the design elements into other solid parts.

WORK FEATURES

To help establish sketch planes and geometric references for solid modeling, you construct work points, work axes, and work planes. Collectively, we call them work features. Whenever a point or a vertex is needed in modeling, you can use a work point instead. The same is true for edges and faces. In essence, vertices and work points, edges and work axes, and planar faces and work planes are interchangeable. For example, you can revolve a profile about an edge, a line, or a work axis. In particular, you can construct a sketch on a work plane when there are no suitable faces on which to establish a sketch plane.

WORK POINTS

Work points are artificial points in a solid part. You can project work points onto a sketch plane for making reference points while sketching, and you can use work points to help construct work axes, work planes, and 3D sketches. There are several places to construct a work point. You can construct a work point

- at the endpoint and midpoint of model edge
- at the intersection of three planar faces or work planes
- at the intersection of a planar face or work plane and a curve (2D sketch, 3D sketch, model edge, or work axis)
- at the intersection of a surface and a curve (2D sketch, 3D sketch, or work axis)
- at the intersection of two edges or work axes
- along an edge or work axis at a point nearest to the tangent point to a cylindrical or elliptical face

To construct a work point, select Work Point on the Part Features Panel Bar or toolbar and select the reference objects. By selecting the midpoint or endpoint of a model edge, you construct a work point there. (See Figure 4.1.)

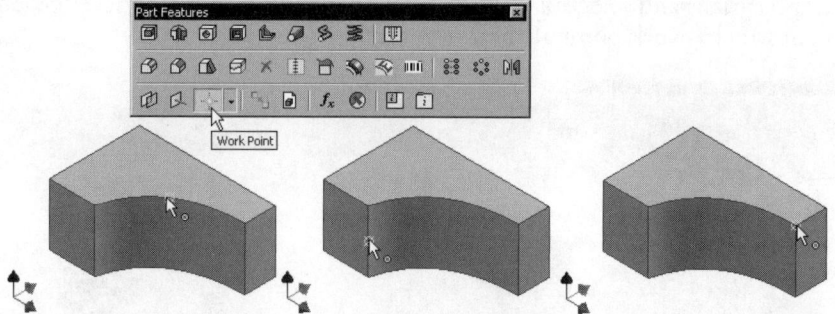

Figure 4.1 *Work points constructed at the midpoint (left and middle) and endpoint (right) of a model edge*

By selecting three planar faces that can be a combination of planar faces, surfaces, and work planes), you construct a work point at their intersection. (See Figure 4.2.)

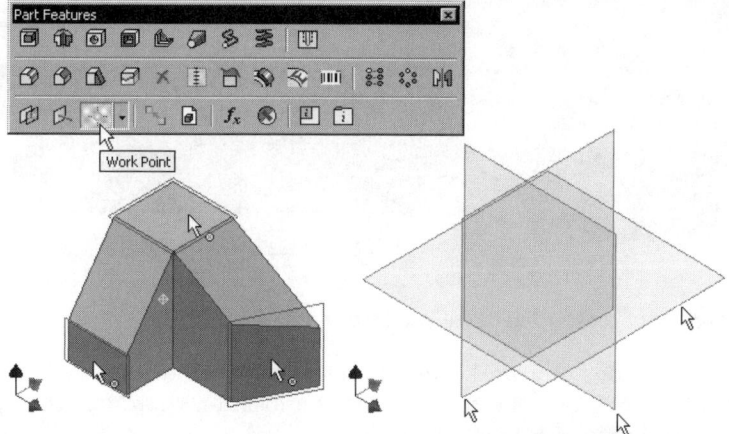

Figure 4.2 *Work point constructed at the intersection of three planar faces*

By selecting a face that can be a planar face or a work plane and selecting a curve that can be a 2D sketch, a 3D sketch, a model edge, or a work axis, you construct a work point at their intersection. (See Figure 4.3.)

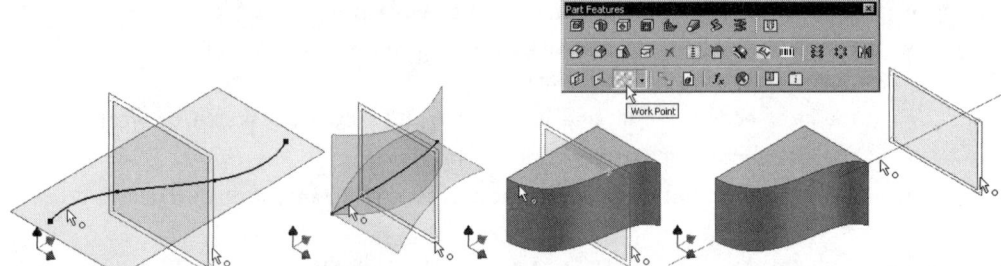

Figure 4.3 *Work points (from left to right) constructed at the intersection of a planar face and a 2D sketch, 3D sketch, model edge, or work axis*

By selecting a surface and selecting a linear sketch element, a linear model edge, or a work axis, you construct a work point at their intersection. (See Figure 4.4.)

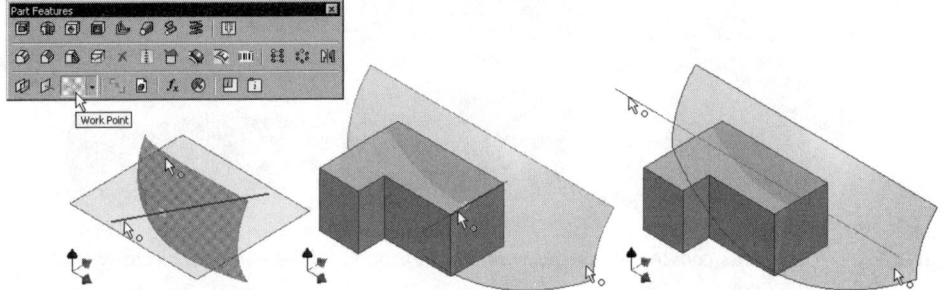

Figure 4.4 *Work points constructed at the intersection of a surface and a linear sketch (left), a linear model edge (middle), and a work axis (right)*

By selecting two intersecting edges of axes, you construct a work point at their intersection. (See Figure 4.5.)

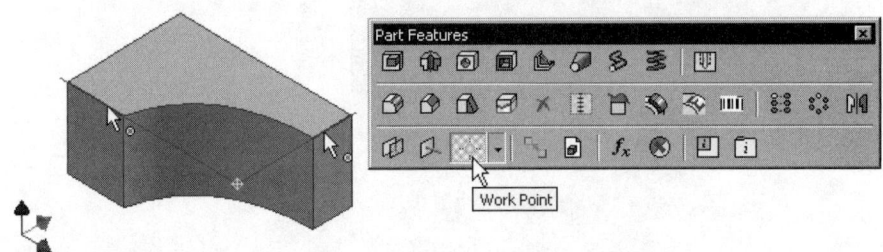

Figure 4.5 *Work point constructed at the intersection of two edges*

By selecting a linear edge or a work axis and selecting a cylindrical or elliptical face, you construct a work point along the edge or axis at a location nearest to the cylindrical or elliptical face. (See Figure 4.6.)

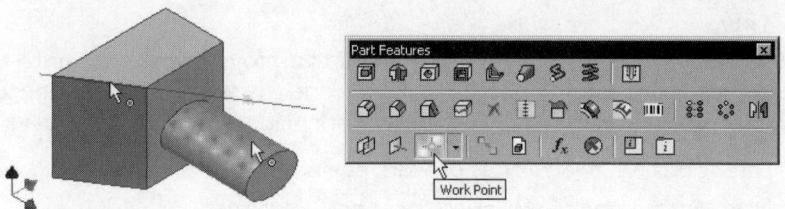

Figure 4.6 *Work point constructed along an edge and nearest to the tangent of an elliptical face*

GROUNDED WORK POINTS

To add flexibility to model construction, you can construct a work point that remains in position in the 3D space regardless of changes to the solid part. Such a work point is called a grounded work point. You can construct a grounded work point in two ways: You can select an existing work point, right-click, and select Ground. You can also construct a grounded work point directly by selecting Grounded Work Point on the Part Features Panel Bar or toolbar and using one of the work point construction methods outlined above. Figure 4.7 shows a grounded work point constructed at the vertex of a solid part.

Figure 4.7 *Construction of a grounded work point and a triad displayed at the grounded work point's location*

When you construct a grounded work point, a triad is displayed at the work point's location. Clicking the triad displays the 3D Move/Rotate dialog box. (See Figure 4.8.) Later if you want to reposition the work point, you can select a grounded work point, right-click, and select 3D Move/Rotate.

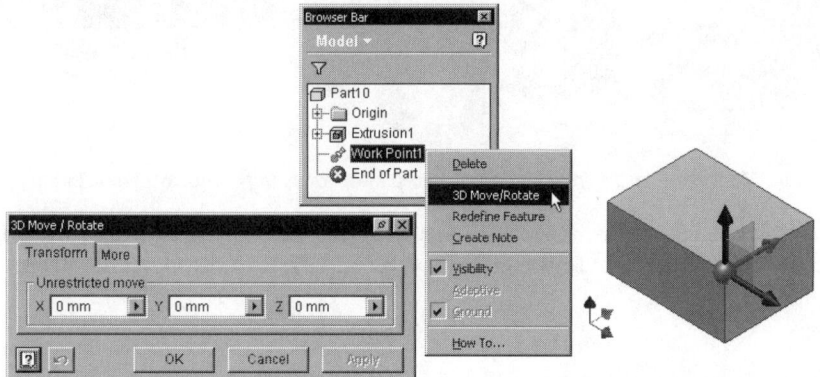

Figure 4.8 *Manipulation of a grounded work point through the triad and the 3D Move/Rotate dialog box*

WORK AXES

Work axes are artificial axes in a solid part. You can project work axes onto the sketch plane as references for sketching purposes, use a work axis as axis of revolution and axis of helical coil, and use work axes to help establish work planes. You can construct a work axis

- passing through two vertices or work points
- at the intersection of two planar faces or work planes
- normal to a planar face or work plane and passing through a vertex or work point
- at the axis of a revolved feature, cylindrical feature, or elliptical feature

To construct a work axis, select Work Axis on the Part Features Panel Bar or toolbar and select reference objects. By selecting two vertices or work points, you construct a work axis passing through them. (See Figure 4.9.)

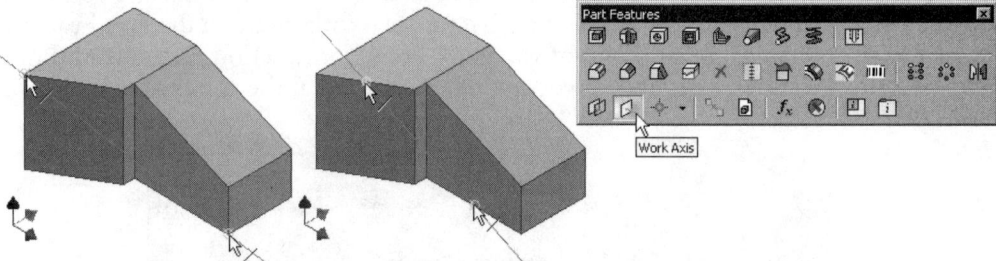

Figure 4.9 *Work axes passing through two vertices (left) and two work points (right)*

By selecting two planar faces or work planes, you construct a work axis at their intersection. (See Figure 4.10.)

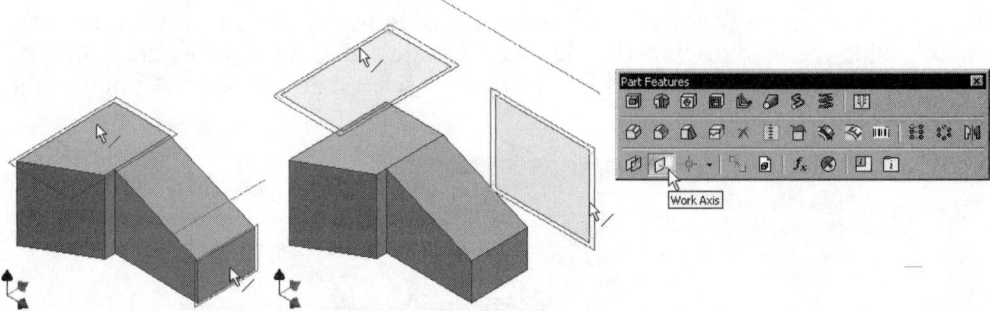

Figure 4.10 *Work axes at the intersection of two planar faces (left) and two work planes (right)*

By selecting a planar face or a work plane and selecting a vertex or a work point, you construct a work axis normal to the face or plane and passing through the vertex or point. (See Figure 4.11.)

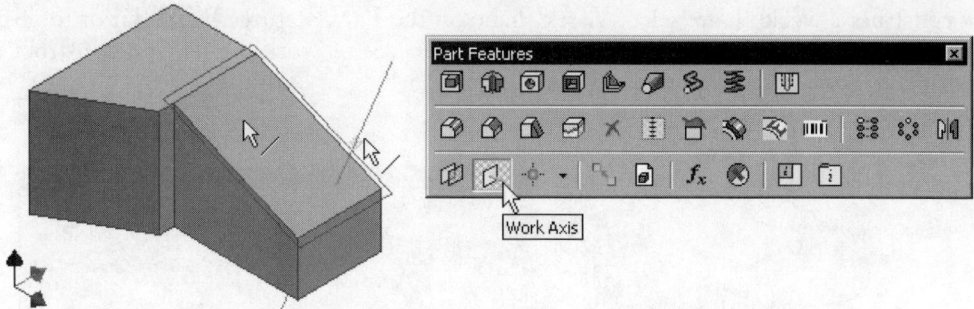

Figure 4.11 *Work axis normal to a model's face and passing through a work point*

By selecting a cylindrical or elliptical face, you construct a work axis along its axis. (See Figure 4.12.)

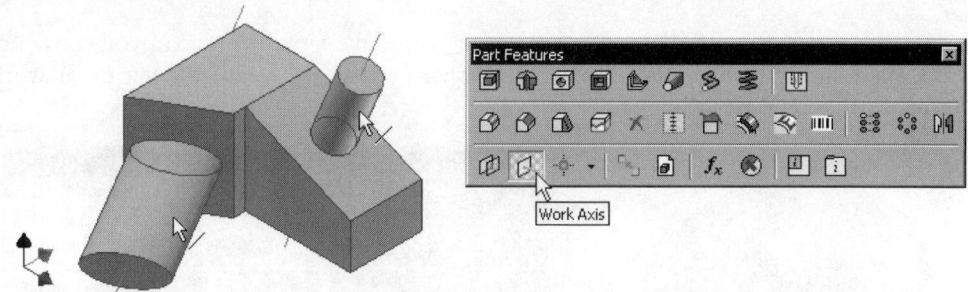

Figure 4.12 *Work axes at the axes of an elliptical feature (left) and a cylindrical feature(right)*

WORK PLANES

Work planes are artificial planes in a solid part. You can use a work plane as sketch plane and termination plane for feature construction, and you can project work planes onto a sketch plane as references. You can construct a work plane

- passing through three vertices or work points
- passing through a vertex or work point and normal to an edge or work axis
- passing through two coplanar edges or work axes
- passing through an edge or a work axis and tangent to a cylindrical or elliptical feature
- parallel to a planar face or work plane and tangent to a cylindrical or elliptical feature
- passing through a vertex or work point and parallel to a face or work plane
- passing through an edge or work axis and at an angle to a face or work plane
- parallel to and offset from a face or work plane
- passing through a spline's control point
- between two parallel faces

To construct a work plane, select Work Plane on the Part Features Panel Bar or toolbar and select reference objects. By selecting three vertices or work points, you construct a work plane passing through them. (See Figure 4.13.)

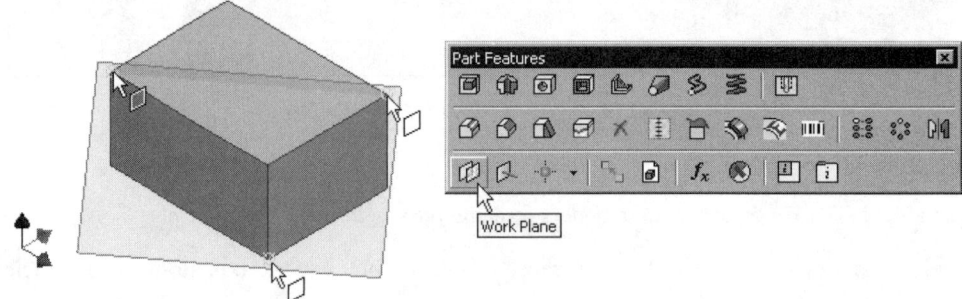

Figure 4.13 *Work plane passing through three vertices of a solid part*

By selecting a vertex or work point and selecting a model edge or work axis, you construct a work plane normal to the edge or work axis and passing through the vertex or work point. (See Figure 4.14.)

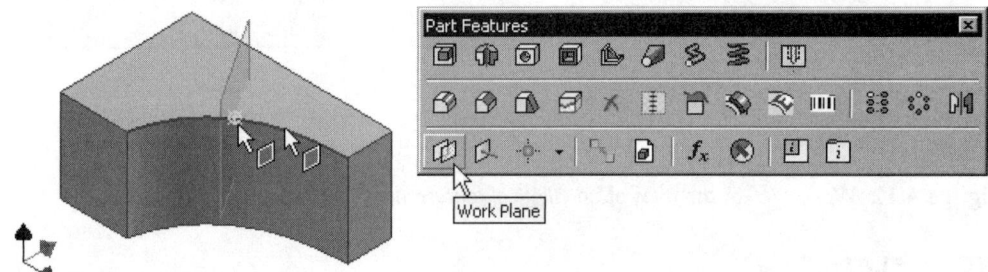

Figure 4.14 *Work plane normal to a model edge and passing through a work point along the edge*

By selecting two linear edges or work axes, you construct a work plane passing through them. (See Figure 4.15.)

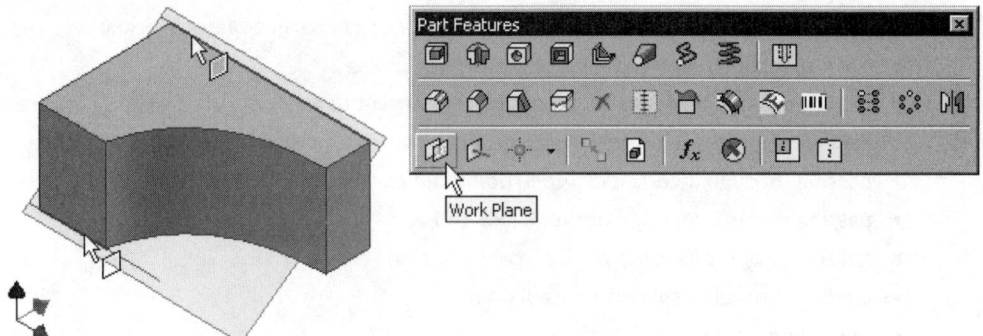

Figure 4.15 *Work plane passing through two coplanar edges*

By selecting a cylindrical or elliptical face and selecting an edge or work axis parallel to the axis of the cylindrical or elliptical feature, you construct a work plane passing through the edge or work axis and tangent to the cylindrical or elliptical face. (See Figure 4.16.)

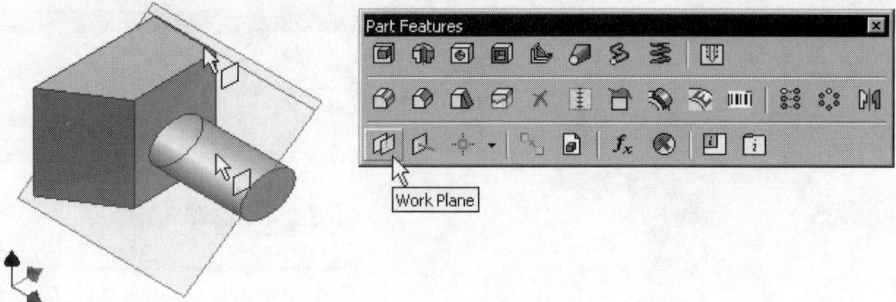

Figure 4.16 *Work plane passing through an edge and tangent to an elliptical face*

By selecting a cylindrical or elliptical face and selecting a planar face or work plane parallel to the axis of the cylindrical or elliptical feature, you construct a work plane tangent to the cylindrical or elliptical face and parallel to the planar face or work plane. (See Figure 4.17.)

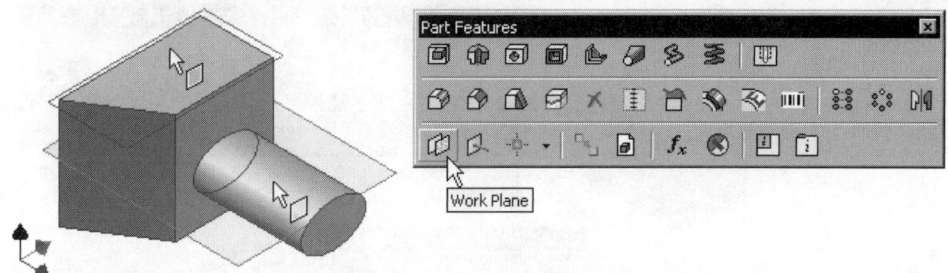

Figure 4.17 *Work plane tangent to an elliptical face and parallel to a face*

By selecting a vertex or work point and selecting a planar face or work plane, you construct a work plane passing through the vertex or work point and parallel to the planar face or work plane. (See Figure 4.18.)

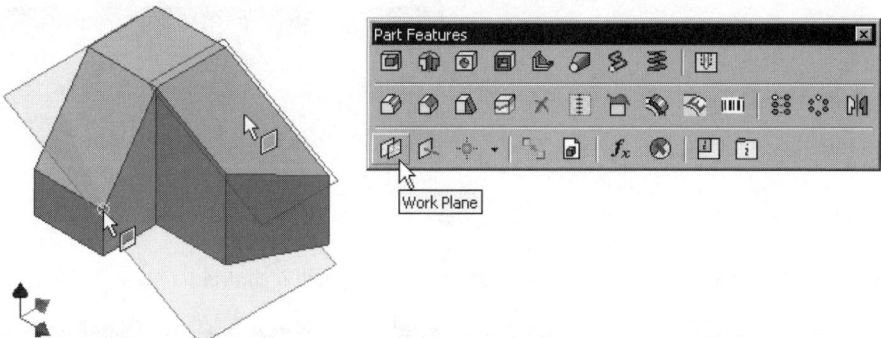

Figure 4.18 *Work plane passing through a vertex and parallel to a planar face*

By selecting an edge or work axis, selecting a planar face or work plane, and specifying an angle, you construct a work plane passing through the edge or work axis and at an angle to the face or work plane. (See Figure 4.19.)

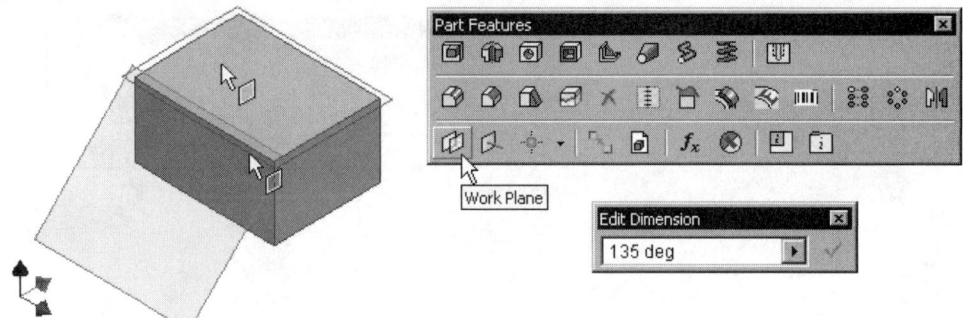

Figure 4.19 *Work plane passing through an edge and at an angle to a face*

By selecting a planar face or a work plane, holding down the left mouse button and dragging it to a new position, and then specifying an offset distance, you construct a work plane offset to a face or work plane. (See Figure 4.20.)

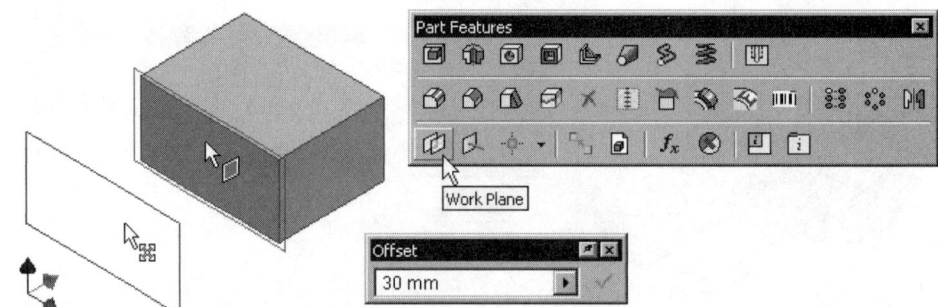

Figure 4.20 *Work plane offsetting from a planar face or work plane*

By selecting a spline and a control point along the spline, you construct a work plane normal to the spline and passing through the control point. (See Figure 4.21.)

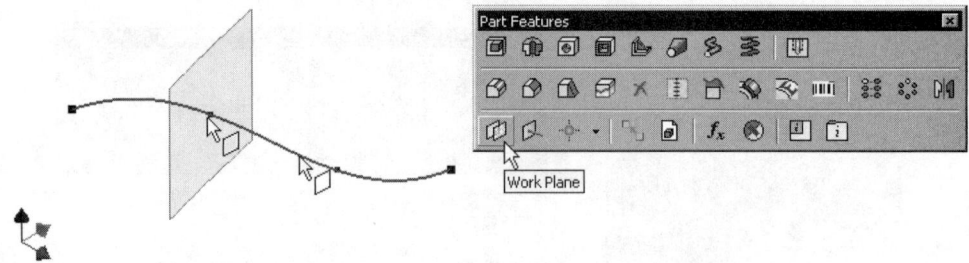

Figure 4.21 *Work plane normal to a spline and passing through a control point*

By selecting two parallel faces, you construct a work plane between them. (See Figure 4.22.)

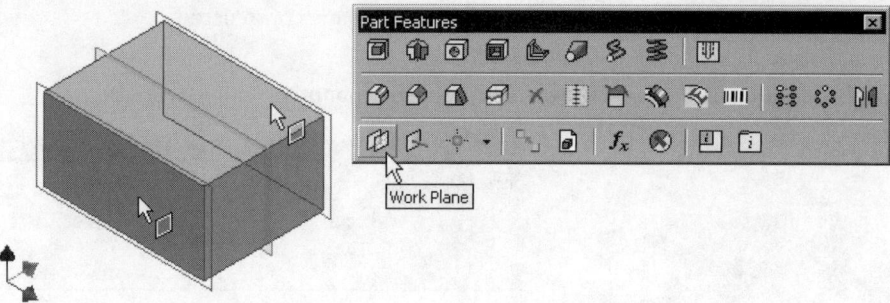

Figure 4.22 *Work plane midway between two parallel planar faces*

IN-LINE WORK FEATURES

To help construct a work feature, you can, in the course of constructing the work feature, construct another work feature. The work feature thus constructed is called an in-line work feature. For example, you can construct three in-line work planes while constructing a work point.

In-line work features can be made hidden or visible, depending on the settings on the Part tab of the Options dialog box. If you select the Auto-hide in-line work features box, the features are hidden automatically after you construct them. If you want to make one visible, you can select it on the Browser Bar, right-click, and select Visibility.

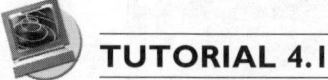

TUTORIAL 4.1

In this tutorial, you will construct two work points while constructing a work axis and use the in-line work points to establish the work axis.

1. Start a new part file. Use the metric template.
2. With reference to Figure 4.23, construct a rectangle measuring 40 mm by 30 mm and extrude the sketch a distance of 20 mm.

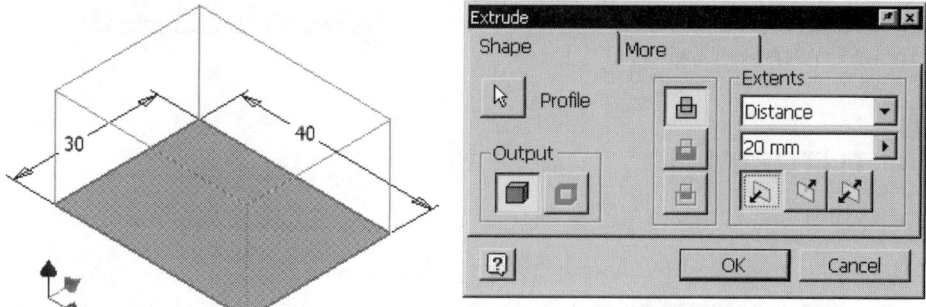

Figure 4.23 *Sketch being extruded*

3. Select Work Axis on the Part Features Panel Bar or toolbar.
4. Right-click and select Create Point.

5. Select midpoint A in Figure 4.24. A work point is constructed.

6. Right-click and select Create Point.

7. Select midpoint B in Figure 4.24 to construct another in-line work point.

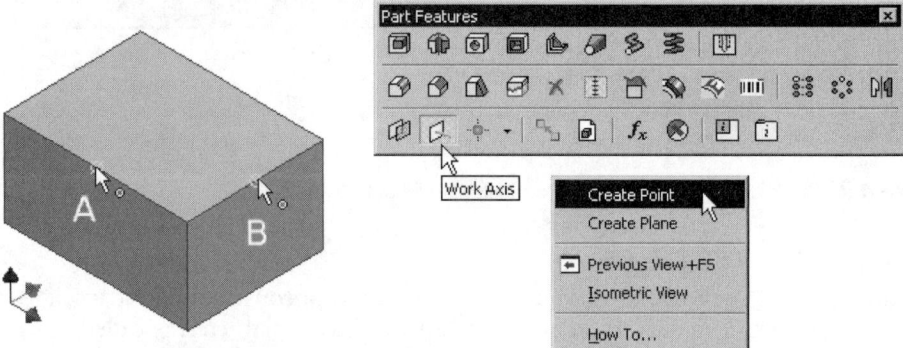

Figure 4.24 *In-line work point being constructed*

As soon as the second in-line work point is constructed, a work axis is automatically constructed to pass through the two in-line work points. If you expand the browser bar, you will find a work axis object with two in-line work points. (See Figure 4.25.) The work axis is complete. Save and close your file (file name: *InlineWork.ipt*).

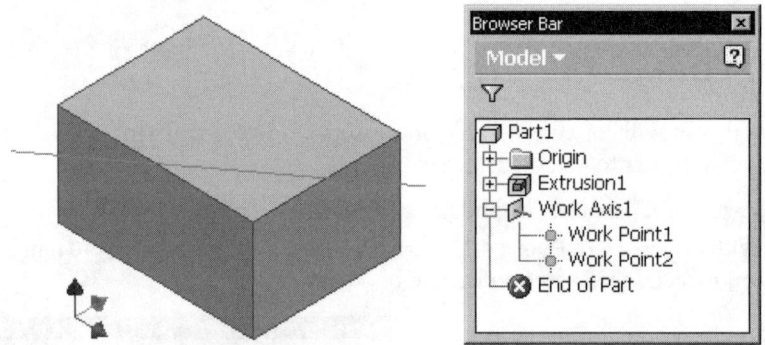

Figure 4.25 *Work axis constructed*

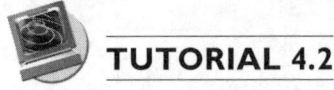 **TUTORIAL 4.2**

In this tutorial, you will construct in-line work features and work features.

1. Start a new part file. Use the metric template.

2. With reference to Figure 4.26, construct a hexagon with its side measuring 12 mm.

3. Extrude the sketch a distance of 30 mm.

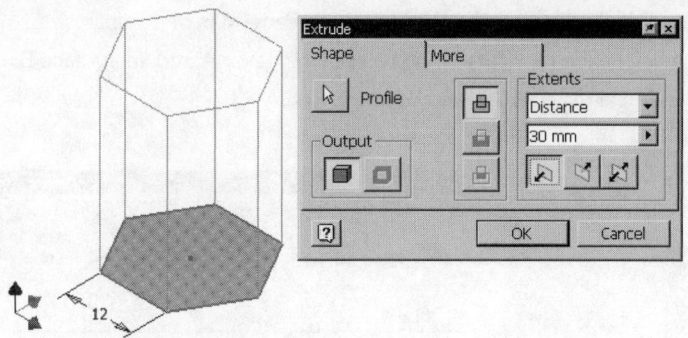

Figure 4.26 *Extruded solid being constructed*

4. Select Work Axis on the Part Features Panel Bar or toolbar.

5. Right-click and select Create Plane.

6. Select face A and B (Figure 4.27) to construct an in-line work plane between the two parallel faces.

7. Right-click and select Create Plane.

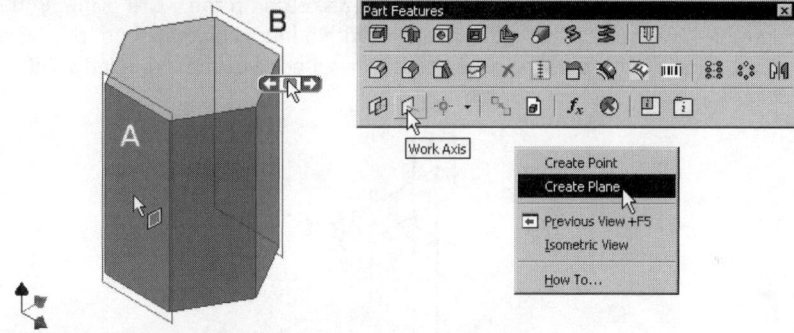

Figure 4.27 *In-line work plane being constructed*

8. Select face A and B in Figure 4.28 to construct another in-line work plane. A work axis is constructed at the intersection of the two in-line work planes.

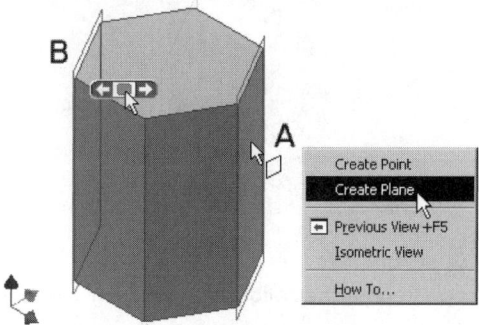

Figure 4.28 *Second in-line work plane being constructed*

9. Select Work Plane on the Part Features Panel Bar or toolbar.

10. With reference to Figure 4.29, select work axis A and select face B.

11. In the Angle dialog box, set the angle to 0.

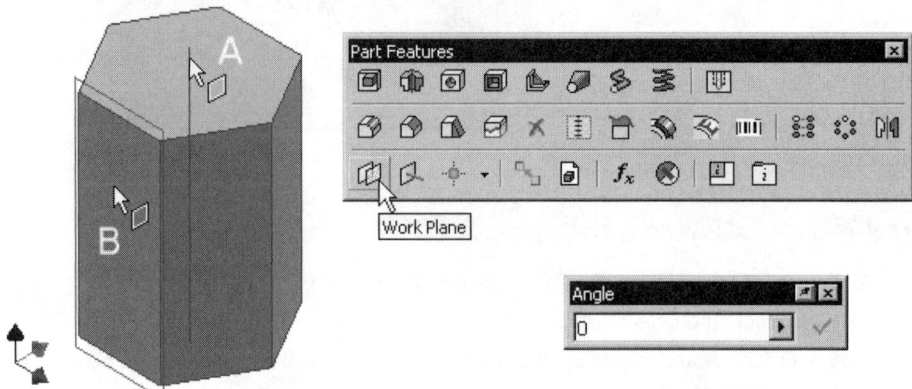

Figure 4.29 *Work plane being constructed*

12. With reference to Figure 4.30, construct a sketch on the work plane. In the sketch, B is a centerline and is collinear with a line projected from the work axis. A is collinear with an edge projected from an edge of the extruded solid.

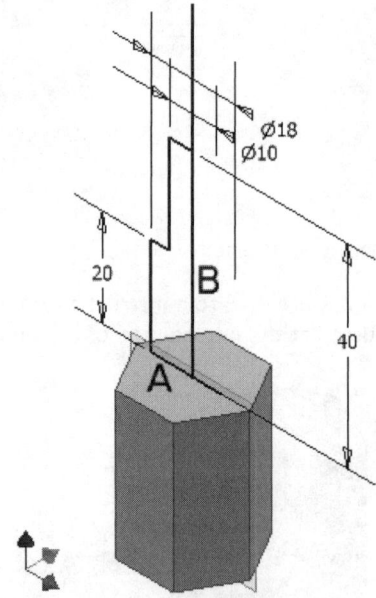

Figure 4.30 *Sketch constructed*

13. Revolve the sketch to join the solid part. (See Figure 4.31.)

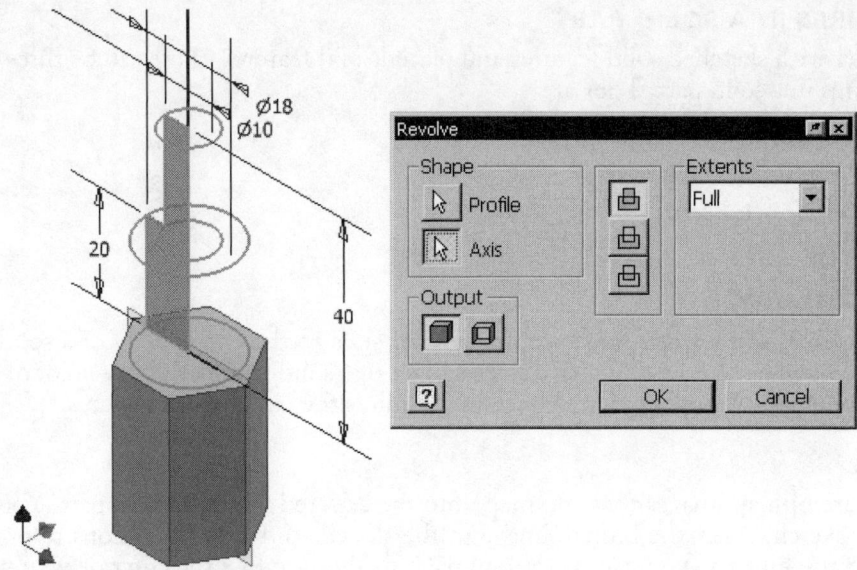

Figure 4.31 *Sketch being revolved*

14. Select Work Plane on the Part Features Panel Bar or toolbar.

15. Select face A (Figure 4.32) and cylindrical face B (Figure 4.32).

A work plane is constructed tangent to the cylindrical face and parallel to a planar face. Now save your file (file name: *Decal.ipt*). You will map a decal on the solid part in the next tutorial.

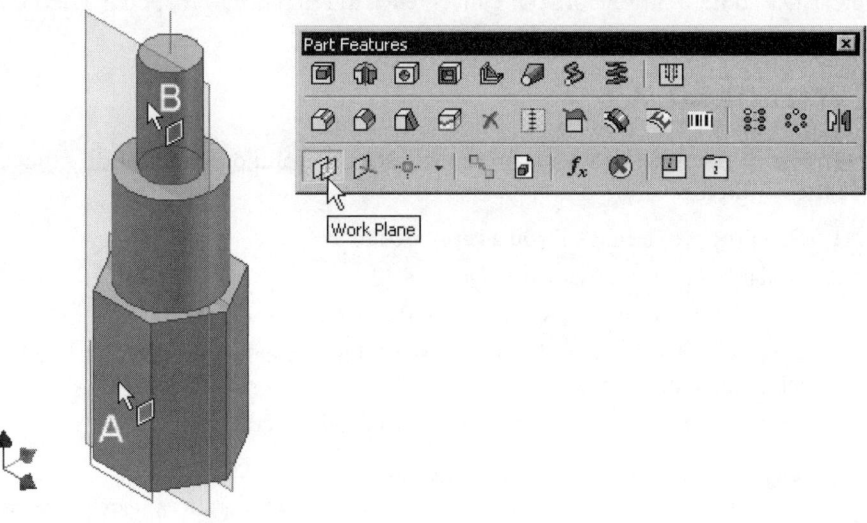

Figure 4.32 *Work plane constructed*

FEATURES IN A SOLID PART

Together with sketched solid features and placed solid features, there can be three kinds of features in a solid part. They are

- sketched solid features
- placed solid features
- work features

APPEARANCE

In Chapter 3, you learned how to apply a color style to the entire body of a solid part. Here, you will learn how to map a decal to a face of a solid and modify the color of a face of the solid part. Mapping a decal on a face requires the use of work features.

DECAL

Decals are bitmap images that you map onto the selected face of a solid part. You construct a sketch, insert the bitmap image in the sketch, and add dimensions to properly position the bitmap. You can wrap the image to the face by projecting or wrapping it perpendicular to the face of the solid. You can control the transparency, rotation, and mirroring of the images.

FACE COLOR

You can set a face's color a unique color, different from the color style that you applied to the entire solid part. Although specifying a face color does not require any sketch or work feature, this topic is placed here simply for your convenience in reference to mapping a decal.

 ## TUTORIAL 4.3

In this tutorial, you will wrap a decal on the surface of a solid part and modify the color of a face of the solid part.

1. Open the file *Decal.ipt*, if you already closed it.
2. Establish a sketch on face A in Figure 4.33.
3. Select Insert Image on the 2D Sketch Panel Bar or toolbar.
4. Select a bitmap from the Inventor 6 sub-folder (*Inventor 6/Textures/Surfaces*) and select the Open button.
5. Select a location on the sketch, right-click, and select Done.

 Note: Inventor may display an alert dialog box warning that the location of the selected bitmap file is not in the active project. You should add the folder to the project or move the bitmap file to a location in the specified project. If you see this dialog box, click OK and continue

6. With reference to Figure 4.33, select an edge of the bitmap and drag the bitmap to reduce its size. (You can use dimensions to govern the location and size of the bitmap.)

7. Select Return on the Inventor Standard toolbar to exit sketching mode.

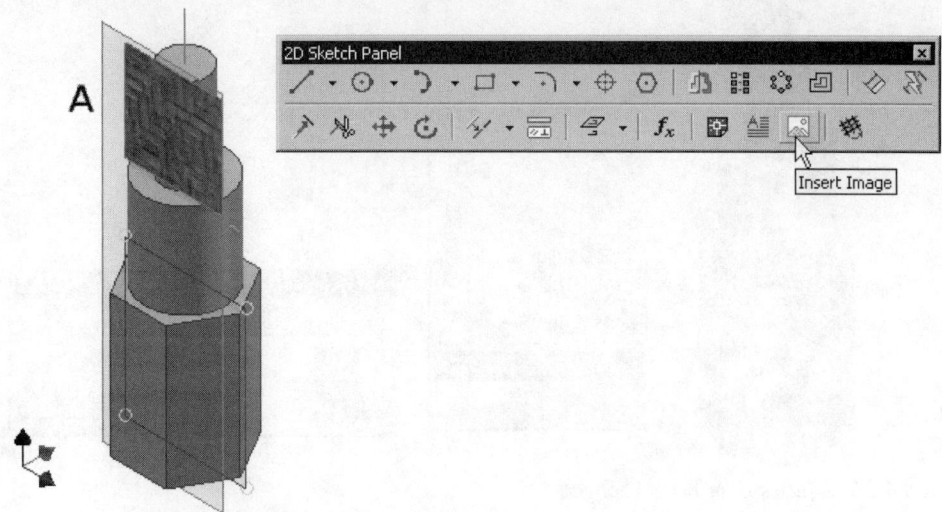

Figure 4.33 *Bitmap positioned*

8. Select Decal on the Part Features Panel Bar or toolbar.
9. Select the image if it is not already selected.
10. Select the Face button and select cylindrical face A in Figure 4.34.

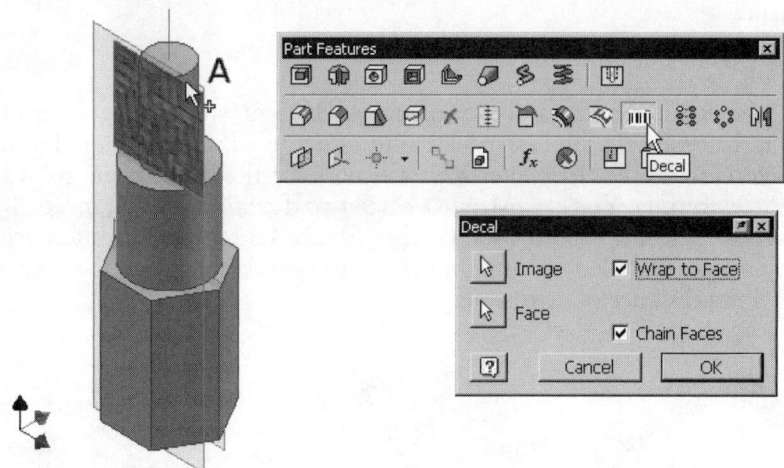

Figure 4.34 *Bitmap being mapped onto a cylindrical face*

11. Select the Wrap to Face box and click the OK button.
12. Select the work features on the Browser Bar, right-click, and deselect Visibility.
13. With reference to Figure 4.35, select face A, right-click, and select Properties.
14. In the Face Properties dialog box, select a color and click the OK button.

A decal is mapped and a face's color is changed. Save and close your file.

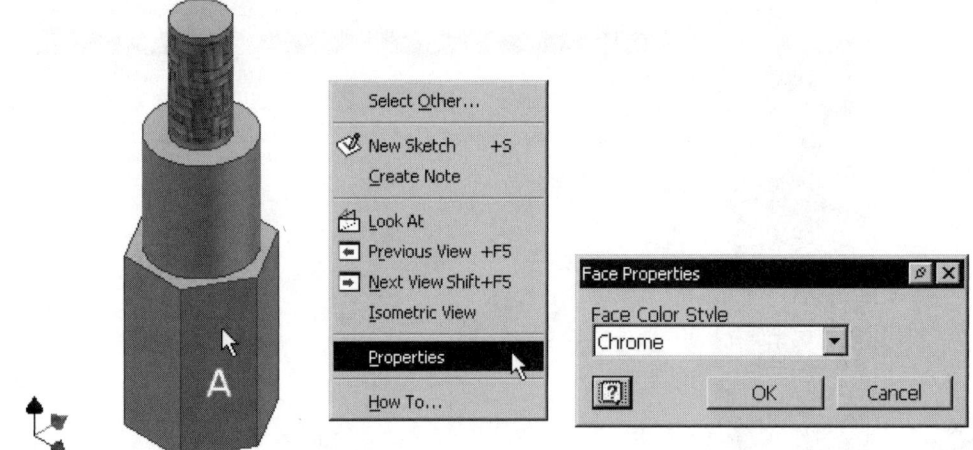

Figure 4.35 *A face's color being changed*

RIB AND EMBOSS FEATURES

Now you will learn how to construct two kinds of sketched solid features: the rib and emboss features. You can regard them as derivatives of the extruded solid feature. . They are explained here because making them requires the use of work features for establishing a sketch plane.

RIB FEATURE

A rib is a triangular or rectangular reinforcing element you add to a component to strengthen it. In essence, a rib is a special kind of extruded solid feature. To construct a rib, you construct an open-loop sketch profile and use the body of the solid to form a closed loop in extruding. You can extrude a rib in two directions until it meets the body of the solid part or extrude it a finite distance. (See Figure 4.36.) The key difference between making a rib and an extruded feature from an open-loop sketch is that the rib enables you to specify a finite distance of extrusion.

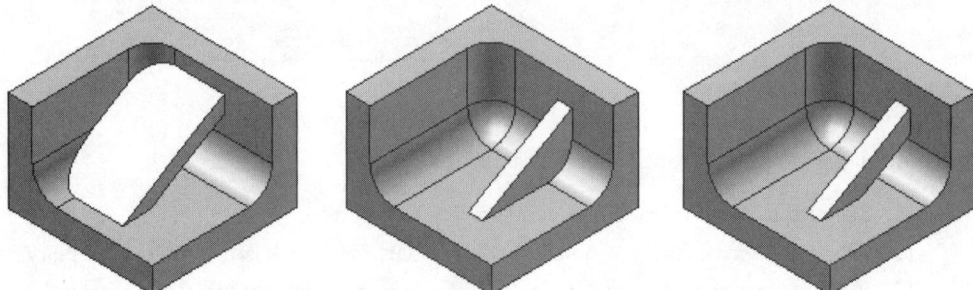

Figure 4.36 *Rib extruded to the body of the solid part in two directions (left and middle) and rib extruded with a finite distance*

TUTORIAL 4.4

In this tutorial, you will learn how to construct three kinds of rib. First you will construct an extruded solid, a shell feature, and a fillet feature.

1. Start a new part file. Use the metric template.
2. Construct a rectangle that measures 30 mm by 30 mm.
3. Extrude the profile a distance of 20 mm. (See Figure 4.37.)

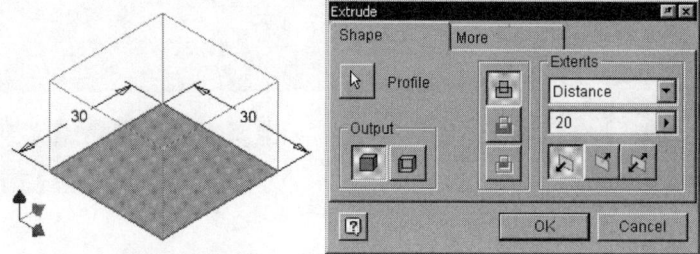

Figure 4.37 *Extruded solid constructed*

4. Select Shell from the Part Features panel bar or toolbar, remove faces A, B, and C, set shell thickness at 5 mm, and select the OK button. (See Figure 4.38.)

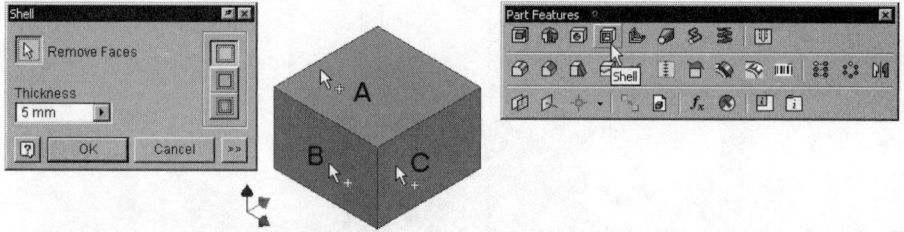

Figure 4.38 *Shell feature being placed*

5. With reference to Figure 4.39, fillet the internal corners. Fillet radius is 5 mm.

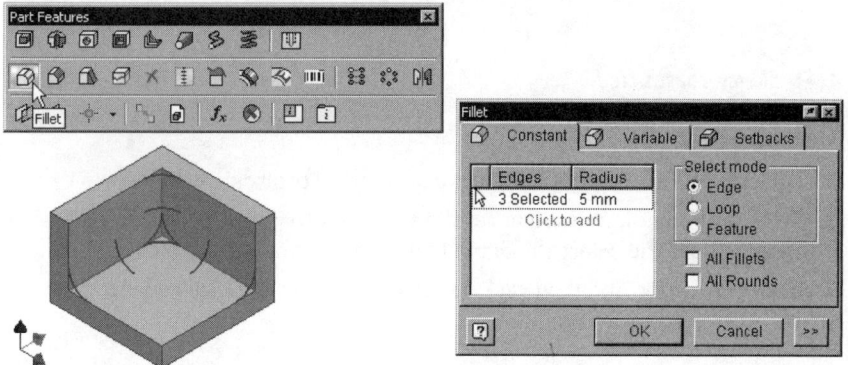

Figure 4.39 *Fillet feature placed*

Now construct a work plane. On the work plane, you will establish a sketch plane and construct a sketch.

6. Construct a work plane that is 10 mm offset from face A indicated in Figure 4.40. (Hold down the mouse button, drag the mouse to a new location, and type **–10** in the Offset dialog box.

7. Construct a sketch on the work plane, right-click and select Slice Graphic.

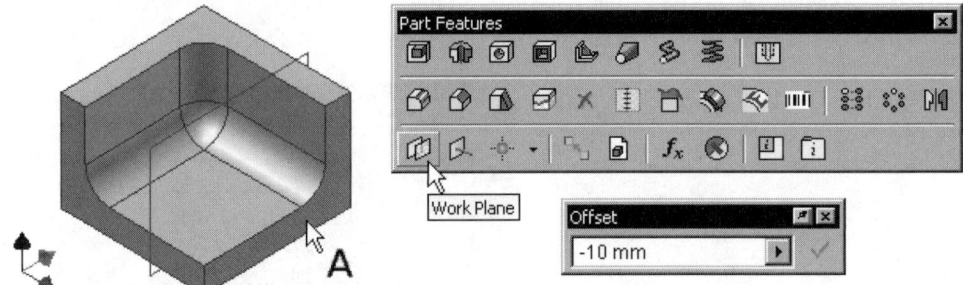

Figure 4.40 *Work plane constructed*

8. With reference to Figure 4.41, construct a line and add dimensions.

9. Select Return on the Inventor Standard toolbar to exit sketch mode.

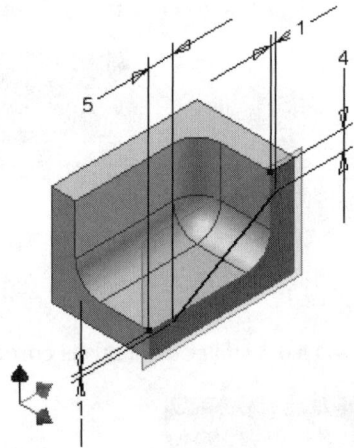

Figure 4.41 *Sketch constructed*

Now construct a rib feature.

10. Select Rib from the Part Features panel bar or toolbar.

11. Select the open-loop sketch, set the thickness to 3 mm, select To Next from the Extents box, and select the From Midplane button in the Rib dialog box.

12. Select Direction and then click to select direction A in Figure 4.42.

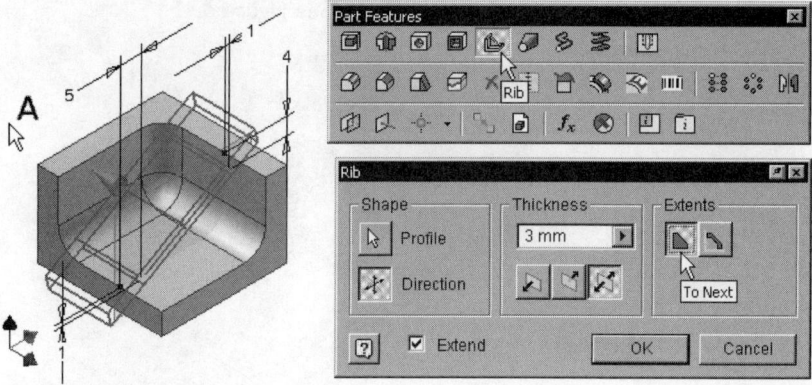

Figure 4.42 *Rib being constructed*

13. Click the OK button. A rib is constructed. (See Figure 4.43.)

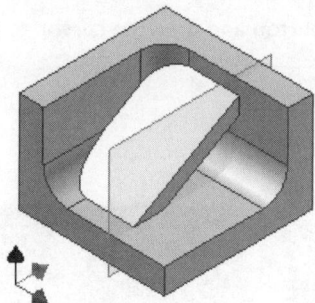

Figure 4.43 *Rib constructed*

Now modify the rib by changing its direction.

14. Select the Rib feature on the Browser Bar, right-click, and select Edit Feature.
15. Select the Direction button and drag the cursor to location A in Figure 4.44.

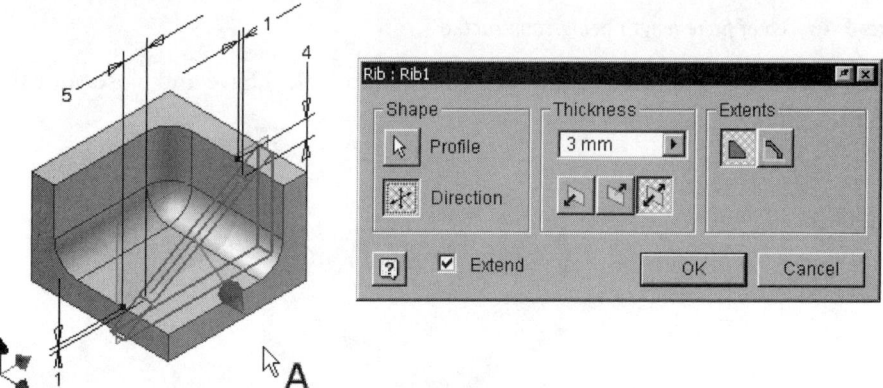

Figure 4.44 *Rib direction being modified*

16. Click the OK button. The rib is modified. (See Figure 4.45.)

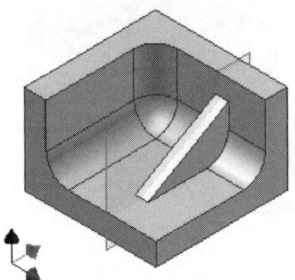

Figure 4.45 *Rib direction modified*

Now modify the rib again.

17. Select the Rib feature on the Browser Bar, right-click, and select Edit Feature.
18. Select Finite in the Extents box and set the distance to 4 mm.
19. Select the Direction button and drag the cursor to location A in Figure 4.46.
20. Click the OK button.

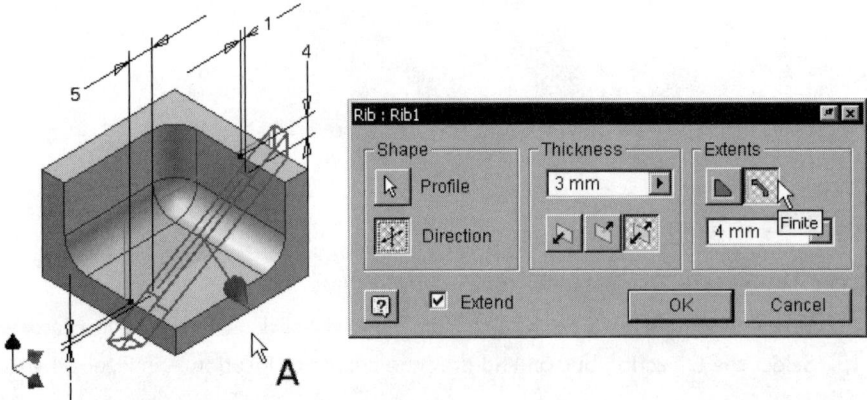

Figure 4.46 *Rib of finite length being constructed*

The part with a rib structure is complete. (See Figure 4.47.) Save and close your file (file name: *Rib.ipt*).

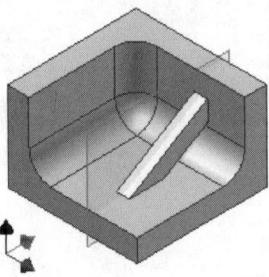

Figure 4.47 *Rib of finite length constructed*

EMBOSS FEATURE

An emboss feature is a profiled feature that is raised from a face of a solid part or is recessed into it. The profiled feature can be a sketch or a text object. Making an embossment is equivalent to extruding it to join or cut the solid part. The key difference between the emboss command and the extrude command is that embossment enables you to wrap the sketch or text profile around a conical or cylindrical face. Note that you can emboss on a flat face, a cylindrical face, and a conical face.

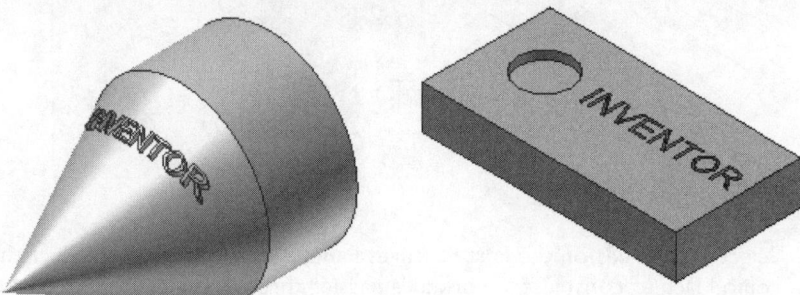

Figure 4.48 *Emboss features*

 **TUTORIAL 4.5**

In this tutorial, you will construct a number of work features for establishing sketch planes and construct emboss features on the sketch planes.

1. With reference to Figure 4.49, construct a sketch.
2. Set the display to an isometric view and revolve the sketch 360° to construct a revolved solid. (See Figure 4.50.)

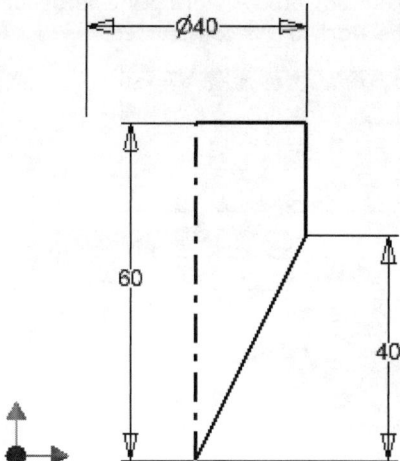

Figure 4.49 *Sketch constructed*

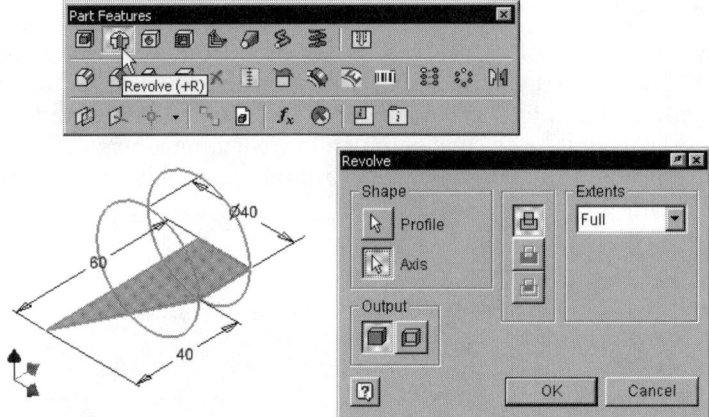

Figure 4.50 *Sketch being revolved*

3. Select Work Axis on the Part Features Panel Bar or toolbar and select the conical face to construct a work axis passing through the centerline of the cone. (See Figure 4.51.)

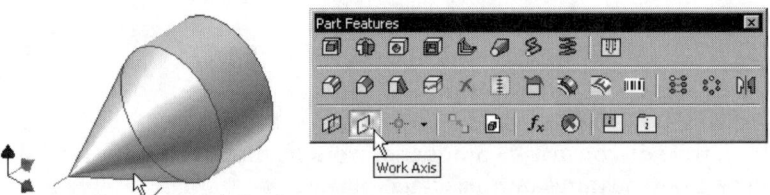

Figure 4.51 *Work axis passing through the axis of the revolved solid constructed*

4. Select Work Plane from the Part Features panel bar or toolbar, select XY plane A (Figure 4.52) on the Browser Bar, select work axis B (Figure 4.52), and specify 90 deg in the Angle dialog box. A work plane perpendicular to the XY plane and passing through the work axis is constructed.

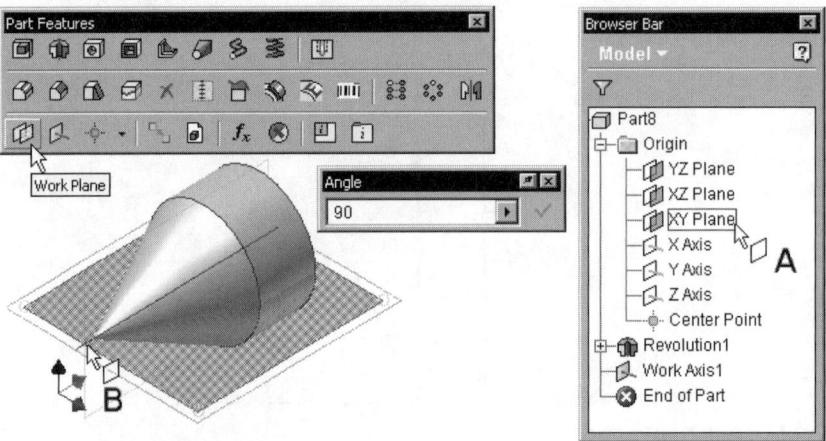

Figure 4.52 *Work plane perpendicular to the XY plane and passing through the work axis constructed*

5. Select Work Point on the Part Features Panel Bar or toolbar, select circular edge A (Figure 4.53), and select work plane B (Figure 4.53) to construct a work point at their intersection.

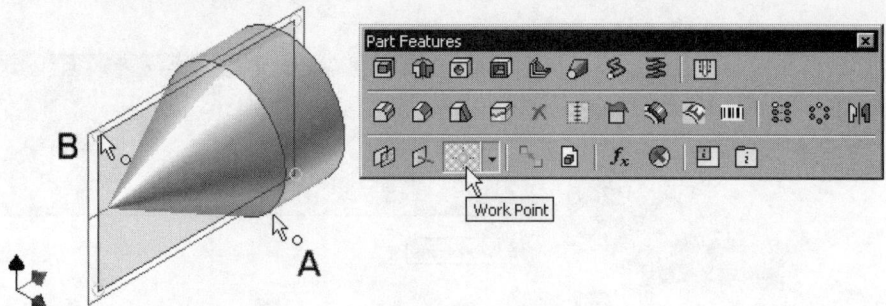

Figure 4.53 *Work point constructed at the intersection between an circular edge and a work plane*

6. Select Work Point on the Part Features Panel Bar or toolbar, select conical face A (Figure 4.54), and select work axis B (Figure 4.54) to construct a work point at their intersection.

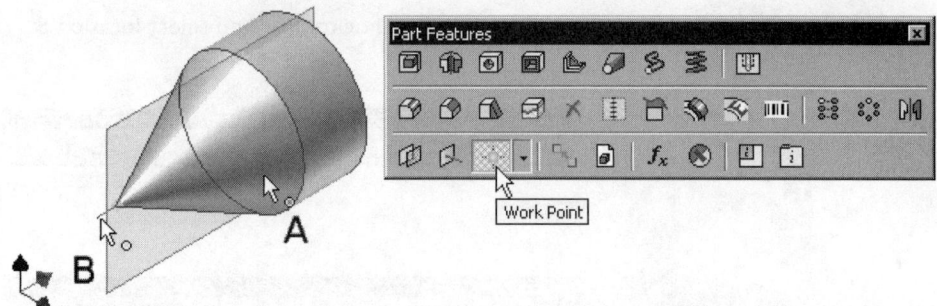

Figure 4.54 *Work point constructed at the intersection between a conical face and a work axis*

7. Select Work Axis on the Part Features Panel Bar or toolbar and select work points A and B (Figure 4.55) to construct a work axis passing through the two work points.

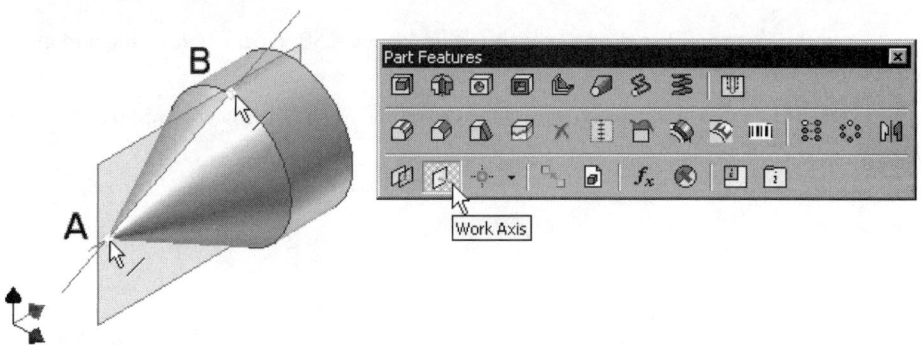

Figure 4.55 *Work axis passing through two work points constructed*

8. Select Work Plane on the Part Features Panel Bar or toolbar, select work axis A (Figure 4.56), and select conical face B (Figure 4.56) to construct a work plane passing through the work axis and tangent to the conical face.

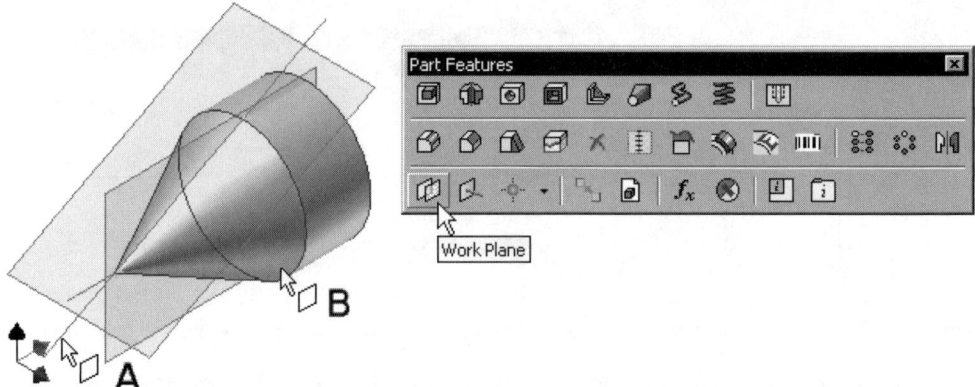

Figure 4.56 *Work plane tangent to the conical face constructed*

9. Establish a sketch on the work plane A (Figure 4.57).

10. Select Create Text from the 2D Sketch Panel or toolbar and select location B (Figure 4.57).

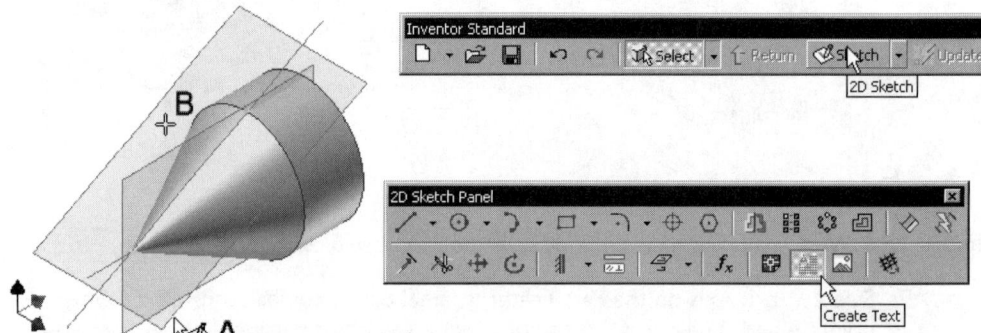

Figure 4.57 *Sketch plane established*

11. In the Format Text dialog box shown in Figure 4.58, type a text string and click the OK button.

Note: In your dialog box (Figure 4-58) change the name of the new component to "emboss". Otherwise, it will be given a part number.

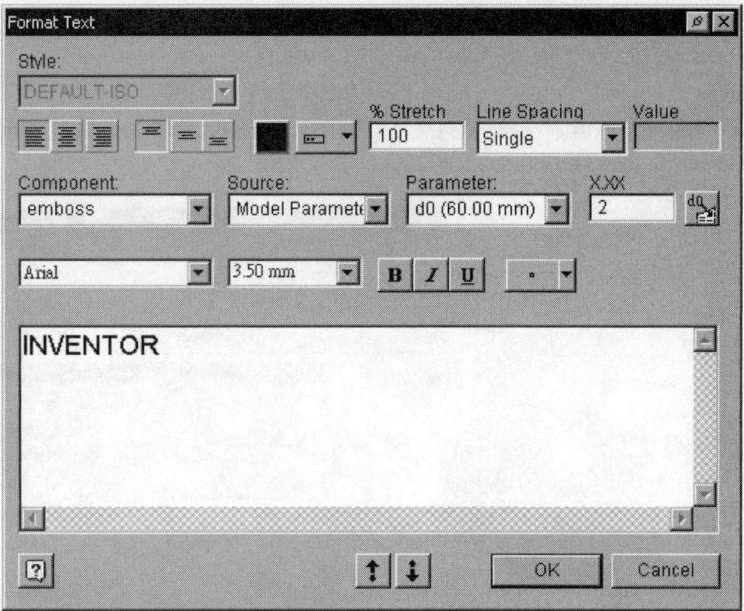

Figure 4.58 *Format Text dialog box*

12. Select the sketch plane, right-click, and select Flip Normal, if the normal direction is incorrect (Figure 4.59.) Otherwise, skip this step.

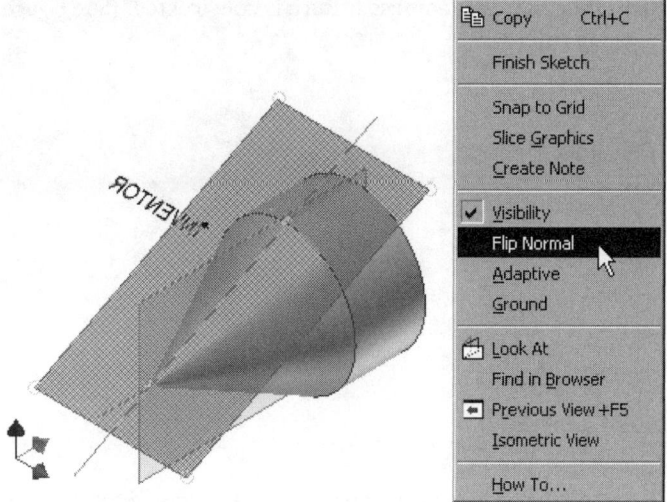

Figure 4.59 *Sketch plane's normal direction being flipped*

13. Position the text in accordance with Figure 4.60.

14. Select Return on the Inventor Standard toolbar to exit sketch mode.

15. Select Emboss on the Part Features Panel Bar or toolbar.

16. In the Emboss dialog box, select the Profile button (if it is not already selected) and select the sketch.

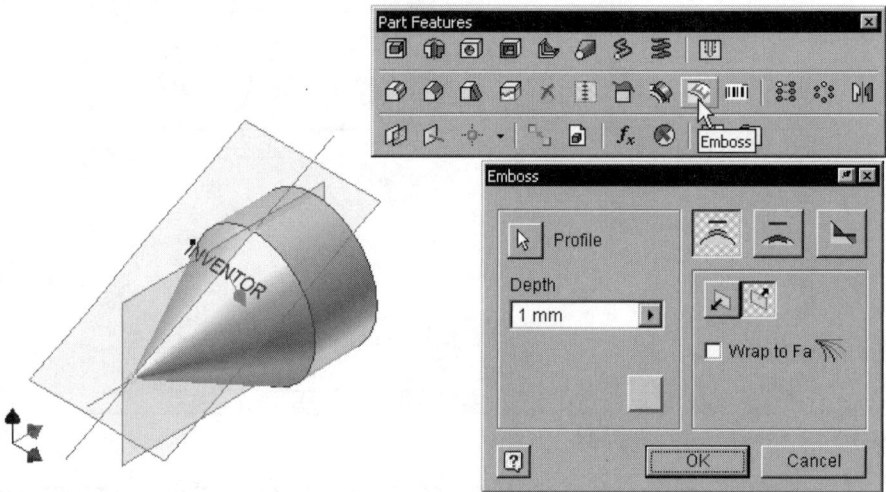

Figure 4.60 *Emboss feature being constructed*

17. Click the OK button. An emboss feature is constructed. (See Figure 4.61.)

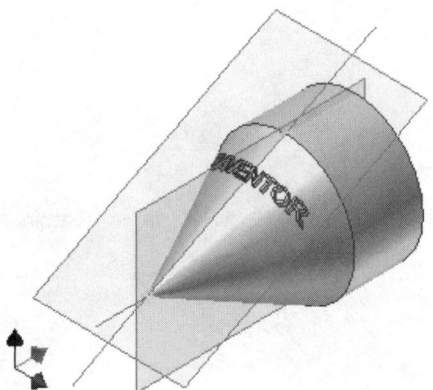

Figure 4.61 *Emboss feature constructed*

To appreciate the difference between embossment and extrusion on a conical face, now undo the emboss command and extrude the sketch.

18. Select Undo from the Edit menu.

19. Select Extrude on the Part Features Panel Bar or toolbar.

20. Select the text sketch, set the extrusion distance to 10 mm, and click the OK button.

An extruded feature is constructed from the sketch. (See Figure 4.62.)

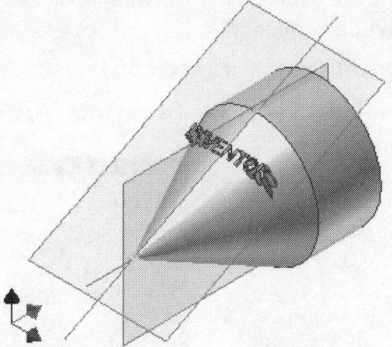

Figure 4.62 *Extruded feature constructed from the text sketch*

Save and close your file (file name: *Emboss1.ipt*).

TUTORIAL 4.6

In this tutorial, you construct an emboss feature on a flat face of a solid part.

1. Start a new part file. Use the metric template.
2. Construct a rectangle measuring 60 mm by 30 mm and extrude it a distance of 10 mm. (See Figure 4.63.)

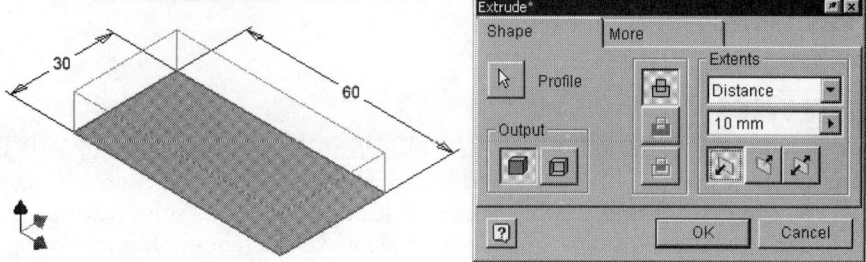

Figure 4.63 *Sketch being extruded*

3. Establish a sketch on face A and construct a circle and a text string. (See Figure 4.64.)

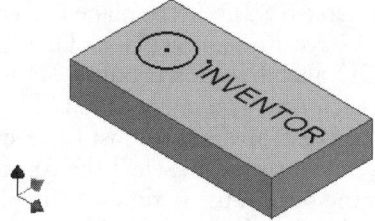

Figure 4.64 *Sketch constructed on a planar face*

4. Extrude the sketch a distance of 2 mm to cut the solid part. (See Figure 4.65.)

5. To observe if there is any difference between extrusion and embossment on a flat face, undo the Extrude command.

6. Construct an embossment. (See Figure 4.66.)

The file is complete. Save and close your file (file name: *Emboss2.ipt*).

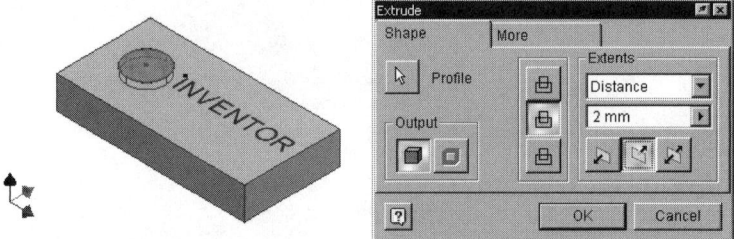

Figure 4.65 *Extruded feature being made*

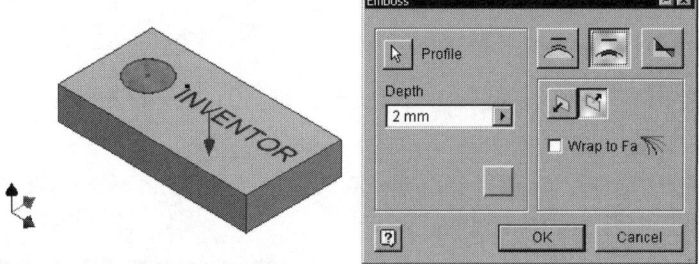

Figure 4.66 *Emboss feature being made*

3D SKETCHES

Sketches are required for making sketched solid features and establishing hole center when placing a hole feature. There are two kinds of sketches: 2D sketches and 3D sketches. Basically, you can use 2D sketches in making all kinds of sketched solid features, and use 3D sketches in sweep and loft features. Because using 3D sketches in loft and sweep solid features expands the repertoire of form and shape, you will learn how to construct 3D sketches before learning how to construct loft and sweep solid features.

A 3D sketch, as its name implies, resides in 3D space rather than on a 2D plane. To construct a 3D sketch, select 3D Sketch on the Inventor Standard toolbar. Unlike 2D sketches, where you need to establish a 2D sketch plane and construct sketch elements on the plane, 3D sketches do not have any sketch plane. Therefore, what you will find immediately after selecting the 3D Sketch button is only a 3D sketch object on the Browser Bar. As we have mentioned, each time you select the 2D Sketch button on the Inventor Standard toolbar, a new sketch is established. It must be re-emphasized that the same is true for making 3D sketches. Each time you select the 3D Sketch button, you set up a new 3D sketch. If you select the button many times, you will end up with many sketch objects. To modify an existing sketch, select the sketch, right-click, and select Edit Sketch, and select Return on the Inventor Standard toolbar to exit sketch mode.

3D SKETCH PANEL BAR AND TOOLBAR

After you select the 3D Sketch button on the Inventor Standard toolbar, the context-sensitive command panel changes to a 3D Sketch Panel Bar. If you prefer to use the toolbar instead of the Panel Bar, select View > Toolbar > 3D Sketch. Figure 4.67 shows the 3D Sketch Panel Bar and toolbar, and Table 4.1 outlines the options on the 3D Sketch Panel Bar or toolbar.

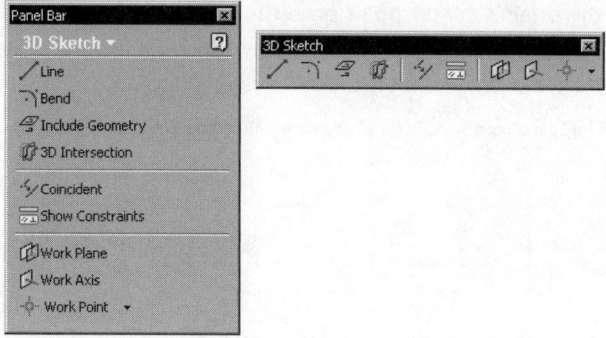

Figure 4.67 *3D Sketch Panel Bar and toolbar*

Table 4.1 *3D Sketch Panel Bar and toolbar options*

Option	Description
Line	Constructs 3D line segments.
Bend	Constructs fillet bends between contiguous 3D sketch elements.
Include Geometry	Includes 2D sketch elements, edges, and vertices in a 3D sketch.
3D Intersection	Constructs a 3D sketch from the intersection of two surfaces.
Coincident	Applies a coincident constraint to an endpoint of a 3D sketch element.
Show Constraints	Displays the constraints applied to select 3D sketch element.
Work Plane	Constructs a work plane.
Work Axis	Constructs a work axis.
Work Point	Constructs a work point and grounded work point.

3D SKETCHING

There are three ways to construct sketch elements of a 3D path:

- You can include elements from 2D sketches you already constructed on planar sketch planes and include edges and vertices of the solid part.
- You can construct an intersection curve between two surfaces.
- You can construct 3D line segments connecting endpoints or vertices of included geometry and work points.

To provide a smooth transition between 3D lines and included geometry, you specify a fillet bend between contiguous 3D sketch elements.

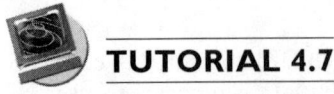

TUTORIAL 4.7

In this tutorial, you will construct two extruded surfaces and construct a 3D sketch from the intersection of the surfaces.

1. Start a new part file. Use the metric template.
2. Project the origin's center point onto the sketch plane.
3. Construct a spline and add dimensions to constrain its shape and location in relation to the projected center point A in Figure 4.68.

Note: The Spline command is on the same flyout as the Line command on the 2D Sketch Panel Bar.

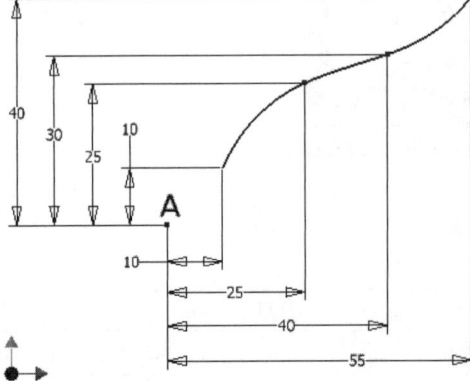

Figure 4.68 *Sketch constructed on the XY plane*

4. Select Return on the Inventor Standard toolbar to exit sketch mode.
5. Set the display to an isometric view.
6. Extrude the sketch a distance of 40 mm to construct a surface feature. (See Figure 4.69.)

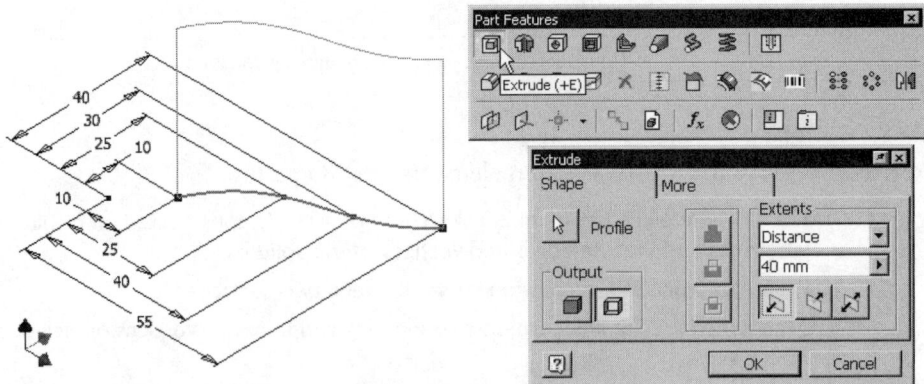

Figure 4.69 *Extruded surface feature being constructed*

7. Select 2D Sketch on the Inventor Standard toolbar, and then select the YZ plane on the Browser Bar to construct a new sketch.

8. With reference to Figure 4.70, construct a spline and add dimensions. Point A is the projected center point from the origin.

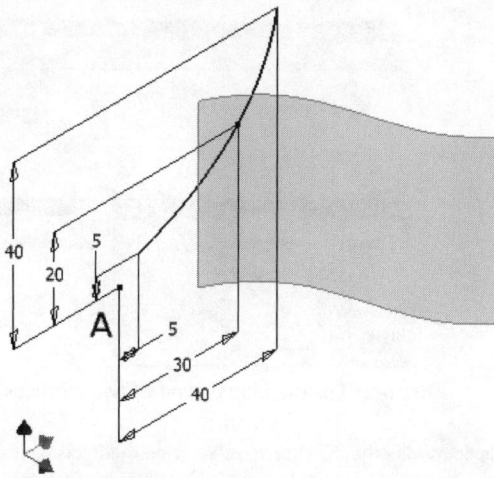

Figure 4.70 *Sketch constructed on the YZ plane*

9. Select Return on the Inventor Standard toolbar to exit sketch mode.

10. Extrude the sketch a distance of 55 mm to construct a surface feature. (See Figure 4.71.)

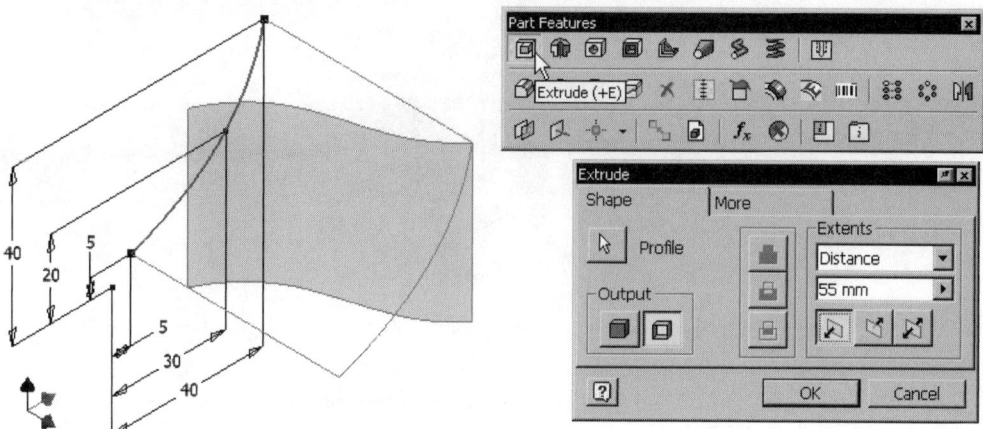

Figure 4.71 *Second extruded surface feature constructed*

11. Select 3D Sketch on the Inventor Standard toolbar to start a 3D sketch.

Note: The 3D Sketch tool is on the same flyout on the Standard toolbar as the 2D Sketch tool.

12. Select 3D Intersection on the 3D Sketch Panel Bar or toolbar.

13. Select surfaces A and B in Figure 4.72 and click the OK button in the 3D Intersection Curve dialog box.

14. Select Return on the Inventor Standard toolbar to exit sketch mode.

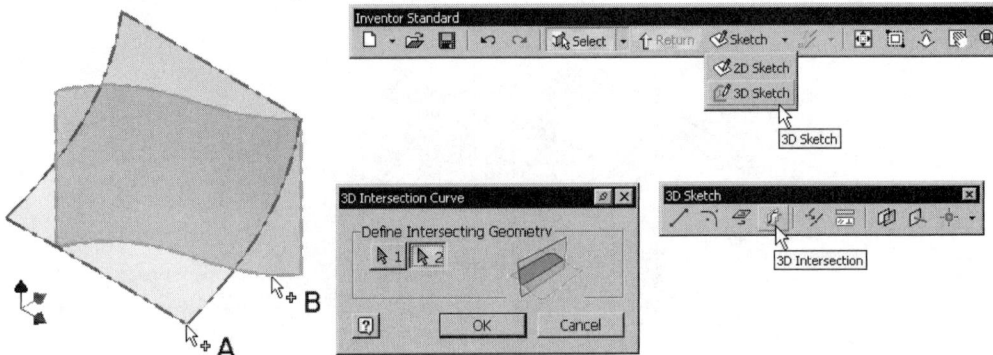

Figure 4.72 *3D curve being constructed at the intersection of two surfaces*

 Tip: If you wish to modify the 3D sketch, select it on the Browser Bar, right-click, and select Edit Sketch. Do not select the 3D Sketch button on the Inventor Standard toolbar because selecting it starts another 3D sketch.

The 3D sketch is complete. (See Figure 4.73.) Now hide the surfaces.

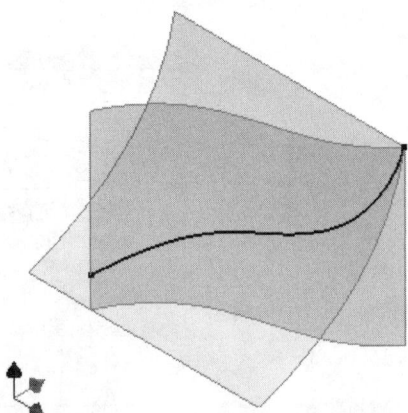

Figure 4.73 *3D sketch constructed from the intersection of two surfaces*

15. Select the surfaces on the Browser Bar (indicated as ExtrusionSrf1 and ExtrusionSrf2), right-click, and deselect Visibility.

The surfaces are hidden. Save your file (file name: *3DSketch1.ipt*).

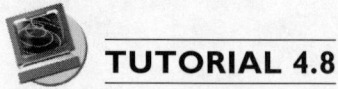

TUTORIAL 4.8

In this tutorial, you will construct a revolved surface, construct an extruded solid extruding to the surface, and construct a 3D sketch from the model edge of the extruded solid.

1. Start a new part file. Use the metric template.
2. With reference to Figure 4.74, construct a sketch (consisting of a spline and a line). Endpoint A of the centerline is coincident with the projected center point.
3. Set the display to an isometric view.

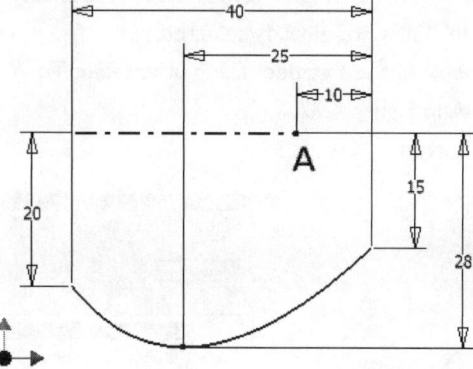

Figure 4.74 *Sketch constructed*

4. Revolve the sketch an angle of 180° to form a revolved surface. (See Figure 4.75.)

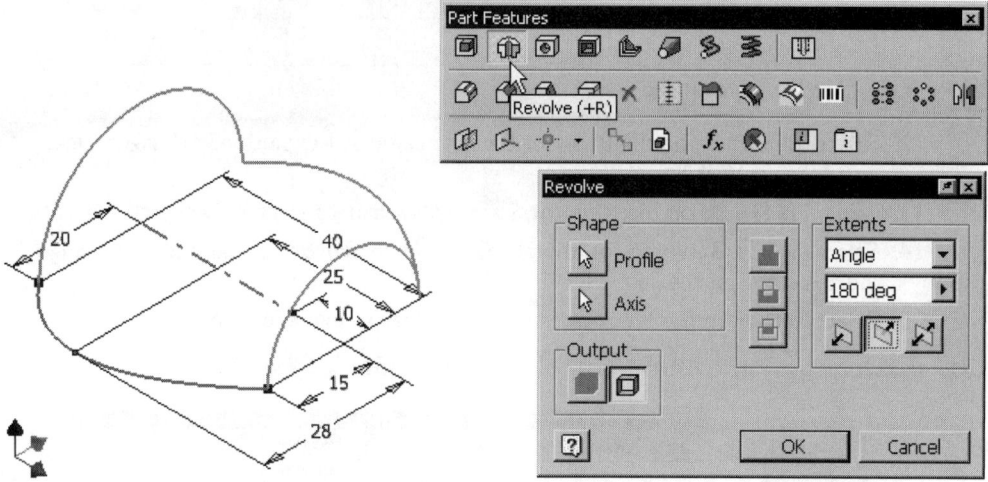

Figure 4.75 *Revolved surface being constructed*

5. Construct a sketch on the XY plane. Point A is the projected center point. (See Figure 4.76.)
6. Select Return on the Inventor Standard toolbar to exit sketch mode.

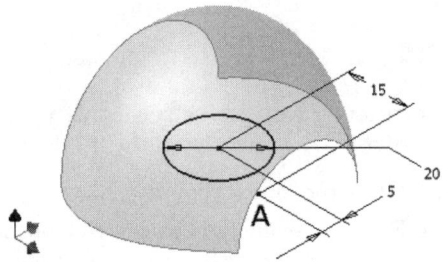

Figure 4.76 *Sketch constructed*

7. Select Extrude on the Part Features Panel Bar or toolbar.

8. Select the circle, if it is not already selected.

9. In the Extents box of the Extrude dialog box, select To.

10. Select surface A in Figure 4.77.

11. Click the OK button.

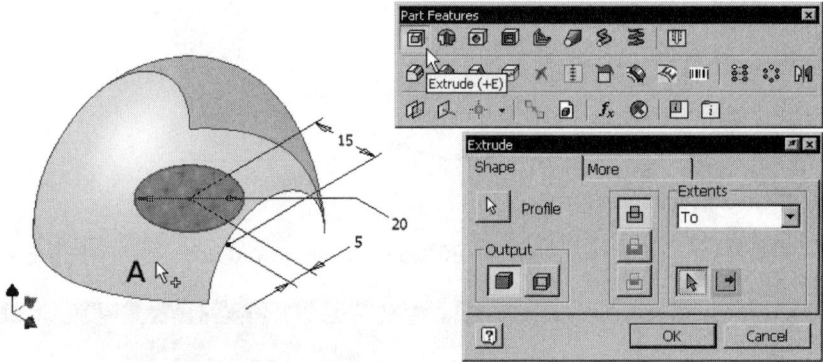

Figure 4.77 *Sketch being extruded*

12. Select the surface on the Browser Bar (indicated as RevolutionSrf1), right-click, and deselect Visibility.

13. Select 3D Sketch on the Inventor Standard toolbar to start a 3D sketch.

14. Select Include Geometry on the 3D Sketch Panel Bar or toolbar and select edge indicated in Figure 4.78.

15. Select Return on the Inventor Standard toolbar to exit sketch mode.

A 3D sketch is constructed. Save your file (file name: *3DSketch2.ipt*).

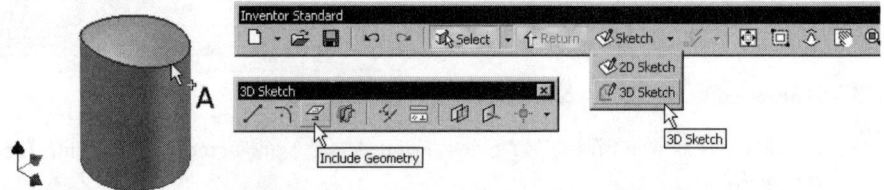

Figure 4.78 *Model edge being included in a 3D sketch*

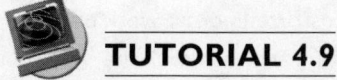

TUTORIAL 4.9

In this tutorial, you will construct a 3D sketch for making a sweep solid feature of a bicycle handlebar shown in Figure 4.79. You will construct three 2D sketches and include their sketch elements in the 3D sketch.

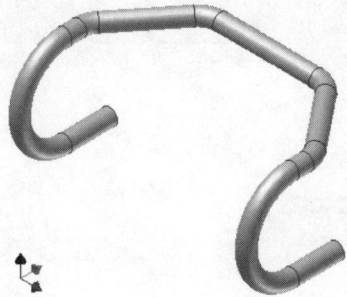

Figure 4.79 *Handlebar of a bicycle*

1. Start a new part file. Use the metric template.
2. If a sketch is already established on the XY plane, select Return on the Inventor Standard toolbar to exit sketch mode, and select the sketch on the Browser Bar, right-click, and select Delete.
3. Select 2D Sketch on the Inventor Standard toolbar, and select YZ Plane on the Browser Bar to establish a sketch plane on the origin's YZ plane.
4. Select Project Geometry and select the origin's center point on the Browser Bar.
5. With reference to Figure 4.80, construct an arc and two horizontal lines tangent to the arc. Point A is the projected center point.

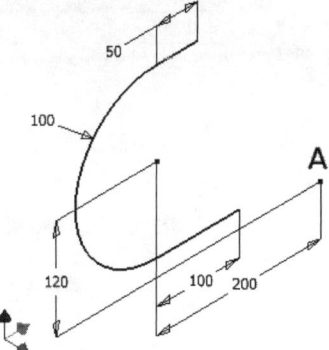

Figure 4.80 *Sketch constructed on the YZ plane*

Now construct a work plane offset a distance of 500 mm from the YZ plane.

6. Select Return on the Inventor Standard toolbar to exit sketch mode.
7. Select Work Plane on the Part Features Panel Bar or toolbar.
8. Select YZ plane on the Browser Bar to highlight it.

9. Select and drag the YZ plane at A in Figure 4.81 to construct an offset work plane.

10. In the Offset dialog box, set the offset value to 500 mm, and then click the Checkmark to construct the work plane.

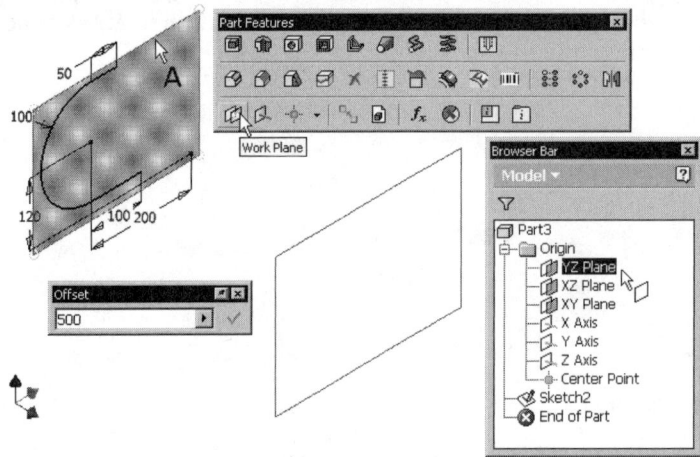

Figure 4.81 *Offset work plane being constructed*

Now project geometry from the first sketch.

11. Select 2D Sketch on the Inventor Standard toolbar and select the new work plane. (See Figure 4.82.)

12. Select Project Geometry on the 2D Sketch Panel Bar or toolbar.

13. Select arc A and lines B and C in Figure 4.82.

14. Select Return on the Inventor Standard toolbar to exit sketch mode.

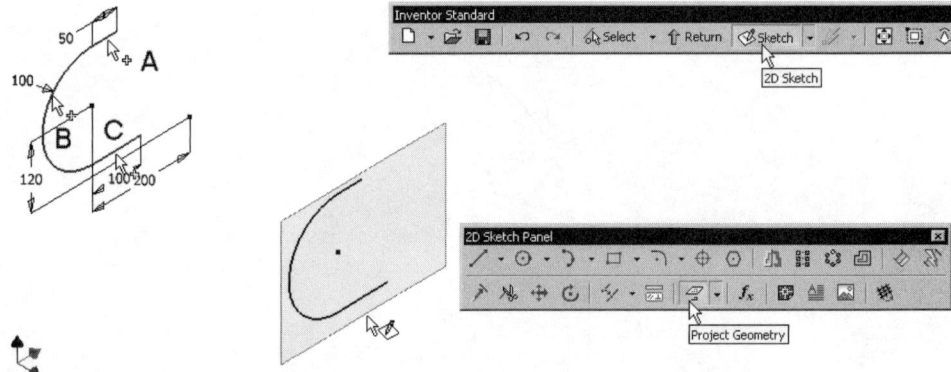

Figure 4.82 *Selected geometry projected*

Now construct a sketch on the XZ plane.

15. Select XZ plane on the Browser Bar, right-click, and select Visibility.

16. Select 2D Sketch on the Inventor Standard toolbar and select the XZ plane.

17. Select Project Geometry and select the origin's center point.

18. With reference to Figure 4.83, construct a line and add dimensions. Point A is the projected center point.

19. Select Return on the Inventor Standard toolbar to exit sketch mode.

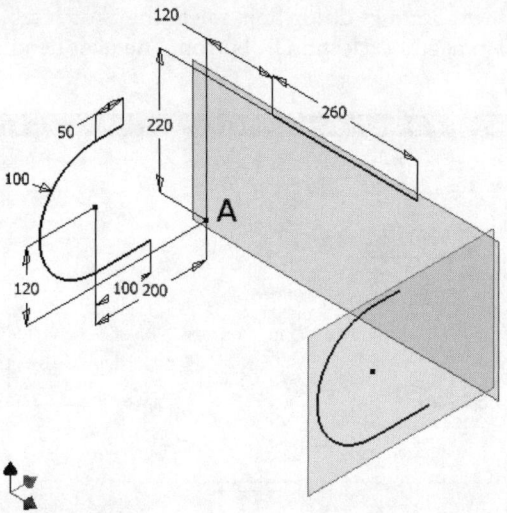

Figure 4.83 *Third sketch constructed*

Now construct a 3D sketch.

20. Select 3D Sketch on the Inventor Standard toolbar.

21. Select Include Geometry on the 3D Sketch Panel Bar or toolbar.

22. Select sketch elements A, B, C, D, E, F, and G in Figure 4.84.

 Note: You should not end the 3D Sketch (do not select Return); remain in sketch mode.

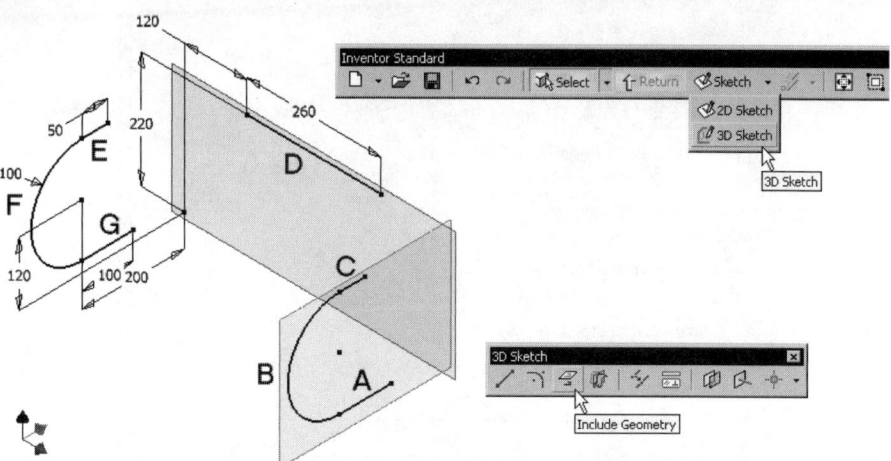

Figure 4.84 *Geometry being included in the 3D sketch*

Now specify a 3D fillet bend of 50 mm to be automatically created while making 3D lines.

23. Select Tools > Document Settings.

24. In the Document Settings dialog box, select the Sketch tab, set Auto-Bend Radius to 50 mm, and click the OK button. The fillet bend radius is set. (See Figure 4.85.)

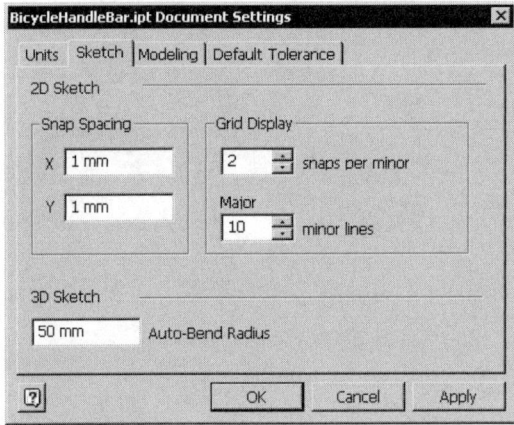

Figure 4.85 *Document Settings dialog box*

25. Select Tools > Application Options.

26. In the Options dialog box, select the Sketch tab, select the Auto-Bend with 3D Line Creation box, and click the OK button. Fillet bends will be placed automatically. (See Figure 4.86.)

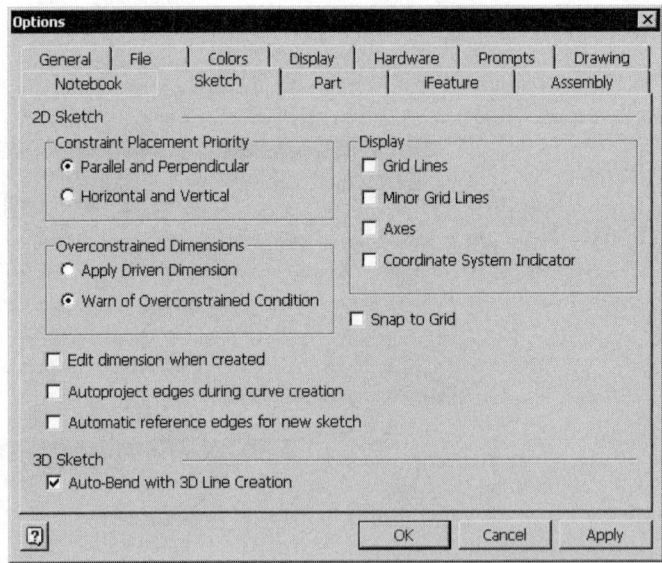

Figure 4.86 *Options dialog box*

Tip: Fillet bends between contiguous 3D line segments can be constructed either automatically, by selecting the Auto-Bend with 3D Line Creation box in the Options dialog box, or manually, by using the Bend command on the 3D Sketch Panel Bar or toolbar.

Now hide the 2D sketches, origin's plane, and work plane.

27. Select the 2D sketches, XZ plane, and the work plane on the Browser Bar one by one, right-click, and deselect Visibility.

Now continue with the 3D sketch by constructing two 3D lines

28. Select the 3D sketch from the browser, right-click, and select Edit Sketch if you exited sketch mode.

29. Select Line on the 3D Sketch Panel Bar or toolbar.

30. Select endpoints A and then B in Figure 4.87, right-click, and select Done.

31. Select Line again on the 3D Sketch Panel Bar or toolbar and then select endpoints C and then D in Figure 4.87, right-click, and select Done.

32. Select Return on the Inventor Standard toolbar to exit sketch mode.

The 3D sketch is complete. Save your file (file name: *BicycleHandleBar.ipt*).

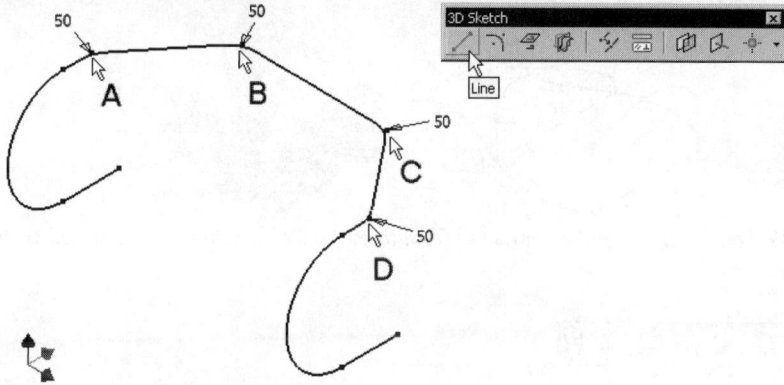

Figure 4.87 *Lines and fillet bends constructed*

LOFT, SWEEP, AND COIL FEATURES

Now you will learn how to construct loft, sweep, and coil features. Both loft and sweep features require more than a sketch. To make a loft solid, you need a number of sketches depicting the cross sections of the feature. To make a sweep solid, you need a sketch to depict the cross section and a sketch to depict the path. The coil feature is a special kind of sweep feature in which a cross section is swept along a helical path.

LOFT FEATURE

Unlike extruded and revolved features that use a single sketch for construction and have a constant cross section, a loft feature builds on multiple sketches and has a variable cross section defined by 2D Sketches or 3D sketches. There are two types of loft features: The first type has two or more cross sections, and the feature is constructed by making a smooth transition from the first cross section through the last cross section. The second type has two or more cross sections together with one or more guiding rails. The use of guide rails in lofting provides better control over how a cross section is transformed to its adjacent cross section. The cross sections are called profile sketches and the guiding rails are called rail sketches.

Prior making a loft feature, you make sketches. The minimum number of sketches required is two. Figure 4.88 shows a loft feature with three profile sketches, and Figure 4.89 shows a loft feature with three profile sketches and three rail sketches.

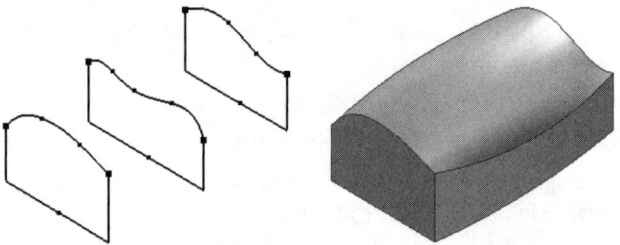

Figure 4.88 *Loft feature (right) constructed from three profile sketches (left)*

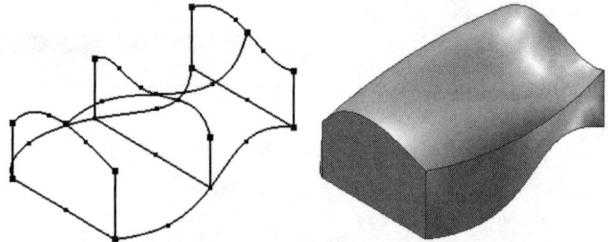

Figure 4.89 *Loft feature (right) constructed from three profile sketches and three rail sketches (left)*

TUTORIAL 4.10

In this tutorial, you will continue to work on a part file that you started in Chapter 3 to construct a number of 2D sketches for making a loft solid feature of the component shown in Figure 4.90.

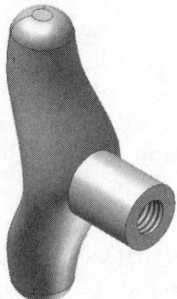

Figure 4.90 *Knob*

In Chapter 3, you already constructed a sketch on the XY plane. Now you will construct four more sketches established on four work planes.

1. Open the file *PunchIndexKnob.ipt* that you created in Chapter 3.

2. Select XY plane on the Browser Bar, right-click, and select Visibility.

3. Select Work Plane on the Part Features Panel Bar or toolbar.

4. Select XY plane, hold down the left mouse button, and drag it from A to B (Figure 4.91).

5. In the Offset dialog box, set the offset distance to 30 mm and select the Checkmark. A work plane offset from the XY plane is constructed. (See Figure 4.91.)

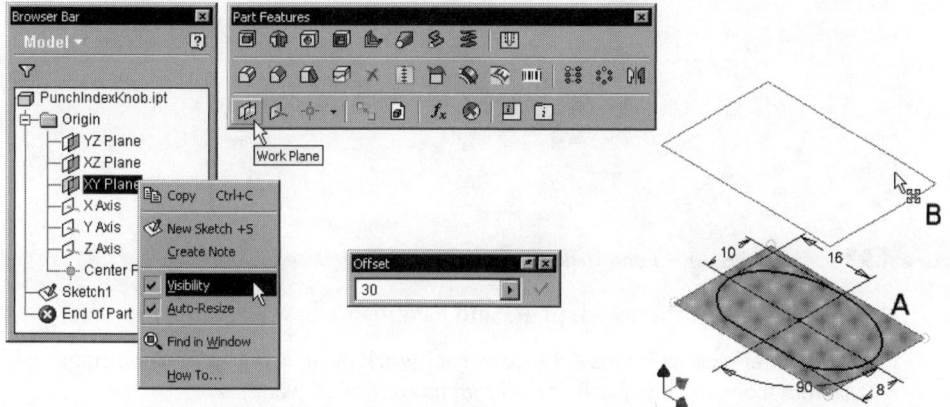

Figure 4.91 *Offset work plane being constructed*

Now construct a sketch on the offset work plane.

6. Select 2D Sketch on the Inventor Standard toolbar and select the work plane to establish a sketch plane.

7. Select Project Geometry on the 2D Sketch Panel bar or toolbar and select line A and center point B in Figure 4.92.

8. Construct a circle.

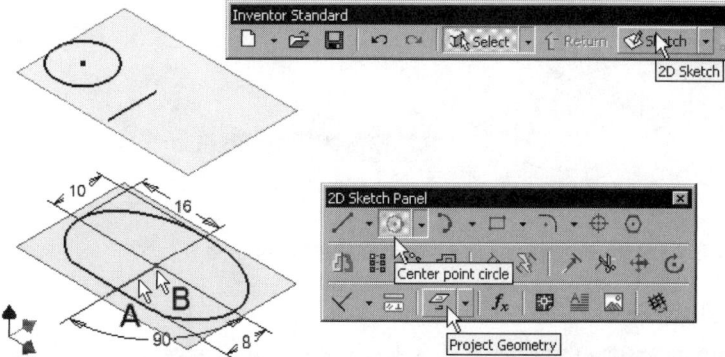

Figure 4.92 *Sketch established, the ellipse's center point projected, and a circle constructed*

9. With reference to Figure 4.93, add dimensions to the sketch. (The diameter of the circle is 20 mm, and its center is 11 mm from the projected geometry.)

10. Select Coincident from the 2D Sketch Panel bar or toolbar and select center point A and line B in Figure 4.93.

11. Select Return on the Inventor Standard toolbar to exit sketch mode.

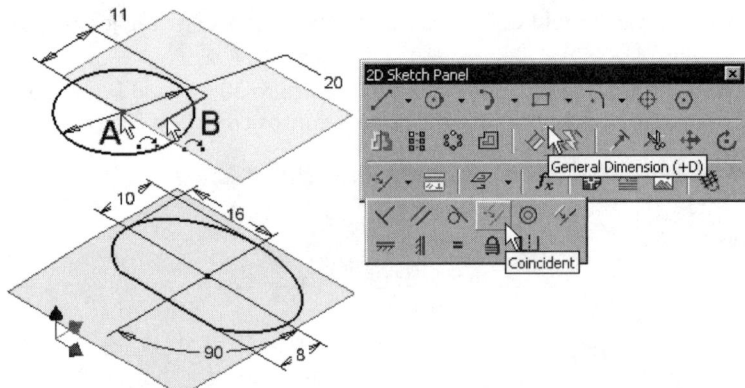

Figure 4.93 *Dimensions added and coincident constraint being placed*

Now construct another work plane and establish a new sketch plane.

12. With reference to Figure 4.94, construct work plane B that offsets a distance of 15 mm from work plane A (or 45 mm from the XY plane).

13. Establish a sketch plane on the new work plane.

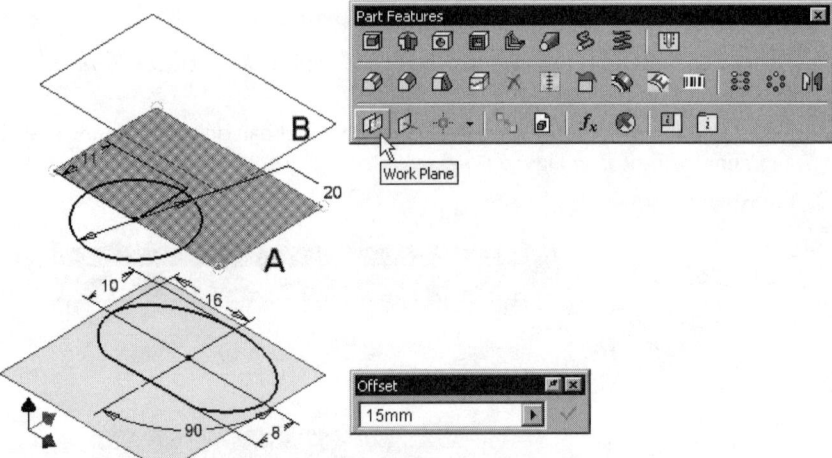

Figure 4.94 *Work plane constructed and sketch plane established*

Now construct a sketch.

14. With reference to Figure 4.95, select Project Geometry on the 2D Sketch Panel bar or toolbar and select center point A.

15. Select Center Point Circle from the 2D Sketch Panel bar or toolbar to construct a circle with the center coincident with project center point B.

16. Add a dimension to the circle. (Diameter = 16 mm)

17. Select Return on the Inventor Standard toolbar.

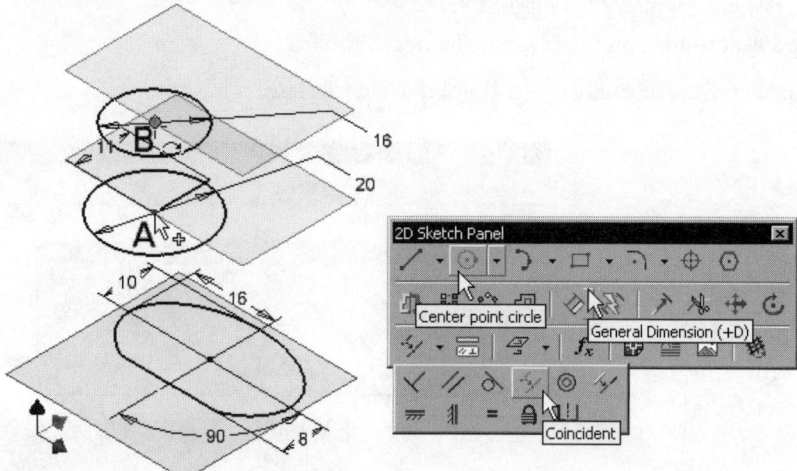

Figure 4.95 *Sketch constructed*

Now construct another work plane and a sketch on the new work plane.

18. With reference to Figure 4.96, construct work plane B that offsets a distance of – 30 mm from the XY plane A, and establish a sketch plane on the work plane.

19. Select Project Geometry on the 2D Sketch Panel Bar or toolbar and select circle C in Figure 4.96.

20. Select Return on the Inventor Standard toolbar.

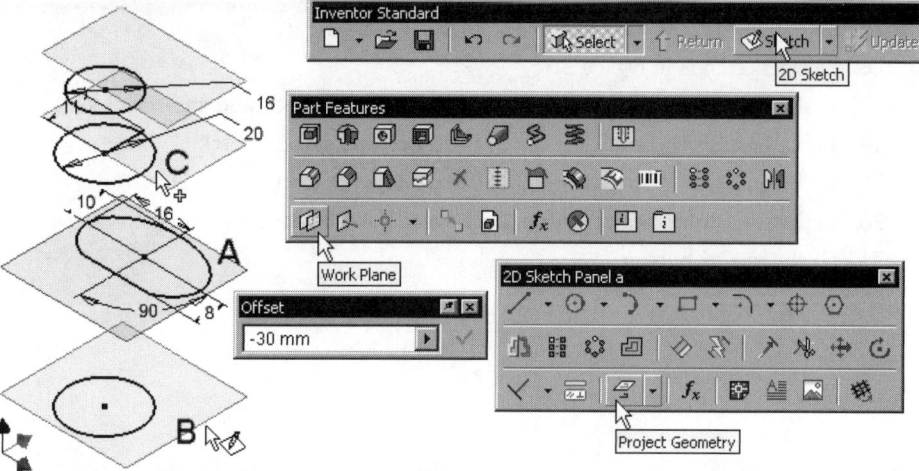

Figure 4.96 *Sketch constructed on a new workplane*

Now construct the fifth sketch.

21. With reference to Figure 4.97, construct work plane A that offsets a distance of 15 mm from work plane B (or –45 mm from the XY plane) and establish a sketch plane on the new work plane.

22. Project geometry C in Figure 4.97.

23. Select Return on the Inventor Standard toolbar.

The sketches for making a loft solid are complete.

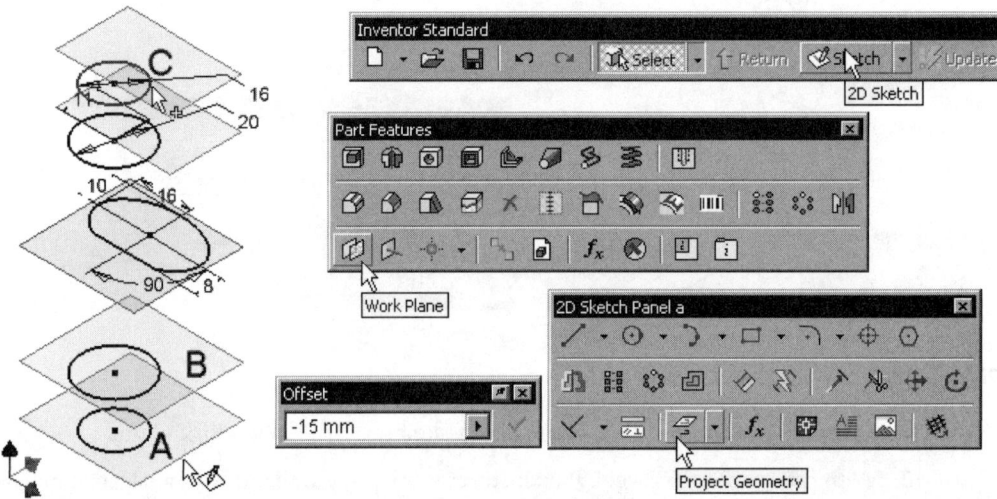

Figure 4.97 *Fifth sketch constructed*

Now construct a loft solid feature from the sketches.

24. Select the XY plane and the work planes on the Browser Bar, right-click, and unselect Visibility.

25. Select Loft on the Part Features Panel Bar or toolbar.

26. Select sketch A in Figure 4.98.

27. In the Loft dialog box, under Sections, click on Click to add and select sketch B.

28. In the Loft dialog box, click on Click to add and select sketch C.

29. In the Loft dialog box, click on Click to add and select sketch D.

30. In the Loft dialog box, click on Click to add and select sketch E.

31. Click the OK button.

The loft solid feature is complete. Save your file.

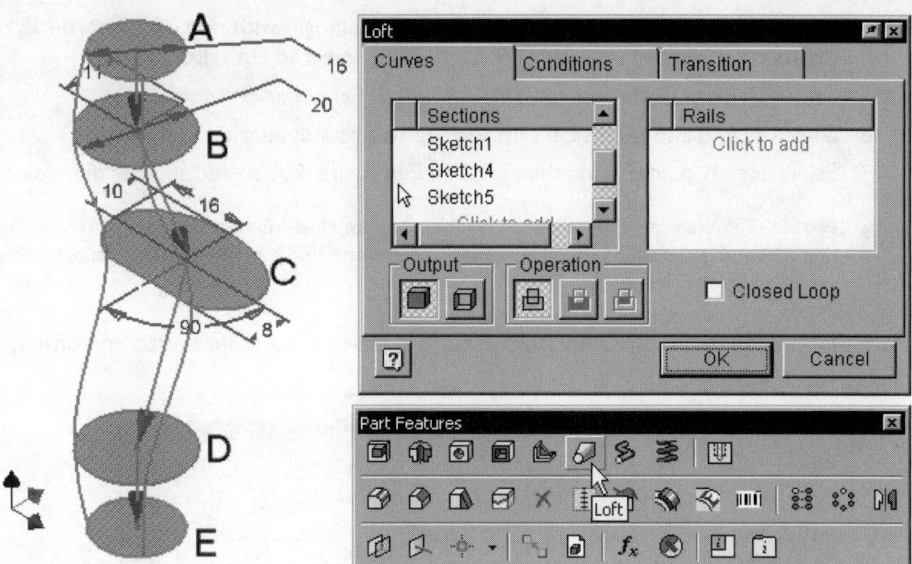

Figure 4.98 *Loft solid being constructed*

Tip: You can use 2D or 3D sketches in making a loft feature.

Project Cut Edges

You can construct a sketch element from the cross section cut by the sketch plane on the solid part. We call this method of sketching projecting cut edges.

TUTORIAL 4.11

In this tutorial, you will continue with the last tutorial by adding a revolved solid feature, a hole feature, and a fillet feature. Now you will construct a revolved feature. While making the revolved feature, you will learn how to project cut edges and use them as sketching elements.

1. Select 2D Sketch on the Inventor Standard toolbar and select XY plane on the Browser Bar.

2. Right-click and select Slice Graphic.

3. Select Project Cut Edges (on the same flyout as Project Geometry) on the 2D Sketch Panel Bar or toolbar. The section cut by the sketch plane across the solid part is projected on the sketch plane.

4. With reference to Figure 4.99, construct a rectangle with one of the corner points coincident with an edge point of the projected cut edge A.

5. Select General Dimension on the 2D Sketch Panel Bar or toolbar.

6. Select edge B and then select location C to add a dimension (38 mm).

7. Select edge B, point D, and then location E in Figure 4.99 to add another dimension.

 Note: This step causes Inventor to display a Create Linear Dimension dialog box warning you that adding this dimension will over-constrain the sketch and prompting you to choose Accept to create a Driven Dimension.

8. In the Create Linear Dimension dialog box, click the Accept button to construct a reference dimension.

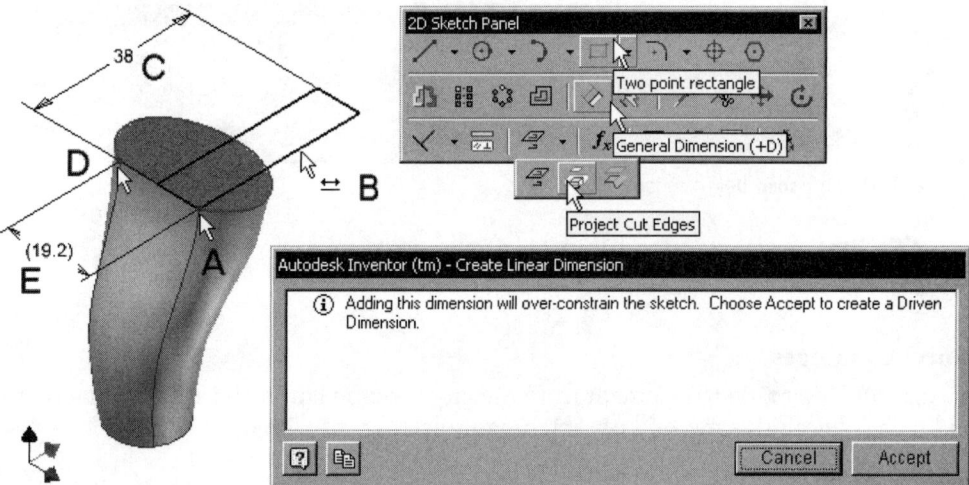

Figure 4.99 *Sketch plane established, graphic sliced, cut edges projected, and rectangle constructed*

Now add a dimension.

9. With reference to Figure 4.100, add a dimension.

10. Select and double-click the last dimension.

11. Select dimension A to copy dimension A's name, and append the string "/2" (meaning divided by 2) in the dimension dialog box. In Figure 4.100, dimension A's name is d22.

12. Select the dimension dialog box's Checkmark.

13. Select Return on the Inventor Standard toolbar.

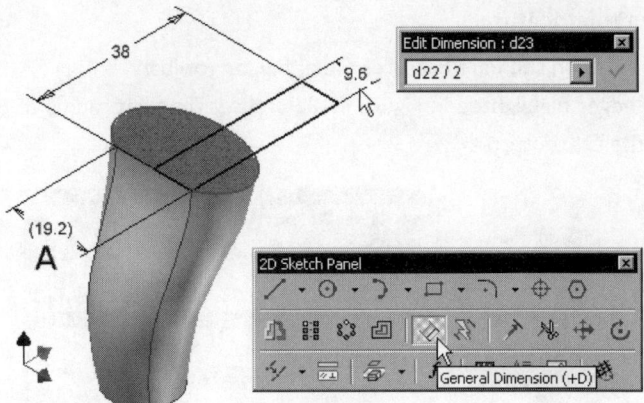

Figure 4.100 *Dimensions being added*

Now revolve the sketch.

14. Select Revolve on the Part Features Panel Bar or toolbar.
15. Select the rectangle as the profile to revolve.
16. In the Revolve dialog box, select Join operation and Full extents.
17. Select the Axis button and select edge A in Figure 4.101.
18. Click the OK button.

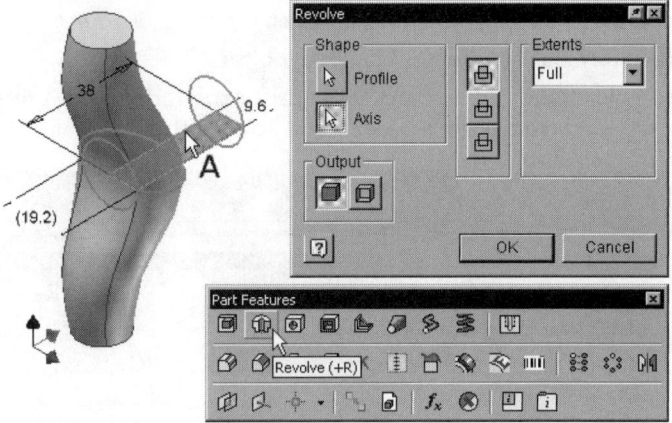

Figure 4.101 *Revolved solid feature being constructed*

Now hide the work planes.

19. Select the work planes, right-click, and unselect Visibility.

Tip: You can hide all the work geometry globally by selecting Work Geometry from the View menu and clearing the appropriate boxes. However, making objects invisible this way is temporary because the work planes will become visible again after you save, close, and reopen the file.

Now add a fillet feature.

20. Select Fillet on the Part Features Panel Bar or toolbar.

21. Select edges highlighted in Figure 4.102 and set the fillet radius to 6 mm.

22. Click the OK button.

Figure 4.102 *Fillet feature being constructed*

Now add a hole feature.

23. Set the display in accordance with Figure 4.103.

24. Select 2D Sketch on the Inventor Standard toolbar and select face A.

25. Select Project Cut Edges on the 2D Sketch Panel Bar or toolbar.

26. Select Return on the Inventor Standard toolbar.

27. Select Hole on the Part Features Panel Bar or toolbar to place a threaded hole (Size = M12, thread length = 18 mm, and hole depth = 20 mm).

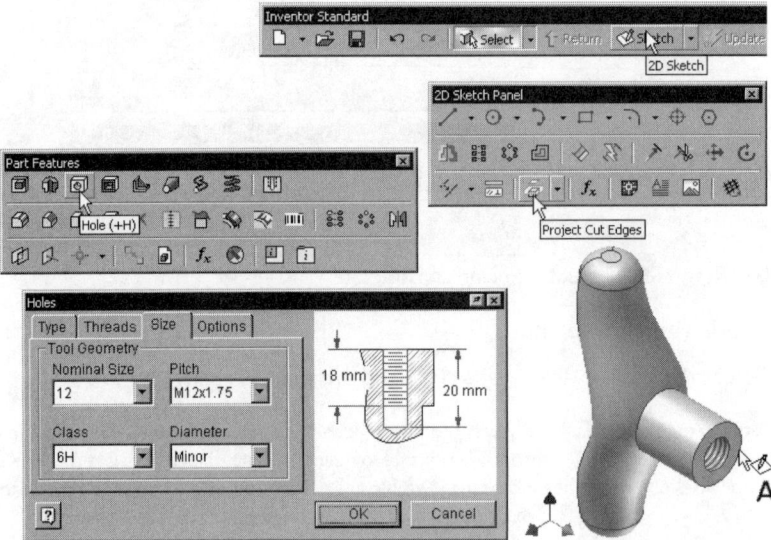

Figure 4.103 *Hole feature being placed*

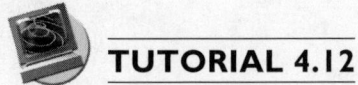

TUTORIAL 4.12

In this tutorial, you will construct a loft feature from three profile sketches and three rail sketches. Figure 4.104 shows the completed solid. You will use this solid part together with some other solid parts that you will construct later in this chapter to form a solid model of a toy car body. Although 2D sketches are used in making this solid, it must be emphasized that both 2D and 3D sketches can be used.

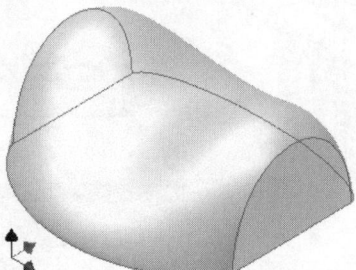

Figure 4.104 *Loft solid*

In this tutorial, you will construct a sketch on the YZ plane.

1. Start a new part file. Use metric template.
2. If a sketch is already established on the XY plane, select Return to exit sketch mode, and then delete the sketch from the Browser.
3. Set the display to an isometric view.
4. Select 2D Sketch on the Inventor Standard toolbar, and select YZ plane on the Browser Bar to construct a new sketch.
5. Project the center point from the solid part's origin.
6. With reference to Figure 4.105, construct a sketch comprising of a line and a spline. Note that point A is the projected center point.
7. Select Return on the Inventor Standard toolbar to exit sketch mode.

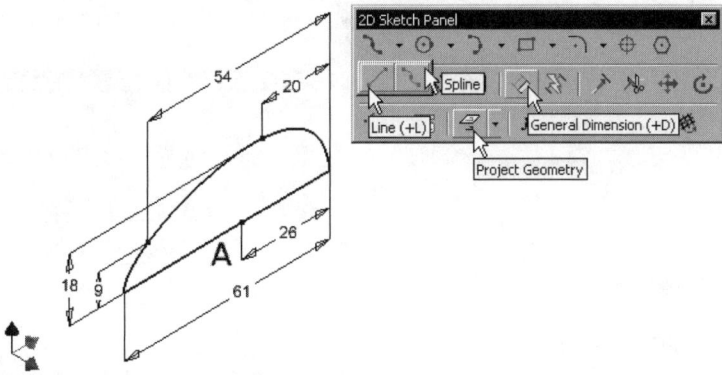

Figure 4.105 *Sketch constructed on the YZ plane*

Now construct a work plane and construct a sketch on it.

8. Select YZ plane on the Browser Bar to highlight it.

9. With reference to Figure 4.106, select Work Plane on the Part Features Panel Bar or toolbar, select the highlighted YZ plane at A, hold down the left mouse button, drag the mouse to location B, type **40** in the Offset dialog box and click the Checkmark to construct a work plane.

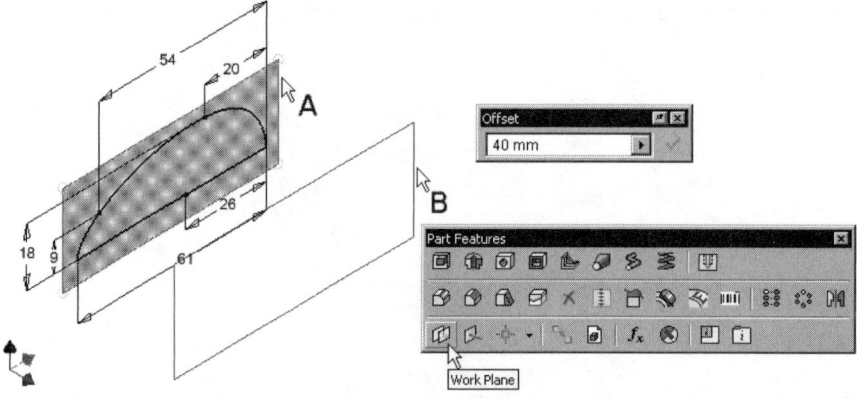

Figure 4.106 *Work plane being constructed*

10. With reference to Figure 4.107, construct a sketch, again using a line and a spline. In the sketch, point A is the projected center point.

11. Select Return on the Inventor Standard toolbar to exit sketch mode.

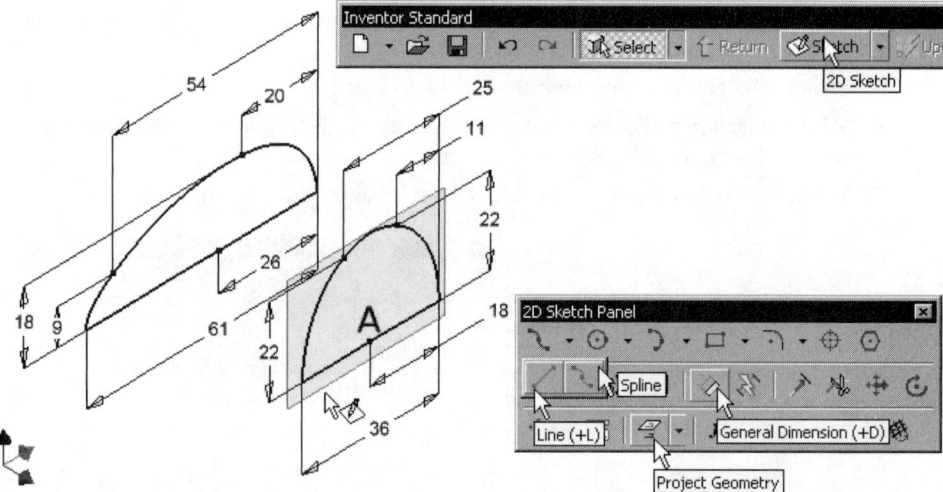

Figure 4.107 *Sketch constructed on the work plane*

12. Highlight the YZ plane from the browser.

13. Construct a work plane offsetting -40 mm from highlighted YZ plane (A in Figure 4.108).

14. Construct a sketch on work plane B.

15. With reference to Figure 4.108, project spline C and line D onto the sketch.

16. Select Return on the Inventor Standard toolbar to exit sketch mode.

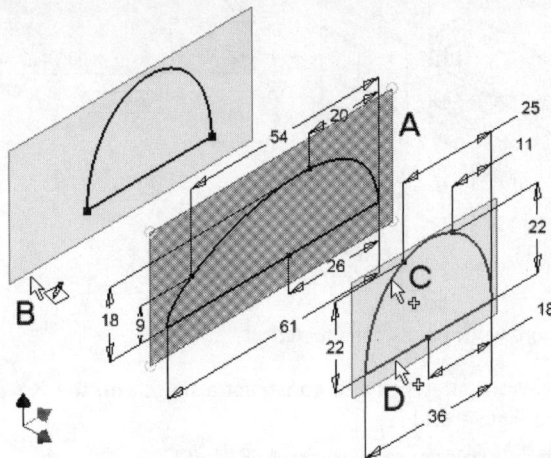

Figure 4.108 *Third sketch constructed*

17. Hide the work planes.

18. Select Work Point on the Part Features Panel Bar or toolbar, select XZ plane (A in Figure 4.109) on the Browser Bar and select curve B to construct a work point.

19. Repeat step 18 twice to construct two more work points at C and D in Figure 4.109.

20. Select Work Point on the Part Features Panel Bar or toolbar and select vertex E to construct a work point.

21. Repeat step 20 twice to construct two more work points at F and G in Figure 4.109.

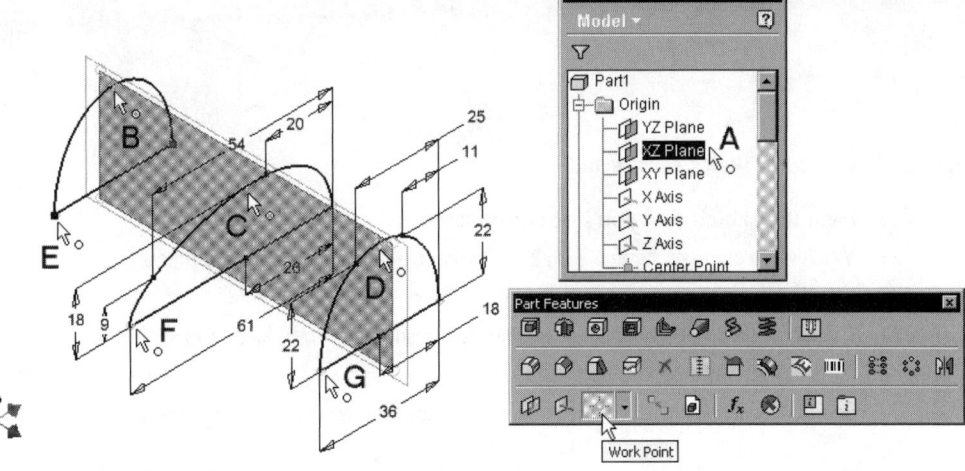

Figure 4.109 *Six work points constructed*

22. Construct three more work points at A, B, and C in Figure 4.110.

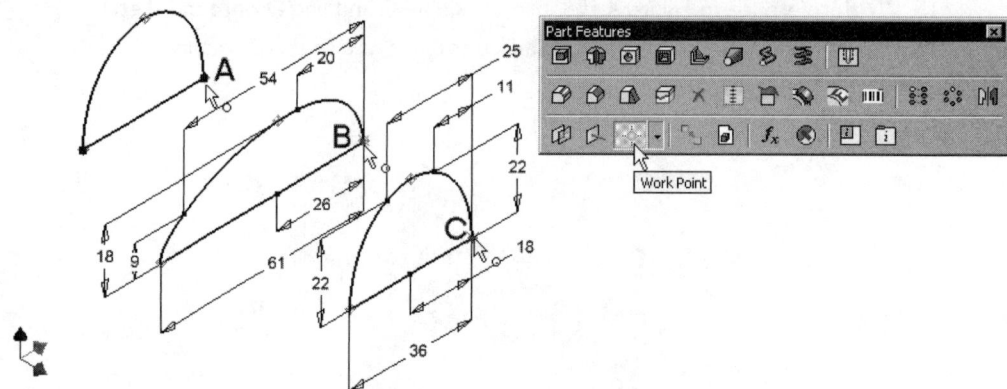

Figure 4.110 *Three more work points constructed*

23. Hide the previous sketches and construct a sketch on the XY plane With reference to Figure 4.111.
24. On the sketch, project work points A, B, and C.
25. Construct a spline with four segments passing through A, B, and C.
26. Add dimensions to the spline.
27. Select Return on the Inventor Standard toolbar to exit sketch mode.

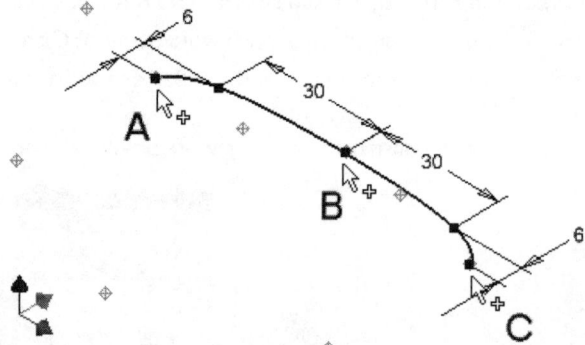

Figure 4.111 *Fourth sketch constructed*

28. Hide the sketch that you just constructed.
29. With reference to Figure 4.112, construct a sketch on the XZ plane.
30. On the sketch, project work points A, B, and C.
31. Construct a spline with four segments passing through A, B, and C.
32. Add dimensions to the spline.
33. Select Return on the Inventor Standard toolbar to exit sketch mode.

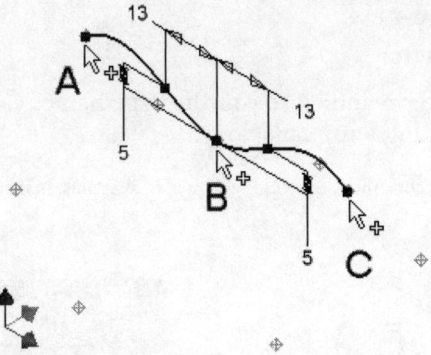

Figure 4.112 *Fifth sketch constructed*

34. Hide the sketch that you just constructed.
35. With reference to Figure 4.113, construct a sketch on the XY plane.
36. On the sketch, project work points A, B, and C.
37. Construct a spline with four segments passing through A, B, and C.
38. Add dimensions to the spline.
39. Select Return on the Inventor Standard toolbar to exit sketch mode.

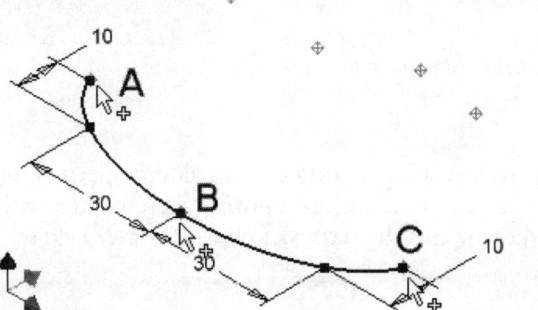

Figure 4.113 *Sixth sketch constructed*

40. Hide all the work points and make all the sketches visible.

 Note: You can select multiple sketches or work points on the Browser Bar by holding down SHIFT or CTRL while selecting them, and then change their visibility all at once, instead of selecting and changing each one individually.

41. Select Loft on the Part Features Panel Bar or toolbar.
42. Click the Sections box in the Loft dialog box and select sketch A in Figure 4.114.
43. Select B in Figure 4.114.
44. Select C in Figure 4.114.
45. Click the Rails box and select D in Figure 4.114.
46. Select E in Figure 4.114.

47. Select F in Figure 4.114.

48. Click the OK button.

A loft solid is complete. Save and close your file (file name: *CarBody1.ipt*). You will use this solid part file for making a toy car's body.

 Tip: The rail sketches must coincide with profile sketches. In other words, they must intersect in 3D space.

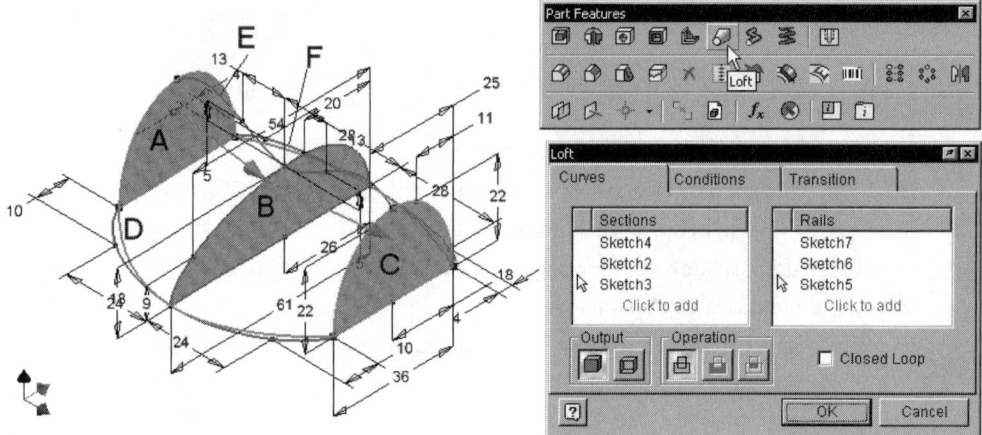

Figure 4.114 *Loft solid being constructed*

SWEEP FEATURE

A sweep feature concerns sweeping a cross section along a path. Therefore, constructing a sweep solid feature requires two sketches: a profile sketch and a path sketch. The profile sketch must be a 2D sketch, and the path sketch can be a 2D sketch or a 3D sketch.

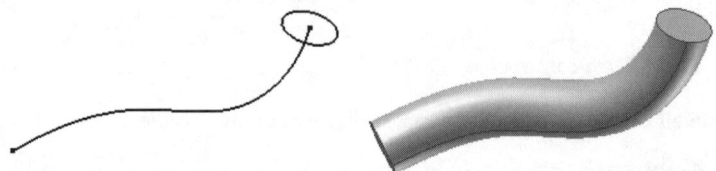

Figure 4.115 *Profile sketch and path sketch (left) and sweep feature constructed from the sketches*

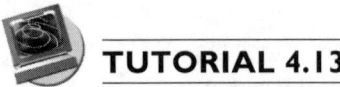 **TUTORIAL 4.13**

In this tutorial, you will use the 3D sketch that you constructed in Tutorial 4.7 as a path, construct a profile sketch, and make a sweep solid from the path and profile.

1. Open the file *3DSketch1.ipt*, if you already closed it.

2. Select Work Plane on the Part Features Panel Bar or toolbar, select the 3D sketch at A, and select endpoint B (Figure 4.116). A work plane is constructed

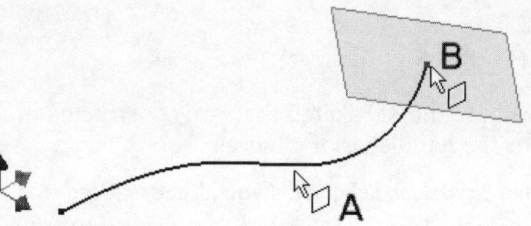

Figure 4.116 *Work plane constructed*

3. Construct a sketch on work plane A in Figure 4.117.
4. With reference to Figure 4.117, construct a sketch. Point B is the projected endpoint of the 3D sketch.
5. Select Return on the Inventor Standard toolbar to exit sketch mode.

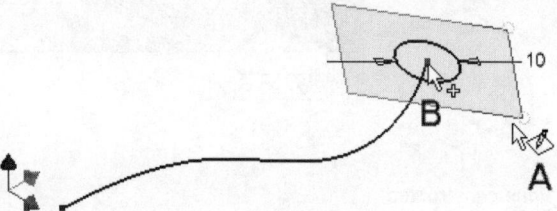

Figure 4.117 *Sketch constructed on a work plane made on the endpoint of the 3D sketch*

6. Select Sweep on the Part Features Panel Bar or toolbar.
7. Select the Profile button, if it is not already selected.
8. Select sketch A in Figure 4.118.
9. Select the Path button and select sketch B in Figure 4.118.
10. Click the OK button.

A sweep solid is constructed. Save and close your file.

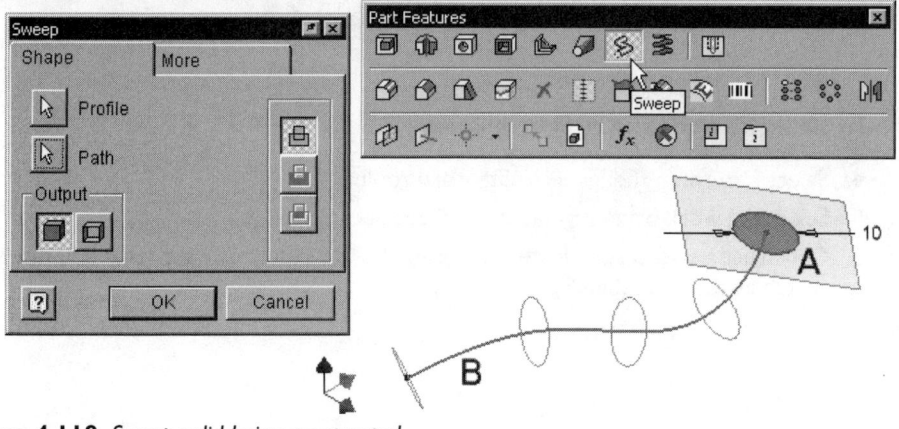

Figure 4.118 *Sweep solid being constructed*

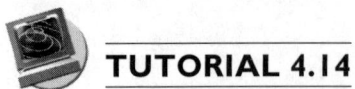

TUTORIAL 4.14

In this tutorial, you will use the 3D sketch that you constructed in Tutorial 4.9 to make a sweep solid feature for the handlebar of a bicycle.

1. Open the file *BicycleHandleBar.ipt*, if you already closed it.
2. Construct a work plane at endpoint A in Figure 4.119

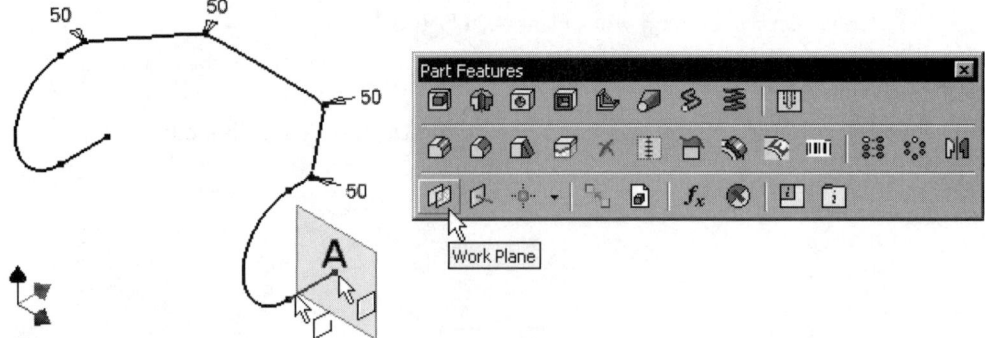

Figure 4.119 *Work plane constructed*

3. With reference to Figure 4.120, establish a sketch plane and construct a sketch.

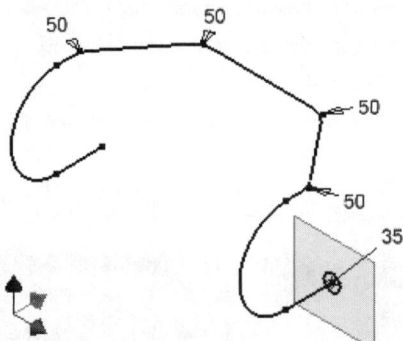

Figure 4.120 *Sketch constructed on the work plane*

4. Select Return on the Inventor Standard toolbar.
5. Select the work plane, right-click, and deselect Visibility.
6. Construct a sweep solid feature by using the 2D sketch as the profile and the 3D sketch as the path. (See Figure 4.121.)

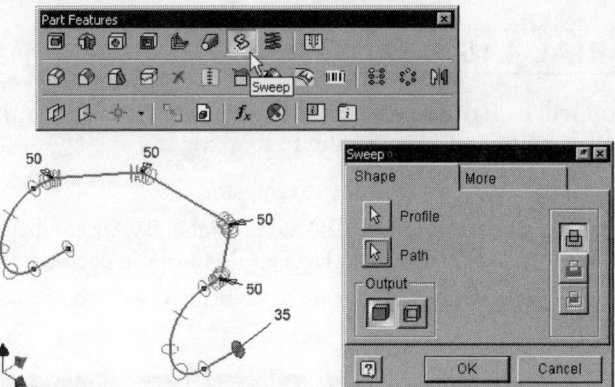

Figure 4.121 *3D sweep solid being constructed*

Now add a shell feature.

7. With reference to Figure 4.122, add a shell feature with a shell thickness of 1 mm and remove faces A and B while shelling.

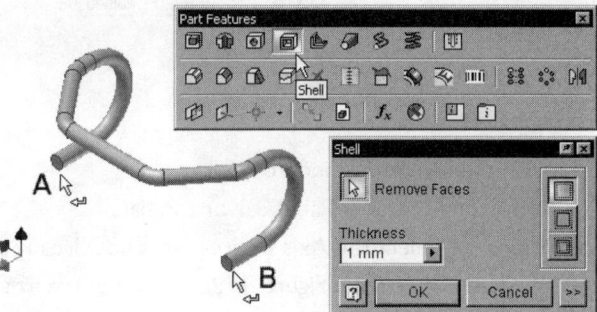

Figure 4.122 *Shell feature placed*

The handlebar is complete. Save and close your file.

COIL FEATURE

A coil solid is a special kind of 3D sweep solid in which the profile sketch is swept along a helical path. To make a coil solid, you construct a 2D sketch to depict the cross section of the coil, specify an axis, and specify the parameters of the helix. The axis can be a line, an edge, or a work axis. Figure 4.123 shows a profile sketch and coil constructed from the profile sketch.

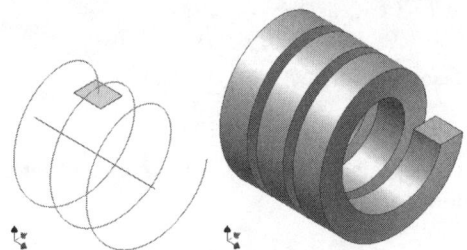

Figure 4.123 *Sketch and coil feature*

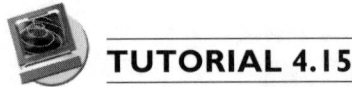

TUTORIAL 4.15

In this tutorial, you will construct a coil solid for making a spring for the punch set. You will use the origin's X axis as the axis of the helix path.

1. Start a new part file. Use the metric template.
2. Select Project Geometry on the 2D Sketch Panel Bar or toolbar and select the origin's X and Y axes. A and B in Figure 4.124 are the projected geometry.
3. Construct a circle with its center point coincident with the projected Y axis and add two dimensions.

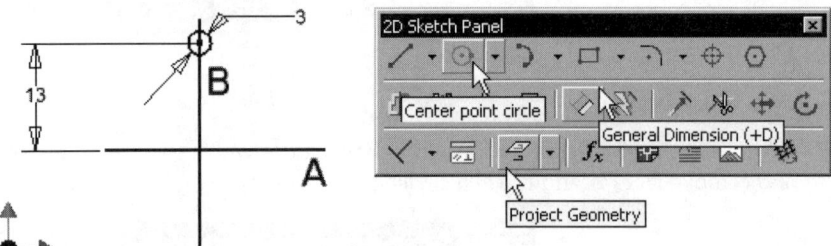

Figure 4.124 *Sketch constructed*

Now construct a coil feature.

4. Select Return on the Inventor Standard toolbar.
5. Select Coil on the Part Features Panel Bar or toolbar.
6. In the Coil dialog box, select the Axis button if it is not already selected.
7. Select the projected X axis (A in Figure 4.125) to use as the axis.

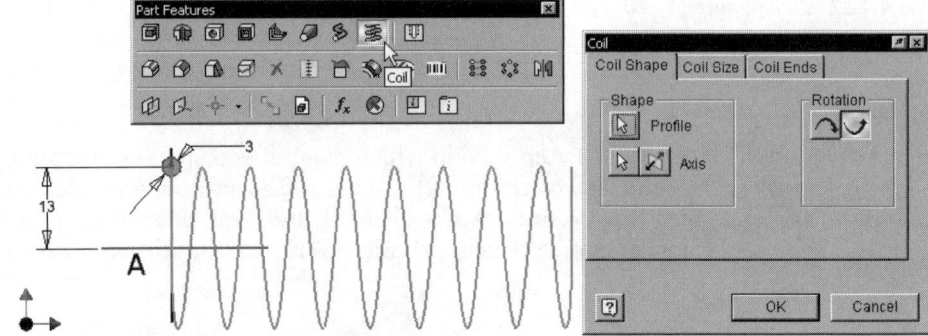

Figure 4.125 *Coil feature being constructed*

8. Select the Coil Size tab. (See Figure 4.126.)
9. On the Coil Size tab, select Pitch and Revolution and set pitch size to 8 mm and revolution number to 8.

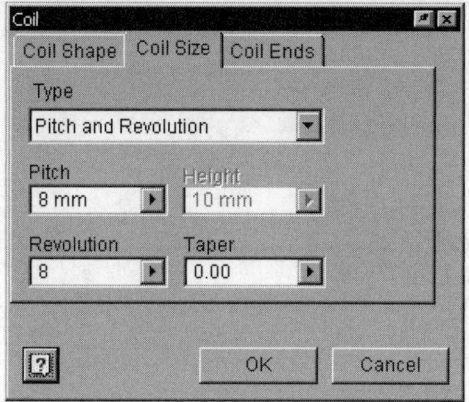

Figure 4.126 *Coil Size tab*

10. Select the Coil Ends tab. (See Figure 4.127.)
11. On the Coil Ends tab, set both start and end to flat, set transition angle to 90 deg, and set flat angle to 90 deg.
12. Click the OK button. A coil feature is constructed.

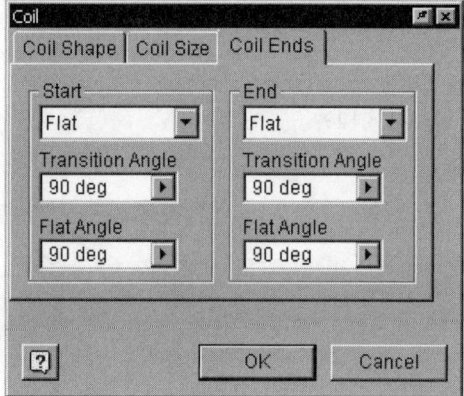

Figure 4.127 *Coil Ends tab*

Now construct a revolved solid feature to intersect with the coil feature.

13. Select 2D Sketch on the Inventor Standard toolbar and select XY plane from the browser.
14. Select Wireframe Display on the Inventor Standard toolbar.
15. Select Project Geometry on the 2D Sketch Panel Bar or toolbar and select the origin's X axis and circular edges A and B in Figure 4.128.
16. Construct a rectangle with its lower edge collinear with the projected X axis and its vertical edges coincident with the center points of the projected edges A and B.
17. Add a dimension.

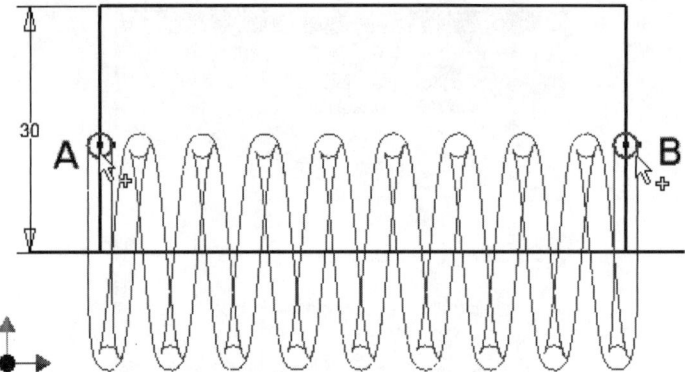

Figure 4.128 *Sketch being constructed*

Now revolve the sketch.

18. Select Return on the Inventor Standard toolbar.

19. Right-click and select Isometric.

20. Select Shaded Display on the Inventor Standard toolbar.

21. Select Revolve on the Part Features Panel Bar or toolbar.

22. In the Revolve dialog box, select the Profile button and then select the rectangle you just constructed, select the Intersect operation, set Extents to full, and select the Axis button, if it is not already selected.

23. Select line A in Figure 4.129.

24. the OK button.

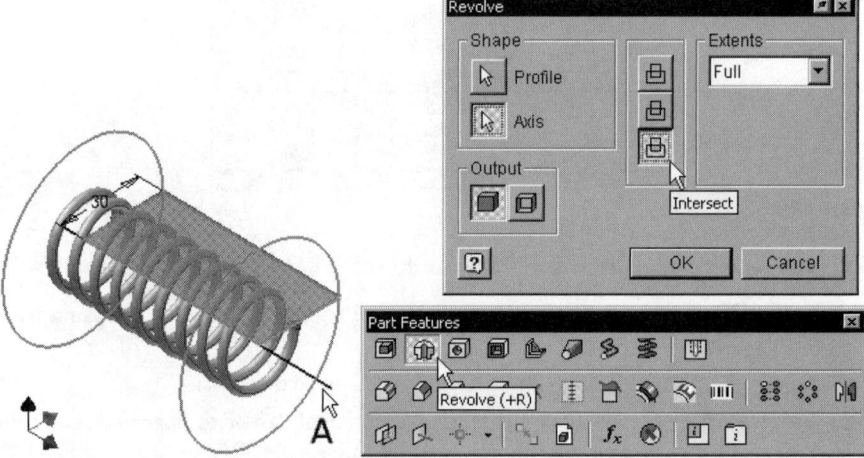

Figure 4.129 *Revolved solid being constructed*

The spring is complete. (See Figure 4.130.) Save and close your file (file name: *PunchSpring.ipt*).

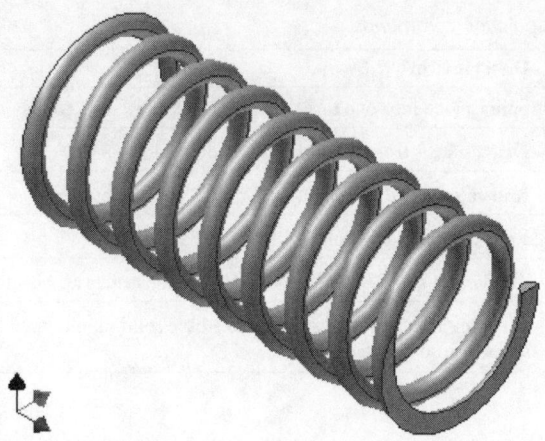

Figure 4.130 *Spring*

SURFACE MANIPULATION

Manipulating surfaces in a solid part increases the flexibility in model construction and expands the repertoire of a model's forms and shapes. You can import surfaces from other CAD applications or construct surfaces by using the four major feature construction methods: extrude, revolve, loft, and sweep. In these operations, if you use a closed-loop sketch or sketches in making these features, you can construct a solid feature or a surface feature. If you use an open-loop sketch or sketches, you construct a surface feature. To construct surfaces, you select the Surface button in the feature construction dialog box.

Although solid features are the main ingredients of a solid part, surfaces play an important role in modeling. Earlier in this chapter, you already learned how to construct a 3D sketch from the intersection of two surfaces, and you also learned how to extrude a solid to terminate at a surface. Beyond these uses, you can use a surface to split a solid into two parts and remove a portion of it.

SURFACE MANIPULATION TOOLS

Commands related to manipulating surfaces in a solid part are highlighted in Figure 4.131 and outlined in Table 4.2.

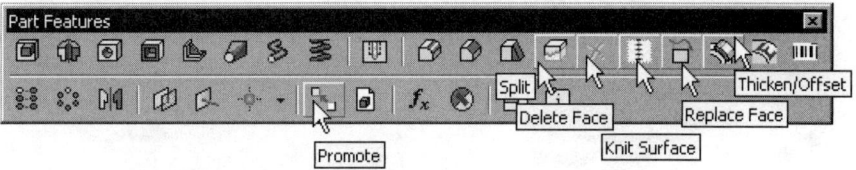

Figure 4.131 *Surface manipulation commands*

Table 4.2 *Surface manipulation commands*

Option	Description
Split	Splits a face into two faces or splits a solid into two solids and removes one of them.
Delete Face	Deletes faces from a set of surfaces or a solid.
Knit Surface	Knits a set of faces into a single quilt.
Replace Face	Replaces a face of a solid with a surface.
Thicken/Offset	Thickens a surface to become a solid or constructs an offset surface.
Promote	Changes a set of IGES surfaces to a base solid or stitches a set of disconnected IGES surfaces into a surface quilt.

SPLIT

You can split a solid part into two portions and remove one of them, and you can split the faces of a solid part into two sets of faces so that you can apply face draft to the faces in two directions. To split a solid part, you can use a work plane, a sketch, or a surface. Figure 4.132 shows a solid part split into two with its upper portion removed by a surface, and Figure 4.133 shows a solid part's faces split into two sets of faces by a plane.

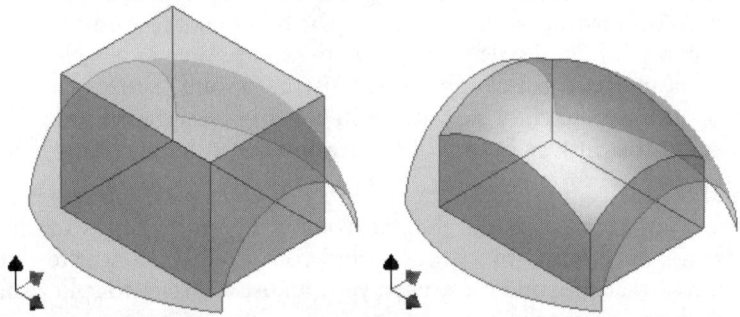

Figure 4.132 *Solid part split and a portion removed by a surface*

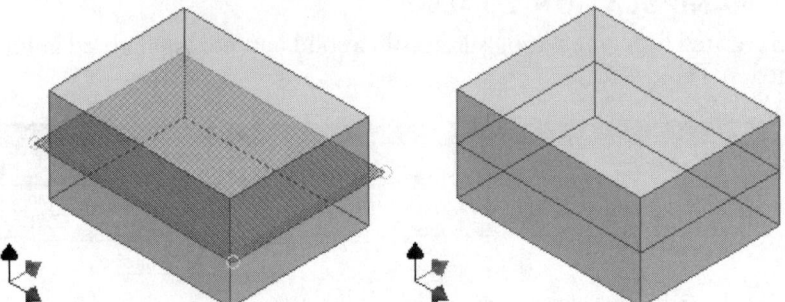

Figure 4.133 *Solid part's faces split into two sets of faces*

The primary purpose for placing a face draft is to facilitate removal of the part from a mold after the component is cast or molded. A mold usually has two halves fitted together along a parting line. You open the mold along the parting line to remove the component part from the mold. A rectangular solid box, for example, has six faces. As you have learned, face draft applies on the entire surface of a selected face. If you want to have a parting line along the waist line of the rectangular block and apply face draft in two directions, you have to split the faces of the solid part into two sets of faces.

TUTORIAL 4.16

This tutorial illustrates how to split a solid part into two and remove a portion after splitting. You will construct an extruded solid part and a revolved surface and use the revolved surface to split the extruded solid.

1. Start a new part file. Use the metric template.
2. With reference to Figure 4.134, construct a rectangle on the default XY plane and extrude it to form an extruded solid.

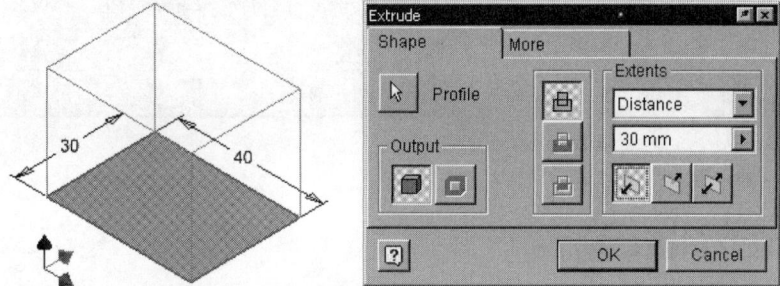

Figure 4.134 *Extruded solid being constructed*

3. Construct a sketch on the XY plane in accordance with Figure 4.135 and revolve it 180° to construct a revolved surface.

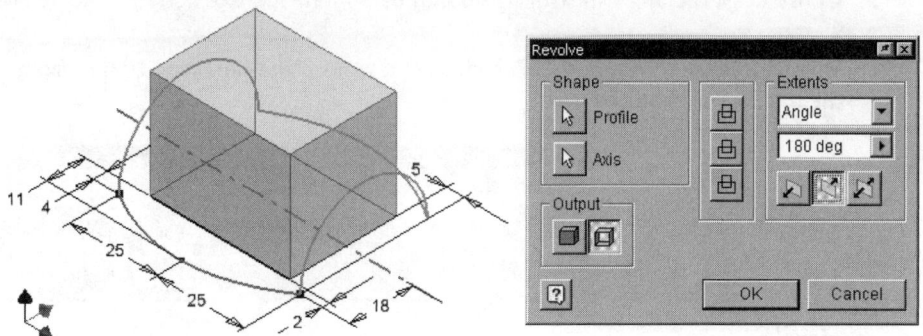

Figure 4.135 *Revolved surface being constructed*

4. Select Split on the Part Features Panel Bar or toolbar.
5. Select Split Part in the Method box of the Split dialog box.
6. Select surface A in Figure 4.136 as the split tool.

7. Select one of the Side to Remove buttons so that the arrowhead indicating the portion to be removed is pointing upward.

8. Click the OK button.

The solid part is split by the surface. Save and close your file (file name: *Split1.ipt*).

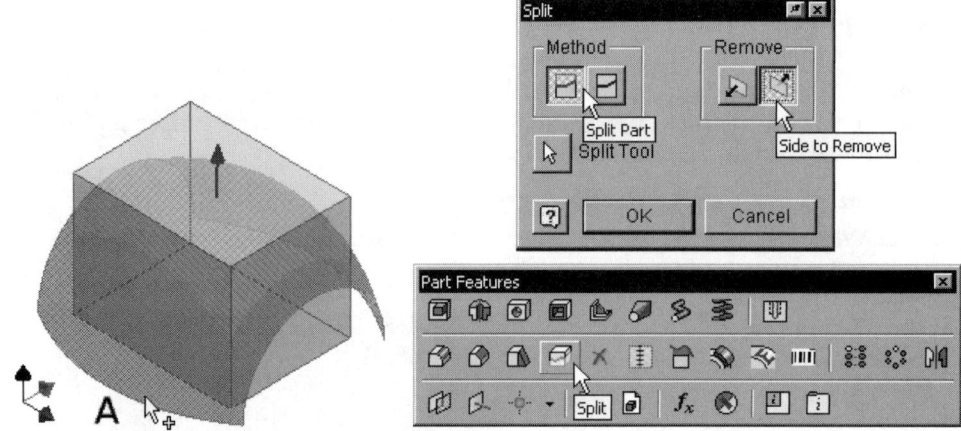

Figure 4.136 *Part being split by a surface*

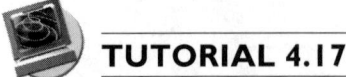

TUTORIAL 4.17

In this tutorial, you will use a work plane to split the faces of a solid part into two sets of surfaces and apply face draft to the faces.

1. Start a new part file. Use the metric template.

2. Construct a rectangle measuring 40 mm by 30 mm and extrude it a distance of 20 mm.

3. With reference to Figure 4.137, construct a work plane offsetting 10 mm from the top face of the solid part.

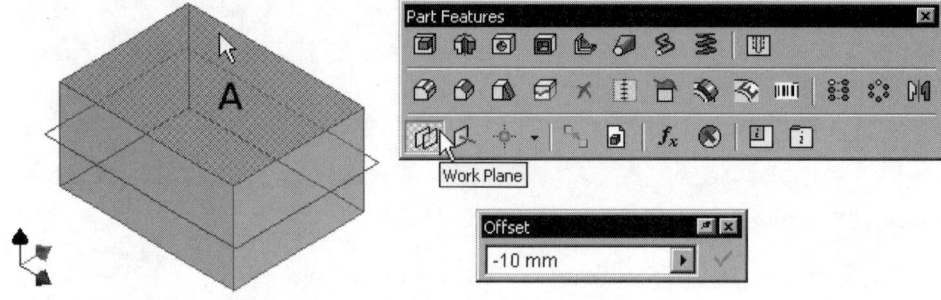

Figure 4.137 *Offset work plane constructed*

4. Select Split on the Part Features Panel Bar or toolbar.

5. In the Split dialog box, select Split Face in the Method box.
6. Select work plane A as the split tool.
7. Select All from the Faces box. (See Figure 4.138.)
8. Click the OK button. All the faces are split.

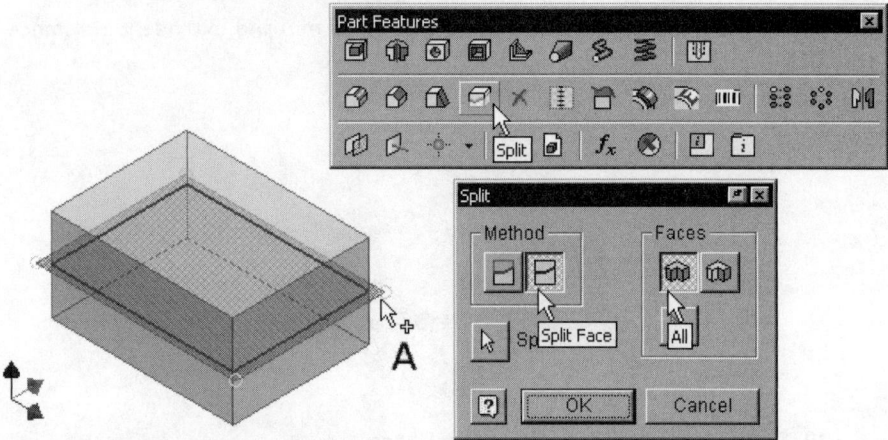

Figure 4.138 *Faces being split*

9. Hide the work plane.
10. Select Face Draft on the Part Features Panel Bar or toolbar.
11. Referring to Figure 4.139, select face A to specify the pull direction.
12. Select faces B, C, D, and E to place a face draft.
13. Specify a draft angle of 2.5 deg.
14. Click the OK button. (See Figure 4.139.)

The solid's faces are split and the face draft is placed. Save and close your file (file name: *Split2.ipt*).

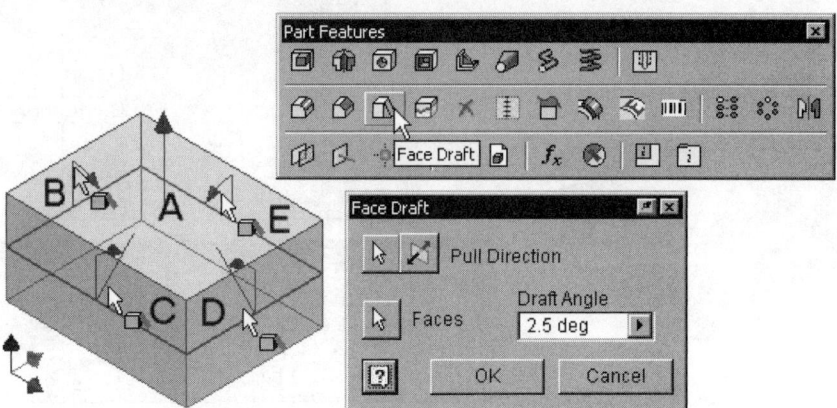

Figure 4.139 *Face draft being placed*

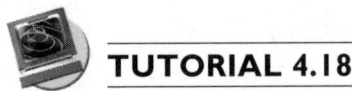

TUTORIAL 4.18

In this tutorial, you will use a sketch to split a solid part and remove a portion of it.

1. Start a new part file. Use the metric template.
2. Construct a rectangle measuring 40 mm by 30 mm and extrude it a distance of 20 mm.
3. Construct a sketch on face A in Figure 4.140.

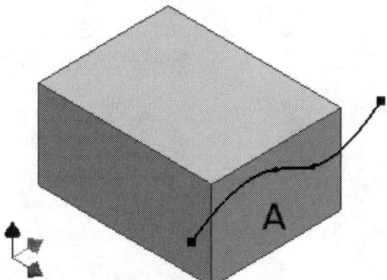

Figure 4.140 *Sketch constructed on a face of a solid part*

4. Select Split on the Part Features Panel Bar or toolbar.
5. Select Split Part in the Method box of the Split dialog box.
6. Select the sketch as the split tool.
7. Select one of the Side to Remove buttons so that the arrowhead indicating the portion to be removed is pointing upward. (See Figure 4.141.)
8. Click the OK button.

The part is split. (See Figure 4.142.) Save and close your file (file name: *Split3.ipt*).

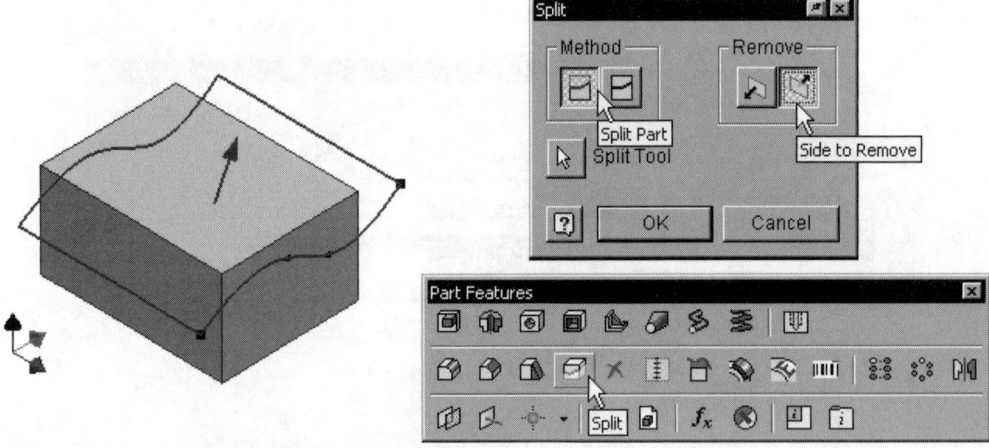

Figure 4.141 *Part being split by the sketch*

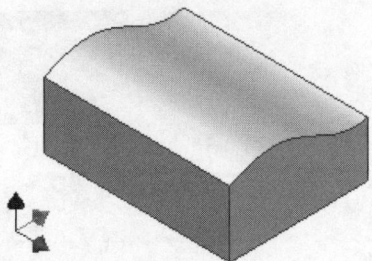

Figure 4.142 *Part split and a portion removed*

DELETE FACE

You can remove faces or lumps of faces from a part file. The prime function of deleting a face is to clean up unwanted faces in a part file. As we have mentioned in Chapter 2, you can open various kinds of file formats, such as an IGES file containing a number of surfaces. If the imported file has some unwanted surfaces, you can delete an unwanted individual face or a set of faces from this kind of imported surface object. Note that if you remove a face from a solid model, it becomes a surface model.

 ## TUTORIAL 4.19

This tutorial illustrates how to delete a face and a lump of faces from a file. You will construct an extruded solid part, place a hole on it, and cut a portion of it away by using another extruded solid. After that, you will remove selected faces and lumps of faces from the solid.

1. Start a new part file. Use the metric template.
2. Construct a rectangle (60 mm by 30 mm) on the default XY plane and extrude it a distance of 10 mm. (See Figure 4.143.)

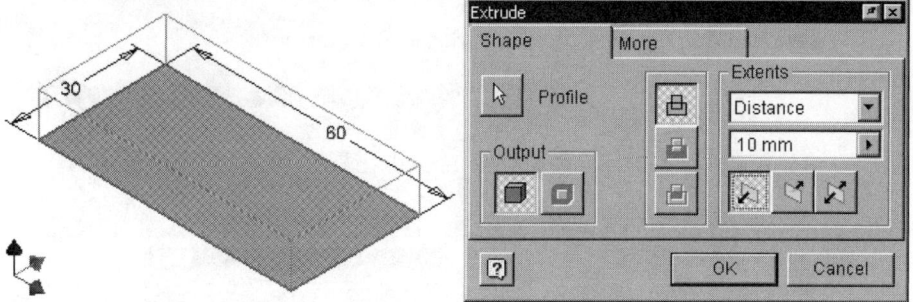

Figure 4.143 *Extruded solid feature being constructed*

3. With reference to Figure 4.144, place two holes of 10 mm diameter.

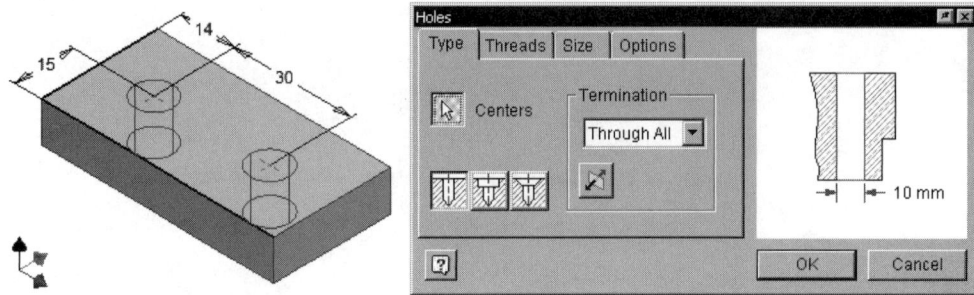

Figure 4.144 *Holes being placed*

4. Construct a sketch in accordance with Figure 4.145 and extrude it to cut through the solid part.

The solid part is cut into two islands of solid.

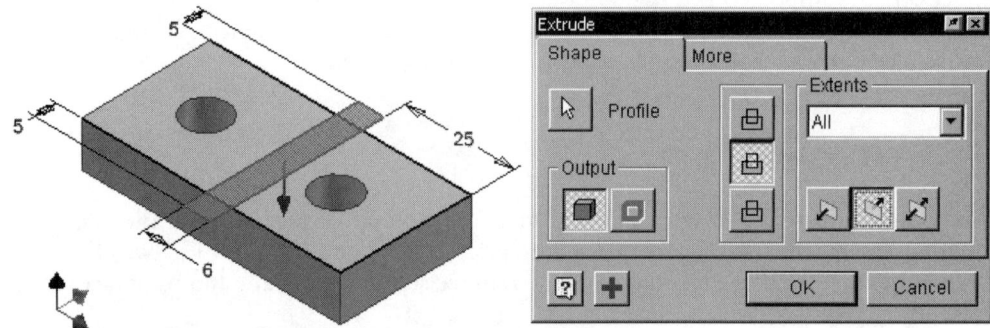

Figure 4.145 *Extruded solid feature being cut through the solid*

5. Select Delete Face on the Part Features Panel Bar or toolbar.
6. In the Delete dialog box, select the Select individual face button, if it is not already selected.
7. Select face A in Figure 4.146 and click the OK button.

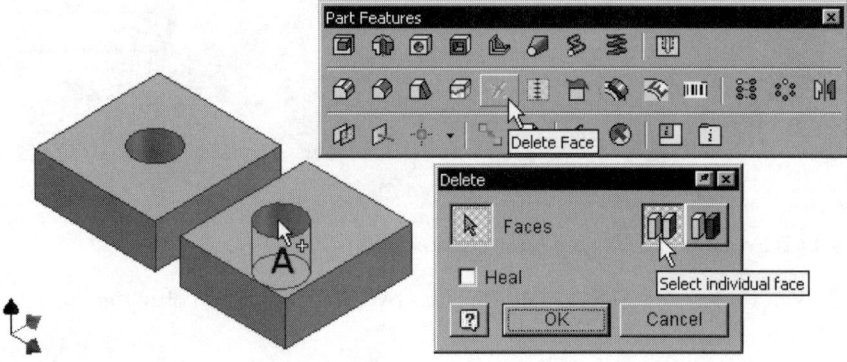

Figure 4.146 *A face being removed*

One of the solid islands is converted to a surface upon removal of a face.

8. Select Delete Face on the Part Features Panel Bar or toolbar.
9. In the Delete dialog box, check the Heal button.
10. Select face A in Figure 4.147 and click the OK button.

Because the opening is healed after removal of a face, the solid island remains as a solid.

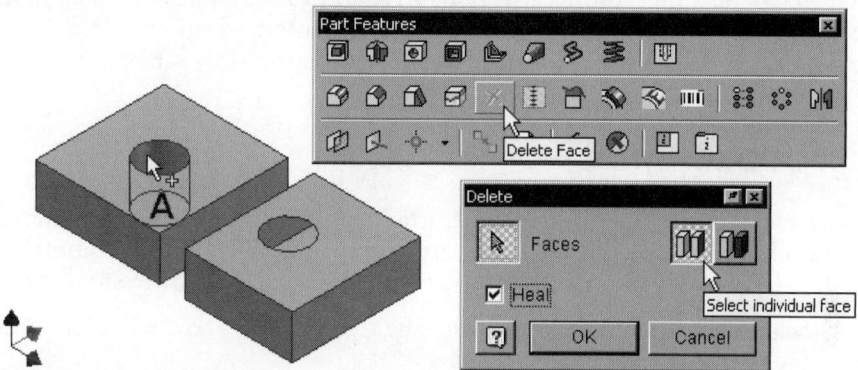

Figure 4.147 *Face being removed and the solid being healed after face removal*

11. Select Delete Face on the Part Features Panel Bar or toolbar.
12. In the Delete dialog box, select the Select lump or void button.
13. Select A in Figure 4.148 and click the OK button.

The lump of surfaces is removed. Save and close your file (file name: *Delete.ipt*).

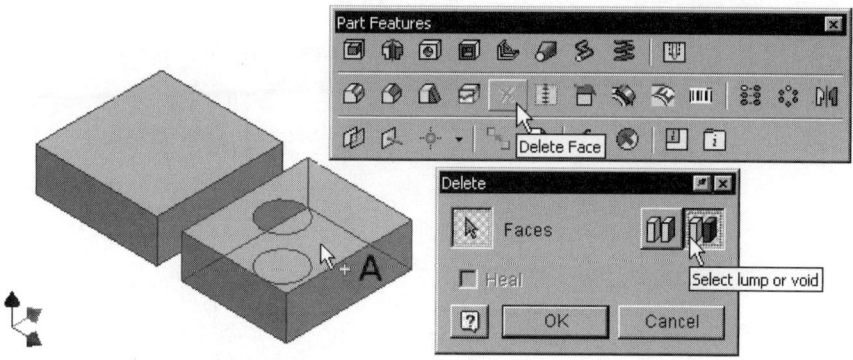

Figure 4.148 *Lump of surfaces being removed*

PROMOTE AND KNIT SURFACE

Promoting refers to repairing any irregularities and inconsistencies of surfaces imported from other CAD applications. Knitting refers to sewing together a set of contiguous surfaces to construct a surface quilt. If the surface quilt encloses a volume without any opening, a solid is formed.

Promoting and knitting are useful for importing face objects from other applications. When you import surfaces from other applications, you might have to knit the surfaces together in a quilt and then promote the quilt to the part environment. Promoting and knitting of imported surfaces can be done automatically by selecting the Auto Stitch and Promote box in the Import Options dialog box.

There are two ways to incorporate into Inventor the surfaces constructed through other computer-aided applications: opening an IGES or ACIS file or inserting an IGES or ACIS file. To open a file, you select IGES or ACIS from the Files of Types pull-down list box in the Open dialog box. To insert a file, select Insert > Import. After you open or insert a file, you will see a set of construction objects on the Browser Bar.

 ## TUTORIAL 4.20

This tutorial will illustrate how faces are knitted together and how a set of knitted faces enclosing a volume is promoted to a solid part. To see how knitting and promoting can be done both manually and automatically, you will first disable AutoStitch and Promote.

1. Select Open from the File menu. In the Open File dialog box, select IGES from the Files of type list box.

2. Select the file *Rev.igs* in the Chapter 4 folder of the CD accompanying this book.

3. In the Open File dialog box, click the Options button to display the Import Options dialog box.

4. Clear the Healer Enabled and AutoStitch and Promote check boxes, and then click OK.

5. In the Open File dialog box, click Open.

6. Reorient the display as shown in Figure 4.149.

A set of surfaces is imported. (See Figure 4.149.) On the Browser Bar, you will find a set of construction objects.

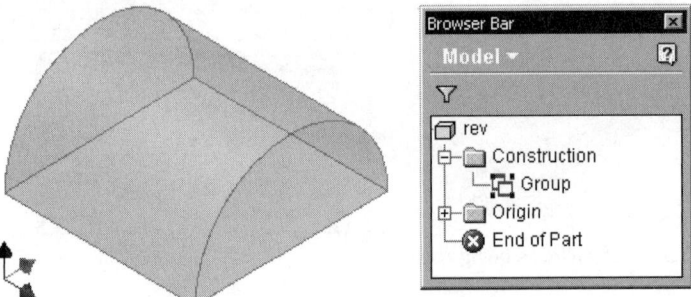

Figure 4.149 *Surfaces imported from an IGES file*

Now promote the construction objects to surfaces.

7. Select Promote on the Part Features Panel Bar or toolbar.

8. Select faces A, B, C, and D in Figure 4.150.
9. In the Promote dialog box, select the Promote as Surface check box, and click the Promote button and then the Done button.

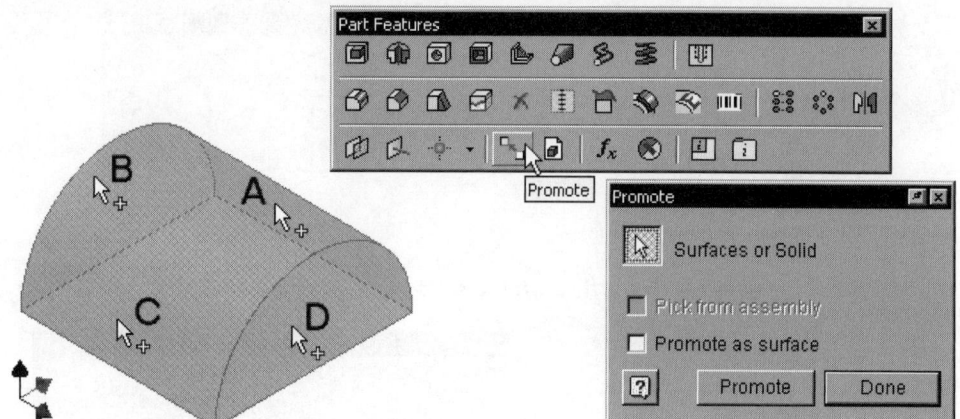

Figure 4.150 *Construction objects being promoted to surfaces*

On the Browser Bar, you will find four surfaces.

10. Select Knit Surface on the Part Features Panel Bar or toolbar.
11. Select surfaces A, B, C, and D in Figure 4.151.
12. Click the OK button in the Knit dialog box.

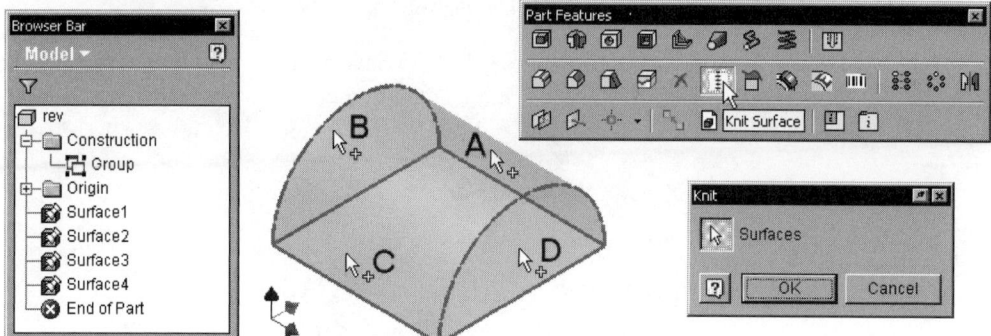

Figure 4.151 *Surfaces being knitted together*

To see that the stitched, closed volume is a solid, construct an extruded solid and cut through the part.

13. Select the surfaces on the Browser Bar, right-click, and deselect Visibility.
14. Select 2D Sketch on the Inventor Standard toolbar and select face A in Figure 4.152.
15. Construct a circle.

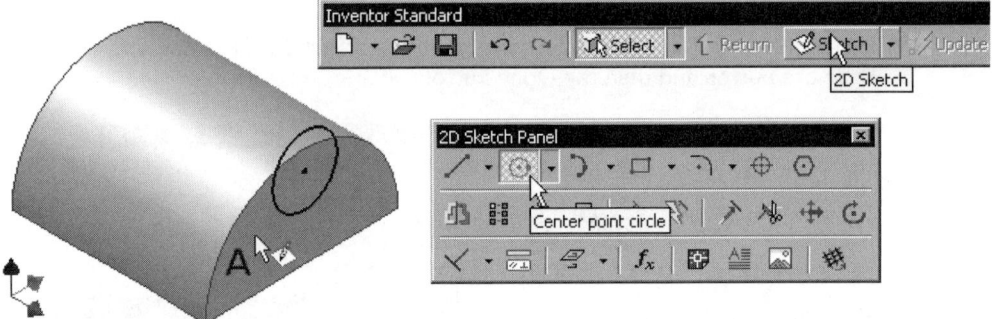

Figure 4.152 *Sketch constructed*

16. Extrude the circle to cut through the solid part. (See Figure 4.153.)

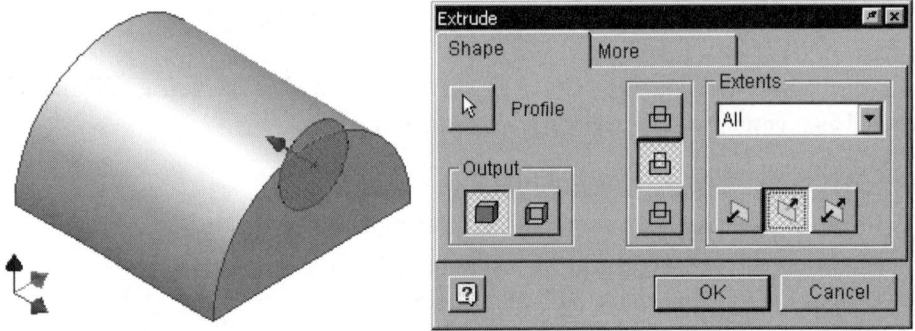

Figure 4.153 *Sketch being extruded to cut through the solid*

The solid part is complete. (See Figure 4.153.) Save and close your file (file name: *Rev1.ipt*).

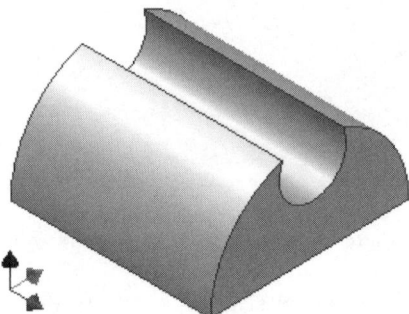

Figure 4.154 *Completed solid part*

Now set knitting and promoting to be automatic and open the same IGES file again.

17. Select File > Open and select the IGES file *Rev.igs* in the Chapter 4 folder of the CD.

18. In the Open File dialog box, click the Options button to display the Import Options dialog box, select the AutoStitch and Promote check box, and click OK.

19. In the Open File dialog box, click Open, and then reorient the part as shown in Figure 4.155.

Construction objects are automatically promoted to surfaces and the surfaces are automatically knitted. As a result, you see a base solid. (See Figure 4.155.)

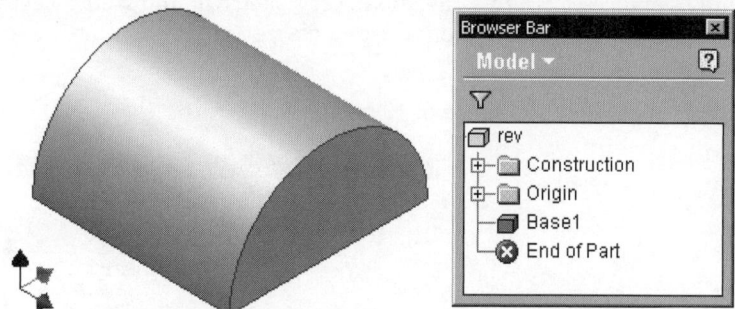

Figure 4.155 *Construction objects automatically promoted and knitted*

20. To see that the imported object is a solid, place a shell feature. (See Figure 4.156.)

Save and close your file (file name: *Rev2.ipt*).

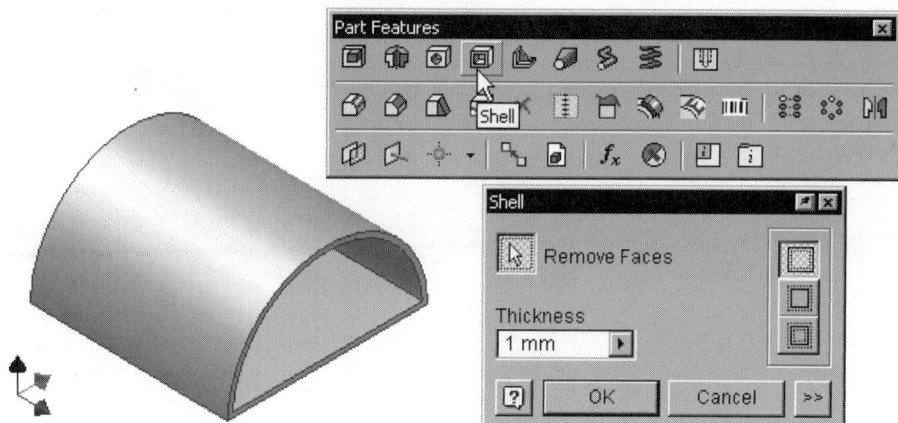

Figure 4.156 *Shell feature placed*

REPLACE FACE

A quick way to construct a solid part with many surfaces is to construct or import surfaces and replace the faces of the solid part by the surfaces. An important point to note for face replacement is that the replaced faces must be larger than or equal to the faces to be replaced.

TUTORIAL 4.21

In this tutorial, you will construct an extruded solid part and a revolved surface and then replace a face of the solid by the surface.

1. Start a new part file. Use the metric template.
2. With reference to Figure 4.157, construct a sketch. In the sketch, center point A is the project center point from the part file's origin.
3. Set the display to an isometric view.
4. Extrude the sketch a distance of 10 mm.

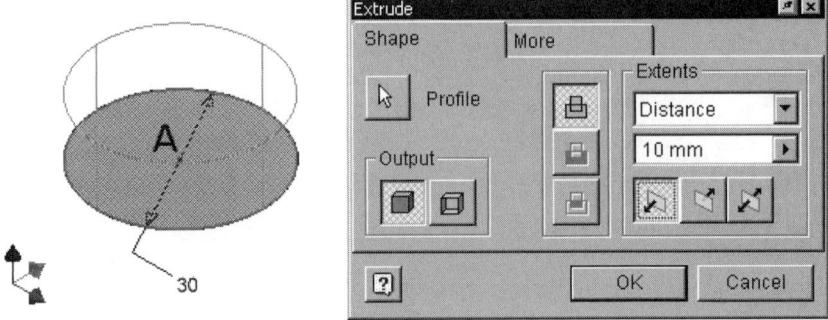

Figure 4.157 *Circle constructed on the XY plane being extruded*

5. Construct a sketch on the XY plane in accordance with Figure 4.158. In the sketch, A is the projected center point.

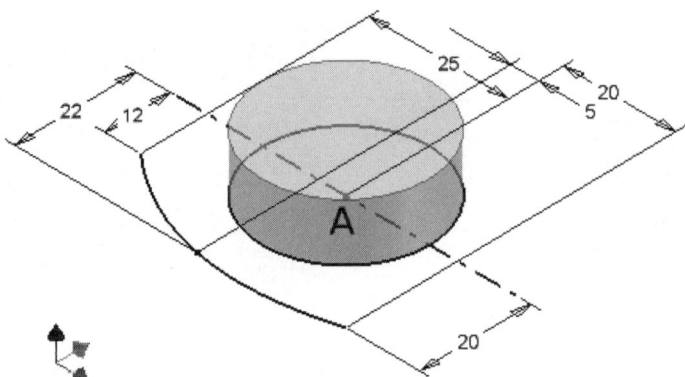

Figure 4.158 *Sketch constructed on the XY plane*

6. Revolve the sketch 180° to construct a revolved surface. (See Figure 4.159.) In the Revolve dialog box, select the Surface button.
7. Select spline A as the profile and centerline B as the axis and click the OK button.

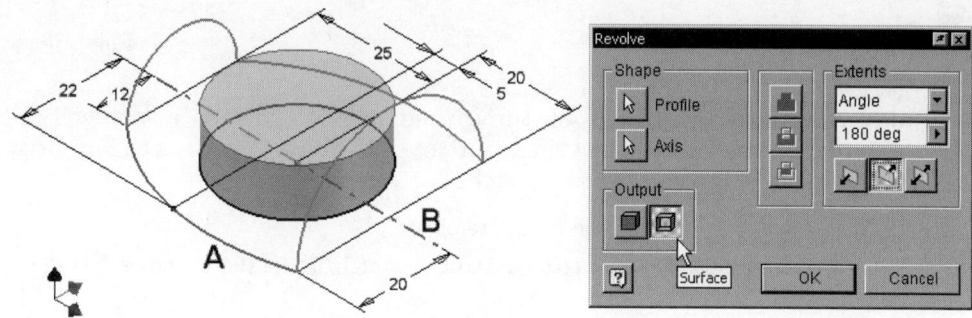

Figure 4.159 *Revolved surface being constructed*

Now replace the top face of the extruded solid by the revolved surface.

8. Select Replace Face on the Part Features Panel Bar or toolbar.
9. In the Replace Face dialog box, select the Existing Faces button and select face A in Figure 4.160.
10. Select the New Faces button and select surface B in Figure 4.160.
11. Click the OK button.

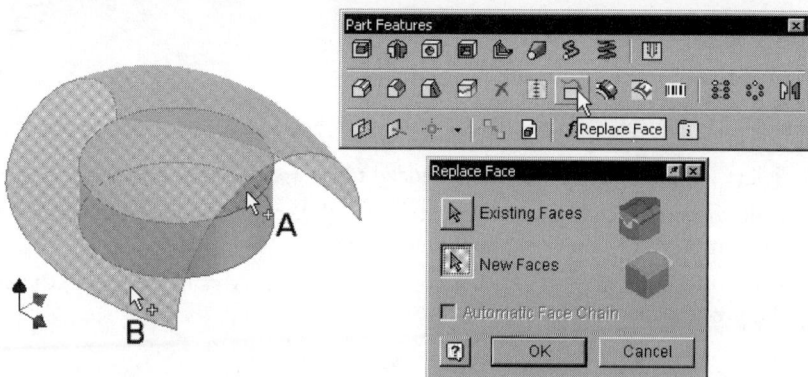

Figure 4.160 *A face being replaced*

A face of the solid part is replaced. (See Figure 4.161.) Save and close your file (file name: *Replace.ipt*). Save another copy of the file as *Offset.ipt*.

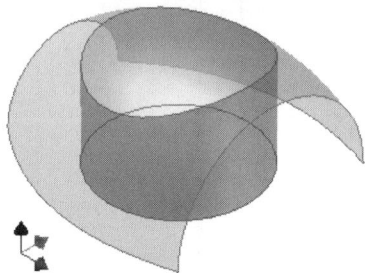

Figure 4.161 *A face replaced*

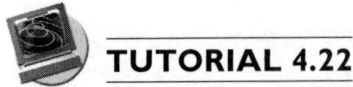

TUTORIAL 4.22

In this tutorial, you will construct a loft surface and an extruded surface. You will then replace faces of the extruded solid by a single surface. First you will construct a loft surface from three profile sketches and two rail sketches.

1. Start a new part file. Use the metric template.

2. With reference to Figure 4.162, construct a sketch. In the sketch, point A is the projected center point of the origin.

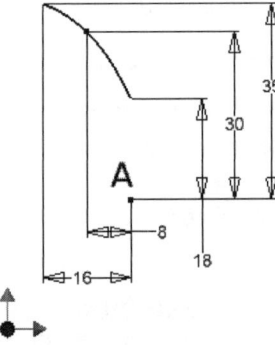

Figure 4.162 *Sketch constructed on the XY plane*

A profile sketch is complete. Now construct the second profile sketch.

3. Select Return on the Inventor Standard toolbar to exit sketch mode.

4. Hide the sketch that you just constructed. (Hiding is done here for clarity in illustration. You can leave the sketch visible.)

5. With reference to Figure 4.163, construct a sketch. Again, point A in the sketch is the projected center point.

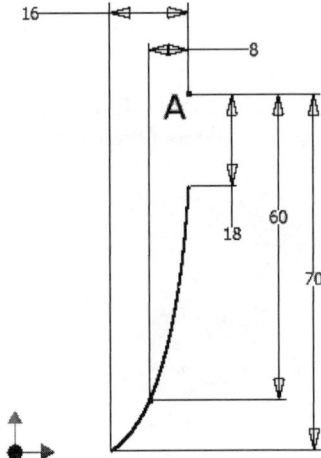

Figure 4.163 *Second sketch constructed*

The second profile sketch is complete. Now construct four work points for establishing the locations for the two rail sketches.

6. Select Return to exit sketch mode.
7. Make the first sketch visible and set the display to an isometric view.
8. Construct four work points A, B, C, and D in Figure 4.164.

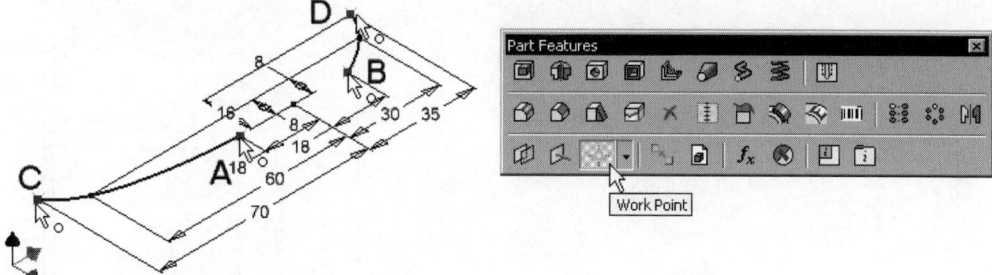

Figure 4.164 *Work points constructed*

Now construct a rail sketch.

9. Hide the first and second sketch. (You can leave them visible.)
10. Construct a sketch on the YZ plane in accordance with Figure 4.165. In the sketch, A and B are the projected points of work points A and B in Figure 4.164.

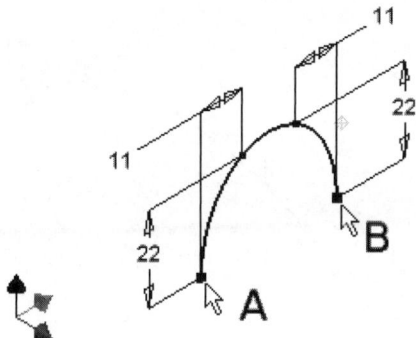

Figure 4.165 *Sketch constructed on the YZ plane*

A rail sketch is complete. Now construct a work plane for sketching another rail sketch.

11. Make the hidden sketches visible.
12. Construct a work plane parallel to the YZ plane and at the work point A in Figure 4.165.

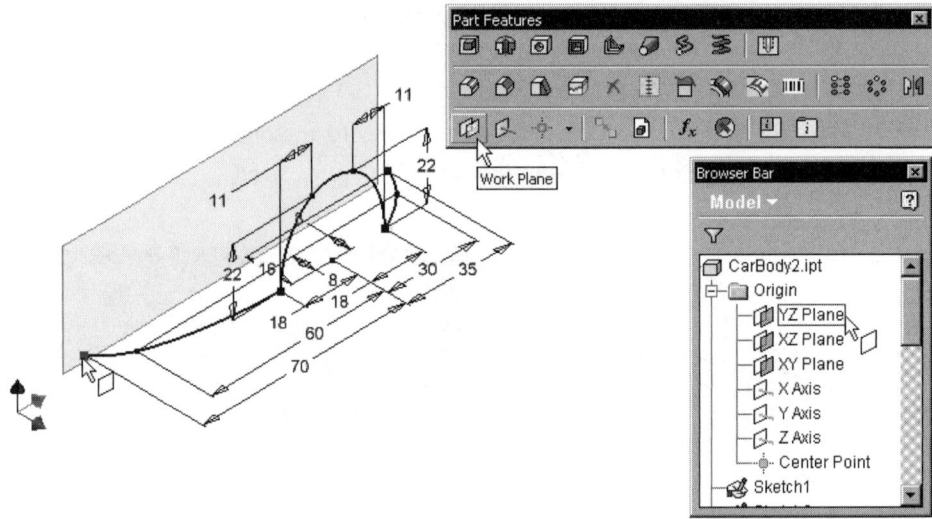

Figure 4.166 *Work plane constructed*

13. Hide all the sketches. (Sketches are hidden here for clarity in illustration. You can leave them visible.)

14. Construct a sketch on the work plane in accordance with Figure 4.166. In the sketch, points A and B are projected from the work points C and D in Figure 4.164.

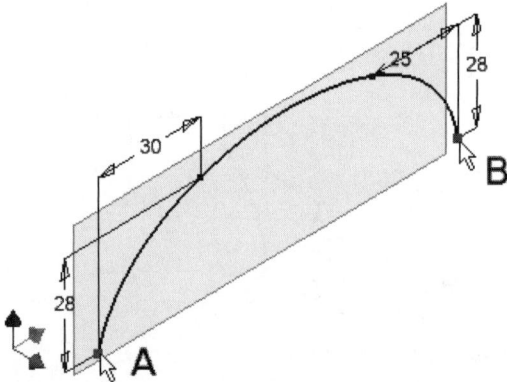

Figure 4.167 *Sketch constructed on the work plane*

The second rail sketch is complete. Now construct two work points for making the third profile sketch.

15. Hide the work plane and make the third sketch visible.

16. Select Work Point on the Part Features Panel Bar or toolbar, select the XZ plane on the Browser Bar, and select curve A in Figure 4.168.

17. Select Work Point on the Part Features Panel Bar or toolbar, select XZ plane on the Browser Bar, and select curve B in Figure 4.168.

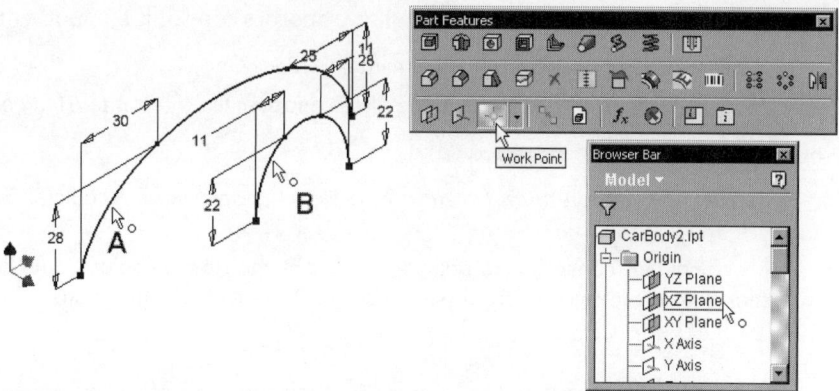

Figure 4.168 *Work points constructed at the intersection of the sketches and the XZ plane*

18. Hide all the sketches. (You can leave them visible.)
19. Construct a sketch on the XZ plane in accordance with Figure 4.169. In the sketch, A and B are projected from work points that you constructed in steps 16 and 17.

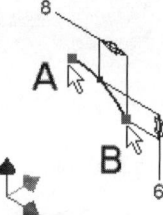

Figure 4.169 *Sketch constructed on the XZ plane*

The sketches are complete. Now construct a loft surface.

20. Exit sketch mode and make all the sketches visible.
21. With reference to Figure 4.170, construct a loft surface using sketches A, B, and C as the profiles and sketches D and E as the rails.

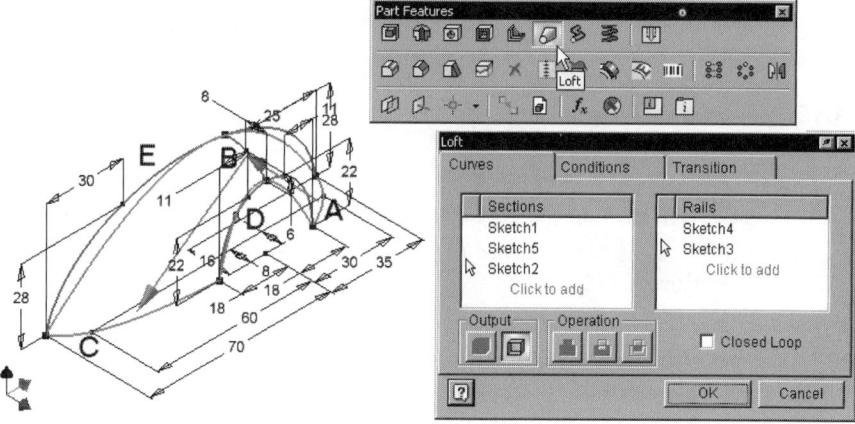

Figure 4.170 *Loft surface being constructed*

A loft surface is complete. Save the file and save another copy of it for another tutorial.

22. Select File > Save to save the file (file name: *CarBody2.ipt*).

23. Select File > Save Copy As to save the file to another file (file name: *Thicken.ipt*).

Now construct a simple extruded solid.

24. Construct a sketch on the XY plane. In the sketch, construct a rectangle. (See Figure 4.171.)

25. Apply coincident constraint to point A and line B and then to point C and line D. (Other than these two constraints, you do not have to add dimensions.)

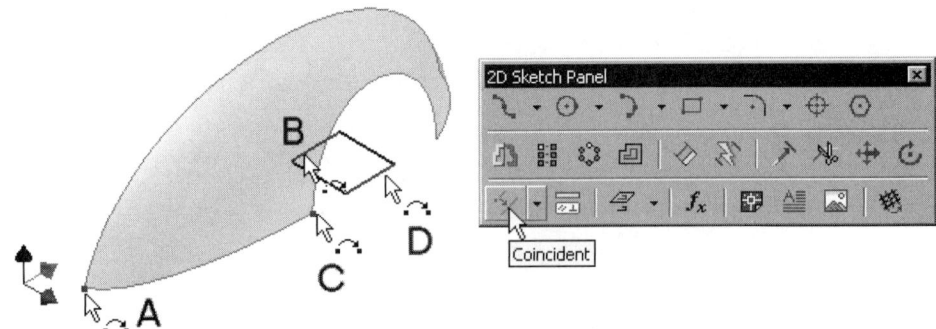

Figure 4.171 *Sketch constructed on the XY plane*

26. Extrude the sketch a distance of 10 mm. (See Figure 4.172.)

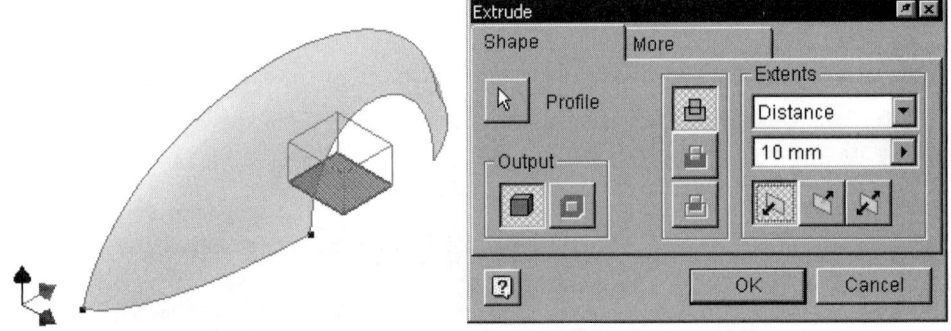

Figure 4.172 *Extruded solid constructed*

Now replace three faces of the extruded solid by the loft surface.

27. Select Replace Face on the Part Features Panel Bar or toolbar.

28. Select the Existing Faces button (if it is not already selected) and select faces A, B, and C in Figure 4.173.

29. Select the New Faces button and select the loft surface.

30. Click the OK button. (See Figure 4.173.)

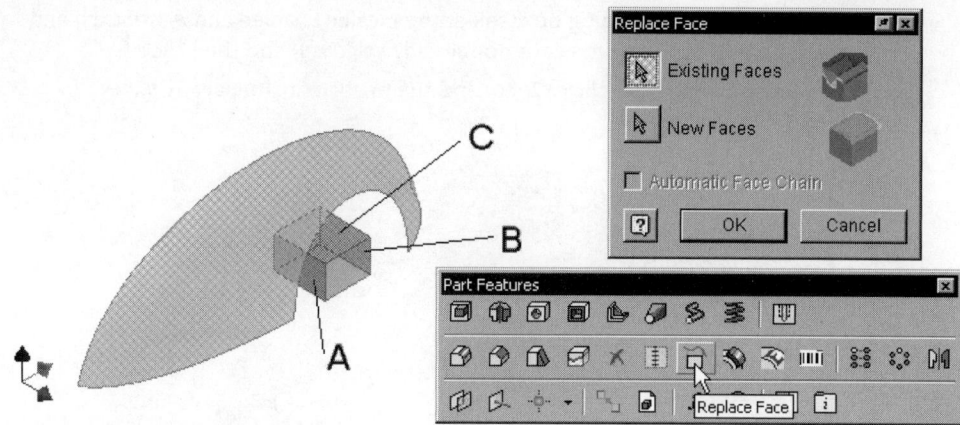

Figure 4.173 *Faces being replaced*

31. Hide the loft surface.

The loft surface replaces three faces of the extruded solid. (See Figure 4.174.) The solid is complete. Save and close your file. You will use this solid part in making a toy car's body.

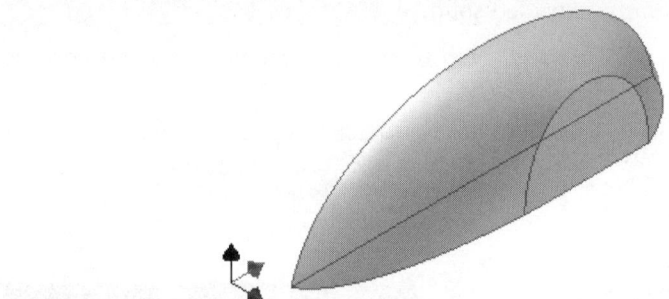

Figure 4.174 *Faces replaced*

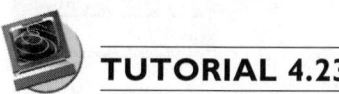 **TUTORIAL 4.23**

In this tutorial, you will import an IGES file that contains a set of surfaces depicting the model of a toy car constructed by other CAD applications. You will construct a simple extruded solid and replace the faces of the solid by the imported surfaces.

Now you will set application options and then import an IGES file.

1. Select File > Open.
2. In the Open dialog box, select IGE from the Files of Type pull-down list and then select the IGES file (file name: *BeetleNext.igs*) from the Chapter 4 folder of the CD accompanying this book.
3. In the Open dialog box, click the Options button.

4. In the Import Options dialog box, select the Healer Enabled and AutoStitch and Promote check boxes (if they are not already selected) and then click OK.

5. In the Open dialog box, click Open, and then reorient the part as shown in Figure 4.175.

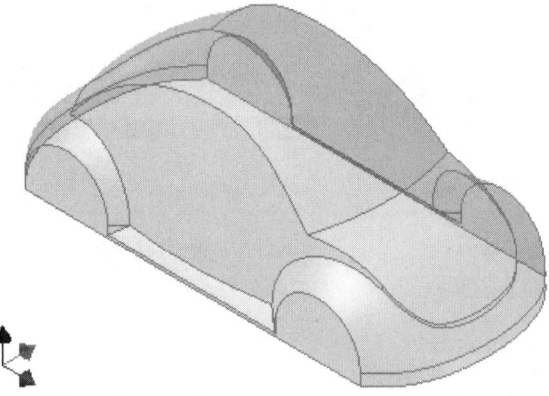

Figure 4.175 *Imported IGES surfaces*

Now construct a simple extruded solid.

6. With reference to Figure 4.176, construct a rectangle and extrude it a distance of 10 mm.

 Note: The dimensions of the rectangle are unimportant.

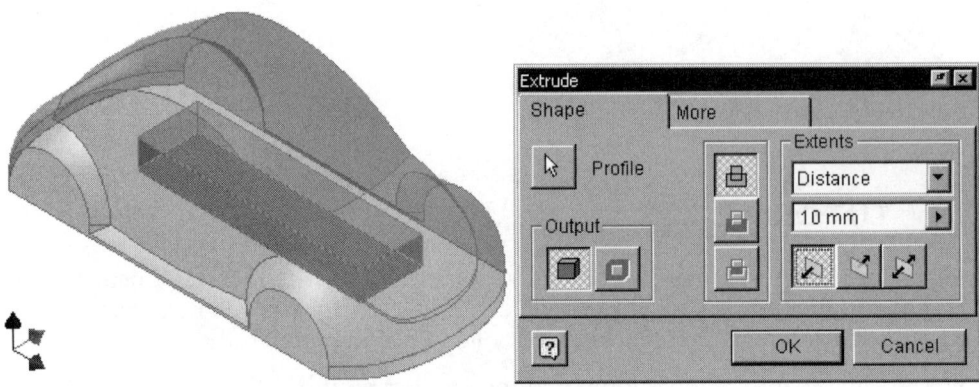

Figure 4.176 *Sketch constructed and being extruded*

Now replace the faces of the solid with the imported IGES surfaces.

7. Select Replace Face on the Part Features Panel Bar or toolbar.

8. Select the Existing Faces button and select the top face and the vertical faces of the extruded solid (all faces except the bottom face).

9. Select the New Faces button and select all the surfaces.
10. Click the OK button. (See Figure 4.177.)

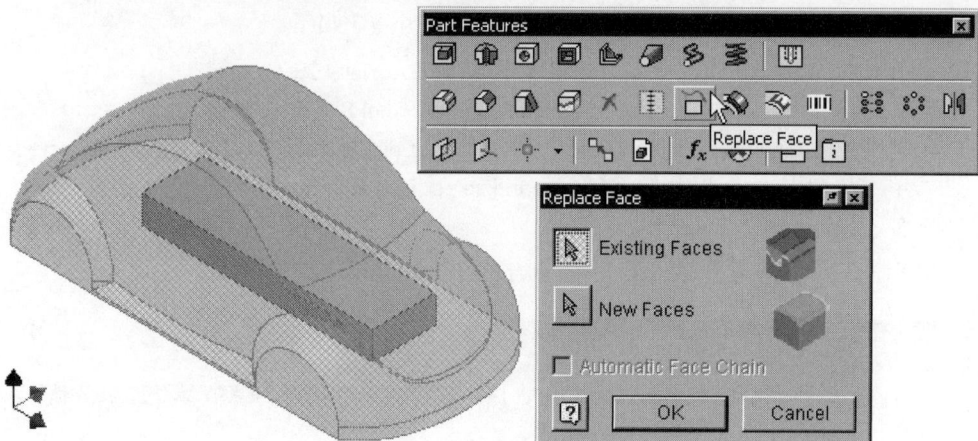

Figure 4.177 *Faces of the solid being replaced by the surfaces*

The solid part is complete. (See Figure 4.178.) Save and close your file.

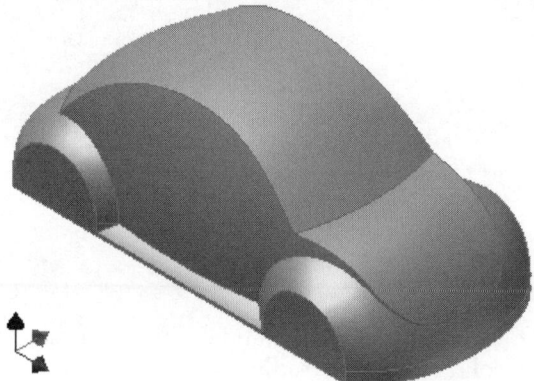

Figure 4.178 *Faces replaced*

THICKEN/OFFSET

You can construct a surface from a surface by offsetting it. The distance of the offset surface from the original surface is constant across the gap between them. A surface constructed in Inventor or an imported surface can be converted to a solid by thickening. The process of thickening is equivalent to making an offset surface from the surface and filling up the volume between the original surface and the offset surface. Thickening can be a quick way to obtain a solid from a surface.

TUTORIAL 4.24

In this tutorial, you will construct a solid by thickening a surface.

1. Open the file *Thicken.ipt* that you saved in Tutorial 4.23.
2. Select Thicken/Offset on the Part Features Panel Bar or toolbar.
3. Select solid part output in the Thicken/Offset dialog box, if it is not already selected.
4. Set thickening distance to 1 mm and select face A in Figure 4.179.
5. Click the OK button.

A solid feature is constructed from the surface. Save and close your file.

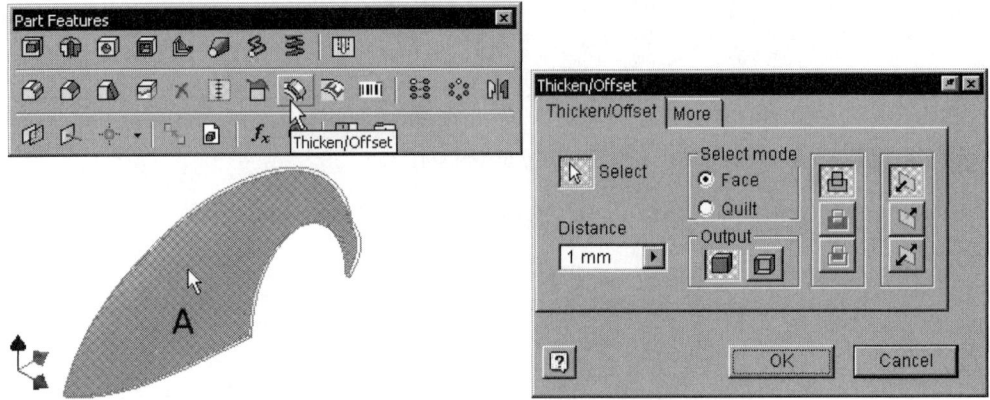

Figure 4.179 *Surface being thickened*

TUTORIAL 4.25

In this tutorial, you will construct a surface and an offset surface from the surface, and construct two extruded solid features extruding to the surfaces.

1. Open the file *Offset.ipt* that you saved in Tutorial 4.22.
2. Select Thicken/Offset on the Part Features Panel Bar or toolbar.
3. Select surface output.
4. Set the offset distance to 3 mm and select surface A in Figure 4.180.
5. Select the flip direction button as necessary to change the direction of offsetting in accordance with Figure 4.180.
6. Click the OK button.

An offset surface is constructed.

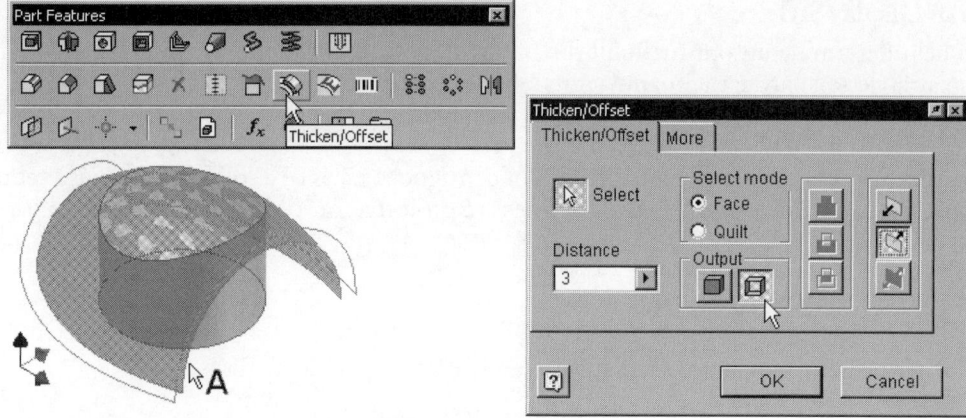

Figure 4.180 *Offset surface being constructed*

Now construct an extruded solid feature.

7. Construct a sketch on the XY plane in accordance with Figure 4.181. In the sketch, the circle A is concentric with the projected edge B.

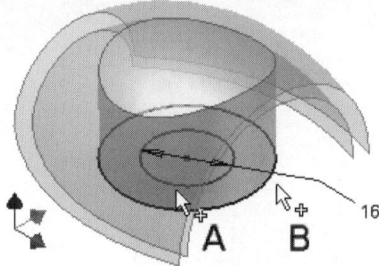

Figure 4.181 *Sketch constructed*

8. Extrude the sketch to the offset surface A. (See Figure 4.182.)

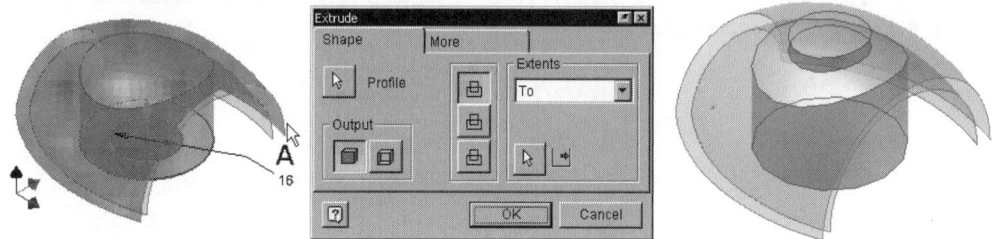

Figure 4.182 *Sketch being extruded to the offset surface (left) and completed solid (right)*

SURFACE ANALYSIS

There are two visual aids to help analyze a surface: one analyzes the draft angle and the other analyzes the surface smoothness.

Draft Display Style

To help determine an optimal pull direction of molded components, you can use draft display style to analyze the manufacturability of the part.

Zebra Display Style

A way to visually inspect the smoothness of contiguous faces of a solid part is to use zebra display style in which parallel strips of lines are projected onto the faces of the solid part.

TUTORIAL 4.26

In this tutorial, you will use draft display and zebra display to analyze a surface.

1. Open the file *CarBody1.ipt* that you constructed.
2. Select Tools > Analyze Faces.
3. In the Analyze Faces dialog box, select Draft and click the OK button. (See Figure 4.183.)

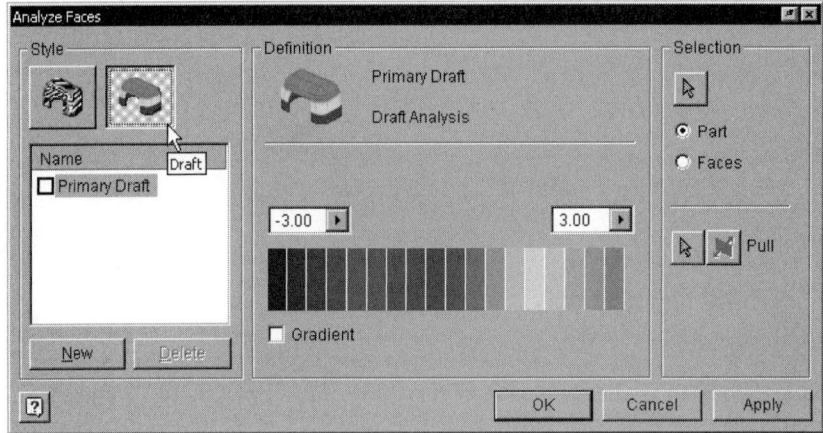

Figure 4.183 *Analyze Faces dialog box*

Draft analysis result is shown in Figure 4.184.

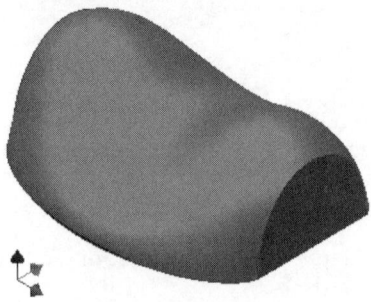

Figure 4.184 *Draft analysis*

4. Select Tools > Analyze Faces.
5. Select the Zebra button and click the OK button.

Zebra analysis is shown in Figure 4.185.

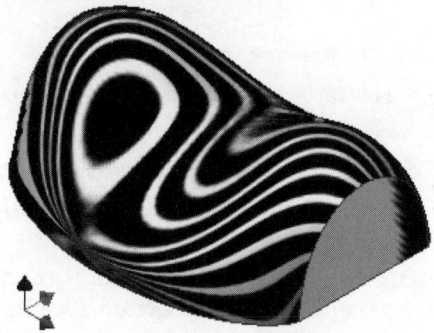

Figure 4.185 *Zebra analysis*

To perform analysis without opening the Analyze Faces dialog box, you can select the Analyze Faces button on the Inventor Standard toolbar. (See Figure 4.186.)

Figure 4.186 *Analyze Faces button on the Inventor Standard toolbar*

Analysis is complete. Close your file.

DERIVED SOLID PART

Sometimes, an existing solid model is a very good starting point for building a new solid model. Of course, you can save a new copy of a solid and continue from there to construct new features. However, if you want to have the features of the new solid to be a mirror copy of another, a scaled copy of it, or to reference an existing solid, you should make a derived solid. By deriving a solid, you can construct a base solid or a set of surfaces that is referenced to a source solid. If the original solid changes, the derived solid changes. You can add features to the derived part.

In addition to deriving a solid part from a solid part, you can also derive a solid part from an assembly. You will learn how to derive a part from an assembly in Chapter 8.

TUTORIAL 4.27

In this tutorial, you will derive a mirrored solid part from a solid part that you constructed in the last tutorial.

1. Start a new part file. Use the metric template.

2. Select Return on the Inventor Standard toolbar, if there is an active default sketch.

3. Set the display to an isometric view.

4. Select Derived Component on the Part Features Panel Bar or toolbar.

5. In the Open dialog box, select the part file (*CarBody2.ipt*) that you constructed in Tutorial 4.23 and select the Open button.

6. In the Derived Part dialog box, select the Mirror Part box, select YZ plane from the pull-down list box, and click the OK button. (See Figure 4.187.)

The derived part is complete. Save and close your file (file name: *CarBody2a.ipt*).

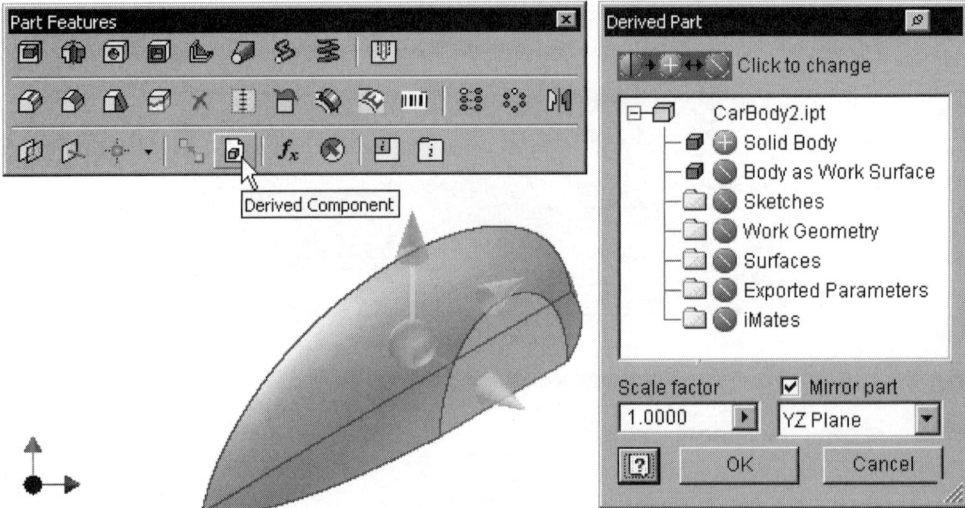

Figure 4.187 *Derived part being constructed*

DESIGN PARAMETERS

As we have mentioned earlier, each dimension that you place in a solid part has a parameter name. By default, the name begins with a letter "d" followed by a number starting from zero. These are called model parameters and they are stored in a parameter table. In addition to model parameters, you can add user parameters, linked parameters, and embedded parameters.

USER PARAMETERS

User parameters are parameters that you add to the table; you specify a parameter name, parameter's unit of measurement, an equation expressing the value of the parameter, a value of the parameter, and a comment about the parameters for easy reference.

TUTORIAL 4.28

In this tutorial, you will learn how to add user parameters to a part file.

1. Open the file *PunchSpring.ipt* that you constructed.
2. Select File > Save Copy As and specify a new file name (*PunchIndexSpring.ipt*).
3. Close the file.
4. Open the file *PunchIndexSpring.ipt*.
5. Select Parameters on the Part Features Panel Bar or toolbar to display the Parameters dialog box.

Because your sequence of dimensioning the solid parts in your file might be different, the equation and value shown in Figure 4.188 might differ from yours. Initially, you will find two tables: Model Parameters and User Parameters. In each table, there are a number of columns:

Parameter Name	The default parameter name begins with the letter d and a number. The first parameter's name is d0 and the second is d1. You can change the name by specifying a name in this column.
Unit	You can use both Imperial size and metric size in a single table.
Equation	You use a simple numeric value or an equation to express the dimension of the parameter.
Nominal Value	This is the numeric value of the equation.
Tol.	If you already specified a dimension tolerance, you can select the dimension to the nominal size, upper limit, or lower limit.
Model Value	This adds the parameter to the custom properties for the model.
Check Box	This adds the parameter to the custom properties for the model.
Comment	You can add textual information.

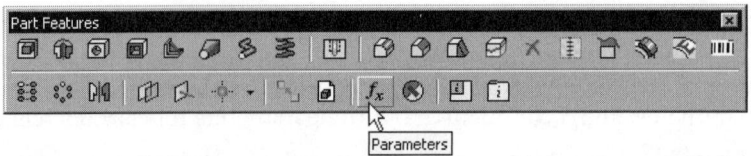

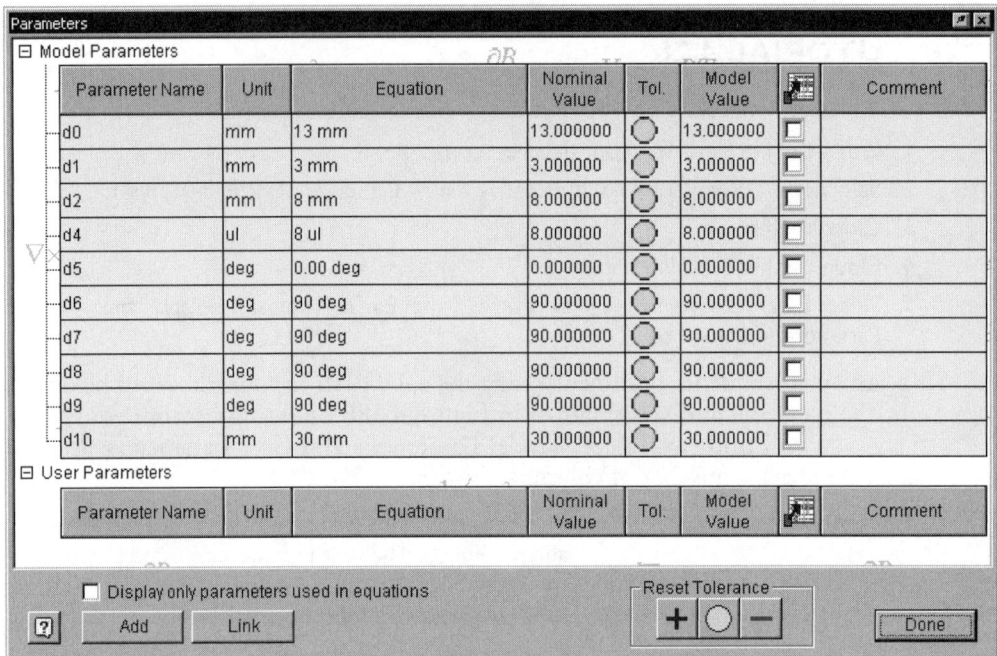

Figure 4.188 *Parameters dialog box*

The model parameters table is a list of parameters that are assigned by Inventor to the solid part when you add dimensions to your solid part. Now you will add an entry to the user parameters.

6. In the Parameters dialog box, select the Add button.

7. In the User Parameter table, type **CoilNumber** in the Parameter Name column. Then click in the Unit column to display the Unit Type dialog box and type **ul** in the Unit Specifier box, or expand Unitless and select Unitless (ul), and then click OK. Then click in the Equation column and type **8 ul**.

8. Select the Add button again.

9. Type **CoilPitch** in the Parameter Name column, accept the default units (mm), and type **6 mm** in the Equation column.

10. In the Parameters dialog box, click the Done button. (See Figure 4.189.) Two parameters are added.

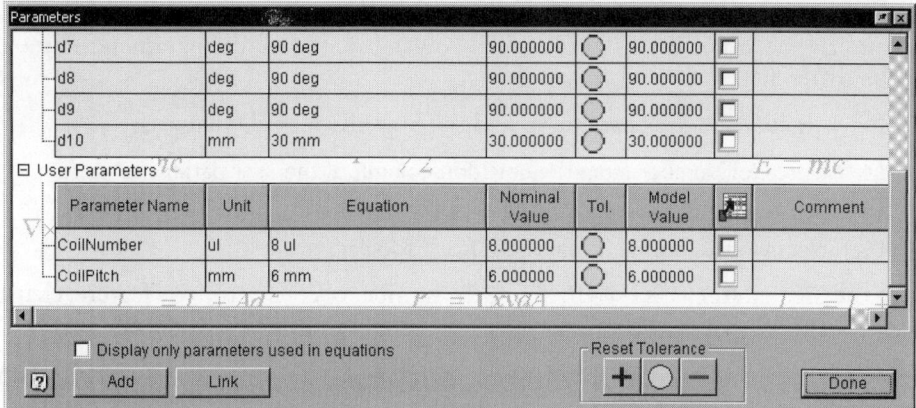

Figure 4.189 *Entries added to User Parameters table*

Now edit the solid part.

11. Select the coil feature on the Browser Bar, right-click, and select Edit Feature.

12. In the Coil dialog box, select the Coil Size tab, type **CoilPitch** in the Pitch box and **CoilNumber** in the Revolution box, and click the OK button. (See Figure 4.190.)

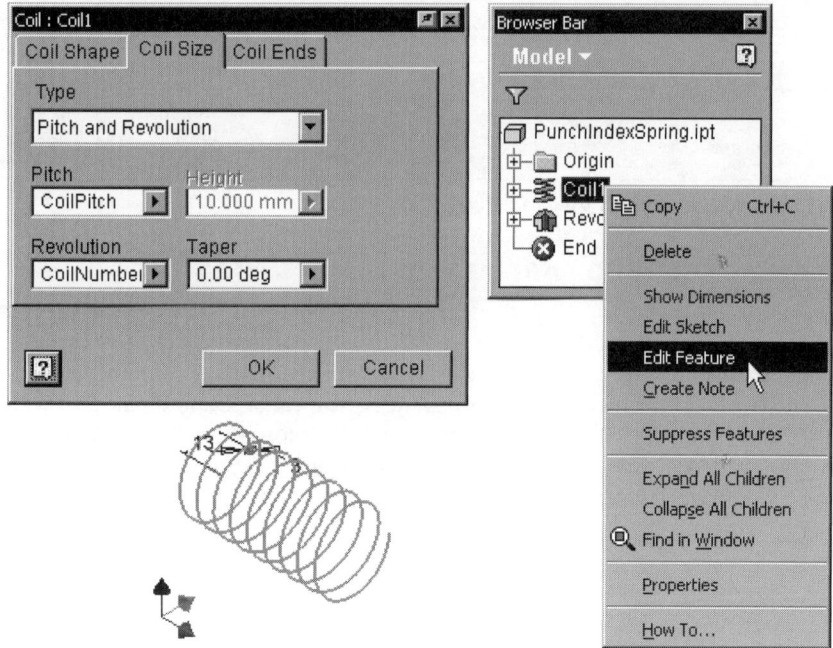

Figure 4.190 *User parameters used to define the number of revolutions of the coil feature*

Now discover the change you made to the parameter table, as a result of editing the part.

13. Select Parameters on the Inventor Standard toolbar to open the parameter table.

In Figure 4.191, the equation of the model parameters, d2, becomes "CoilPitch" and d4, becomes "CoilNumber."(In your table, the model parameter name may be different.)

Now modify the user parameter and see how the model changes.

14. Select CoilNumber from the Equation column of the user parameter table.

15. Change the equation to 7 ul and click the Done button.

16. Select Update on the Inventor Standard toolbar.

Because the user parameter is used to define the number of coils of the coil feature, changing the user parameter causes the coil number to change as well. Save your file.

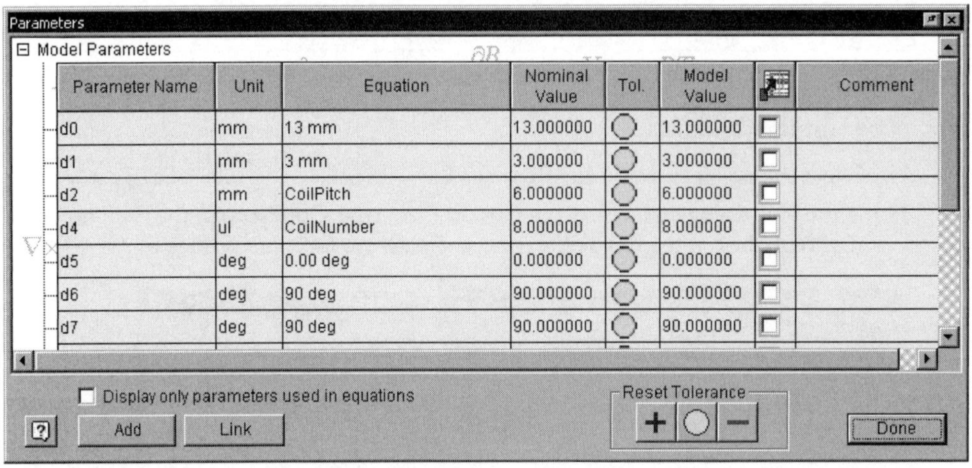

Figure 4.191 *User parameters used in defining the solid part*

LINKED AND EMBEDDED PARAMETERS

To link or embed a collection of parameters, you can use an Excel spreadsheet. Parameters in a linked Excel spreadsheet are called linked parameters, and parameters in an embedded spreadsheet are called embedded parameters. Linked parameters link to the Excel spreadsheet. If you modify the spreadsheet, the parameters change as well. However, embedded parameters imported from the Excel spreadsheet do not maintain any link with the original Excel spreadsheet, and changes in the Excel spreadsheet will not affect the embedded parameters.

Using an Excel Spreadsheet

To enter a set of parameters collectively, you can use an Excel spreadsheet. In the spreadsheet, construct two, three, or four columns.

The first column is the parameter name, the second the value, the third the units of size, and the fourth the comment. If you do not specify the third column, the default units of size will be used. You should have Excel properly installed on your computer.

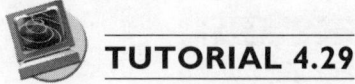

TUTORIAL 4.29

In this tutorial, you will construct two spreadsheets. You will link one spreadsheet and embed the second spreadsheet.

1. Start a new spreadsheet and fill in the sheet as shown in Figure 4.192.

2. Save your file (file name: *Book1.xls*).

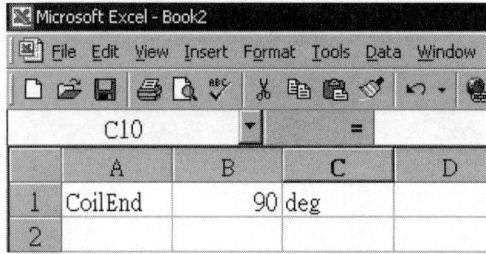

Figure 4.192 *Book1 spreadsheet*

3. Start another new spreadsheet and fill in the sheet as shown in Figure 4.193.

4. Save and close your file (file name: *Book2.xls*).

Figure 4.193 *Book2 spreadsheet*

When you link a spreadsheet, the data in the spreadsheet is not saved in the solid part file. Every time you open a solid part file, the data in the linked spreadsheet is imported. As a result, you always get the latest version of the spreadsheet in your solid part. Now link an Excel spreadsheet to an Inventor file.

5. In Inventor, open the file *PunchIndexSpring.ipt*, if you already closed it.

6. Select Parameters on the Inventor Standard toolbar.

7. Click the Link button.

8. In the Open dialog box, select the Link option, select the file *Book1.xls*, and click the Open button. (See Figure 4.194.) A spreadsheet file is linked to the solid part file. Do not close the Parameters dialog box yet.

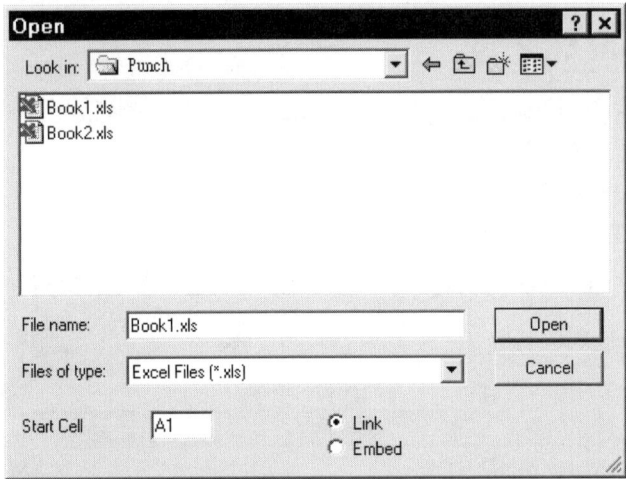

Figure 4.194 *Book1.xls linked to the solid part*

Another way of using a spreadsheet is to embed the data from the spreadsheet in the solid part file. After embedding, there will be no more connection between your solid part file and the spreadsheet. Now embed an Excel spreadsheet in an Inventor file.

9. In the Parameters dialog box, click the Link button again. Then, in the Open dialog box, select Book2.xls, select the Embed option, and then click Open. (See Figure 4.195.)

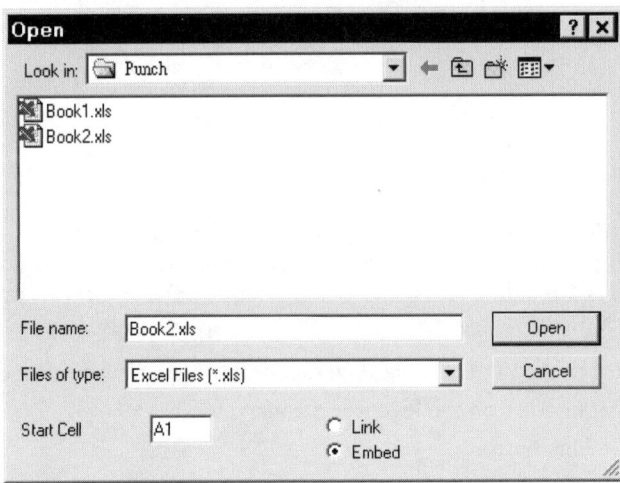

Figure 4.195 *Book2.xls embedded*

10. Now the parameter box has two more parameter tables. (See Figure 4.196.) In the Parameters dialog box, click the Done button.

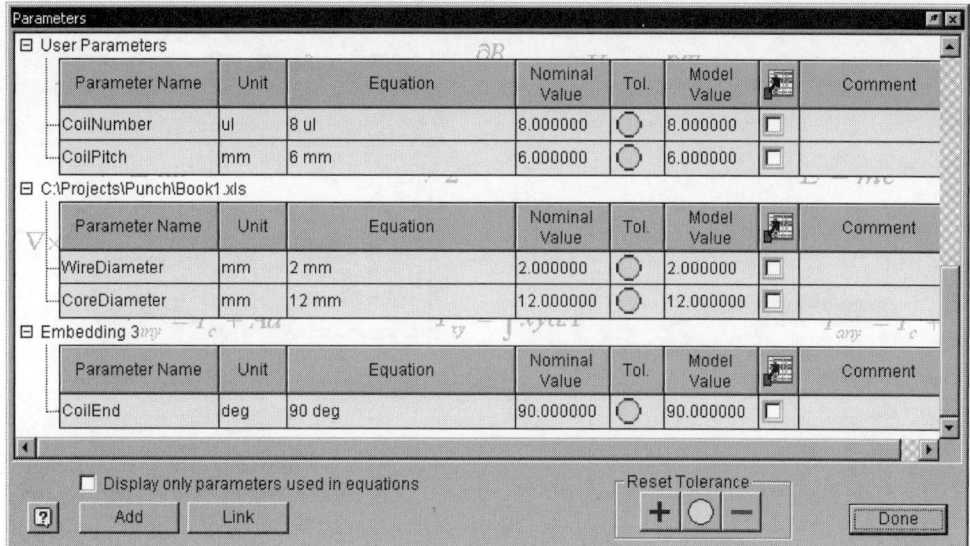

Figure 4.196 *A spreadsheet linked and a spreadsheet embedded*

A spreadsheet is linked and a spreadsheet is embedded. They are listed on the Browser Bar. (See Figure 4.197.)

Figure 4.197 *Objects listed on the Browser Bar*

Now use the parameters in both the linked table and the embedded table to dimension your solid part.

11. Select the sketch for making the coil feature on the Browser Bar, right-click, and select Edit Sketch.

12. Double-click the diameter of the circle. In the Edit Dimension dialog box type **WireDiameter**, and then click the Checkmark. Then, double-click the 13mm dimension (the distance to the center of the circle), in the Edit Dimension dialog box type **CoreDiameter/2+WireDiameter**, and then click the Checkmark. (See Figure 4.198.).

13. Select Return on the Inventor Standard toolbar.

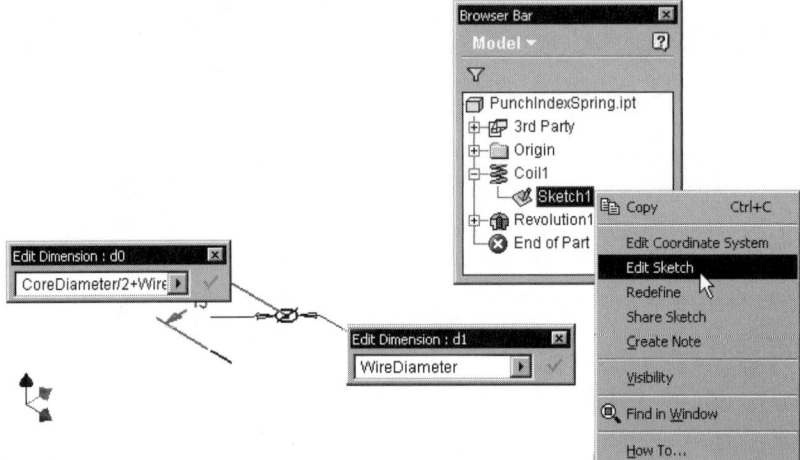

Figure 4.198 *Parameters in a the linked spreadsheet used*

14. Select the coil feature on the Browser Bar, right-click, and select Edit Feature.
15. In the Coil dialog box, select the Coil Ends tab.
16. Change the transition angle and flat angle of the start and end of the coil to **CoilEnd** and click the OK button. (See Figure 4.199.)

The solid part is modified. Save your file.

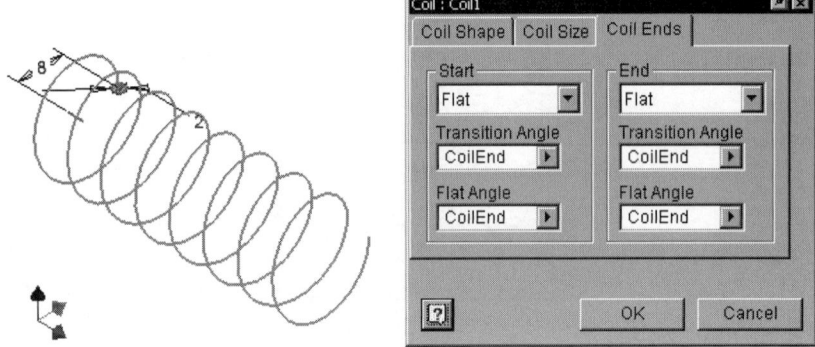

Figure 4.199 *Parameter in an embedded spreadsheet used*

Now link a spreadsheet to another solid part file so that parameters in two solid parts are controlled by a single spreadsheet file.

17. Open the file *PunchIndex.ipt*.
18. Select Parameters on the Inventor Standard toolbar.
19. In the Parameters dialog box, click the Link button.

20. In the Open file dialog box, select the spreadsheet *Book1.xls*, select the Link option, and click the OK button.

21. Click the Done button. The parameters in the spreadsheet *Book1.xls* are linked to the solid part.

22. Select the sketch of the revolved solid feature from the browser, right-click, and select Edit Sketch.

23. Double-click the dimension indicated in Figure 4.200 and change it to **CoreDiameter**.

24. Select Return on the Inventor Standard toolbar.

The solid part is modified. Save and close your file.

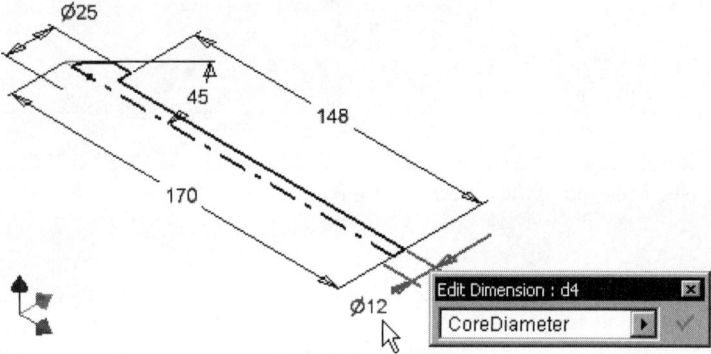

Figure 4.200 *Parameter in a linked spreadsheet used*

If you change the values in the spreadsheet *Book1.xls* and then save your changes, you can then click the Update button to change the part accordingly, because *Book1.xls* is linked to the part. But if you change the values in *Book2.xls*, the part will not change, because that spreadsheet is embedded rather than linked.

DESIGN NOTEBOOK

To record your design intention and other design information for later retrieval, you can maintain a design notebook in your Inventor file. In the notebook, you keep graphical and textual records of your design.

SYSTEM SETTING

The Notebook tab of the Options dialog box let you control the display of the design notebook icon, keep or delete design notes for deleted objects, and manipulate color elements in the design notebook. To access the Notebook tab of the Options dialog box, select Application Options from the Tools menu. (See Figure 4.201.)

The Notebook tab has three areas: Display in model, History, and Color. In the Display in model area, select the Note icons box to display the notebook icon in the model, and select the Note text icon to display the text of a design note when the cursor pauses over a note symbol. In the History area, select the Keep notes on deleted objects box to keep the design notes attached to deleted objects. In the Color area, select colors for the design notebook.

Accept the default and click the OK button.

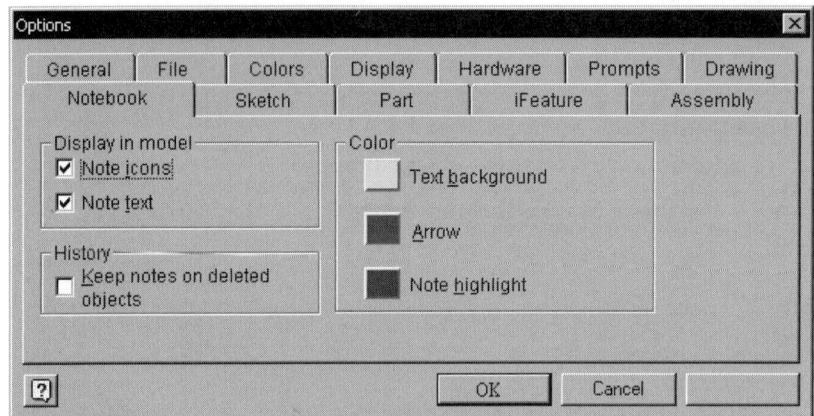

Figure 4.201 *Notebook tab of the Options dialog box*

 TUTORIAL 4.30

In this tutorial, you will create a design notebook in a solid part file.

1. Open the file *PunchIndexSpring.ipt*, if you already closed it.
2. Select the coil feature on the Browser Bar, right-click, and select Create Note. (See Figures 4.202 and 4.203.)

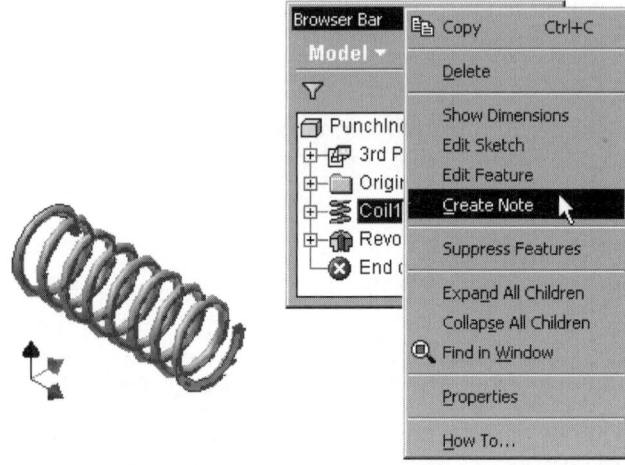

Figure 4.202 *Design notebook being constructed*

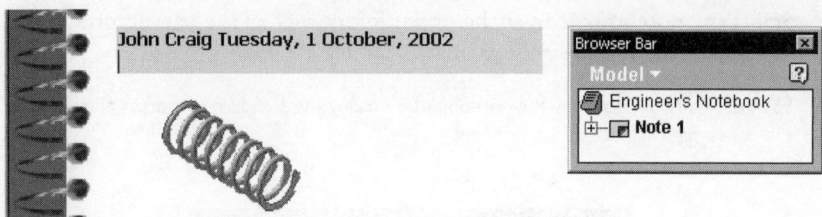

Figure 4.203 *Design notebook constructed*

Now add an arrow and a comment in the design notebook.

3. Select Arrow on the Notebook Panel toolbar, select points A, B, and C in Figure 4.204, right-click, and select Done to construct an arrow.

4. Right-click and select Insert Comment.

5. Select point D and E in Figure 4.104 to describe a rectangular box for placing the comment.

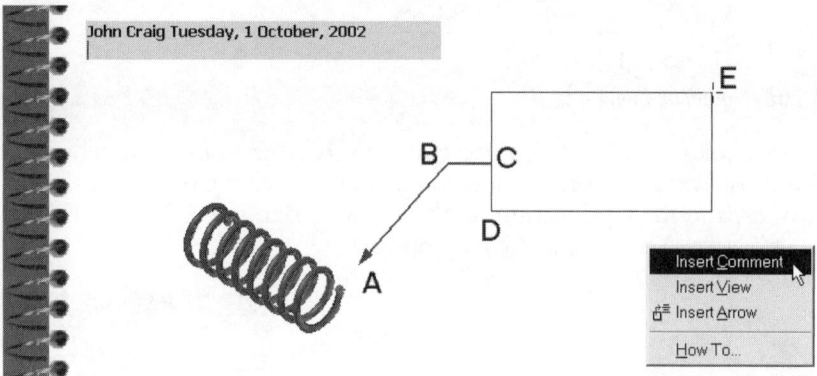

Figure 4.204 *Comment being inserted*

6. In the inserted comment area, type the notes that you want to include. (See Figure 4.205.)

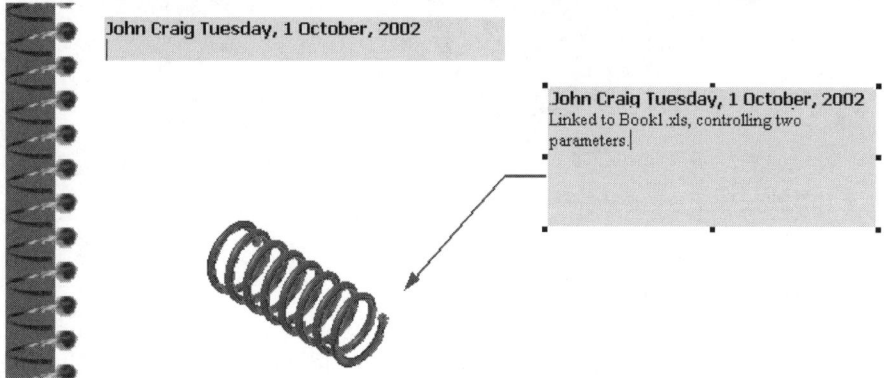

Figure 4.205 *Comment typed*

7. Select the notebook icon at the upper left corner of the user interface and select Close to close the design notebook. (See Figure 4.206.)

Note: You can also close the notebook by clicking the standard Windows close button (the X) in the upper right corner of the window.

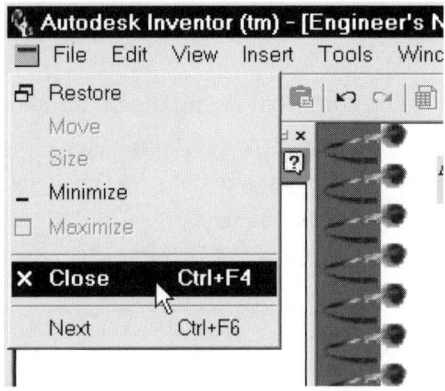

Figure 4.206 *Notebook being closed*

A design notebook is constructed, and a comment is inserted in the notebook. (See Figure 4.207.) If you want to open the notebook, double-click the notebook icon in the user interface, or expand the Coil feature on the Browser Bar, select Note 1, right-click and select Display Note. Now save and close your file.

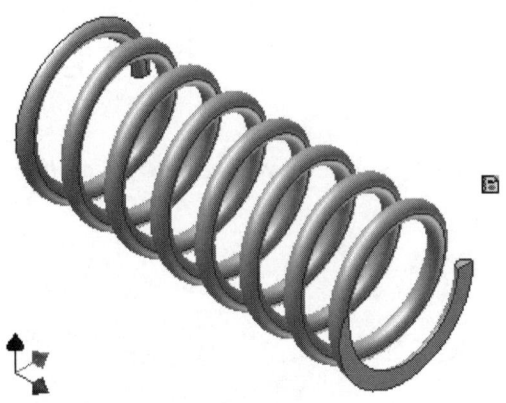
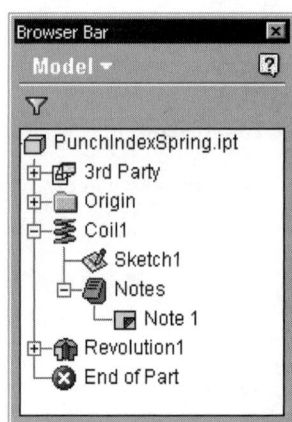

Figure 4.207 *Notebook constructed*

FAMILY OF PARTS — iPARTS

An iPart is a family of similar solid parts having different parameters and properties. To construct an iPart, you first construct a solid part and then use the iPart author to construct an embedded spreadsheet, the iPart table. In the iPart table, you set up a number of iParts and specify feature parameters, properties, and suppression information about the iParts that will be accessible to downstream users. There are two kinds of iParts: standard and custom. A standard iPart does not allow any modification. A custom iPart enables a user to modify some of the parameters. To use an iPart, you open an iPart and select a family member from the iPart table.

TUTORIAL 4.31

Now you will construct an iPart from the bolt that you constructed in Chapter 3. First you will rename the parameters of the solid part to make them more meaningful to downstream users.

1. Open the file *PunchBolt.ipt*.
2. Select Parameters on the Inventor Standard toolbar. (See Figure 4.208.)

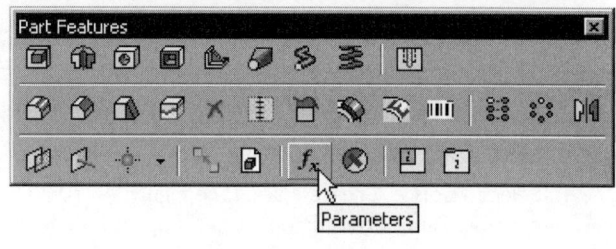

Figure 4.208 *Bolt*

3. Change the parameter names (first column of the model parameters table) as shown in Table 4.3. (See Figure 4.209.) When you are finished, click Done.

Table 4.3 *Parameters*

Dimension	Parameter Name
Hexagon's half width	HexWidth
Hexagonal extruded solid's height	HexHeight
Diameter of the shank	Diameter
Total length of the bolt	Length
Threaded length	Thread

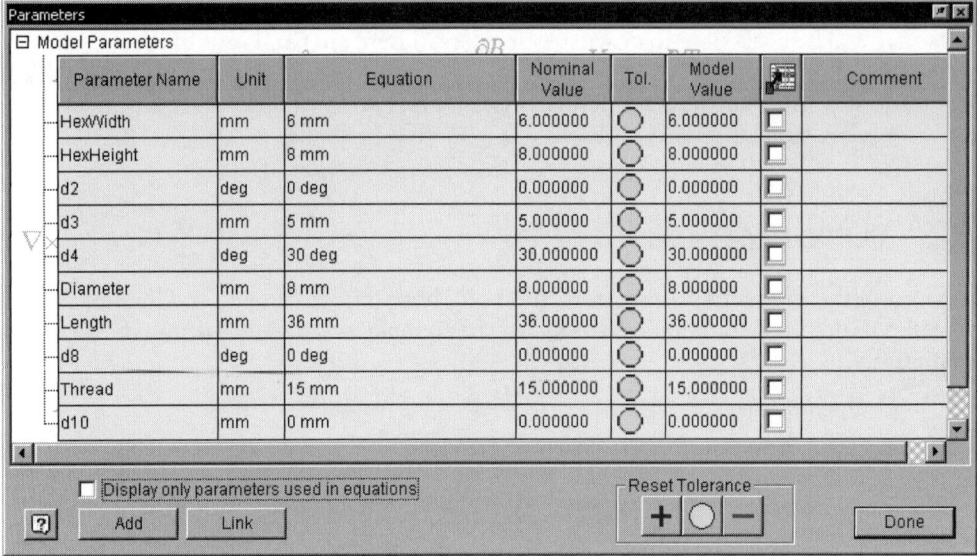

Figure 4.209 *Parameter's name changed*

Now construct an iPart table.

4. Select Tools > Create iPart. (See Figure 4.210.) The iPart Author dialog box has six tabs: Parameters, Properties, Suppression, iMates, Threads, and Other. (You will learn about iMate in Chapter 7.)

5. On the Parameters tab, select the parameters HexWidth, HexHeight, Diameter, Length, and Thread one by one and select the >> button to put them in the list box, if they are not already put in the box.

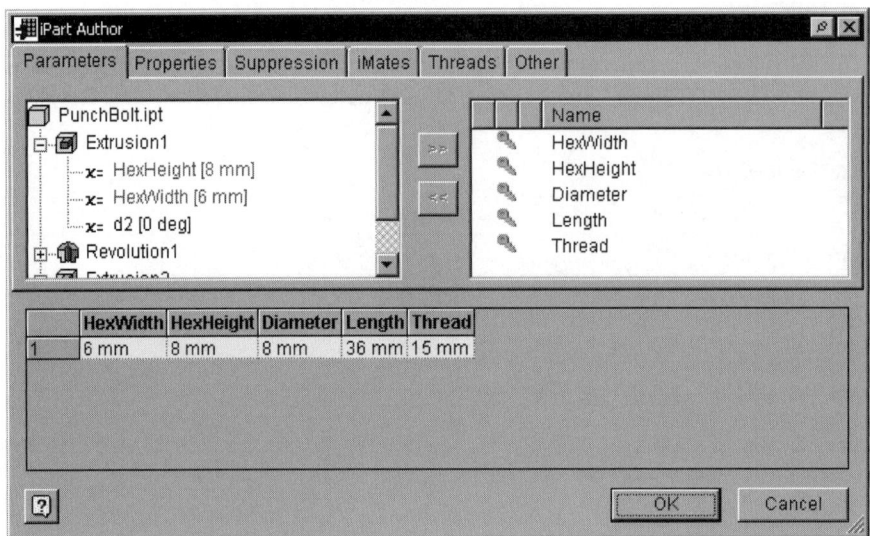

Figure 4.210 *Parameters tab*

Now include the material property in the iPart table.

6. Select the Properties tab. Here you will find three kinds of properties: summary, project, and physical.

7. Select Material from Physical and select the >> button. (See Figure 4.211.)

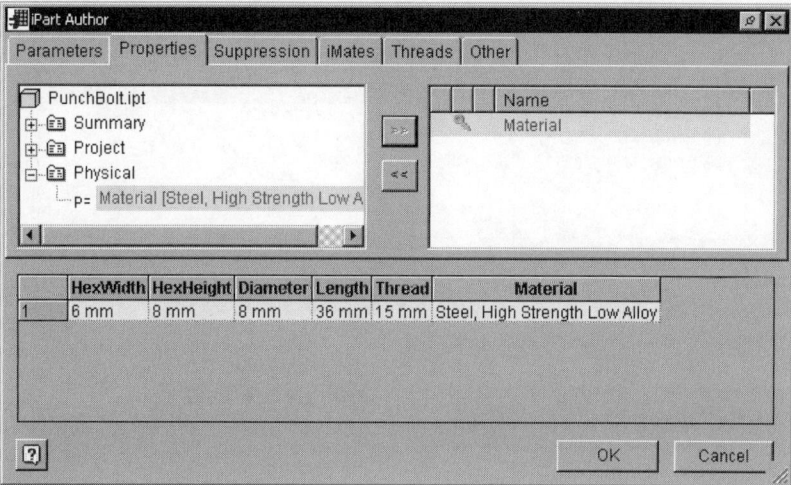

Figure 4.211 *Properties tab*

Now select a feature that will become suppressible.

8. Select the Suppression tab.

9. Select the thread feature and then the >> button. This feature will be suppressible by the user. (See Figure 4.212.)

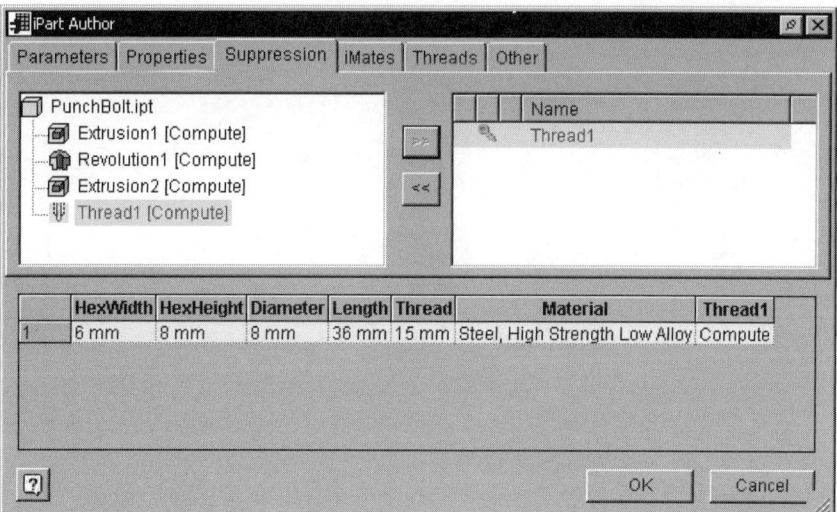

Figure 4.212 *Suppression tab*

Now make the thread feature suppressible and add a row in the iPart table.

10. In the iPart table, select the Thread cell, right-click, and select Custom Parameter Column.

11. Select any column, right-click, and select Insert Row. (See Figure 4.213.)

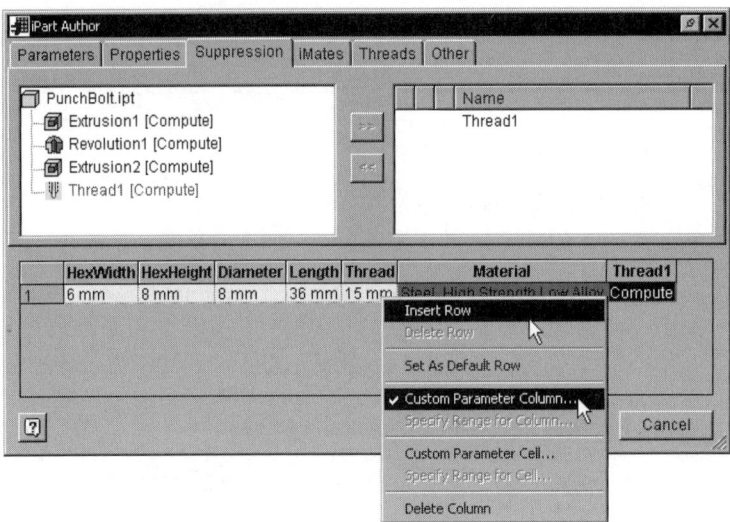

Figure 4.213 *Thread feature made suppressible*

Now change the values in the iPart table.

12. With reference to Figure 4.214, set columns Length and Thread to be custom parameter columns and set the values in second row: HexWidth = 4 mm, HexHeight = 4 mm, and Diameter = 6 mm.

13. Click the OK button.

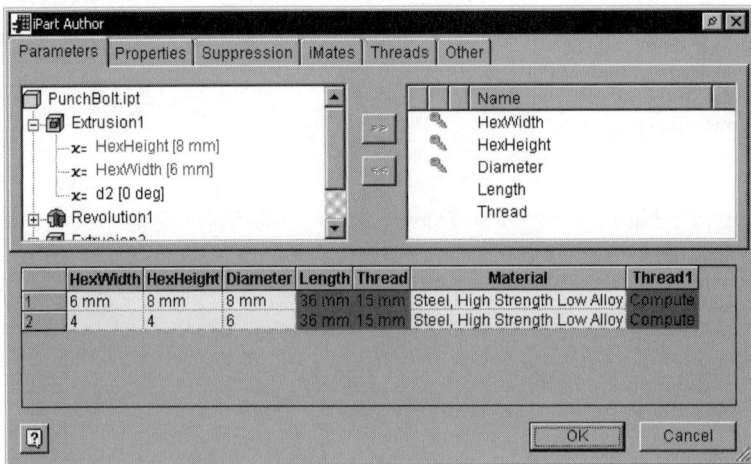

Figure 4.214 *iPart table modified*

An iPart table is constructed. In the table, there are two iParts. When you insert this solid part in an assembly, you can choose either part 1 or part 2 (with different parameter values), and you can decide the length of the bolt and suppress the thread feature. Save and close your file.

USER-DEFINED CUSTOMIZED FEATURES — iFEATURES

Unlike iParts, which are standard solid parts that you use in new assemblies, iFeatures are standard features of a solid part that you store in a library folder (design catalog) and insert into your new solid parts. You can consider iFeatures as custom-made placed solid features. For example, you can construct an iFeature in the form of a slot, and when you insert the iFeature in a new solid part, you get a slot.

There are two major kinds of iFeatures: unconsumed sketches and sketched features. Sketched features are further divided into three categories: join feature, cut feature, and intersect feature. When you insert a sketched feature into a solid part, a join element joins to the part, a cut element cuts the part; and an element feature intersects with the part. In addition, you can construct a composite iFeature consisting of a number of features.

SYSTEM SETTINGS

To access system settings related to managing iFeatures, select the iFeature tab of the Options dialog box. (See Figure 4.215.)

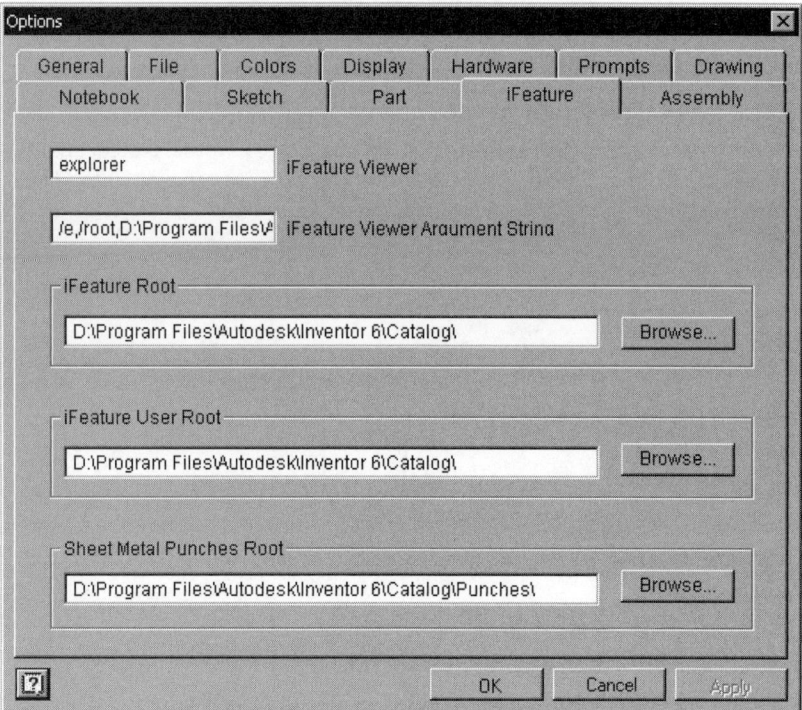

Figure 4.215 *iFeature tab of the Options dialog box*

The iFeature tab of the Options dialog box has five fields. Their functions are shown in Table 4.4.

Table 4.4 *iFeature tab of the Options dialog box*

iFeature Viewer	Specifies the viewer application used for viewing iFeatures.
iFeature Viewer Argument String	Sets the viewer command line argument.
iFeature Root	Sets the location of iFeature files used by the View Catalog dialog box
iFeature User Root	Sets the location of iFeature files used by both the Create iFeature and Insert iFeature dialog boxes.
Sheet Metal Punches Root	Sets the location of iFeature files used by the sheet metal Punch Tool dialog box.

SKETCH iFEATURE

Sketching complex shapes is laborious and can take quite some time. To reuse these sketches, you can put them in a folder (called design catalog) as iFeatures and insert a selected iFeature sketch on a plane of a solid part for making sketched solid features.

 TUTORIAL 4.32

Now you will learn how to construct a 2D sketch, use it as a design element, and insert the element into another solid part file.

1. Start a new part file. Use metric template.
2. With reference to Figure 4.216, construct a sketch. In the sketch, arcs A, B, and C are equal, lines D and E are equal, and the arcs are tangent to the lines.
3. Select Return on the Inventor Standard toolbar.

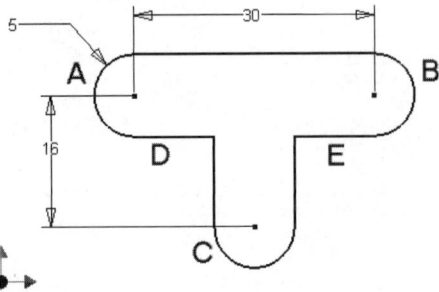

Figure 4.216 *Sketch constructed*

Now construct an iFeature from the sketch.

4. Select Extract iFeature from the Tools menu.
5. Select the sketch on the Browser Bar to put it in the iFeature dialog box.

6. Select the parameters of the sketch one by one and select the >> button to place each one in the Size Parameters box. (See Figure 4.217.)

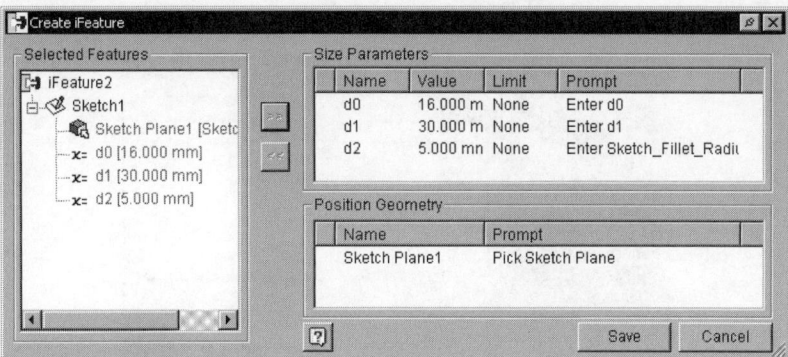

Figure 4.217 *Create iFeature dialog box*

Now save the iFeature in a catalog folder.

7. Click the Save button.

8. In the Save As dialog box, select a folder and specify a file name (*SketchiFeature.ide*). The extension of an iFeature is .ide.

The iFeature is complete. Save and close your file. (file name: *SketchiFeature.ipt*).

Now use the sketch iFeature in solid part modeling.

9. Start a new part file. Use the metric template.

10. Select Return on the Inventor Standard toolbar to exit sketch mode, if there is a default sketch.

11. Select Insert iFeature on the Part Features Panel Bar or toolbar.

12. In the Insert iFeature dialog box, select the Browse button and select the iFeature file (*SketchiFeature.ide*). (See Figure 4.218.)

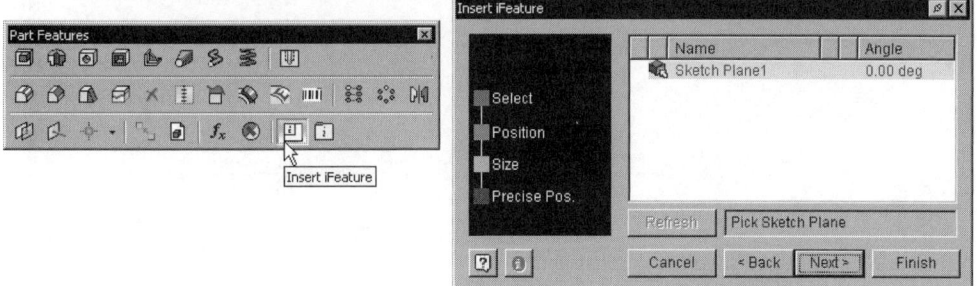

Figure 4.218 *iFeature file selected*

13. Select the XY plane on the Browser Bar to specify a plane for the iFeature sketch.

14. Click the Next button.

15. Modify the parameters if necessary.

16. Click the Finish button. (See Figure 4.219.)

The iFeature in the form of a sketch is inserted. You can use this sketch for making sketched solid features. Save and close your file (file name: *SketchPart.ipt*).

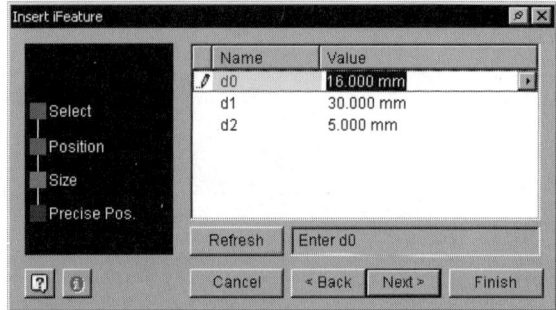

Figure 4.219 *Sketch parameters being modified*

JOIN iFEATURE

A join iFeature will join to the solid part into which it is inserted. To construct a join iFeature, you construct a sketched solid feature that joins a solid part and save the feature in the design catalog.

 **TUTORIAL 4.33**

Now you will learn how to construct a join iFeature and use it in part modeling.

1. Start a new part file. Use the metric template.

2. Construct an extruded solid feature in accordance with Figure 4.220.

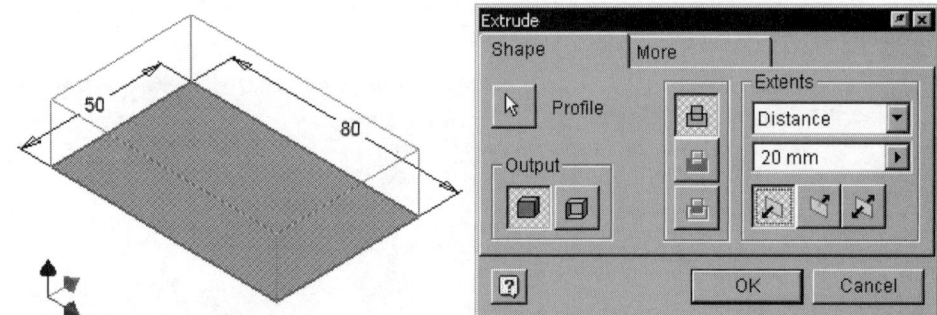

Figure 4.220 *Extruded solid being constructed*

3. With reference to Figure 4.221, construct a sketch, add two dimensions, and extrude the sketch to join the solid part.

4. Save your file (file name: *iFeaturePart.ipt*).

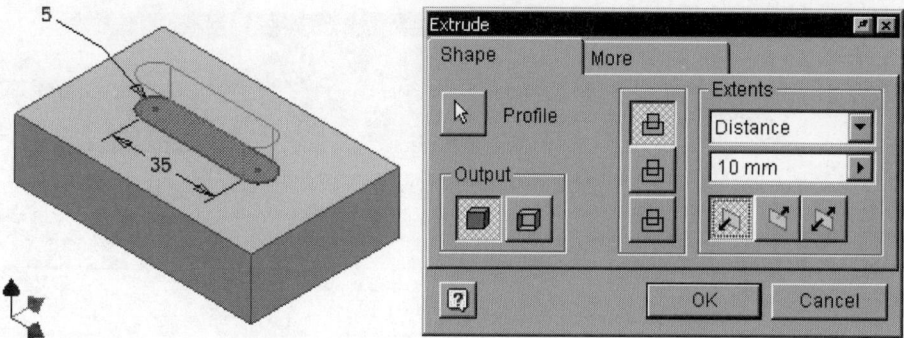

Figure 4.221 *Second sketch being extruded*

Now construct an iFeature from the extruded (join) feature.

5. Select Extract iFeature from Tools menu.

6. Select the second extruded feature to use it as an iFeature.

7. In the Create iFeature dialog box, select the parameters one by one and select the >> button to put them in the Size Parameters box.

8. Select the Save button and specify a file name (*JoiniFeature.ide*).

Because the source feature that you use to construct the iFeature joins to the source solid part, the iFeature will join to the solid part when you insert it in a part.

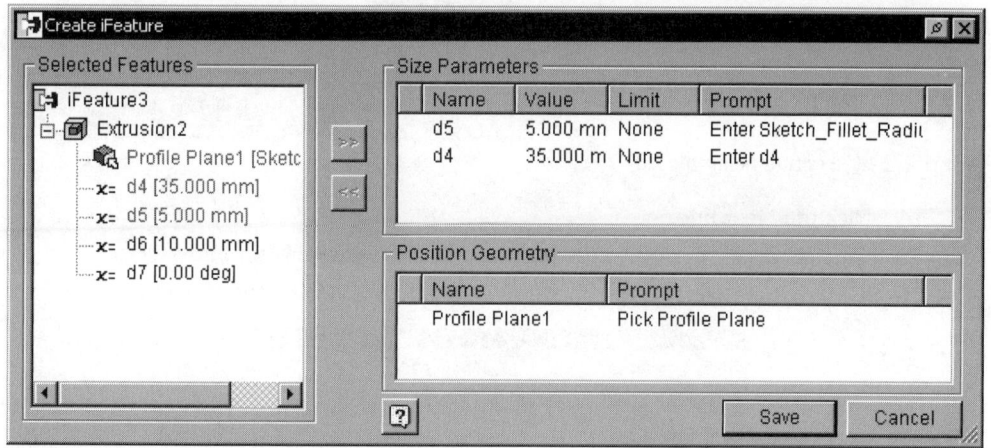

Figure 4.222 *Feature and its parameters selected*

Now insert the iFeature in a solid part.

9. Start a new part file. Use the metric template.

10. With reference to Figure 4.223, construct a rectangular sketch 100 mm by 100 mm and extrude it a distance of 30 mm.

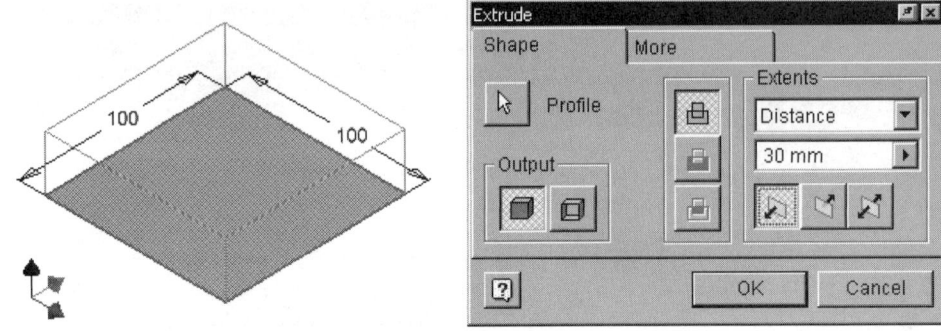

Figure 4.223 *Solid part being constructed*

11. Select Insert iFeature on the Part Features Panel Bar or toolbar.

12. Select the Browse button and select the iFeature file (*JoiniFeature.ide*).

13. Select face A in Figure 4.224 to specify a profile plane for the iFeature.

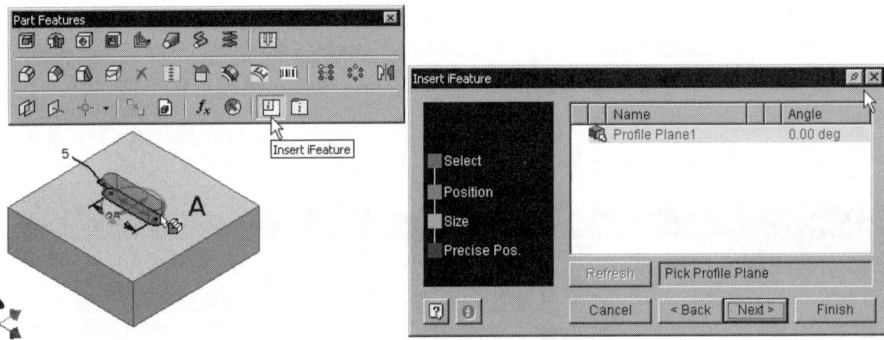

Figure 4.224 *iFeature being placed*

14. Click the Next button.

15. Make necessary change to the parameters and click the Next button.

16. Select Activate Sketch Edit Immediately and click the Finish button. (See Figure 4.225.)

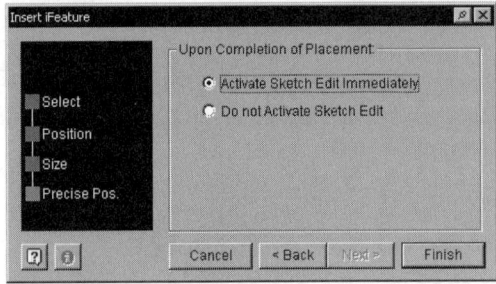

Figure 4.225 *Activating sketch edit*

17. With reference to Figure 4.226, add two dimensions.

18. Select Return on the Inventor Standard toolbar.

The iFeature is inserted and joined to the solid part. Save your file (file name: *PartwithiFeature1.ipt*).

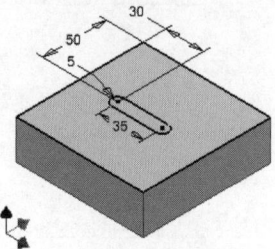

Figure 4.226 *Dimensions added to the sketch of the iFeature*

CUT iFEATURE

When you insert a cut iFeature into a solid part, it will cut the solid part to form a cavity. To construct a cut iFeature, you construct a sketched solid feature that cuts a solid part and store the iFeature in the design catalog.

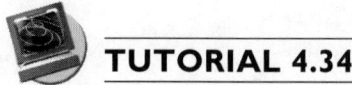

TUTORIAL 4.34

In this tutorial, you will learn how to construct an iFeature that cuts a solid part.

1. Open the file *iFeaturePart.ipt*, if you already closed it.
2. Select the second extruded feature on the Browser Bar, right-click, and select Edit Feature.
3. In the Extrude dialog box, select the Cut button and click the OK button. (See Figure 4.227.) Now you have a cut feature.

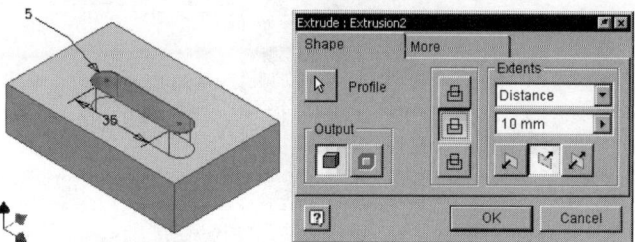

Figure 4.227 *Join operation changed to cut operation*

Now construct an iFeature from the extruded (cut) feature.

4. Select Extract iFeature from Tools menu.
5. Select the extruded (cut) feature indicated in Figure 4.228 to use it as an iFeature.
6. In the Create iFeature dialog box, select the parameters one by one and select the >> button to put them in the Size Parameters box.
7. Select the Save button and specify a file name (*CutiFeature.ide*).

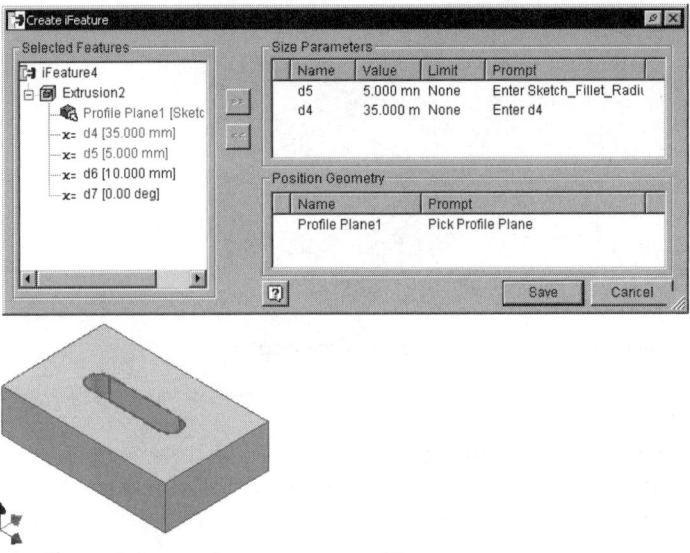

Figure 4.228 *Cut feature being used to construct an iFeature*

Now insert the iFeature in a solid part.

8. Open the file *PartwithiFeature.ipt*, if you already closed it.
9. Select Insert iFeature on the Part Features Panel Bar or toolbar.
10. Select the Browse button and select the iFeature file (*CutiFeature.ide*).
11. Select face A in Figure 4.229 to specify a profile plane for the iFeature.

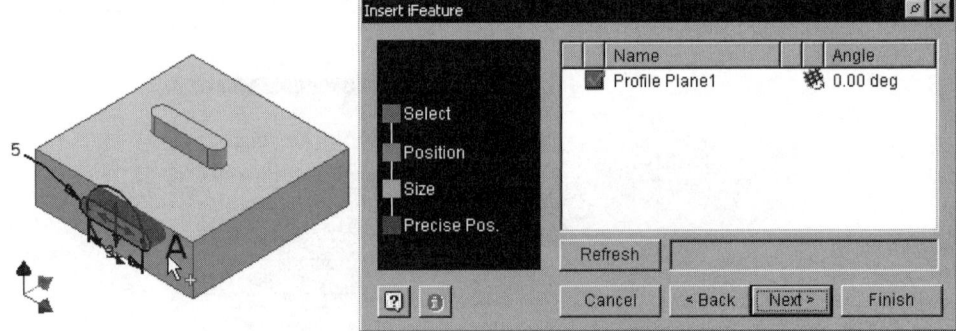

Figure 4.229 *iFeature being inserted*

12. Click the Next button.
13. Make necessary changes to the parameters and click the Next button.
14. Select Activate Sketch Edit Immediately and click the Finish button.
15. With reference to Figure 4.230, add two dimensions.
16. Select Return on the Inventor Standard toolbar.

The iFeature is inserted and joined to the solid part. Save your file (file name: *PartwithiFeature1.ipt*).

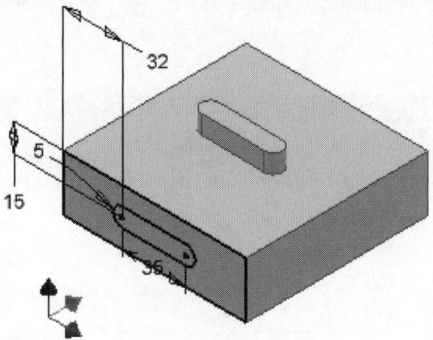

Figure 4.230 *Sketch being modified*

INTERSECT iFEATURE

When you insert an intersect iFeature into a solid part, it will intersect with the solid part. To construct an intersect iFeature, you construct a sketched solid feature that intersects with a solid part and save the iFeature in the design catalog.

TUTORIAL 4.35

In this tutorial, you will construct an intersect iFeature.

1. Open the file *iFeaturePart.ipt*, if you already closed it.
2. Select the second extruded feature on the Browser Bar, right-click, and select Edit Feature.
3. In the Extrude dialog box, select the Intersect button and click the OK button. (See Figure 4.231.) Now you have an intersect feature.

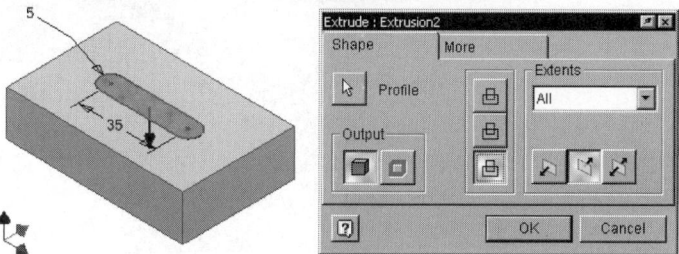

Figure 4.231 *Cut operation being changed to intersect operation*

Now construct an iFeature from the extrude (intersect) feature.

4. Select Extract iFeature from the Tools menu.
5. Select the extruded (intersect) feature indicated in Figure 4.232 to use it as an iFeature.

6. In the Create iFeature dialog box, select the parameters one by one and select the >> button to put them in the Size Parameters box.

7. Select the Save button and specify a file name (*IntersectiFeature.ide*).

Save and close the file.

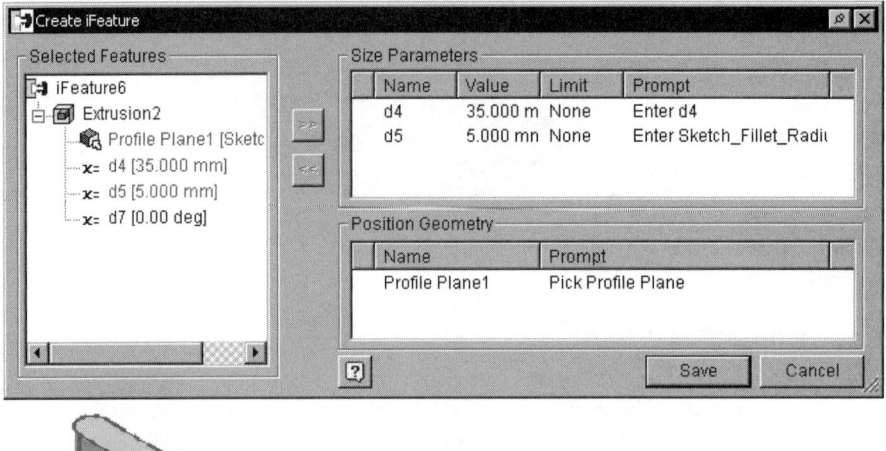

Figure 4.232 *Intersect iFeature being constructed*

Now you will insert the iFeature in a solid part.

8. Start a new part file. Use the metric template.

9. With reference to Figure 4.233, construct a polygon (six-sided) and extrude it to form a solid part.

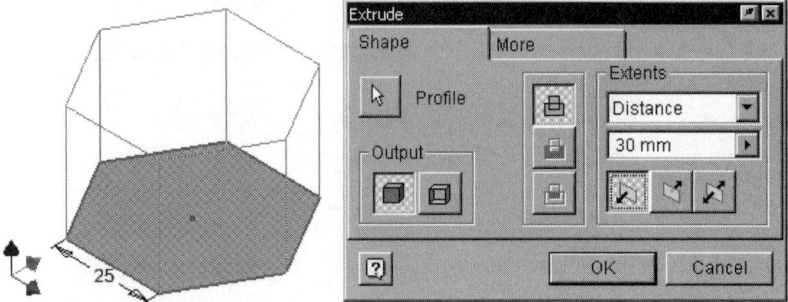

Figure 4.233 *Hexagon extruded*

10. Select Insert iFeature on the Part Features Panel Bar or toolbar.

11. Select the Browse button and select the iFeature file (*IntersectiFeature.ide*).

12. Select the face indicated in Figure 4.234 to specify a profile plane for the iFeature.

13. In the Insert iFeature dialog box, change the value in the angle column to 90 deg and click the Next button.

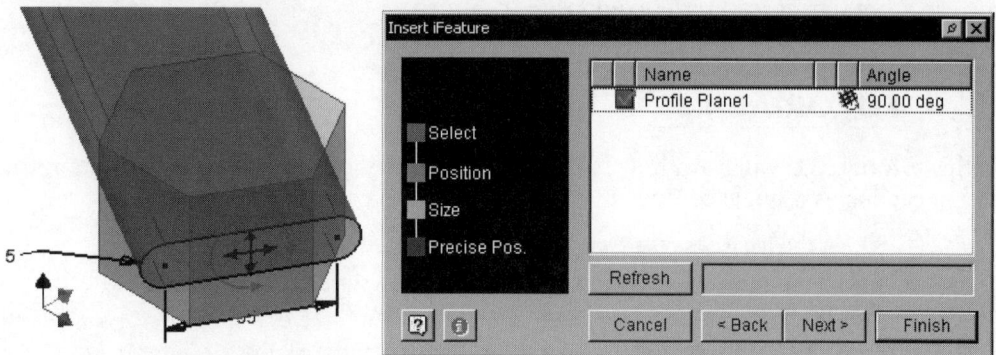

Figure 4.234 *Plane selected for placing an iFeature*

14. Change the extrusion depth of the iFeature to 100mm and click the Next button.
15. Select Activate Sketch Edit Immediately and click the Finish button.
16. With reference to Figure 4.235, add two dimensions.
17. Select Return on the Inventor Standard toolbar.

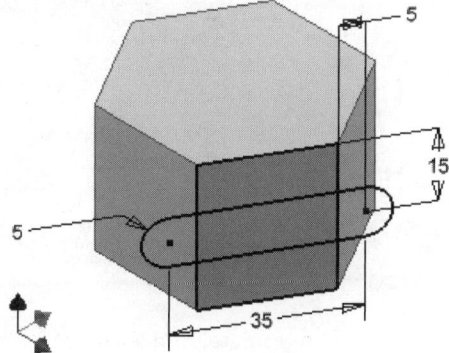

Figure 4.235 *Sketch edit activated*

The iFeature is inserted and joined to the solid part. (See Figure 4.236.) Save and close your file (file name: *PartwithiFeature2.ipt*).

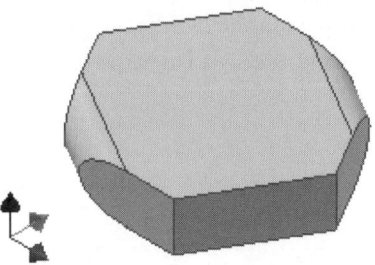

Figure 4.236 *Inserted iFeature intersected with the solid part*

COMPOSITE iFEATURE

Besides putting individual features into the design catalog as join, cut, or intersect iFeatures, you can store sets of features as composite iFeatures.

TUTORIAL 4.36

In this tutorial, you will learn how to construct a number of features and put them in the design catalog as a single composite iFeature.

1. Open the file *iFeaturePart.ipt*, if you already closed it.
2. Select the second extruded feature on the Browser Bar, right-click, and select Edit Feature.
3. In the Extrude dialog box, select the cut button and click the OK button. Now you changed the intersect feature to a cut feature.
4. With reference to Figure 4.237, construct a circle, add a dimension (12 mm), and extrude the sketch a distance of 3 mm.

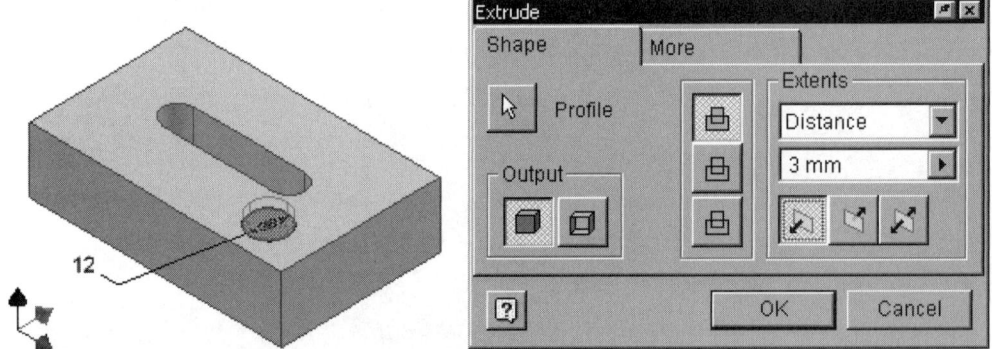

Figure 4.237 *Sketch being extruded*

5. Place a through hole of 5 mm diameter and a fillet of 3 mm.
6. Select Extract iFeature from the Tools menu.
7. Select the extruded feature, hole feature, and the fillet feature indicated in Figure 4.238 to use them collectively as a composite iFeature.
8. In the Create iFeature dialog box, select the parameters one by one and select the >> button to put them in the Size Parameters box.
9. Click the Save button and specify a file name (CompoundiFeature.ide).

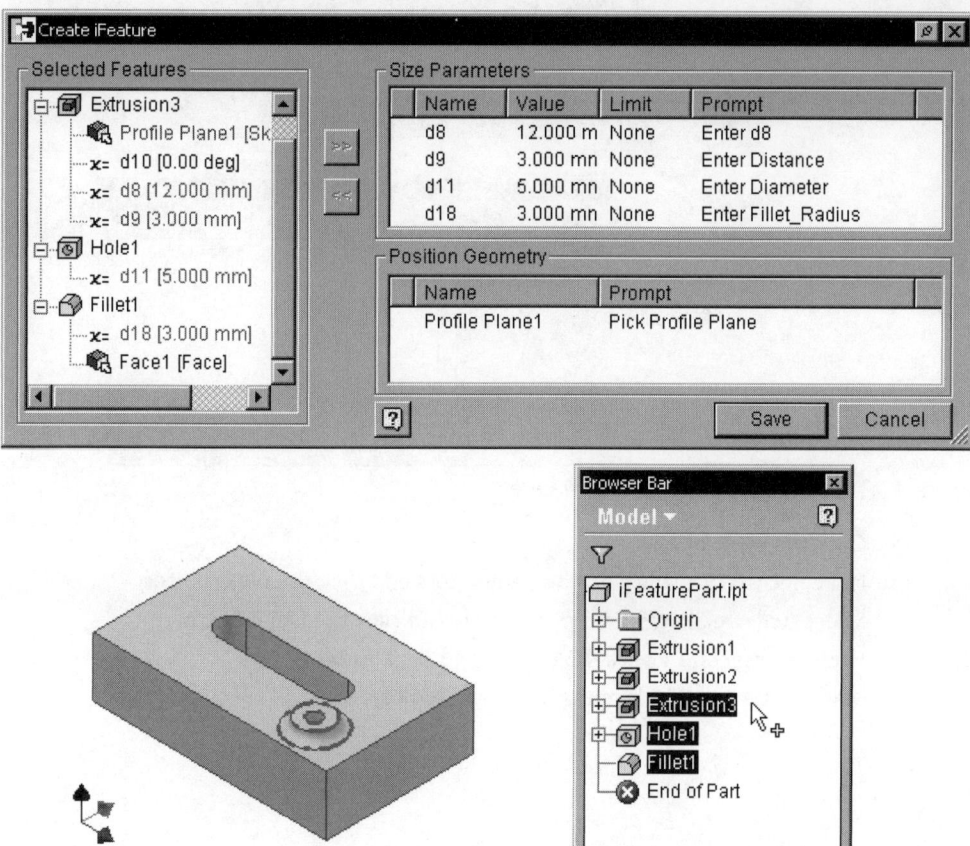

Figure 4.238 *Compound iFeature being constructed*

Now insert the iFeature in a solid part.

10. Open the file *PartwithiFeature1.ipt*, if you already closed it.

11. Select Insert iFeature on the Part Features Panel Bar or toolbar.

12. Select the Browse button and select the iFeature file (*CompoundiFeature.ide*).

13. Select the face indicated in Figure 4.239 to specify a profile plane for the iFeature.

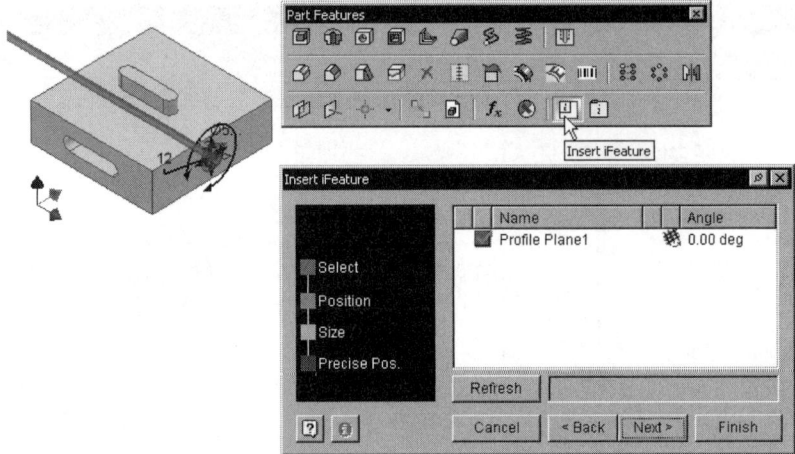

Figure 4.239 *Compound iFeature being inserted*

14. Click the Next button.
15. Make necessary changes to the parameters and click the Next button.
16. Select Activate Sketch Edit Immediately and click the Finish button.
17. With reference to Figure 4.240, add two dimensions.
18. Select Return on the Inventor Standard toolbar.

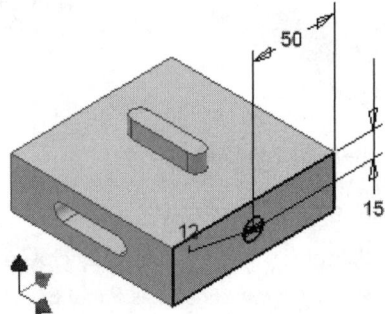

Figure 4.240 *Dimensions added*

The compound iFeature is inserted. (See Figure 4.241.) Save and close your files.

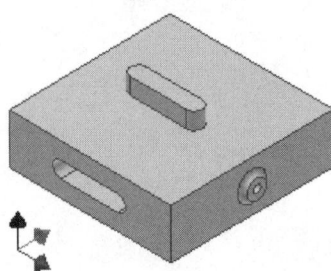

Figure 4.241 *Compound iFeature inserted*

SUMMARY

You use work features to help construct features in a solid part. There are three kinds of work features—work plane, work axis, and work point. Sketches are needed in making sketched features. There are two kinds of sketches—2D sketches and 3D sketches. From sketches, you can construct sketched solid features or sketched surface features.

Solid features are the main ingredients of the solid part. To expand the repertoire of form and shape, you incorporate surfaces in a solid part in various ways. You can replace a face of a solid with a surface, and you can thicken a surface to become a solid. You can knit a set of contiguous surfaces to form a single quilt. If the quilt encloses a volume without any opening, you can promote it to a solid part.

There are four basic kinds of solid and surface features—extrude, revolve, loft, and sweep. You use a single 2D sketch for making an extruded feature or a revolved feature. A loft feature has a number of cross sections and, optionally, a number of rails. In making a loft feature, you use a number of 2D or 3D sketches. A sweep feature has a cross section and a path. To make a sweep solid, you use a 2D sketch as the cross section and use a 2D sketch or a 3D sketch as the path along which the cross section sweeps.

In addition to the four basic kinds of features, there are rib, emboss, and coil solid features. Both the rib and the emboss features are special kinds of extruded solid features. A rib feature derives from an open-loop 2D sketch to form a rib or web. An embossment consists of a 2D sketch that can incorporate text profiles, and it can wrap around a cylindrical or conical face. A coil feature is a special kind of sweep feature. You construct a 2D sketch depicting the cross section and specify a helical path.

If you want to construct a solid part by referencing an existing solid part (optionally scaled or mirrored) and using the referenced solid part as a foundation for making additional solid features, you derive a solid part. Any modification made to the original solid part will be reflected in the derived part.

Dimensions that control the size and shape of a feature are called design parameters. There are four kinds of design parameters—model, user, linked, and embedded. Model parameters are parameters that are assigned by Inventor to a dimension of the solid part. User, linked, and embedded parameters are parameters that you add to the parameter table. User parameters are parameters that you add individually. Linked and embedded parameters are sets of parameters that you add collectively by using an Excel spreadsheet. You can use these parameters to control the model parameter. By linking a set of solid parts to a single Excel spreadsheet, you control the dimensions of the solid parts collectively.

While designing, you can record what you want to be retrieved later in a design notebook. The design notebook gives clues and information to you and other designers.

To use a design in various circumstances (with different parameters and properties), you construct a family of parts from the solid part. A family of parts is called an iPart. It has an embedded spreadsheet, the iPart table, maintaining the parameters, properties, and suppression information of the individual members of the iPart.

To enhance design efficacy and efficiency, you construct customized features for inserting (or placing) in a solid part. These features are called iFeatures. You can set up a design catalog (a catalog of iFeatures) by exporting selected sketches and solid features as design elements, and then you can insert iFeatures in new solid parts. There are four kinds of iFeatures—sketch iFeature, join iFeature, cut iFeature, and intersect iFeature. An iFeature that is joined to, cut from, or intersected with the solid part in the source solid part file will join, cut, or intersect the new solid part.

REVIEW QUESTIONS

1. Use sketches to explain the ways you can construct work planes, work axes, and work points.

2. Besides extruded and revolved solids, what kinds of features can you construct by making sketches? With the aid of sketches, describe how these features are constructed.

3. Differentiate between a face split feature and a face draft feature. Use examples to illustrate.

4. What kind of surface features can you construct from sketches? How will you use surface features in solid modeling?

5. Explain how can you derive a solid part from an existing solid part.

6. Explain the four kinds of design parameters in a solid part file. State the difference between linking and embedding a parameter spreadsheet.

7. What object can you include in your file to store textual design information?

8. What is meant by an iPart? How do you construct an iPart?

9. What is meant by an iFeature? State the four kinds of design elements and depict how a design catalog can be constructed.

CHAPTER 5

Sheet Metal Modeling

OBJECTIVES

This chapter explains the key concepts of sheet metal modeling, introduces the set of tools for constructing sheet metal parts, and outlines the ways to construct a sheet metal part and a flat pattern from a sheet metal part. This chapter also explains how to convert a 3D solid part to a 3D sheet metal part. After studying this chapter, you should be able to

- Describe the key concepts of sheet metal modeling
- Use sheet metal tools to construct 3D sheet metal parts
- Develop a 2D flat pattern from a 3D sheet metal part
- Convert 3D solid parts to 3D sheet metal parts

OVERVIEW

A sheet metal part is a special kind of 3D solid part. In reality, you make a 3D sheet metal object from a piece of 2D sheet metal of uniform thickness by cutting out a flat pattern and folding it into the final shape of the part. When you design a sheet metal part, you think about how you will manufacture the 3D part from a flat sheet, round off the joints between the faces of a sheet metal part, and provide recesses at the joints of the faces or the bend corners. Furthermore, you must think about how you will unfold the 3D part into a 2D flat pattern for manufacturing purposes. To manage these requirements, you need a special set of computer modeling tools that enable you to construct various kinds of bends and seams and to unfold the 3D sheet metal part into a 2D flat pattern.

SHEET METAL CONCEPTS

To manufacture a 3D object, you use material cutting or material forming processes. Material cutting processes involve the removal of material: You start with a piece of material that is larger than the overall size of the final shape of the 3D object. Then you use various machining processes to remove unwanted material from it. Typically, you use turning, milling, etc. Material forming processes involve the deformation of material to change its shape to the final appearance of the product. Sheet metal work is a kind of material forming process. You cut out a 2D flat pattern from a large sheet of material that has a uniform thickness. Then you bend the 2D flat pattern into a 3D complex object. To design a 3D sheet metal part, you should, in addition to thinking about the functional requirement of the part, think about the bend radii of the faces and relieves at the bend of the faces. Most importantly, you should think about how to unfold the 3D part to make a 2D flat pattern. Figure 5.1 shows a 3D sheet metal part and its flat pattern.

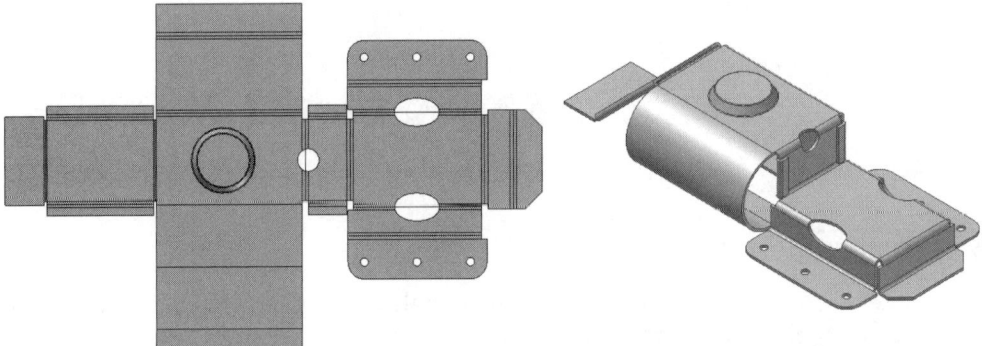

Figure 5.1 *3D sheet metal part and its flat pattern*

To construct a 3D sheet metal part in the computer, you can use either the building block approach or the conversion approach. Using the building block approach, you construct the elements of the sheet metal part one by one. Using the conversion approach, you first use the ordinary set of solid modeling tools to construct a 3D solid part (with a uniform thickness) and convert the 3D solid part to a 3D sheet metal part. Of the two methods, you should use the building block approach, because the set of sheet metal tools is far easier to use than the ordinary set of tools in constructing 3D sheet metal parts, particularly for making seams, hems, and joints. You should use the conversion approach only if you already have a 3D solid part that can readily be unfolded into a flat sheet.

BUILDING BLOCK APPROACH

This is the preferred way to construct a 3D sheet metal part. You critically analyze the sheet metal part to identify the number of individual sheet metal elements, determine the shape and size of the elements, and construct the elements one by one. Using the sheet metal modeling tools, you can construct some of the elements automatically or semi-automatically, such as a bend or a joint between two faces and a hem at the edge of a sheet metal face. For example, rounded joints between flat faces are done automatically each time you add a new face to the sheet metal part, and you can place joints and hems semi-automatically by selecting the element from the menu and specifying the parameters.

Before starting to design a sheet metal part, you should have general knowledge about the following kinds of elements specific to sheet metal modeling: face, flange, contour flange, cut, fold, project flat pattern, bend, corner seam, and hem.

Face

A face is a piece of flat sheet. In essence, it is an extruded solid feature that extrudes a sketch a distance equal to the thickness of the sheet metal. To construct a face, you determine the thickness of the sheet metal and other settings, use the sketching tool to construct a sketch, and use the face tool to extrude the sketch to form a face. As in making an extruded solid feature for an ordinary solid part, the sketch can be any shape. Figure 5.2 shows a sketch and a face constructed from the sketch.

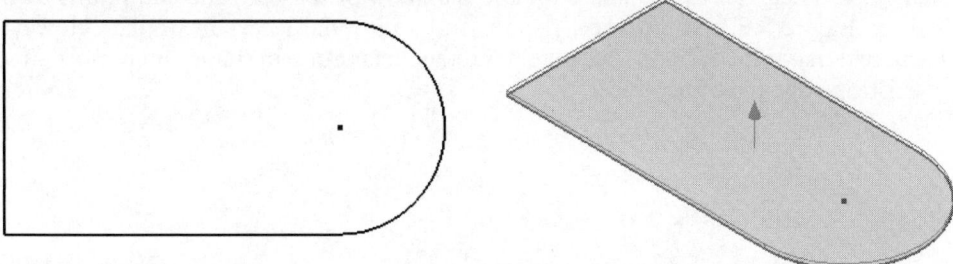

Figure 5.2 *Sketch and a face constructed from the sketch*

Similar to working on solid parts, we call the first sketched feature (normally a face) of a sheet metal part the base feature. To construct an additional face, you set up a sketch plane and construct a sketch. If one of the additional face's edges shares an edge of an existing face, bends and notches will be constructed automatically at the contiguous edge. (See Figure 5.3.)

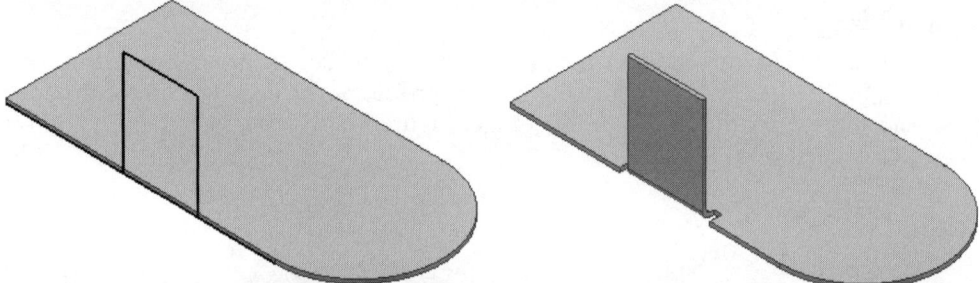

Figure 5.3 *Rounded bends and notches added automatically at the joint of new faces*

To manage design needs, you can set each individual bend to have its own unfold and bend relief options. Figure 5.4 shows bends with different settings.

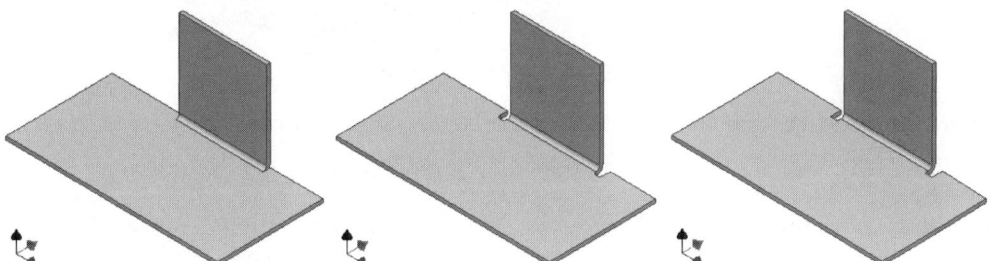

Figure 5.4 *Bends with different relief settings: (from left to right) none, straight, and round*

Flange

A flange is a rectangular flat face attached to a straight edge of an existing face of a sheet metal part. You select an edge and specify the length of the flange. While constructing the flange, you can specify distances for the flange to offset from the end points of the edge. (See Figure 5.5.) At a glance, flanges and additional faces are similar. However, there are two major differences: A flange has to be rectangular in shape and it must attach to an existing edge from the outset.

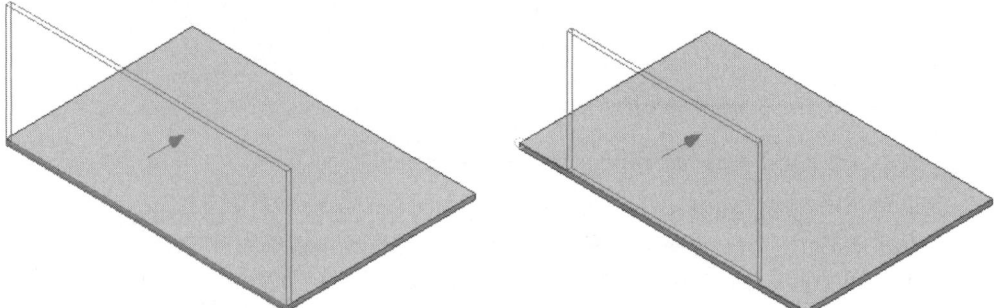

Figure 5.5 *Full length flange (left) and flange offset from the end points of the edge (right)*

Contour Flange

A contour flange is a curved sheet metal element. To make a contour flange, you construct an open-loop sketch to depict the general profile of the cross section of the sheet metal part and specify the length of the flange. Figure 5.6 shows a sketch and a contour flange constructed from the sketch.

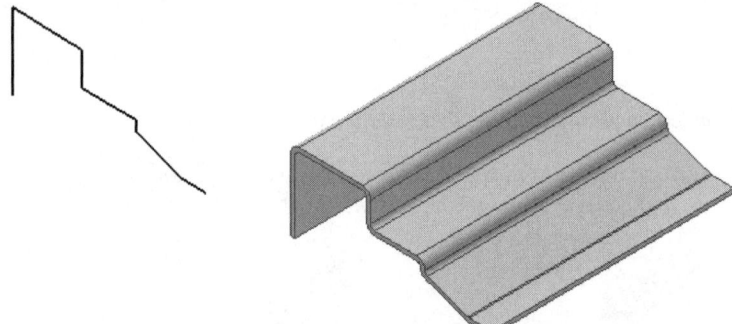

Figure 5.6 *Sketch depicting the contour (left) and the contour flange constructed from the sketch (right)*

Cut

In contrast to adding faces and flanges, you remove unwanted portions of a sheet metal by cutting. You construct a sketch on a sketch plane and use the sketch to cut an opening. As in making a face from a sketch, a cut can be any shape. Figure 5.7 shows a sketch and a cut constructed from the sketch.

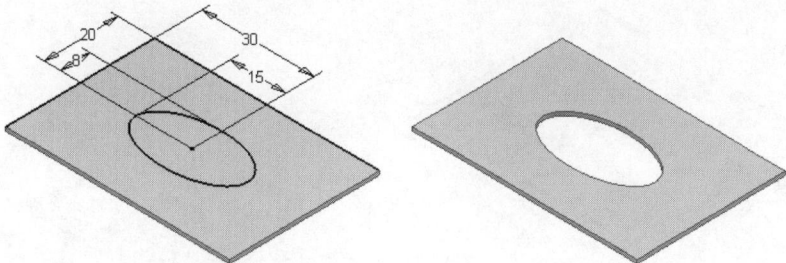

Figure 5.7 *Sketch and cut made on a face*

Fold

To fold a flat face, you construct a sketch line on a face of the sheet metal part and fold the sheet metal around the fold line. Figure 5.8 shows a fold line constructed on a sheet metal face and the folded sheet metal part.

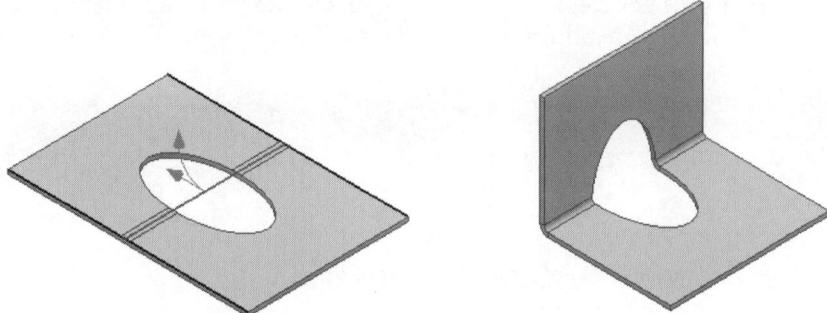

Figure 5.8 *Fold line on a face (left) and the folded part (right)*

Project Flat Pattern

If you already constructed a set of contiguous faces and you want to construct a cut across them as if the cut were made before the sheet metal was unfolded along the fold lines, you project a flat pattern from the contiguous faces and make a cut on the projected faces. Figure 5.9 shows a set of contiguous faces and the projected flat pattern of the faces, and Figure 5.10 shows a sketch constructed on the projected flat pattern and the cut made from the sketch.

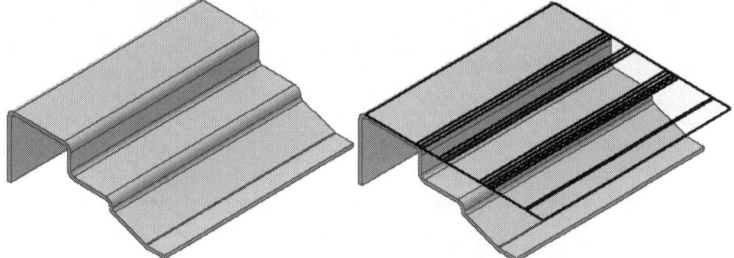

Figure 5.9 *Contiguous faces and a projected flat pattern of the faces*

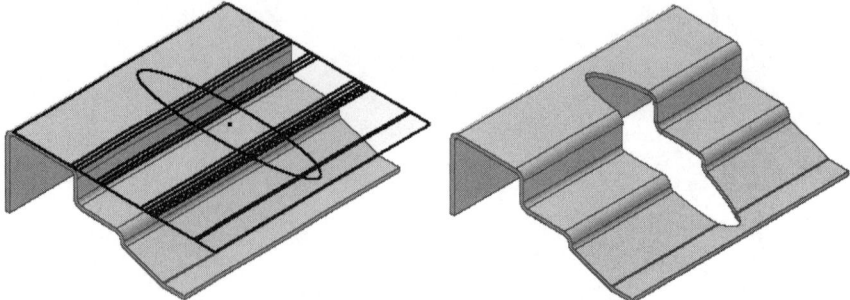

Figure 5.10 *Sketch constructed on the projected flat pattern and a cut made from the sketch*

Bend

You can connect two parallel edges from a pair of disjoint faces by making a bend feature between them. Depending on the relative positions of the faces, you can construct a fixed edge bend, a 45 degree bend, a full radius bend, or a 90 degree bend. Figure 5.11 shows the four ways of making a bend feature between to disjoint faces.

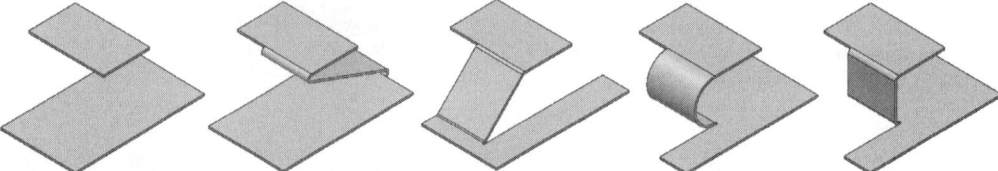

Figure 5.11 *Two individual faces (left) and four kinds of bends*

Corner Seam

To modify how three faces meet at a corner, you place a corner seam feature. Figure 5.12 shows various kinds of corner seams.

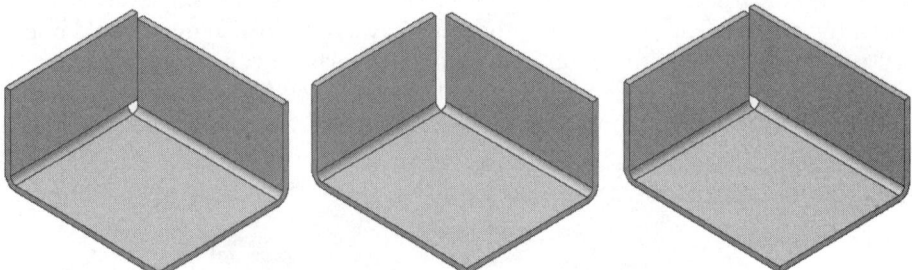

Figure 5.12 *Various kinds of corner seams*

Hem

Edges of thin sheet metal part can be hazardous. To round off the edges, you apply a hem. There are four kinds of hems: single, teardrop, rolled, and double. Figure 5.13 shows various kinds of hems.

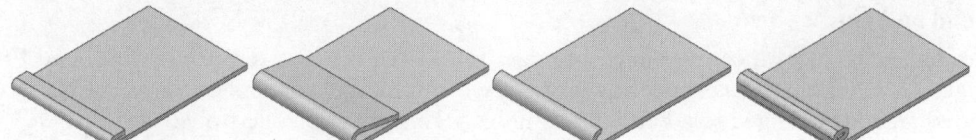

Figure 5.13 *Different kinds of hems: single, teardrop, rolled, and double (left to right)*

FLAT PATTERN

Perhaps the most important aspect regarding the design of a sheet metal part is to make a flat pattern (development) of the part so that you can cut the component from a flat sheet and then fold the flat sheet along the fold lines. Because a material will deform upon bending, the dimensions of a sheet metal after it is folded will change slightly. You can measure the flat pattern to determine the change. See Figure 5.14 shows a sheet metal component and a flat pattern generated from the 3D part.

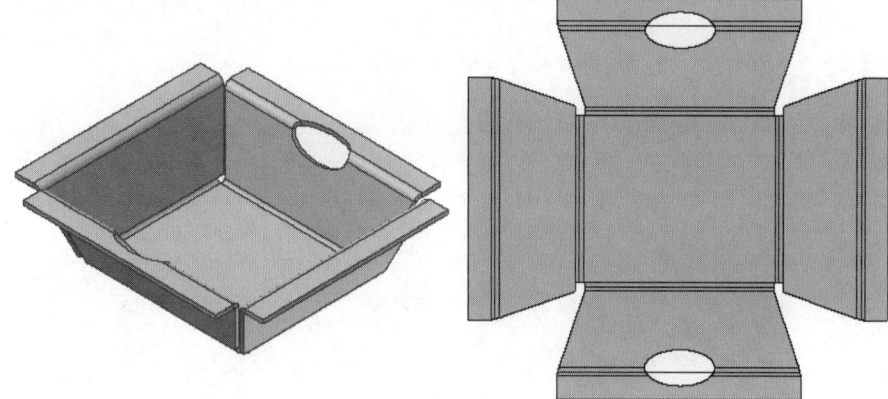

Figure 5.14 *Sheet metal part (left) and its flat pattern*

CONVERSION APPROACH

To convert a solid part to a sheet metal part, you bend all the sharp edges and rip an edge at each corner so that you can unfold the component to a flat pattern. Figure 5.15 shows a thin shell solid part converted to a sheet metal part with edges bent and ripped.

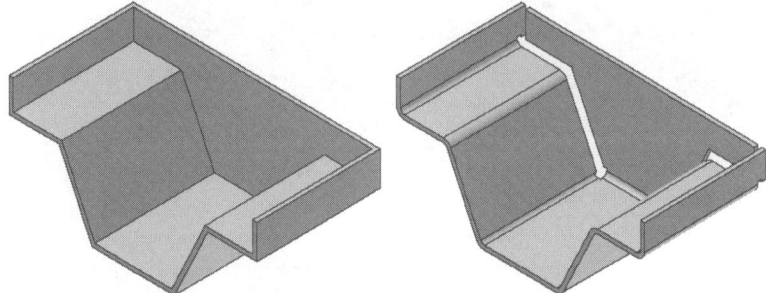

Figure 5.15 *Solid part (left) and converted sheet metal part (right)*

Bend and Corner Seam

To make two faces meeting along an edge unfoldable into a flat pattern, you bend the edge. To make three faces meeting at a corner unfoldable, you rip an edge. Figure 5.16 shows an edge changed to a bend, and Figure 5.17 shows an edge ripped at a corner.

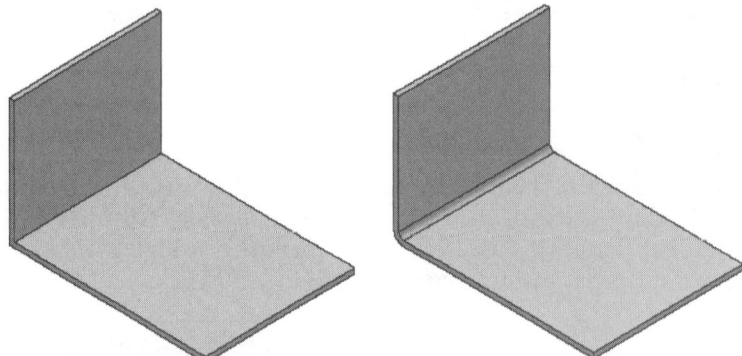

Figure 5.16 *Sharp edge (left) changed to a bend (right)*

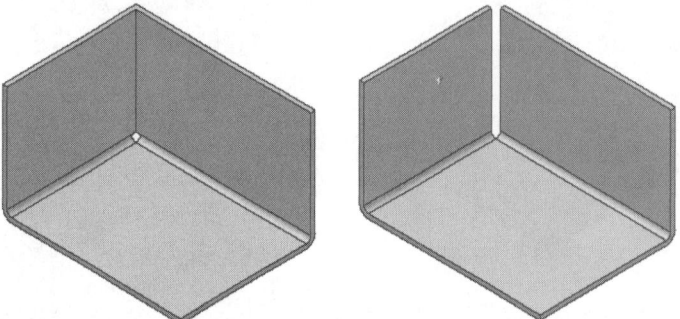

Figure 5.17 *An edge at a corner ripped*

To modify how you unfold three faces meeting at a corner, you convert a corner seam to a bend and a bend to a corner seam. (See Figure 5.18.)

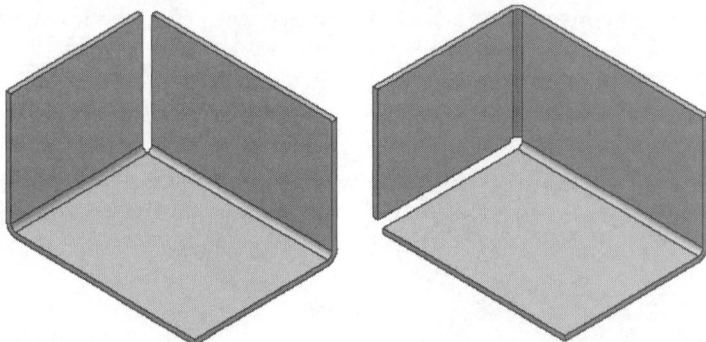

Figure 5.18 *Original corner (left) and modified corner (right)*

USER INTERFACE

Now you will familiarize yourself with the sheet metal modeling user interface. To start a sheet metal part file, follow these steps:

1. Start a new assembly file by selecting New from the File menu.
2. In the New dialog box, select the Default tab.
3. Select the *Sheetmetal.ipt* template and click the OK button.

After you start a new sheet metal part file, the user interface will be similar to an ordinary solid part file. (See Figure 5.19.)

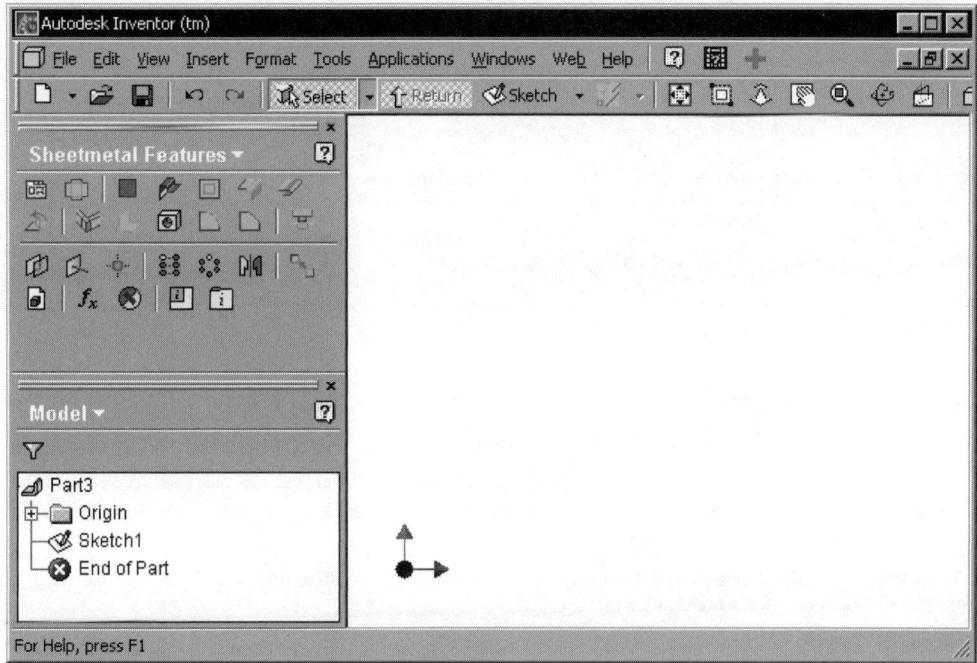

Figure 5.19 *Sheet metal modeling user interface*

SHEETMETAL FEATURES PANEL BAR AND TOOLBAR

After you finished the sketch and select the Return button on the Inventor Standard toolbar, the Sheetmetal Features Panel Bar will be displayed. To view the accompanying Sheetmetal Features Toolbar, select View > Toolbars > Sheetmetal Features.

Basically, constructing a 3D sheet metal part is similar to constructing the 3D solid part that you learned about in earlier chapters. You start your design by making a sketch, which you extrude to form a face of the sheet metal. Then you elaborate your design by adding more faces to it or by cutting holes on it. Like other 3D solid parts, a sheet metal part has four kinds of features: sketched solid features, placed solid features, work features, and pattern features. Because of the unique characteristics of the sheet metal part, the sheet metal modeling tool set is slightly different from the set of tools that you learned about earlier. Figure 5.20 shows the Sheetmetal Features Panel Bar and toolbar.

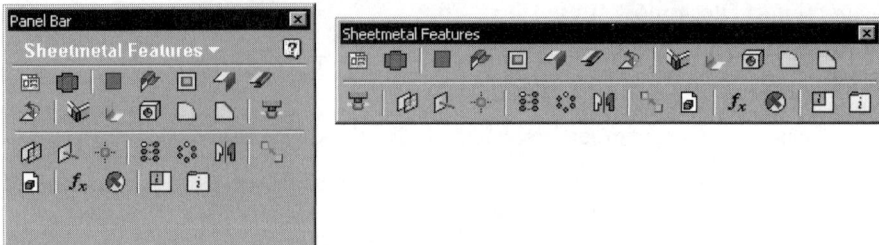

Figure 5.20 *Sheetmetal Features Panel Bar and toolbar*

The Sheetmetal Features Panel Bar and toolbar have twenty-six buttons, the first fourteen of which are sheet metal tools. Table 5.1 describes the choices.

Table 5.1 *Sheetmetal Features Panel Bar and toolbar options*

Option	Description	Feature Type	Tool
Styles	Sets sheet metal thickness and bend parameters.	—	Sheet metal tool
Flat Pattern	Flattens a 3D sheet metal part into a 2D flat pattern.	—	Sheet metal tool
Face	Extrudes a sketch to form a sheet metal face.	Sketched feature	Sheet metal tool
Contour Flange	Constructs a contour flange from an open-loop sketch.	Sketched feature	Sheet metal tool
Cut	Extrudes a sketch to cut an opening in a sheet metal face.	Sketched feature	Sheet metal tool
Flange	Constructs a sheet metal flange at the edge of a sheet metal face.	Placed feature	Sheet metal tool
Hem	Constructs a folded hem along an edge of a sheet metal part	Placed feature	Sheet metal tool
Fold	Folds an existing sheet metal part along a fold line.	Sketched feature	Sheet metal tool
Corner Seam	Treats the construct between two disjoint sheet metal faces.	Placed feature	Sheet metal tool
Bend	Constructs a sheet metal bend at the intersection of two non-parallel sheet metal faces.	Placed feature	Sheet metal tool
Hole	Constructs a hole on a sheet metal face.	Placed feature	General tool
Corner Round	Rounds off the corner of a sheet metal face.	Placed feature	General tool
Corner Chamfer	Bevels the corner of a sheet metal face.	Placed feature	General tool
Punch Tool	Inserts a sheet metal iFeature	—	Sheet metal tool

Table 5.1 *Sheetmetal Features Panel Bar and toolbar options (continued)*

Option	Description	Feature Type	Tool
Work Plane	Constructs a work plane.	Work feature	General tool
Work Axis	Constructs a work axis.	Work feature	General tool
Work Points	Constructs a work point.	Work feature	General tool
Rectangular Pattern	Places a rectangular pattern of features.	Placed feature	General tool
Circular Pattern	Places a circular pattern of features.	Placed feature	General tool
Mirror Feature	Place a mirror feature.	Placed feature	General tool
Promote	Changes a set of IGES surfaces to a base solid or stitches a set of disconnected IGES surfaces into a surface quilt.	—	General tool
Derived Component	Derives a solid part from another solid part or from an assembly.	—	General tool
Parameters	Sets model parameters.	—	General tool
Create iMate	Constructs an assembly interface to selected features of a component.	—	General tool
Insert iFeature	Inserts an iFeature.	—	General tool
View Catalog	Displays the iFeature catalog.	—	General tool

OTHER SHEET METAL MODELING TOOLS

The set of sheet metal modeling tools enables you not only to construct specific sheet metal elements (face, flange, contour flange, cut, fold, project flat pattern, bend, corner seam, and hem), but also to construct other features such as iFeature, punch tool, hole, corner round, corner chamfer, work features, rectangular pattern, circular pattern, and mirror features.

iFeatures

Sweep and loft features in sheet metal design will not show up properly in the flat pattern. Typical examples are louvers, dimples, and countersink holes. You should construct these features as iFeatures (design elements) and insert the iFeature in the sheet metal part. This way, the features will show up as 3D objects in the flat pattern. Figure 5.21 shows a square emboss feature inserted as an iFeature.

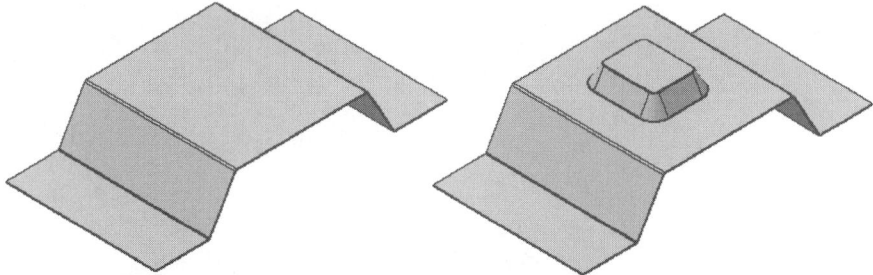

Figure 5.21 *Sheet metal part with inserted iFeature*

Punch Tool

Using the punch tool, you can insert specific sheet metal iFeatures in a sheet metal part by first constructing a hole center on a face of the sheet metal and then placing the iFeature by referencing to the center hole. Figure 5.22 shows a special iFeature in the form of a punch tool inserted in a sheet metal part.

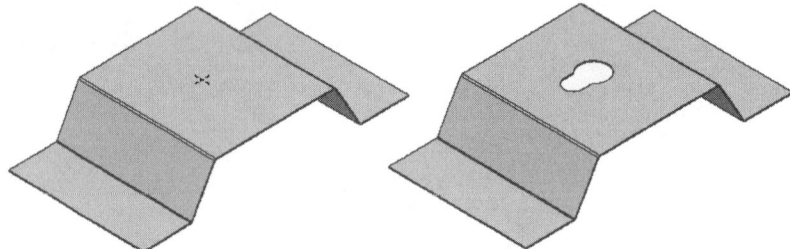

Figure 5.22 *Sheet metal part with punch feature placed*

Hole

If you want to make a circular cutting on a sheet metal part, you do not have to make a sketch. Instead, place a hole feature. Figure 5.23 shows a hole feature placed on a sheet metal part.

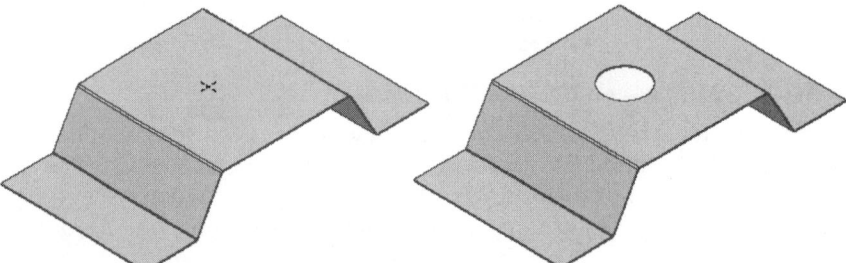

Figure 5.23 *Hole feature placed*

Corner Round and Corner Chamfer

To modify the sharp corners of a face, you add corner rounds and corner chamfers. (See Figure 5.24.)

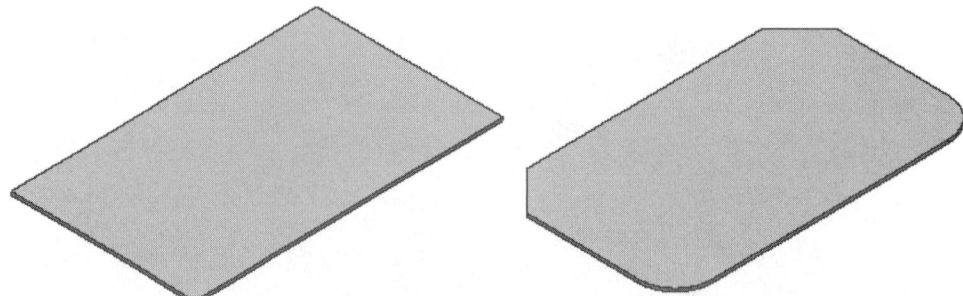

Figure 5.24 *Rectangular face (left) and corner round and corner chamfer features placed (right)*

Work Features

To help establish geometric references, you construct work features (work planes, work axes, and work points) in much the same way as you do in making ordinary 3D solid parts.

Rectangular Pattern and Circular Pattern

To repeat features in a rectangular or circular array, you place rectangular or circular patterns.

Mirror Feature

To construct a mirror copy of an existing feature, you place a mirror feature.

SHEET METAL CONSTRUCTION

Now you will learn how to construct a sheet metal component by using various methods.

BUILDING BLOCK APPROACH

To make a sheet metal part by using the building block approach, you critically analyze the component to determine the kind of features to construct and make them one by one.

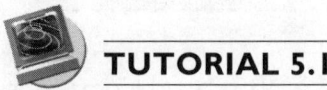 **TUTORIAL 5.1**

In this tutorial, you will construct a sheet metal component by using building block approach.

1. Start a new sheetmetal file. Use the sheetmetal metric template.
2. With reference to Figure 5.25, construct a sketch on the XY plane. In the sketch, construct a rectangle that measures 50 mm by 80 mm.

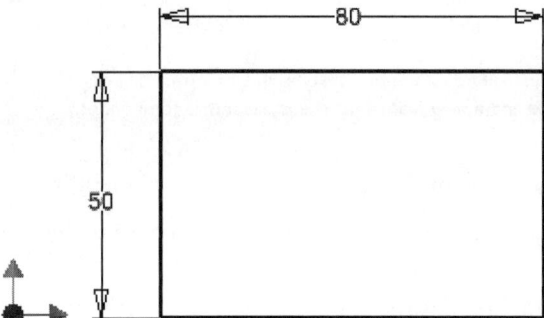

Figure 5.25 *Sketch constructed*

Now exit sketch mode and use the sheet metal modeling tool.

3. Select Return on the Inventor Standard toolbar.

The Sheetmetal Features Panel Bar should be displayed. If not, you can either move the cursor over the Panel Bar area, right-click, and select Sheetmetal or you can select Applications > Sheetmetal.

4. Select Styles on the Sheetmetal Features Panel Bar or toolbar. (See Figure 5.26.)

Sheet Metal Styles

Style List

Default
Default

Active Style

Default

| Sheet | Bend | Corner |

Sheet

Material   Thickness
Default   1 mm

Flat Pattern

Unfold Method   Unfold Method Value
Linear   KFactor

Modify List

Save   Delete   New   Done

Figure 5.26 *Sheetmetal Styles dialog box*

The Sheet Metal Styles dialog box has three tabs: Sheet, Bend, and Corner. On the Sheet tab, you select a material, set the sheet thickness, and determine how the flat pattern is evaluated. On the Bend tab, you set the bend's radius, minimum remnant, transition, relief shape (if any), relief width, and relief depth. On the Corner tab, you set the corner's relief shape and relief size.

5. Select the Sheet tab, if it is not already selected.

6. Set the sheet thickness to 1 mm and click the Save and Done buttons.

 Tip: You can maintain a number of styles in a sheetmetal file, and you can override default settings each time you add a new feature to the sheet metal component.

Now construct a face of the sheet metal component from the sketch.

7. Set the display to an isometric view and select Face on the Sheetmetal Features Panel Bar or toolbar. (See Figure 5.27.)

The Face dialog box has three tabs: Shape, Unfold Options, and Relief Options. On the Shape tab, you set the direction of extrusion (offset), set bend radius, and select bend edges. On the Unfold Options tab, you override default unfold settings (unfold method, bend transition, and unfold method value) specified in the Sheet Metal Styles dialog box. On the Relief Options tab, you override default bend relief settings (relief shape, minimum remnant, relief width, and relief depth) specified in the Sheet Metal Styles Styles dialog box.

8. Select the Shape tab, if it is not already selected.

9. Select or deselect the Offset button to change the direction of extrusion in accordance with Figure 5.27 and click the OK button.

A sheet metal face is constructed from the sketch.

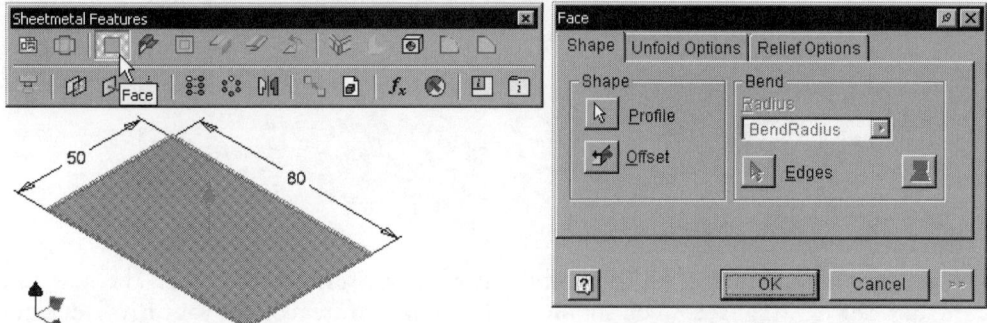

Figure 5.27 *Sketch being extruded to form a face of the sheet metal component*

Now construct a work plane, establish a sketch on the work plane, construct a sketch, and make a face from the sketch.

10. Construct a work plane that offsets a distance of –10 mm from vertical face A in Figure 5.28. (You might have to zoom in to select the face.)

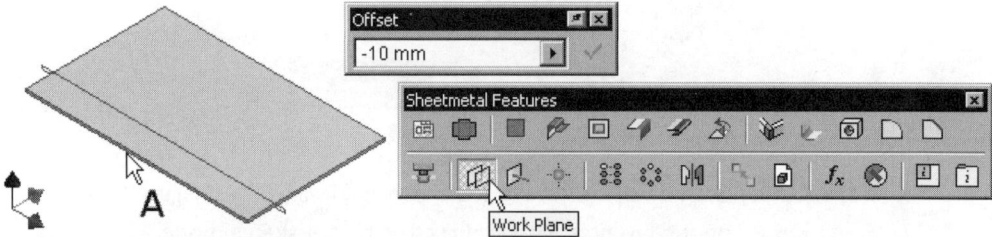

Figure 5.28 *Work plane offset from an edge's face being constructed*

11. With reference to Figure 5.29, establish a sketch on the work plane and construct a sketch on the work plane.

12. Select Return on the Inventor Standard toolbar.

13. Select Face on the Sheetmetal Features Panel Bar or toolbar and select the sketch, if it is not already selected.

14. On the Shape tab of the Face dialog box, select the Edges button and the Extend Bend Aligned to Side Faces button, select edge A in Figure 5.29, and click the OK button.

Tip: If you do not check the Edges button and select an edge, the face will simply be added to the face without any bend and relief cuts.

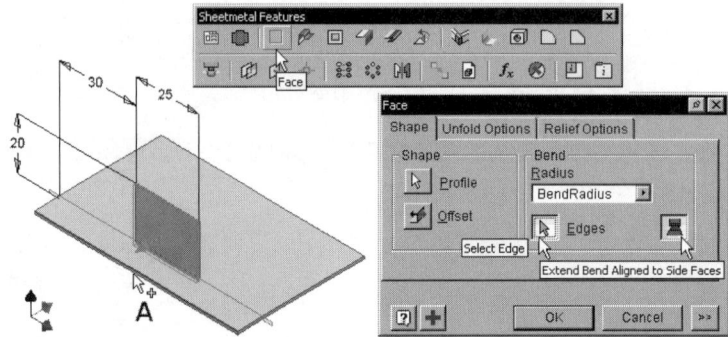

Figure 5.29 *Face being constructed*

A face is constructed, and a bend is automatically included between the new face and the existing face. (You should find a Bend feature icon on your Browser Bar.) Now construct another sheet metal face on an offset work plane and then construct a bend to join the face to the main body of the sheet metal component.

15. Construct a work plane that offsets 40 mm from face A in Figure 5.30.

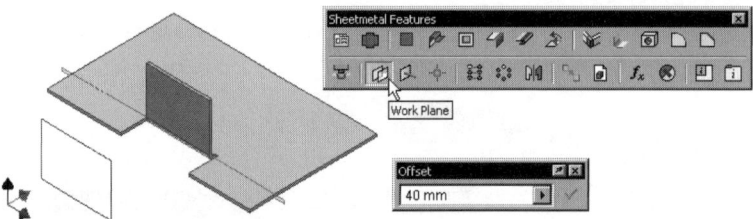

Figure 5.30 *Offset work plane constructed*

16. Construct a sketch on the offset work plane in accordance with Figure 5.31.

17. Select Return on the Inventor Standard toolbar to exit sketch mode.

18. Select Face on the Sheetmetal Features Panel Bar or toolbar to construct a face from the sketch.

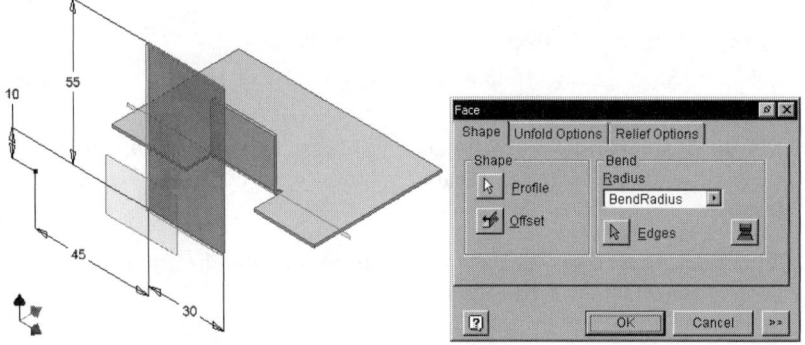

Figure 5.31 *Face being constructed*

19. Hide the work planes.

20. Select Bend on the Sheetmetal Features Panel Bar or toolbar.

21. Select edges A and B in Figure 5.32.

22. In the Bend dialog box, select Fixed Edges and click the OK button.

A bend is constructed, connecting two sheet metal faces.

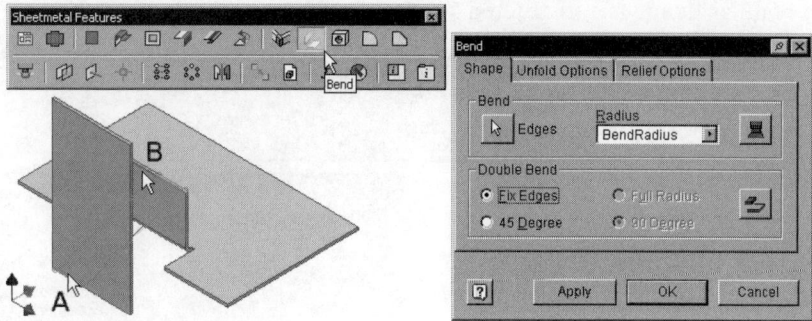

Figure 5.32 *Bend being constructed*

Now construct flanges along the edges of the sheet metal component.

23. Select Flange on the Sheetmetal Features Panel Bar or toolbar.

24. Select edge A in Figure 5.33.

25. In the Flange dialog box, set the distance to 25 mm and angle to 90 deg.

26. Select the >> button to expand the dialog box.

27. Select Width in the Type pull-down list box.

28. Set offset value to 0 and width value to 50 mm.

29. Click the Select Start Point button, select vertex B, and then click the OK button.

A flange with specified width is constructed.

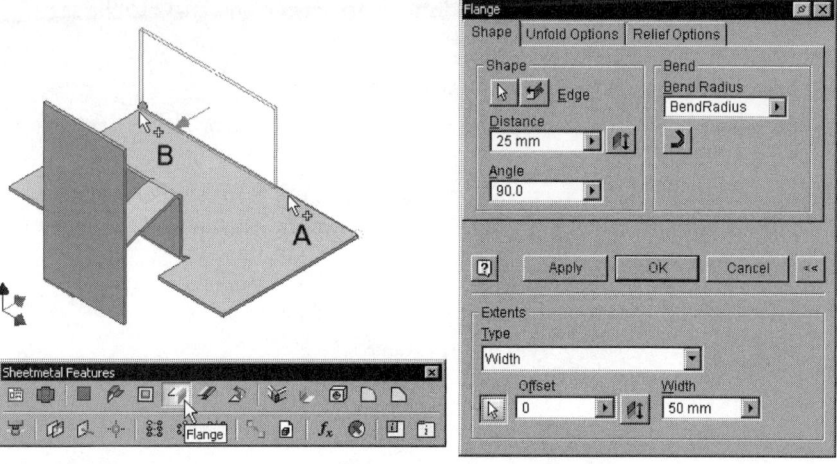

Figure 5.33 *Flange of specified width being constructed*

Now construct another flange.

30. Select Flange on the Sheetmetal Features Panel Bar or toolbar.

31. Select edge A in Figure 5.34.

32. In the Flange dialog box, set the angle to 60 deg and click the OK button.

An angular flange is constructed.

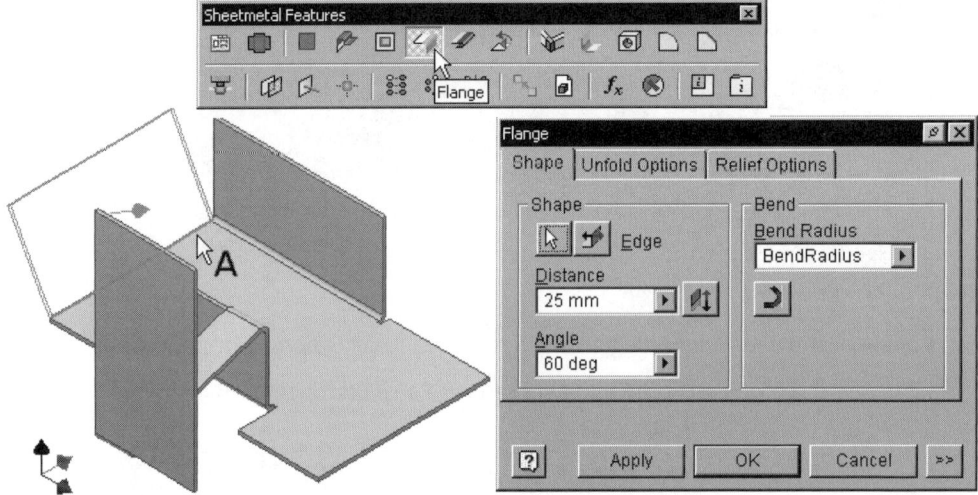

Figure 5.34 *Angular flange being constructed*

Now construct an offset flange.

33. Select Flange on the Sheetmetal Features Panel Bar or toolbar.

34. Select edge A in Figure 5.35. If necessary, click the Flip Direction and/or Flip Offset buttons so that the preview image looks like Figure 5.35.

35. In the Flange dialog box, set the distance to 25 mm and angle to 90 deg.

36. Select the >> button.

37. Select Offset in the Type pull-down list box.

38. Set offset1 to –30 mm, click the Select Start Point button, and select vertex B in Figure 5.35.

39. Set offset2 to 0. The vertex C (indicated in Figure 5.35) should already be highlighted; otherwise click the Select Endpoint button and select vertex C.

40. Click the OK button.

An offset flange is constructed.

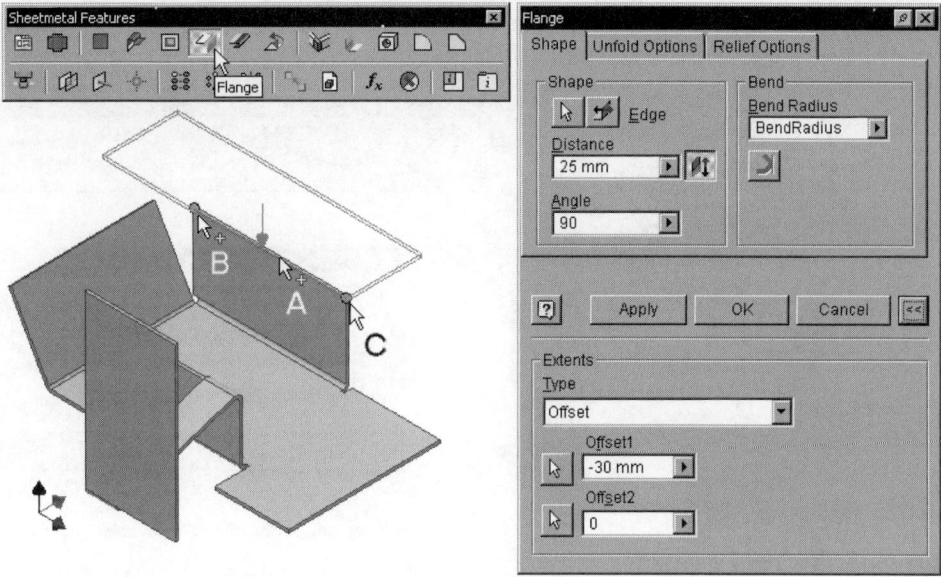

Figure 5.35 *Offset flange being constructed*

Now construct a face and then a bend to connect the face to the main body of the sheet metal component.

41. Set up a sketch on face A in Figure 5.36 and construct a rectangle.

42. Add a dimension and apply a collinear constraint to edges B and C, edges D and E, and edges F and G.

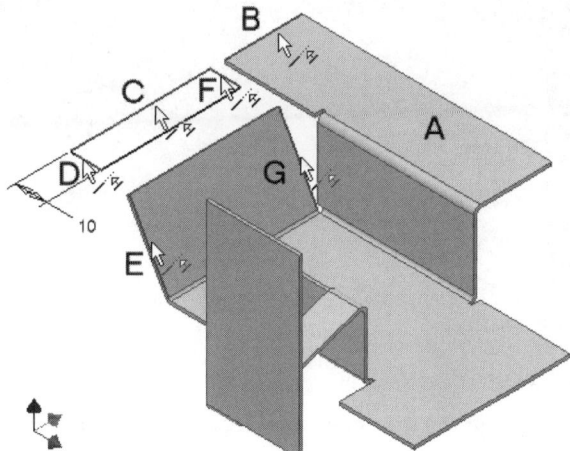

Figure 5.36 *Sketch being constructed*

43. Select Return on the Inventor Standard toolbar to exit sketch mode.

44. Construct a face from the sketch. (See Figure 5.37.)

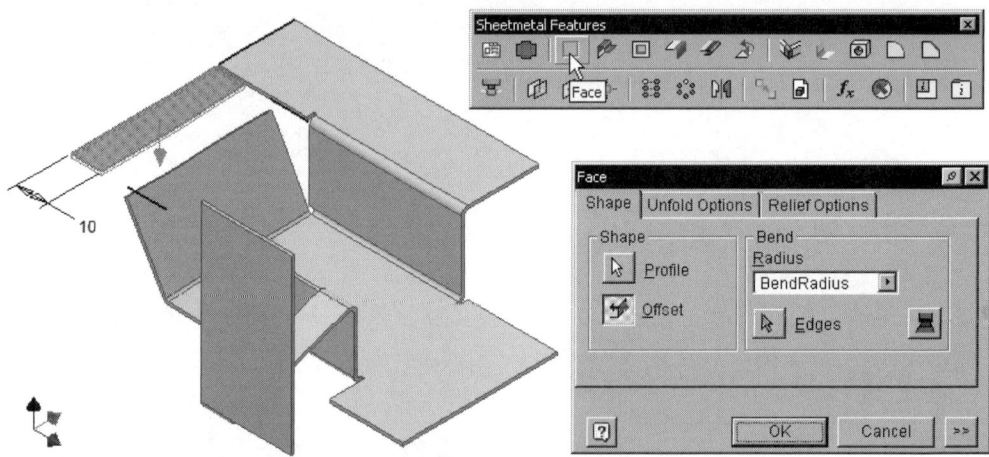

Figure 5.37 *Face being constructed*

45. Construct a bend to connect edges A and B in Figure 5.38.

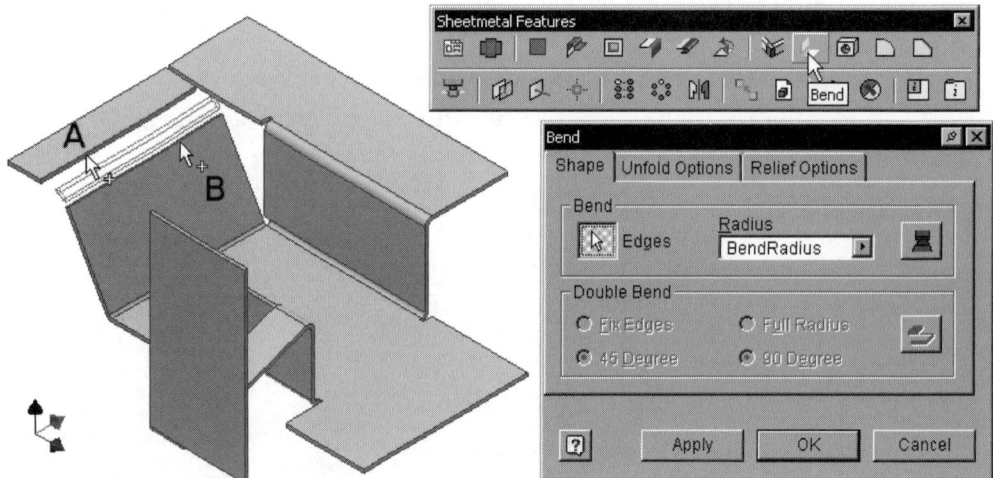

Figure 5.38 *Bend being constructed*

Now construct a corner seam.

46. Select Corner Seam on the Sheetmetal Features Panel Bar or toolbar.

47. Select edge A and then edge B in Figure 5.39.

48. In the Corner Seam dialog box, select the Overlap button and click the OK button.

A corner seam is constructed.

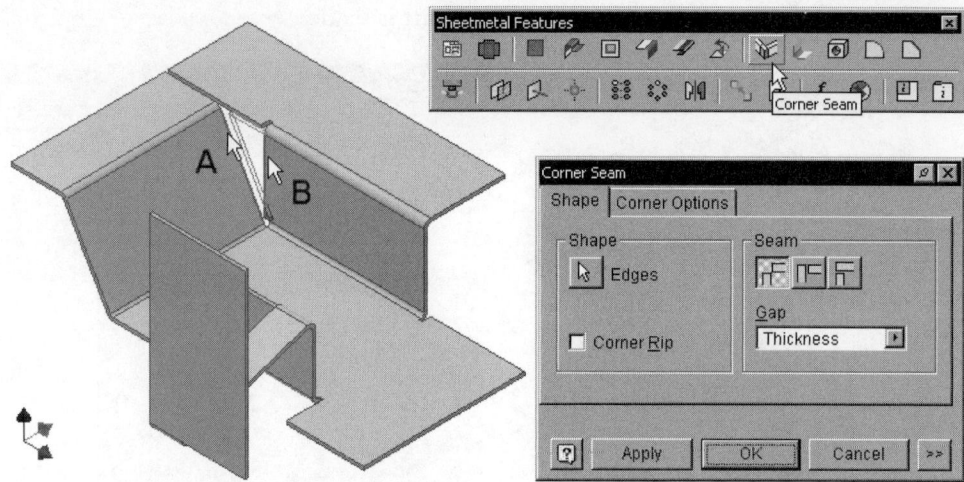

Figure 5.39 *Corner seam being constructed*

Now construct a number of cuts on the sheet metal component. The first cut will cut a distance equal to the thickness of the sheet metal, the second cut will cut through the sheet metal, and the third cut will cut across the bend.

49. Establish a sketch on face A in Figure 5.40 and construct a circle.

50. Select Cut on the Sheetmetal Features Panel Bar or toolbar.

51. In the Cut dialog box, select Distance and Thickness in the Extents box and click the OK button.

A cut is made on the sheet metal component.

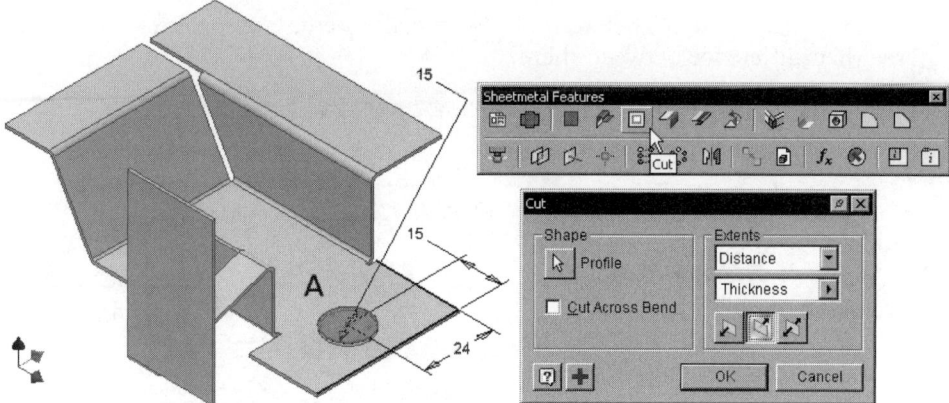

Figure 5.40 *A cut being made on the sheet metal*

52. Establish a sketch on face A (in Figure 5.41), construct a circle, and then click Return on the Inventor Standard toolbar.

53. Select Cut on the Sheetmetal Features Panel Bar or toolbar.

54. In the Cut dialog box, select All in the Extents box and click the OK button.

An opening cut through the sheet metal part is made.

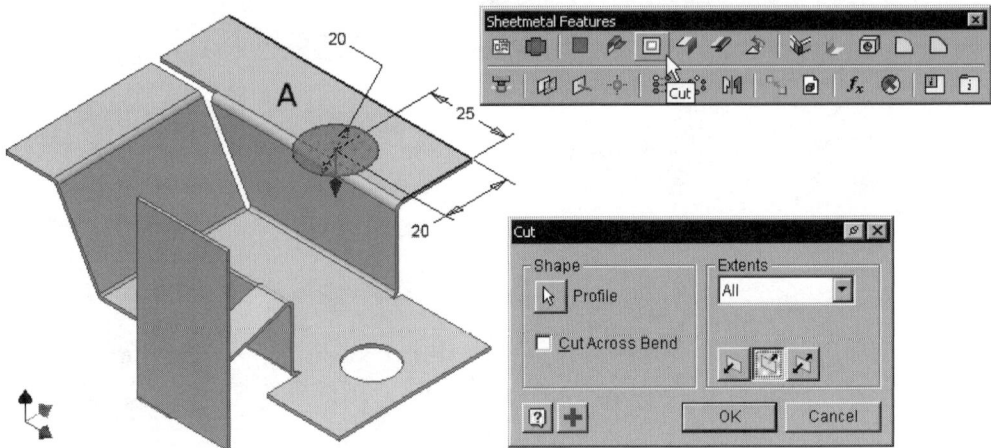

Figure 5.41 *A cut is being made to cut through the sheet metal component*

Now construct a cut to cut across a bend.

55. Set up a sketch on face A (indicated in Figure 5.42) and construct a circle, but do not end the sketch.

56. Select Project Flat Pattern on the 2D Sketch Panel Bar or toolbar (located on the same flyout as Project Geometry), select face B to project the face to the sketch plane, and then click Return on the Inventor Standard toolbar.

57. Select Cut on the Sheetmetal Features Panel Bar or toolbar.

58. In the Cut dialog box, select the Cut Across Bend box and click the OK button.

A cut is made as if cut before the sheet metal is bent. Compare the three cuts to see their difference between them.

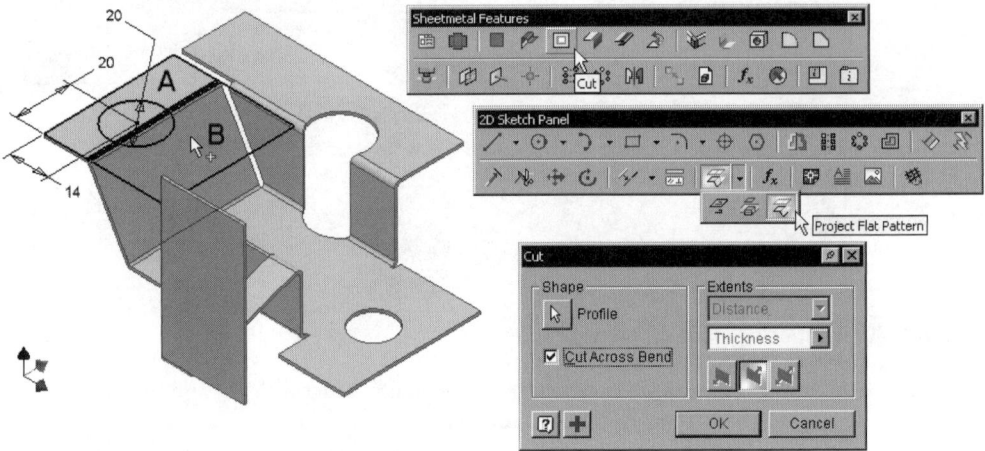

Figure 5.42 *A cut being made across a bend*

Now construct a sketch line and fold the sheet metal along the line.

59. Set up a sketch on face A in Figure 5.43 and construct a line.

60. Add a dimension and apply a coincident constraint to the end points of the line, and then click Return on the Inventor Standard toolbar.

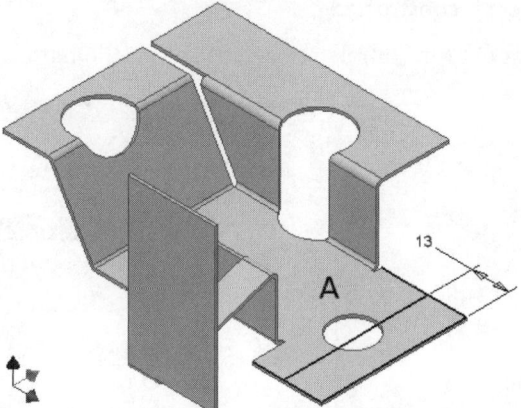

Figure 5.43 *A line constructed*

61. Select Fold on the Sheetmetal Features Panel Bar or toolbar.

62. Select the line as the fold line.

63. In the Fold dialog box, select or deselect the Flip Side and Flip Direction buttons such that the side to fold and the direction of fold are in accordance with Figure 5.44.

64. Click the OK button. A fold is made.

Tip: Making a cut first and then folding the sheet metal component is equivalent to cutting across a bend.

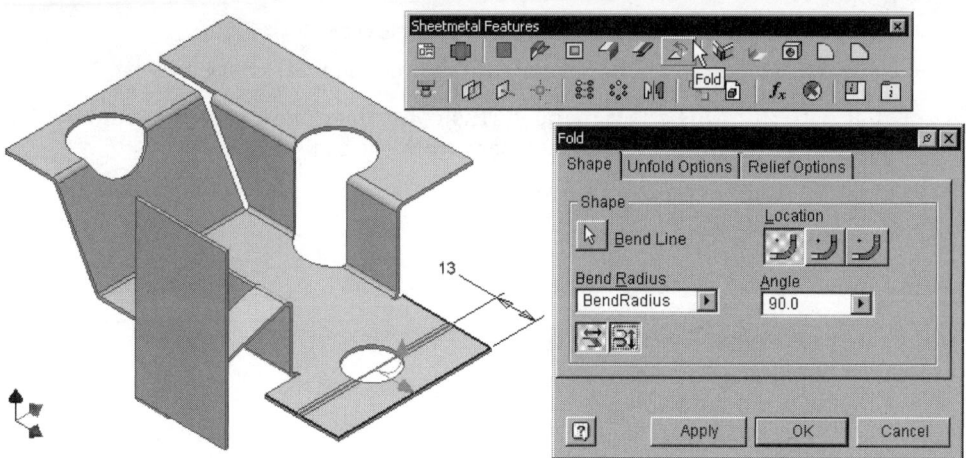

Figure 5.44 *Sheet metal component being folded along a line*

Now construct a hem.

65. Select Hem on the Sheetmetal Features Panel Bar or toolbar.

66. Select edge A in Figure 5.45.

67. In the Hem dialog box, select Single in the Type pull-down box and click the OK button. A hem is constructed.

The sheet metal component is complete. Save your file (file name: *SheetMetalA.ipt*).

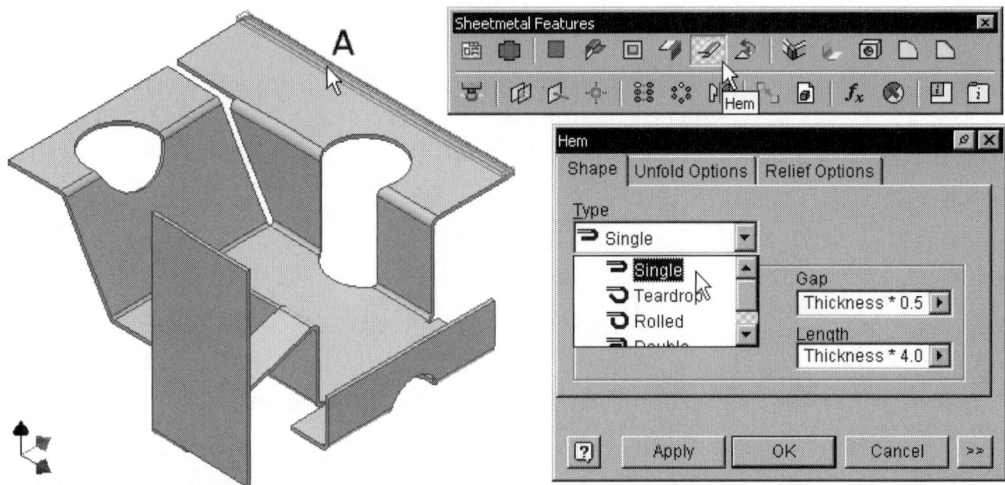

Figure 5.45 *A hem being constructed*

TUTORIAL 5.2

In this tutorial, you will construct a flat pattern from the sheet metal component.

1. Open the file *SheetMetalA.ipt*, if you already closed it.

2. Select Flat Pattern on the Sheetmetal Features Panel Bar or toolbar.

A flat pattern of the sheet metal part is constructed in a separate window. (See Figure 5.46.) Now take a measurement from the flat pattern.

3. Select Tools > Measure Distance.

4. Select A and B in Figure 5.46.

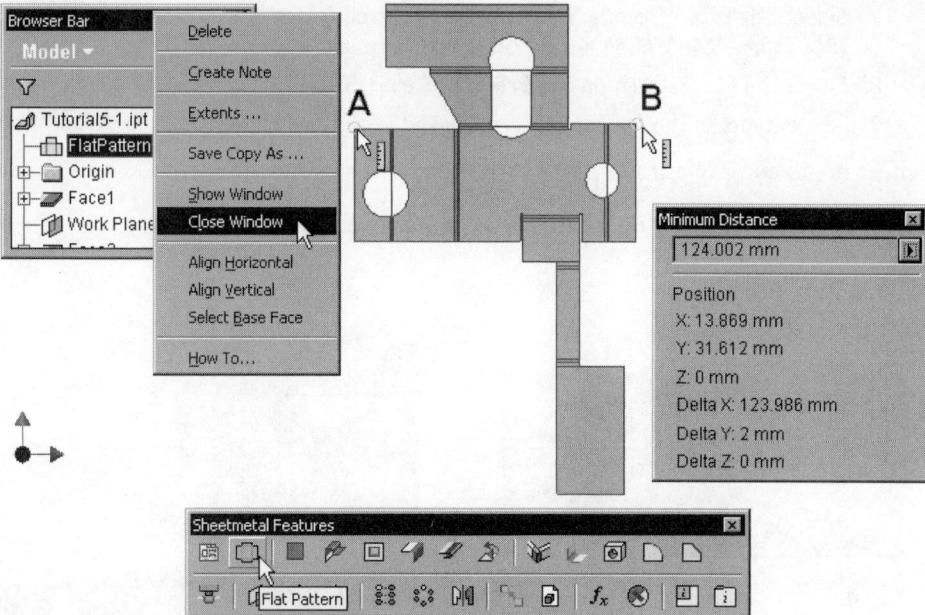

Figure 5.46 *Flat pattern constructed*

Distance between these vertices is displayed. To close the flat pattern window, select Flat Pattern on the Browser Bar, right-click, and select Close Window. You can also simply close it by clicking the standard Windows close button (the X in the upper-right corner of the window). Now align the flat pattern.

5. Select Flat Pattern on the Browser Bar, right-click, and select Align Horizontal.
6. Select edge A in Figure 5.47.

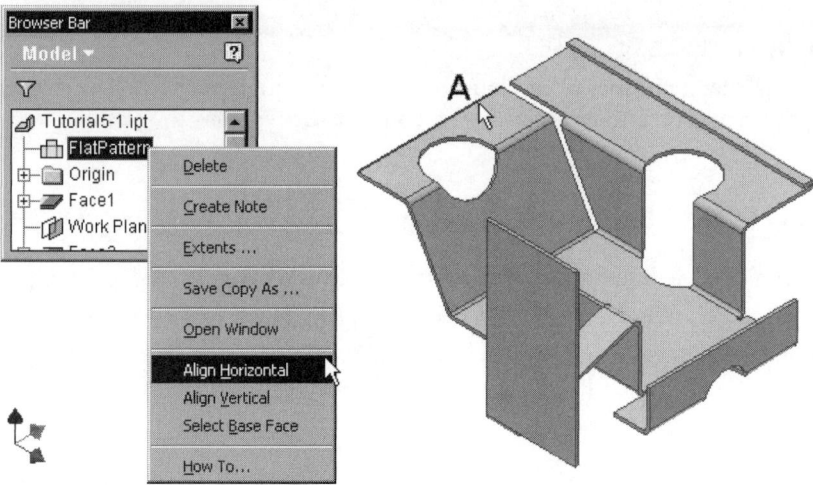

Figure 5.47 *Flat pattern being aligned*

7. Select Flat Pattern on the Browser Bar, right-click, and select Open Window. (See Figure 5.48.) The orientation of the flat pattern is changed.

8. Select the Flat Pattern on the Browser Bar, right-click, and select Extents.

9. The extents of the flat pattern is displayed in a dialog box.

The sheet metal component is complete. Save and close your file.

 Tip: You can align a flat pattern vertically or horizontally and you can select a face of the sheet metal as the base face.

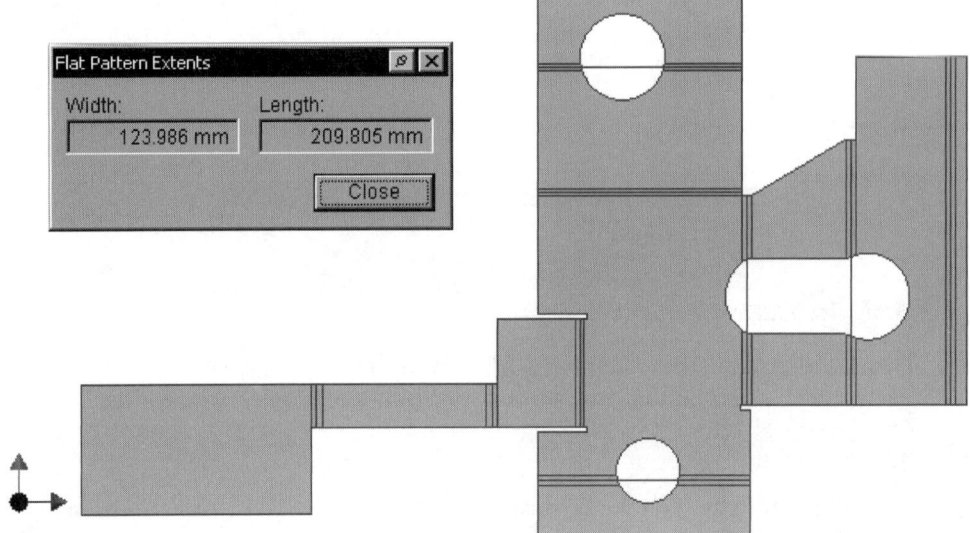

Figure 5.48 *Flat pattern aligned and extents displayed*

CONTOUR FLANGE

A way to construct a number of full length flanges in a single step is to construct a sketch depicting the cross section and construct a contour flange from the sketch.

 **TUTORIAL 5.3**

In this tutorial, you will learn how to construct contour flanges.

1. Start a new sheetmetal file. Use the metric template.

2. With reference to Figure 5.49, construct a sketch to depict the contour of a flange.

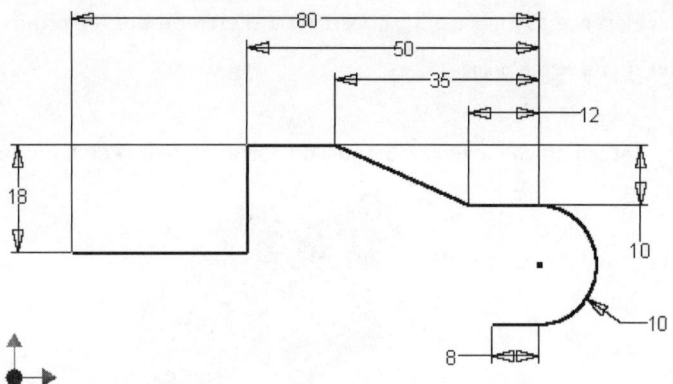

Figure 5.49 *Sketch depicting the contour of a flange*

Now construct contour flanges.

3. Select Return on the Inventor Standard toolbar to exit sketch mode.

4. Set the display to an isometric view.

5. Select Styles on the Sheetmetal Features Panel Bar or toolbar and set the sheet metal thickness to 1 mm.

6. Select Contour Flange on the Sheetmetal Features Panel Bar or toolbar.

7. In the Contour Flange dialog box, set the distance to 50 mm, click the Profile button, select the sketch, and then click the OK button. (See Figure 5.50.)

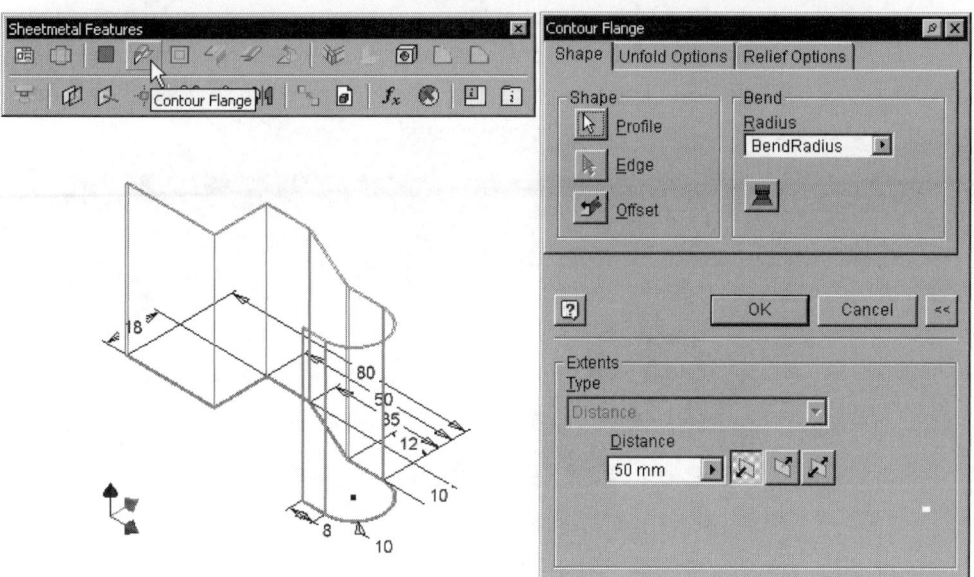

Figure 5.50 *Contour flange being constructed*

8. With reference to Figure 5.51, construct a sketch on the XY plane.

 Note: The sketch is a line.

9. Select Return on the Inventor Standard toolbar to exit sketch mode.

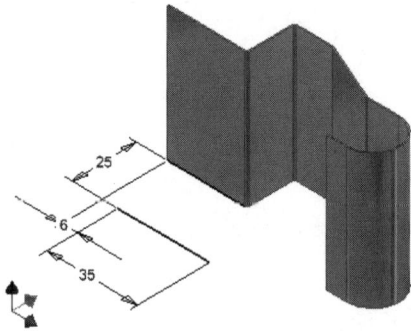

Figure 5.51 *Sketch constructed*

10. Select Contour Flange on the Sheetmetal Features Panel Bar or toolbar.
11. In the Contour Flange dialog box, select the Profile button (if it is not already selected) and select A in Figure 5.52.
12. Select the Edge button and select edge B in Figure 5.52.
13. Select the >> button.
14. Select Offset in the Type pull-down list box.
15. Click the Select Start Point button, select vertex C, and then set Offset1 to 10mm.
16. Click the Select End Point button, select vertex D, and then set Offset2 to 5 mm.
17. Click the OK button. A flange is constructed.

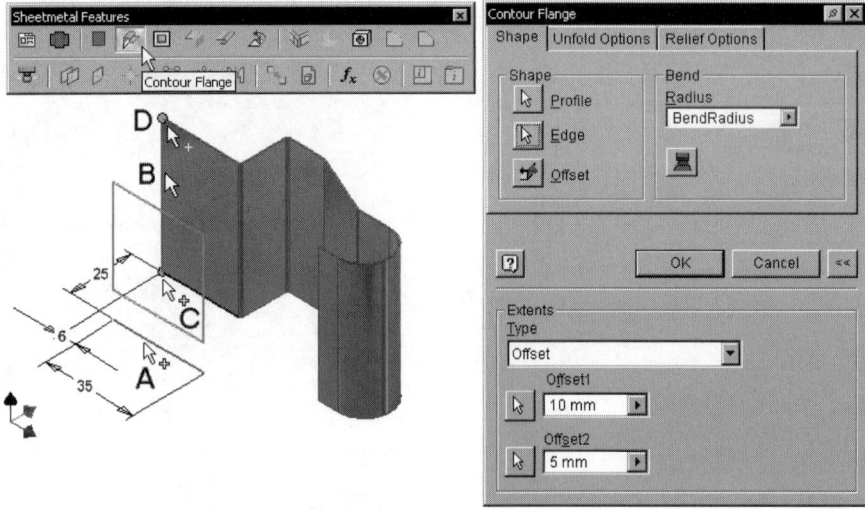

Figure 5.52 *Flange being constructed*

18. Select Bend on the Sheetmetal Features Panel Bar or toolbar.
19. Select edges A and B in Figure 5.53.
20. In the Bend dialog box, select Full Radius and click the OK button. A bend is constructed.

The sheet metal component is complete. Save and close your file (file name: *SheetmetalB.ipt*).

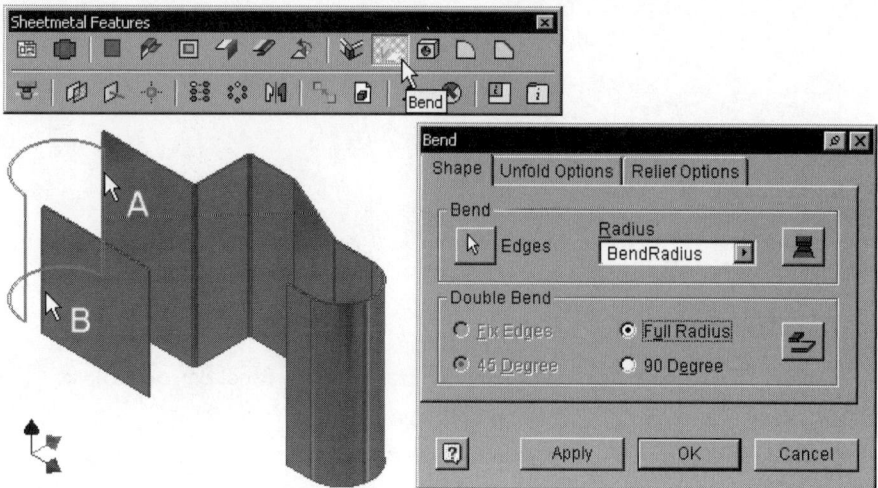

Figure 5.53 *Bend being constructed*

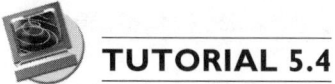

TUTORIAL 5.4

In this tutorial, you will learn how to construct a contour flange joining an edge of a sheet metal component.

1. Start a new sheetmetal part file. Use the metric template.
2. With reference to Figure 5.54, construct a sketch on the XY plane and make a face from the sketch. The thickness of the sheet metal is 1 mm.

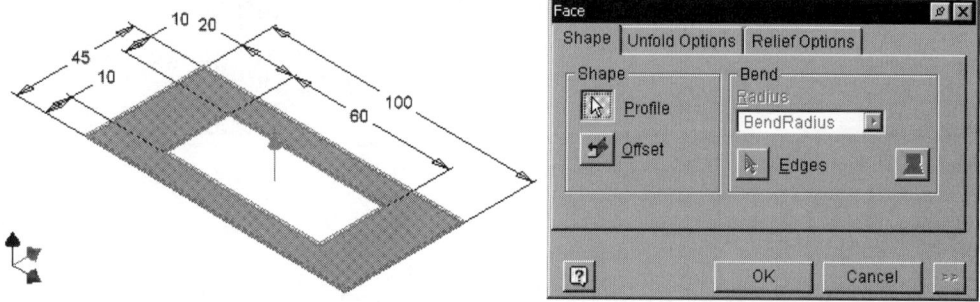

Figure 5.54 *Face being constructed*

Now construct a flange.

3. Construct a sketch on the XZ plane in accordance with Figure 5.55. In the sketch, construct two lines and add four dimensions.

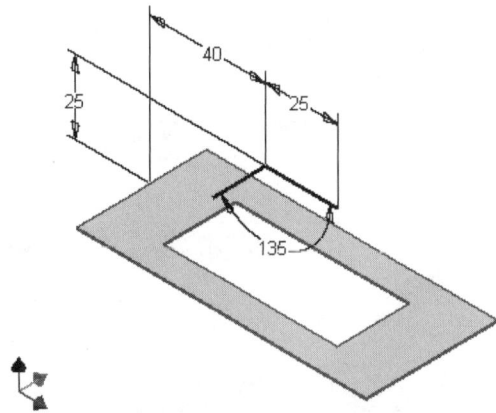

Figure 5.55 *Sketch constructed on XZ plane*

4. Select Contour Flange on the Sheetmetal Features Panel Bar or toolbar.
5. Select A as the profile and B as the edge. (See Figure 5.56.)
6. Click the OK button. A contour flange is constructed.

The sheet metal component is complete. Save and close your file (file name: *SheetmetalC.ipt*).

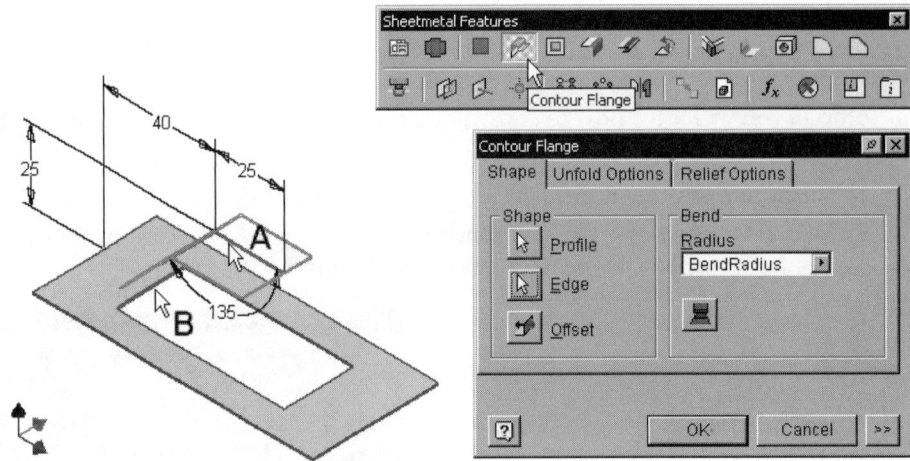

Figure 5.56 *Contour flange being constructed*

CONVERSION APPROACH

If you already have a solid part that can be unfolded to a flat pattern, you use the conversion approach to construct a sheet metal part.

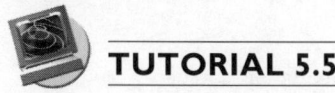

TUTORIAL 5.5

In this tutorial, you will construct a solid part and convert it to a sheet metal component.

1. Start a new part file. Use the metric template.
2. With reference to Figure 5.57, construct a sketch and extrude the sketch a distance of 30 mm.

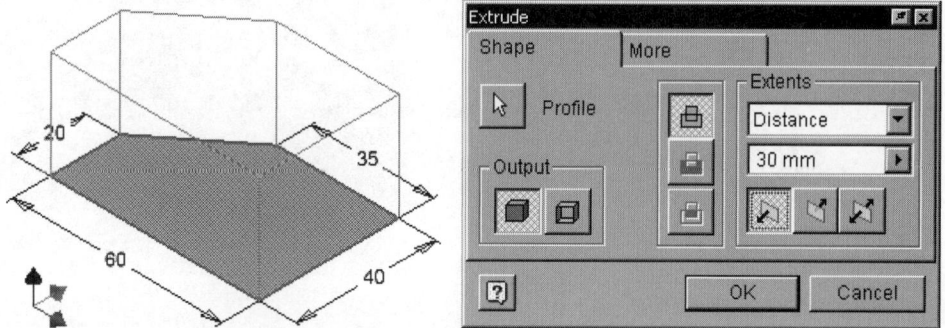

Figure 5.57 *Sketch being extruded*

3. Place a shell feature with a shell thickness of 1 mm. Remove faces A, B, and C in Figure 5.58.

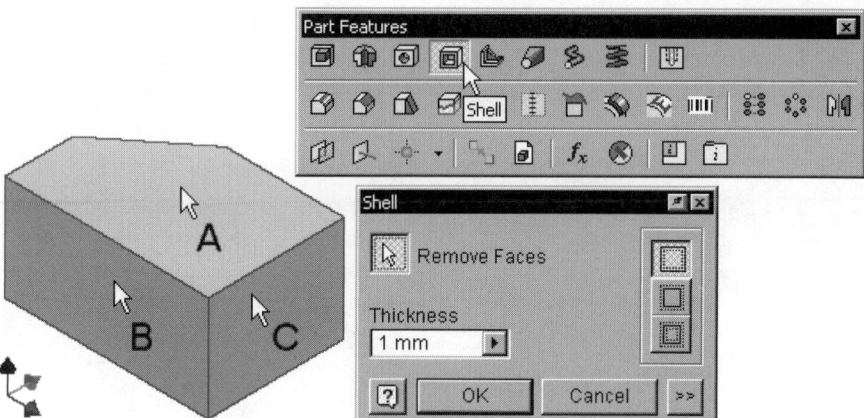

Figure 5.58 *Shell feature being placed*

Now convert the component to a sheet metal component.

4. Select Applications > Sheet Metal and then select Styles on the Sheetmetal Features Panel Bar or toolbar.
5. In the Sheet Metal Styles dialog box, set the thickness to 1 mm and click the Save button and then the Done button.

Tip: Shell thickness must equal to sheet metal thickness.

Now rip two edges.

6. Select Corner Seam on the Sheetmetal Features Panel Bar or toolbar.

7. In the Corner Seam dialog box, select the Corner Rip check box.

8. Select edges A and B in Figure 5.59 and click the OK button. Two edges are ripped.

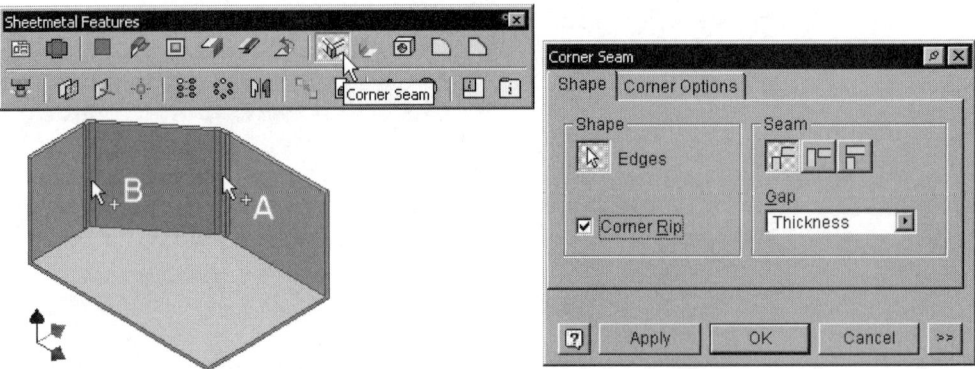

Figure 5.59 *Edges being ripped*

Now add three bends.

9. Select Bend on the Sheetmetal Features Panel Bar or toolbar.

10. Select edge A in Figure 5.60 and click the Apply button.

11. Select edge B in Figure 5.60 and click the Apply button.

12. Select edge C in Figure 5.60 and click the OK button. Three edges are converted to bends.

The sheet metal component is complete. Save and close your file (file name: *SheetmetalD.ipt*).

Figure 5.60 *Edges being converted to bends*

DESIGN ELEMENTS AND PUNCH TOOLS

In Chapter 4, you learned how to construct a design catalog and incorporate design elements in your new design. Here in a sheet metal model, you can use a design element as well.

TUTORIAL 5.6

In this tutorial, you will construct a sheet metal face and insert a design element and cut a slot by using the punch tool.

1. Start a new sheetmetal file. Use the metric template.
2. With reference to Figure 5.61, construct a sketch and make a face from the sketch. The thickness of the sheet metal is 1 mm.

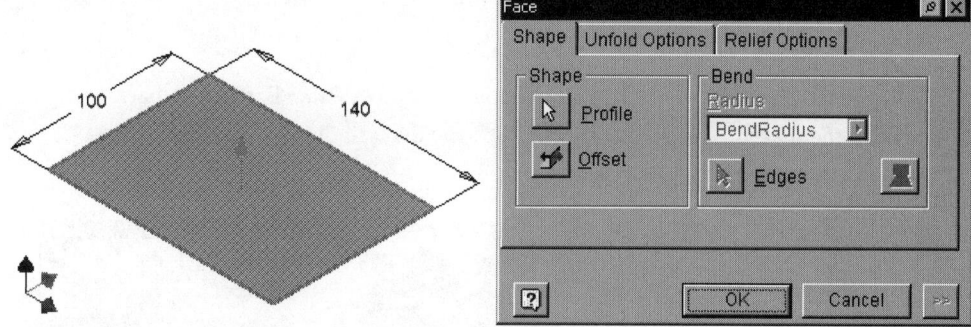

Figure 5.61 *Sketch constructed and a face being made from the sketch*

Now use the punch tool to cut an opening. First you will construct a center point for locating the punch tool.

3. Set up a sketch on face A in Figure 5.62 and construct a center point.

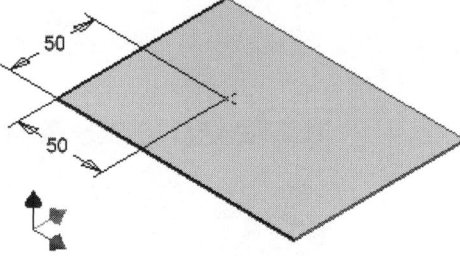

Figure 5.62 *Second sketch constructed*

4. Select Punch Tool on the Sheetmetal Features Panel Bar or toolbar. (See Figure 5.63.)
5. In the Punch Tool dialog box, select a punch tool (D-Sub connector2.ide), click the Next button, and then click the OK button. An opening is cut.

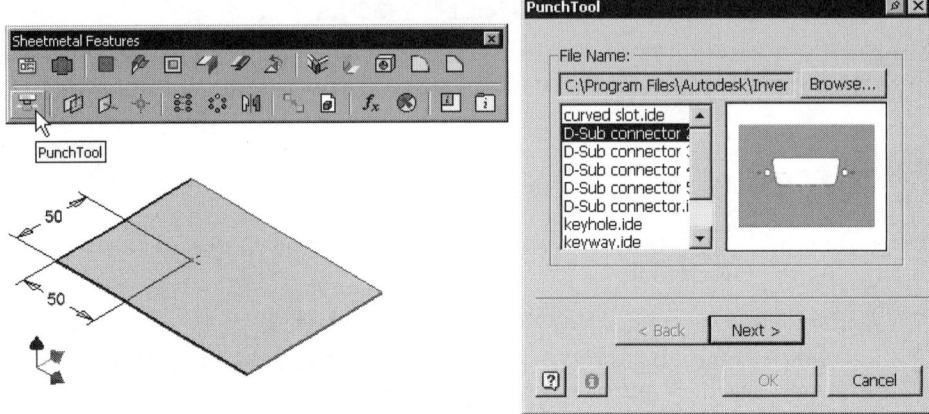

Figure 5.63 *Punch tool being used to cut an opening*

Now insert an iFeature.

6. Select Insert iFeature on the Sheetmetal Features Panel Bar or toolbar.

7. In the Insert iFeature dialog box, click the Browse button. (See Figure 5.64.)

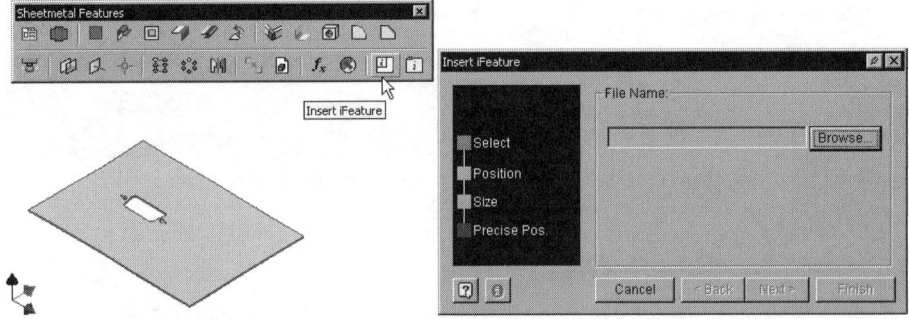

Figure 5.64 *Insert iFeature being activated*

8. In the Open dialog box, select the Punch folder, select the Round Emboss, and then click Open.

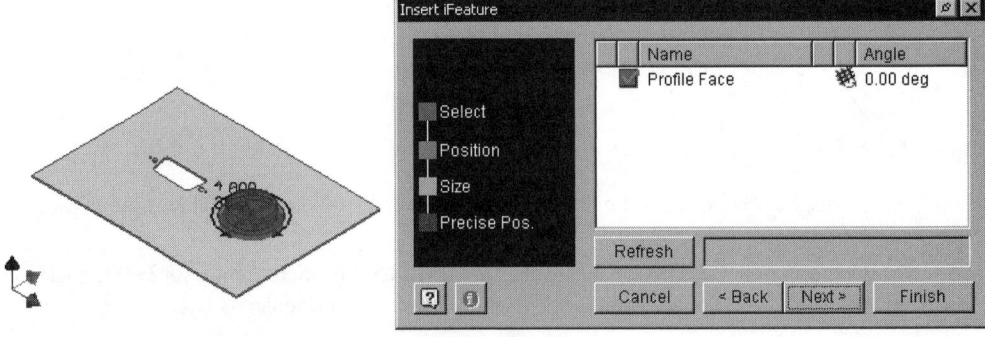

Figure 5.65 *iFeature selected from the catalog*

9. Click to select the top face of the sheet metal part, and then click Next twice.

10. Select Activate Sketch Edit Immediately and click the Finish button.

11. With reference to Figure 5.66, add two dimensions.

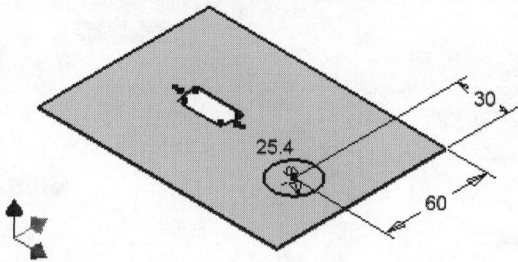

Figure 5.66 *Dimensions added to position the iFeature*

12. Select Update on the Inventor Standard toolbar. An iFeature is inserted. (See Figure 5.67.)

The sheet metal component is complete. Save and close your file (file name: *SheetmetalE.ipt*).

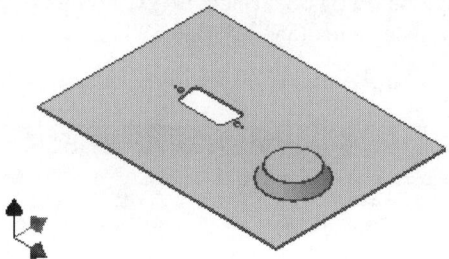

Figure 5.67 *iFeature inserted*

EXPORT FLAT PATTERN FROM MODEL

You can export the flat pattern of a sheet metal part in SAT, DWG, and DXF formats. If you output DWG or DXF formats, you can put the bend lines and bend tangent lines on separate layers. In the tutorial that follows, you will save the flat pattern of a sheet metal part to AutoCAD format.

 **TUTORIAL 5.7**

In this tutorial, you will export the flat pattern of a sheet metal component to AutoCAD format.

1. Open the file *SheetmetalA.ipt*.

2. Select Flat Pattern on the Browser Bar, right-click, and select Save Copy As. (See Figure 5.68.)

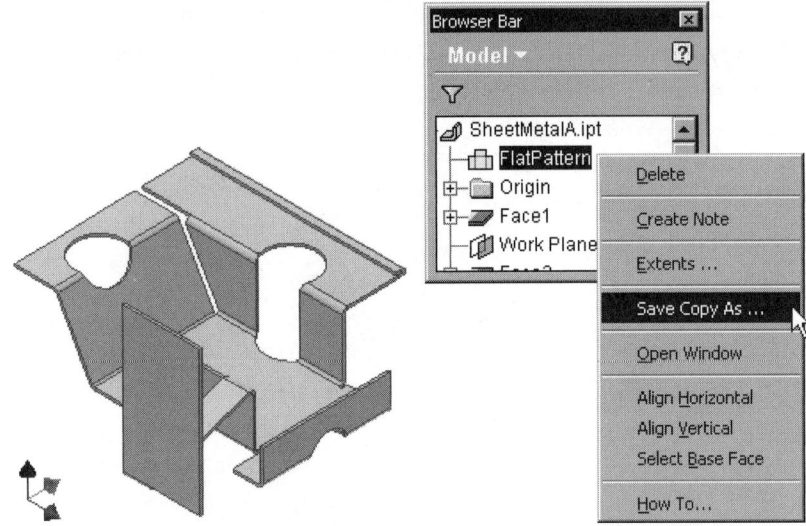

Figure 5.68 *Flat pattern selected on the Browser Bar*

3. In the Save Copy As dialog box, select DWG in the Save as type pull-down list box and specify a file name. (See Figure 5.69.)

The flat pattern is saved as an AutoCAD drawing. Close your file.

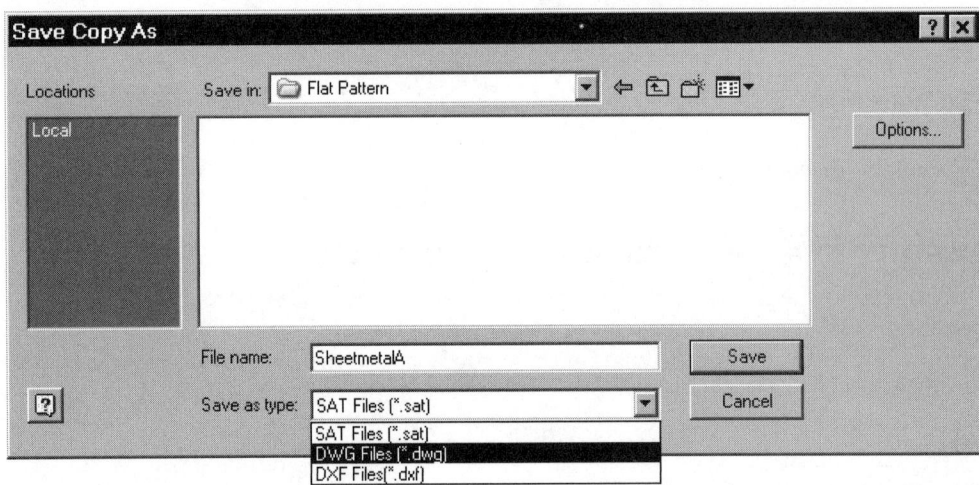

Figure 5.69 *Flat pattern being saved as an AutoCAD drawing*

SUMMARY

Sheet metal parts are a special kind of 3D solid object. Basically, a sheet metal part is a 3D object composed of a number of faces of uniform thickness that are joined together at rounded bends. To manage practical manufacturing needs, you need to place relief cuts at the joints, and you should design the sheet metal part in such a way that it can be developed into a flat pattern.

To construct a sheet metal part, you need to use a special set of tools as well as the general tools. Using the tools specific to sheet metal modeling, you construct sheet metal elements such as face, flange, contour flange, cut, fold, bend, corner seam, and hem. Using the general tools, you construct general and special iFeature, hole, corner round, corner chamfer, work features, rectangular pattern, circular pattern, and mirror features. To develop the sheet metal part for manufacturing, you use the flat pattern tool. Before making the bends in a sheet metal part, you determine the sheet metal settings. While you make a model, bends and relieves are placed automatically on the sheet metal part. Another way of making a sheet metal component is to first make a solid part and convert it to a sheet metal by shelling, ripping, and adding bends.

Similar to an ordinary 3D solid part, a sheet metal part can consist of three kinds of features: sketched solid features, placed solid features, and work features.

REVIEW QUESTIONS

1. How does a sheet metal part differ from an ordinary 3D solid part?

2. State the similarities and differences between the 3D part modeling tool set and the 3D sheet metal modeling tool set.

3. What are the two approaches to making a sheet metal part?

4. What kinds of features are specific to sheet metal modeling? Briefly explain each of them.

CHAPTER 6

Assembly Modeling I

OBJECTIVES

The aims of this chapter are to explain the key concepts of assembly modeling, to delineate the three design approaches to constructing an assembly, and to let you practice placing components (including library components), constructing components, and manipulating components in an assembly. This chapter also explains how to place an assembly pattern and construct a parts list. After studying this chapter, you should be able to

- Describe the key concepts of assembly modeling
- State the three design approaches in making an assembly
- Place existing components and library components in an assembly
- Construct new components in an assembly
- Establish an assembly pattern
- Manipulate components in an assembly
- Construct a parts list

OVERVIEW

An assembly is a collection of components put together properly to form a device that serves a purpose. To construct an assembly in the computer, you use an assembly file. Because you already used part files to construct the individual 3D solids and to store the information about them, you will use the assembly file for two main purposes: linking the components together and keeping the information regarding how the components are put together.

To construct an assembly to link to a set of components, you use the bottom-up, top-down, or the hybrid approach. Using one of these approaches, you construct an assembly file by placing in the assembly components that you already constructed, by constructing new components while you work on the assembly, or by placing existing components and constructing new components simultaneously.

Initially, components that you link to an assembly are free to be translated in 3D space. To establish proper spatial relationships between them, you manipulate them by applying assembly constraints to selected pair of faces, edges, or vertices. In circumstances where a component in an assembly is repeated in a rectangular array or circular array, or is assembled in multiple copies with a feature pattern of another component, you set up an assembly pattern. To facilitate design using standard parts, you incorporate a shared content link and insert the standards parts from the uniform resource locator. Using an assembly, you generate a bill of materials for use in database management and in constructing a parts list in an engineering drawing.

ASSEMBLY MODELING CONCEPTS

With the exception of very simple objects, such as a ruler, most objects have more than one part put together to form a useful whole. The set of parts put together is called an assembly. When you design the parts for an assembly, the relative dimensions and positions of parts, and the way they fit together, are crucial. You need to know whether there is any interference among the mating parts. If there is interference, you need to find out where it occurs; then you can eliminate it. Besides fitting the parts together, you also need to validate relative motions and check clearances if there are moving parts in the assembly. Moreover, you should critically evaluate the parts and the assembly as a whole to ensure that the assembly functions correctly in accordance with the design intent. To shorten the design lead time, you construct virtual assemblies in the computer to validate the integrity of the parts and the assembly.

COMPONENTS

For complex devices that have many parts, it is common practice to organize the parts into a number of smaller sub-assemblies such that each sub-assembly has fewer parts. Therefore, an assembly set consists of an assembly file and a number of part files or an assembly file together with a number of sub-assembly files and part files. Collectively calling the individual parts or sub-assemblies as components, you can define an assembly in the computer as a data set containing information about a collection of components linked to the assembly and about how the components are assembled together. Figure 6.1 shows an assembly of a model car, a set of components.

Figure 6.1 *An assembly of components*

PART FILE AND ASSEMBLY FILE

To construct an assembly in the computer, you start a new assembly file and connect a set of relevant part files and/or assembly files (sub-assemblies) to the assembly file. The part files store the information about the 3D objects, and the assembly files store the information about how you assemble the 3D objects together. (See Table 6.1.) Because the part files hold the definitions of the solid parts and link to the assembly file, any change you make to the part files will be incorporated in the assembly automatically.

Table 6.1 *Information*

File	Information
Part File	Definition and information about individual 3D solid part
Assembly File	Information about the locations of the linked components and how the linked components are assembled together

CONSTRUCTING AN ASSEMBLY

Constructing an assembly in the computer involves two major tasks: gathering a set of components in an assembly file and assembling the components by applying appropriate assembly constraints. How you gather a set of components in an assembly depends on which design approach you take, and how you apply assembly constraints depends on the shapes of the features of the components in the assembly.

DESIGN APPROACHES

There are three design approaches: bottom-up, top-down, (bottom being the parts, and top being the assembly) and hybrid. In a bottom-up approach, you construct all the component parts and then assemble them in an assembly file. In a top-down approach, you start an assembly file and construct the individual component parts while you are doing the assembly. The hybrid approach is a combination of the bottom-up and top-down approaches.

The Bottom-Up Approach

When you already have a good idea on the size and shape of the components of an assembly or you are working as a team on an assembly, you use the bottom-up approach. Through parametric solid modeling methods, you construct all the parts to appropriate sizes and shapes that best describe the components of the assembly. Then you start an assembly file and place the parts in the assembly. In the assembly, you align the components together by applying assembly constraints. After putting all the parts together, you analyze and make necessary changes to the parts.

The Top-Down Approach

Sometimes you have a concept in your mind, but you do not have any concrete ideas about the component parts. You use the top-down approach—you start an assembly file and design some component parts there. From the preliminary component parts, you improvise. The main advantages in using this approach are that you see other parts while working on an individual part, and you can continuously switch from designing one part to another.

The Hybrid Approach

In reality, you seldom use one approach alone. You use the bottom-up approach for standard component parts and new parts that you already know about, and you use the top-down approach to figure out new component parts with reference to the other component parts. This combined approach is the hybrid approach.

PART MODELING MODE AND ASSEMBLY MODELING MODE

No matter which design approach you use to construct an assembly, you can switch from assembly modeling mode to part modeling mode (and vice versa) whenever you feel it is appropriate. Selecting a component and double-clicking switches from assembly modeling mode to part modeling mode, working on the selected parts. When you are working in part modeling mode, the other components in the assembly are dimmed, indicating that they are unavailable. (Opacity of inactive component in an assembly depends on the Component Opacity setting in the Assembly tab of the Options dialog box.) To return from part modeling mode to assembly modeling mode, you select the Return button on the Inventor Standard toolbar. In assembly modeling mode, you use the assembly modeling tools to construct an assembly.

COMPONENTS IN AN ASSEMBLY

In an assembly file, the components are free to be translated in three linear directions and three rotational directions. You move and rotate them as if you were manipulating a real object. To impose restriction to the movements and to align a component properly with another component in the assembly, you apply assembly constraints.

Degrees of Freedom

Initially, the component parts (except the first component) that you place or create in an assembly are free to be translated in the 3D space, in three linear directions and three rotational directions. These free translations are called the six degrees of freedom (DOF). The DOF of a component is represented by a DOF symbol (see Figure 6.2). You discover the DOF of the objects in an assembly by selecting Degrees of Freedom from the View menu.

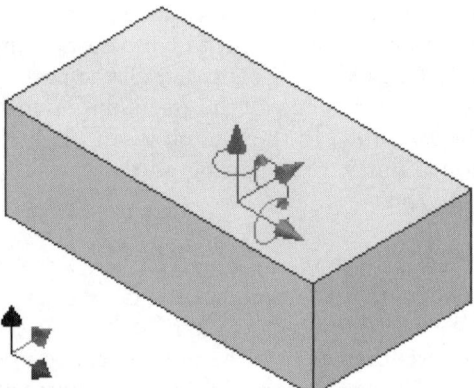

Figure 6.2 *Six degrees of freedom of a free object*

Grounding

By default, the first component part that you place or create in the assembly is fixed in the 3D space. We call a fixed object a grounded object; it has no degree of freedom—you cannot move it. To free a grounded object, you select the object, right-click, and deselect Grounded. (See Figure 6.3.) On the other hand, if you want to fix an object in 3D space, you ground the object by right-clicking and selecting Grounded.

Translation of Objects in 3D Space

Unless a component part is grounded or constrained, it is free to be translated in the 3D space. To translate the component parts in an assembly to their appropriate locations, you move or rotate them. Note that moving or rotating a component does not affect a component's DOF. You simply put it in a new position and new orientation.

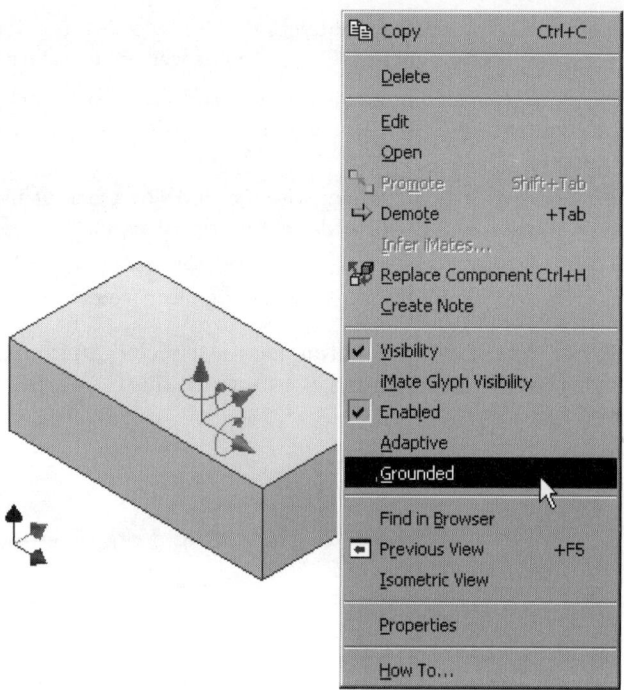

Figure 6.3 *Grounded selected*

APPLYING ASSEMBLY CONSTRAINTS

To restrict the movement of a component in 3D and to align it with another component in the assembly, you apply assembly constraints to selected faces, edges, and vertices of parts and the origin (with three work axes along the X, Y, and Z directions) of the part file or the assembly file.

There are four kinds of assembly constraints: mate, angle, tangent, and insert. The kind of constraint to be applied to a pair of components depends on the design intent of the assembly and the function and shape of the individual components.

Mate Constraint

A mate constraint causes two selected objects (face, edge, or vertex) to mate or flush at a specified offset distance. You can mate a face to a face, an edge to an edge, a point to a point, an edge to a face, a vertex to a face, and a vertex to an edge. If you mate a face to a face, you have to decide how the faces face each other, in the same direction or in the opposite directions.

By mating a face of a component with six DOFs to a face of another component, you remove three DOFs (two degrees of rotation freedom and one degree of linear freedom).

By mating an edge of a component with six DOFs to an edge of another component, you remove four DOFs (two degrees of rotation freedom and two degrees of linear freedom).

By mating a vertex of a component with six DOFs to a vertex of another component, you remove three DOFs (three degrees of linear freedom).

By mating an edge of a component with six DOFs to a face of another component, you remove two DOFs (one degree of rotation freedom and one degree of linear freedom).

By mating a vertex of a component with six DOFs to a face of another component, you remove one DOF (one degree of linear freedom).

By mating a vertex of a component with six DOFs to an edge of another component, you remove two DOFs (two degrees of linear freedom).

Figure 6.4 shows (from top to bottom) mating face A to face B, mating edge A to edge B, mating vertex A to vertex B, mating edge A to face B, mating vertex A to face B, and mating vertex A to edge B.

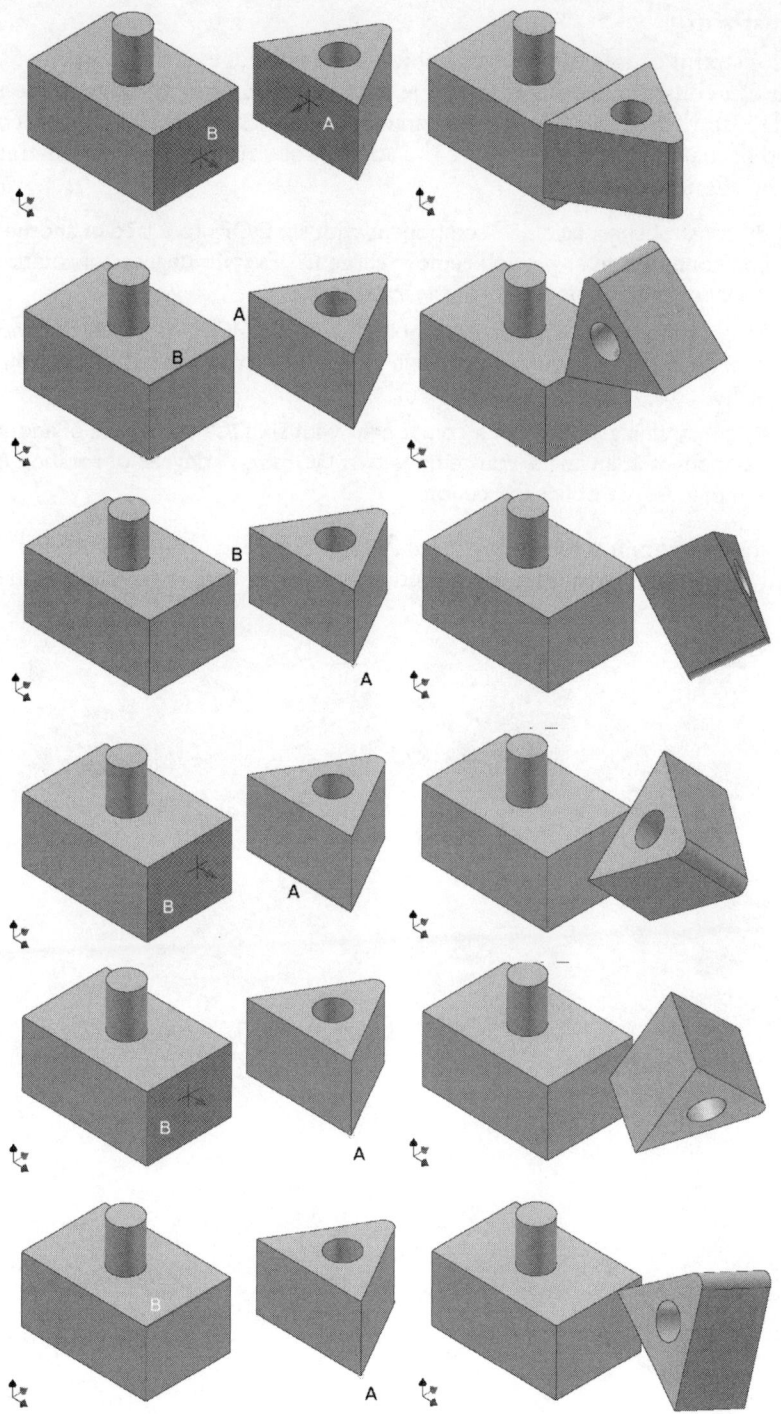

Figure 6.4 *Mate constraints*

Angle Constraint

An angle constraint causes two selected objects (face or edge) to be aligned at a specified angle. You can constrain a face at an angle with another face, an edge at an angle with another edge, and a face at an angle with an edge. You can consider an angle constraint a special kind of mate constraint, because the angle constraint causes the constrained component to be aligned at an angle.

> By constraining a face of a component with six DOFs to a face of another component at an angle, you remove three DOFs (two degrees of rotation freedom and one degree of linear freedom).

> By constraining an edge of a component with six DOFs to an edge of another component at an angle, you remove four DOFs (two degrees of rotation freedom and two degrees of linear freedom).

> By constraining an edge of a component with six DOFs to a face of another component at an angle, you remove two DOFs (one degree of rotation freedom and one degree of linear freedom).

Figure 6.5 shows (from top to bottom) face A constrained at an angle with face B, edge A constrained at an angle with edge B, and edge A constrained at an angle with face B.

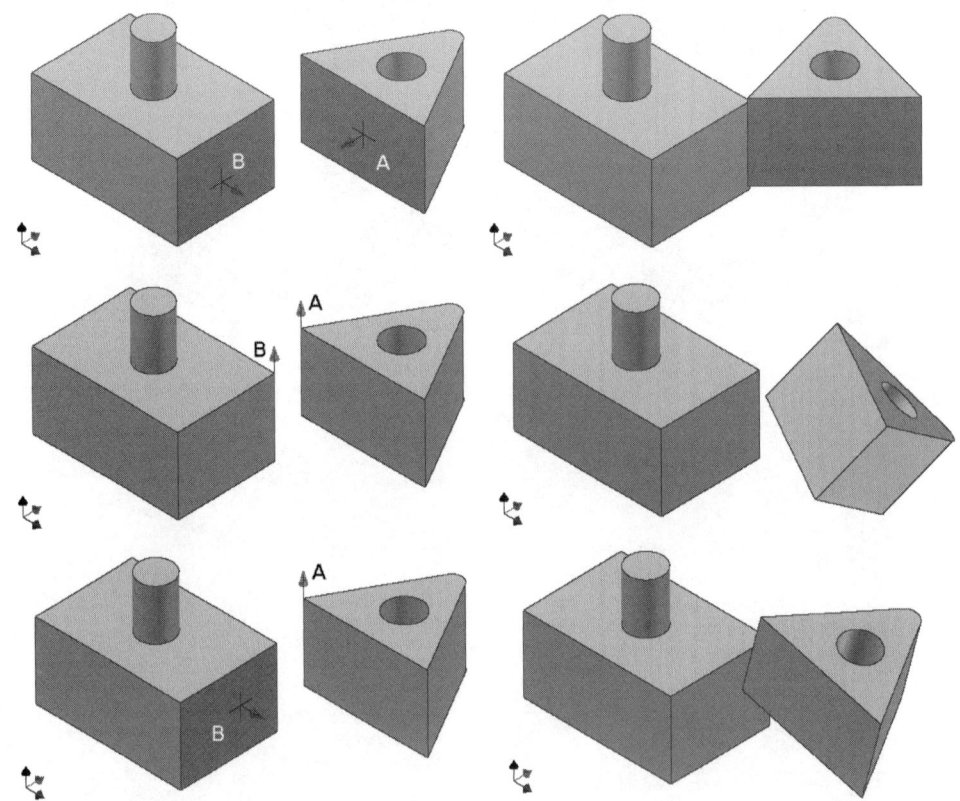

Figure 6.5 *Angle constraints*

Tangent Constraint

A tangent constraint causes selected faces, planes, cylinders, spheres, and cones to make contact at their tangential point and at a specified offset distance; you select faces.

> By applying tangent constraint to a cylindrical face of a component with six DOFs to a face of another component, you remove two DOFs (one degree of rotation freedom and one degree of linear freedom).

> By applying tangent constraint to a spherical face of a component with six DOFs to a face of another component, you remove one DOFs (one degree of linear freedom).

Figure 6.6 shows (from top to bottom) a cylindrical face constrained to be tangent to a planar face and a spherical face constrained to be tangent to another spherical face.

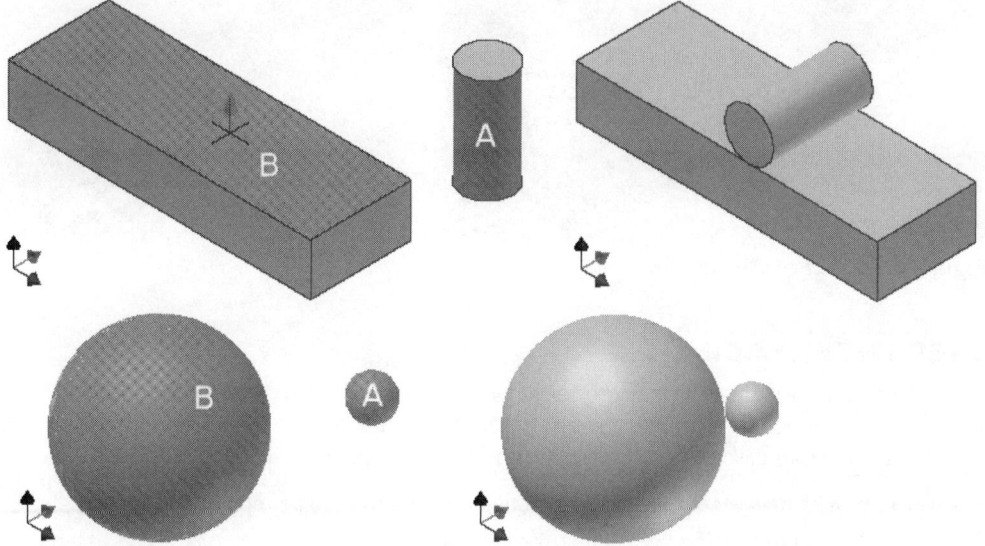

Figure 6.6 *Tangent constraints*

Insert Constraint

An insert constraint causes selected circular edges to align face to face and concentrically at a specified distance—you select circular edges. An insert constraint is a combination of two mate constraints: mating an axis of a cylindrical object with the axis of another cylindrical object and mating the end face of the cylindrical object with the end face of another cylindrical object.

Figure 6.7 shows (from top to bottom) a cylindrical object inserted in another cylindrical object in an opposed direction, and a cylindrical object inserted in another cylindrical object in an aligned direction.

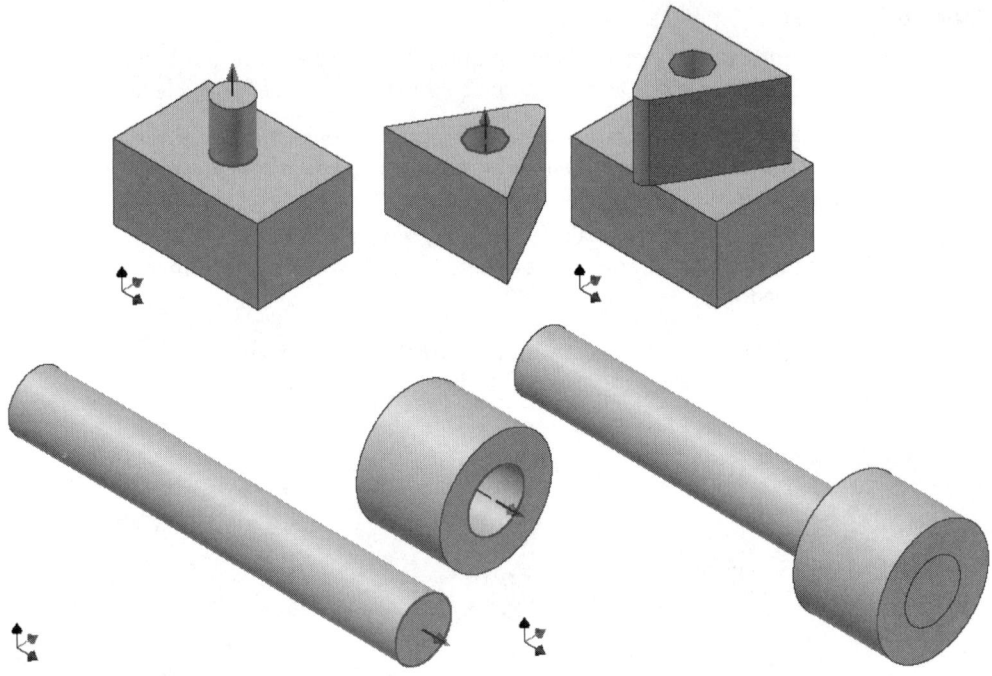

Figure 6.7 *Insert constraints*

USER INTERFACE

To construct an assembly of components, you use an assembly file. To familiarize yourself with Autodesk Inventor's assembly modeling user interface, you will start an assembly file. To start an assembly file, use the following steps. (See Figure 6.8.)

1. Start a new assembly file by selecting New from the File menu.
2. In the New dialog box, select the Default tab.
3. Select the *Standard.iam* template and click the OK button.

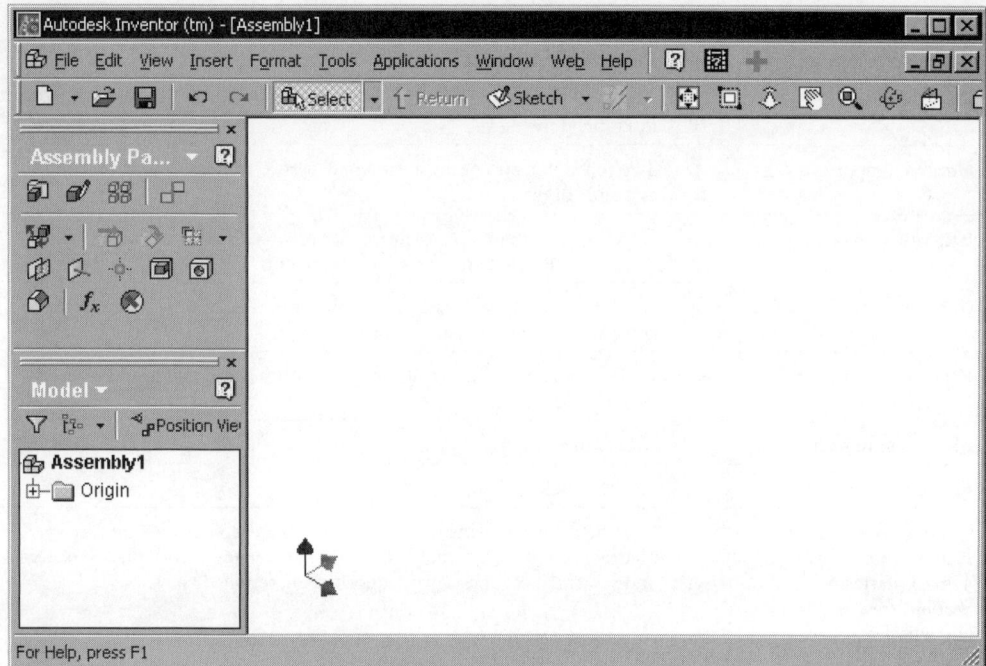

Figure 6.8 *New assembly file started*

ASSEMBLY PANEL BAR AND TOOLBAR

The user interface of an assembly file is very similar to that of a part file, except that the Assembly Panel Bar is in place rather than the Sketch or Features Panel Bar. In addition to the Assembly Panel Bar, tools for constructing an assembly are available on the Assembly Panel toolbar. To display the Assembly Panel toolbar, select View > Toolbar > Assembly Panel or select Tools > Customize, select Assembly Panel from the Toolbars tab and then select the Show button. Figure 6.9 shows the Assembly Panel Bar and toolbar.

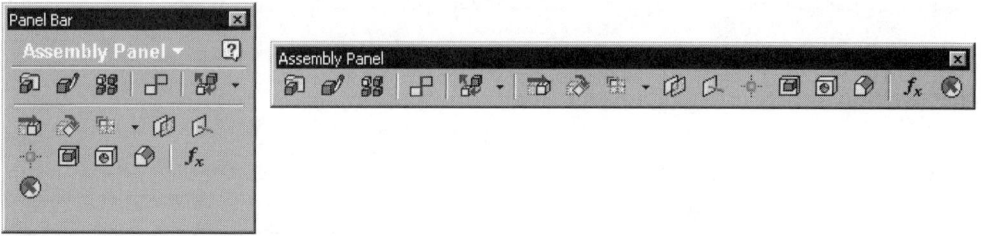

Figure 6.9 *Assembly Panel Bar and toolbar*

The Assembly Panel Bar and toolbar have sixteen button areas. Table 6.2 describes the choices:

Table 6.2 *Assembly Panel Bar and toolbar options*

Button	Function
Place Component	The bottom-up approach; places existing parts or sub-assemblies in the assembly drawing.
Create Component	The top-down approach; starts a new part or a new sub-assembly in the assembly drawing.
Pattern Component	Repeats a component in a rectangular or circular array or constructs an associative assembly pattern of a component.
Constraint	Applies assembly constraints, motion constraint, or transitional surface constraint to selected pair of objects.
Replace Component/Replace All	Replaces a component or all occurrence of the component with another component.
Move Component	Moves to a new location.
Rotate Component	Rotates a selected object in 3D space.
Quarter Section View/ Half Section View/ Three Quarter Section View/ Unsectioned View	Sets the display to a quarter section view, half section view, or three quarter section view through selected sketch planes, work planes, or faces and resets the section view to a normal unsectioned view.
Work Plane	Constructs a work plane in the context of an assembly.
Work Axis	Constructs a work axis in the context of an assembly.
Work Point	Constructs a work point in the context of an assembly.
Extrude	Constructs an extruded solid feature in the context of an assembly.
Hole	Constructs a hole in the context of an assembly.
Chamfer	Constructs a chamfer in the context of an assembly.
Parameters	Sets model parameters in the context of an assembly.
Create iMate	Constructs an assembly interface in the context of an assembly.

SHORTCUT KEYS

The assembly shortcut keys are listed in Table 6.3 below:

Table 6.3 *Assembly modeling shortcut keys*

Shortcut Key	Functions
P	Places component in an assembly.
C	Constructs component in an assembly.
ALT + Drag	Adds mate constraint.

SELECTION PRIORITY

Depending on which kind of objects are selected in the Selection Priority pull-down list of the Inventor Standard toolbar shown in Figure 6.10, placing the cursor on an assembly will select an assembly (or sub-assembly), part (in the assembly or in an sub-assembly), a feature of a part, a face of a part, or a sketch of a sketched feature of a part.

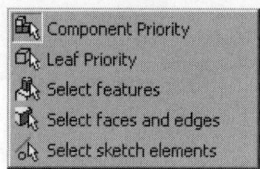

Figure 6.10 *Selection priority in an assembly file*

ASSEMBLY BROWSER BAR

The assembly Browser Bar has two panels, the Model panel and the Library panel. Like the Browser Bar of a part file, the Model panel of the assembly Browser Bar shows objects in the assembly in a hierarchy, and initially it has an origin that consists of the YZ, XZ, and XY planes, X, Y, and Z axes, and the center point. After you place or create a component in the assembly, the component's origin will be displayed on the Browser Bar. If you apply assembly constraints to the components, the constraint will also appear here.

The Model panel of the assembly Browser Bar shown in Figure 6.11 displays two components placed or constructed in an assembly with an insert constraint applied to them.

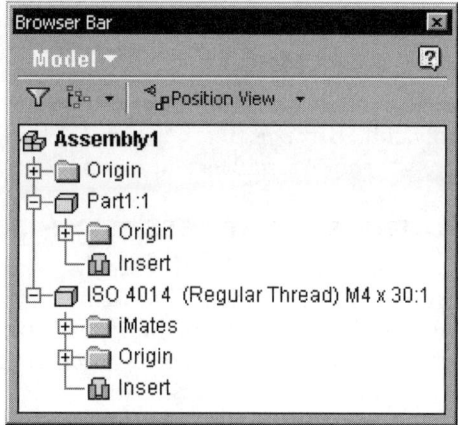

Figure 6.11 *Assembly Browser Bar showing the Model panel*

In the Library panel, you will find various kinds of standard components that you can place in your assembly. (See Figure 6.12.)

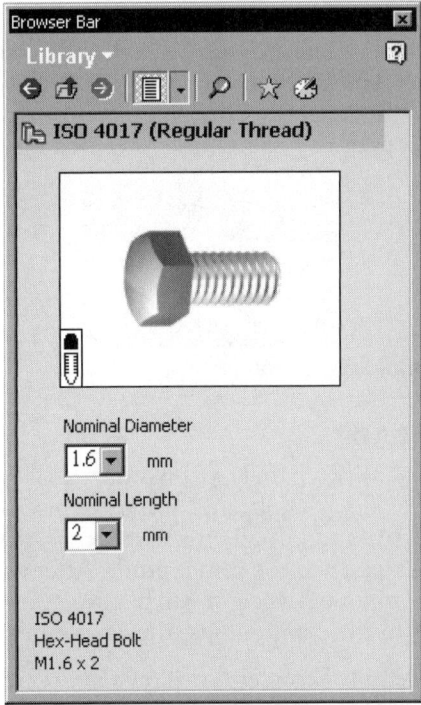

Figure 6.12 *Assembly Browser Bar showing Library panel*

HIERARCHY OF AN ASSEMBLY

The hierarchy of an assembly refers to the sequence of placement or construction of components in the assembly. The Browser Bar in Figure 6.13 shows a typical hierarchy of an assembly in which there are sub-assemblies and component parts.

RE-ORDERING COMPONENTS IN THE ASSEMBLY HIERARCHY

A component that is placed or constructed earlier in the assembly will be placed in a higher position in the hierarchy. When you apply assembly constraints to a pair of objects, an object located lower in the hierarchy will be translated toward the object located higher in the hierarchy unless it is grounded. To re-order the hierarchy, you select the object on the Browser Bar and drag it to a new position.

RESTRUCTURING COMPONENTS IN THE ASSEMBLY HIERARCHY

In a large assembly that has many parts organized into a number of sub-assemblies, you can reorganize a component in the assembly into a sub-assembly or reorganize a component in a sub-assembly into the assembly. (See Figure 6.14.)

REPLACING COMPONENTS

Sometimes you need to replace a component in an assembly with another component. You might use a placeholder component in the initial design stage and replace it with a standard component in the final assembly, or you might replace a component from one vendor with one from another. During replacement, you can replace a single instance of the component or all occurrence of the component. After replacement, assembly constraints might have to be re-applied if the replacement component has a different shape than the original component.

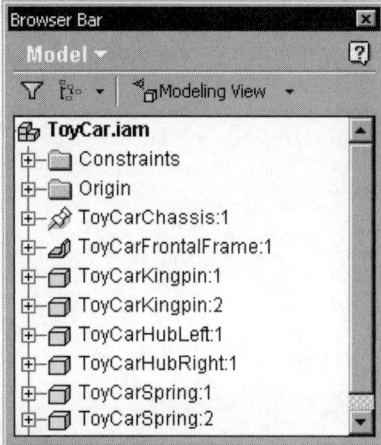

Figure 6.13 *Assembly hierarchy*

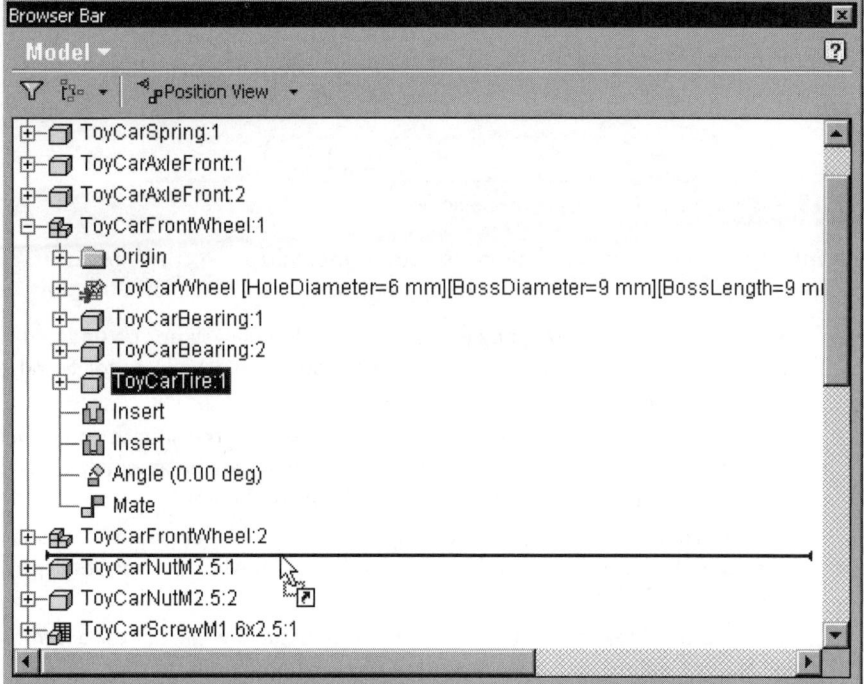

Figure 6.14 *Restructuring components in an assembly*

SYSTEM OPTIONS

Before you use the assembly modeling tools to construct assemblies of solid parts, you should spend some time studying the related system settings. To set assembly options, select Application Options from the Tools menu. (See Figure 6.15.)

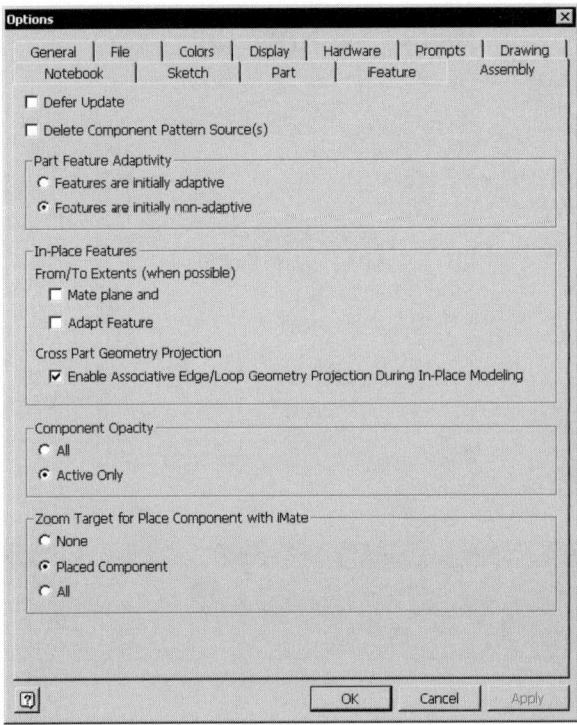

Figure 6.15 *Assembly tab of the Options dialog box*

The Assembly tab of the Options dialog box lets you set various settings regarding assembly modeling. The functions of the check boxes on this tab are explained below:

Defer Update If selected, any change in the assembly will not be updated automatically. You need to select the Update button on the Inventor Standard toolbar to update the assembly manually.

Delete Component Pattern Source(s) If selected, the source component of an assembly pattern will also be deleted when you delete an assembly pattern. (You will learn about assembly patterns in the next chapter.)

Part Feature Adaptivity This area specifies whether new features of solid parts constructed in the assembly are adaptive. (You will learn about adaptive technology in Chapter 9.)

Mate Plane If selected, whenever you construct a new part in the assembly, the new feature will mate to a selected face of a component in the assembly.

Adapt Feature If selected, whenever you construct a new part in the assembly, the new feature will adapt in size to the feature size of a selected face of a selected component in the assembly.

Enable Associative Edge/Loop Geometry Projection During In-Place Modeling If selected, whenever you construct a new part in the assembly, a reference sketch projected from the geometry of a selected feature of a selected part in the assembly will be constructed.

Component Opacity This area determines whether all components or only the active component will be displayed as opaque when an assembly cross section is displayed.

Zoom Target for Place Component with iMate This area specifies zoom behavior for the graphics window when components are placed with iMates.

THE BOTTOM-UP APPROACH

Now you will learn how to construct an assembly by using the bottom-up approach. In the bottom-up approach, you construct a number of 3D solid parts. Then you start an assembly file and place the components in the assembly file.

You select Place Component on the Assembly Panel Bar or toolbar and select a solid part or assembly from the Open dialog box. In the Open dialog box, you select a part file or assembly file and place the file in the assembly file. If you select an assembly, it becomes a sub-assembly in the current assembly. After you place components (solid part files or assembly files), you translate them to their appropriate location and apply an assembly constraint to put them together.

PLACE COMPONENT

The first task in constructing an assembly with the bottom-up approach is to place the components that you already constructed in an assembly file. To facilitate subsequent placement of assembly constraints, you might have to relocate them to their appropriate position.

 TUTORIAL 6.1

In this tutorial, you will construct the assembly of the punch set, which consists of fourteen components: PunchBolt, PunchHandle, PunchBush, PunchIndexBush, PunchHinge, PunchPin, PunchIndex, PunchIndexKnob, PunchColumn, PunchBase, PunchTable, PunchPost, PunchSpring, and PunchIndexSpring (see Figure 6.16). Because you already constructed these components in Chapters 3 and 4, you will place them in an assembly file and add assembly constraints to put them together.

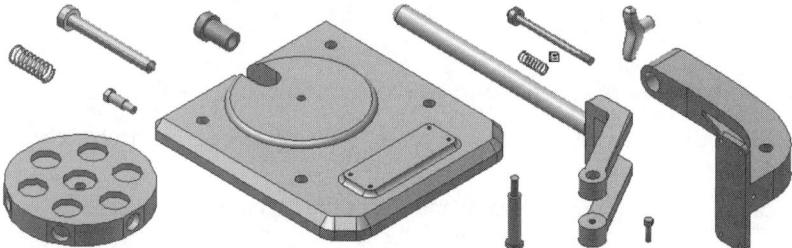

Figure 6.16 *Components of the punch set*

1. Select New from the File menu or select New on the Inventor Standard toolbar.

2. In the New dialog box, select the Metric tab.

3. Select Standard.iam and click the OK button.

4. Select Place Component on the Assembly Panel Bar or toolbar. (Alternatively, you can place the cursor in the blank space of the graphics area, right-click, and select Place Component.)

5. In the Open dialog box, select the file *PunchBase.ipt* from the *Punch* folder where you saved the components in Chapters 3 and 4, and click Open.

The display now changes to an isometric view automatically, and the selected component is placed in the assembly with the cursor at the selected component. If you select another location in the graphics area, you will place another copy of the selected component in the assembly.

6. Right-click and select Done.

By default, the first component placed or created in the assembly is grounded. Therefore, you find a pushpin symbol next to the PunchBase on the Browser Bar shown in Figure 6.17. It denotes that the object is grounded.

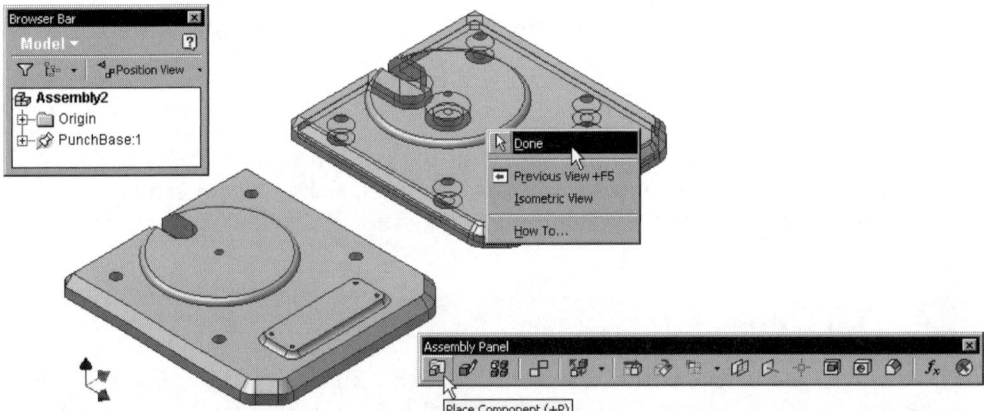

Figure 6.17 *First component being placed in the assembly*

 Tip: A grounded component has no degree of freedom at all. To remove the grounding effect, select the component, either in the graphics screen or on the Browser Bar, right-click, and deselect Grounded. On the other hand, you can ground a component by selecting it, right-clicking, and selecting Grounded.

Now add two more components to the assembly and display the DOF symbols.

7. Move the cursor over the blank space of the graphics area, right-click, and select Place Component.

8. Select the file (PunchColumn) and select the Open button.

9. Select a location in the blank space of the graphics area to place the component. (See Figure 6.18.)

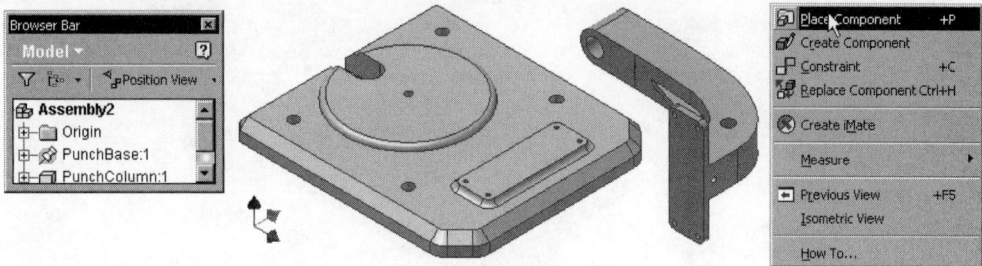

Figure 6.18 *Second component placed in the assembly*

10. Right-click and select Done. Another component is placed.

11. Repeat steps 4 through 7 to place the file (PunchTable) in the assembly.

12. Select View > Degrees of Freedom. The DOF symbols are displayed. (See Figure 6.19.)

 Note: The grounded component does not have any symbol because it is not free to rotate or be translated, and the other two components are free to rotate about three axes and be translated along three axes.

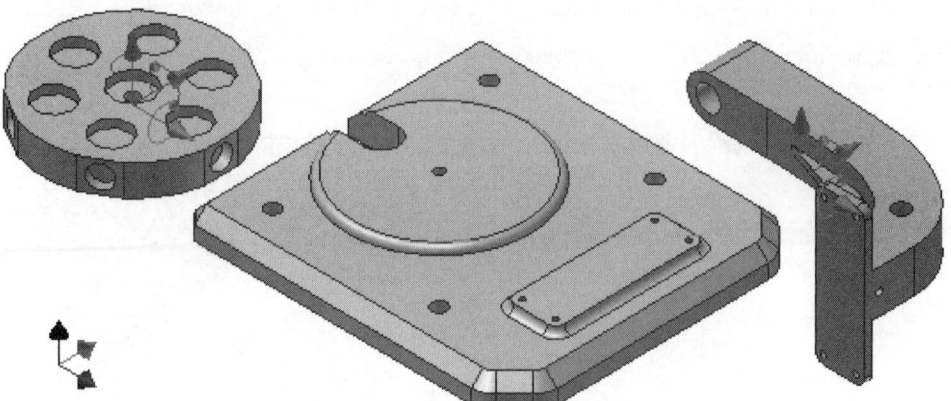

Figure 6.19 *Third component placed in the assembly and DOF symbols displayed*

Now translate the components by moving and rotating.

13. Select Move Component on the Assembly Panel Bar or toolbar.

14. Select the punch column (PunchColumn), hold down the left mouse button, and drag the cursor to a new position. The selected component is moved. (See Figure 6.20.) When you are finished, right-click and select Done.

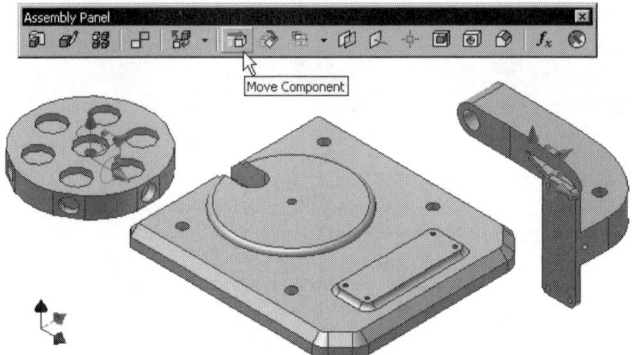

Figure 6.20 *Component being moved*

15. Select Rotate Component on the Assembly Panel Bar or toolbar.

16. Select the table (PunchTable).

17. Select a point inside the rotate symbol and drag to make a free rotation or select the horizontal or the vertical axis of the rotate symbol and drag to rotate horizontally or vertically. (See Figure 6.21.) When you are finished, right-click and select Done.

Now hide the DOF symbols.

18. From the View menu, deselect the Degrees of Freedom checkmark.

Three components are placed and their positions are changed. Save your file (file name: *Punch.iam*).

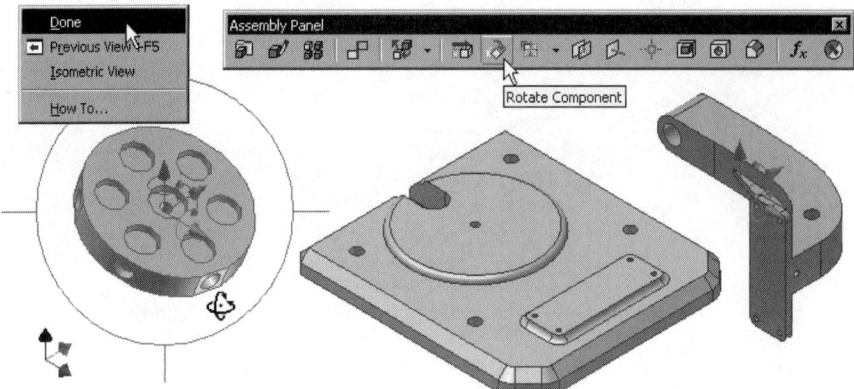

Figure 6.21 *Component being rotated*

APPLY ASSEMBLY CONSTRAINT

Components in an assembly, except the grounded ones, are free to be translated along three axes and rotate about three axes. To restrict their relative movement and to align them to their proper position and orientation, you apply assembly constraints to selected pairs of geometric references on the components.

After you apply an assembly constraint to an assembly, you will find an icon on the Browser Bar depicting the type of assembly constraints (mate, angle, tangent, or insert). If you move the cursor over the icon, a tooltip will be displayed telling you the constraint type, offset value, direction, and part names affected by the constraint. To modify a constraint's offset value, you can double-click it on the Browser Bar and modify the offset value in the Edit Dimension dialog box. If you want to modify the type of constraint, select the icon on the Browser Bar, right-click, and select Edit. If you want to determine the other half of the constraint, select the icon on the Browser Bar, right-click, and select Other Half. (Other half means the matching part participating in the assembly constraint.)

TUTORIAL 6.2

In this tutorial, you will learn how to apply assembly constraints to components of an assembly.

1. Open the assembly file *Punch.iam*, if you already closed it.

2. Select Constraint on the Assembly Panel Bar or toolbar. (You can also move the cursor over the blank space of the graphics area, right-click, and select Constraint.)

Tip: Checking the Show Preview button causes the selected component to be translated to its constraint position. For the sake of clarity, Show Preview is disabled in the tutorials in this book.

3. In the Place Constraint dialog box, select the Mate button, if it is not already selected, and clear the Show Preview button.

4. Select faces A and B in Figure 6.22 and click the Apply button.

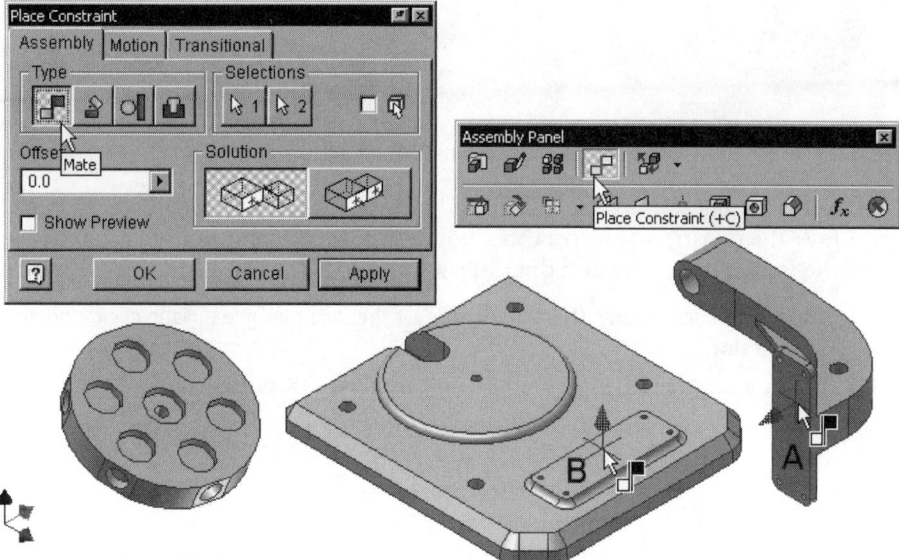

Figure 6.22 *Mate constraint being placed*

You will hear an audio signal denoting that the selected faces are constrained. The bottom face of the column is mated to a face of the base. Now place two more mate constraints to properly position the column of the punch set.

5. Select axes A and B in Figure 6.23 and click the OK button.

Tip: To select a feature, move the cursor to a position near the feature, pause a moment to wait for the selection dialog box to appear, select either the right or left arrow to cycle through the possible features, and select the center button of the selection dialog box when the required feature is highlighted. If the required feature is not highlighted after you cycle through all the possible features, move the cursor away from the feature and move it back to the feature to try again.

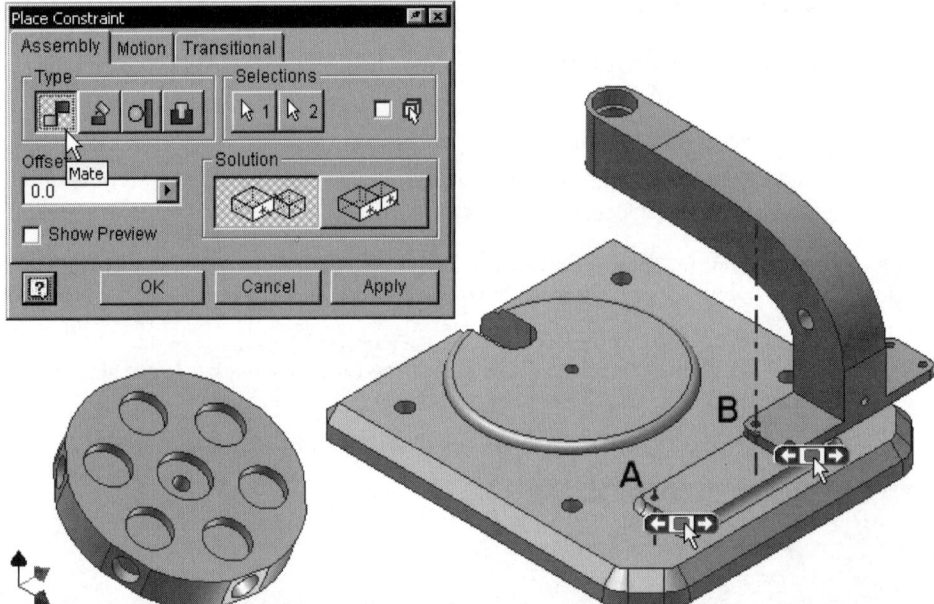

Figure 6.23 *Mate constraint being applied to constrain an axis to an axis*

Two mate constraints (face to face and axis to axis) are applied. To get a better idea of how the constraints restrict the movement of the column in relation to the base of the punch set, select and drag the column to a new location. (See Figure 6.24.)

6. Move the cursor over the blank space of the graphics area, right-click, and select Constraint.

7. Select axes A and B in Figure 6.24 and click the OK button.

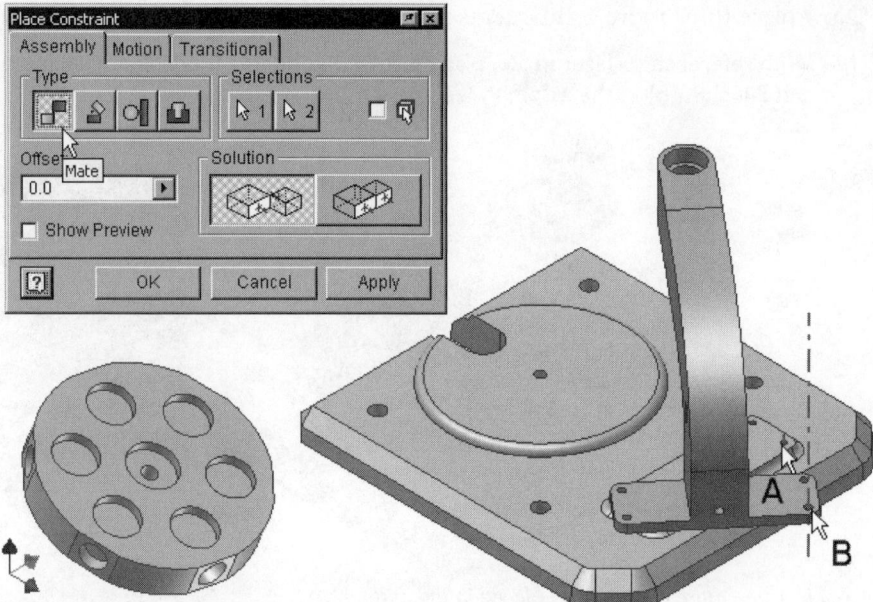

Figure 6.24 *Mate constraint being applied to constrain two axes*

Now apply an insert constraint.

8. Rotate the punch table.
9. Select Constraint on the Assembly Panel Bar or toolbar.
10. Select the Insert button in the Place Constraint dialog box.
11. Select circular edges A and B in Figure 6.25 and click the OK button.

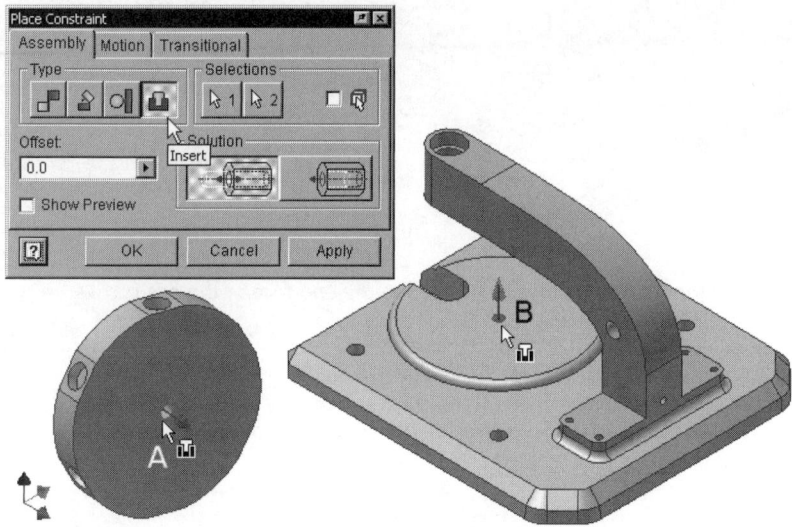

Figure 6.25 *Insert constraint being applied to circular edges*

Now place three more components in the assembly.

12. With reference to Figure 6.26, place three components (PunchPin, PunchSpring, and PunchBush) in the assembly.

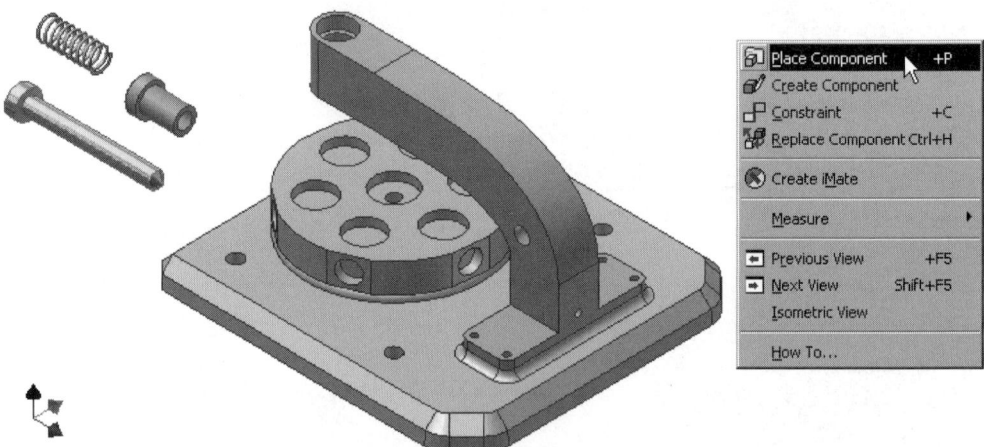

Figure 6.26 *Three more components placed in the assembly*

Now place assembly constraints to assemble the three components together.

13. Select Constraint on the Assembly Panel Bar or toolbar.

14. Select end face A of the spring and end face B of the bush in Figure 6.27 and click the Apply button.

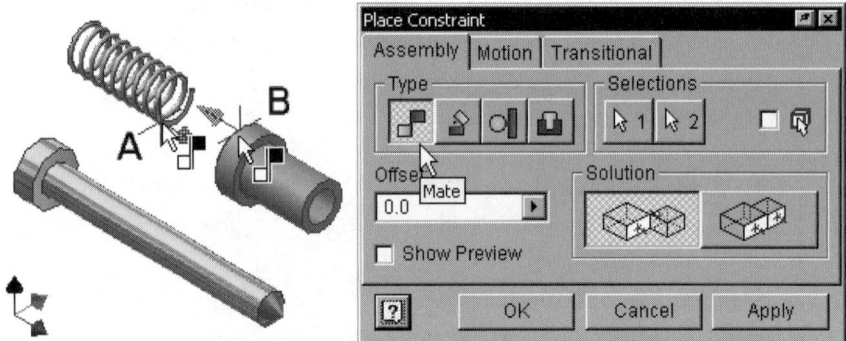

Figure 6.27 *End face of the spring being mated to the end face of the bush*

15. Select the X axis of the spring on the Browser Bar (A in Figure 6.28), select axis B in Figure 6.28, and click the Apply button.

Note: The axis of the helical path of the spring is coincident with the X axis.

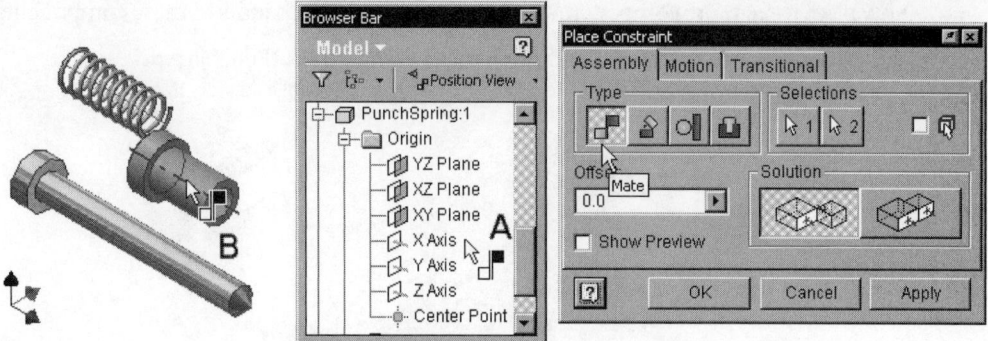

Figure 6.28 *X axis of the spring being mated to the axis of the bush*

16. Select end face A of the pin and end face B of the spring in Figure 6.29 and click the Apply button.

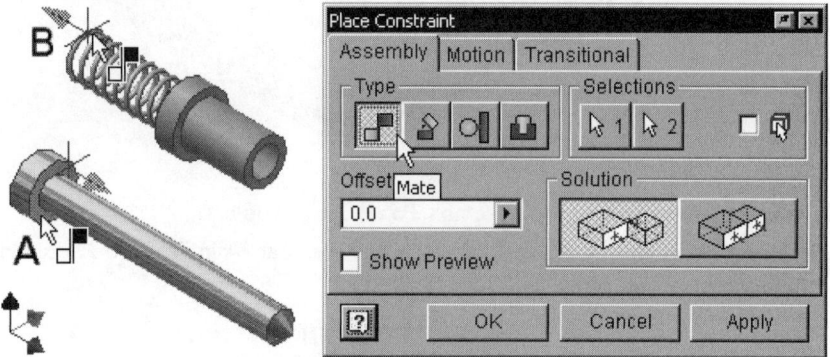

Figure 6.29 *End face of the spring being mated to the end face of the pin*

17. Select axes A and B in Figure 6.30 and click the OK button.

The three components (PunchPin, PunchSpring, and PunchBush) are assembled.

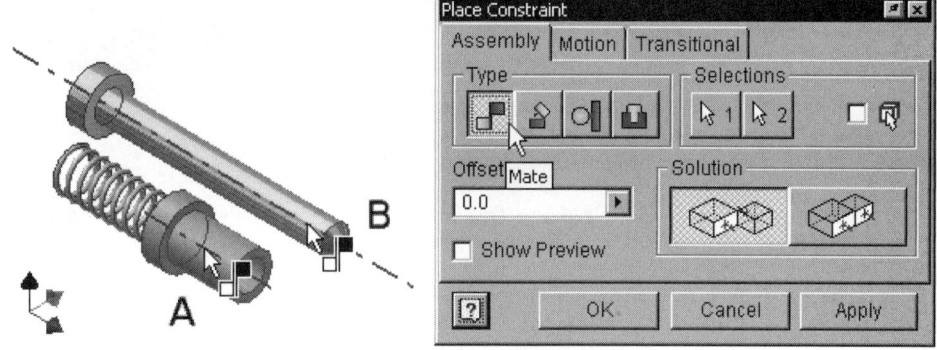

Figure 6.30 *Axis of the pin being mated to the axis of the bush*

Now place three more components and apply assembly constraints to these components.

18. With reference to Figure 6.31, place three components (PunchIndex, PunchIndexSpring, and PunchIndexKnob) in the assembly.

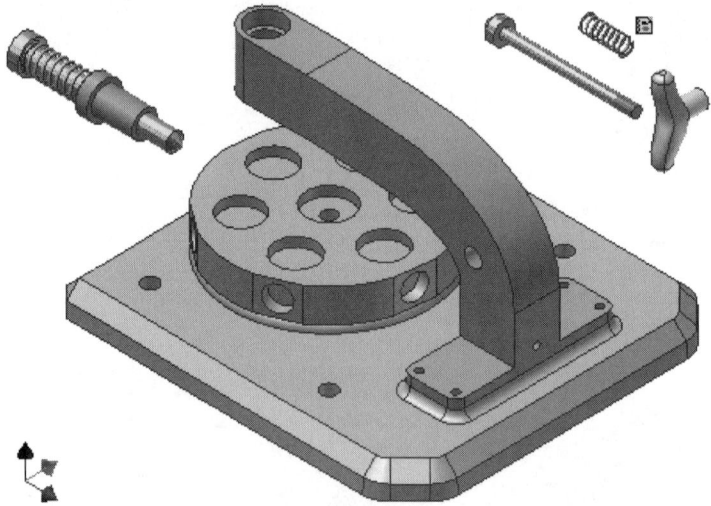

Figure 6.31 *Three components placed in the assembly*

19. Select Constraint on the Assembly Panel Bar or toolbar.

20. Select the X axis of the spring on the Browser Bar (A in Figure 6.32), select axis B in Figure 6.32, and click the Apply button.

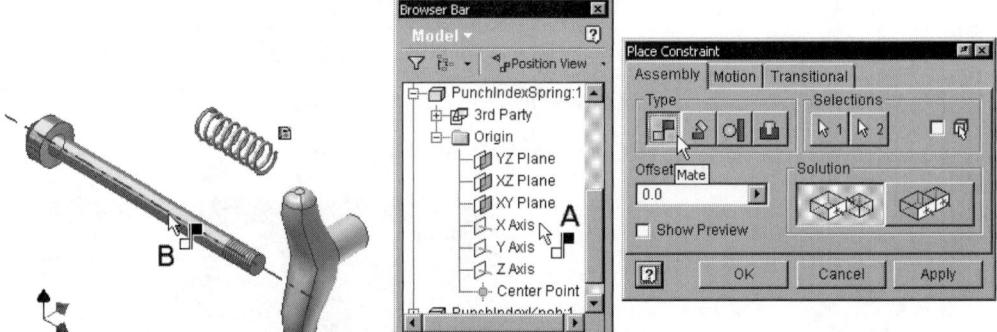

Figure 6.32 *Axis of the spring being mated to the axis of the index pin*

21. Select end face A of the pin and end face B of the spring in Figure 6.33 and click the Apply button.

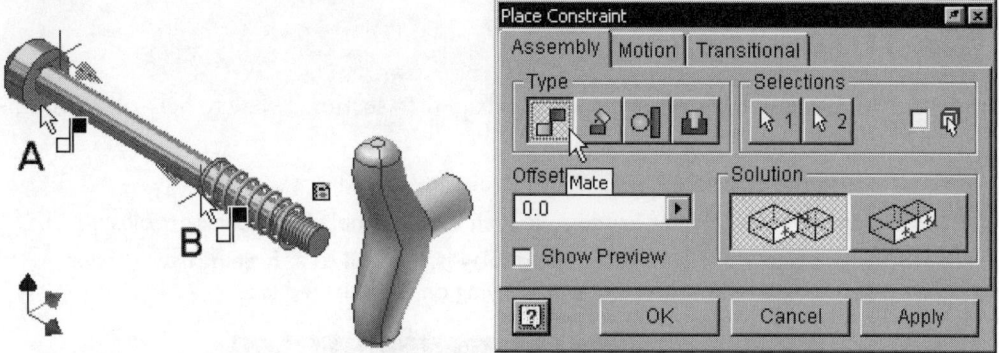

Figure 6.33 *End face of the spring being mated to the end face of the index pin*

22. Select the Insert button in the Place Constraint dialog box and set the offset value to –12 mm.

23. Select circular edges A and B in Figure 6.34 and click the OK button.

Three components (PunchIndex, PunchIndexSpring, and PunchIndexKnob) are assembled. Save your file.

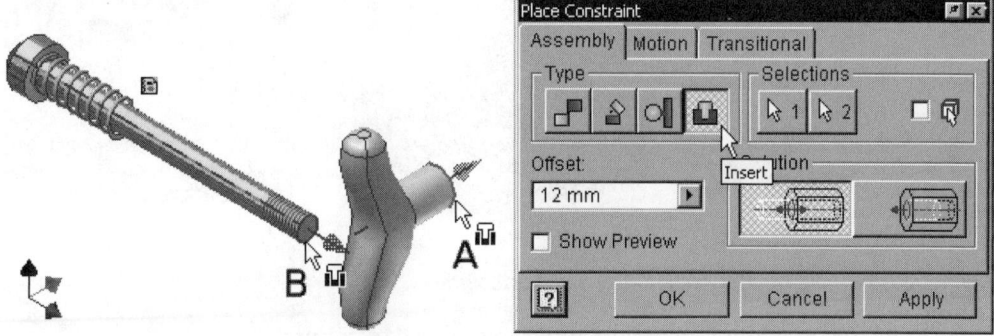

Figure 6.34 *Insert constraint being applied to circular edges of the index pin and the knob*

DISPLAY SECTION VIEW

While you construct an assembly, you have to select objects from individual components. To facilitate selection, you can set the display to either shaded mode or wireframe mode. In addition, you can set the display to three kinds of section view: quarter section, half section, and three quarter section. To set the display to a half section view, select Half Section View on the Assembly Panel Bar or toolbar and select a plane, work plane, or face to define the section plane. To set the display to a quarter or three quarter section view, select Quarter Section View or Three Quarter Section View on the Assembly Panel Bar or toolbar. Then select two mutually perpendicular planes, work planes, or faces to specify the section planes.

TUTORIAL 6.3

In this tutorial, you will learn how to set the display to sectional view to help place assembly constraints.

1. Open the assembly *Punch.iam*, if you already closed it.
2. Select Three Quarter Section View on the Assembly Panel Bar or toolbar.
3. Select face A in Figure 6.35. The display is changed to a three quarter section section view with the cutting plane lying on the selected face.

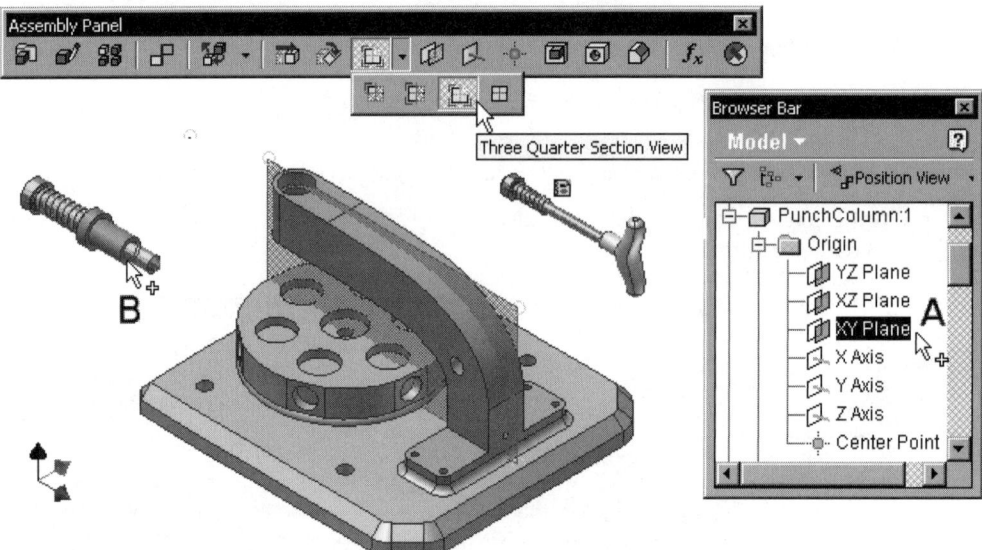

Figure 6.35 *Three quarter section view being established*

Now place a component and apply assembly constraints.

4. Place the component (PunchPost) into the assembly. (See Figure 6.36.)

 Tip: Because the display is set to three quarter section, you will not see the placed component if you put it in the sectioned zone. If you do not find the placed component in the graphics area but it appears on the Browser Bar, you have probably placed it in the sectioned zone. The placed component will reappear after you end the section view display.

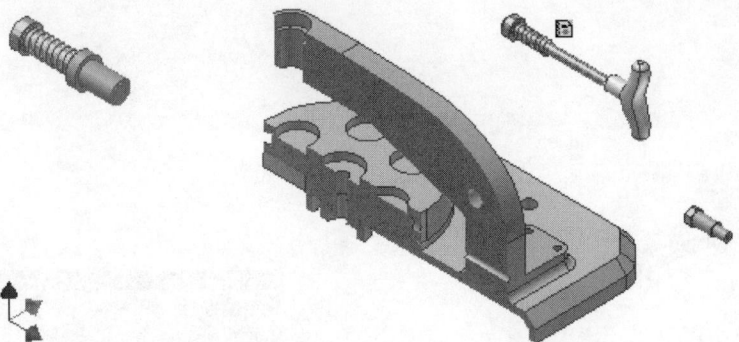

Figure 6.36 *Display set to three quarter section and a component placed (outside the sectioned zone)*

5. Select Constraint on the Assembly Panel Bar or toolbar.
6. Select the Insert button in the Place Constraint dialog box.
7. Select circular edges A and B in Figure 6.37 and click the Apply button.

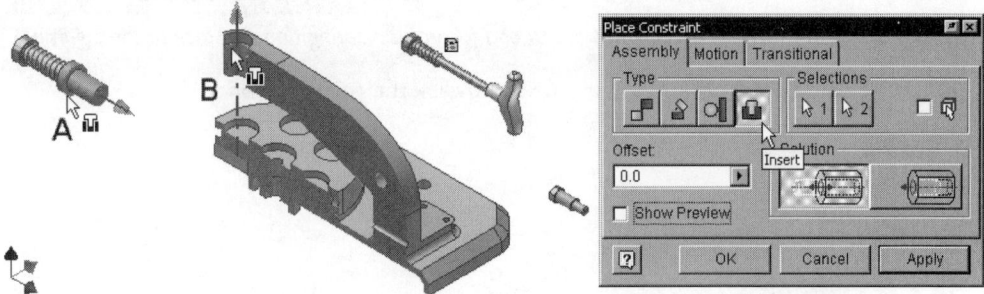

Figure 6.37 *Insert constraint being applied to circular edges of the post and the table*

8. Select circular edges A and B in Figure 6.38 and click the OK button.

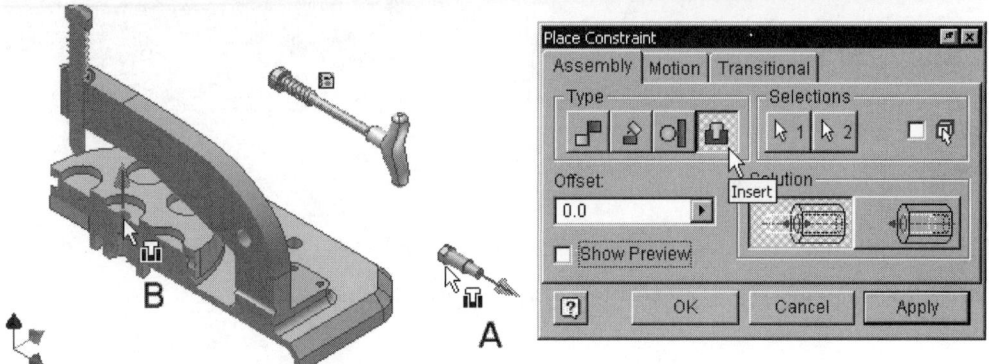

Figure 6.38 *Insert constraint being applied to circular edges of the pin and the column*

9. Select the Mate button in the Place Constraint dialog box.
10. Select end face A of the index spring and face B in Figure 6.39 and click the Apply button.

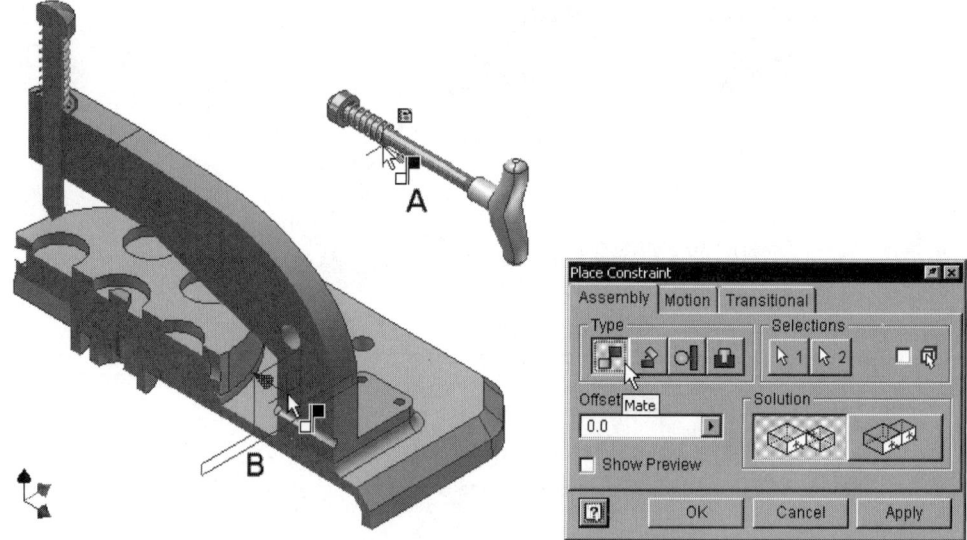

Figure 6.39 *Mate constraint being applied to end face of the spring and end face of the column*

11. Select axes A and B in Figure 6.40 and click the Apply button.

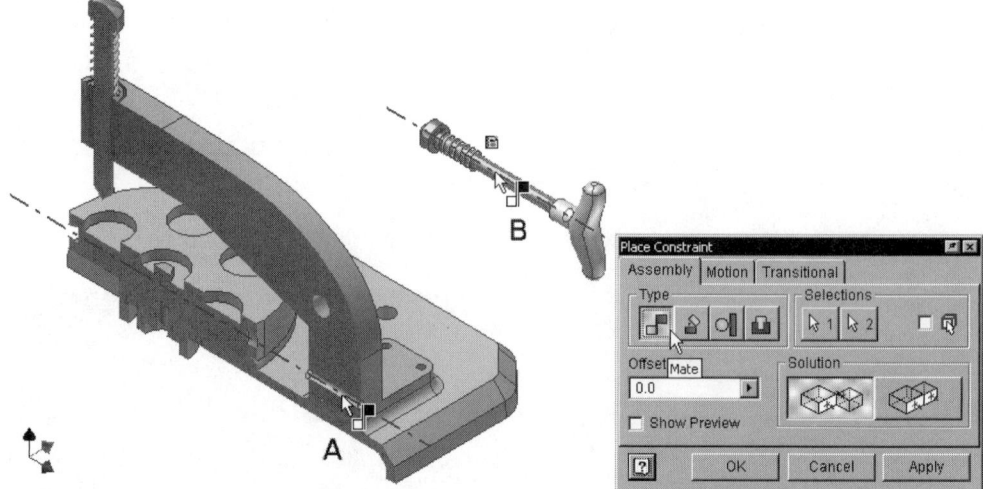

Figure 6.40 *Mate constraint being applied to axis of the index pin and axis of the column*

12. Select axes A and B in Figure 6.41 and click the OK button.

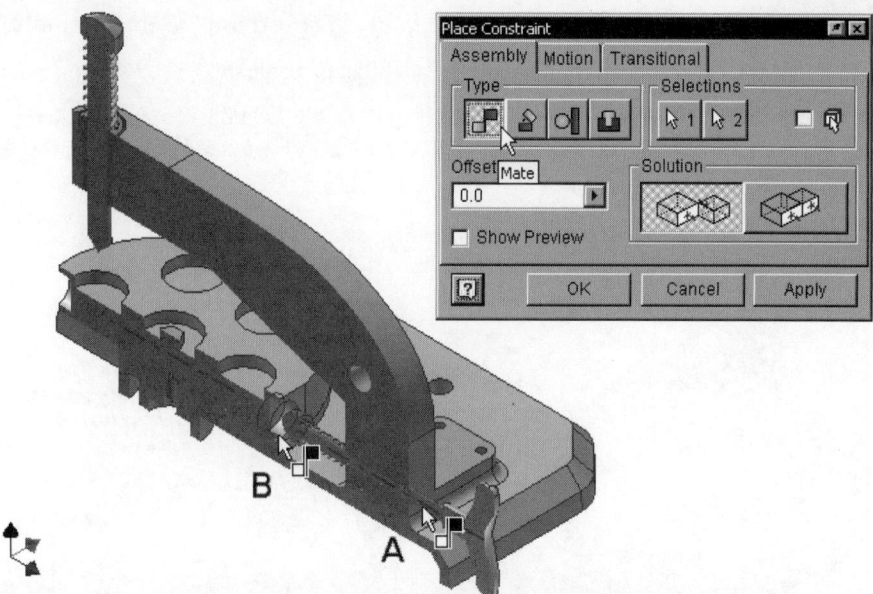

Figure 6.41 *Mate constraint being applied to axis of the table and the index*

Now you will end section view display.

13. Select End Section View on the Assembly Panel Bar or toolbar.

14. Place two more components (PunchHandle and PunchHinge) in the assembly. (See Figure 6.42.)

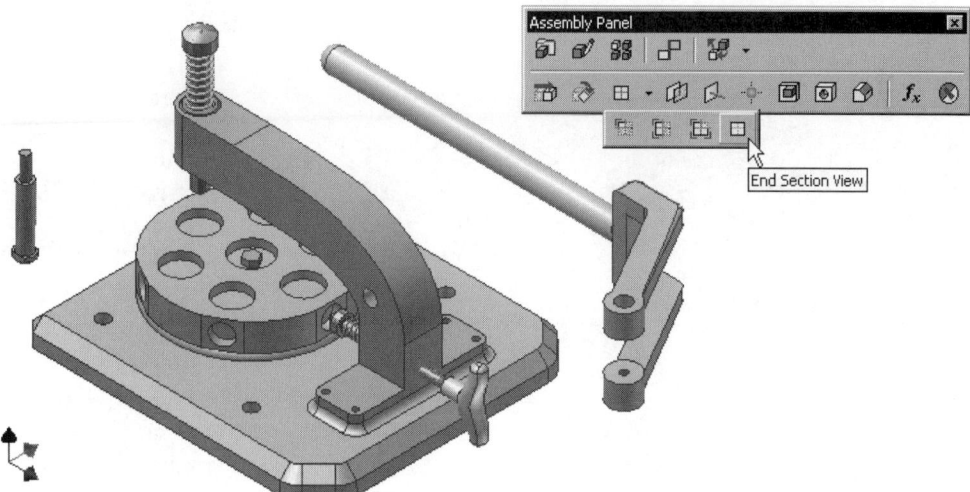

Figure 6.42 *Section view ended and two more components placed*

Now apply a tangent constraint and two insert constraints to the assembly.

15. Select Constraint on the Assembly Panel Bar or toolbar.
16. Select the Tangent button in the Place Constraint dialog box.
17. Select spherical face A and cylindrical face B in Figure 6.43 and click the Apply button.

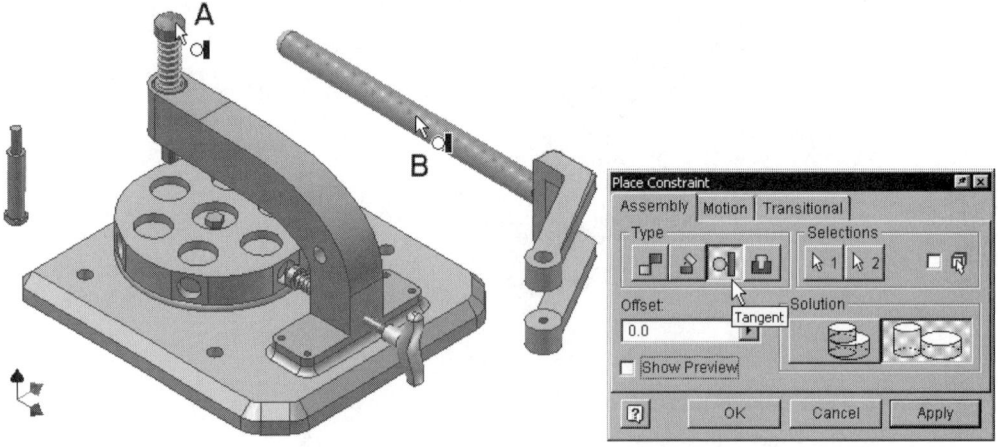

Figure 6.43 *Tangent constraint being applied to spherical face and planar face*

18. Select the Insert button in the Place Constraint dialog box.
19. Select edges A and B in Figure 6.44 and click the Apply button.

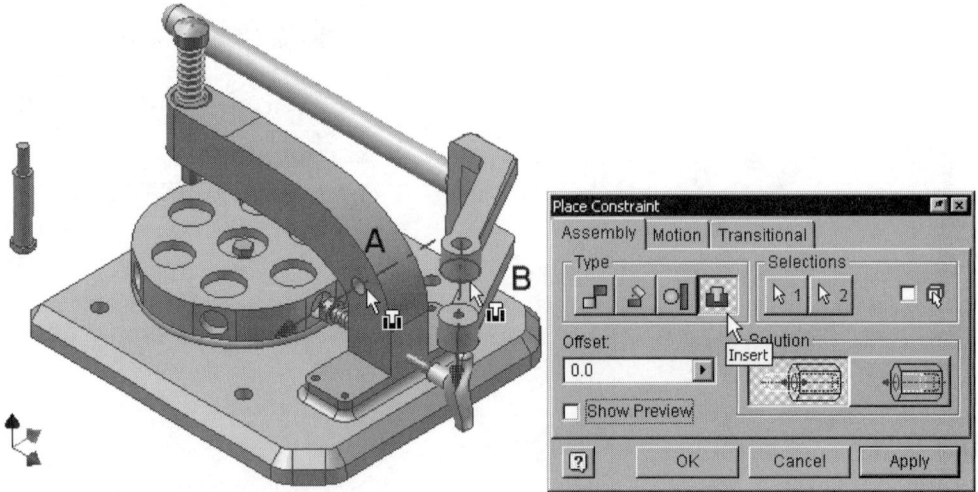

Figure 6.44 *Insert constraint being applied to circular edges*

20. Select edges A and B in Figure 6.45 and click the OK button.

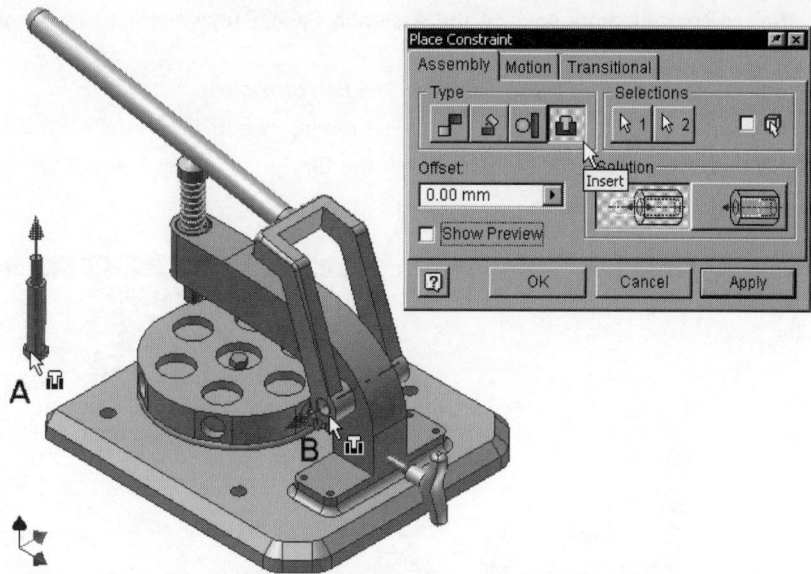

Figure 6.45 *Insert constraint being applied to circular edges*

Now add another component to the assembly.

21. With reference to Figure 6.46, place the file (PunchIndexBush) in the assembly.

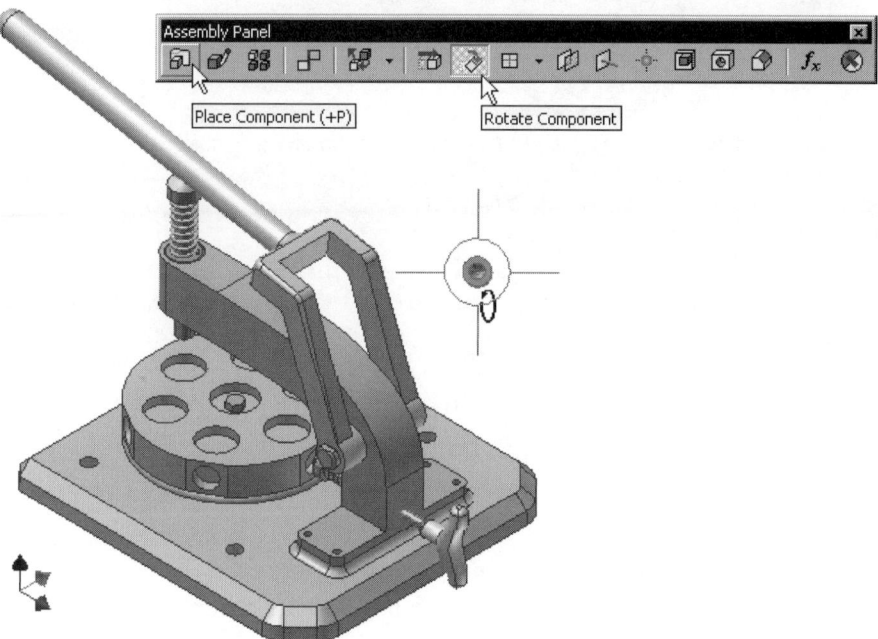

Figure 6.46 *Index bush placed and rotated*

22. Select Rotate Component on the Assembly Panel Bar or toolbar, select the index bush, and drag it to rotate it.

23. Select Constraint on the Assembly Panel Bar or toolbar.

24. In the Place Constraint dialog box, select the Insert button and the Aligned button.

25. Select circular edges A and B and click the OK button. (See Figure 6.47.)

The components are assembled. Save your file.

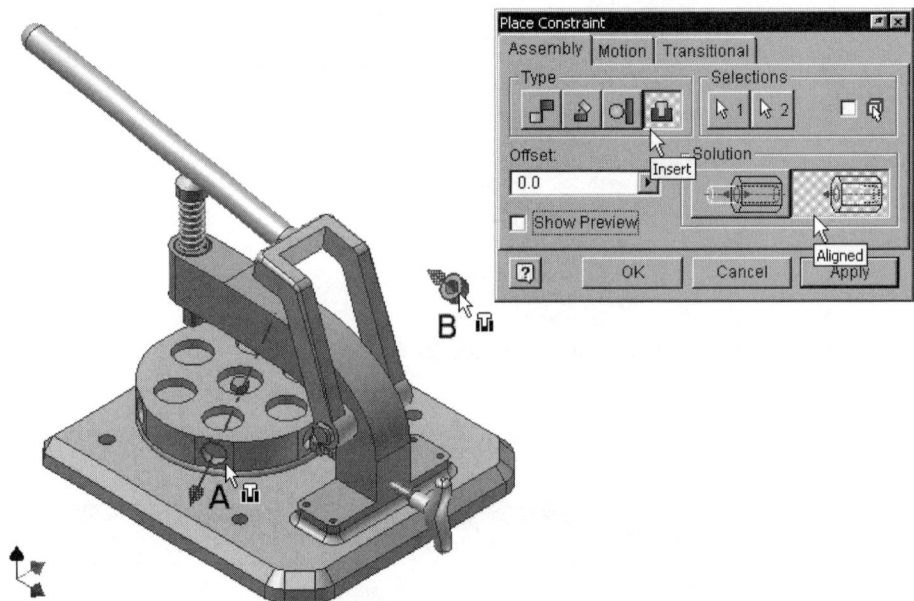

Figure 6.47 *Insert constraint being applied to the index bush and the table*

PLACE iPART

An iPart is a family of parts. Placing an iPart concerns generating a solid part from the iPart in accordance with the specification saved in the iPart. To place an iPart in an assembly, you select it from the folder where you saved the iPart, specify a file name for the solid part that is generated from the iPart, and select a location for the generated part.

 **TUTORIAL 6.4**

In this tutorial, you will place an iPart in the assembly.

1. Open the assembly file *Punch.iam*, if you already closed it.

2. Select Place Component on the Assembly Panel Bar or toolbar and open the file (PunchBolt).

3. In the Place Custom iPart dialog box, select the Table tab and select the first row. (See Figure 6.48.)

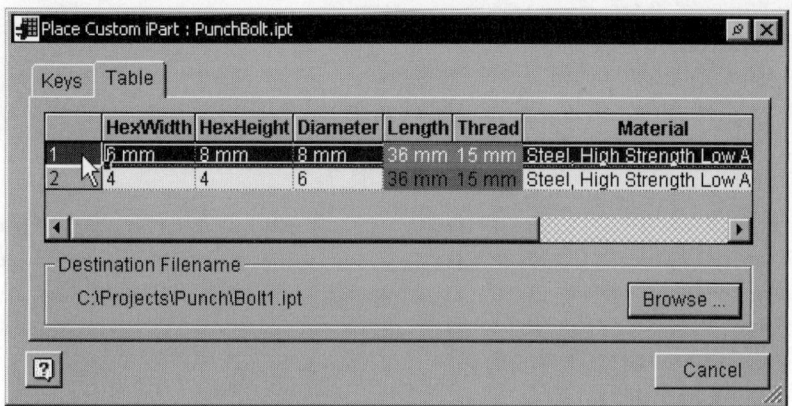

Figure 6.48 *Place Custom iPart dialog box*

4. Select the Browse button, specify a file name (Bolt1) for the solid part generated from the iPart, and then select Open.

5. Select a location in the graphics area to place the component, right-click, and select Done.

6. With reference to Figure 6.49, apply an insert constraint to circular edges A and B.

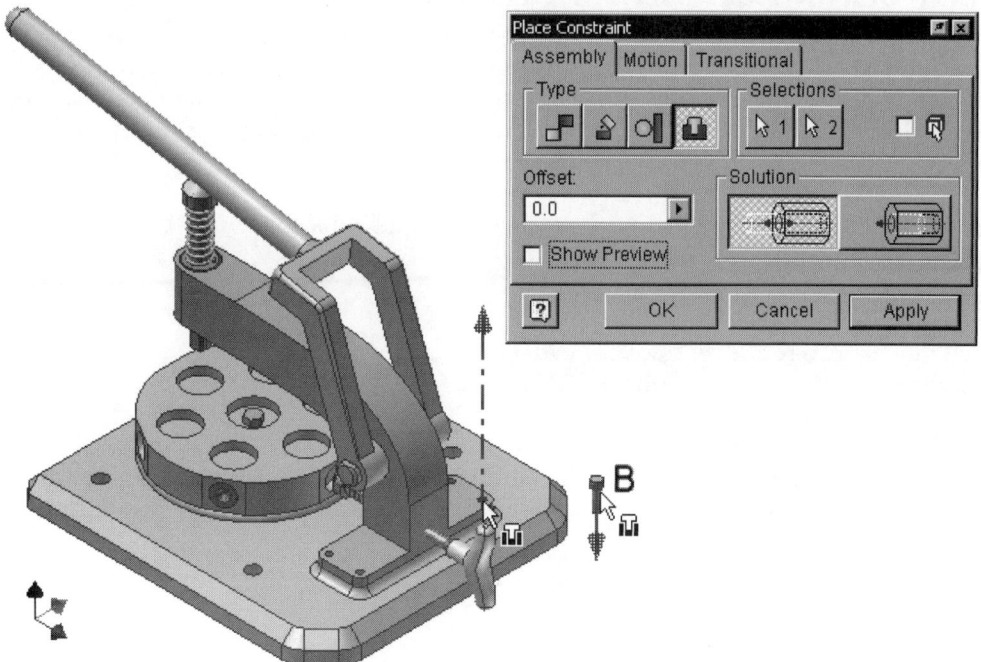

Figure 6.49 *Insert constraint being applied*

The iPart is placed and assembled. (See Figure 6.46.) Save your file.

CONSTRUCT ASSEMBLY PATTERN

When constructing an array of components in an assembly, it is very time consuming to place the components one by one and apply assembly constraints to each component individually. To save time, you put a component in the assembly, apply assembly constraints to properly position it, and construct an assembly pattern.

RECTANGULAR, CIRCULAR, AND ASSOCIATIVE ASSEMBLY PATTERNS

There are three kinds of assembly patterns: rectangular, circular, and associative. In a rectangular assembly pattern, you repeat selected components of an assembly in a rectangular pattern by specifying the number of instances and distance between instances in two specified directions. In a circular assembly pattern, you repeat selected components of an assembly in a circular pattern by specifying an axis, number of instances, and angle between instances. These two kinds of pattern are very similar to the kinds of patterns you constructed in a solid part.

In an associative assembly pattern, you repeat selected components of an assembly by referencing the components to a pattern feature of a part in the assembly. The assembly pattern constructed this way is associative to the part's pattern feature (rectangular or circular). When you change the part's pattern feature, the assembly pattern will change accordingly.

MANIPULATING AN ASSEMBLY PATTERN

You can manipulate instances of an assembly pattern individually in two ways: you suppress an instance of an assembly pattern and you separate an instance from the assembly pattern while it remains in place in the assembly. After separation, you can retain a copy of the original source instance of the pattern even if you delete the assembly pattern.

Component Pattern Replace

You can replace components in an associated assembly in a single step by replacing the source component.

Component Pattern Restructure

You can restructure a set of assembly pattern components into or out of a sub-assembly by selecting the pattern on the Browser Bar and dragging it to a new location in the hierarchy.

Component Pattern Visibility

In terms of visibility, you can treat the entire assembly pattern as an single object. You select the pattern on the Browser Bar, right-click, and select or deselect Visibility.

 **TUTORIAL 6.5**

In this tutorial, you will place two associative assembly patterns in the punch set.

1. Open the assembly file *Punch.iam*, if you already closed it.
2. Select Pattern Component on the Assembly Panel Bar or toolbar. (See Figure 6.50.)

3. Select the bolt on the Browser Bar as the component to array.

4. Select the Associative tab, if it is not already selected.

5. Select the Associative Feature Pattern button, select feature pattern A in Figure 6.50, and click the OK button.

 Tip: Prior to placing an associative component pattern, you need to place the source component on the source feature of the feature pattern.

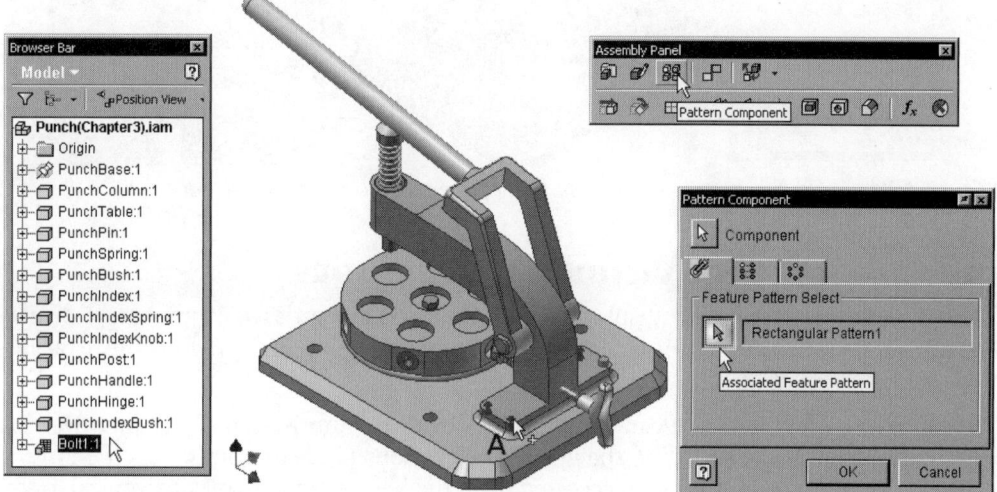

Figure 6.50 *Associative pattern component being placed on a rectangular feature pattern*

6. Select Pattern Component on the Assembly Panel Bar or toolbar again.

7. Select the index bush on the Browser Bar as the component to array.

8. Select the Associative tab, if it is not already selected.

9. Select the Associative Feature Pattern button, select feature pattern A in Figure 6.51, and click the OK button.

 Tip: You do not have to place the source component of an associative pattern component on the source feature of a circular pattern if the feature pattern is fitted in 360 deg.

The assembly pattern components are complete. Save your file.

386

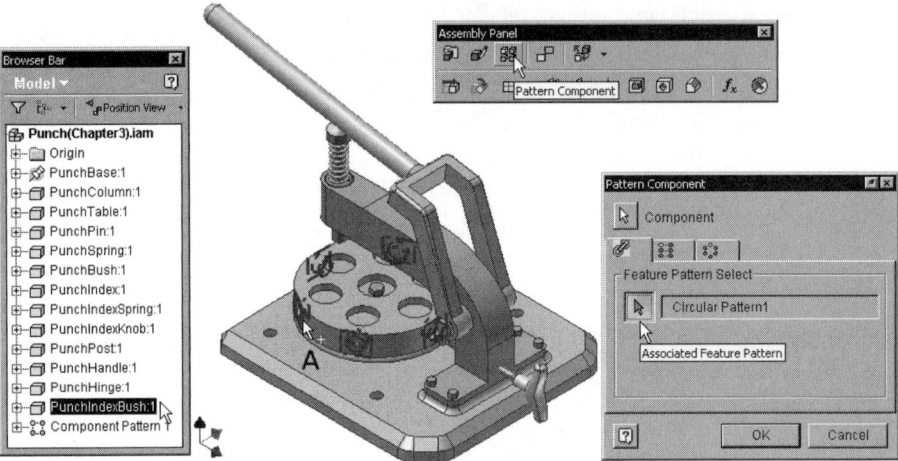

Figure 6.51 *Associative pattern component being placed on a circular feature pattern*

DESIGN VIEW AND VISIBILITY OVERRIDES

Design views are saved view display configurations. You can save a number of display configurations, such as an overall view and a number of zoom-in views looking at various details of the assembly.

Along with design view, you can save visibility override information. Visibility override information concerns overriding the visibility of various kinds of objects specified in the Object Visibility cascading menu, accessed from the View menu. Typically, you can override the visibility of origin planes, origin, axes, origin points, work planes, work axes, work points, sketches, welds, and weld symbols. These objects might have been left visible in their source part file. Instead of opening these files and turning them off, you can turn them off in the assembly. It must be noted that turning them off in the assembly does not affect their mode of visibility in their source part files. If they are visible in the source file, they will remain visible in their source file.

 TUTORIAL 6.6

In this tutorial, you will save a design view of the punch assembly. In the design view, you will turn off all the work features of the source part files.

1. Open the assembly file *Punch.iam*, if you already closed it.
2. Select View > Object Visibility and then deselect User Work Planes. Repeat this process again to deselect User Work Axes and User Work Points.

Now the work features are overridden.

3. Select View > Design Views.
4. In the Design Views dialog box, type the name you want to assign to the design view in the Design View box, and then click OK.

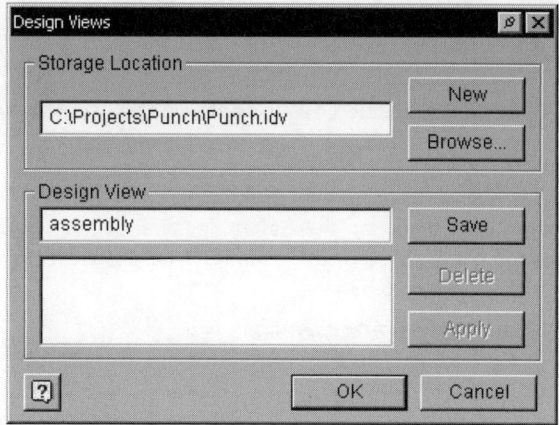

Figure 6.52 *Design Views dialog box*

Now the work features of all the part files will be turned off automatically in the assembly whenever you retrieve this design view in the future. The assembly is complete. Save and close your file.

THE TOP-DOWN APPROACH

In a top-down design approach, you start an assembly without having any component parts ready for assembling. Instead, you construct parts or sub-assemblies while working in the assembly environment.

Using the top-down approach, you select Create Component on the Assembly Panel Bar or toolbar and specify a new solid part or a new assembly. In the Create In-Place Component dialog box, you specify a part file or an assembly file that you will construct in the assembly. After you start a component part, you can proceed to constructing the component part in a way similar to what you learned in Chapters 3, 4, and 5.

The advantages of constructing components in the assembly environment are that you see other parts and you make use of the features of other parts to construct the solid.

WORKING MODE

Once you have constructed a part, there is no difference as to whether it is a placed part or a created part, because you can work in assembly modeling mode or part modeling mode in an assembly, regardless of which approach you use to construct the assembly.

Whichever approach you use, you can display a section view, construct an assembly pattern, save design views, and incorporate a shared content link. You can also construct an iPart by using the top-down approach.

CONSTRUCT COMPONENT

The first task to do to construct an assembly by using top-down approach is to start a part file in an assembly file and construct a solid part. Then you continue to construct other parts in the assembly and put the parts together by applying assembly constraints.

TUTORIAL 6.7

In this tutorial, you will construct the components of an assembly of a pair of pliers by using the top-down design approach.

1. Start a new assembly file. Use the metric template.
2. Select Create Component on the Assembly Panel Bar or toolbar, or move the cursor over the blank space of the graphics area, right-click, and select Create Component. (See Figure 6.53.)

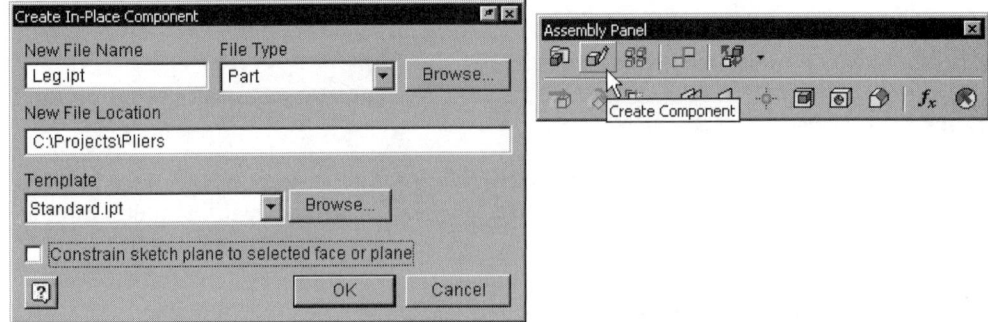

Figure 6.53 *Create In-Place Component dialog box*

In the Create In-Place Component dialog box, you select a file type (part file or assembly file), specify a file name and file location, and select a template. If you select an assembly file, you will make a sub-assembly.

3. In the Create In-Place Component dialog box, type the file name (*leg.ipt*) in the New File Name box. and select Part from the File Type pull-down list box.
4. Click the Browse button adjacent to the Template box in the Create In-Place Component dialog box. In the Open Template dialog box, select the Metric tab, select the metric *Standard.ipt* template, and then click OK.
5. In the Create In-Place Component dialog box, clear the Constrain sketch plane to selected face or plane check box, so that you don't place an unnecessary constraint on the new solid part you are about to construct.
6. Click OK to close the Create In-Place Component dialog box.

A part file is started. On the Browser Bar shown in Figure 6.54, you will find the assembly icon and its origin dimmed, denoting that assembly commands are unavailable at the moment and you are now working in part modeling mode.

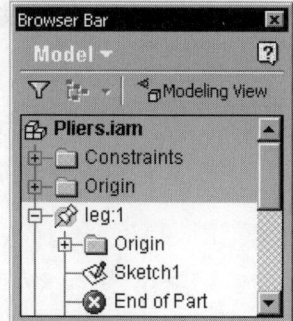

Figure 6.54 *Browser bar with the assembly and its origin dimmed*

Now construct a solid part. To reiterate, the way to construct a solid part while working in the top-down approach is identical to working in the bottom-up approach.

7. With reference to Figure 6.55, construct a sketch on the XY plane of the solid part file.

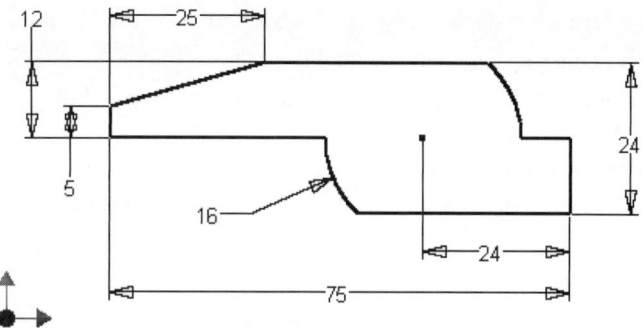

Figure 6.55 *Sketch constructed on the XY plane*

8. Select Return on the Inventor Standard toolbar to exit sketch mode.

Tip: After sketching and selecting the Return button once, you exit sketch mode. If you select Return button again, you exit part modeling mode and return to assembly modeling mode. To return to part modeling mode, select the part on the Browser Bar or on the graphics screen, and double-click.

9. Set the display to an isometric view.
10. Extrude the sketch a distance of 16 mm from mid-plane. (See Figure 6.56.)

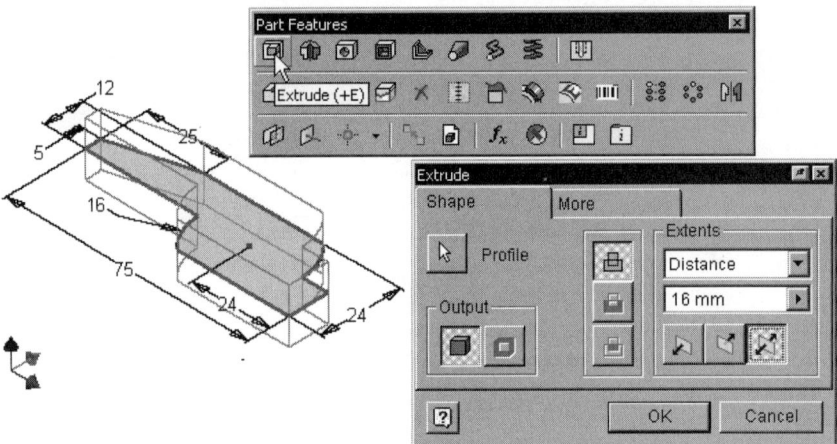

Figure 6.56 *Sketch being extruded from mid-plane*

Now construct a sweep solid to join the base solid feature.

11. With reference to Figure 6.57, establish a sketch on the XY plane

12. Project geometry from edges A and B.

13. Construct a line C and a spline D, with one end point of the spline tangent to the line.

14. Add dimensions to the sketch.

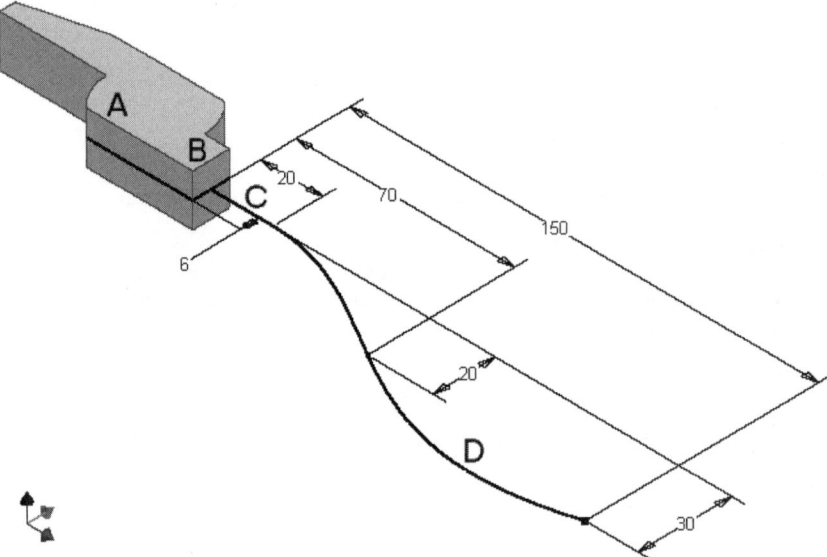

Figure 6.57 *Sketch constructed on the XY plane*

15. Construct another sketch on the vertical face indicated in Figure 6.58.

16. Project geometry from edges A and B and line C.

17. Construct an ellipse with center at the projected geometry from line C and tangent to projected geometry from edges A and B.

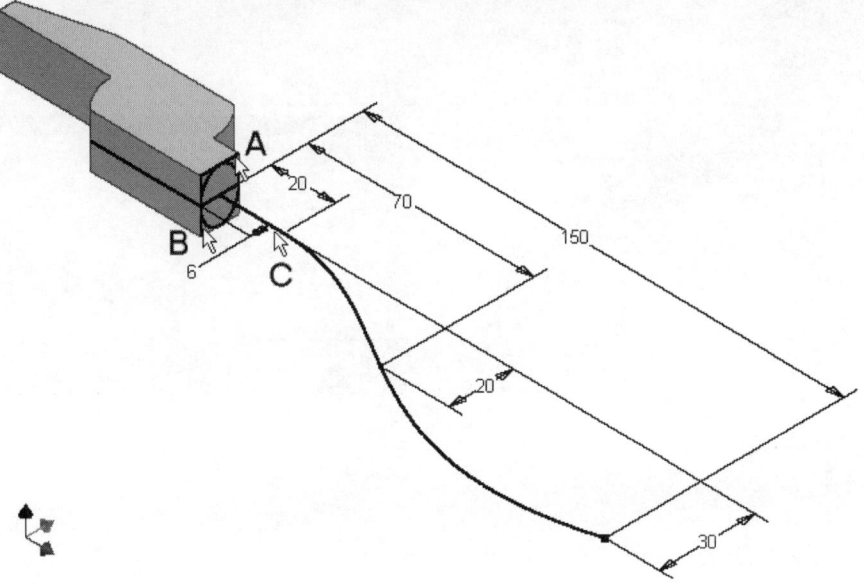

Figure 6.58 *Sketch constructed on a face of the solid*

18. Construct a sweep solid in accordance with Figure 6.59 and join the feature to the base solid.

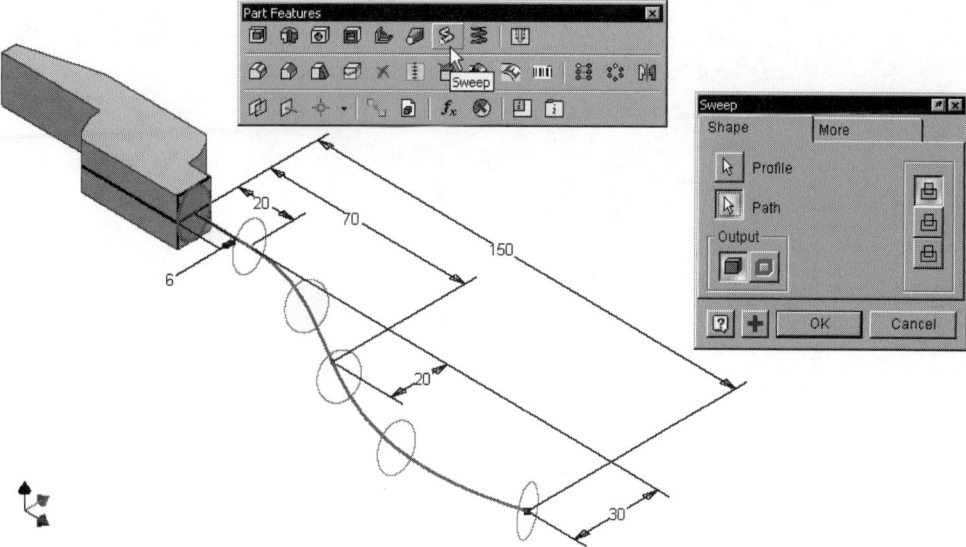

Figure 6.59 *Sweep solid feature being constructed*

19. With reference to Figure 6.60, construct a fillet feature.

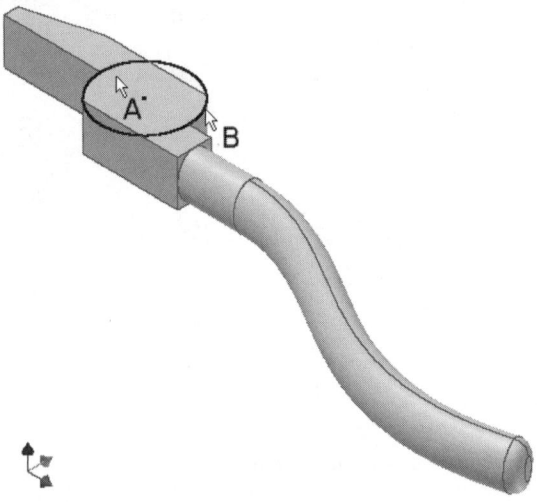

Figure 6.60 *Circle constructed on a face of the solid part*

Now construct an extruded solid feature to cut the part and place a countersink hole.

20. With reference to Figure 6.61, establish a sketch on face A and construct a circle.

21. Project edge B.

22. Add a coincident constraint to center of the circle and the projected center of B, and add an equal constraint to the circle and the project edge B.

Figure 6.61 *Circle being extruded to cut the solid*

24. Extrude the circle a distance of 8 mm to cut the solid part. (See Figure 6.62.)

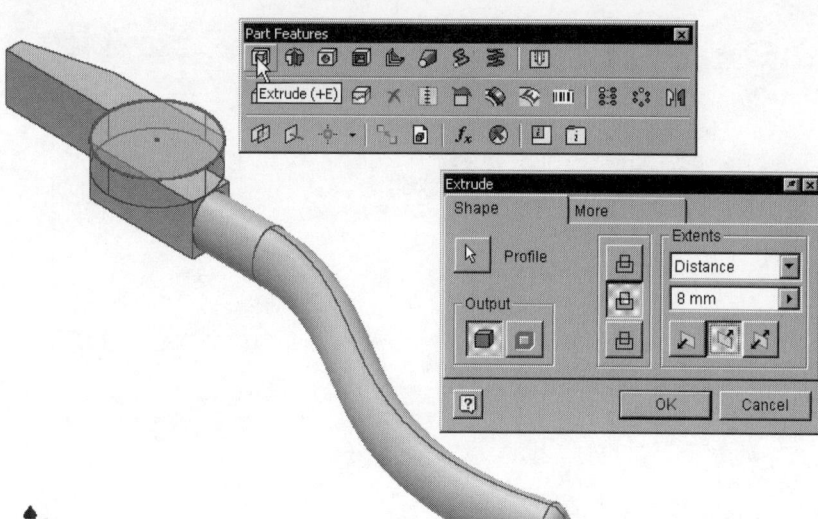

Figure 6.62 *Circle being extruded to cut the solid*

25. With reference to Figure 6.63, rotate the display and place a countersink through hole (8 mm hole diameter and 12 mm countersink diameter) on the solid part.

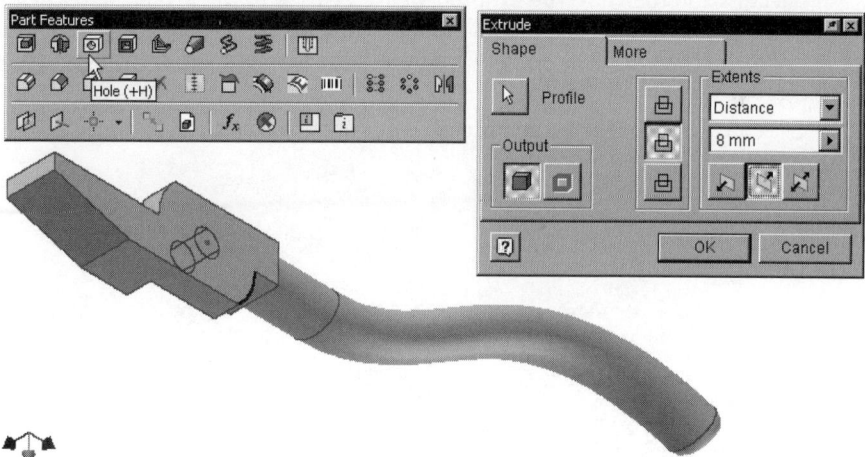

Figure 6.63 *Countersink through hole being placed*

Now add an instance of the solid part that you constructed.

26. Select Return on the Inventor Standard toolbar to exit part modeling mode.

27. Select the solid part on the Browser Bar or the graphics area, right-click, and select Copy.

28. Move the cursor over the blank space of the graphics area, right-click, and select Paste.

An additional instance of the solid part is placed in the assembly. (See Figure 6.64.)

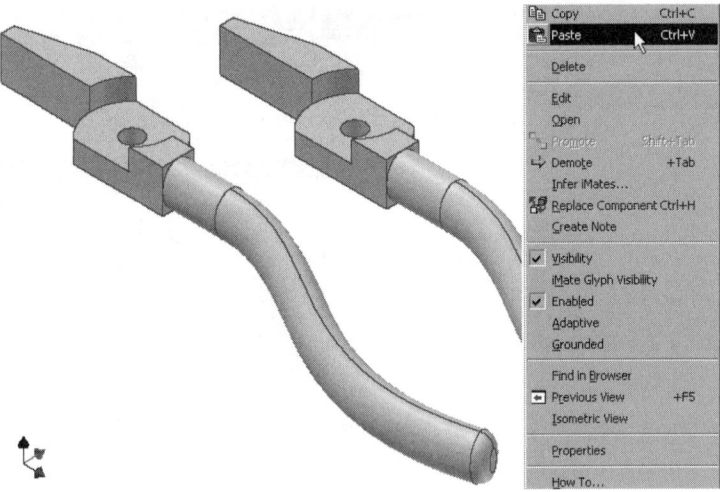

Figure 6.64 *An instance pasted*

Now construct another solid part in the assembly.

29. Select Create Component on the Assembly Panel Bar or toolbar.

30. In the Create-In-Place Component dialog box, specify a part file name (*rivet.ipt*), select Part in the File Type drop-down list, browse to the proper folder (*Pliers*) in which to create the part, Browse to select the metric *Standard.ipt* template file, and then click OK to close the Create-In-Place Component dialog box and begin creating the new part.

31. Click anywhere in the assembly and then, with reference to Figure 6.65, construct a sketch of the new part.

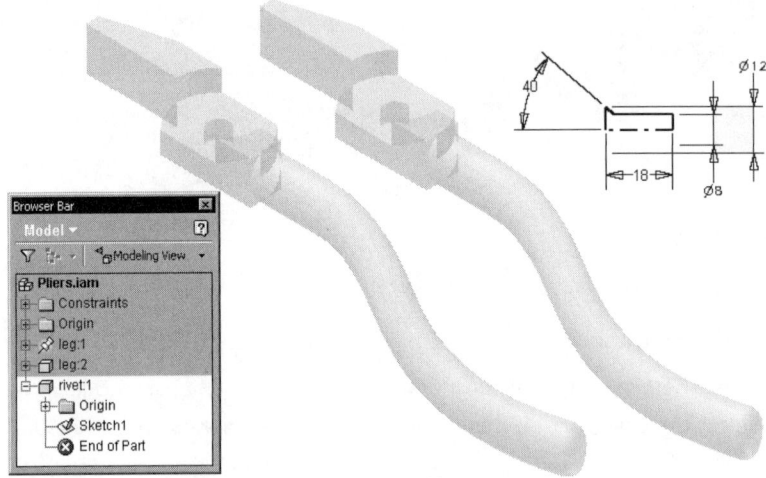

Figure 6.65 *Sketch constructed on a solid part in the assembly*

Tip: On the Browser Bar, the assembly icon (depicting the assembly) and the part file icons (depicting the instances of last solid part that you constructed) are unavailable.

32. Select Return on the Inventor Standard toolbar to exit sketch mode.

33. Revolve the solid part to form a revolved solid. (See Figure 6.66.) The solid part is complete.

34. Select Return on the Inventor Standard toolbar to return to assembly modeling mode.

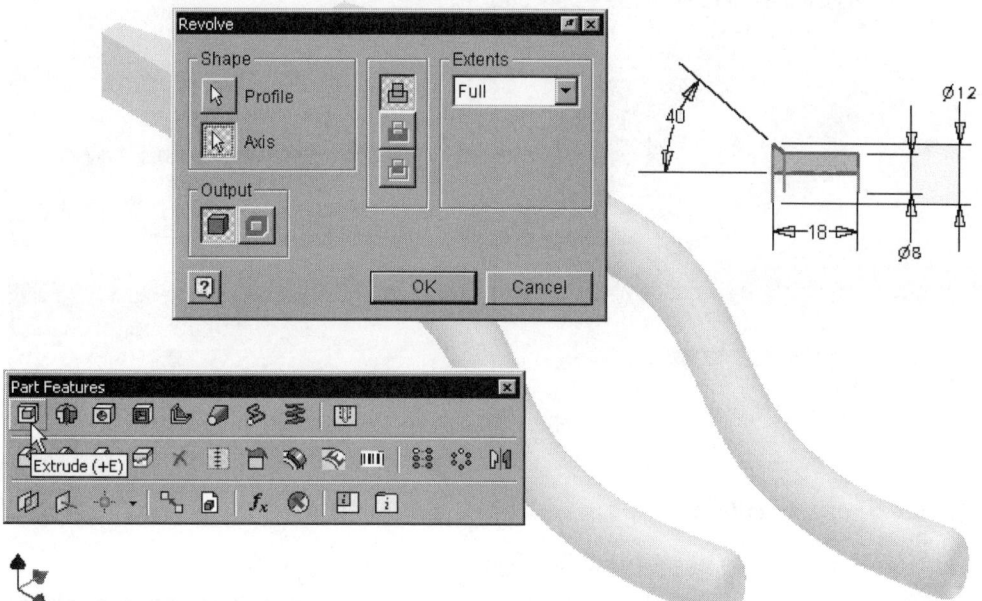

Figure 6.66 *Sketch being revolved*

The parts for the assembly are complete. Save your file (file name: *Pliers.iam*).

 Note: Inventor will display a dialog box asking if you want to save both of the parts you just created. Click OK to save the parts.

 Tip: Now you have three files: an assembly file, *Pliers.iam*, linking to two solid part files, *leg.ipt* and *rivet.ipt*.

As you have seen, the way to construct a part in assembly mode is virtually the same as when you work in part modeling mode. The difference is that you see the other parts in the assembly.

PLACE ASSEMBLY CONSTRAINT

Regardless of which design approach you use to construct an assembly, you use the same set of assembly constraints (mate, angle, tangent, and insert) to assemble the components together.

TUTORIAL 6.8

In this tutorial, you will complete the pliers assembly by applying assembly constraints to the components.

1. Open the assembly file *Pliers.iam*, if you already closed it.
2. Select Constraint on the Assembly Panel Bar or toolbar.
3. Select Insert in the Place Constraint dialog box, select circular edges A and B in Figure 6.67, and click the Apply button.

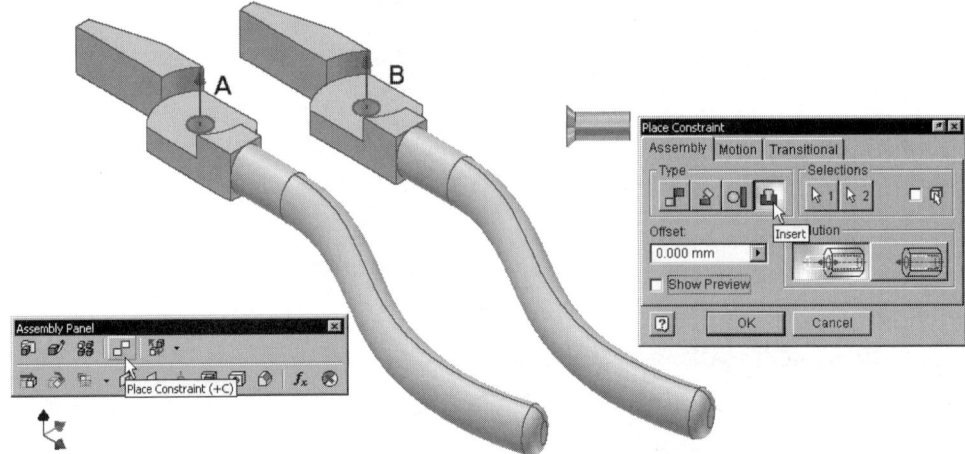

Figure 6.67 *Insert constraint being applied to a pair of circular edges*

4. Select circular edges A and B in Figure 6.68 and click the Apply button.

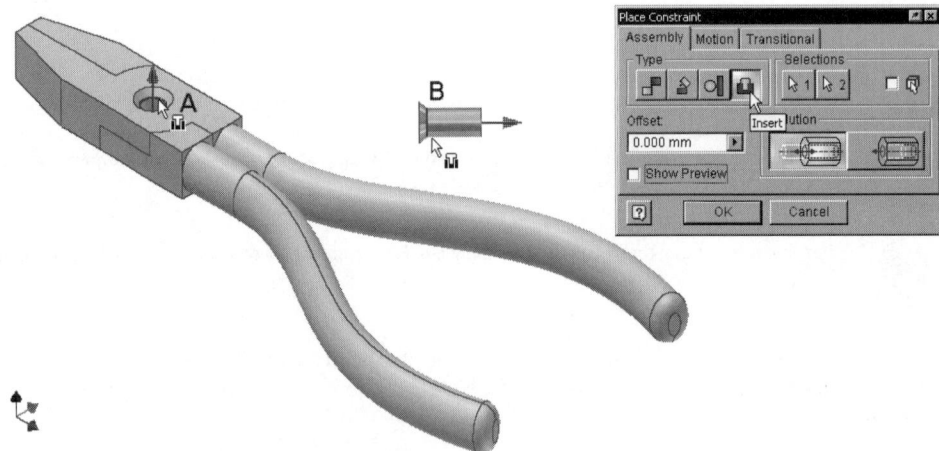

Figure 6.68 *Insert constraint being applied*

5. In the Place Constraint dialog box, select Angle. Specify an angle of 180 deg., select edges A and B (indicated in Figure 6.69), and then click OK.

The assembly is complete. Save and close your file.

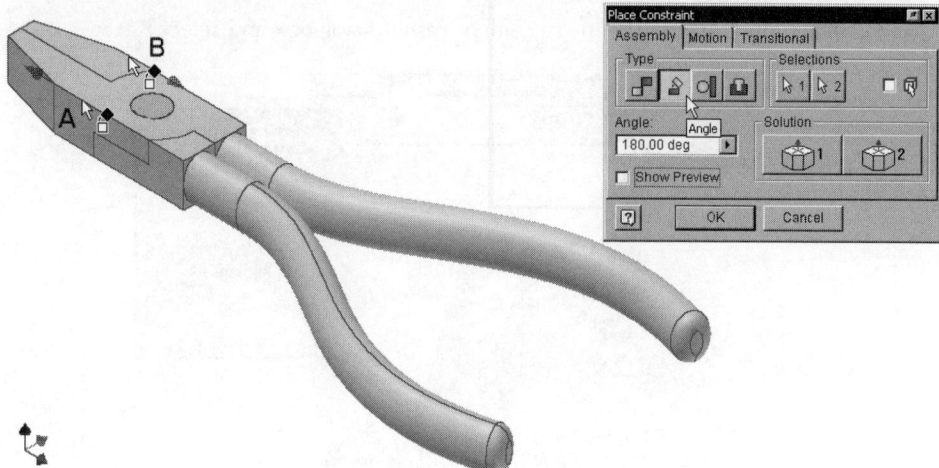

Figure 6.69 *Angle constraint being applied*

HYBRID APPROACH

In reality, you seldom use the bottom-up approach or top-down approach alone. It is normal practice to place components that you already constructed in an assembly and construct new components to fit the other components in the assembly. Using both the bottom-up and the top-down approaches, you use the hybrid approach.

To reiterate, there is no difference in the way to treat the solid part files and the assembly file once the assembly is constructed.

MEASURING AND COMPONENT CONSTRUCTION

While editing and constructing parts in assembly mode, you see all the parts put together and you can determine the distance between selected geometry of the parts selected. No matter which approach you use to construct an assembly, you can measure distance, angle, loop, and area. As when working in part modeling mode, you can access the measurement tools from the Tools menu.

TUTORIAL 6.9

In this tutorial, you will open the punch assembly, take a measurement of a component of the assembly, and construct a solid part in the assembly. Because you already used the bottom-up approach to place components you constructed and you now add new components in the assembly while working in assembly mode, you are, in fact, using a hybrid approach.

1. Open the assembly file *Punch.iam*.
2. Select Tools > Measure Distance.

3. Select circular edge A in Figure 6.70.

4. The diameter of the circular edge (45 mm) is displayed.

5. Select the arrow of the measurement result dialog box and select Reset.

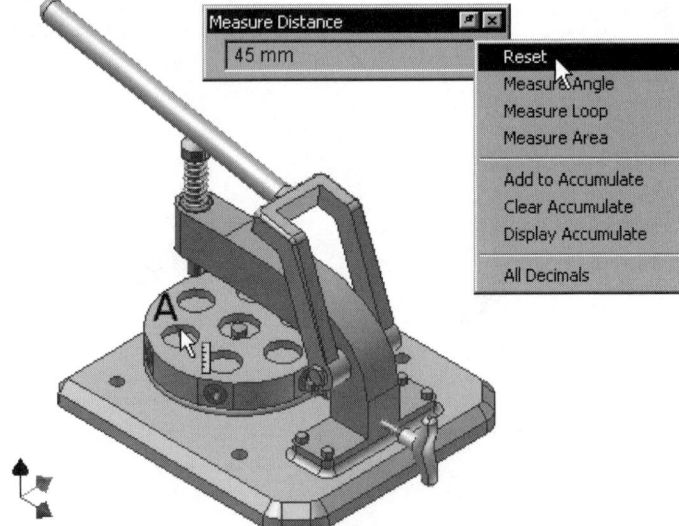

Figure 6.70 *Diameter of a component measured*

6. Move the cursor over circular edge A in Figure 6.71, pause a moment to wait for the selection dialog box to be displayed, select the left or right arrow until the center of the circular edge A is highlighted, and select the center button of the selection dialog box.

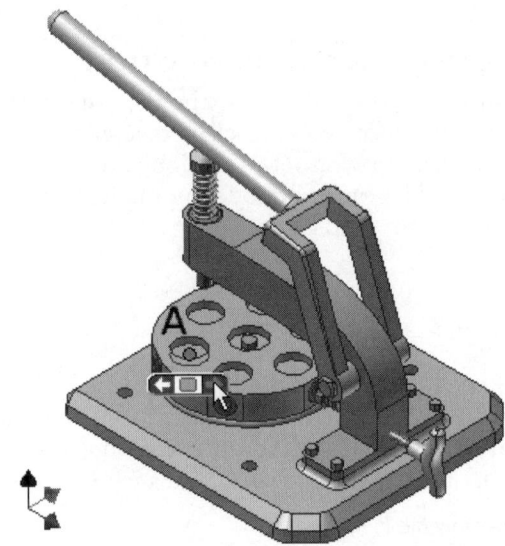

Figure 6.71 *Center of the circular edge selected*

7. Select circular edge A in Figure 6.72.

8. The distance between the two circular edges (10 mm) is displayed.

9. Select the X mark of the measurement dialog box to close it.

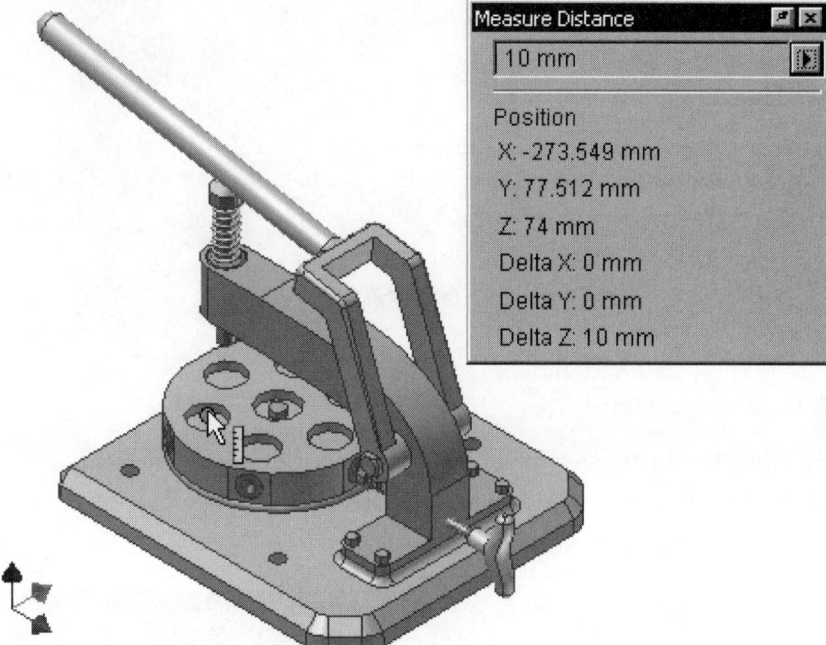

Figure 6.72 *Distance between circular edges measured*

Now construct a solid part in the assembly.

10. Move the cursor over the blank space of the graphics area, right-click, and select Create Component.

11. Specify a file name (*PunchDie.ipt*), put it in the proper folder (*Punch*), and be sure to use the metric part template. Click OK, and then click in the assembly to begin a new sketch.

12. Select Parameters on the Part Features Panel Bar or toolbar.

13. In the Parameters dialog box, select the Add button to add a user parameter (Parameter name: Size; Unit: mm; and Equation: 3 mm).

14. With reference to Figure 6.73, construct a sketch.

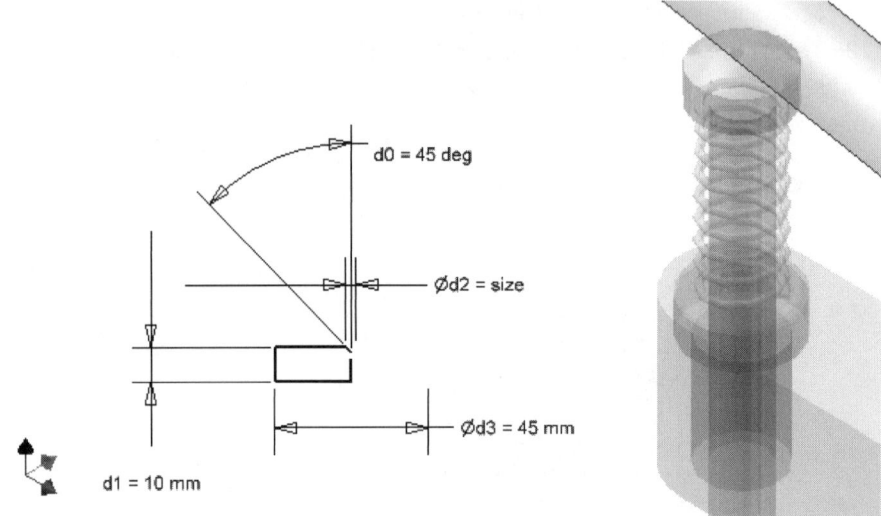

Figure 6.73 *Sketch constructed*

15. Select Return on the Inventor Standard toolbar to exit sketch mode.

16. Revolve the sketch to form a revolved solid. (See Figure 6.74.)

A solid part is constructed. Save your files.

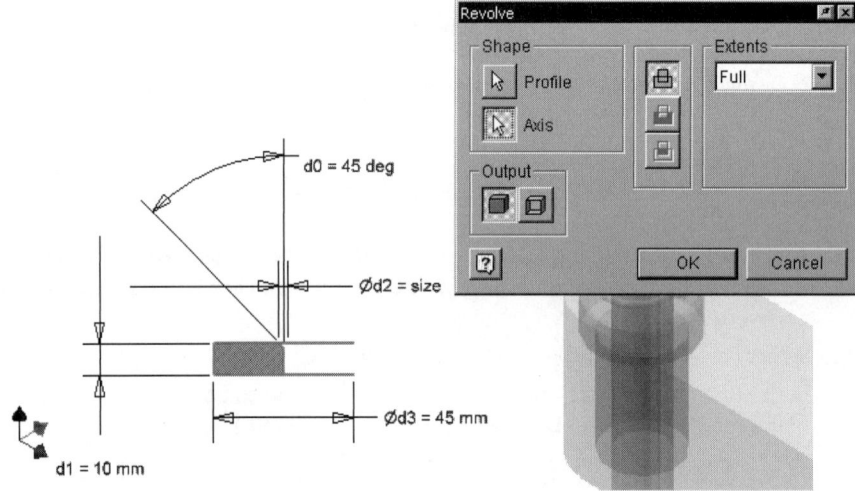

Figure 6.74 *Sketch revolved*

CONSTRUCTING AN iPART IN ASSEMBLY MODE

As well as using the bottom-up approach to construct an iPart and place a family member of the iPart into the assembly, you can construct an iPart in the assembly and use copy and paste to place a family member of the iPart in the assembly while working in assembly modeling mode.

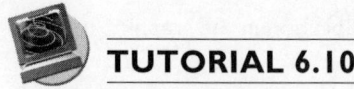

TUTORIAL 6.10

In this tutorial, you will construct an iPart from the solid part in the assembly and insert a number of parts from the family of parts. Now you will construct an iPart table.

1. Open the assembly file *Punch.iam*, if you already closed it.
2. Double-click the component Die.ipt, if you have already exited part modeling mode.
3. Select Tools > Create iPart.
4. On the Parameters tab, select the parameter (size) and select the >> button.
5. Insert five more rows in the iPart table (a total of six rows).
6. Change the value of the parameter (size) in the rows as follows: 3 mm, 4 mm, 5 mm, 6 mm, 7 mm, and 8 mm.
7. Click the OK button. (See Figure 6.75.) An iPart is constructed in the assembly.
8. Select Return on the Inventor Standard toolbar to return to assembly modeling mode.

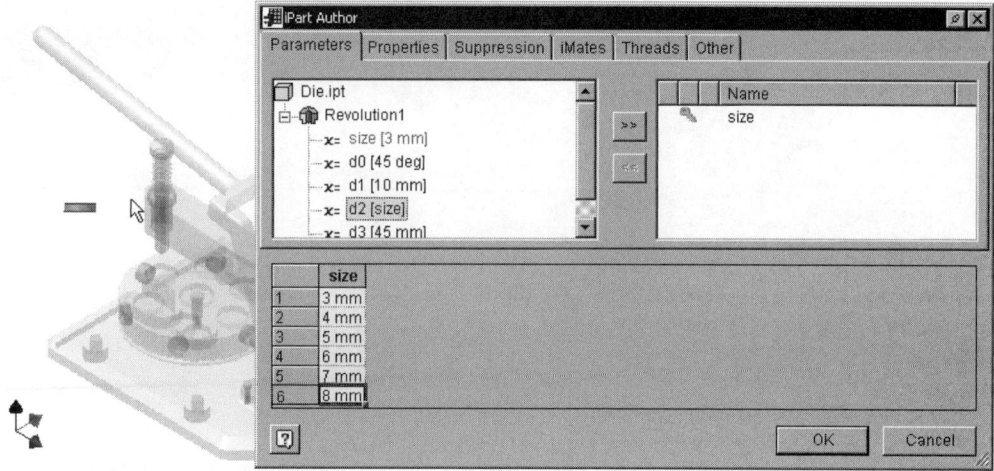

Figure 6.75 *iPart table*

An iPart with six versions is constructed. Now place five more instances of the iPart in the assembly.

9. Select the solid part (iPart that you constructed) on the Browser Bar or in the graphics area, right-click, and select Copy.
10. Move the cursor over a blank area of the graphics area, right-click, and select Paste.
11. Select the Table tab in the Place Standard iPart dialog box, select the second row of the iPart table, and select a location in the graphics area. A family member from the second row of the iPart table in placed in the assembly. After you place the new part, Inventor displays a dialog box asking if you want to save changes to the PunchDie and its dependents. Click OK.

12. Select the third row of the iPart table and select a location in the graphics area to place the third iPart. Again, save your changes.

13. Select the fourth row of the iPart table and select a location in the graphics area to place the fourth iPart.

14. Select the fifth row of the iPart table and select a location in the graphics area to place the fifth iPart.

15. Select the sixth row of the iPart table and select a location in the graphics area to place the sixth iPart.

16. Right-click and select Done. Five additional iParts are placed. (See Figure 6.76.)

Altogether, you should have six solid parts from the family of iParts.

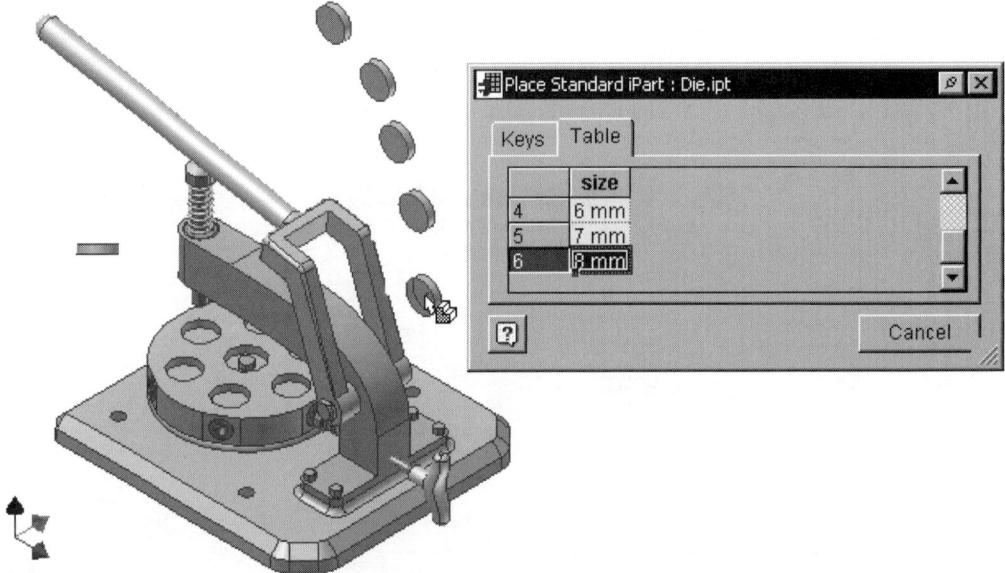

Figure 6.76 *iParts placed in the assembly*

17. With reference to Figure 6.77, apply insert constraints to the iParts one by one to assemble them in the table of the punch set.

The assembly is complete. Save your files.

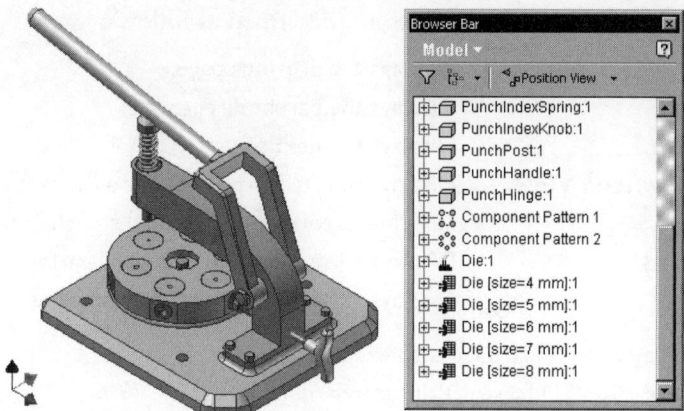

Figure 6.77 *iParts assembled*

CONTENT LIBRARY

It is common engineering practice to use standard components in our design. Quite often, we have to construct individual solid part files for these components and put them in our assembly to help evaluate our design. To speed up our design work, we can use standard components from the content library that is accessible on the assembly Browser Bar. In an assembly file, the browser has two panels: the Model panel and the Library panel. The Model panel depicts the objects in the assembly, and the Library panel lists standard components. The content of the library can reside in your computer as you installed Inventor, or it can be provided by library servers.

TUTORIAL 6.11

In this tutorial, you will select the standard components from the content library and place them into your assembly.

1. Open the assembly file *Punch.iam*, if you already closed it.

2. Select Library from the Browser Bar.

If this is the first time you have opened the Library panel, it displays a standard catalog. (See Figure 6.78.)

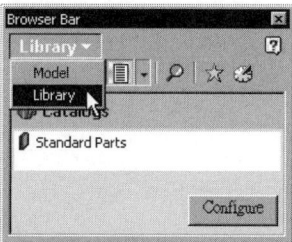

Figure 6.78 *Library panel of the Browser Bar*

The Library panel has seven buttons, described as follows:

Back	Displays the previous page.
Parent	Displays the parent directory.
Forward	Displays the next page.
List View/Icon View	Displays the standards in either a list or a set of icons.
Search	Searches according to specified search criteria.
Favorites	Displays a list of favorite standard parts.
History	Displays a list of recently retrieved parts.

3. Double-click Standard Parts to expand.
4. Double-click ISO to select this standard. (See Figure 6.79.)
5. Double-click Screws and Threaded Bolts.
6. Double-click Hex Head Types.
7. Double-click ISO 4014 (Regular Thread).
8. In the ISO 4014 (Regular Thread), set nominal diameter to 18 mm and nominal length to 70 mm.

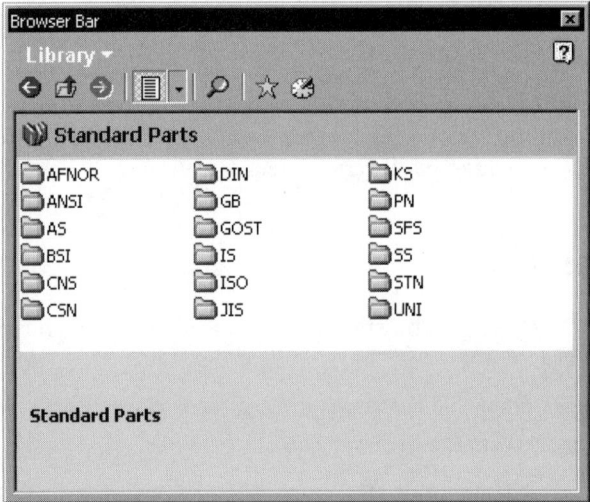

Figure 6.79 *Various standards*

9. Moving the cursor over the graphics inside the panel, you will see an iDrop symbol. (See Figure 6.80.)

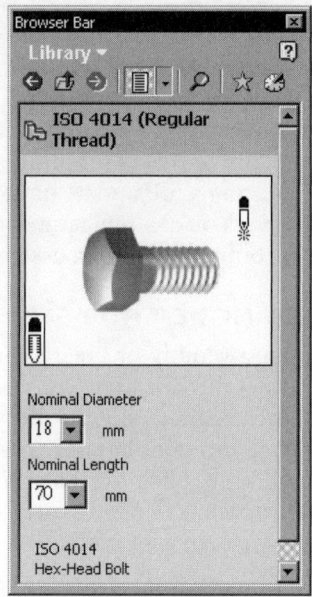

Figure 6.80 *Regular thread*

10. Hold down the left mouse button, drag the cursor to the graphics screen, and release the mouse button.

11. Place three more copies of the standard part in the assembly and then right-click and select Done.

12. With reference to Figure 6.81, apply insert constraints to properly assemble the screws.

The assembly is complete. Save and close your file.

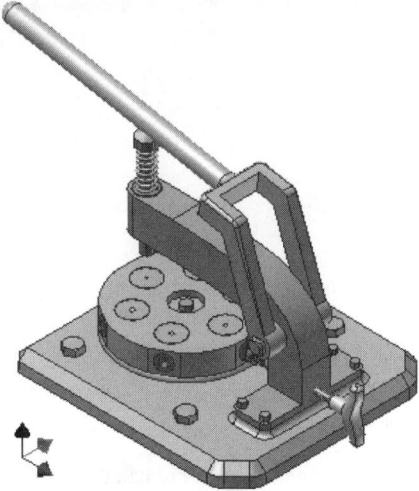

Figure 6.81 *Standard parts constrained*

MANIPULATING COMPONENTS

No matter which way you construct an assembly, you can toggle between part modeling mode (to modify the solid part) and assembly mode (to change the way the parts are assembled together). To find out how components fit together in the assembly, you check interference between the components. After you modify a part, you can update all the instances of the solid part by performing a full update or you can perform a local update in which only the active part is updated. You can replace a component with another component, and you can restructure the components in an assembly.

TOGGLING PART MODELING MODE AND ASSEMBLY MODELING MODE

After you place components in an assembly or create components in an assembly, the component part files link to the assembly file. The component parts will be displayed as objects on the assembly file Browser Bar. To edit a solid part, you select the part on the Browser Bar or in the graphics area and double-click to switch to part modeling mode. To return to assembly mode, you double-click the assembly on the Browser Bar or select Return on the Inventor Standard toolbar. To translate parts in the 3D space or apply assembly constraints to parts, you work in assembly mode. To construct or edit a part in an assembly, you work in part modeling mode.

TUTORIAL 6.12

In this tutorial, you will edit and modify the parts of the pliers assembly by working in part modeling and assembly modes.

1. Open the assembly file *Pliers.iam*.

2. Select one of the legs on the Browser Bar or in the graphics screen and double-click to toggle to part modeling mode. The other instances in the assembly are dimmed. (See Figure 6.82.)

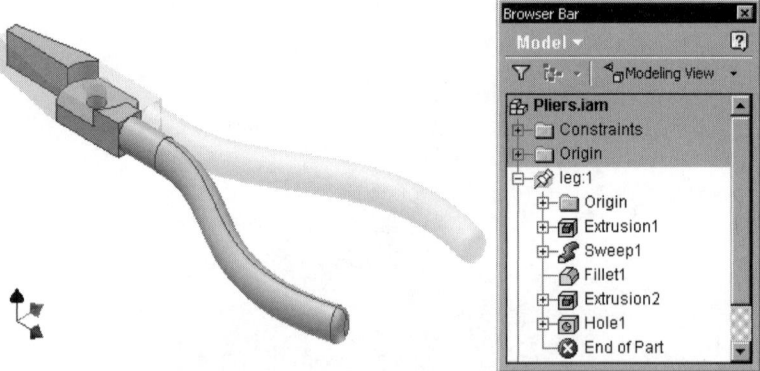

Figure 6.82 *Part modeling mode activated*

3. Select the extruded feature indicated in Figure 6.83 on the Browser Bar and select Show Dimensions.

4. Select the dimension indicated in Figure 6.83, double-click, and change it to 95 mm.

5. Select Return on the Inventor Standard toolbar to return to assembly mode.

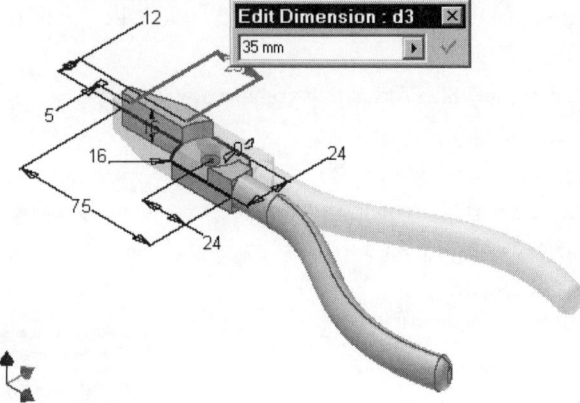

Figure 6.83 *Part being modified*

The part is changed and the assembly is updated. Save your files.

FULL UPDATE AND LOCAL UPDATE

When you edit the components of an assembly, changes are held in the memory of your computer. To incorporate changes to the components, you select Update on the Inventor Standard toolbar. By default, when you select Update on the Inventor Standard toolbar, the selection is Full Update. By selecting Full Update, you update all files and spreadsheets in one action. If you want to update only the active component or the components in the active sub-assembly, select Local Update. Figure 6.84 shows the Full Update and Local Update buttons on the Inventor Standard toolbar.

Both update buttons reflect the update condition of the part and assembly. If they are up-to-date, the buttons are unavailable. If the buttons are active, selection of them is necessary to bring the assembly up-to-date.

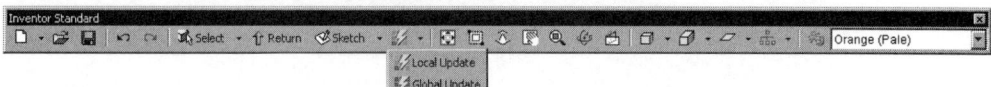

Figure 6.84 *Full update and local update*

CHECKING INTERFERENCE AND DIMENSION SIZE

Because it is difficult to tell from the display whether there is any interference among the parts in an assembly, you will let the computer find it out for you.

In reality, components can never be made to their exact dimensions, and we specify tolerances to the dimensions. Before we check interference among the components, we need to specify tolerances and to decide to use the maximum size, nominal size, or the minimum size for evaluation.

TUTORIAL 6.13

In this tutorial, you will carry out interference checking on the pliers assembly twice. First you will check them in their nominal size and then you will check them with the rivet's diameter set to its maximum size.

1. Open the assembly file *Pliers.iam*, if you already closed it.
2. Select Tools > Analyze Interference.
3. Select the rivet (A in Figure 6.85).
4. Select Define Set #2 button in the Interference Analysis dialog box.
5. Select B and C in Figure 6.85.
6. Click the OK button.

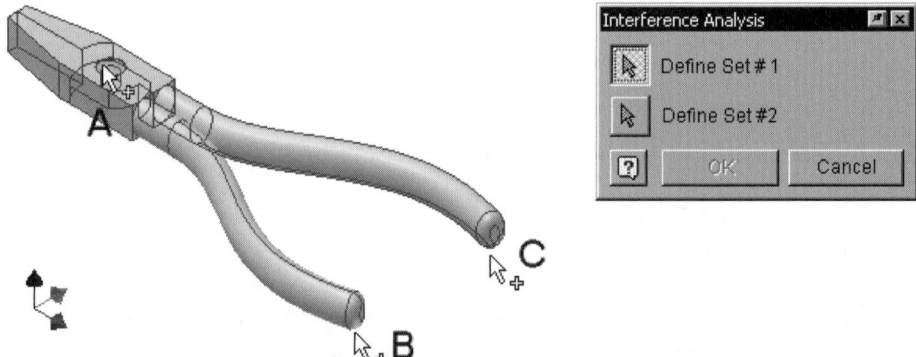

Figure 6.85 *Analyzing interference*

The dialog box shown in Figure 6.86 indicates that there is no interference. If there is any interference, a message will indicate where the interference occurs.

7. Save your files.

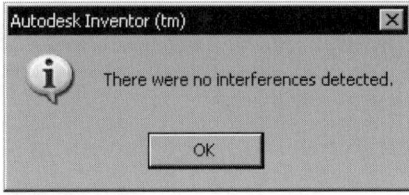

Figure 6.86 *Result of interference analysis*

8. Select the rivet on the Browser Bar and double-click to activate part modeling mode and work on the rivet.
9. Select the revolved feature on the Browser Bar, right-click, and select Edit Sketch.
10. Select the dimension indicated in Figure 6.87, right-click, and select Dimension Properties.

11. In the Dimension Properties dialog box, select the + sign in the Evaluated Size box, set tolerance to symmetric and 0.1 mm, and then click OK.

12. Select Update on the Inventor Standard toolbar to update the change.

Now evaluate interference again.

13. Select Return on the Inventor Standard toolbar to return to assembly modeling mode.

14. Select Tools > Analyze Interference.

15. Select Define Set #1 (if it is not already selected) and select the rivet on the Browser Bar.

16. Select Define Set #2 and select the legs one by one on the Browser Bar.

17. Click the OK button.

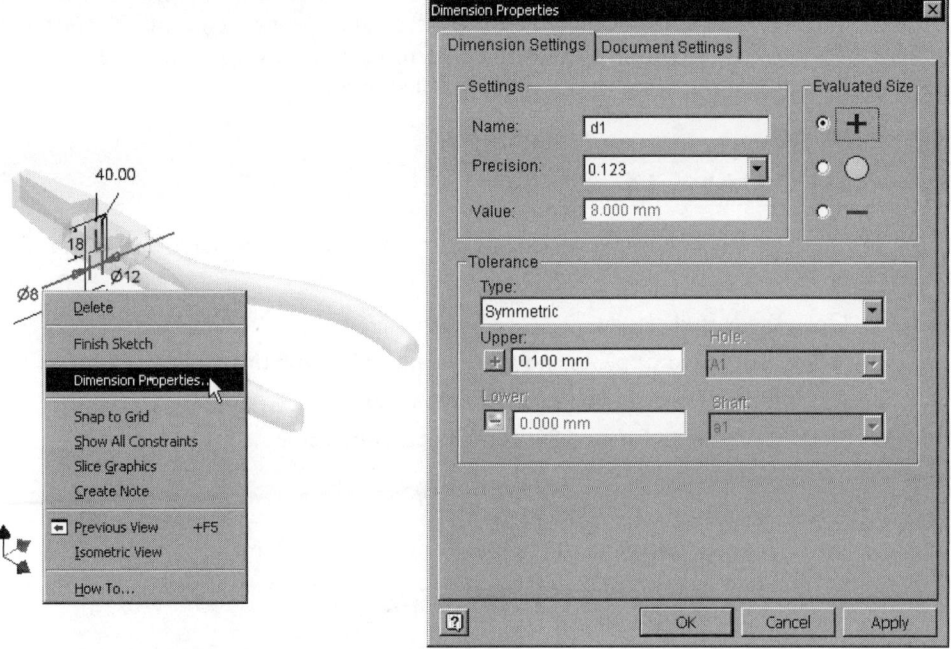

Figure 6.87 *A dimension's properties changed*

Evaluation result indicates that there is interference. (See Figure 6.88)

 Note: The amount of interference depends on the tolerance setting and may, therefore, differ from your result.

To ensure that there is no interference between the rivet and the leg, you should specify tolerance to the leg's hole, modify the dimensions of the rivet and/or the leg's hole, and check interference again. Save your file when finished.

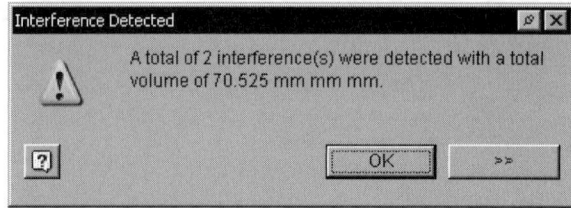

Figure 6.88 *Evaluation result after setting tolerance value in the rivet*

REPLACING A COMPONENT IN AN ASSEMBLY

In the process of designing a new product, you might want to replace a component part in an assembly with another component. To save the effort needed to remove a component and re-link another component, you use a replacement. You can replace a component in an assembly with a new component and yet preserve some or all of the assembly constraints in the replaced components, if the constraints are still relevant. Note that the instances of a pattern of components cannot be replaced. During replacement, you can replace an instance of the component or all the instances.

 TUTORIAL 6.14

In this tutorial, you will construct a solid part and use it to replace a component in an assembly.

1. Open the part file *leg.ipt.*
2. Select File > Save Copy As and specify a new file name (*newleg.ipt*).
3. Close the file *leg.ipt* and open the file *newleg.ipt.*
4. With reference to Figure 6.89, construct a sketch.

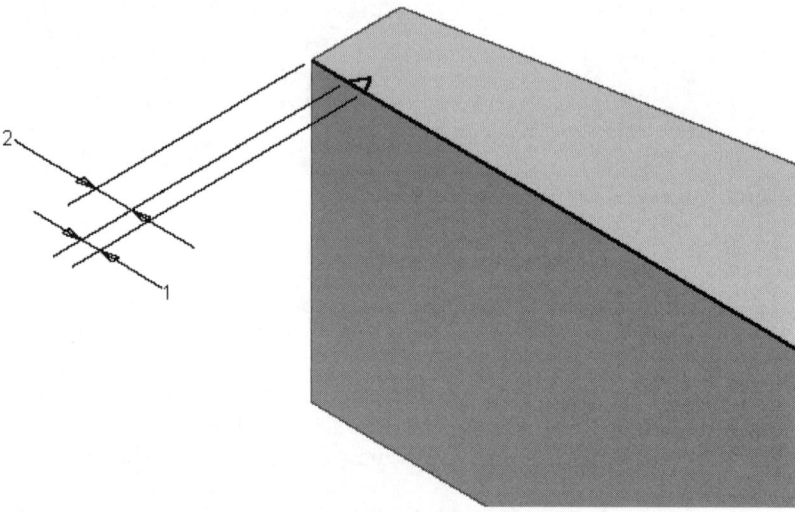

Figure 6.89 *Sketch constructed on a face of the solid part*

5. Extrude the sketch to cut through the solid part and construct a rectangular pattern of the extruded solid feature (count = 10 and space = 3 mm). (See Figure 6.90.)

6. The solid part is complete. (See Figure 6.90.) Save and close your file.

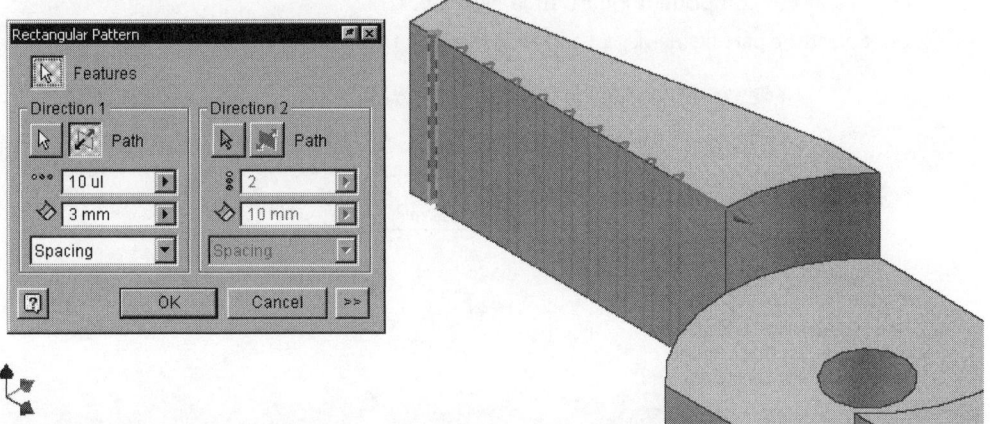

Figure 6.90 *Sketch extruded and the extruded feature being patterned*

Now replace a component in an assembly.

7. Open the assembly file *Pliers.iam*, if you already closed it.

8. Select Replace on the Assembly Panel Bar or toolbar.

9. Select the component indicated in Figure 6.91.

10. In the Open dialog box, select the part file *newleg.ipt* and click the Open button. The selected component is replaced.

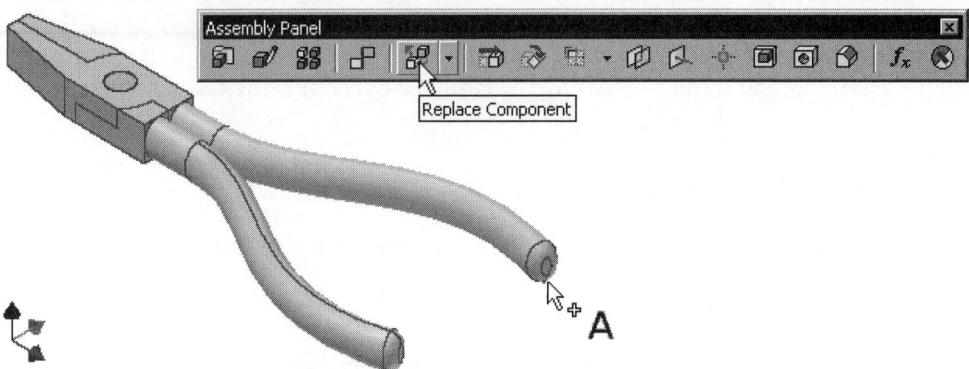

Figure 6.91 *An instance replaced*

Now undo the last command and learn how to replace all the instances of a component.

11. Select Edit > Undo Replace Component.

12. Select Replace All on the Assembly Panel Bar or toolbar.

 Note: Replace All is on the same flyout as Replace.

13. Select the component indicated in Figure 6.92.
14. Select the part file *newleg.ipt* and click the Open button. All the instances are replaced.

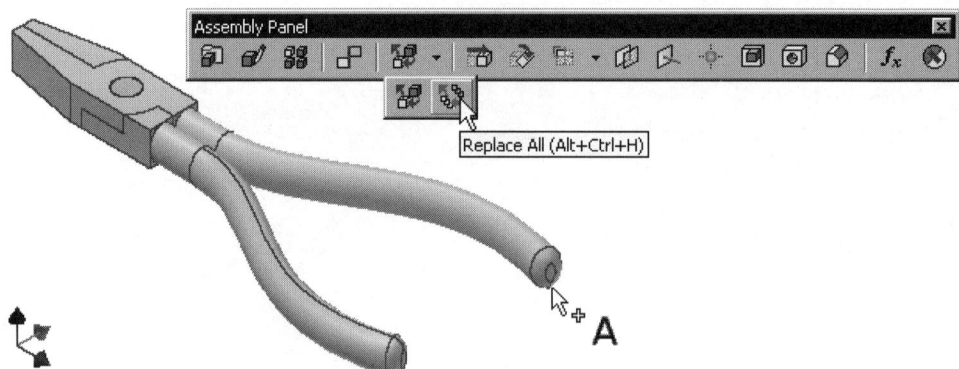

Figure 6.92 *All the instances replaced*

Save and close your file.

RESTRUCTURING COMPONENTS IN THE ASSEMBLY HIERARCHY

Objects (components or sub-assemblies) placed or created in an assembly form a hierarchy in the browser. The hierarchy refers to the sequence of placement or creation. In a large assembly that has many component parts, it is common practice to organize the components into sub-assemblies of smaller number of components. If you want to reorganize a component in an assembly into a sub-assembly, you select the part on the Browser Bar, right-click, and select Demote. If you want to move a component in a sub-assembly into the assembly, you select the part on the Browser Bar, right-click, and select Promote.

 TUTORIAL 6.15

Now you will learn how to restructure the components in an assembly to organize them into sub-assemblies.

1. Open the assembly file *Punch.iam*.
2. Select File > Save Copy As and specify a file name (*PunchReorganize.iam*).
3. Close the file *Punch.iam* and open the file *PunchReorganize.iam*.
4. Select the part file (PunchIndex) on the Browser Bar, right-click, and select Demote. (See Figure 6.93.)

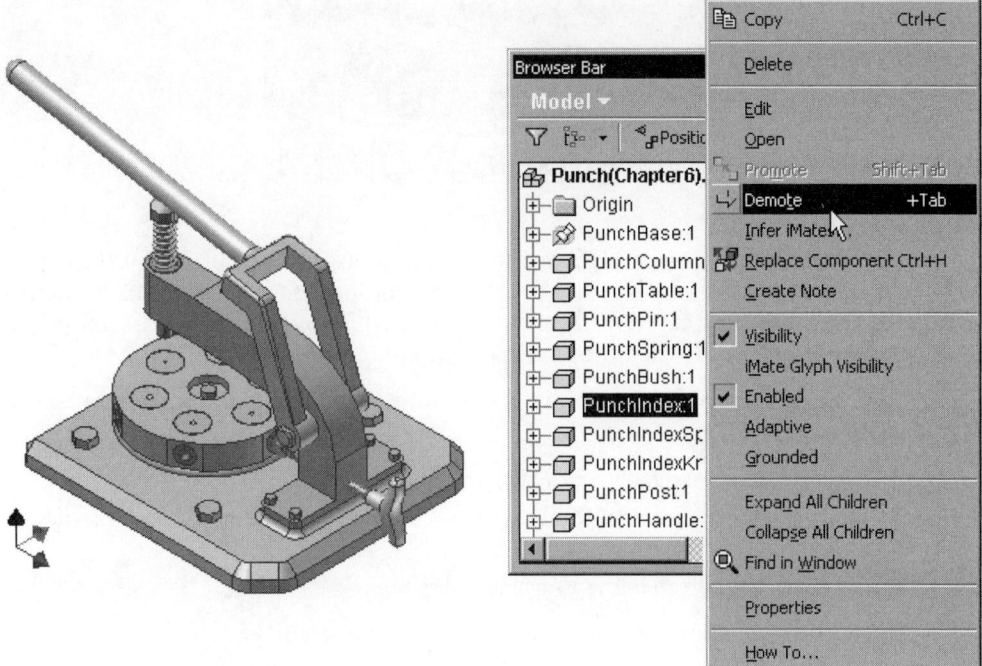

Figure 6.93 *Demoting a component*

5. In the Create In-Place Component dialog box, specify a file name (*Index.iam*) and click the OK button. (See Figure 6.94.)

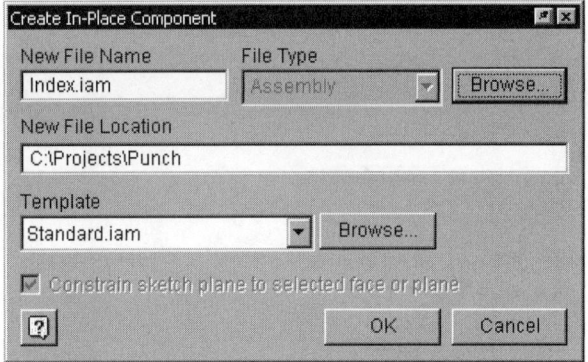

Figure 6.94 *Sub-assembly file being constructed for the demoted component*

 Note: Assembly constraints that you applied might be removed as a result of demoting.

6. In the Warning dialog box, click the Yes button. (See Figure 6.95.)

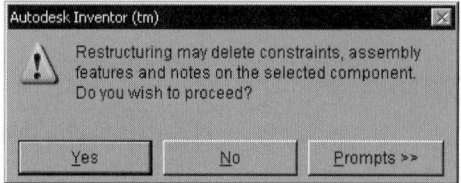

Figure 6.95 *Warning dialog box*

After demoting, an assembly is constructed, this assembly is placed in the current assembly as a sub-assembly, and the selected component is removed from the current assembly and placed in the new assembly. Consequently, the assembly constraints placed on the demoted component are removed.

Now you will learn how to move a component from the main assembly to the sub-assembly.

7. Expand the sub-assembly on the Browser Bar.
8. Select the file (PunchIndexKnob) on the Browser Bar and drag it under the sub-assembly. (See Figure 6.96.)
9. Click the OK button in the Warning dialog box.

Again, the restructured component's assembly constraints are removed because it is now moved from the current assembly to the sub-assembly.

10. Now complete the assembly by moving the file *PunchIndexSpring.ipt* to the sub-assembly and reapply the necessary assembly constraints to the components.

Save and close your files.

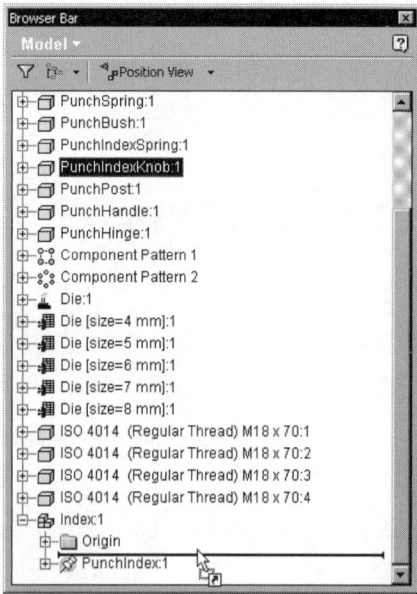

Figure 6.96 *Component being reorganized*

CENTER OF GRAVITY

You can determine the center of gravity of an assembly by displaying the center of gravity glyph.

TUTORIAL 6.16

In this tutorial, you will manipulate the center of gravity glyph.

1. Open the file *Punch.iam*, if you already closed it.

2. Select View > Center of Gravity.

If an assembly component is out of date and needs updating, an error message will be displayed. (See Figure 6.97.)

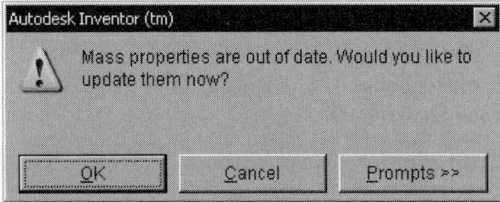

Figure 6.97 *Warning message*

3. Click the OK button to update the assembly. A center of gravity glyph is displayed.

4. On the Inventor Standard toolbar, click the arrow adjacent to the Select button, and then choose Select Features from the flyout.

5. Move the cursor over the glyph to display the coordinates of the center of gravity. (Figure 6.98.)

Note: The coordinate values may differ from yours.

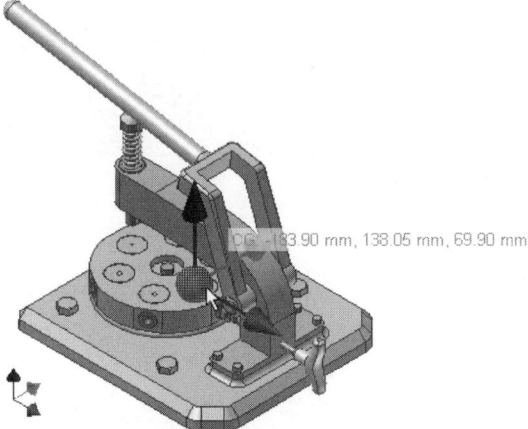

Figure 6.98 *Center of gravity glyph and its coordinates displayed*

BILL OF MATERIALS

An assembly is a collection of components. To summarize the statistics about the components in an assembly, you construct a bill of materials. Using the bill of materials, you construct a parts list in the engineering drawing.

 ## TUTORIAL 6.17

In this tutorial, you will construct a bill of materials for an assembly. A bill of materials is a table detailing the properties and textual information regarding the component parts of an assembly.

1. Open the assembly file *Punch.iam*.
2. Select Tools > Bill of Materials.
3. In the Bill of Materials dialog box, select the >> button.
4. Set the width of the columns and the alignment of the name and data. Click the OK button. (See Figure 6.99.)

The bill of materials is complete. Save your file.

	ITEM	QTY	PART NUMBER	DESCRIPTION
	1	1	PunchBase	
	2	1	PunchColumn	
	3	1	PunchTable	
	4	1	PunchPin	
	5	1	PunchSpring	
	6	1	PunchBush	
	7	1	PunchIndex	
	8	1	PunchIndexSpring	
	9	1	PunchIndexKnob	
	10	1	PunchPost	
	11	1	PunchHandle	

Figure 6.99 *Bill of materials dialog box*

PROPERTIES

You already learned about setting iProperties in Chapter 3. Now you will set properties in assembly modeling mode.

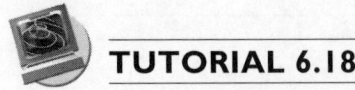

TUTORIAL 6.18

1. Select the solid part (PunchBase), right-click, and select Properties.
2. As in the part modeling Properties dialog box, there are seven tabs. In addition to the six tabs in part modeling mode, there is an Occurrence tab. You set how the solid part occurs in the current assembly.
3. Select As Material in the Color Style pull-down list box and click the OK button to close the Properties dialog box. (See Figure 6.100.)

Save and close your file.

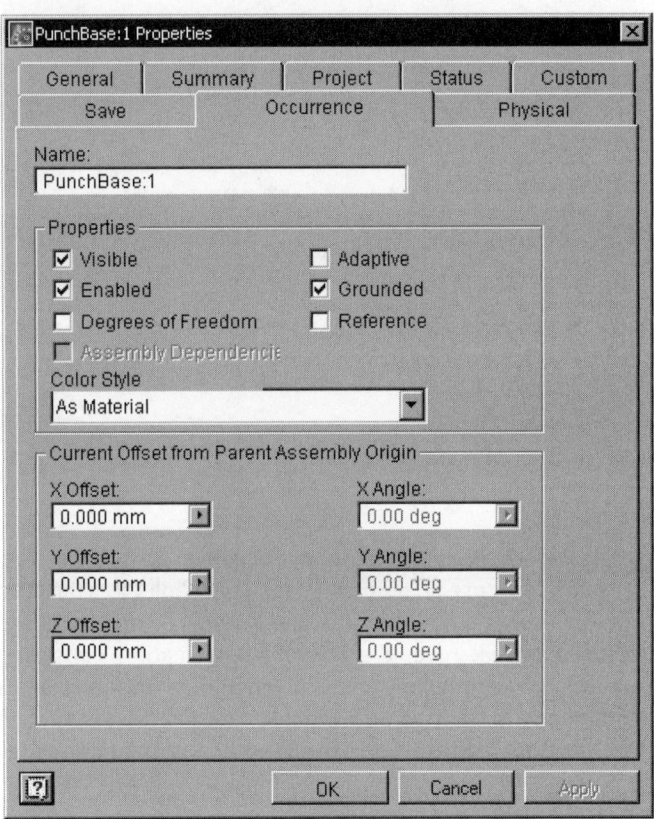

Figure 6.100 *Occurrence tab of the Properties dialog box in assembly modeling mode*

ENGINEERING DRAWING OUTPUT

Now you have completed two assemblies. If you wish to learn how to produce engineering drawings from them, you may proceed to Chapter 10. Figure 6.101 shows an engineering drawing for the assembly of pair of pliers.

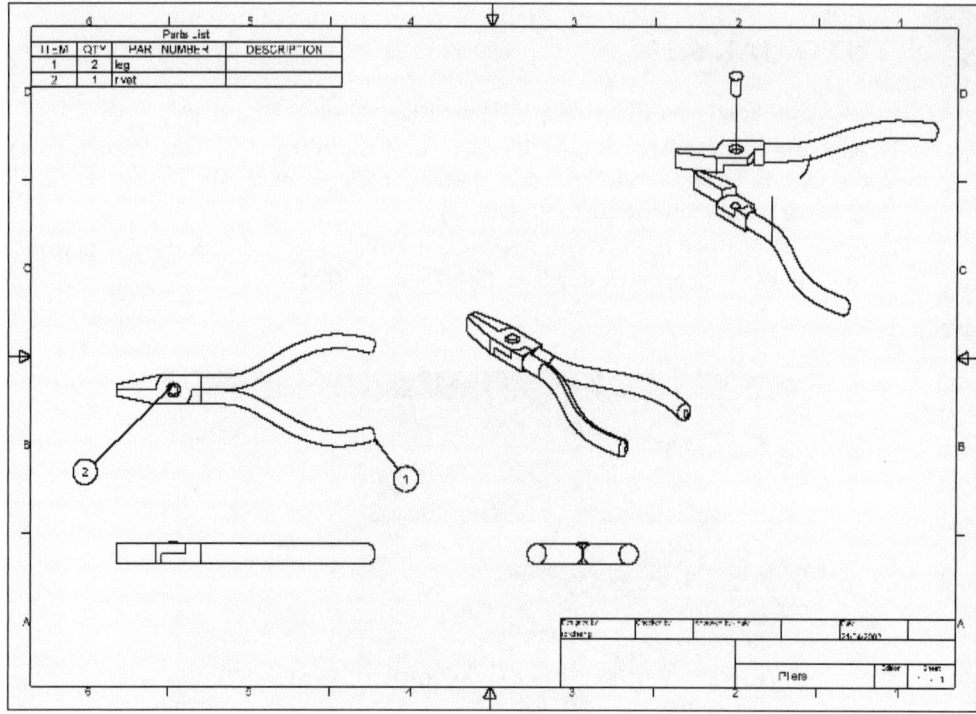

Figure 6.101 *Engineering drawing of the pliers assembly*

SUMMARY

An assembly is a collection of components put together properly to form a device. To construct an assembly, you use an assembly file, which links to a set of solid part files or sub-assembly files. In essence, definitions of the individual solid parts are stored in the part files, and an assembly file saves the assembly information, consisting of the location of the solid parts (or sub-assemblies) and the way the components are assembled. Because the components of an assembly are not saved in the assembly file, every time you open an assembly drawing, the latest versions of the solid parts are loaded.

To design an assembly, you can use three approaches: bottom-up, top-down, and hybrid. In a bottom-up approach, you construct a set of solid parts and place the components in an assembly. In a top-down approach, you start an assembly file and construct the components in the assembly. In a hybrid approach, you place some existing components in the assembly file and construct some new components while working in the assembly environment.

To speed up the design process, you can use standard parts available from the content library by selecting them and dragging them to your assembly.

In 3D space, an object in the assembly can be translated along three linear directions and rotated about three axes. The freedom of movement is called the six degrees of freedom (DOF). To restrict the movement of an object in 3D space, you can fix it by grounding. To maintain a proper positional relationship among a set of components in an assembly, you apply assembly constraints.

An assembly pattern is a set of instances of a component in an assembly. You can suppress or separate each instance individually. There are three kinds of assembly patterns: rectangular, circular, and associative. In an associative assembly pattern, the pattern corresponds to a feature pattern of another component in the assembly.

After you use assembly constraints to put all the components together, you can check interference between the components and find a way to eliminate it.

For an assembly of large numbers of components, it is common practice to organize the components into sub-assemblies of smaller number of components. You can reorganize the components in an assembly into sub-assemblies and vice versa. You can also replace a component with another component.

A parts list (bill of materials) details the number of different components in an assembly.

REVIEW QUESTIONS

1. Differentiate between a solid part file and an assembly file in terms of the data stored in the files.

2. Briefly describe the three approaches in constructing an assembly.

3. What are the six degrees of freedom of an object?

4. What kinds of assembly constraints can you apply to selected features of a pair of components?

5. What are the three kinds of assembly patterns? Briefly explain how to construct each of them.

6. Illustrate, with the aid of sketches, how components of an assembly can be organized.

7. What is the purpose of incorporating shared content link in an assembly file?

8. List the attributes that you can include in a bill of materials.

CHAPTER

Assembly Modeling II

OBJECTIVES

The aims of this chapter are to delineate advanced assembly constraint methods (the snap and go and iMate techniques), and to detail ways to simulate mechanical motion of the components in an assembly (by placing a transitional surface constraint, varying a parameter of an assembly constraint, and setting up motion relationships between components). This chapter also explains how to construct assembly features and make a welded structure. After studying this chapter, you should be able to

- Use the snap and go technique
- Construct an iMate and a composite iMate
- Apply a transitional surface constraint
- Drive a constraint (vary the parameters of an assembly constraint)
- Set relative motions between components in a system
- Construct assembly features and welded structures

OVERVIEW

Constructing an assembly consists of two major tasks: linking a set of components to the assembly file and applying assembly constraints to the components in the assembly to define proper geometric spatial relationships. For complex assemblies with many components, placing assembly constraints can be a tedious job. To place assembly constraints in a semi-automatic way, you use the snap and go technique, which is a series of dragging and dropping actions. To further automate the process of applying assembly constraints in a large assembly, you use the iMate technique to pre-define an assembly interface in individual components to specify how it will be assembled with other components, and you select the interface and drag it to a corresponding component in the assembly.

After you properly constrain the components in an assembly, you can animate the motion of individual or linked members of the assembly to illustrate mechanical motions. To keep a face of a component in contact with a set of contiguous faces of another component during motion simulation, you apply a transitional surface constraint. To animate mechanical motion, you vary the parameter of an assembly constraint in the assembly. To simulate motions among components that are not linked by any assembly constraint, you set up a motion relationship between them.

In an assembly, you can construct assembly work features to help establish sketch planes and geometric references in an assembly, and you can construct two solid features that are specific to the assembly such as a hole drilled through a set of assembled components. Moreover, you can construct a welded structure from a set of assembled parts.

SNAP AND GO

Snap and go is an assembly modeling technique whereby you apply assembly constraints semi-automatically to components of an assembly without having to use the Place Constraint command. You select a feature of a component in an assembly, press and hold down ALT, and drag the component to a feature of another component in the assembly. You will hear an audio signal after you drag the component to a constrained position. With the release of the mouse button, you place an assembly constraint.

In essence, the kind of assembly constraint applied this way depends on the geometry type of the selected features. (See first and second columns of Table 7.1.) By pressing SPACE after you release ALT (still holding down the left mouse button), you toggle between mate and flush, inside tangent and outside tangent, and direction of insert. (See third column of Table 7.1.)

Table 7.1 *Constraint applied with snap and go*

Geometry type of selected feature	Assembly constraints to be applied	Releasing ALT, and pressing SPACE after dragging the selected component
Planar face, linear edge, or axis	Mate or flush constraint	Toggles between mate and flush constraint.
Cylindrical face	Tangent constraint	Toggles between tangent constraint inside or outside the selected cylindrical face.
Circular edge	Insert constraint	Changes the direction of the insert constraint.

To change the kind of constraint in the middle of using snap and go, you release ALT and press a shortcut key. Table 7.2 below lists the shortcut keys and the constraints corresponding to the shortcut keys.

Table 7.2 *Snap and go shortcut keys*

Shortcut	Action
M or 1	Changes to a mate constraint.
A or 2	Changes to an angle constraint.
T or 3	Changes to a tangent constraint.
I or 4	Changes to an insert constraint.
R or 5	Changes to a rotation motion constraint.
S or 6	Changes to a translation constraint.
X or 8	Changes to a transitional constraint.

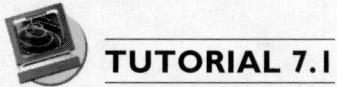

TUTORIAL 7.1

In this tutorial, you will construct an assembly of a cam set to learn how to use snap and go in assembling components of an assembly. Figure 7.1 shows the completed assembly.

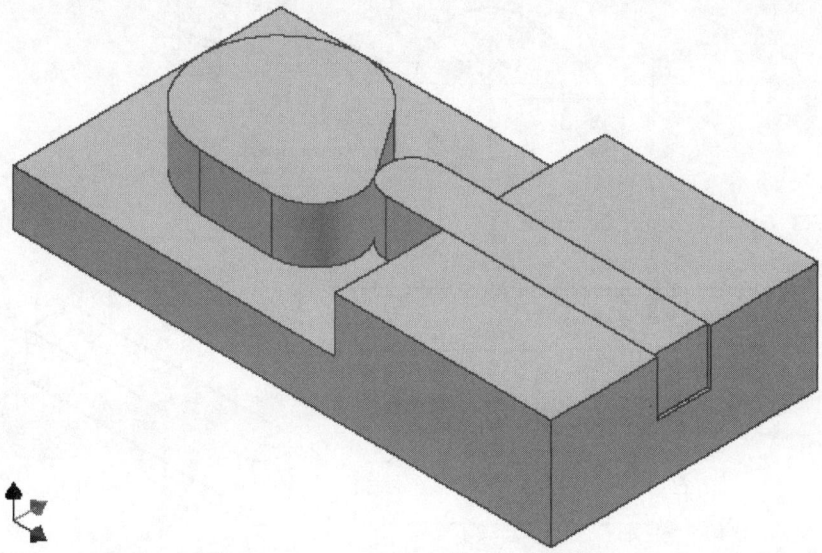

Figure 7.1 *Cam set assembly*

1. With reference to Figures 7.2 through 7.4, construct three solid parts. (You can open the part files for these components on the CD accompanying this book.)

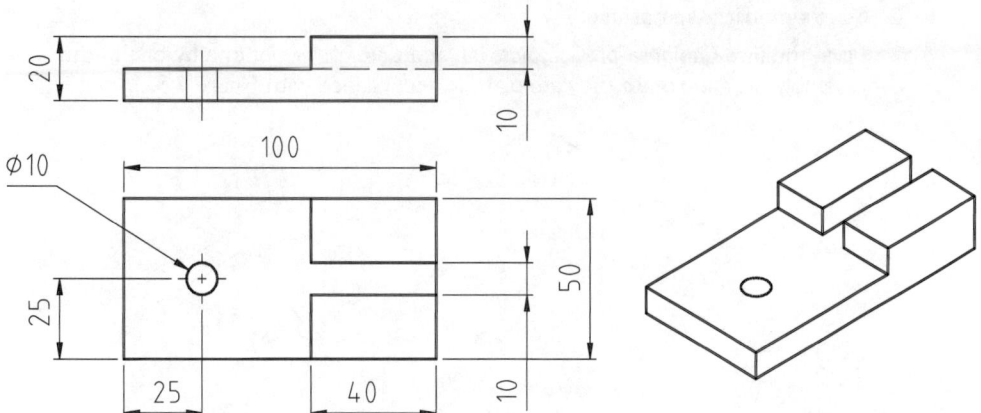

Figure 7.2 *Base unit (file name: Cambase.ipt)*

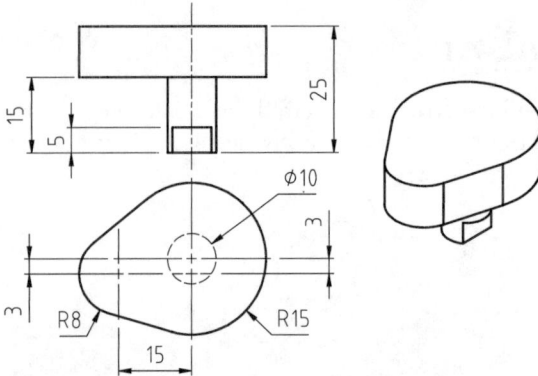

Figure 7.3 *Cam (file name: Camplate.ipt)*

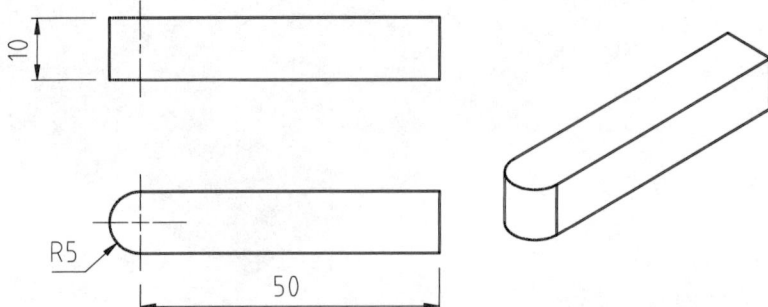

Figure 7.4 *Follower (file name: Camfollower.ipt)*

Now place the components in an assembly.

2. Start a metric assembly file.

3. Place the files *Cambase.ipt*, *Camplate.ipt*, and *Camfollower.ipt* one by one in the assembly file and rotate the cam plate in accordance with Figure 7.5.

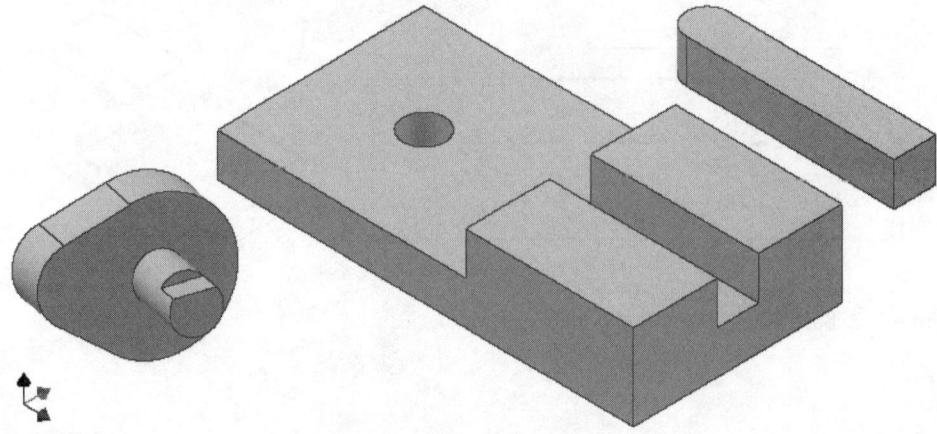

Figure 7.5 *Three components placed and a component rotated*

Now use snap and go to place an insert constraint.

4. Set the display to a hidden edge display by selecting Hidden Edge Display on the Inventor Standard toolbar.

5. Hold down ALT, select circular edge A in Figure 7.6, and hold down the left mouse button.

 Note: The insert constraint symbol is displayed at the cursor because you selected a circular edge.

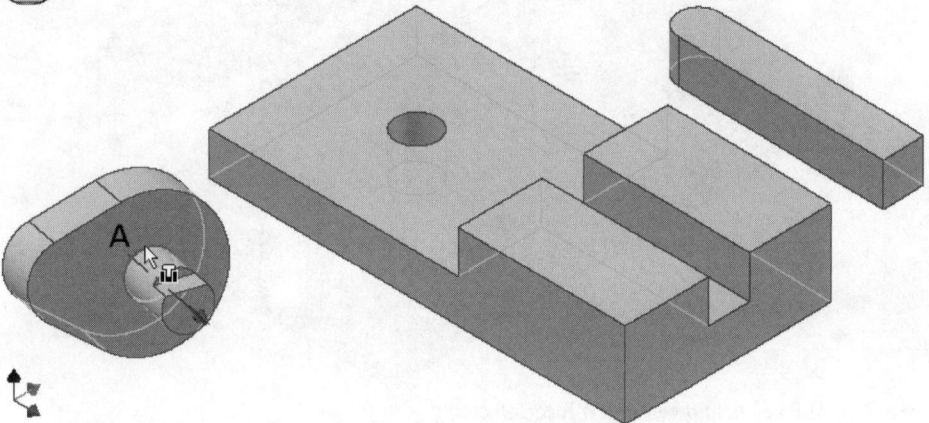

Figure 7.6 *ALT key held down and a circular edge selected*

6. Release ALT (do not release the left mouse button) and drag the component to the circular edge A in Figure 7.7.

7. Release the left mouse button.

Upon releasing the mouse button, you will hear an audio signal confirming that the assembly constraint is applied. An insert constraint is applied between the two selected circular edges.

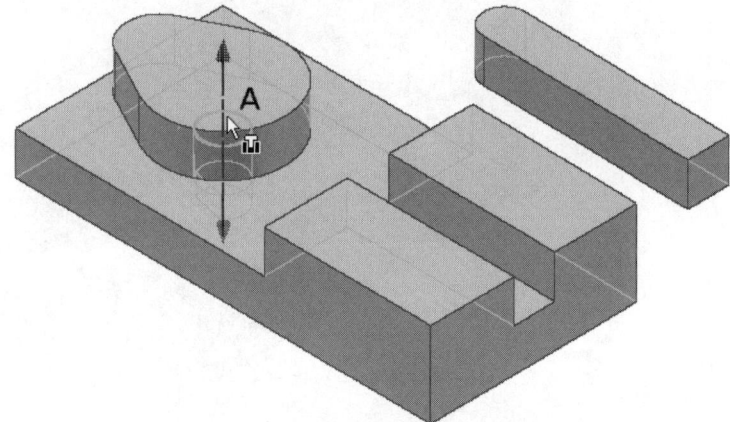

Figure 7.7 *Selected circular edge dragged to another circular edge*

Now use snap and go to place a mate constraint.

8. Hold down ALT, select face A in Figure 7.8, and hold down the left mouse button.

 Note: A mate constraint symbol is displayed at the cursor because you selected a face.

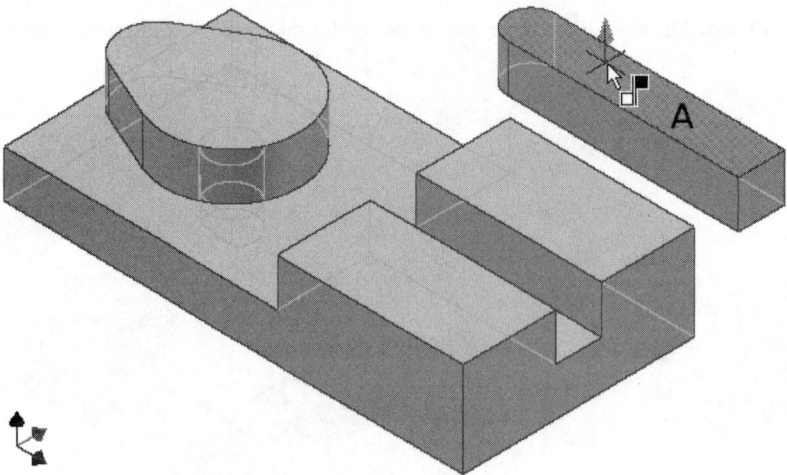

Figure 7.8 *ALT key held down and a face selected*

9. Release ALT (still holding down the left mouse button) and drag the selected component to face A in Figure 7.9.

10. Release the left mouse button.

A mate constraint is placed between the two selected faces.

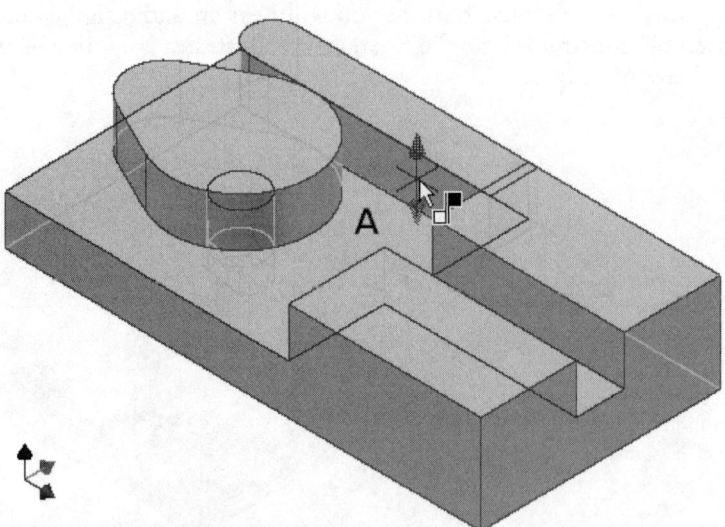

Figure 7.9 *Selected face dragged to another face*

Now place another mate constraint by using the snap and go technique.

11. Hold down ALT, select face A in Figure 7.10, and hold down the left mouse button. A mate constraint symbol is displayed at the cursor.

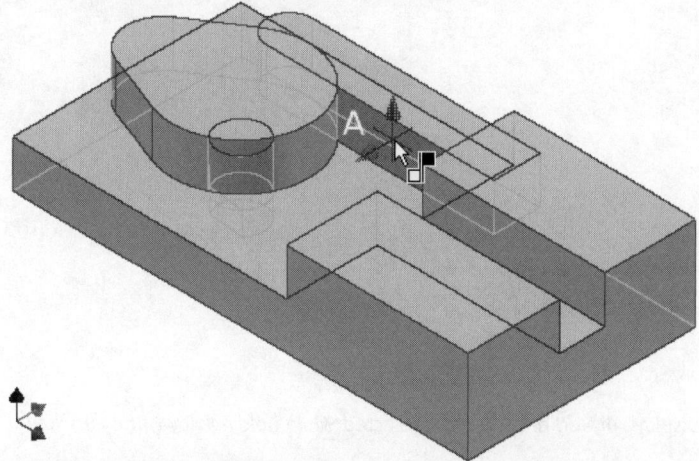

Figure 7.10 ALT *key held down and a face selected*

12. Release ALT (still holding down the left mouse button) and drag the selected component to face A in Figure 7.11.

13. Wait a moment for the selection dialog box to appear.

14. Release the left mouse button and select the left or right button in the selection dialog box until the vertical face A is highlighted.

15. Select the center button in the selection dialog box.

A mate constraint is placed.

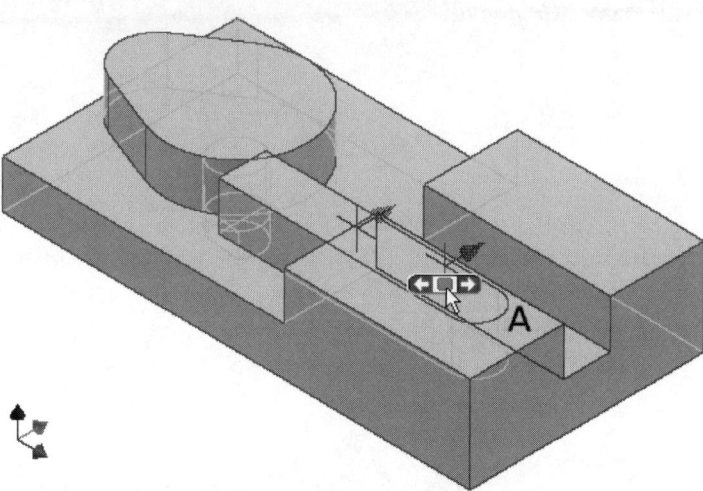

Figure 7.11 *Selected face dragged to a vertical face*

Now place an angle constraint.

16. With reference to Figure 7.12, rotate the display and change the display to shaded display.

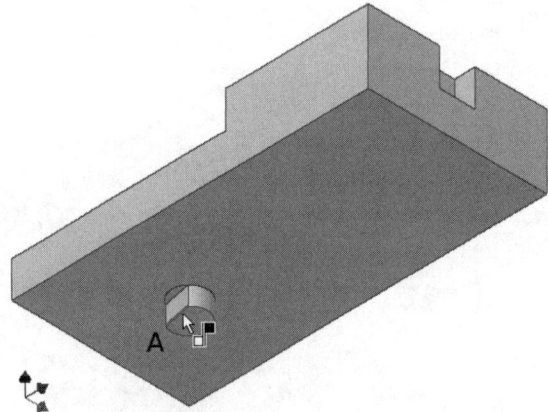

Figure 7.12 *Display rotated and an edge selected while holding down the ALT key*

17. Hold down ALT, select edge A in Figure 7.12, and hold down the left mouse button. A mate constraint symbol is displayed at the cursor.

18. Release ALT.

19. Press A to change to an angle constraint and select edge A in Figure 7.13. An angle constraint symbol is displayed.

20. Release the left mouse button.

An angle constraint is placed.

21. Set the display to an isometric view.

Save your file (file name: *Cam.iam*).

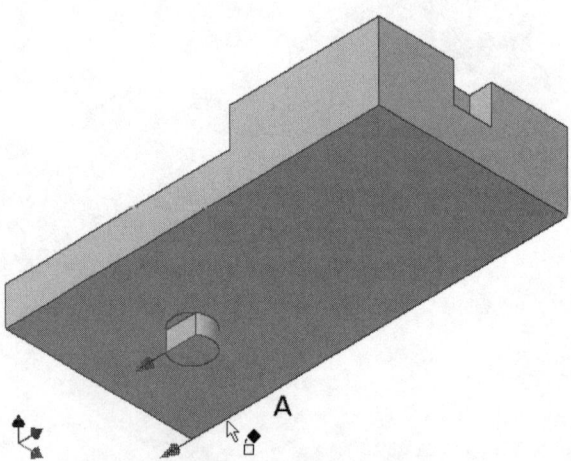

Figure 7.13 *Changed to an angle constraint and an edge selected*

iMATE

Very often, you apply the same kinds of assembly constraint to a component when you place it in any assembly. For example, you use bolts and nuts frequently and repeatedly in assemblies, and you always constrain them in the same way by using the insert constraint. Although you can use the snap and go technique to speed up the assembly constraint placement process, in a very complex assembly with many similar components like bolts and nuts, the task of adding assembly constraints can be very tedious.

To save time, you pre-define an assembly interface in the component to specify how you will constrain it in an assembly. In the assembly, you simply select the assembly interface of one component and drag it to the assembly interface of another component to assemble them together. We call the assembly interface an iMate.

In essence, the pre-defined constraint interface, iMate, is half of a constraint pair in an assembly. Therefore, the iMate command is quite similar to the Place Constraint command, except that you apply constraint to a single component in part modeling mode rather than to two components in assembly mode.

COMPOSITE iMATE

If you already constructed a number of iMate constraints in a component, you can group them together to form a composite iMate constraint by selecting them on the browser, right-clicking, and selecting Composite iMate. As a result, you can apply several assembly constraints grouped under the composite iMate in a single step.

iMATE IN iPART

As we have explained in Chapter 4, you can construct a family of parts from a solid part, an iPart. In defining an iPart, you can incorporate an iMate.

iMATE PUBLISH

To reuse assembly constraint information that you previously applied to components in the context of an assembly, you can extract such data to convert it to an iMate. We call this process of converting constraints placed in an assembly to an iMate as iMate publishing.

 ## TUTORIAL 7.2

To appreciate how to use the iMate technique in making an assembly, you will delete the cam follower from the cam assembly, construct an iMate in the cam follower and the cam base, place the modified cam follower in the assembly, and use the iMate in the assembly.

1. Open the assembly file *Cam.iam*, if you already closed it.
2. Select the cam follower and press DELETE to delete the component from the assembly. (See Figure 7.14.)

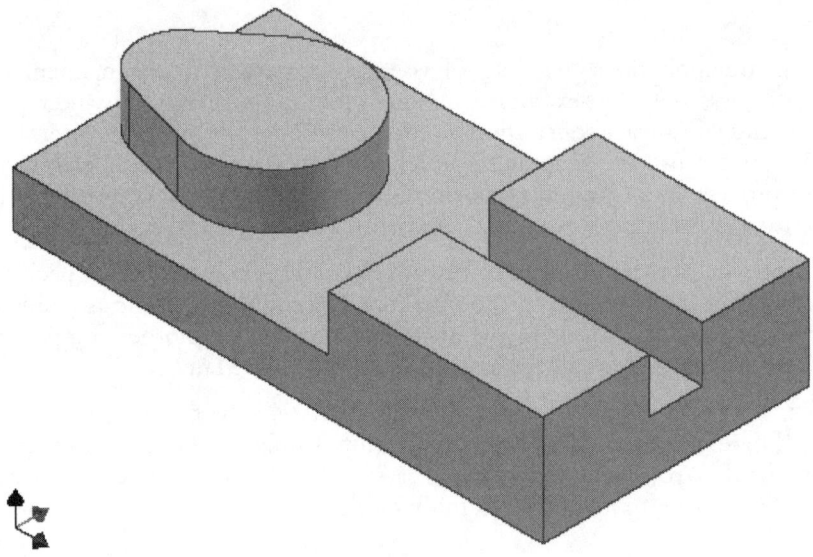

Figure 7.14 *Cam follower deleted*

Now construct two iMates in the cam follower.

3. Open the part file *Camfollower.ipt*.
4. Select Tools > Create iMate or select Add Interface from the Part Features toolbar
5. In the Create iMate dialog box, select the Mate button.
6. Select face A in Figure 7.15 and click the Apply button.

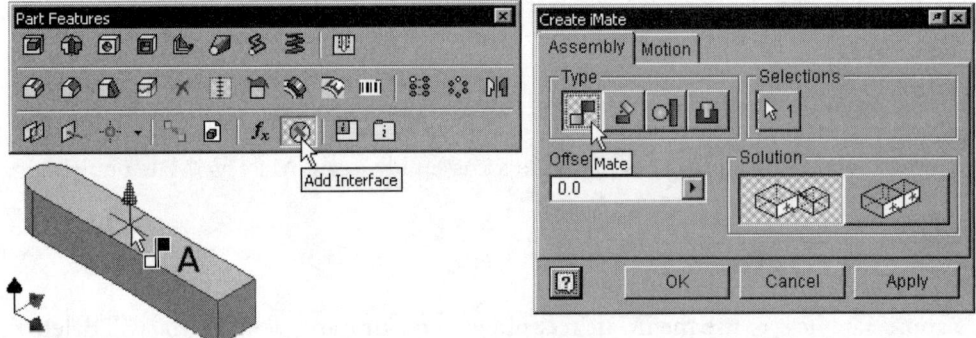

Figure 7.15 *An iMate being constructed on the horizontal face*

7. With reference to Figure 7.16, select face A and click the OK button.

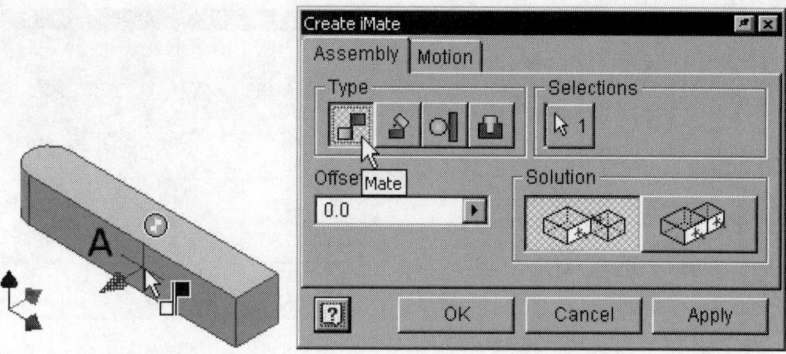

Figure 7.16 *Second iMate being constructed on the vertical face*

Now combine two iMates into a single, composite iMate.

8. Select the + sign of the iMate icon on the Browser Bar to expand it.

9. Hold down CTRL and select the iMate icons.

10. Right-click and select Create Composite. (See Figure 7.17.)

The selected iMates are combined into a single composite iMate.

11. Save and close your file.

 Tip: When two components with composite iMates are assembled, the first iMate of a component will match the first iMate of the other component. To reorder an iMate in a composite iMate, select it on the Browser Bar and drag it to a new position.

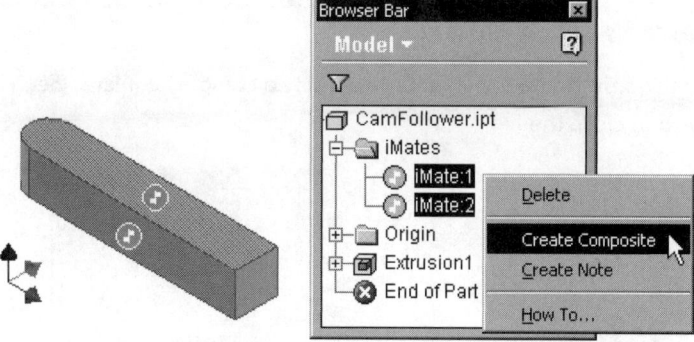

Figure 7.17 *Composite iMate constructed*

Now construct a composite iMate for the base of the cam set assembly.

12. Open the file Cambase.ipt.

13. With reference to Figures 7.18 and 7.19, construct two iMate constraints. Because it is essential that the sequence of iMate constraints in one composite iMate match the sequence of iMate constraints in another composite iMate, here you should apply iMate to the horizontal face first and then to the vertical face.

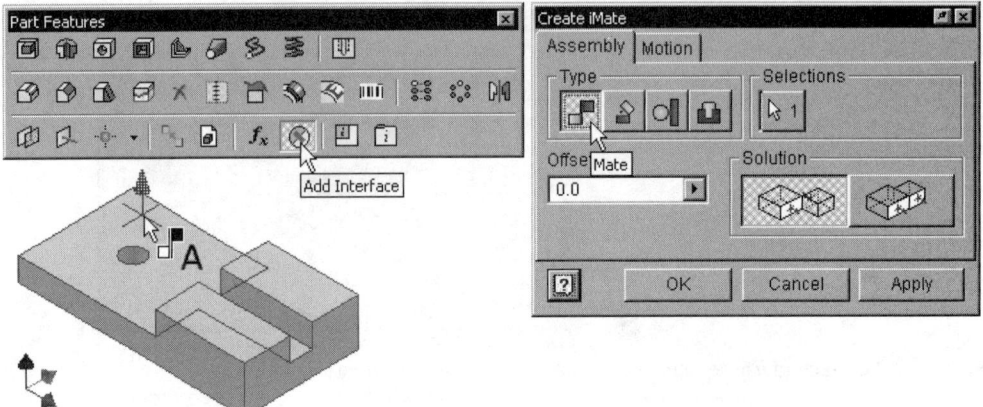

Figure 7.18 *iMate placed on a horizontal face*

Figure 7.19 *iMate placed on a vertical face*

14. Combine the iMate constraints into a single composite iMate. (See Figure 7.20.)

15. Save and close the file.

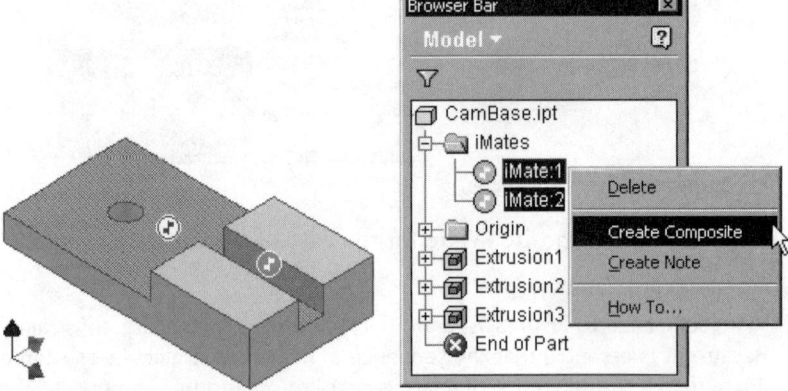

Figure 7.20 *iMates combined*

Now place the cam follower in the cam assembly and use iMate.

16. Place the cam base in the assembly, select the cam base and the cam follower one by one, right-click, and select iMate Glyph Visibility to turn on the iMate symbol.

17. Hold down ALT, select the iMate symbol of the follower, and drag it to the iMate symbol of the base component. (See Figure 7.21.)

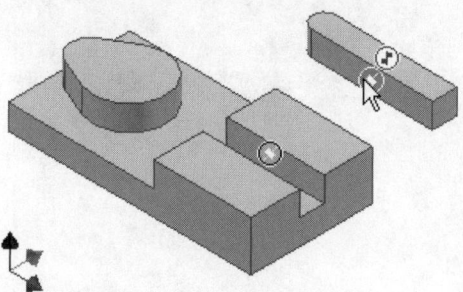

Figure 7.21 ALT *key held down and composite iMate icon selected*

18. Release the left mouse button.

The follower is assembled to the base. (See Figure 7.22.) Save your file.

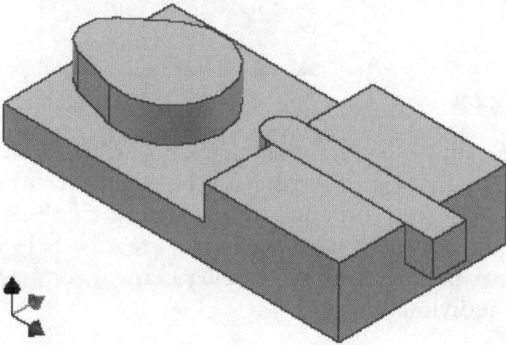

Figure 7.22 *Cam follower assembled*

TUTORIAL 7.3

In this tutorial, you will publish an iMate. In other words, you will convert the insert constraint that you applied to the cam unit to an iMate.

1. Open the assembly file *Cam.iam*, if you already closed it.

2. Select the cam plate, right-click, and select Infer iMates. (See Figure 7.23.)

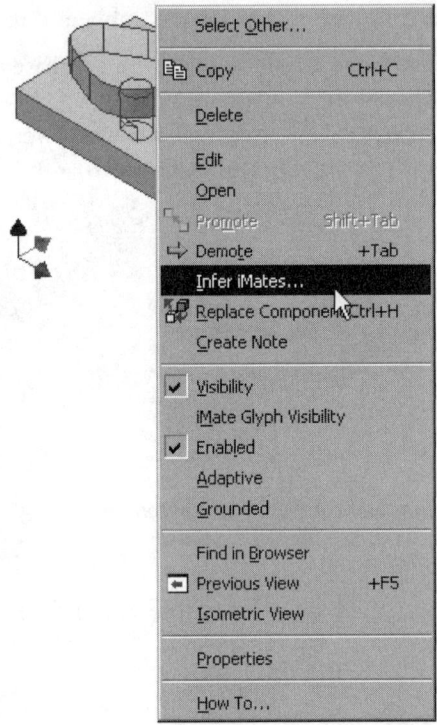

Figure 7.23 *Right-click menu*

The Infer iMates dialog box has two check boxes, Selected Occurrence Only and Create Composite iMates. Selecting the Selected Occurrence Only box applies the iMate to the selected component. If you clear this box, all the occurrences of the selected component will have the iMate created. Selecting the Create Composite iMates box puts all the iMates into a single composite iMate. Otherwise, you get a set of individual iMates.

3. In the Infer iMates dialog box, select both boxes and click the OK button. (See Figure 7.24.)

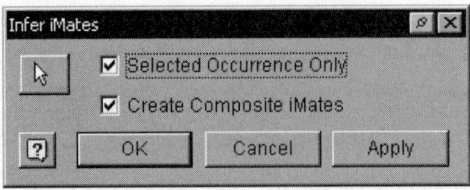

Figure 7.24 *Infer iMate dialog box*

4. To appreciate the change to the cam plate, open the *CamPlate.ipt*. In the part file, you will see the iMate created. (See Figure 7.25.)

Close the part file.

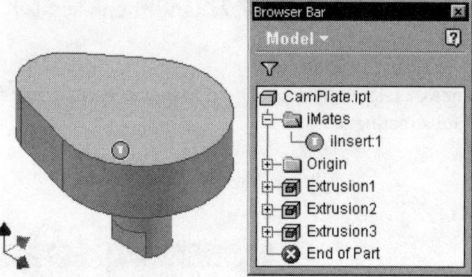

Figure 7.25 *iMate published in the part file*

TRANSITIONAL SURFACE CONSTRAINT

Using a tangent constraint, you keep two faces in contact. If you want to keep a set of contiguous faces of a component in contact with a face of another component in an assembly, you apply a transitional surface constraint. To apply a transitional surface constraint, select one of the faces of a set of contiguous faces of a component and a face of another component. A transitional surface constraint is particularly useful in simulating cam and follower motion, in which you constrain the cylindrical face of a follower to a cam plate that has a number of faces. (You will learn about motion simulation in Tutorial 7.5.)

 ## TUTORIAL 7.4

In this tutorial, you will apply a transition surface constraint to the cam and cam follower of a cam set assembly.

1. Open the assembly file *Cam.iam*, if you already closed it.
2. Select Constraint on the Assembly Panel Bar or toolbar.
3. Select the Transitional tab in the Place Constraint dialog box.
4. With reference to Figure 7.26, select the cylindrical face of the follower.

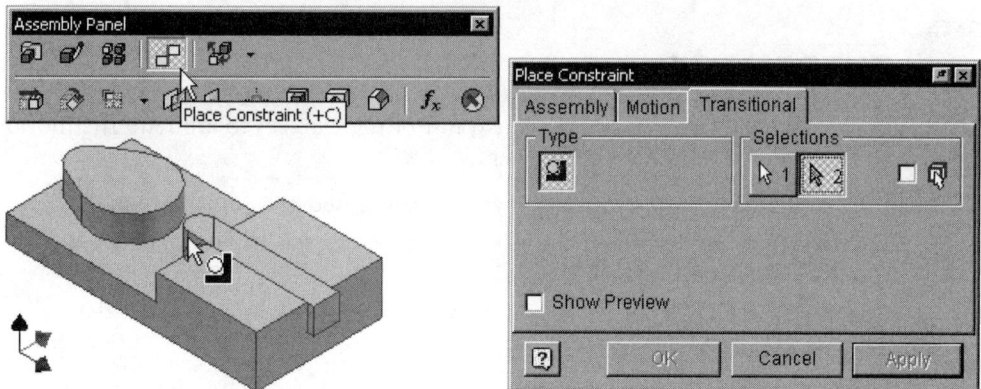

Figure 7.26 *Single face selected*

5. Select the face indicated in Figure 7.27, and then click OK. A transitional constraint is applied.

 Tip: The sequence of selection is crucial: You select the single face first and then select a face from the set of contiguous mating faces.

Save your file.

Figure 7.27 *One of the set of contiguous faces selected*

CONSTRAINT DRIVING

To help visualize and evaluate design, you set the component(s) of an assembly in motion by manipulating the animation parameters of an assembly constraint that you apply to a component of an assembly. We call this kind of motion simulation constraint driving. To drive a constraint, you select an assembly constraint on the Browser Bar, right-click, and select Drive Constraint to use the Drive Constraint command. In the Drive Constraint dialog box, you set the animation parameters.

In an assembly with components properly assembled, driving a constraint will animate other related components as well.

 ## TUTORIAL 7.5

In this tutorial, you will drive the angle constraint of the cam set to animate the motion between the cam plate and the cam follower.

1. Open the assembly file *Cam.iam*, if you already closed it.
2. On the Browser Bar, select the insert constraint for the drrive, right-click, and select Drive Constraint.
3. In the Drive Constraint dialog box, set the End Angle to 360 and select the Forward button. (See Figure 7.28.)

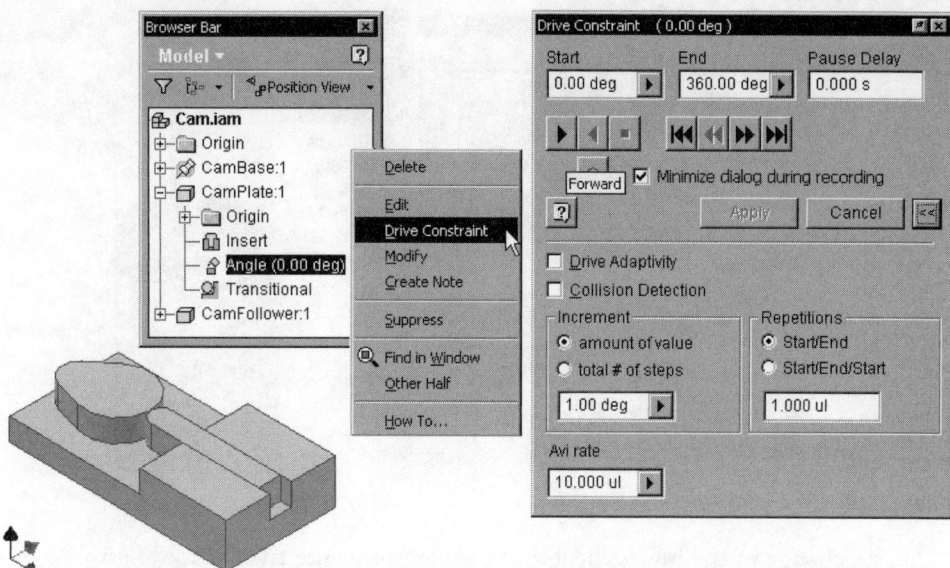

Figure 7.28 *Angle constraint being driven to animate cam plate and cam follower motion*

The change of the angle constraint's value from 0 deg to 360 deg causes the cam plate to rotate. Because there is a transitional constraint keeping the cylindrical face of the cam follower in contact with the set of contiguous faces of the cam plate, the follower is translated in a linear direction. If there are other components connected to the cam plate or cam follower, they will be translated as well. To save the animation in an AVI file, select the Record button in the Drive Constraint dialog box, specify a file name, and select the Play button.

Now terminate the constraint driving and close the file.

    4. Click the Cancel button in the Drive Constraint dialog box.

The cam assembly is complete. Save and close your file.

 TUTORIAL 7.6

To further familiarize yourself with constraint driving, you will drive the constraints of the pliers assembly.

    I. Open the assembly file *Pliers.iam*.

    2. Select the insert constraint on the Browser Bar, right-click, and select Drive Constraint.

    3. In the Drive Constraint dialog box, set the End distance to 10 mm and select the Forward button. (See Figure 7.29.)

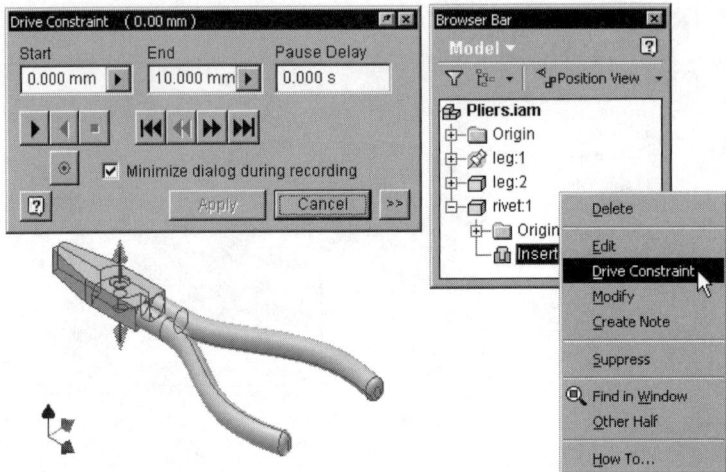

Figure 7.29 *Insert constraint being driven*

The change in the insert constraint's value causes the rivet to move up.

4. Click Cancel in the Drive Constraint dialog box.

5. On the Browser Bar, select the angle constraint for leg 2, right-click, and select Drive Constraint.

6. In the Drive Constraint dialog box, set the End Angle to 190 deg and select the Forward button. (See Figure 7.30.)

The change in the angle constraint's value causes a leg of the pliers to rotate about the axis of the rivet.

7. Click Cancel in the Drive Constraint dialog box.

Now close your file.

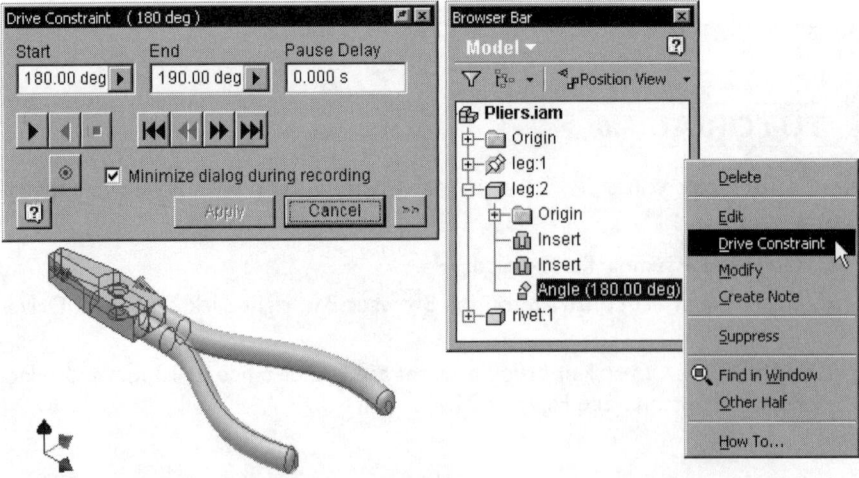

Figure 7.30 *Angle constraint being driven*

SET RELATIVE MOTION

In a properly constrained mechanical linkage system, you can use constraint driving to simulate mechanical motion. To cause a set of related components (like gears, friction wheels, and chain sprockets) in an assembly to move simultaneously according to a pre-defined speed ratio, you set up relative motions between the components.

COMPOUND GEAR SYSTEM

In a compound gear system, the rotation of a shaft will cause another shaft to rotate at a speed that is determined by the gear ratio. To simulate this kind of motion, you need to set up relative motions between the components in the gear system assembly.

 TUTORIAL 7.7

In this tutorial, you will construct a friction drive assembly and learn how to set up relative motions between the components in the assembly. Figure 7.31 shows the assembly with the upper casing exploded, with the standard engineering components such as bearings, seals, and fasteners omitted for the sake of simplicity in illustration.

The mechanism has five component parts: lower casing, upper casing, input shaft, intermediate shaft, and output shaft.

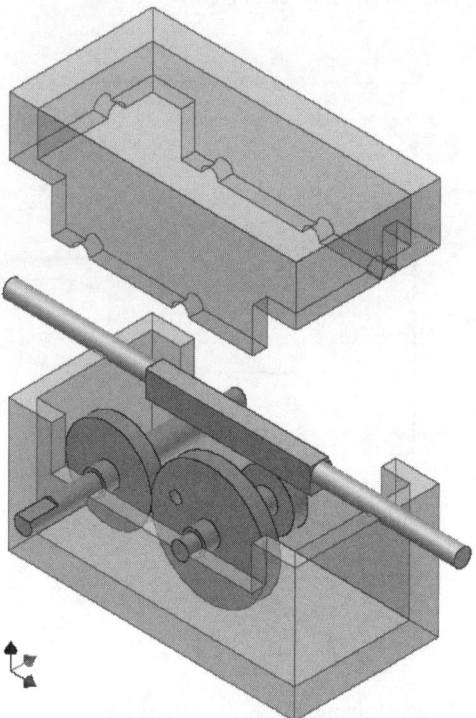

Figure 7.31 *Friction drive mechanism with the upper casing exploded*

1. Construct the part files as shown in Figures 7.32 through 7.36. (You can open the part files of these components on the CD accompanying this book.)

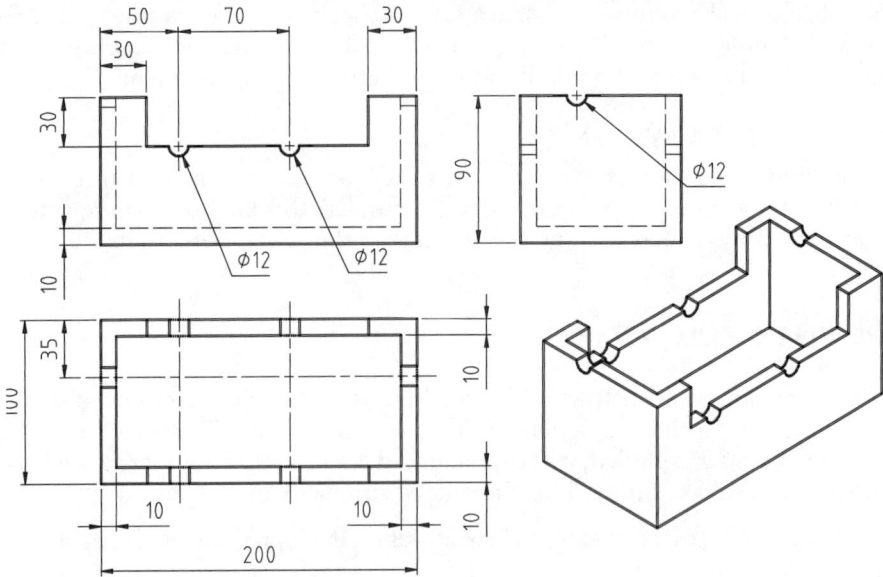

Figure 7.32 *Lower casing (file name: LowerCasing.ipt)*

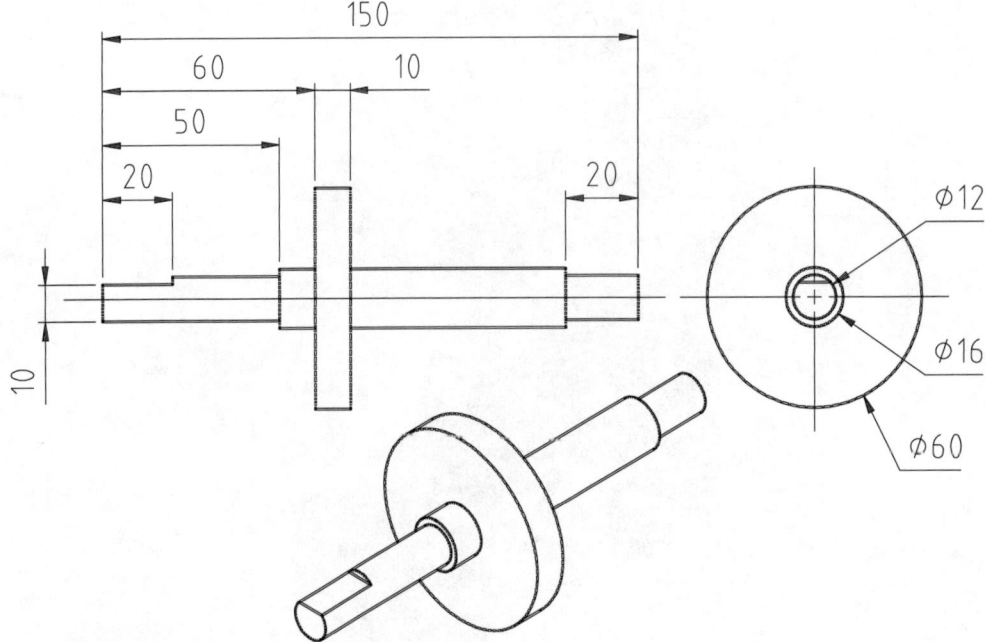

Figure 7.33 *Input shaft (file name: Input.ipt)*

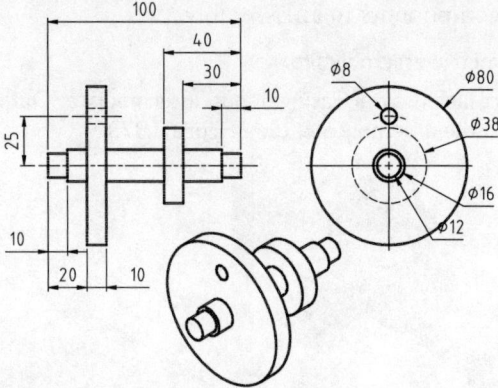

Figure 7.34 *Intermediate shaft (file name: Intermediate.ipt)*

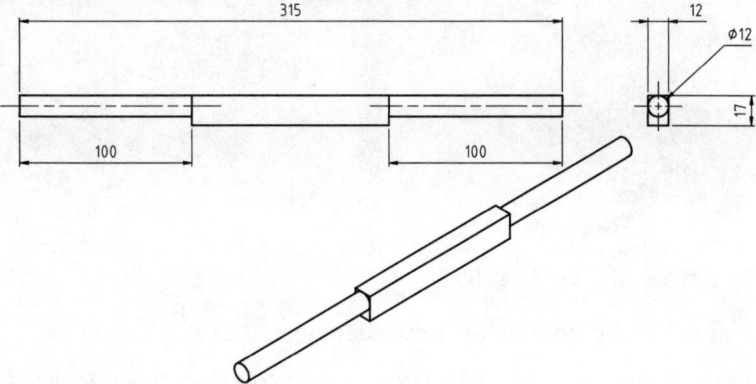

Figure 7.35 *Output (file name: Output.ipt)*

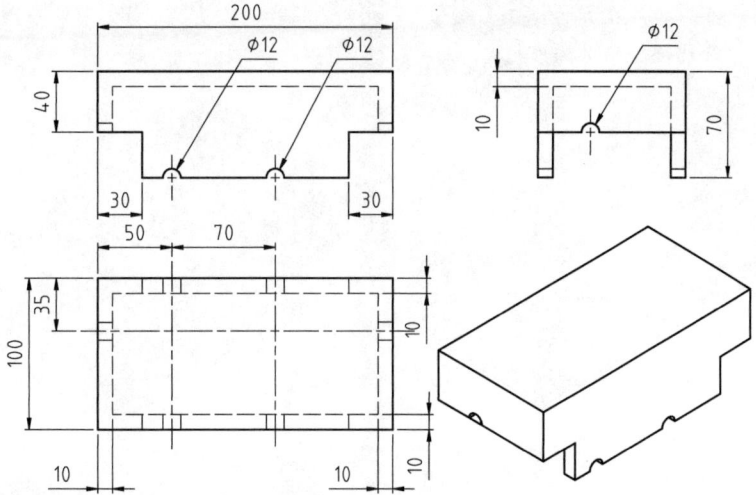

Figure 7.36 *Upper casing (file name: UpperCasing.ipt)*

Now place the components in an assembly.

2. Start a new metric assembly file.

3. Place the part files (LowerCasing, Input, Intermediate, Output, and UpperCasing) one by one in the assembly file. (See Figure 7.37.)

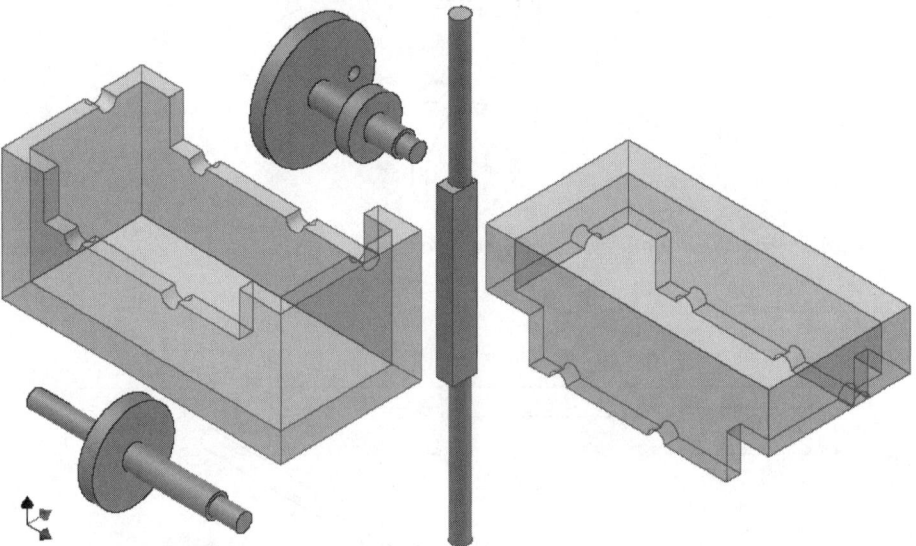

Figure 7.37 *Parts placed in the assembly*

Now add assembly constraints to the components.

4. With reference to Figure 7.38, place an insert constraint to locate the input shaft.

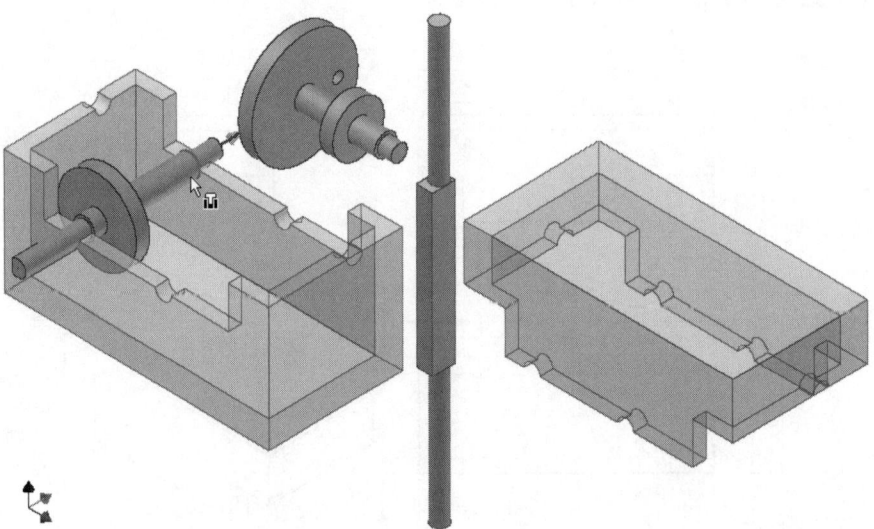

Figure 7.38 *Insert constraint being placed*

5. Add an angle constraint to the input shaft. (See Figure 7.39.)

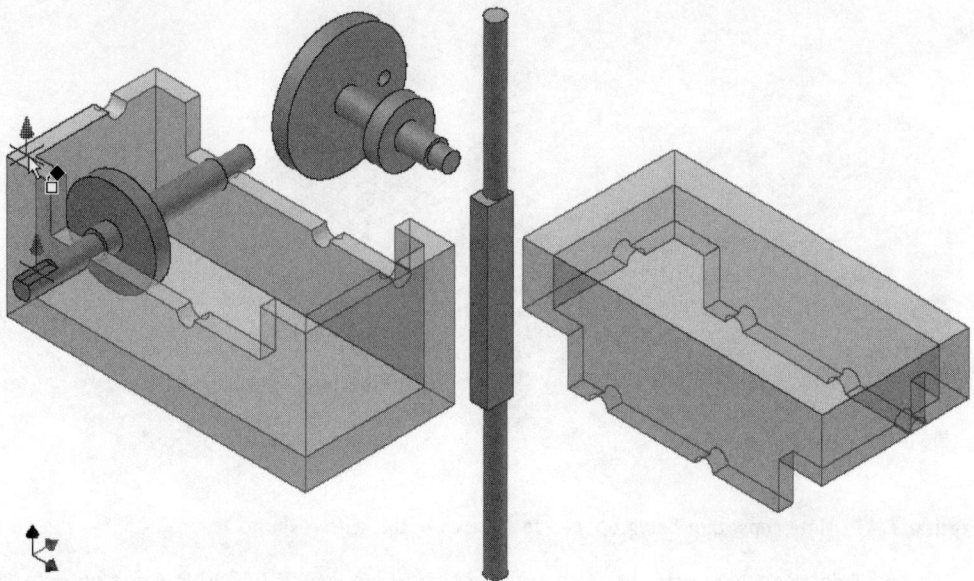

Figure 7.39 *Angle constraint being placed*

Now the input shaft has no DOF. If you drive the angle constraint, it will rotate.

6. Add an insert constraint to locate the intermediate shaft. (See Figure 7.40.) Now the intermediate shaft has one DOF (rotation) remaining.

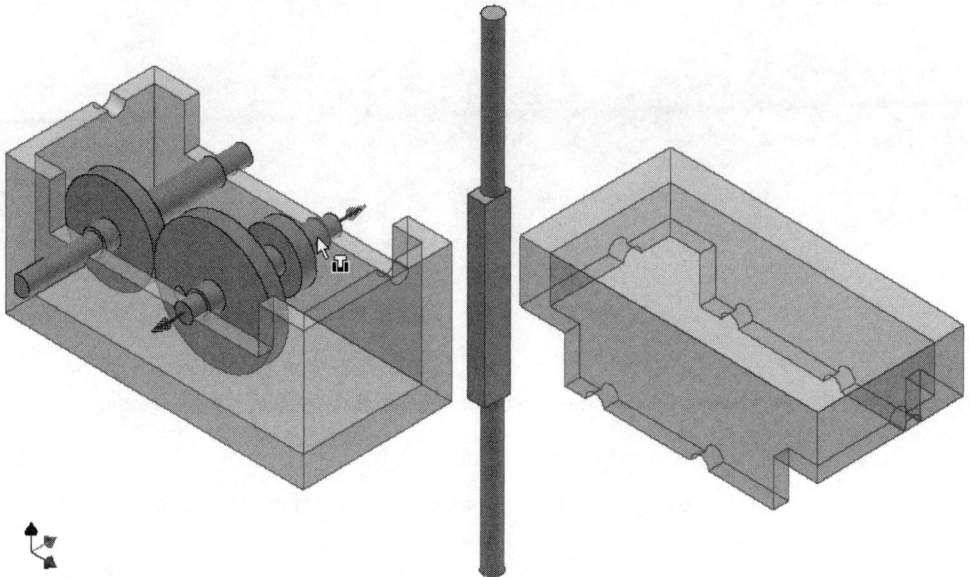

Figure 7.40 *Insert constraint applied to the intermediate shaft*

7. Add a mate constraint to locate the axis of the output shaft. (See Figure 7.41.)

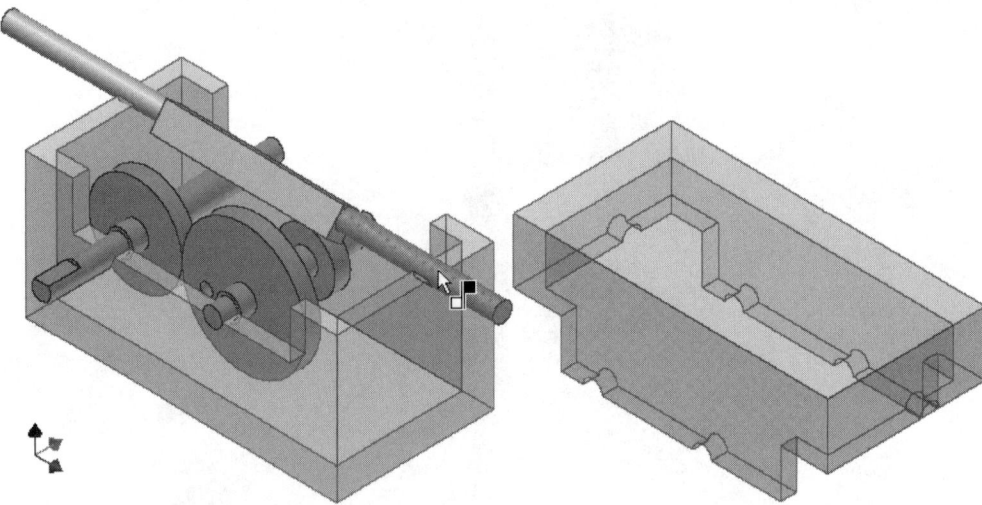

Figure 7.41 *Mate constraint being applied to the axis of the output shaft*

8. Add an angle constraint to the output shaft. (See Figure 7.42.) Now the output shaft has one DOF (linear translation) remaining.

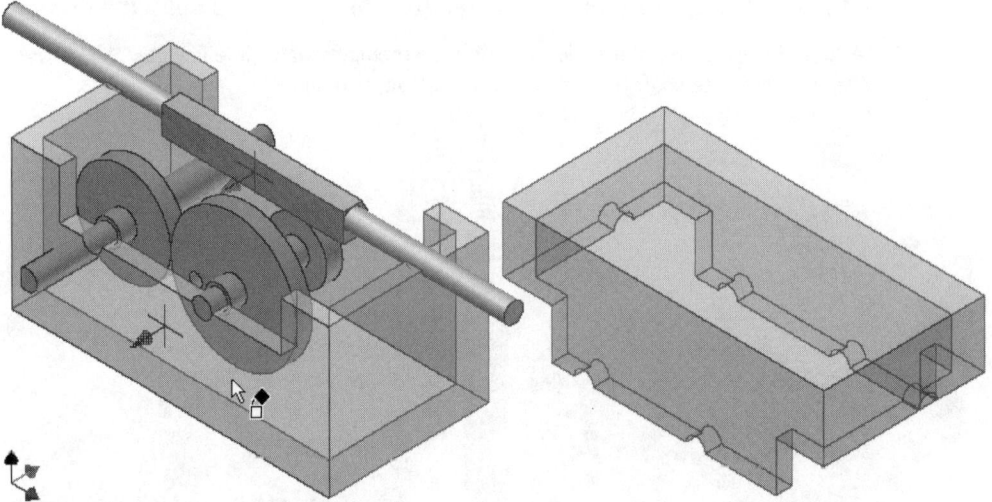

Figure 7.42 *Angle constraint being applied to the output shaft to prevent it from rotating*

Now you will set the upper casing to be invisible.

9. With reference to Figure 7.43, place assembly constraints to properly locate the upper casing.

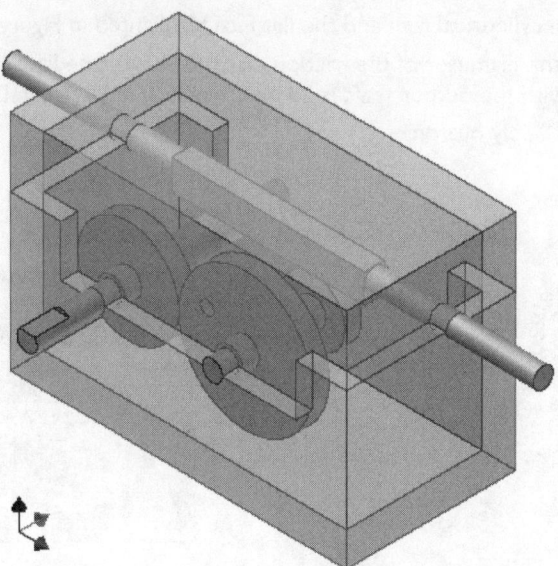

Figure 7.43 *Upper casing assembled*

10. Select the upper casing, right-click, and deselect Visibility. (See Figure 7.44.)

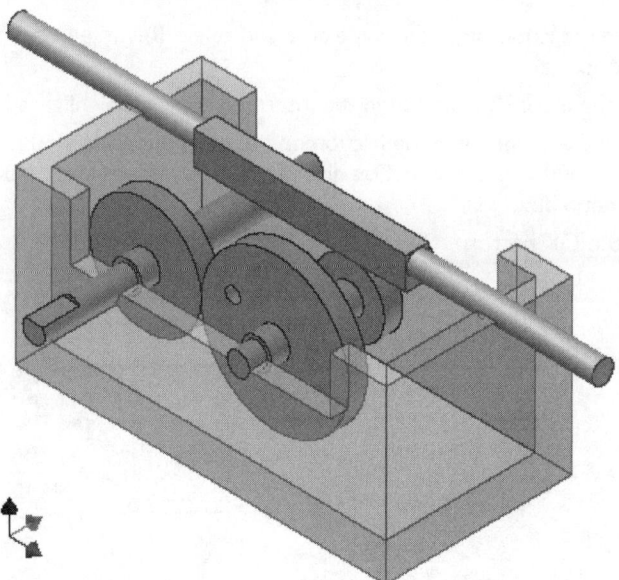

Figure 7.44 *Upper casing invisible*

Now set relative motions between the components.

11. Select Constraint on the Assembly Panel Bar or toolbar.

12. In the Place Constraint dialog box, select the Motion tab, select Rotation-Translation in the Type area, and select Forward in the Solution area.

13. Select the cylindrical face and the flat face highlighted in Figure 7.45.

14. Because the diameter of the friction disc of the intermediate shaft that is in contact with the output shaft is 38 mm, type **22/7*38** in the Distance box.

15. Click the Apply button.

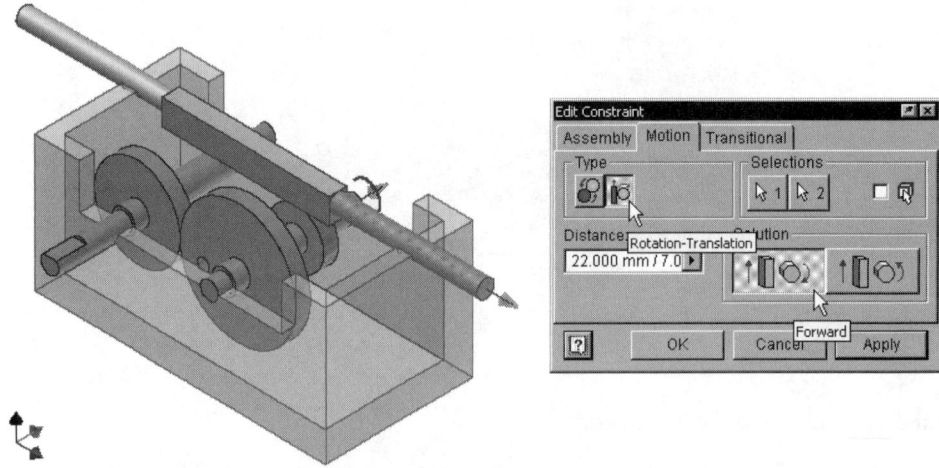

Figure 7.45 *Rotation-Translation motion being set between the intermediate shaft and the output shaft*

16. Now select Rotation in the Type area and select Reverse in the Solution area of the Motion tab.

17. Select the input shaft and then the intermediate shaft highlighted in Figure 7.46.

18. Because the diameter of the friction disc of the input shaft is 60 mm and the diameter of the meshing friction disc of the intermediate shaft is 80 mm, type **30/40** in the Ratio box.

19. Click the OK button.

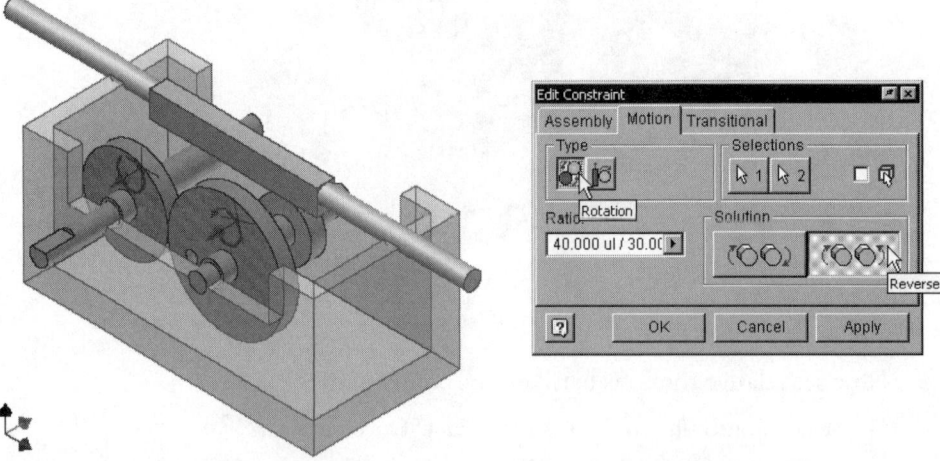

Figure 7.46 *Rotational motion being set between the input shaft and the intermediate shaft*

Now drive the angle constraint of the input shaft to animate the rotational motion of the input shaft. Because there is a relative motion between the input shaft and the intermediate shaft and a relative motion between the intermediate shaft and the output shaft, the output shaft will be translated in a linear direction.

20. Select the angle constraint of the input shaft highlighted in Figure 7.47, right-click, and select Drive Constraint.

21. In the Drive Constraint dialog box, set End to 360 degrees and select the >> button to expand the dialog box.

22. Set total # of steps to 10.

23. Select the Forward button.

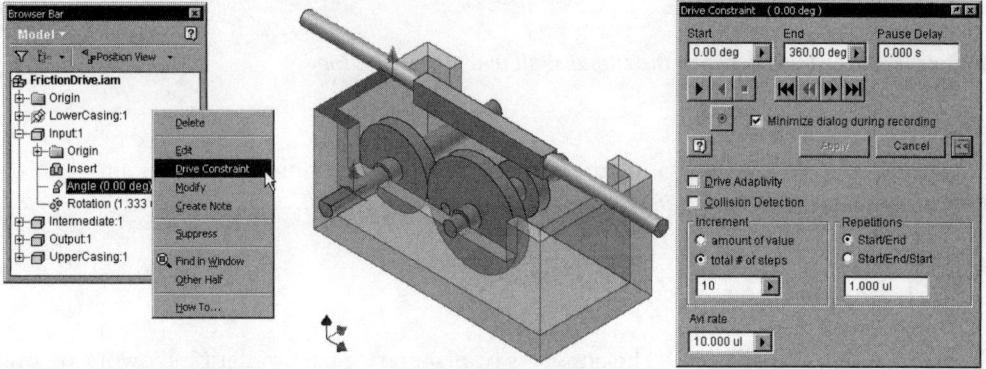

Figure 7.47 *Angle constraint of the input shaft being driven*

After you select the Forward button, the output shaft will be translated to the right. You will find that the output shaft interferes with the casing.

24. Select the Reverse button to reverse the mechanism.

25. To check whether there is any interference among the components, select Collision Detection in the Drive Constraint dialog box.

26. Select the Forward button again.

The mechanism stops when a face of the output shaft collides. (See Figure 7.48.)

27. Select the Reverse button and click the Cancel button.

Save and close your file (file name: *FrictionDrive.iam*).

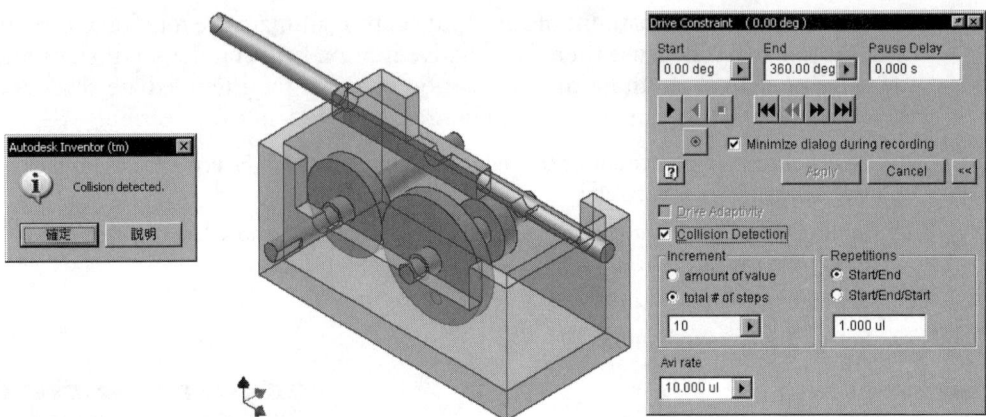

Figure 7.48 *Collision between the output shaft and the lower casing*

PLANETARY GEAR SET

By setting relative motions among components of a mechanism, you can simulate the motion of a set of planetary gears.

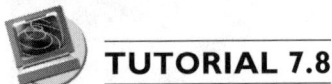

TUTORIAL 7.8

In this tutorial, you will learn the concepts of planetary gear simulation by working on a planetary friction wheel system. Figure 7.49 shows the completed assembly.

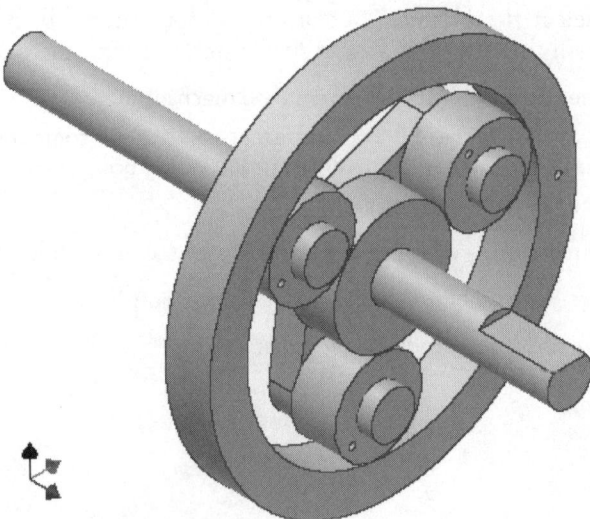

Figure 7.49 *Planetary wheels assembly*

1. With reference to Figures 7.50 through 7.53, construct four part files. (You can open the part files of these components on the CD accompanying this book.)

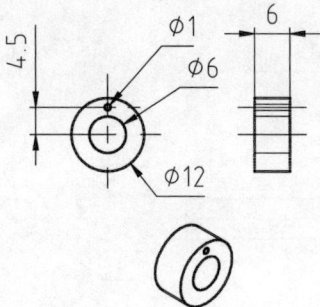

Figure 7.50 *Planet wheel (file name: PlanetWheel.ipt)*

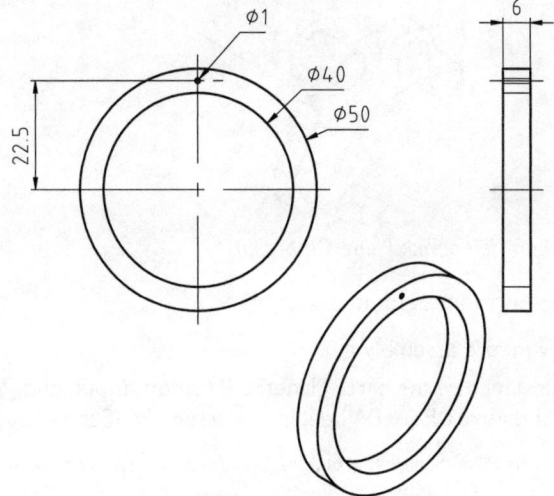

Figure 7.51 *Ring wheel (file name: PlanetRing.ipt)*

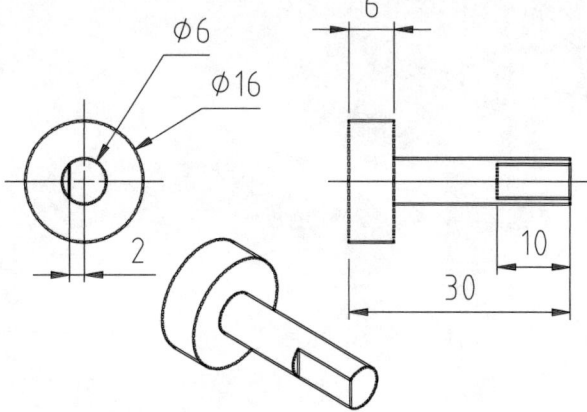

Figure 7.52 *Sun wheel (file name: PlanetSun.ipt)*

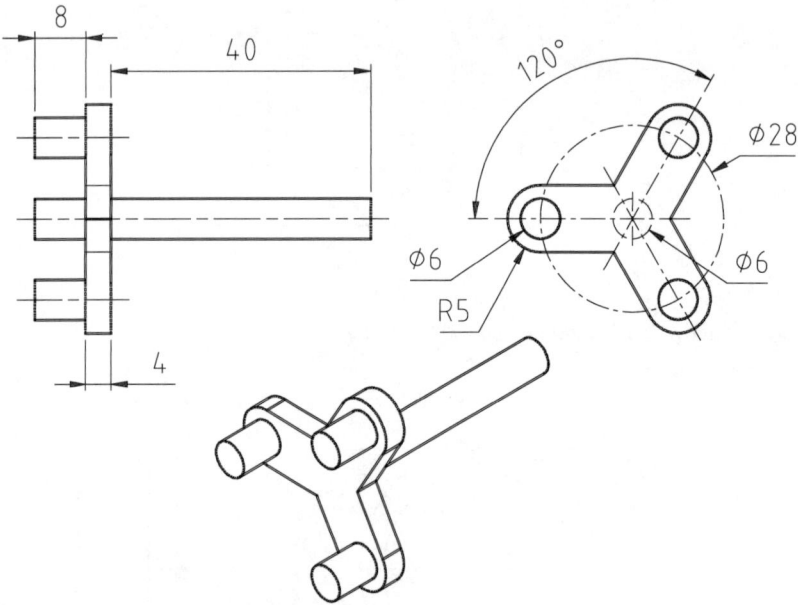

Figure 7.53 *Output shaft (file name: PlanetOutput.ipt)*

Now put the components in an assembly.

2. Start a new metric assembly file.

3. Place one instance of the parts PlanetSun, PlanetOutput, and PlanetRing and three instances of the part PlanetWheel in the assembly. (See Figure 7.54.)

 Note: The part PlanetSun.ipt is grounded because it is the first component placed in the assembly.

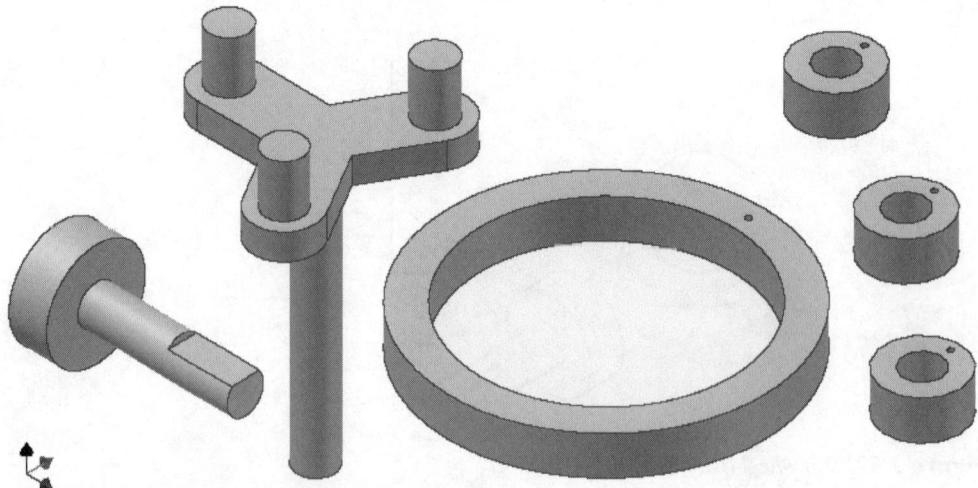

Figure 7.54 *Components placed in the assembly*

Now place assembly constraints to assemble the components.

4. Apply insert constraints on the planet wheels and the output shaft. (See Figure 7.55.)

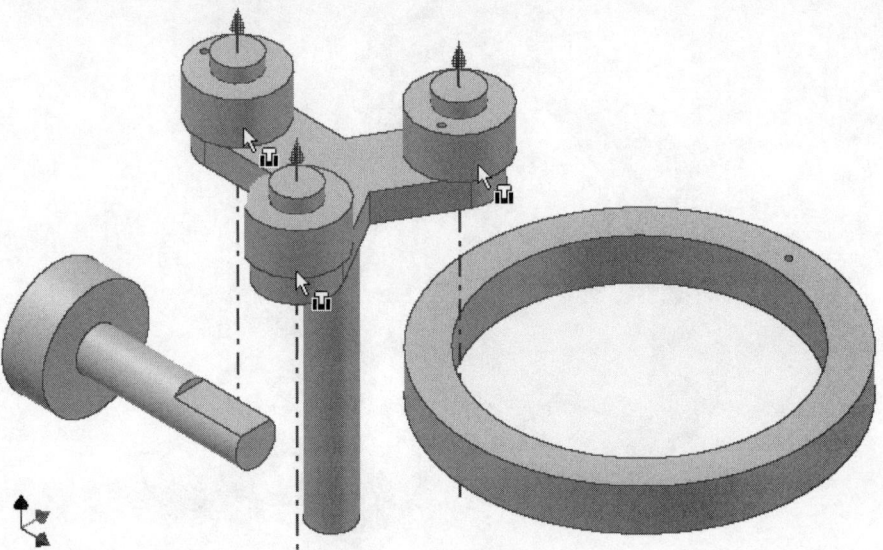

Figure 7.55 *Insert constraints being applied to the planet wheels and the output shaft*

5. Add insert constraints to assemble the ring wheel to the sun wheel. (See Figure 7.56.)

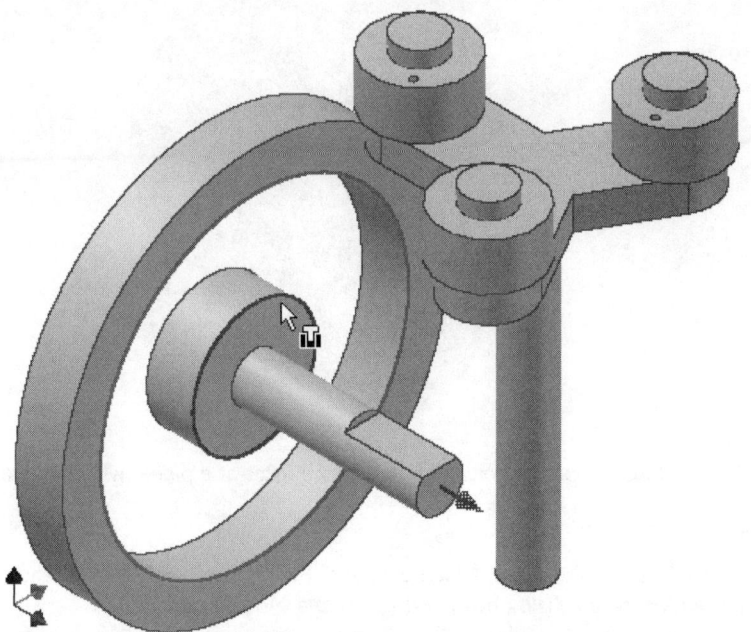

Figure 7.56 *Insert constraints applied to the ring wheel and the sun wheel*

6. Add a mate constraint to the axes of the sun wheel and the output shaft. (See Figure 7.57.)

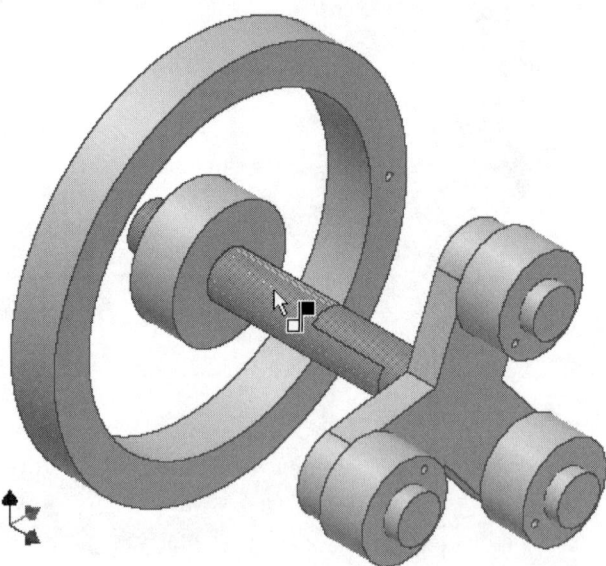

Figure 7.57 *Mate constraint being applied to the axes of the sun wheel and the output shaft*

7. Add a mate (flush) constraint to the faces of a planet wheel and the ring wheel. (See Figure 7.58.)

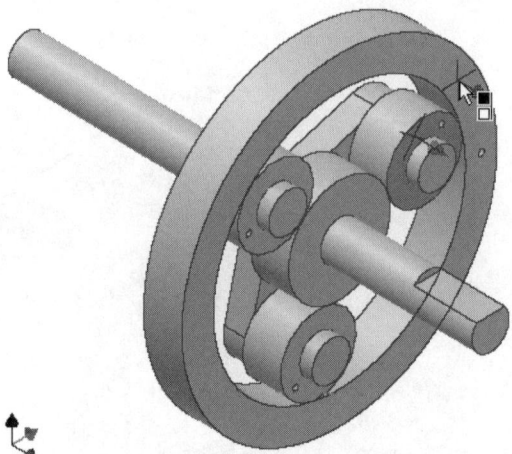

Figure 7.58 *Mate (flush) constraint being applied to the faces of a planet wheel and the ring wheel*

Now apply motion constraint.

8. Select Constraint on the Assembly Panel Bar or toolbar (if you already closed the Place Constraint dialog box) and select the Motion tab.

9. Select Rotation in the Type box and select Forward in the Solution box.

10. Set the ratio to 12/40 and select the planet wheel and then the ring wheel. (See Figure 7.59.)

 Note: The sequence of selecting components is related to the ratio.

11. Click the Apply button. Motion between the planet wheel and the ring wheel is set.

Figure 7.59 *Relative motion between the planet wheel and the ring wheel being set*

12. Refer to Figure 7.60, select Reverse in the Solution box and set the ratio to 12/16.
13. Select the planet wheel and then the sun wheel.
14. Click the OK button.

Motion relation is set. Test the motion by selecting and rotating the ring wheel. If the motion is correct, save and close your file (file name: *Epicyclic.iam*).

Figure 7.60 *Relative motion between the planet wheel and the sun wheel being set*

ASSEMBLY FEATURES

In the context of an assembly, you can construct work features and solid features. In addition, you can construct a welded structure.

To help locate components in an assembly, you construct assembly work features. In part files, work features serve mainly as references for making solid features. In assembly files, we use work features as references mainly for the assembly of components.

Quite often, we have to perform machining processes on a set of component parts after they are assembled. For example, we use match drilling to ensure that holes on two components are perfectly matched to each other. In match drilling, we clamp the parts together and drill a hole through them. As engineering designers, we need to tell people to drill the hole not in the context of making the part but only after the parts are properly assembled. There are also occasions when we need to weld two parts together and perform post-weld machining operations. Again, the machining operations must be done after the parts are assembled and welded together.

One of the ways to put several component parts together permanently is to deploy welding processes. To ensure good penetration in the welding of heavy components that have a substantial thickness, we usually pre-treat the weld joints by cutting 'V' or 'U' grooves.

ASSEMBLY WORK FEATURES

When constructing work features in a top-down approach, you must not confuse work features constructed in part modeling mode with work features constructed in assembly modeling mode. If you activate part modeling and construct work features, the work features are ordinary work features residing in part files. If you work in assembly modeling mode and construct work features, you construct assembly work features. Methods to construct work features in assembly modeling mode are the same as those you learned in Chapter 4 to construct work features in part modeling mode. You can construct assembly work features relative to features of the components placed in the assembly or relative to the three basic planes (XY, XZ, and YZ), three basic axes, and center point of the origin of the assembly.

ASSEMBLY SOLID FEATURES

To depict machining operations that are to be carried out after the parts are assembled, we construct assembly solid features instead of part solid features, because assembly solid features affect only the solid parts when they are viewed in the context of an assembly. For example, when you construct a hole in the context of an assembly, the hole is illustrated in the assembly but there is no hole in the part file. The same is true when you construct an extruded cut feature or a chamfer feature in an assembly. They are exhibited only in the assembly, not in the solid parts. Altogether, there are three kinds of assembly solid features: extruded cut feature, hole feature, and chamfer feature.

WELDING ENVIRONMENT

Welding is a special kind of assembly in which a number of parts are welded together to form a single component. To depict a welded structure in the computer, we put the parts together in an assembly and convert the assembly to a weldment. In the weldment, we perform preparation work by constructing extruded cuts, holes, and chamfers as may be necessary. After preparing the joints, we construct weldments. After welding, we can add machining features. Both preparation and machining can be extruded cut, hole, and chamfer features.

TUTORIAL 7.9

In this tutorial, you will construct assembly solid features to depict machining processes that are to be carried out after the components are assembled.

1. Start a new part file. Use the metric template.
2. With reference to Figure 7.61, construct a rectangle (60 mm by 30 mm) and extrude it a distance of 10 mm.
3. Save and close the file (file name: *AssemblyFeature.ipt*).

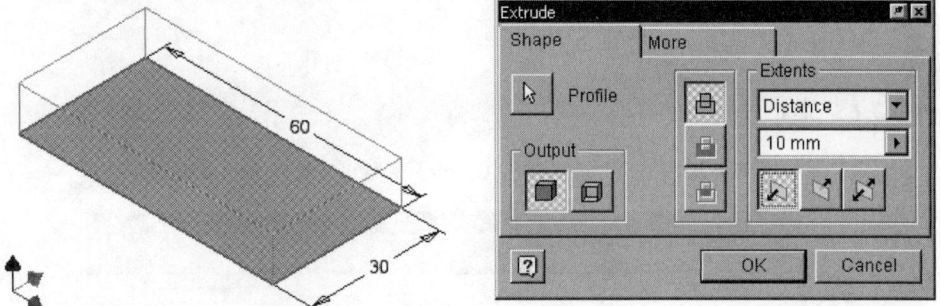

Figure 7.61 *Sketch being extruded*

Now place the part in an assembly and construct solid features in the assembly.

4. Start a new assembly file. Use the metric template.
5. Place two copies of the part file *AssemblyFeature.ipt* in the assembly. (See Figure 7.62.)
6. Add a mate constraint to faces A and B in Figure 7.62.
7. Add flush constraints to faces C and D and then E and F in Figure 7.62.

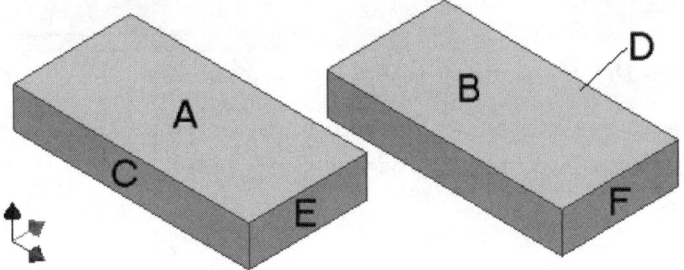

Figure 7.62 *Parts placed in an assembly*

8. Select 2D Sketch on the Inventor Standard toolbar and select face A in Figure 7.63 to establish a sketch plane.

9. Project edges B and C, a center point, and add dimensions to the center point in accordance with Figure 7.63.

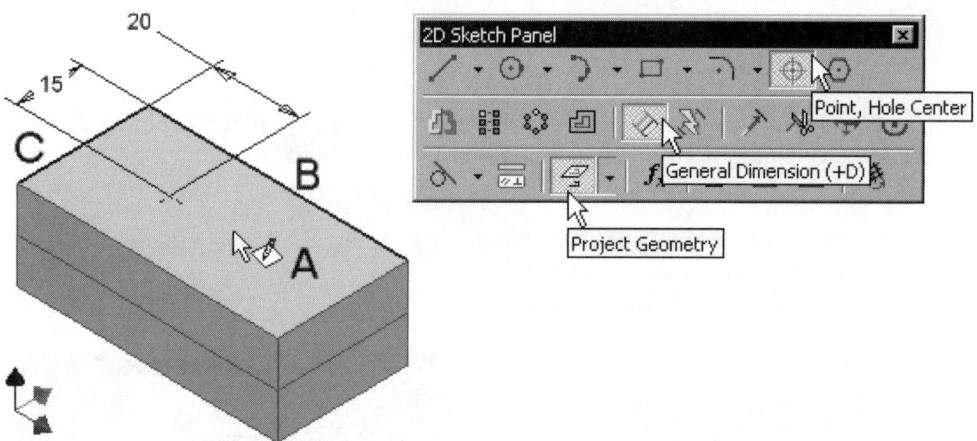

Figure 7.63 *Sketch constructed*

10. Place a through hole cutting through all the solid parts. (See Figure 7.64.)

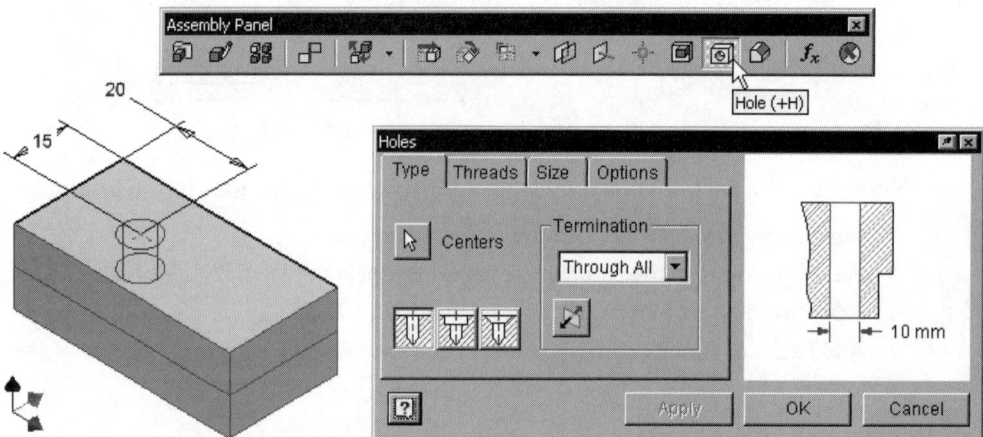

Figure 7.64 *Through hole placed in the context of the assembly*

11. Construct a sketch on face A in accordance with Figure 7.65.

12. Extrude the sketch to cut through all the solid parts.

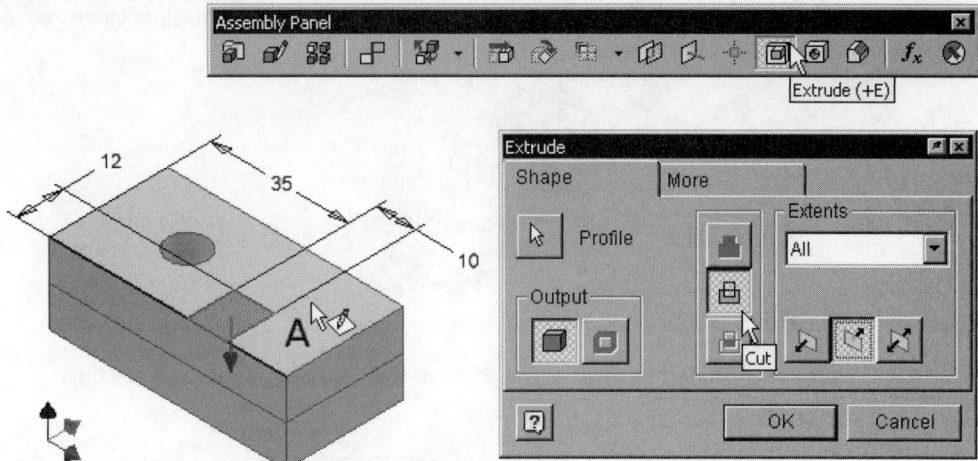

Figure 7.65 *Extruded solid feature constructed in the context of the assembly*

13. With reference to Figure 7.66, place a chamfer feature.

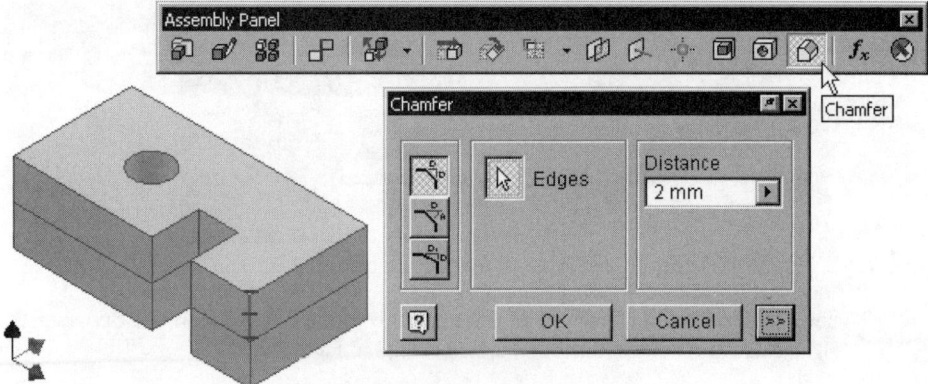

Figure 7.66 *Chamfer feature placed in the context of the assembly*

14. Open the part file *AssemblyFeature.ipt*.

You will see that no changes have been made to the original part file, because the hole, extruded feature, and chamfer were all added in the context of the assembly.

Now place one more copy of the part file in the assembly and include it as a participant in the assembly solid features.

15. Place one more copy of the solid part in the assembly by selecting one of the solid parts, right-clicking, selecting Copy, right-clicking, and selecting Paste.

Because the assembly solid features do not affect the solid part file, there is no hole, extruded cut, or chamfer in the pasted solid part.

16. With reference to Figure 7.67, apply a mate and two flush constraints to properly locate the solid part.

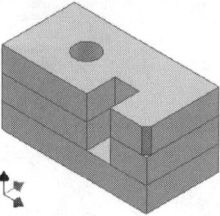

Figure 7.67 *Part copied and assembled*

17. Select the extruded feature on the Browser Bar, right-click, and select Add Participant.

18. Select solid part A in Figure 7.68.

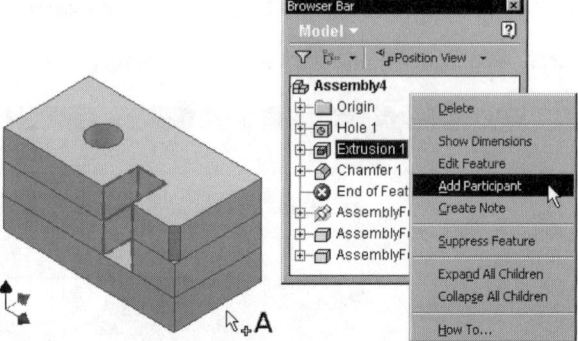

Figure 7.68 *Participant to the assembly extruded feature being added*

19. Select the second participant of the assembly extruded solid on the Browser Bar, right-click, and select Remove Participant. (See Figure 7.69.)

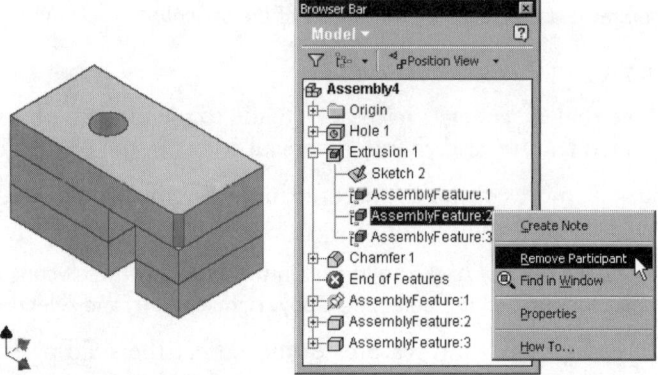

Figure 7.69 *Participant removed*

The assembly is complete. Save and close your file (file name: *AssemblyFeature.iam*).

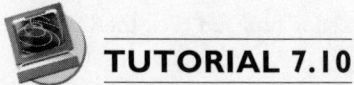

TUTORIAL 7.10

In this tutorial, you will construct an assembly of parts, convert the assembly to a welded structure, construct preparations, construct weldments, and construct post-welding machining features.

1. Start a new metric assembly file.
2. Place three copies of the part file *AssemblyFeature.ipt* that you constructed in the last tutorial.
3. With reference to Figure 7.70, apply mate and flush constraints to put the components together.

As we have said, the assembly solid features that you constructed in the previous tutorial apply only to the assembly, but not to the part file. Therefore, in the solid part file, you will not find any solid features that you constructed in the assembly.

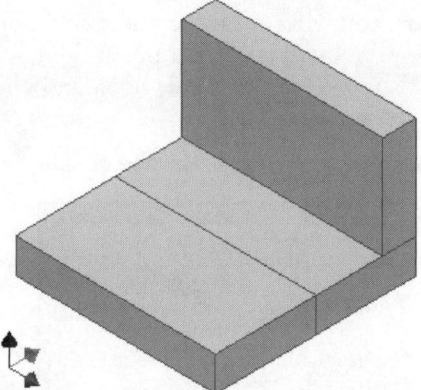

Figure 7.70 *Components assembled*

Now convert the assembly to a welded structure.

4. Select Application > Weldment.

Because conversion from an assembly to a welded structure is irreversible, a warning message appears.

5. Click Yes in the Autodesk Inventor dialog box shown in Figure 7.71.

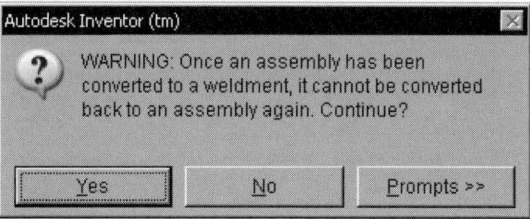

Figure 7.71 *Warning message*

6. Select ISO standard in the Convert to Weldment dialog box and click the OK button. (See Figure 7.72.)

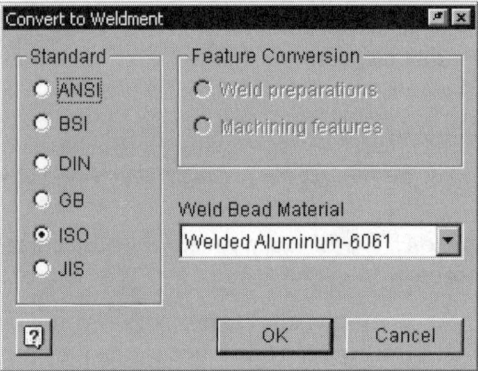

Figure 7.72 *Convert to Weldment dialog box*

On the Browser Bar, you will find three additional objects: Preparations, Welds, and Machining. Now you will construct pre-welding preparation features. There are three kinds of features, extruded cut, hole, and chamfer. Here you will construct an extruded cut and chamfer.

7. Select Preparations on the Browser Bar and double-click. (See Figure 7.73.)

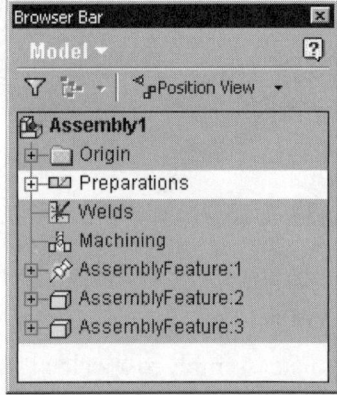

Figure 7.73 *Browser bar with three additional objects and Preparations activated*

Now construct a sketch and make an extruded cut from the sketch.

8. Select 2D Sketch on the Inventor Standard toolbar and select face A in Figure 7.74.

9. Construct a sketch. In the sketch construct an arc, a horizontal line, and two vertical lines.

 Note: The two vertical lines are on either side of the joint between the two parts and form legs attached to the arc, and the horizontal line connects across the top of the two vertical lines.

10. Apply a tangent constraint to the arc and the two vertical lines.

11. Apply a collinear constraint to the horizontal line and horizontal edge B.

12. Apply a coincident constraint to the center of the arc and vertical edge C.

13. Add two dimensions (3 mm for the length of the arc and 4 mm for the radius of the arc).

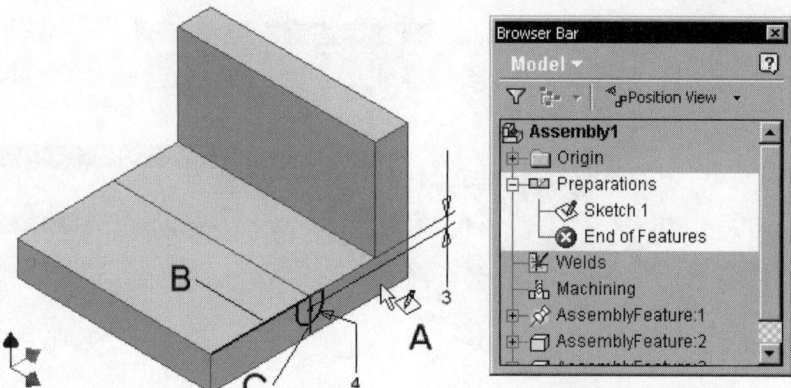

Figure 7.74 *Sketch constructed for construction of weldment preparation features*

14. Select Return to exit sketch mode.

Tip: Care must be taken not to select the Return button more than once. Otherwise, you exit the preparation. If you exit the preparation, select it on the Browser Bar and double-click.

15. Select Extrude on the Weldment Features Panel Bar or toolbar.

16. In the Extrude dialog box, select Profile, if it is not already selected, and select All in the Extents box.

17. Select the profile highlighted in Figure 7.75.

18. Click the OK button.

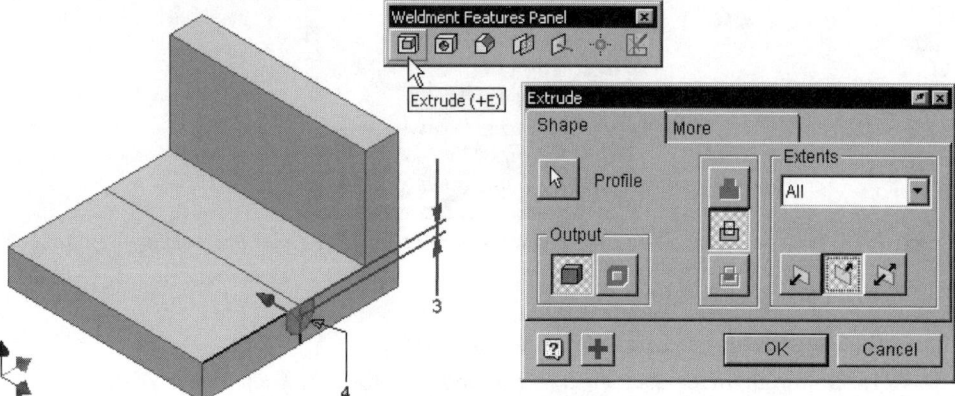

Figure 7.75 *Pre-welding extruded cut being constructed*

Now you will construct a pre-welding chamfer.

19. With reference to Figure 7.76, select Chamfer on the Weldment Features Panel Bar or toolbar, select edge A, specify 5 mm distance in the Chamfer dialog box, and click the OK button.

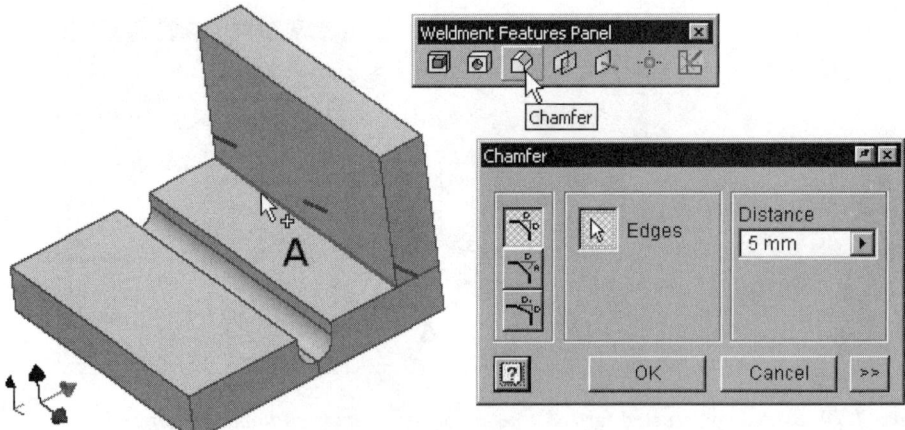

Figure 7.76 *Pre-welding chamfer being constructed*

Now you will see an extruded feature and a chamfer feature on the Browser Bar. (See Figure 7.77.)

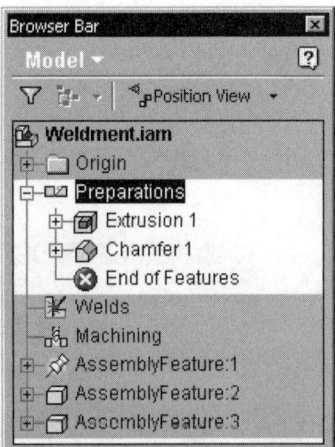

Figure 7.77 *Browser bar showing extruded and chamfer features under pre-welding preparations*

Now construct weldments.

20. Select Return to exit weldment preparation.

21. On the Browser Bar, select Welds and double-click. Then, in the Weldment Features Panel Bar, click Weld.

22. In the Weld Feature dialog box shown in Figure 7.78, select the Cosmetic Weld button, select the Arrow Side Symbol button and select U Butt Weld.

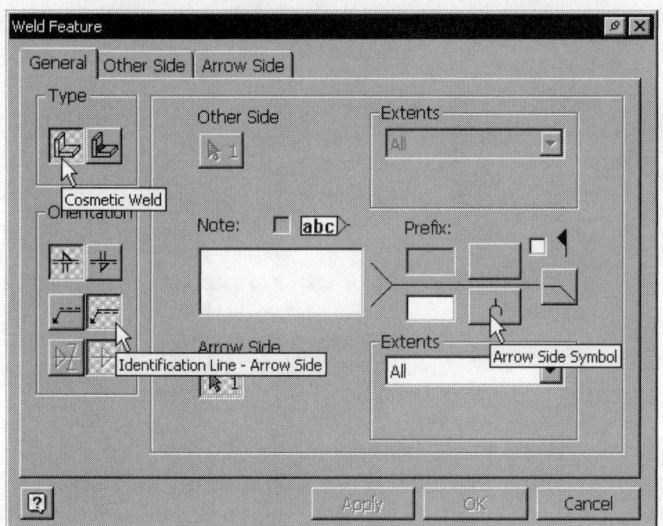

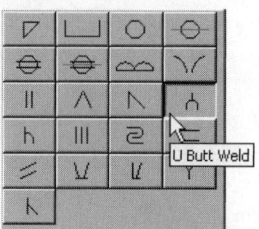

Figure 7.78 *Weld Feature dialog box*

23. Select the Identification Line-Arrow side button and select edge A in Figure 7.79.
24. Click the Apply button. A U bevel butt weld symbol is placed.
25. Select Arrow Side button and select Bevel Butt Weld.
26. Select edge B in Figure 7.79.
27. Click the OK button.

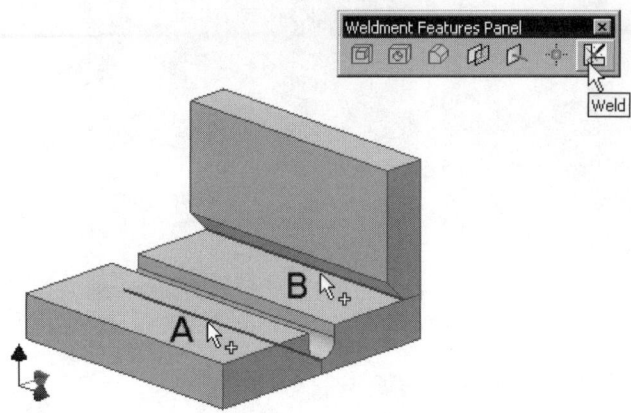

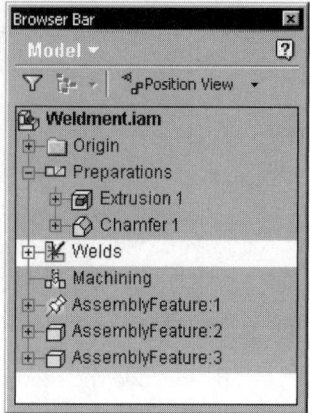

Figure 7.79 *Weldments being placed*

Two weldment symbols are placed. (See Figure 7.80.)

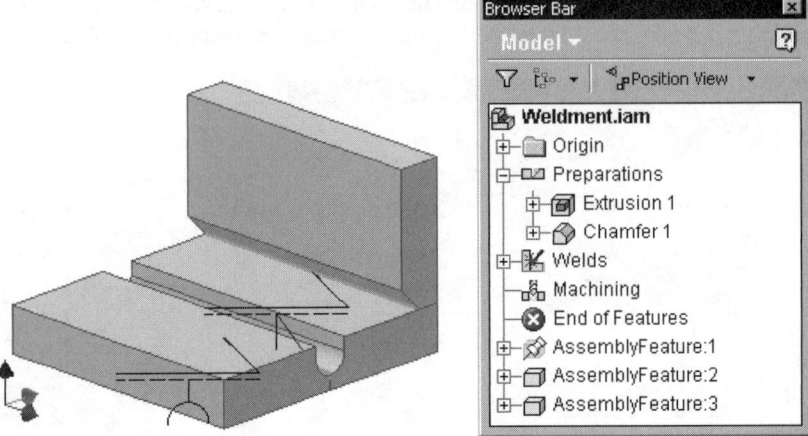

Figure 7.80 *Weldment symbols placed*

Now perform post-weldment machining. Valid operations are extruded cuts, hole cuts, and chamfers. To add such features, double-click Machining on the Browser Bar and proceed to make the features. Here you will add an extruded cut on the welded structure.

28. Select Machining on the Browser Bar and double-click.

29. Set up a sketch on face A in Figure 7.81. In the sketch, vertex of the rectangle C is coincident with vertex B.

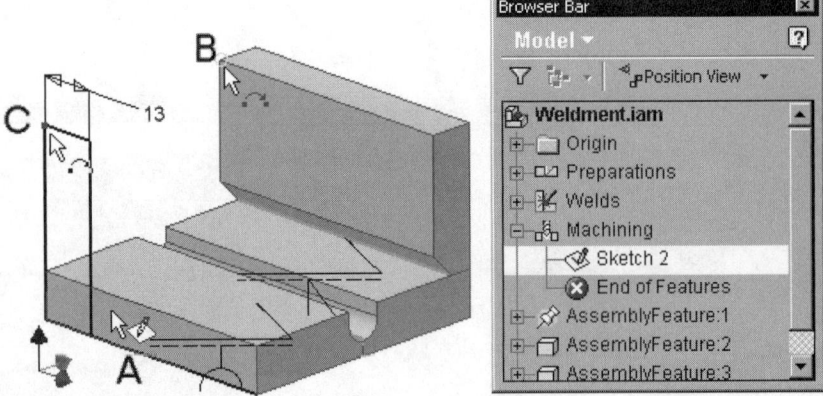

Figure 7.81 *Sketch constructed*

30. Extrude the sketch to cut through the solid part. (See Figure 7.82.)

The welded structure, consisting of preparations, welds, and machining, is complete. Save and close your file (file name: *Weldment1.iam*).

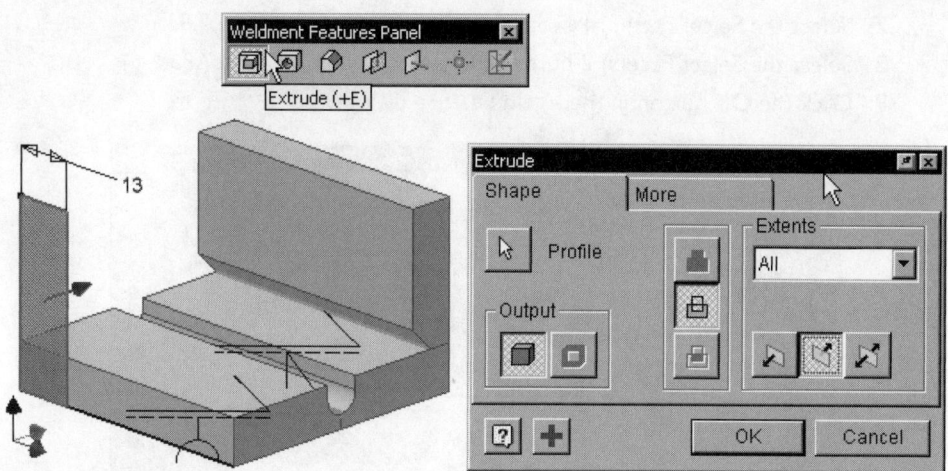

Figure 7.82 *Post-weld extrusion being constructed*

TUTORIAL 7.11

In this tutorial, you will construct a fillet weld bead on a welded structure.

1. Start a new metric assembly file, place two copies of the part file *AssemblyFeature.ipt*, and apply assembly constraints to put them together.

2. Select Applications > Weldment.

3. Click the OK buttons in the Warning message dialog box and the Convert to Weldment dialog box.

4. Select Welds on the Browser Bar and double-click. (See Figure 7.83.)

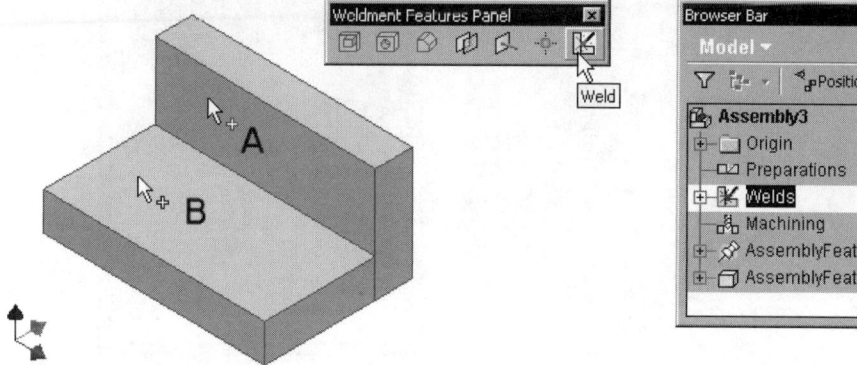

Figure 7.83 *Weld selected*

5. Select Weld on the Weldment Features Panel Bar or toolbar.

6. In the Weld Feature dialog box, select the Fillet Weld button and select the Identification Line-Arrow Side button. (See Figure 7.84.)

7. Select the Select Face(s) 1 button and select face A in Figure 7.83.
8. Select the Select Face(s) 2 button and select face B in Figure 7.83.
9. Click the OK button in the Weld Feature dialog box.

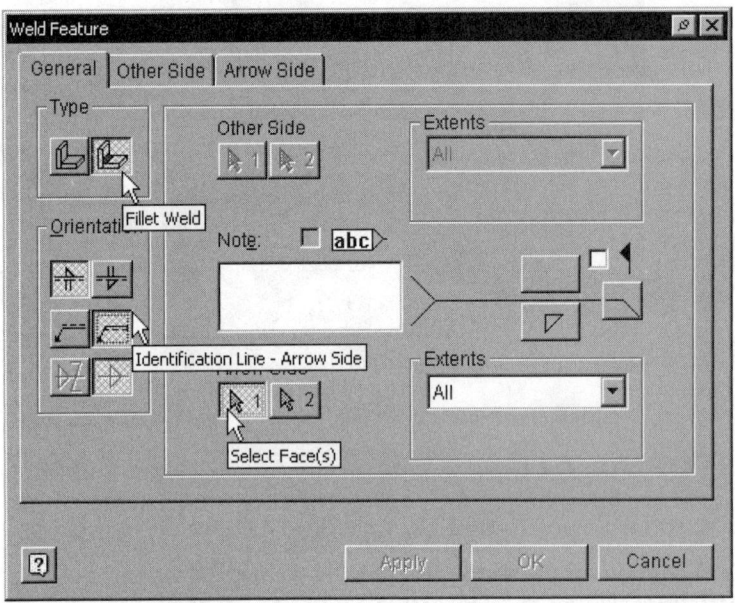

Figure 7.84 *Weld Feature dialog box*

A fillet weld is constructed. (See Figure 7.85.) Save and close your file (file name: *Weldment2.ipt*).

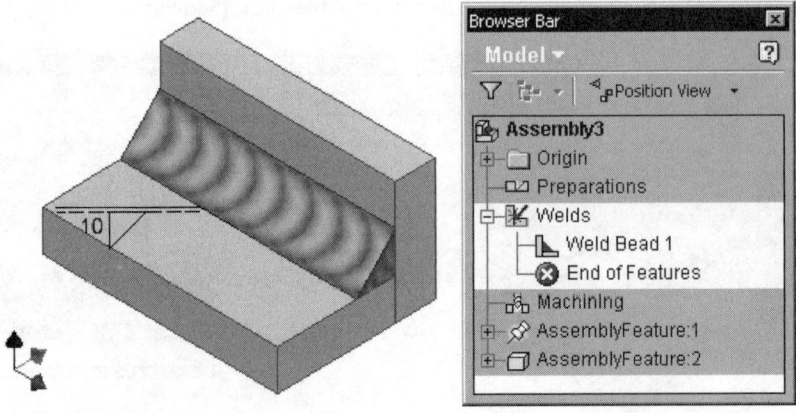

Figure 7.85 *Fillet weld placed*

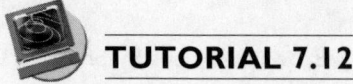

TUTORIAL 7.12

In this tutorial, you will construct a set of assembly work features in an assembly file as references for making the assembly of a bicycle frame. Figure 7.86 shows the completed bicycle frame, and Figure 7.87 shows the work features that you will construct.

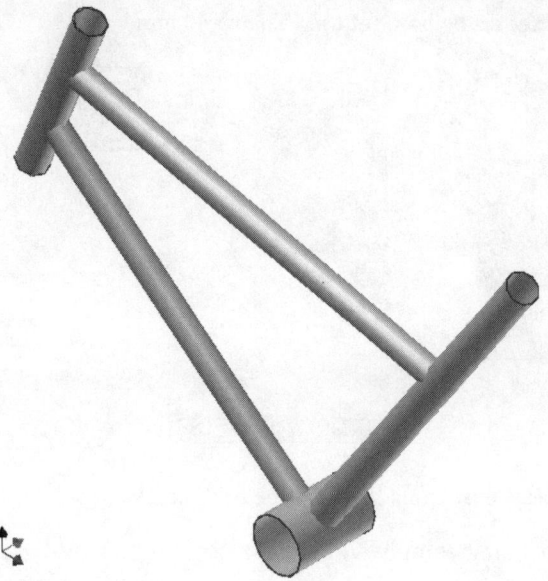

Figure 7.86   *Bicycle frame*

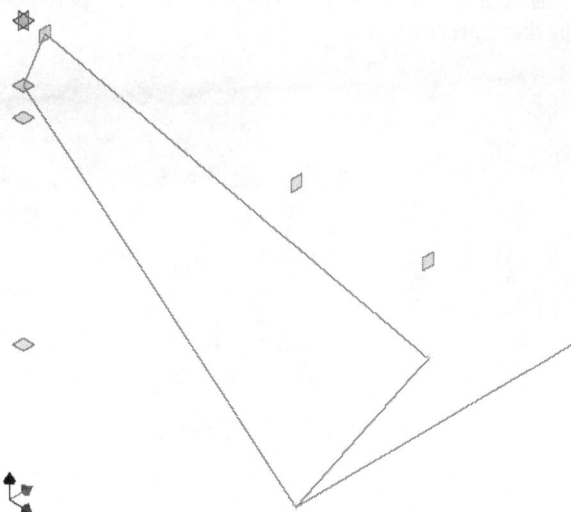

Figure 7.87   *Work features for making the assembly of the bicycle frame*

1. Start a new metric assembly file.

2. Set the display to an isometric view.

3. Select the XY plane, YZ plane, and XZ plane on the Browser Bar one by one, right-click, and select Visibility.

4. Select the Work Plane on the Assembly Panel Bar or toolbar.

5. Select the YZ plane, hold down the left mouse button, and drag to a location indicated in Figure 7.88.

6. In the Offset dialog box, set the value to 40 mm.

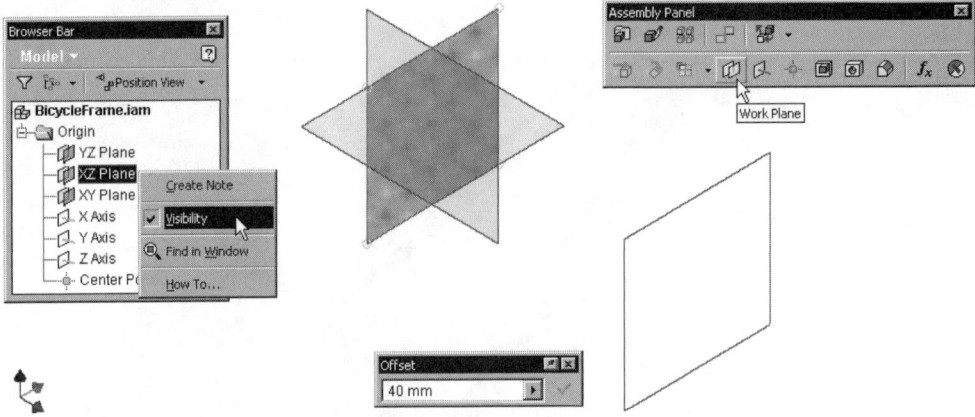

Figure 7.88 *Work plane offset from the YZ plane being constructed*

Now construct an assembly work point.

7. Select Work Point on the Assembly Panel Bar or toolbar.

8. Select planes A, B, and C in Figure 7.89. A work point D (Figure 7.89) is constructed at their intersection.

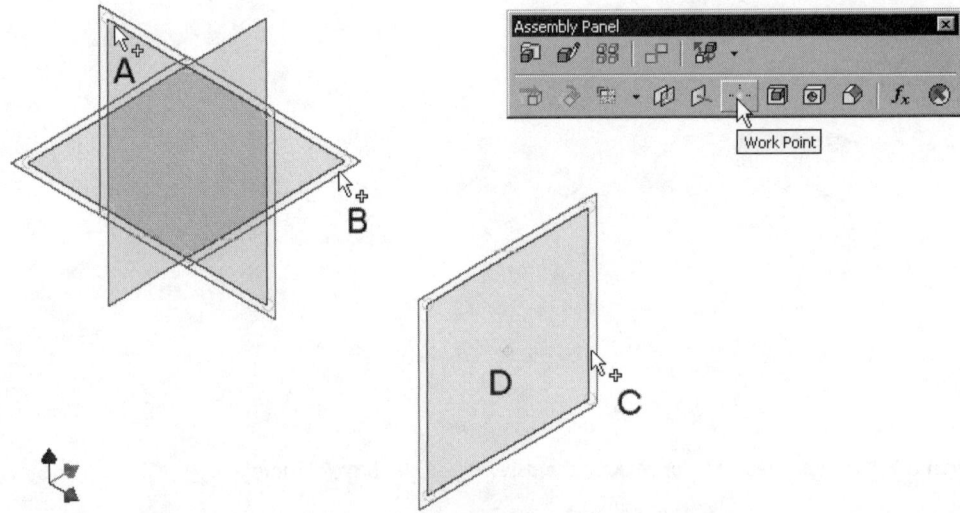

Figure 7.89 *Work point at the intersection of three planes being constructed*

Now construct an offset work plane.

9. Select Work Plane on the Assembly Panel Bar or toolbar.

10. Select the XY plane, hold down the left mouse button, and drag to a location indicated in Figure 7.90.

11. In the Offset dialog box, set the value to −100 mm.

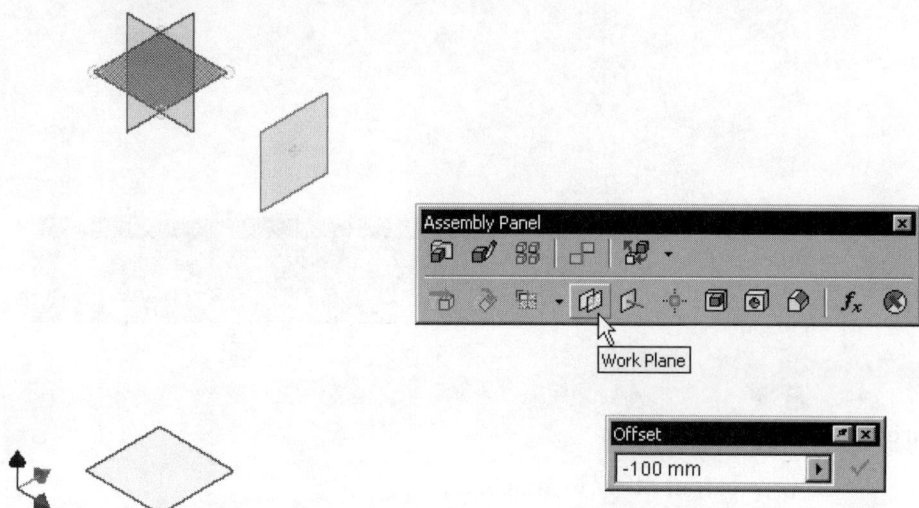

Figure 7.90 *Work plane offset from the XY plane being constructed*

Now construct a work point.

12. Select Work Point on the Assembly Panel Bar or toolbar.

13. Select work plane A in Figure 7.91 and select the YZ plane and XZ plane on the Browser Bar. A work point is constructed at their intersection.

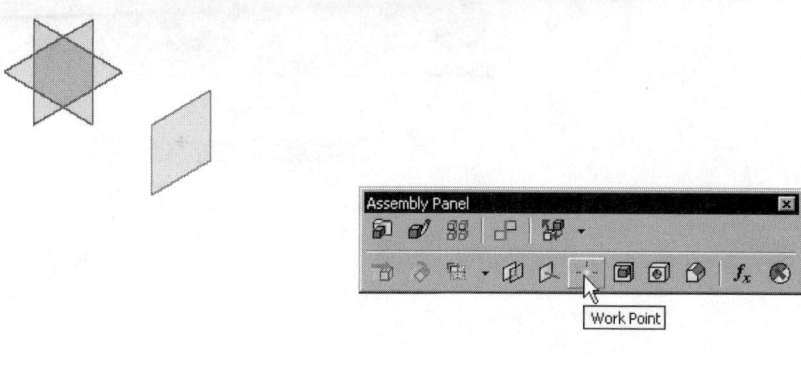

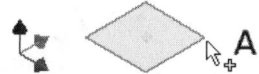

Figure 7.91 *Work point at the intersection of three planes being constructed*

Now construct an assembly work axis joining two work points.

14. Select Work Axis on the Assembly Panel Bar or toolbar.
15. Select work points A and B in Figure 7.92.

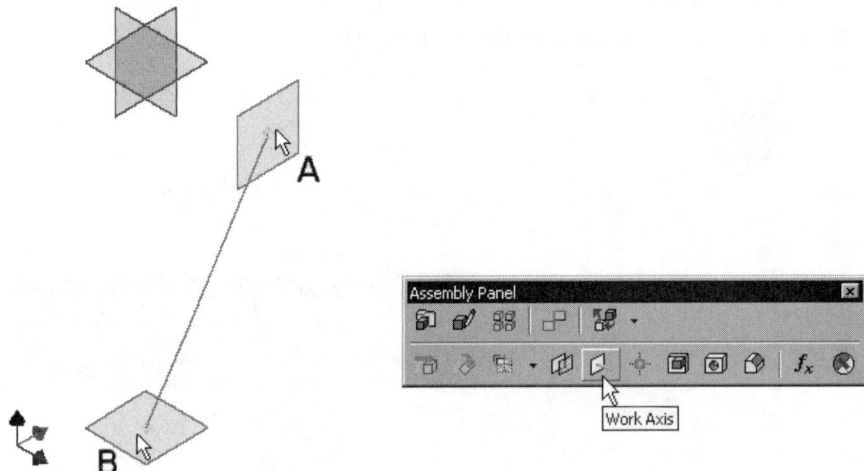

Figure 7.92 *Work axis constructed*

Now construct four offset work planes.

16. Select Work Plane on the Assembly Panel Bar or toolbar.
17. Select the YZ plane, hold down the left mouse button, and drag to location A in Figure 7.93.
18. In the Offset dialog box, set the value to 740 mm.

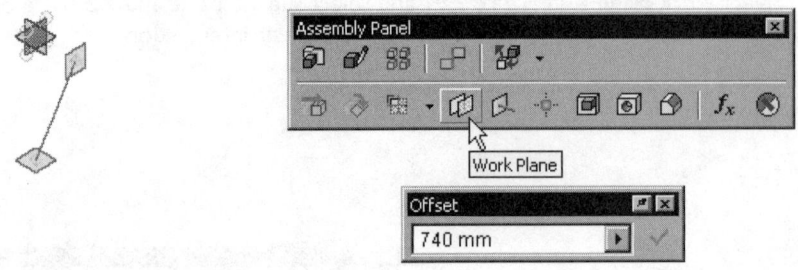

Figure 7.93 *Work plane offset from the YZ plane constructed*

19. Select Work Plane on the Assembly Panel Bar or toolbar.
20. Select the YZ plane, hold down the left mouse button, and drag to location A in Figure 7.94.
21. In the Offset dialog box, set the value to 500 mm.

Figure 7.94 *Another work plane offset from the YZ plane constructed*

22. Select Work Plane on the Assembly Panel Bar or toolbar.
23. Select the XY plane, hold down the left mouse button, and drag to location A in Figure 7.95.
24. In the Offset dialog box, set the value to −150 mm.

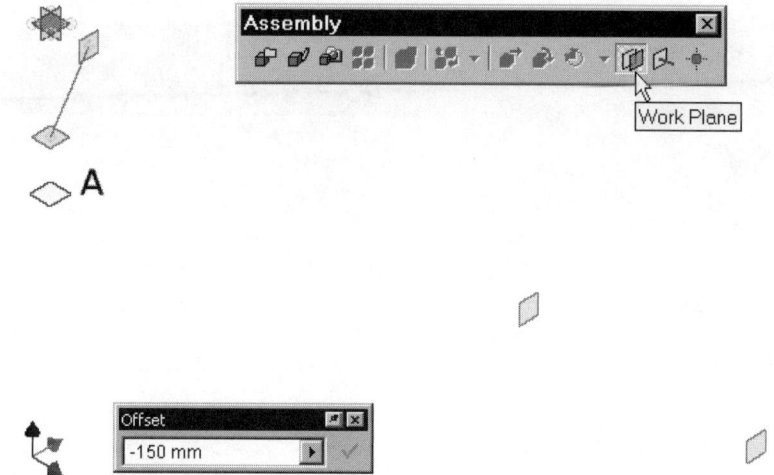

Figure 7.95 *Work plane offset from the XY plane constructed*

25. Select Work Plane on the Assembly Panel Bar or toolbar.

26. Select the XY plane, hold down the left mouse button, and drag to location A in Figure 7.96.

27. In the Offset dialog box, set the value to –500 mm.

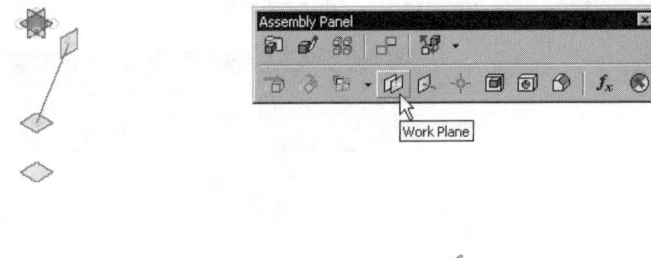

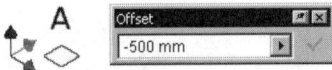

Figure 7.96 *Another work plane offset from the XY plane constructed*

Now you will construct two work points.

28. Select Work Point on the Assembly Panel Bar or toolbar.

29. Select the XZ plane and planes A and B in Figure 7.97.

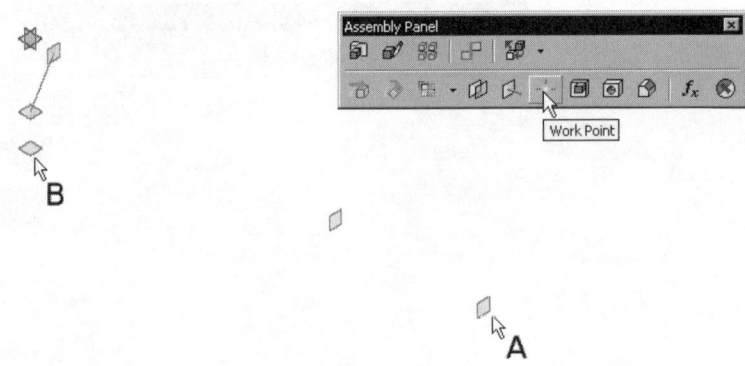

Figure 7.97 *Work point constructed*

30. Select Work Point on the Assembly Panel Bar or toolbar.
31. Select the XZ plane and planes A and B in Figure 7.98.

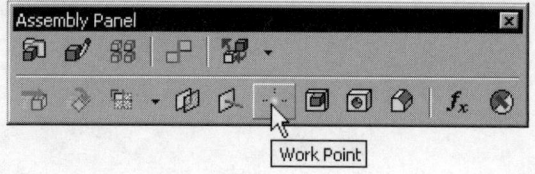

A

B

Figure 7.98 *Work point constructed*

Now you will construct four work axes.

32. Select Work Axis on the Assembly Panel Bar or toolbar.
33. Select work points A and B in Figure 7.99.
34. Select Work Axis on the Assembly Panel Bar or toolbar.
35. Select work points B and C in Figure 7.99.
36. Select Work Axis on the Assembly Panel Bar or toolbar.
37. Select work points C and D in Figure 7.99.

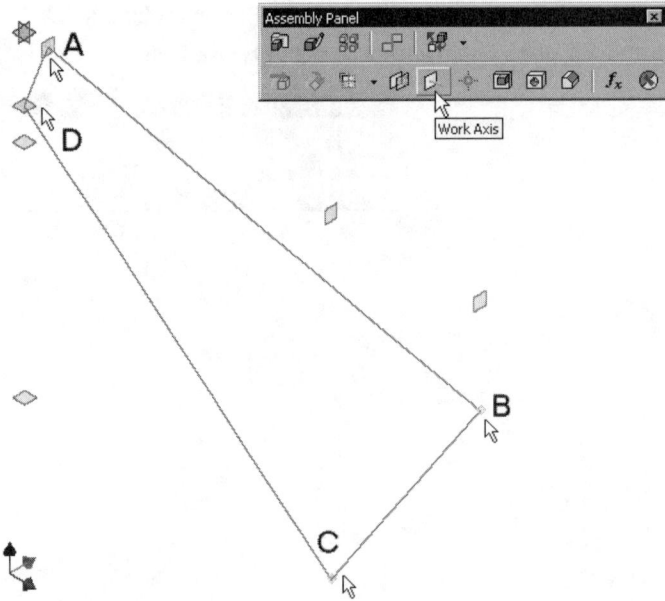

Figure 7.99 *Work axes constructed*

38. Select Work Axis on the Assembly Panel Bar or toolbar.
39. Select the XZ plane on the Browser Bar and work point A in Figure 7.100.

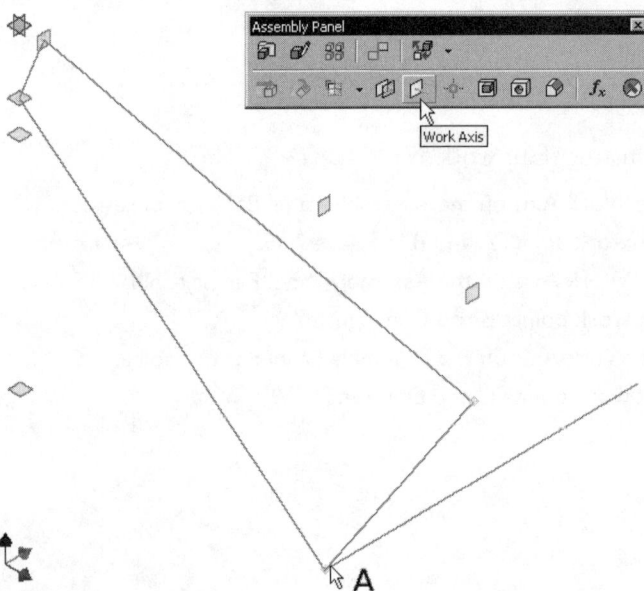

Figure 7.100 *Work axis constructed*

The work features are complete. Save your file (file name: *BicycleFrame.iam*, folder name: *Bicycle*).

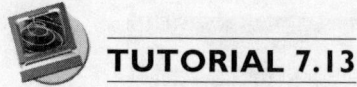

TUTORIAL 7.13

In this tutorial, you will use the work features you constructed in the previous tutorial as a framework for assembling components of the bicycle frame. To continue, you will construct three component parts of the bicycle frame.

1. Open the assembly file *BicycleFrame.iam*, if you already closed it.

2. Select Create Component on the Assembly Panel Bar or toolbar.

3. In the Create In-Place Component dialog box, specify the new part file name (*Frame01.ipt*), browse to save the file in the folder (*Bicycle*), and use the metric *Standard.ipt* template file. Click OK to close the dialog box and begin to sketch the new part.

4. Construct a circle with a diameter of 50 mm and extrude it a distance of 300 mm. (See Figure 7.101.)

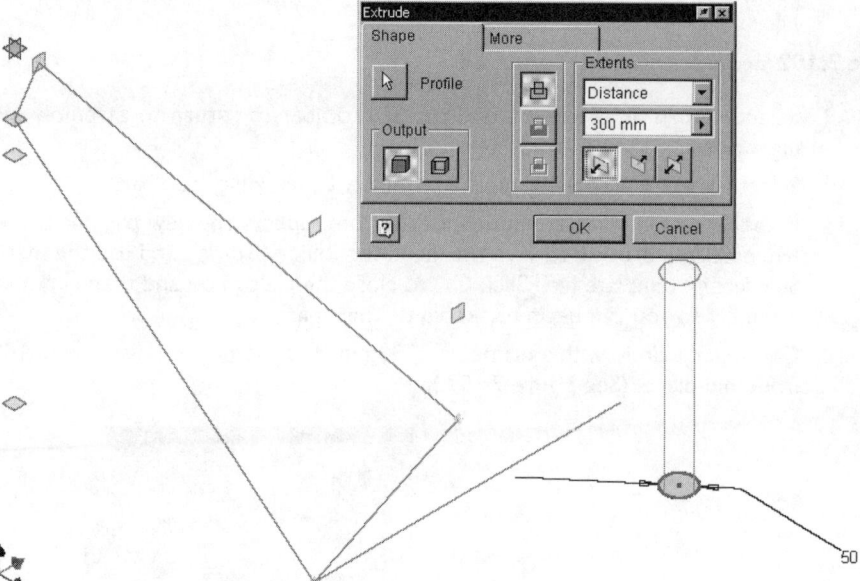

Figure 7.101 *A part constructed in the assembly and a circle being extruded*

5. Select Return on the Inventor Standard toolbar to return to assembly modeling mode.

6. Select Create Component on the Assembly Panel Bar or toolbar.

7. In the Create In-Place Component dialog box, specify the new part file name *Frame02.ipt*, browse to save the file in the folder *Bicycle*, and use the metric *Standard .ipt* template file. Click OK to close the dialog box and then click in the assembly so you can begin to sketch the new part.

8. Construct a circle with a diameter of 50 mm and extrude it a distance of 700 mm. (See Figure 7.102.)

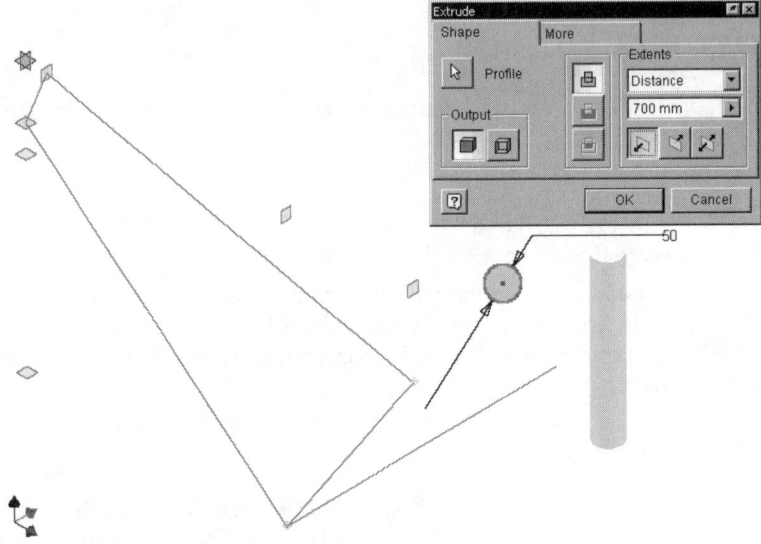

Figure 7.102 *Second component constructed*

9. Select Return on the Inventor Standard toolbar to return to assembly modeling mode.

10. Select Create Component on the Assembly Panel Bar or toolbar.

11. In the Create In-Place Component dialog box, specify the new part file name (*Frame03.ipt*), browse to save the file in the folder (*Bicycle*), and use the metric *Standard.ipt* template file. Click OK to close the dialog box and then click in the assembly so you can begin to sketch the new part.

12. Construct a circle with a diameter of 90 mm and extrude it a distance of 140 mm from mid-plane. (See Figure 7.103.)

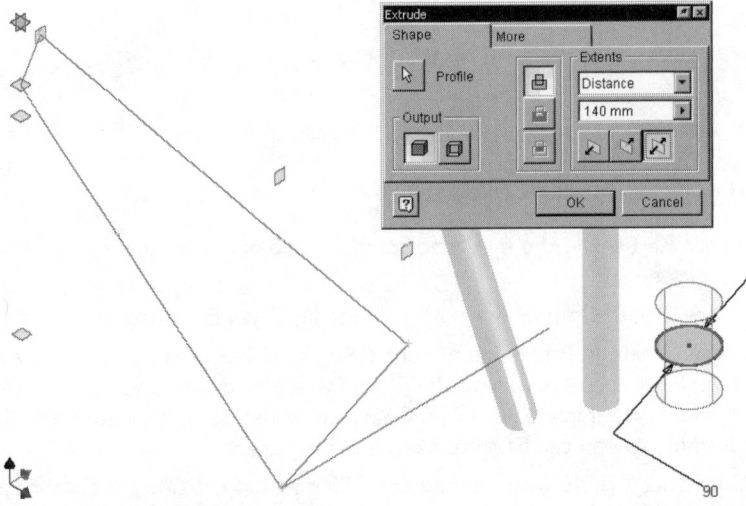

Figure 7.103 *Third component constructed*

Now assemble the components to the framework of the bicycle established by the work features.

13. Select Return on the Inventor Standard toolbar to return to assembly modeling mode.

14. Select Frame01 on the browse bar, right-click, and deselect Grounded.

15. With reference to Figure 7.104, apply a mate constraint to axis A and work axis B.

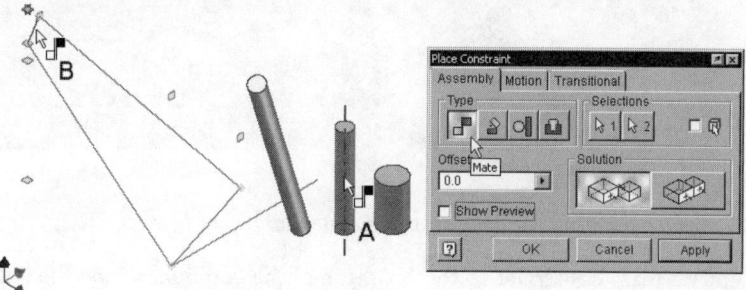

Figure 7.104 *Axis of a component being mated to a work axis*

16. Apply a mate constraint to face B and work point A at an offset distance of –100 mm in accordance with Figure 7.105.

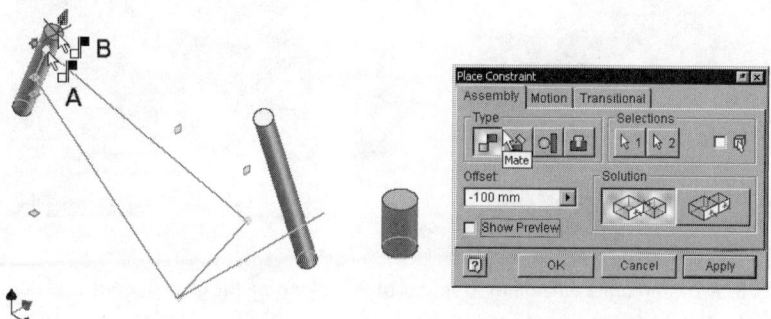

Figure 7.105 *Face of a component being mated to a work point at an offset distance*

17. Apply a mate constraint to axis A of a component and work axis B in Figure 7.106.

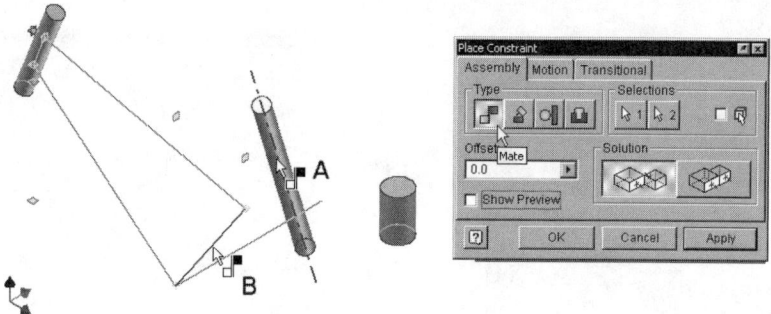

Figure 7.106 *Axis of a component being mated to a work axis*

18. Apply a mate constraint to the center point A of the component and to work point B in Figure 7.107.

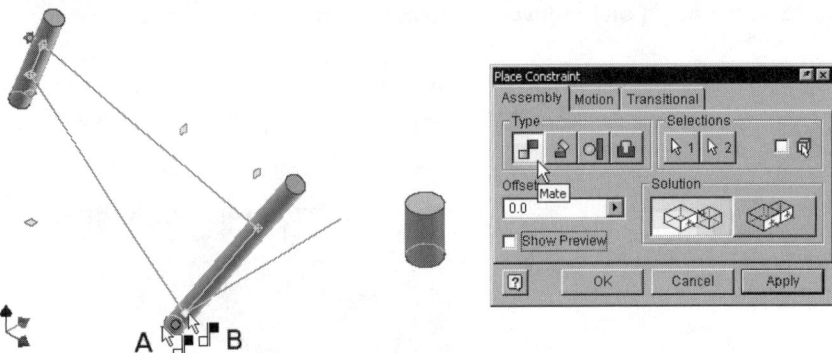

Figure 7.107 *Center point of a component being mated to a work point*

19. Apply a mate constraint to the XY plane of the component and the XZ plane of the assembly. (See Figure 7.108.) Select the planes on the Browser Bar.

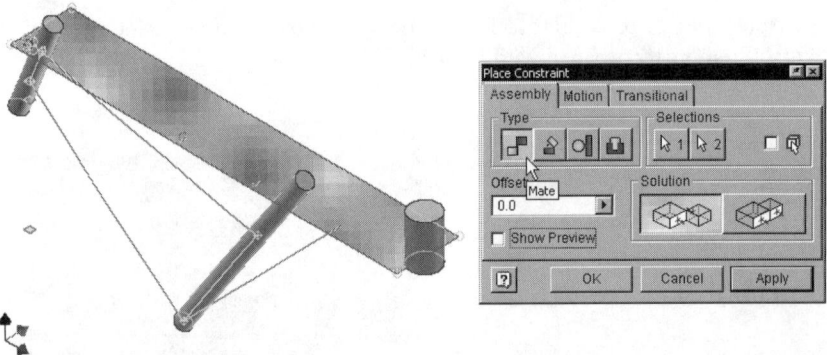

Figure 7.108 *A mate constraint being applied to XY plane of the component and XZ plane of the assembly*

20. Apply a mate constraint to axis A of the component and work axis B in Figure 7.109.

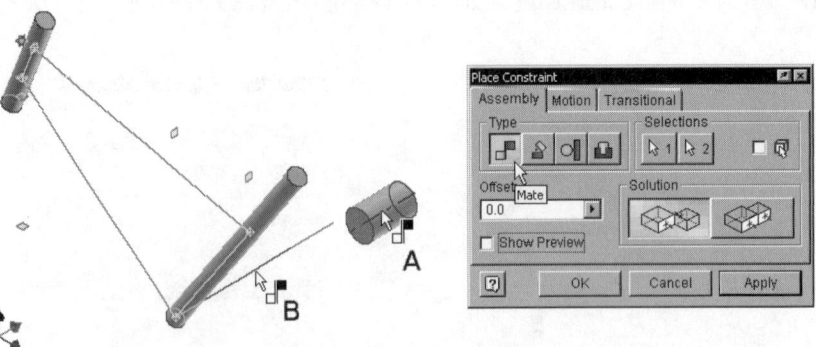

Figure 7.109 *Axis of the component being mated to a work axis*

21. Select View > Object Visibility and clear the Origin Planes and User Work Planes boxes one by one to hide the planes. (See Figure 7.110.)

Three components of the bicycle frame assembly are complete. You will complete the bicycle frame by using adaptive technology and deriving parts from the assembly in Chapter 8. Now save and close your files.

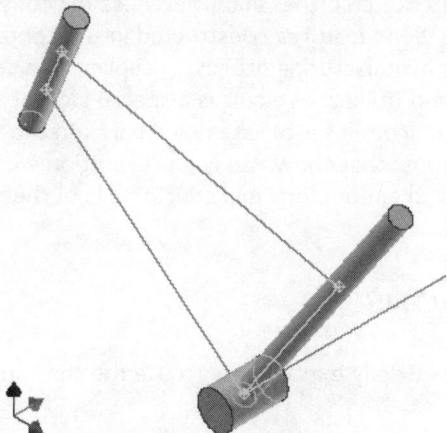

Figure 7.110 *Origin planes and work planes hidden*

SUMMARY

The snap and go technique is a semi-automatic way to apply assembly constraints to components of an assembly by selecting a feature of a component, holding down ALT, dragging the feature to a feature of another component in the assembly, and releasing the mouse button. The kind of constraint applied this way is determined by the geometry type of the selected component as well as the shortcut key you use during the course of applying assembly constraints

iMate is half of a constraint pair. It is a pre-constructed assembly constraint placed in a component indicating the kind of assembly constraint to be applied when the component is placed in an assembly. In an assembly, you select and drag the iMate interface of one component to the iMate interface of another component. With a number of iMates defined, you can combine them together to form a composite iMate. To specify the intended constraint relationship between a cylindrical face and a set of contiguous faces, you apply a transitional constraint. A transitional surface constraint keeps the set of constrained contiguous faces in contact with a face of another component as you translate a component along the open degrees of freedom. It is particularly useful in simulating motion between a cam plate and a cam follower.

Drive constraint is a technique whereby a parameter of an assembly constraint in an assembly is varied to cause a relative motion among the components in the assembly. This technique is particularly useful in simulating the motion of a set of linkages.

To simulate complex mechanical motions in mechanical drive systems such as a belt drive, friction drive, and gear drive, you can set up motion relationships between the members of the assembly.

In the context of an assembly, you can construct work features and solid features. Assembly work features, unlike work features in part files, are objects in the assembly file that help you set up sketch planes and references for constructing solid parts in an assembly environment. Solid features constructed in the context of an assembly can best be used to depict manufacturing processes applied to a set of assembled components, such as milling and drilling. Welding is a special kind of assembly in which a single component is produced from a set of assembled parts. In the welding environment, you can construct features that show welding preparation work, and you can add cosmetic weld seams and annotations depicting details of the weld.

REVIEW QUESTIONS

1. What is snap and go? Briefly explain how to use the snap and go technique in assembly modeling.

2. Explain how to construct an iMate and a composite iMate in a component.

3. Illustrate how mechanical motions can be animated in a mechanism by varying the parameters of an assembly constraint.

4. Briefly explain how relative motion between the components in an assembly can be set and simulated.

5. Give an example where a transitional surface constraint is needed.

6. How do assembly features differ from features constructed in the part modeling environment?

7. What kind of features can you construct in the welding environment?

8. Can the welding environment be reverted to the assembly environment?

Advanced Part Modeling Techniques

OBJECTIVES

The aims of this chapter are to enable you to gain an in-depth understanding about adaptive technology, to delineate the way to derive a solid part from an assembly of simpler parts, to detail the use of 2D layout drawings in mechanism design, and to promote copied surfaces and solid features. This chapter also outlines how to design collaboratively. After studying this chapter, you should be able to

- Apply adaptive technology in design
- Derive a solid part from an assembly
- Promote copied surfaces and solid features in the context of an assembly
- Use 2D layout drawings to design mechanisms
- Design collaboratively with other design team members

OVERVIEW

This chapter will address various advanced modeling techniques.

In an assembly, the dimensions of corresponding parts have to match in order to function properly. To maintain this kind of relationship, you use adaptive technology to make the size and shape of a feature of one solid to adapt to the size and shape of a feature of another solid part.

The steps to making a feature-based solid are sequential. You construct the solid features of a solid part one by one and combine them in a sequential way. To overcome this limitation, you construct the features in two or more solid part files, put them in an assembly, and derive a solid part from the assembly. While deriving, you combine the solid parts together by joining and cutting.

In the context of an assembly, you can copy surface and solid features between solid parts. Although the copied features are not associative with the source part, you can redefine the copied features to manage any future changes.

During the initial stage of designing a mechanism, you can construct a 2D layout drawing outlining the linkages in terms of line sketches. Using the sketches, you evaluate the validity of the mechanism and improve the design further. When you are satisfied with the mechanism, you build solid features on the 2D layout drawing.

You can work with a team of designers collaboratively on a design project by using various strategies. You set up project paths, share files, reserve files, and use NetMeeting to communicate and to work with others on the same file in real time.

ADAPTIVE TECHNOLOGY

In the context of an assembly, it is very common that the size and shape of a feature of a component are required to match the size and shape of the feature of another component. To help mandate this kind of relationship among components in an assembly, you use adaptive technology, a solid modeling technique whereby the parameters of a feature of a solid part automatically change to adapt to the parameters of the feature of another solid part in the assembly. You can apply adaptive technology to sketched features and work features of a solid part in the root assembly and the sub-assembles. You can also apply it to projected and fixed geometry.

UNDER-CONSTRAINED ADAPTIVE SKETCH

A way to construct an adaptive sketched solid feature is to construct it from a sketch that is under-constrained, either geometrically or dimensionally. For example, if you want to set the diameter of a hole to adapt to the diameter of a shaft, you must not dimension the hole's diameter. Because you must add a dimension in a hole feature, you should construct a circle without diameter dimension and extrude it to cut the solid part to obtain a hole. You set the sketch and the sketched solid feature to be adaptive in part modeling mode and set the solid part adaptive in the assembly. After you apply an assembly constraint to the cylindrical faces of the hole and shaft, the hole's diameter will change to adapt. If you later modify the shaft's diameter, the hole's diameter will also change. In making a cylindrical solid feature, you can either extrude a circle or revolve a rectangle. If you extrude a circle, you can leave the diameter un-dimensioned but you must set the extrusion length. If, however, you revolve a rectangle, you can leave both the length and width of the rectangle without defined dimension, allowing the diameter and length to be adaptive.

 ## TUTORIAL 8.1

In Chapter 7, you constructed three components of a bicycle frame and assembled them with reference to a set of assembly work features. Now you will use adaptive technology to construct two more components. Because the length of these two components depends on other parameters in the assembly, you will not assign dimension to the length of the components (making them under-constrained), but set the feature adaptive, and apply an assembly constraint to help establish the length.

1. Open the file *BicycleFrame.iam*.
2. If you did not hide the work features in the previous chapter, select the origin planes and work planes on the Browser Bar, right-click, and deselect Visibility to hide them.

3. Select Create Component on the Assembly Panel Bar or toolbar to construct a solid part.

4. Specify a file name (*Frame04.ipt*), make sure it will be saved in the proper folder (*Bicycle*), clear the Constrain sketch plane to selected face or plane check box if selected, use the metric *Standard.ipt* template, and then click OK.

5. With reference to Figure 8.1, construct a rectangle, change a line's linestyle to Centerline, and add a dimension to specify the diameter as 40 mm. (Do not dimension the length.)

 Tip: A sketch needs to be under-constrained before you can set it adaptive.

Now revolve the under-constrained sketch to a revolved solid.

6. Select Return on the Inventor Standard toolbar. (Do not select this button more than once. Otherwise, you return to assembly modeling mode instead of part modeling mode.)

7. Select Revolve on the Part Features Panel Bar or toolbar.

8. In the Revolve dialog box, click the OK button. A revolved solid feature is constructed. (See Figure 8.2.)

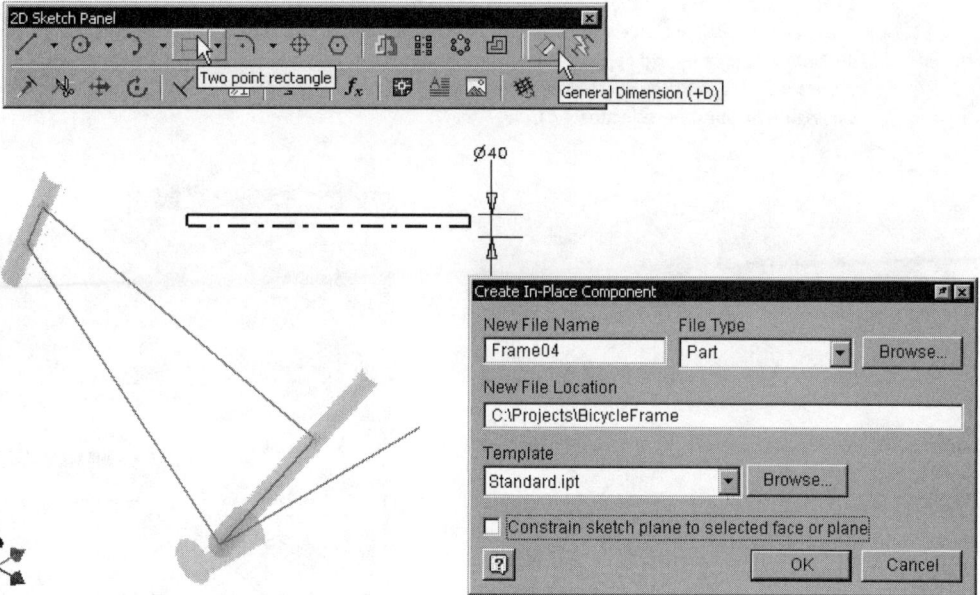

Figure 8.1 *Under-constrained sketch constructed*

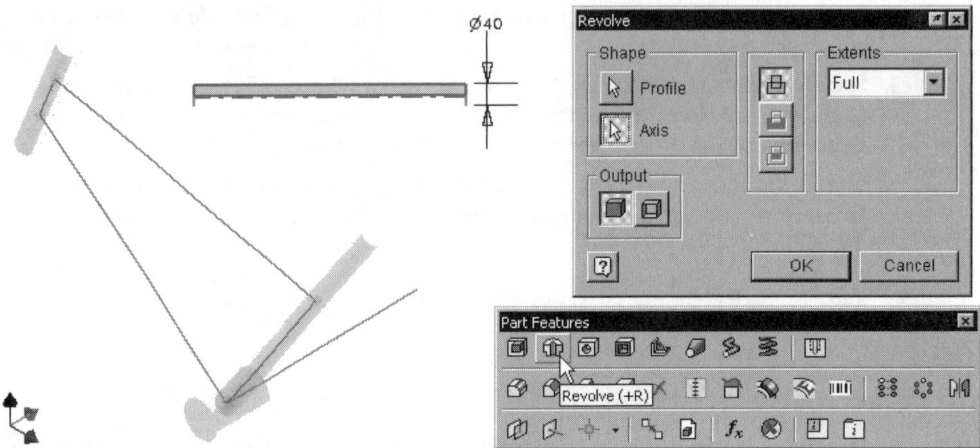

Figure 8.2 *Under-constrained sketch being revolved*

Now set the feature and the component to be adaptive.

9. Select the revolved feature on the Browser Bar, right-click, and select Adaptive. (See Figure 8.3.)

 Tip: Because you used the top-down approach to construct a component, setting a feature adaptive in part modeling mode will automatically set the solid part adaptive in the assembly. If you use the bottom-up or hybrid approach to place a component, you have to first set the feature adaptive in the solid part and then set the component adaptive in the assembly by selecting it on the Browser Bar, right-clicking, and selecting Adaptive.

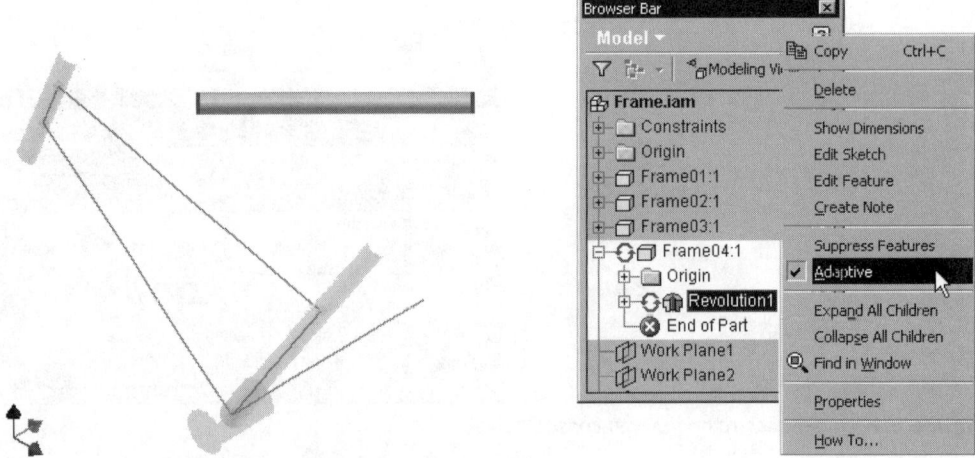

Figure 8.3 *Under-constrained sketched feature set adaptive*

Now return to assembly modeling mode and place assembly constraints.

10. Select Return on the Inventor Standard toolbar to return to assembly modeling mode.

11. Set the display to Hidden Edge Display.

12. Select Constraint on the Assembly Panel Bar or toolbar.

13. Select the component's axis A and work axis B in Figure 8.4, and click the Apply button in the Place Constraint dialog box. A mate constraint is placed.

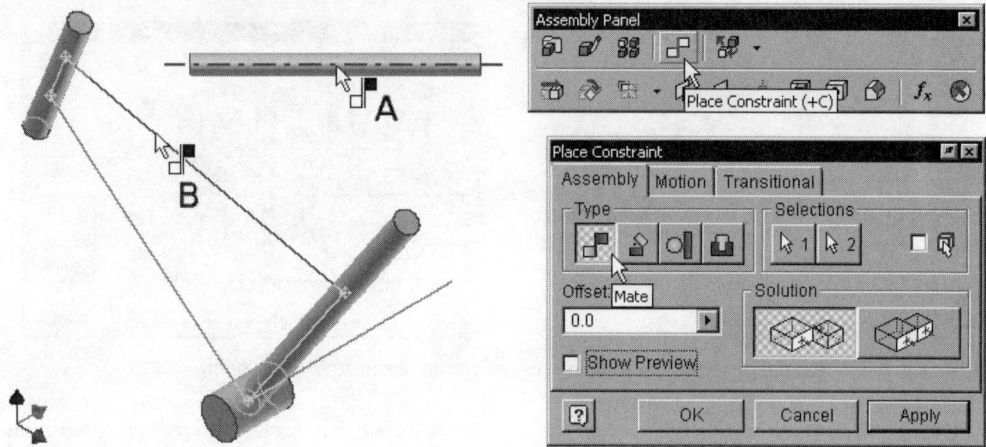

Figure 8.4 *Component's axis being mated to a work axis of the assembly*

Now place a mate constraint to position the center point of the adaptive solid part in reference to a work point.

14. Select center point A and work point B in Figure 8.5 and click the Apply button in the Place Constraint dialog box.

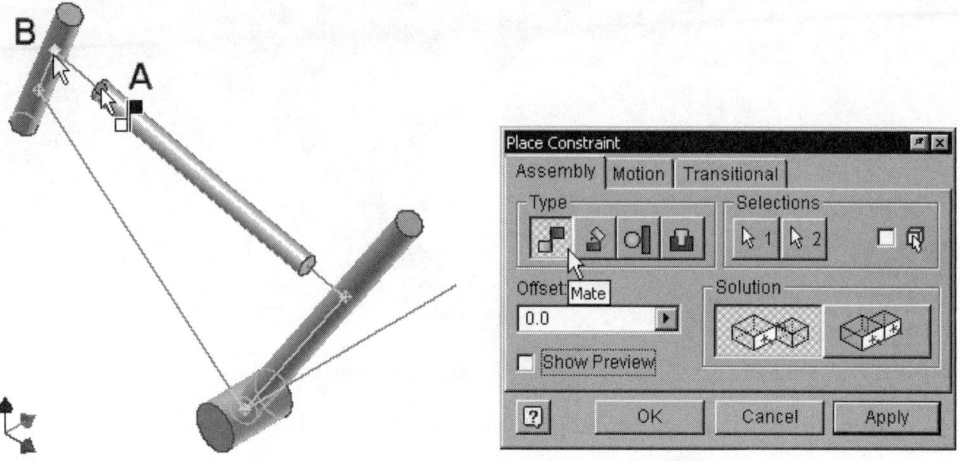

Figure 8.5 *Component's face being mated to the assembly's work point*

Now add another assembly constraint.

15. Select center point A and work point B in Figure 8.6 and click the OK button in the Place Constraint dialog box.

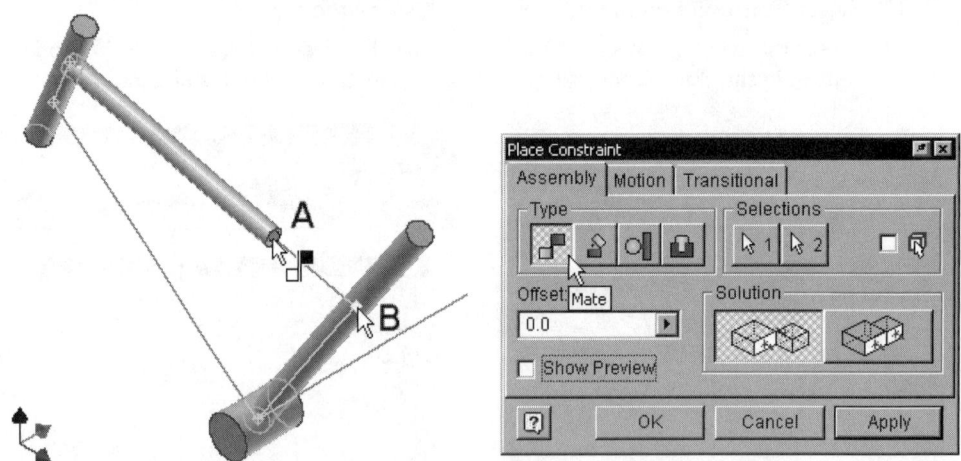

Figure 8.6 *Component's center point being mated to the assembly's work point*

 Note: After you click the Apply or OK button, the adaptive feature changes in length (because the length is not dimensioned) to adapt to the locations of the mated work points. See Figure 8.7 and compare it with Figure 8.8.

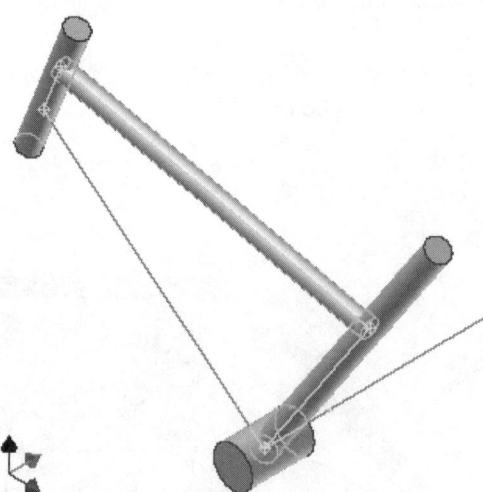

Figure 8.7 *Parameter of the under-constrained sketched feature changed*

Now construct another solid part with an adaptive sketched solid feature.

16. Repeat Steps 3 through 15 to construct another solid part with an under-constrained adaptive sketched solid feature (file name: *Frame05.ipt*), and add assembly constraints to assemble it. (See Figure 8.8.)

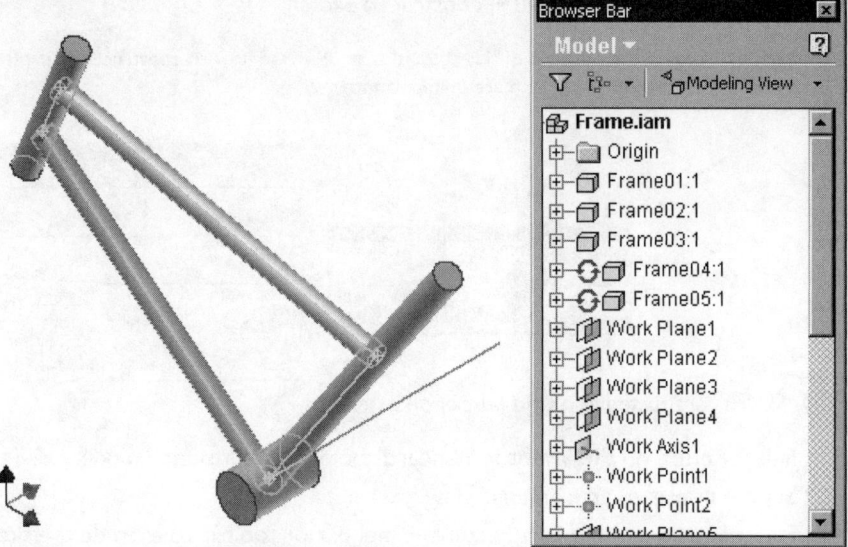

Figure 8.8 *Second adaptive component constructed*

The basic components of the bicycle frame are complete. (See Figure 8.1.) You will further elaborate the design by deriving solid parts from the assembly later in this chapter. Now save and close all your files.

SKETCH ADAPTIVE TO A FACE OF ANOTHER COMPONENT

In the context of an assembly, you can construct a sketch of a solid part on the face of another component. You can also associatively reference an edge or a connected loop of edges to the other component part in an assembly. A sketch constructed this way is adaptive to the edges of the referenced solid part.

 **TUTORIAL 8.2**

To learn how to construct adaptive sketch on the face of another part, you will use the top-down approach to construct an assembly of two component parts. First you will construct a solid part in an assembly.

1. Start a new assembly file. Use the metric template.
2. Select Create Component on the Assembly Panel Bar or toolbar.
3. In the Create In-Place Component dialog box, specify a file name (*AdaptiveSketch1.ipt*).

 Note: The part should be created using the metric Standard.ipt template.

4. With reference to Figure 8.9, construct a sketch.

 Note: Because the purpose of this tutorial is to illustrate how to construct an adaptive sketch, the shape and size of this sketch are unimportant.

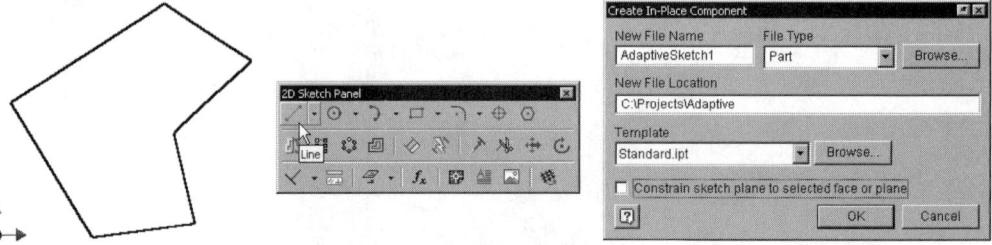

Figure 8.9 *Sketch constructed in a solid part of an assembly*

5. Select Return on the Inventor Standard toolbar to exit sketch mode.

6. Set the display to an isometric view.

7. Select Extrude on the Part Features Panel Bar or toolbar to extrude the sketch a distance of 10 mm. (See Figure 8.10.)

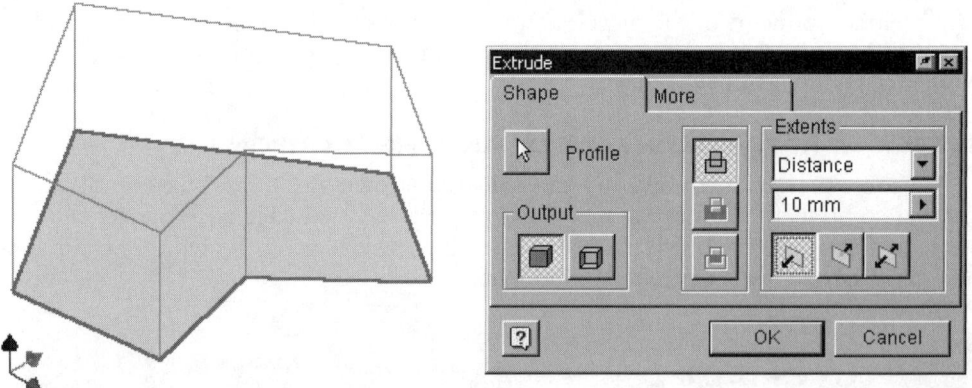

Figure 8.10 *Sketch being extruded*

Now construct another solid part in the assembly.

8. Select Return on the Inventor Standard toolbar to return to assembly modeling mode.

9. Select Create Component on the Assembly Panel Bar or toolbar.

10. In the Create In-Place Component dialog box, select the Constrain sketch plane to selected face or plane check box, specify the new part file name (*AdaptiveSketch2.ipt*), make sure that it will be saved in the same folder as the previous part, select the metric *Standard.ipt* template, and then click OK.

11. Select face A in Figure 8.11 to specify a plane.

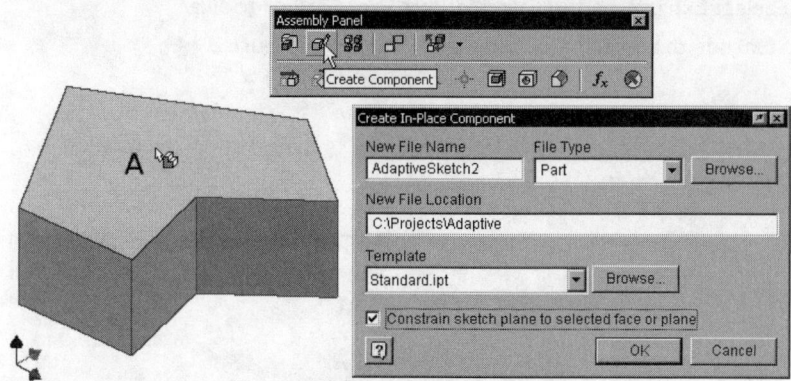

Figure 8.11 *Second solid part in the assembly being constructed*

Now construct an adaptive sketch by projecting the edges of another solid part of the assembly to the current sketch plane.

12. Select Project Geometry on the 2D Sketch Panel Bar or toolbar.

13. Select edges A, B, and C in Figure 8.12.

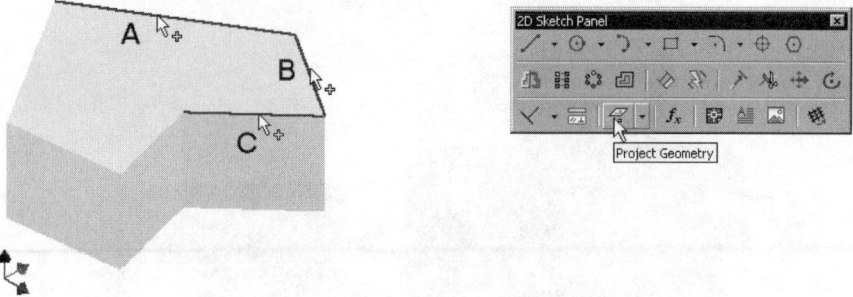

Figure 8.12 *Edges of another solid part in the assembly being projected*

Now complete the sketch and extrude the sketch to a solid.

14. Referring to Figure 8.13, construct a line AB.

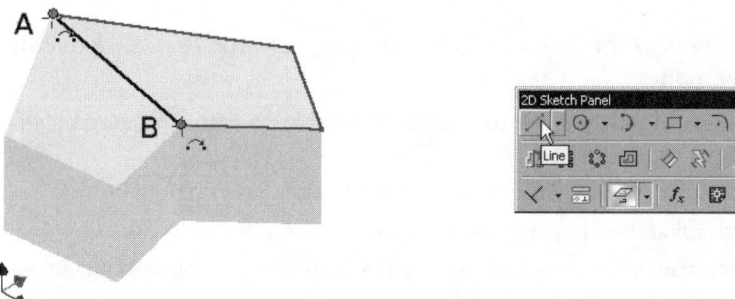

Figure 8.13 *Adaptive sketch constructed*

15. Select Return on the Inventor Standard toolbar to exit sketch mode.
16. Select Extrude on the Part Features Panel Bar or toolbar.
17. Extrude the sketch a distance of 10 mm. (See Figure 8.14.)

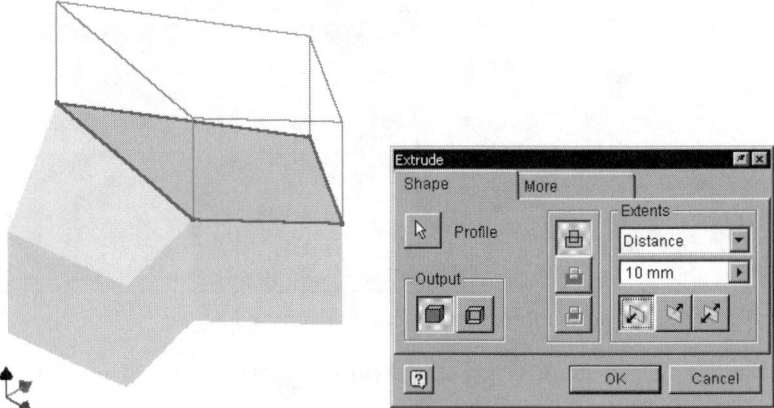

Figure 8.14 *Adapted sketch being extruded*

A solid part with its sketch referenced to the edges of another solid part is constructed. (See Figure 8.15.)

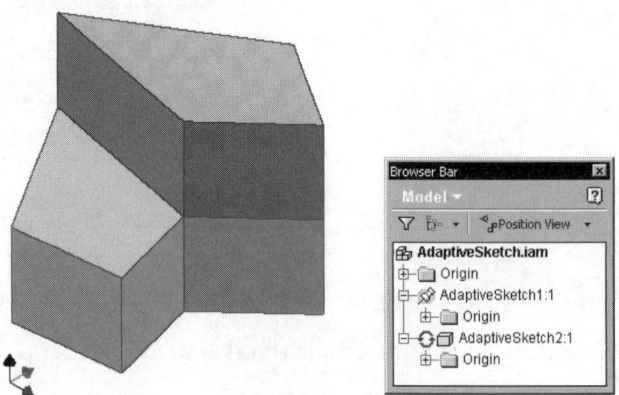

Figure 8.15 *Solid part constructed from an adaptive sketch*

To appreciate how the second component adapts to the first component, modify the first component.

18. Select Return on the Inventor Standard toolbar to return to assembly modeling mode.
19. Select the part (AdaptiveSketch1) on the Browser Bar and double-click.
20. Select the sketch of the solid part, right-click, and select Edit Sketch.
21. With reference to Figure 8.16, modify the sketch's shape by selecting its vertices and dragging them to new positions.

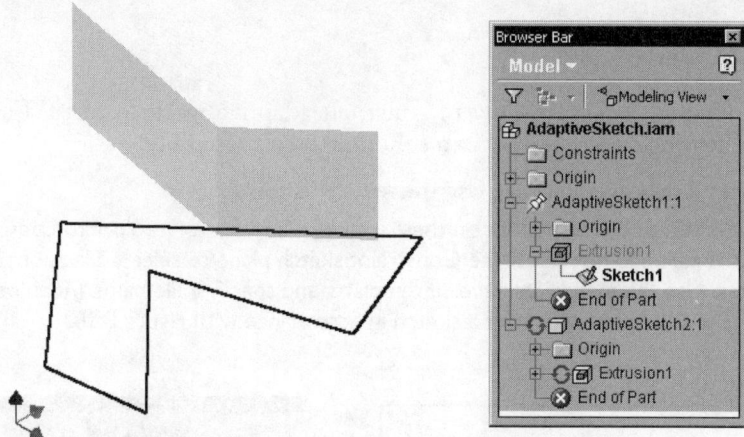

Figure 8.16 *Sketch modified*

22. Right-click and select Finish Sketch.

23. Select the Update button on the Inventor Standard toolbar.

The solid part is updated and, as you can see, the other component part is also modified. (See Figure 8.17.) Save and close your files (file name of the assembly: *AdaptiveSketch.iam*).

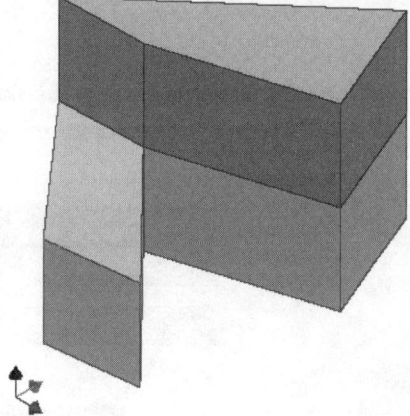

Figure 8.17 *Parts modified*

ADAPTIVE WORK FEATURES

You can apply adaptive technology to work features as well as to sketched features. Adaptive work features are work features constructed in part modeling mode and referenced to features of other components in the context of an assembly in a top-down design approach. However, they are not assembly work features, because they reside in solid part files. The location of an adaptive work feature is referenced to the geometry of other component parts in the assembly. If the referenced geometry changes, the adaptive work features adapt to the change.

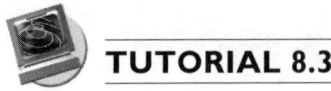

TUTORIAL 8.3

In this tutorial, you will learn how to construct adaptive work features. First you will construct an extruded solid feature in a solid part of an assembly.

1. Start a new assembly file. Use the metric template.

2. Select Create Component on the Assembly Panel Bar or toolbar to construct a solid part. Do not select the Constraint sketch plane to selected face or plane check box. Use the metric *Standard.ipt* template and specify a file name (*AdaptiveWF1.ipt*). In the solid part, construct a sketch in accordance with Figure 8.18.

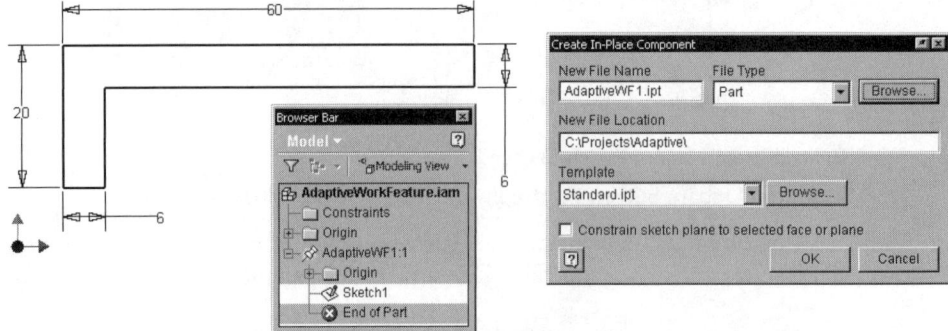

Figure 8.18 *Sketch constructed in a solid part of an assembly*

3. Set the display to an isometric view and extrude the sketch a distance of 30 mm.

4. With reference to Figure 8.19, construct a sketch and extrude it to cut through the solid.

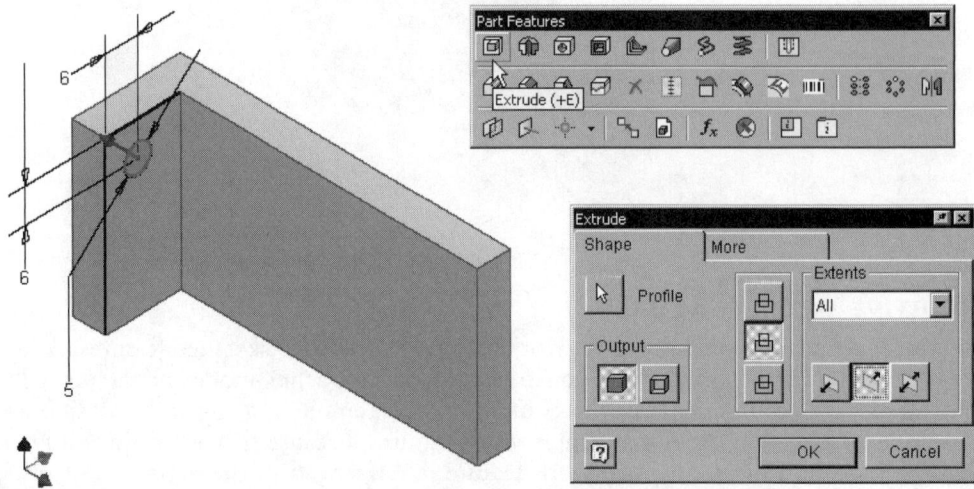

Figure 8.19 *Display set to isometric and a sketch constructed on a face of the solid*

5. Construct another sketch in accordance with Figure 8.20 and extrude it to cut through the solid part. The solid part is complete.

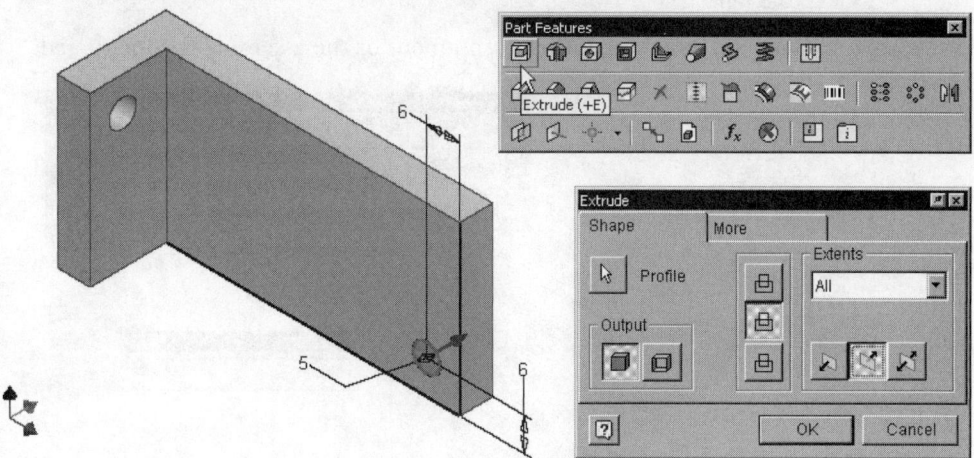

Figure 8.20 *Sketch constructed on a horizontal face of the solid part*

Now construct another solid part in the assembly.

6. Select Return on the Inventor Standard toolbar to return to assembly modeling mode.

7. Select Create Component on the Assembly Panel Bar or toolbar.

8. In the Create In-Place Component dialog box, specify a file name (*AdaptiveWF2.ipt*), clear the Constrain sketch plane to selected face or plane check box, be sure to use the metric *Standard.ipt* template, and then click OK. (See Figure 8.21.)

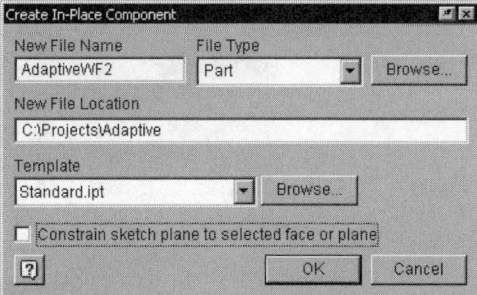

Figure 8.21 *Second component being started in the assembly*

Now construct adaptive work planes, work axes, and work points.

9. Select a point in the graphics area to locate the solid part.

10. Select Return on the Inventor Standard toolbar to exit sketch mode, select the new sketch on the Browser Bar, right-click, and select Delete to remove the new sketch.

11. Select Work Plane on the Part Features Panel Bar or toolbar.

12. Select the face highlighted in Figure 8.22, hold down the left mouse button, and drag to construct an offset work plane. In the Offset dialog box, set the offset value to 5 mm.

A work plane adaptive to another component of the assembly is constructed.

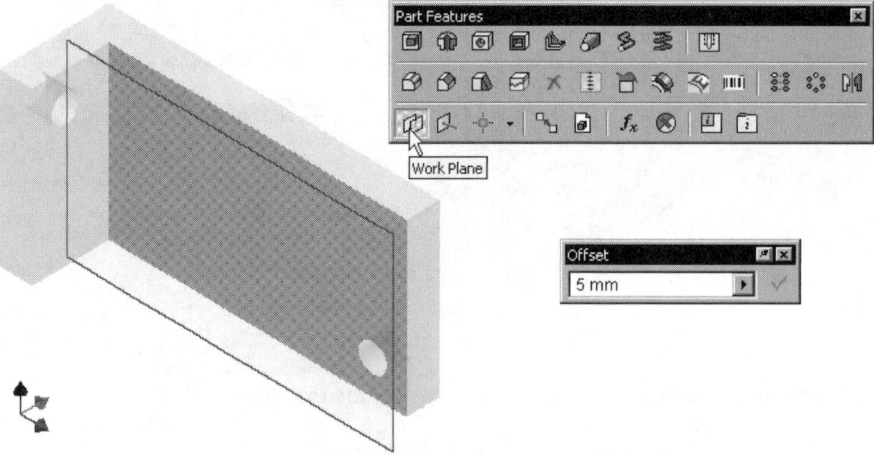

Figure 8.22 *Adaptive work plane being constructed*

 Tip: Constructing a work feature with reference to other solid parts in the top-down approach makes the work feature an adaptive work feature.

Now construct three more adaptive work planes.

13. Select Work Plane on the Part Feature Panel Bar or toolbar. Then select the face highlighted in Figure 8.23 and drag to construct a work plane that offsets a distance of 5 mm. Another adaptive work plane is constructed.

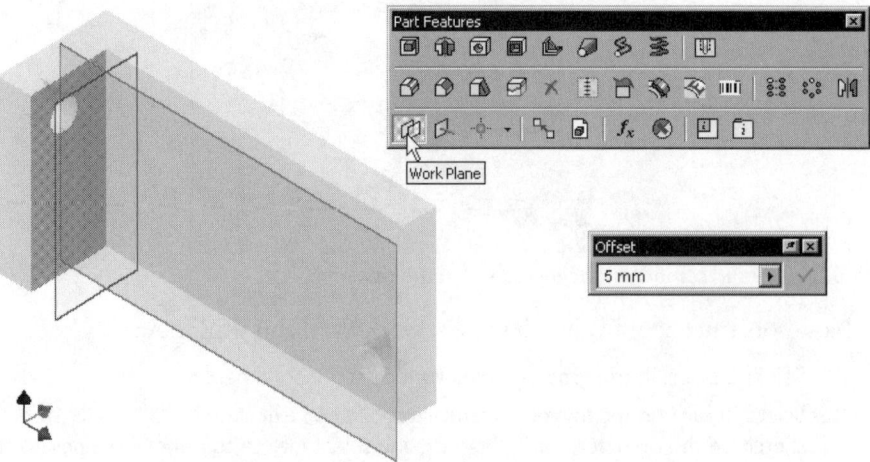

Figure 8.23 *Second adaptive work plane being constructed*

14. Rotate the display in accordance with Figure 8.24. Then construct the third adaptive work plane accordingly. Offset distance is 5 mm.

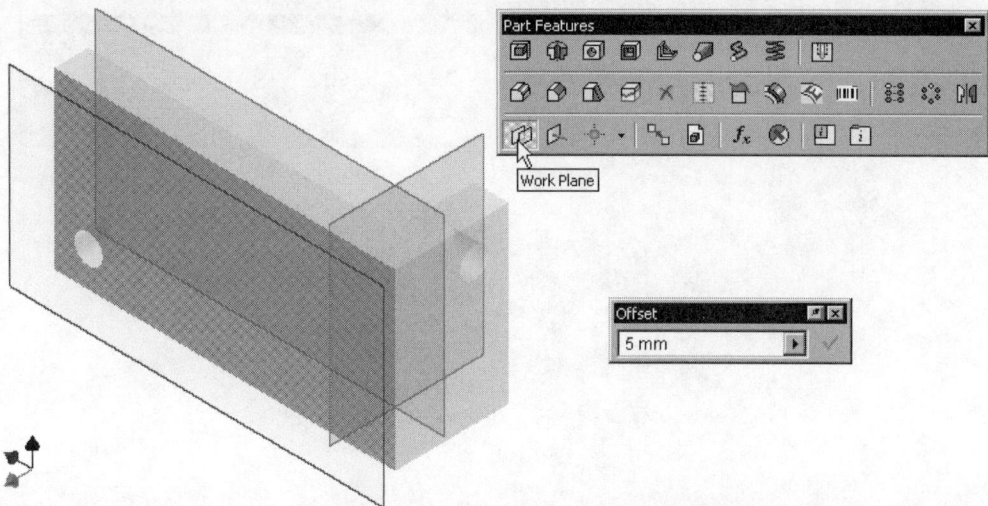

Figure 8.24 *Display rotated and third adaptive work plane being constructed*

15. Construct the fourth adaptive work plane in accordance with Figure 8.25. The offset distance is also 5 mm.

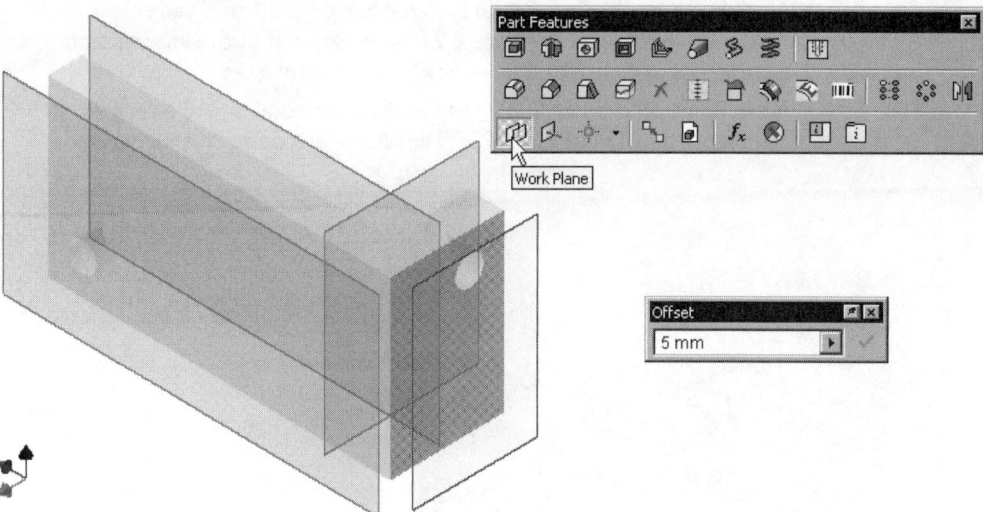

Figure 8.25 *Fourth adaptive work plane being constructed*

Now construct two adaptive work axes.

16. Select Work Axis on the Part Features Panel Bar or toolbar and select axes A and B in Figure 8.26 one by one to construct two work axes.

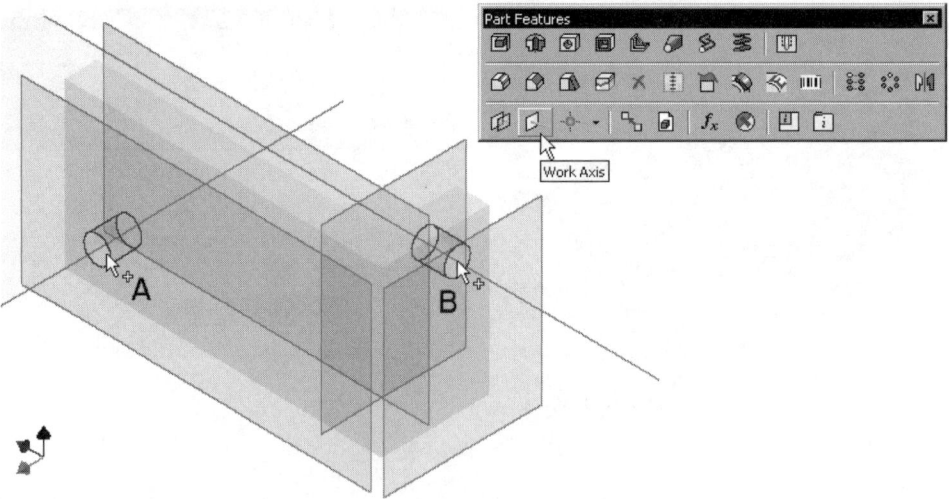

Figure 8.26 *Display set to isometric and two adaptive work axes constructed*

Now construct four work points at the intersections of the work axes and the work planes.

17. Select Work Point on the Part Features Panel Bar or toolbar. Then select work plane A and the work axis B in Figure 8.27. A work point at the intersection of an adaptive work axis and an adaptive work plane is constructed.

18. Select Work Point on the Part Features Panel Bar or toolbar and select work plane C and work axis D in Figure 8.27. Another work point is constructed.

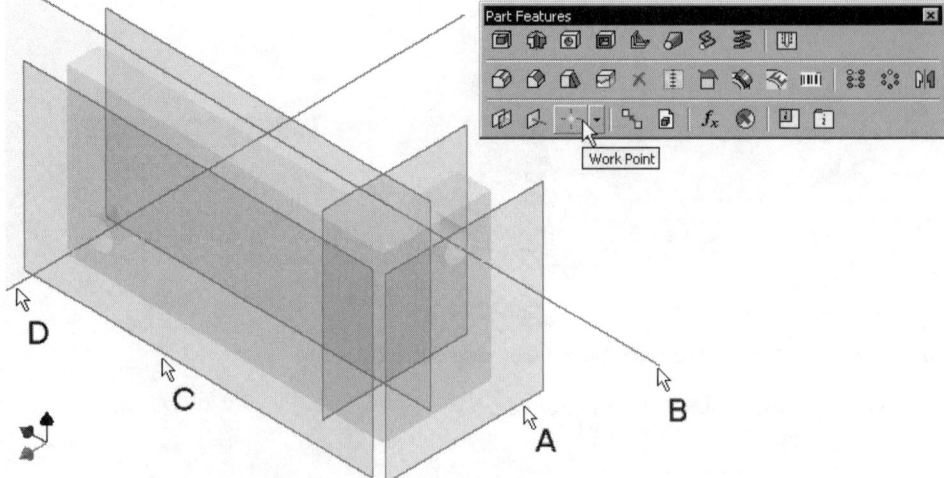

Figure 8.27 *Adaptive work point constructed*

19. Set the display to isometric. Then similarly, construct two more work points at the intersection of the work axis A and work plane B and the intersection of work axis C and work plane D.

A set of adaptive work features is constructed. (See Figure 8.28.)

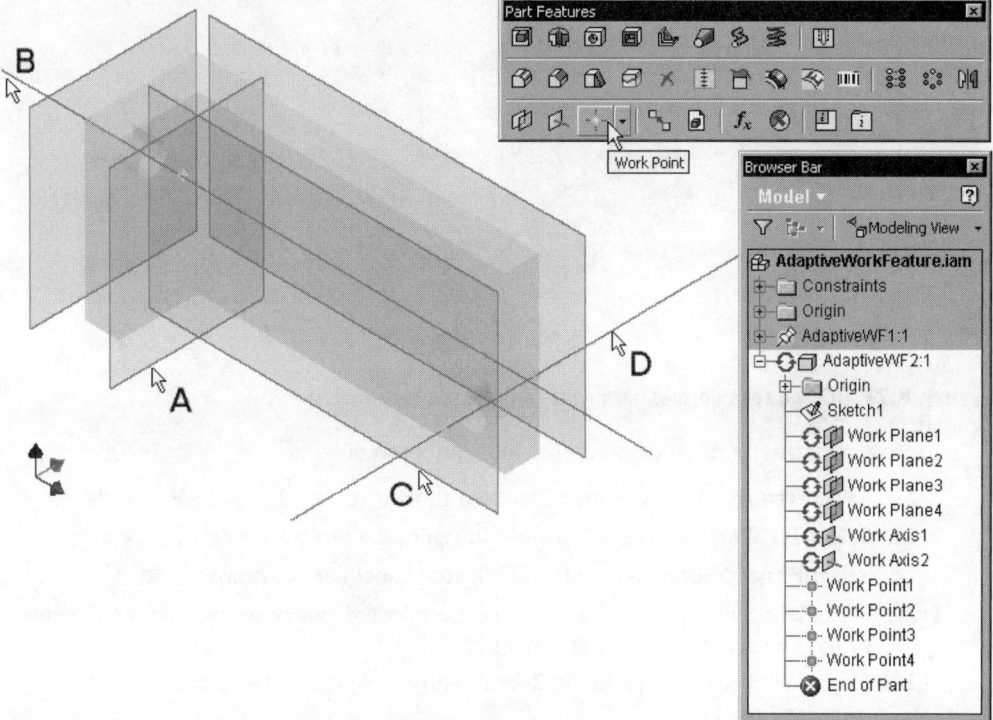

Figure 8.28 *Three more adaptive work points constructed*

Now use the adaptive work points as vertices to construct a 3D sweep path.

20. Select 3D Sketch on the Inventor Standard toolbar to start a 3D sketch.

 Note: 3D Sketch is on the same flyout as the 2D Sketch button.

21. Select Line on the 3D Sketch Panel Bar or toolbar and select the work points indicated in Figure 8.29 one by one to construct a set of 3D lines with tangent corner bends (radius = 5 mm). A 3D sketch is constructed.

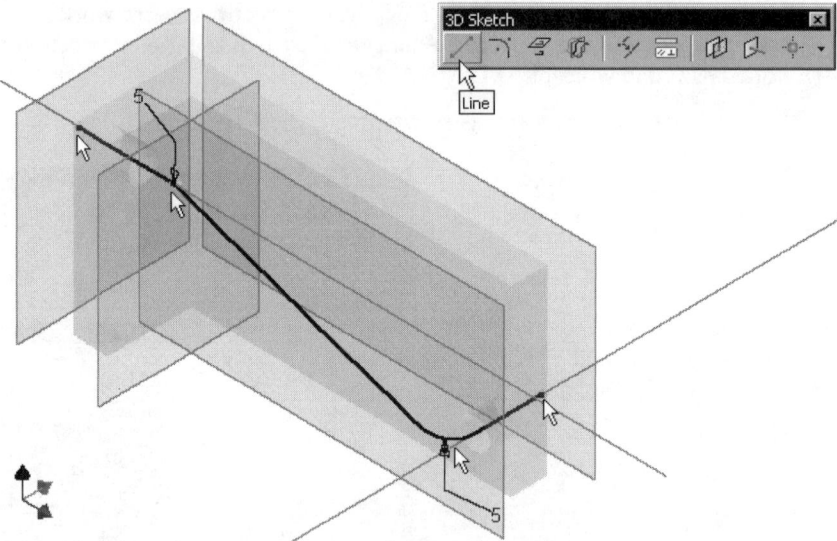

Figure 8.29 *3D line segments with tangent corner bend constructed*

Now construct a profile sketch for making a sweep solid.

22. Select Return on the Inventor Standard toolbar to exit 3D sketching mode.

23. Select 2D Sketch and select plane A in Figure 8.30 to start a new 2D sketch.

24. Select Project Geometry on the 2D Sketch Panel Bar or toolbar.

25. Select axis B in Figure 8.30 to project the selected axis onto the current sketch plane. The projected geometry is a point.

26. Construct a circle (diameter = 3 mm) with the center point located on the projected geometry.

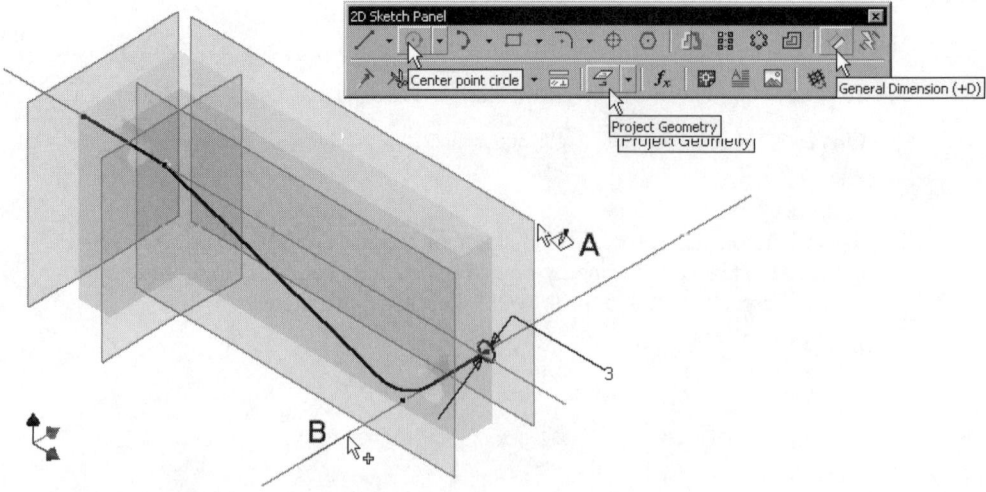

Figure 8.30 *2D sketch constructed*

Now construct a 3D sweep solid.

27. Exit sketch mode and select Sweep on the Part Features Panel Bar or toolbar to construct a sweep solid. (See Figure 8.31.)

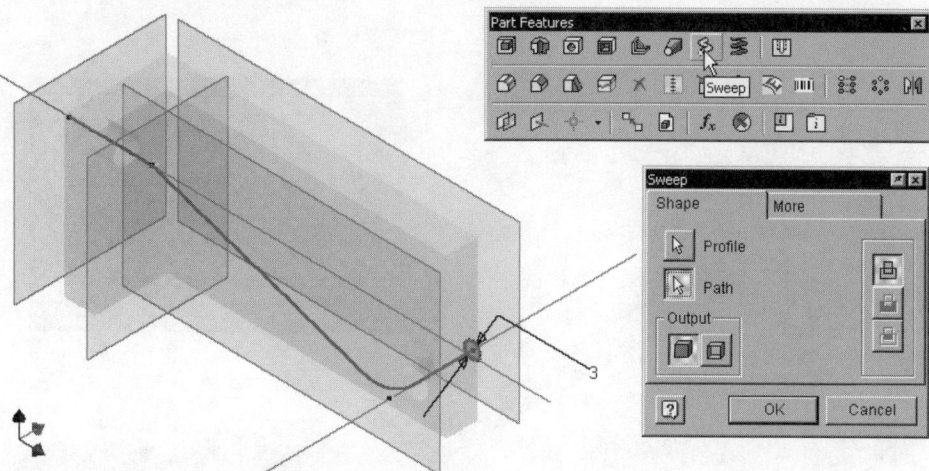

Figure 8.31 *Sweep solid being constructed*

Now hide the work features.

28. Select the work planes, work axes, and work points one by one, right-click, and deselect Visibility to hide them. (See Figure 8.32.)

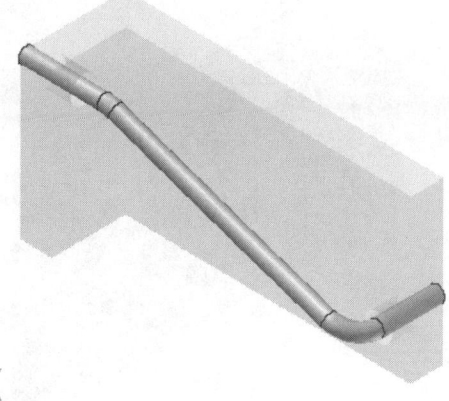

Figure 8.32 *Work features hidden*

The sweep solid with adaptive work features is complete. To appreciate how the adaptive work features help to establish a relationship between the solid parts, edit a component part of the assembly.

29. Double-click the solid part (AdaptiveWF1) on the Browser Bar to activate part modeling mode of the selected part.

30. Select the first extruded feature on the Browser Bar, right-click, and select Edit Feature.

31. Change the extrusion distance to 40 mm. (See Figure 8.33.)

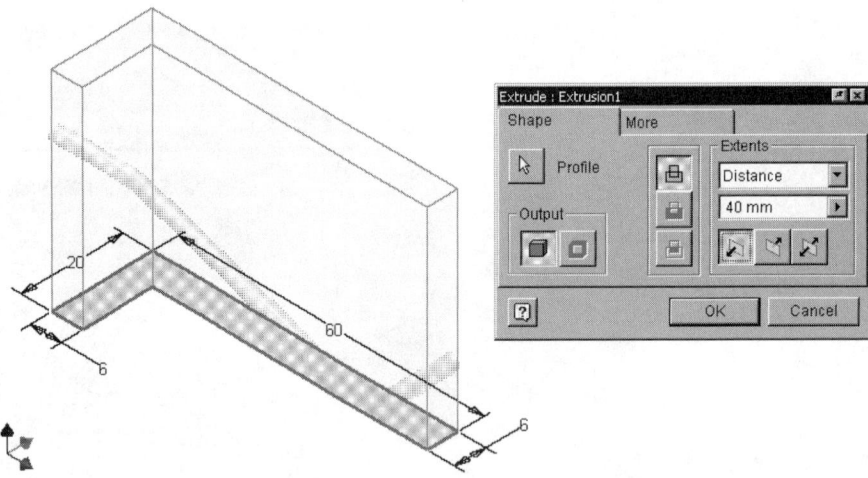

Figure 8.33 *Extrusion distance being modified*

32. Double-click the assembly file on the Browser Bar. The other component adapts to the change. (See Figure 8.34.)

The assembly is complete. Save and close your files (file name of the assembly: *AdaptiveWorkFeature.iam.*).

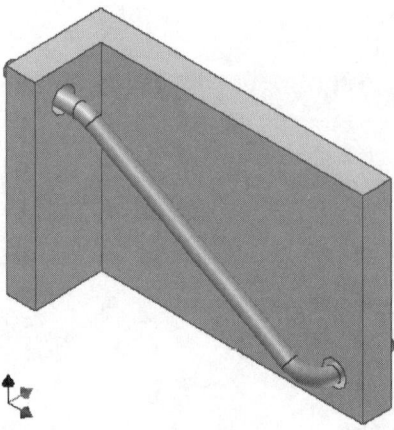

Figure 8.34 *Assembly updated*

ADAPTIVE PARTS IN A SUB-ASSEMBLIES

You can set sub-assemblies to be adaptive in the root assembly. If you mark a sub-assembly to be adaptive, the assembly will not be treated as a rigid part in the root assembly. You can set adaptive parts in a sub-assembly to be adaptive in the root assembly.

TUTORIAL 8.4

To learn how to set sub-assemblies as adaptive, you will construct an assembly with two components: a bush and a sub-assembly that you constructed in Tutorial 8.3.

First you will construct a solid part.

1. Start a new part file. Use the metric template.
2. Construct a sketch in accordance with Figure 8.35.

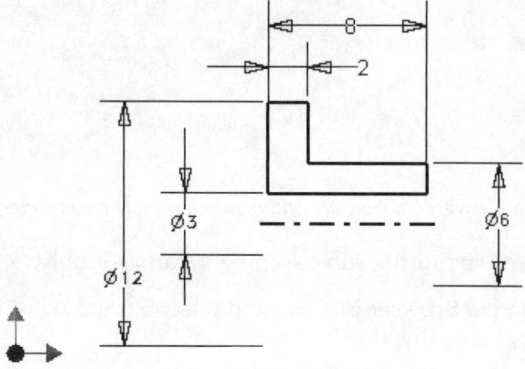

Figure 8.35 *Sketch constructed*

3. Revolve the sketch 360 degrees. (See Figure 8.36.) The part file is complete.
4. Save and close the file (file name: *Bush.ipt*).

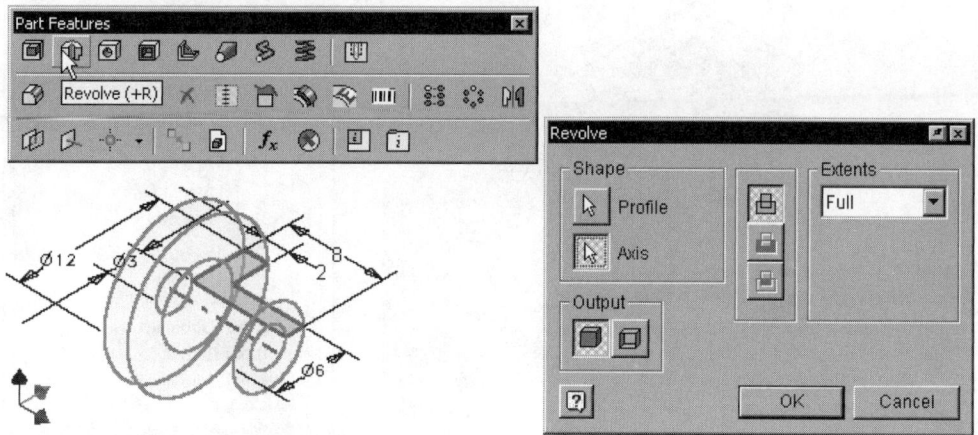

Figure 8.36 *Sketch being revolved*

Now construct an assembly. In the assembly, you will place a solid part and a sub-assembly.

5. Start a new assembly file. Use the metric template.

6. Place two components (*AdaptiveWorkFeature.iam* and *Bush.ipt*) in the assembly. (See Figure 8.37.)

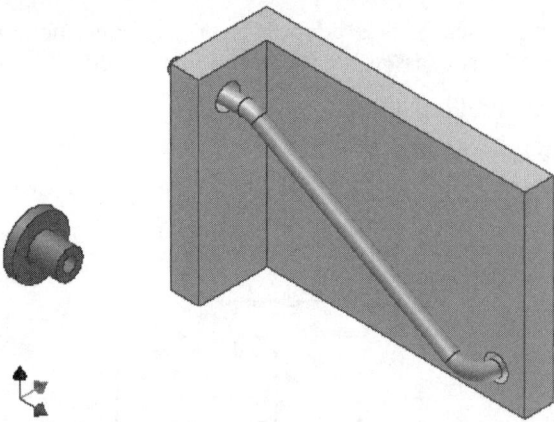

Figure 8.37 *New assembly with a solid part and a sub-assembly constructed*

Now edit a solid part in the sub-assembly of an assembly.

7. Right-click in the Browser Bar area and select Expand All to expand the items in the assembly hierarchy.

8. Double-click the part file (AdaptWF1) from the sub-assembly to activate part modeling mode. (See Figure 8.38.)

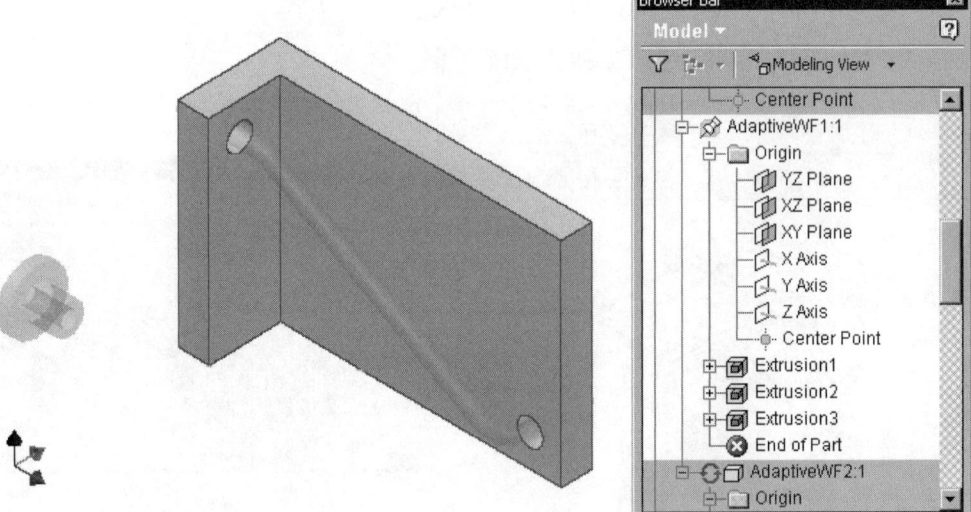

Figure 8.38 *Part in the sub-assembly activated*

Now delete a dimension in the sketches of sketched features to make them under-constrained.

9. Refer to Figure 8.39. Select Sketch 2, right-click and select Edit Sketch, delete the diameter dimension, and then select Return. Repeat this sequence, referring to Figure 8.40, to delete the other diameter dimension.

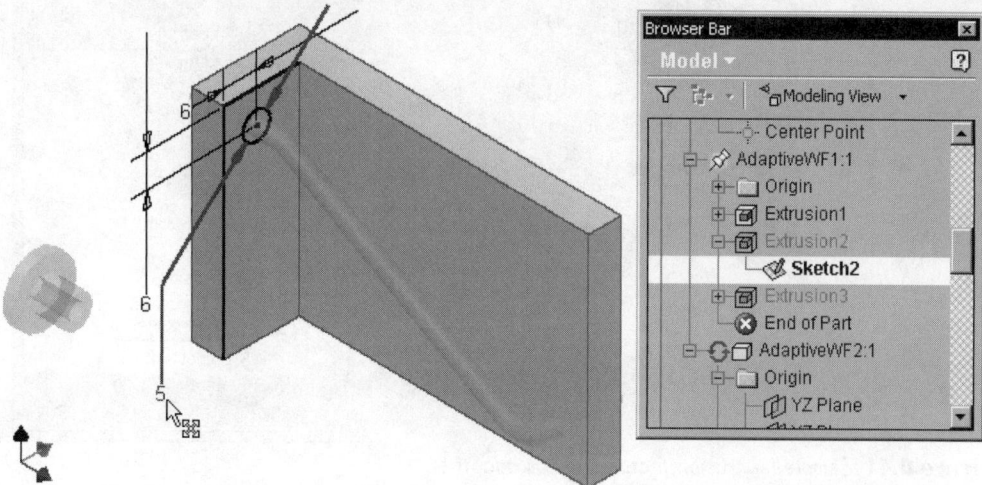

Figure 8.39 *Diameter dimension being deleted*

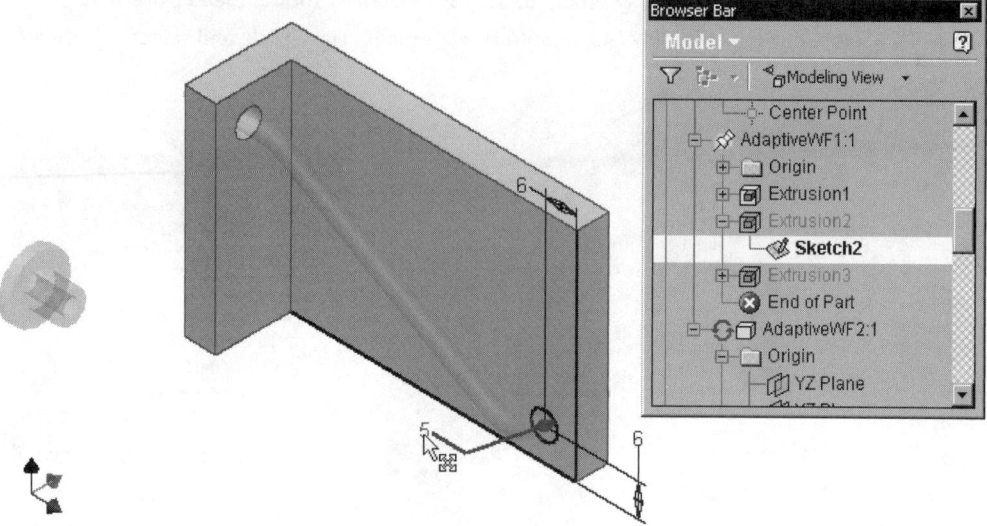

Figure 8.40 *Diameter dimension being deleted*

Now set the under-constrained features to be adaptive.

10. Referring to Figure 8.41, select the extrusion features one by one, right-click, and select Adaptive.

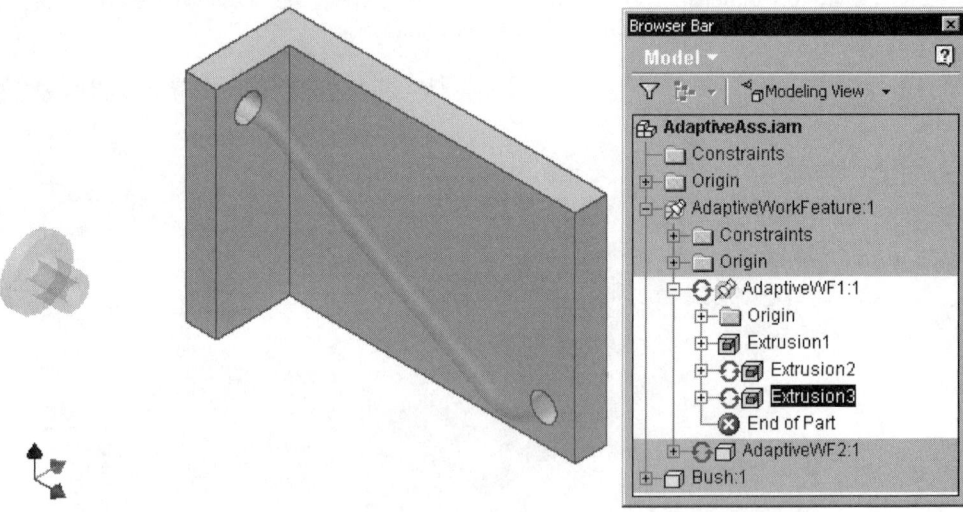

Figure 8.41 *Selected extrusion features set as adaptive*

Now place assembly constraints.

11. Double-click the root assembly to activate assembly mode. (See Figure 8.42.)

12. Select the sub-assembly (*AdaptiveWorkFeature.iam*), right-click, and select Adaptive.

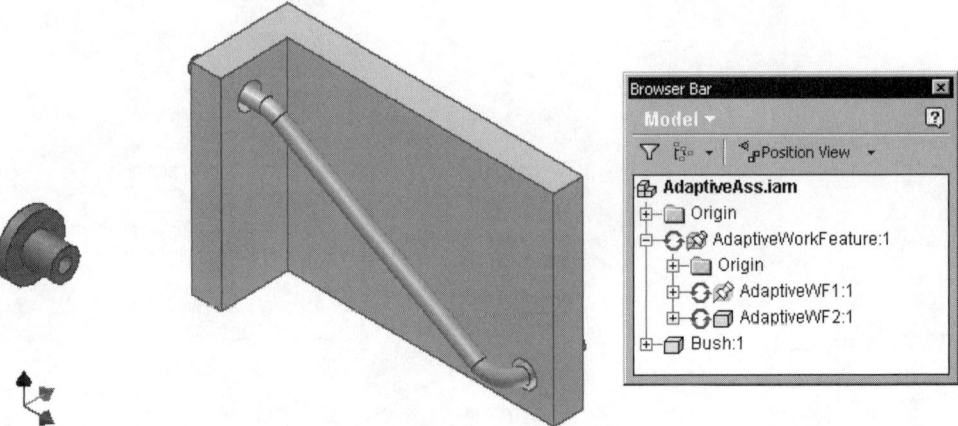

Figure 8.42 *Subassembly marked as adaptive*

Now apply assembly constraints to mate the bush to a component in the sub-assembly.

13. Apply a mate constraint to the cylindrical faces A and B (not the axes) highlighted in Figure 8.43.

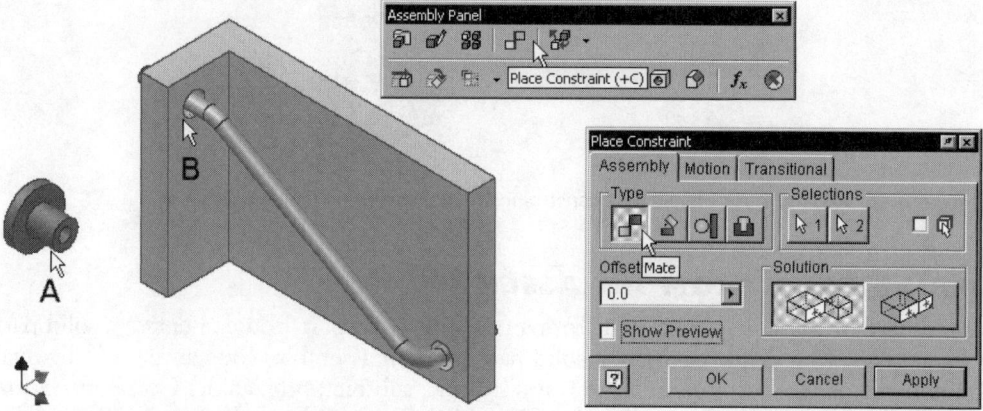

Figure 8.43 *Cylindrical faces being mated*

14. Click the Apply button. Note the change in size of the cylindrical face of the solid part of a sub-assembly.

15. Select the faces highlighted in Figure 8.44 and click the OK button to complete the mating.

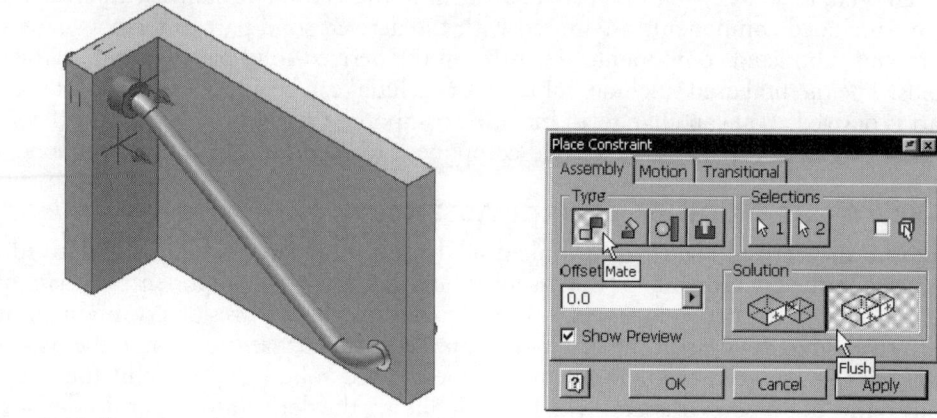

Figure 8.44 *Flush mate being placed*

16. Place another instance of the bush.

17. Repeat steps 13 and 14 above to apply assembly constraints.

The assembly is complete. (See Figure 8.45.) Save and close your file (file name: *AdaptAss.iam*).

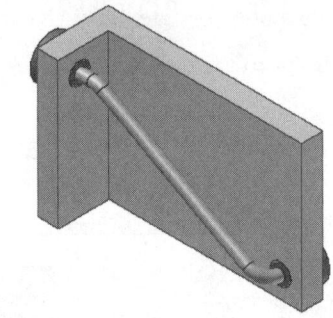

Figure 8.45 *Another bush placed and constrained*

DERIVE PART FROM AN ASSEMBLY

As well as deriving a new solid part from an existing solid part, you can derive a solid part from an assembly. While deriving a solid part from an assembly, you can treat individual components of the assembly in three ways: include, subtract, or exclude. Components are included in, subtracted from, or excluded from the derived solid part. Unlike deriving a solid part from another solid part, you cannot mirror or scale the components from the source.

COMBINATION OF COMPONENTS IN THE DERIVED PART

By using different combinations of include, subtract, and exclude, you can derive a number of different parts from a single assembly. Except for exclude, in which the components are ignored, include and subtract are equivalent to the union and subtract Boolean operations. Included components are united with the derived solid part to form a single solid part, and subtracted components are cut from the derived solid part to form cavities and voids. The method used (include, subtract, or exclude) can be edited any time after a solid part is derived. You can change an included component to become a subtracted component, and you can change an excluded component to become an included component.

LINK TO THE ORIGINAL SOURCE ASSEMBLY

Because the derived solid part is linked to the assembly from which it is derived, any changes in the components of the assembly and the way the components are assembled are reflected in the derived part. If you want to modify the features of a component of the derived solid part, you edit the original component. If you want to change the positional relationships between the components of the derived solid part, you edit the assembly constraints of the source assembly. Although linking the derived solid part to the original source assembly is advantageous, you can de-link the derived solid part from the source assembly if you want the derived solid part to become an independent solid part and not affected by the change in the source assembly. However, it must be realized that de-linking a derived part from the source assembly is irreversible.

CONSTRUCTION OF COMPLEX SOLID PART IN A NON-LINEAR WAY

Deriving a solid part from an assembly adds flexibility to the modeling of complex solid parts. The normal way to make a solid part is to construct its solid features one by one and combine them as you construct them. We call the first sketched feature the base solid feature. Each subsequent sketched solid feature has to be combined with the base solid feature as they are being constructed. This way of modeling a solid part is called the linear approach. The major disadvantage of this approach is that you cannot first combine sketched solid features into two or more sets of features and then combine the sets of features into a single part. For example, you cannot join feature A with B and feature C with D and then cut the combined C and D features from the combined A and B features. As a result, the final form and shape of the solid part is restricted. To overcome this drawback, you can first use the linear approach to construct a number of solid parts of simpler shapes, put the solid parts together in an assembly, and derive a new part from the assembly.

TUTORIAL 8.5

In this tutorial, you will construct a number of derived solid parts from the bicycle assembly.

1. Start a new part file. Use the metric template.
2. Select Return on the Inventor Standard toolbar, if there is a default sketch being activated.
3. Select Derive Component on the Part Features Panel Bar or toolbar.
4. In the Open dialog box, select the assembly file *BicycleFrame.iam* that you constructed in Tutorial 8.1.
5. In the Derived Assembly dialog box, select Frame01 twice to change its state to exclude. (See Figure 8.46.)
6. Select Frame03 once to change its state to subtract.
7. Select Frame04 and Frame05 one by one twice to change their state to exclude.
8. Click the OK button.

A solid part is derived from the assembly. In the derived part, Frame02 is included, Frame01, Frame04, and Frame05 are ignored, and Frame03 is subtracted.

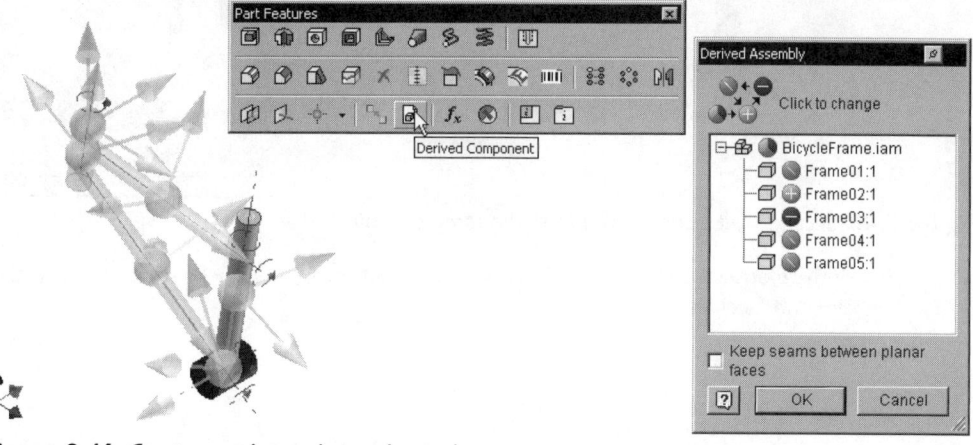

Figure 8.46 *Components' state being changed*

Now place a shell feature.

9. Select Shell on the Part Features Panel Bar or toolbar.

10. Select the faces at both ends of the cylindrical object indicated in Figure 8.47.

11. In the Shell dialog box, set thickness to 2 mm and click the OK button.

12. The solid part is complete. Save and close the file (file name: *Frame02a.ipt*).

Figure 8.47 *Shell feature being placed*

Now construct another derived solid part from the same assembly.

13. Repeat Steps 1 through 4 to construct a solid part.

14. In the Derived Assembly dialog box, subtract Frame01 and Frame02, exclude Frame03 and Frame05, include Frame04, and click the OK button. (See Figure 8.48.)

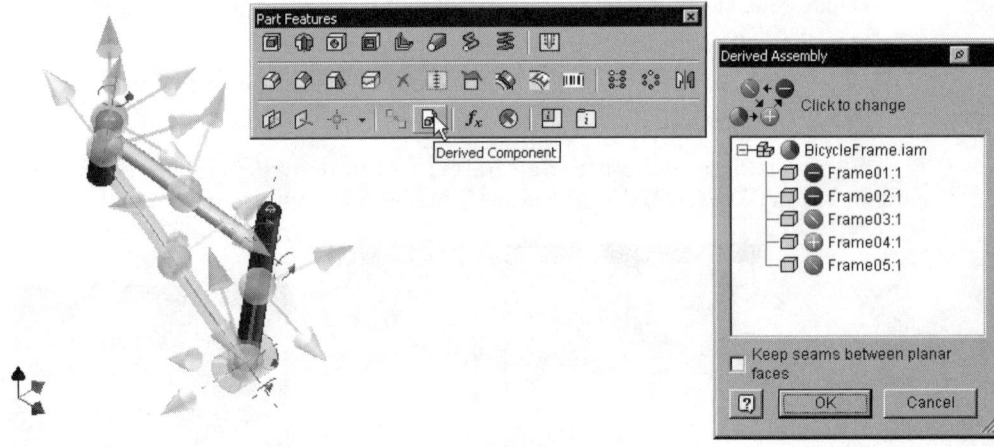

Figure 8.48 *Second solid being derived from the same assembly*

15. With reference to Figure 8.49, place a shell feature with a thickness of 2 mm with two end faces removed.

16. The second solid part derived from the same assembly is complete. Save and close your file (file name: *Frame04a.ipt*).

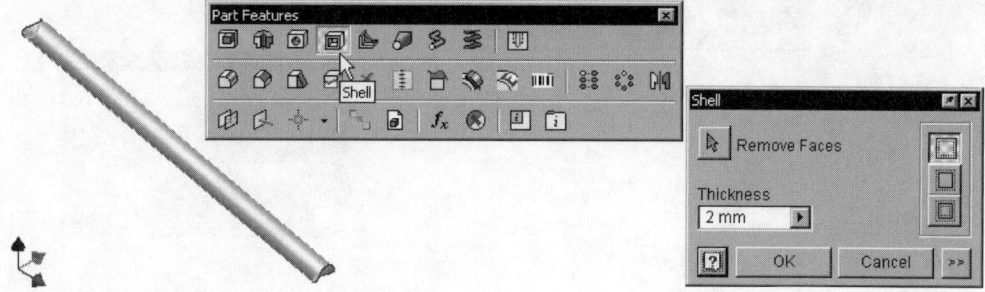

Figure 8.49 *Shell feature being placed*

Now derive the third component from the same assembly.

17. Repeat Steps 1 through 4.

18. In the Derived Assembly dialog box, subtract Frame01 and Frame03, exclude Frame02 and Frame04, include Frame05, and click the OK button. (See Figure 8.50.)

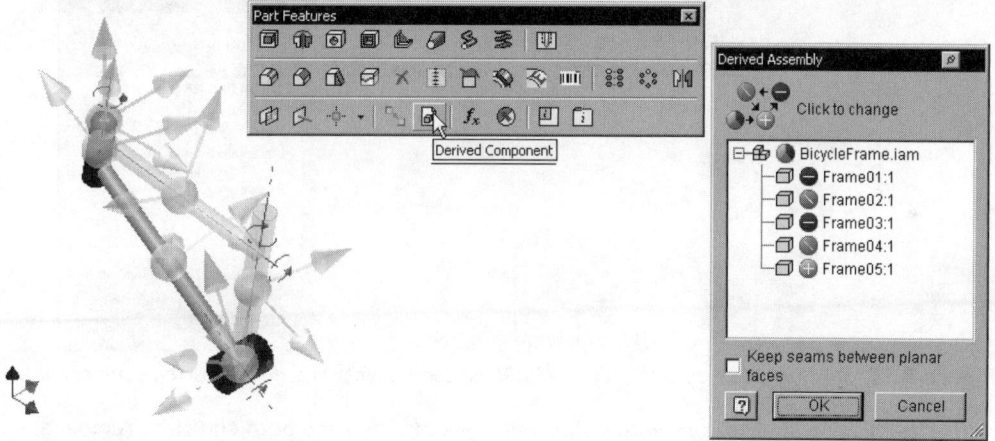

Figure 8.50 *Third component being derived*

19. Add a shell feature with a thickness of 2 mm and two end faces removed. (See Figure 8.51.)

20. The third derived solid part is complete. Save and close the file. (File name: *Frame05a.ipt*)

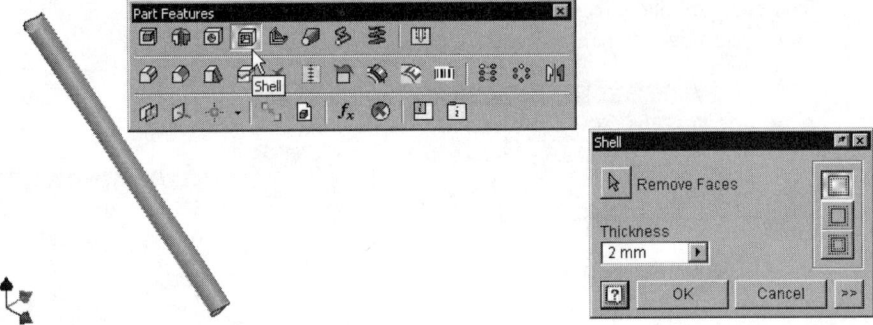

Figure 8.51 *Shell feature placed*

Now derive the fourth solid part from the assembly.

21. Repeat Steps 1 through 4.

22. In the Derived Assembly dialog box, include Frame01, exclude all other components, and click the OK button. (See Figure 8.52.)

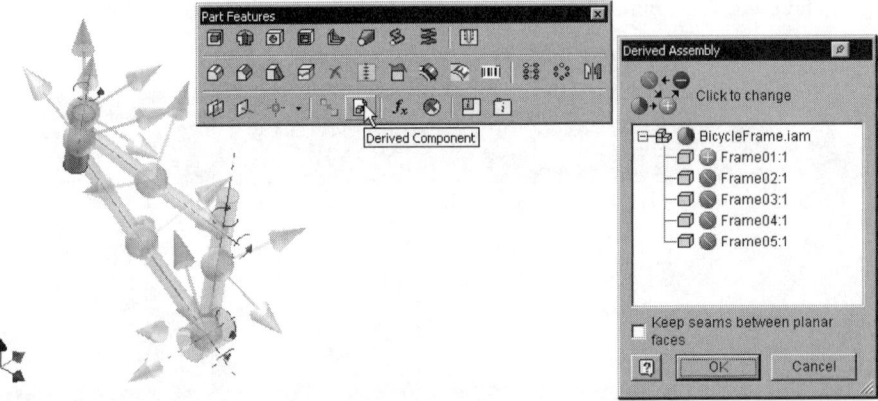

Figure 8.52 *Solid part derived with only one included components and no subtracted component*

23. Place a shell feature with a shell thickness of 2 mm and both end faces removed. (See Figure 8.53.)

24. The fourth derived component is complete. Save and close your file (file name: *Frame01a.ipt*).

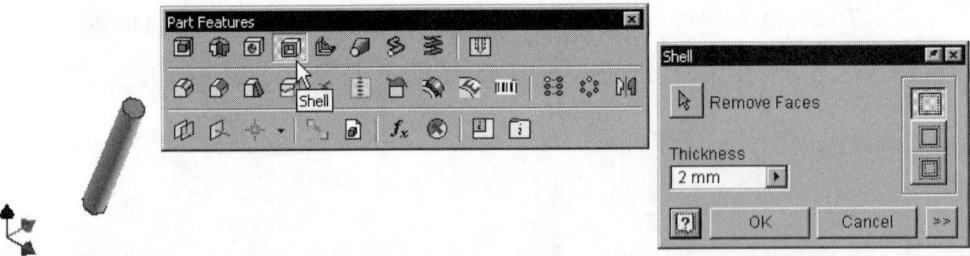

Figure 8.53 *Shell feature placed*

In the last derived solid part, you included only one component and excluded all other components. This is equivalent to deriving from the included solid part. For comparison purposes, now derive a solid part from a single solid part.

25. Start a new part file. Use the metric template.
26. Select Return on the Inventor Standard toolbar, if there is a default sketch.
27. Select Derive Component on the Part Features Panel Bar or toolbar.
28. In the Open dialog box, select the file *Frame03.ipt*.
29. In the Derived Part dialog box, click the OK button. (See Figure 8.54.)

Figure 8.54 *Solid part being derived from another solid part*

30. Select Shell on the Part Features Panel Bar or toolbar. (See Figure 8.55.)
31. Select both end faces of the derived component.
32. In the Shell dialog box, set shell thickness to 2 mm and click the OK button.
33. The solid part derived from another solid part is complete. Save and close your file (file name: *Frame03a.ipt*).

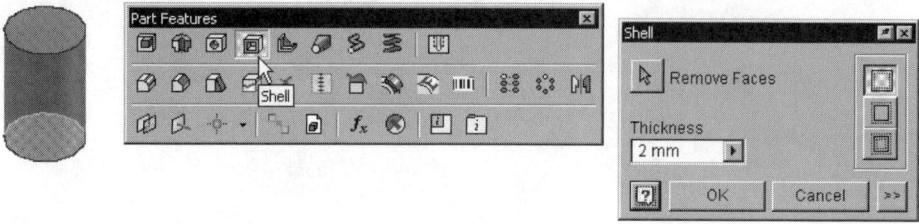

Figure 8.55 *Solid part derived from another solid part*

Now you have five derived solid parts. To complete the bicycle frame assembly, which is a set of tubular components, instead of a set of solid rods with interferences, construct another assembly.

34. Start a new assembly file. Use the metric template.

35. Place the assembly file *BicycleFrame.iam* in the assembly. The placed assembly becomes a sub-assembly.

36. Select the components and the work planes of the sub-assembly one by one on the Browser Bar, right-click, and deselect Visibility. (Do not activate the subassembly. Otherwise, you modify the sub-assembly.)

The components and work planes are hidden. Now you have a skeleton of work axes. (See Figure 8.56.)

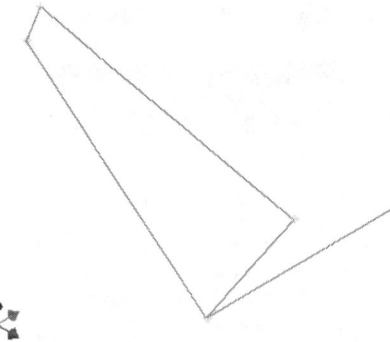

Figure 8.56 *Sub-assembly's component hidden*

Now place the derived solid parts in the assembly and add assembly constraint to properly locate them.

37. Place the part files (Frame01a, Frame02a, Frame03a, Frame04a, and Frame05a) in the assembly. (See Figure 8.57.)

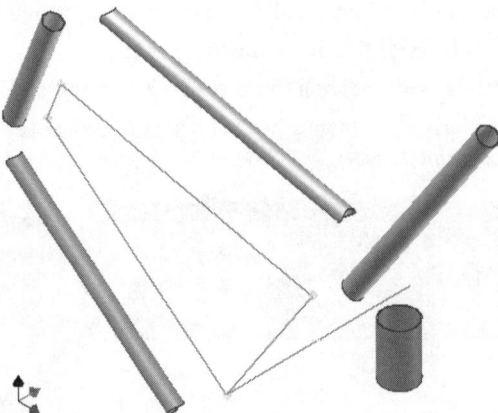

Figure 8.57 *Part files placed in the assembly*

38. Using the work features of the sub-assembly as references, add assembly constraints to assemble the part files in accordance with Figure 8.58.

The final assembly is complete. If you wish to modify the size of the bicycle frame, open the assembly file *BicycleFrame.iam* and change the parameters of the work planes.

Now save and close your file (file name: *Frame.iam*).

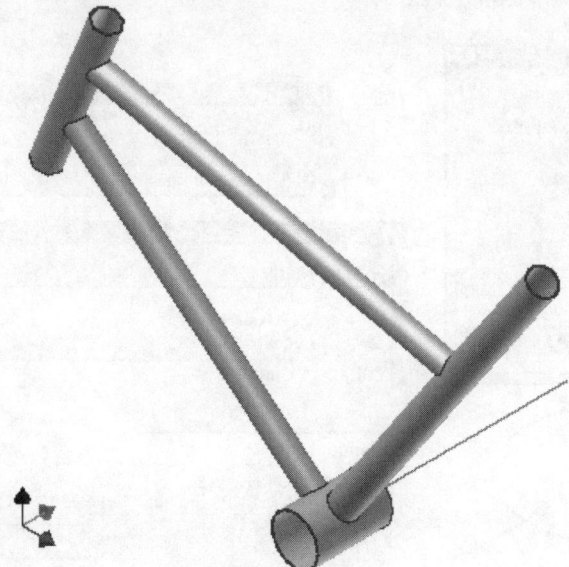

Figure 8.58 *Completed bicycle frame assembly*

PROMOTE FROM ASSEMBLY

In designing a solid part by using a top-down approach, you can copy surfaces and solid features between solid parts. To perform this copying task, you use the Promote tool. Although the copied objects are un-associated and grounded, you can use them in modeling processes that you learned in other chapters.

Typically, you use the Promote tool to help construct a feature that matches the contour and profile of the feature of another solid part. Because surfaces or solids promoted are grounded and un-associated, you need to properly position the solid parts prior to promoting.

 ## TUTORIAL 8.6

In this tutorial, you will promote a surface from another solid part in the context of an assembly and use the promoted surface as an extrusion termination surface.

Now you will use top-down approach to construct an assembly of three 2D layout parts.

1. Start a new assembly file. Use the metric template
2. Select Create Component on the Assembly Panel Bar or toolbar.
3. In the Create In-Place Component dialog box, specify a file name (*Promote.ipt*).

 Note: Use the metric Standard.ipt template.

4. Set the display to an isometric view and construct a sketch on the XY plane of the part file. (See Figure 8.59.)

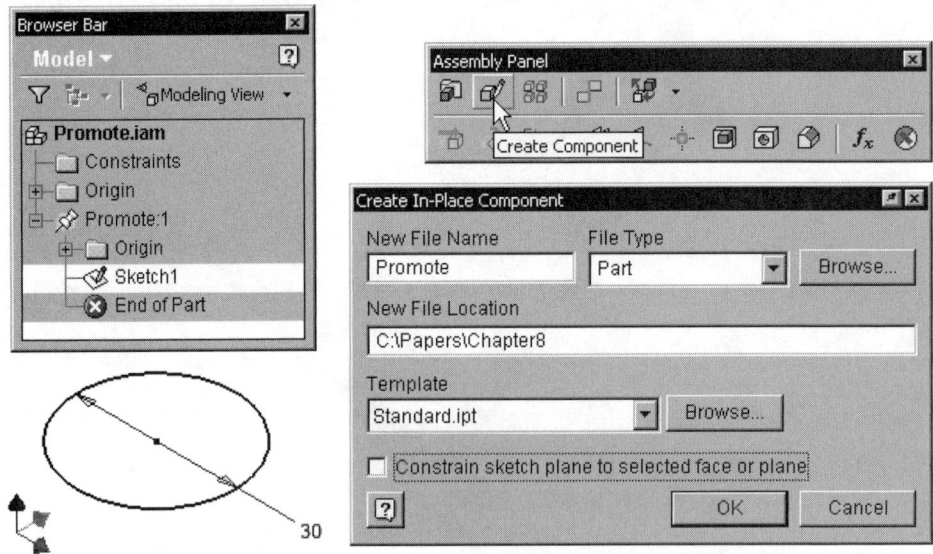

Figure 8.59 *Sketch of a solid part constructed in the context of an assembly*

5. Select Return on the Inventor Standard toolbar to exit sketch mode.
6. Select Return on the Inventor Standard toolbar to return to assembly modeling mode.
7. Select Place Component on the Assembly Panel Bar or toolbar.
8. Select the file *Replace.ipt* that you constructed in Chapter 4, select a location on the screen to position it (Figure 8.60), right-click, and select Done.

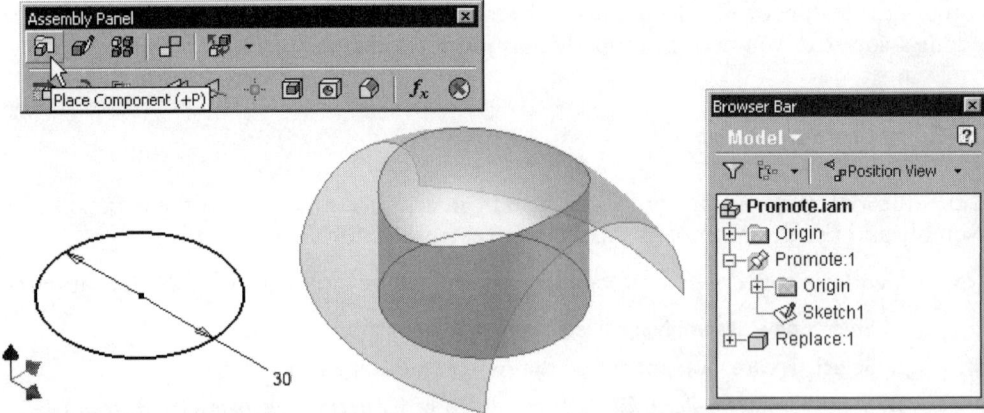

Figure 8.60 *Component placed in the assembly*

Assembly constraints can be applied not only to solid features but also to sketches. Now apply an assembly constraint to a sketch of a component.

9. Select Constraint on the Assembly Panel Bar or toolbar.

10. Select face A in Figure 8.61 and select the XY plane of the part file (*Promote.ipt*).

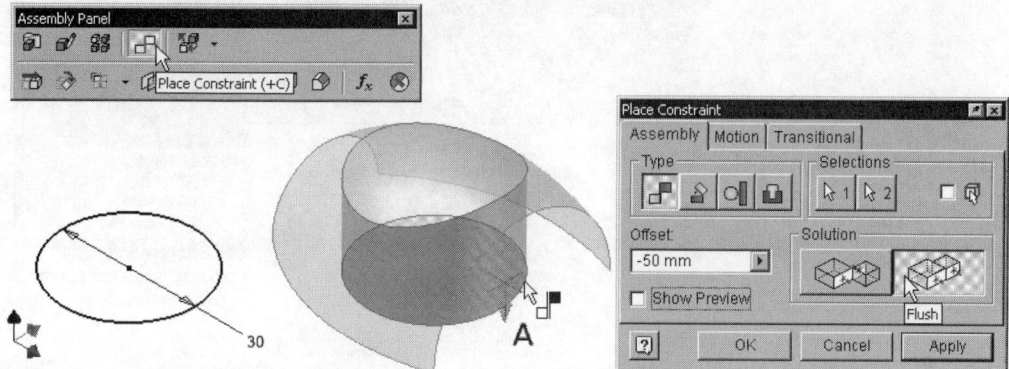

Figure 8.61 *Mate constraint applied to the axis of the sketch of a component and axis of another component*

11. In the Place Constraint dialog box, select Flush, set the offset to –50 mm, and click the Apply button.

12. Select circular edge A, select cylindrical face B and click the OK button. (See Figure 8.62.)

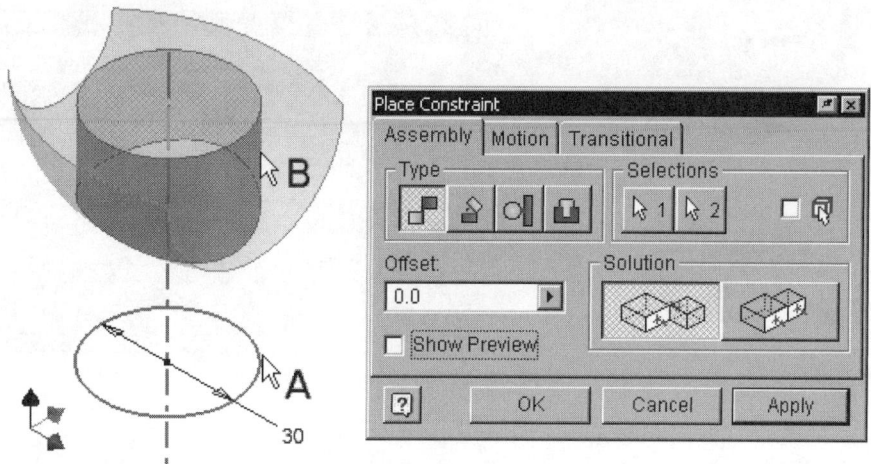

Figure 8.62 *Second assembly constraint applied*

The components are properly assembled. Now you will promote a surface.

13. Double-click the part file (*Promote.ipt*) on the Browser Bar to activate part modeling mode.

14. Select Promote on the Part Features panel or toolbar.

15. In the Promote dialog box, select the Pick from assembly and Promote as surface check boxes, if they are not already selected.

16. Select A and B in Figure 8.63 and click the Promote and Done buttons.

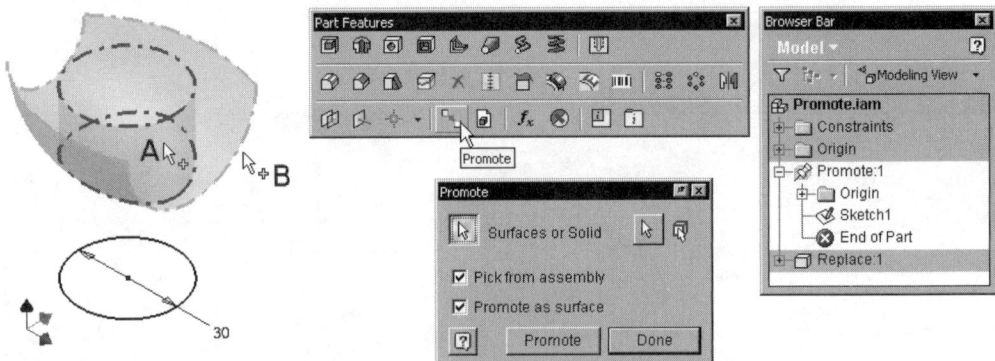

Figure 8.63 *Surface being promoted from one solid part to another part*

Now use the promoted surface as an extrusion termination surface.

17. Extrude the sketch to meet the promoted surface (A in Figure 8.64).

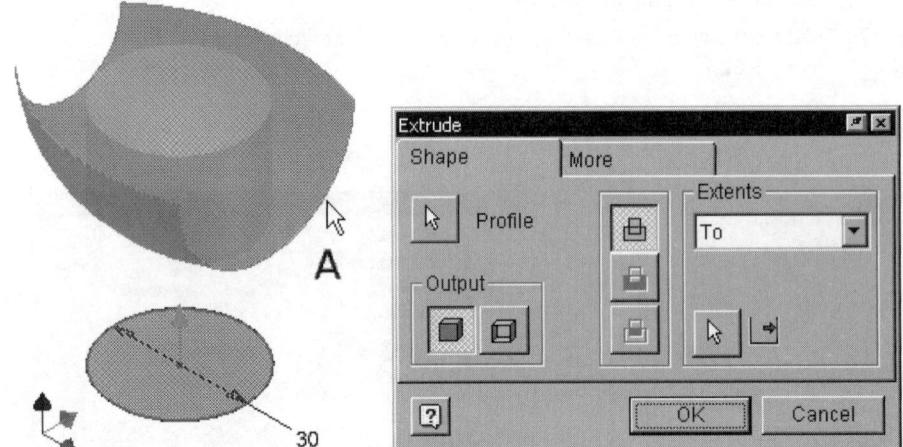

Figure 8.64 *Sketch be extruded to terminate at the promoted surface*

18. Hide the promoted surface. (See Figure 8.65.)

19. Select Return on the Inventor Standard toolbar to exit part modeling mode.

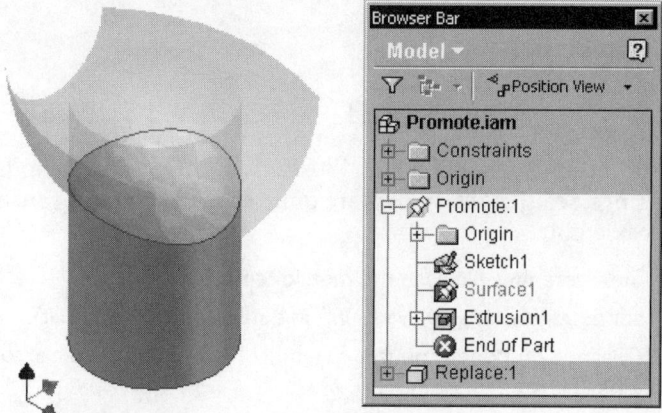

Figure 8.65 *Promoted surface hidden*

To complete the assembly, activate the other component, hide the surface there, and return to the assembly. (See Figure 8.66.)

Save and close your file. (file name: *Promote.iam*).

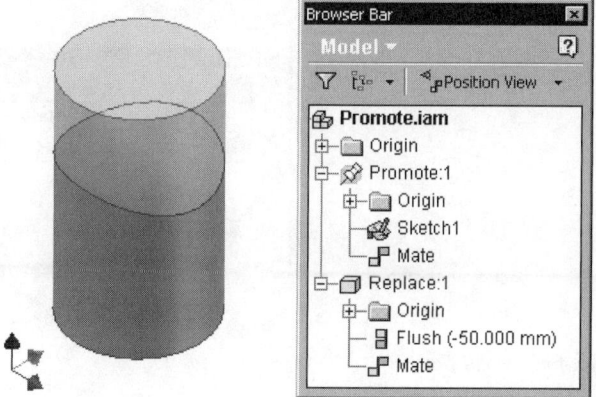

Figure 8.66 *Completed assembly*

2D DESIGN LAYOUT

When you start to design a mechanism, you have an idea of a set of linkages in your mind, but you do not know yet how the mechanism works. Before spending time designing the details of the component parts, you can quickly construct a set of 2D sketches to depict the linkages and assemble the 2D sketches together to form the mechanism.

To align the 2D component parts together, you assign assembly constraints to selected sketched elements. Because there are no solid features in the 2D sketches, you might need to construct work features for the purpose of placing assembly constraints. After validating and evaluating the simplified mechanism, you construct 3D solid models from the sketches.

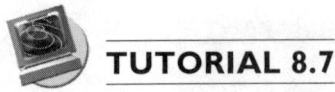

TUTORIAL 8.7

In this tutorial, you will use top-down approach to construct an assembly of three 2D layout parts. You will apply assembly constraints to the sketch entities to assemble the parts together. For the sake of simplicity in illustration, the standard engineering components such as bearings, seals, and fasteners are omitted. First you will construct a sketch in a solid part of an assembly.

1. Start a new assembly file. Use the metric template.

2. Save your assembly file (*2Dlayout.iam*) in a new folder (*2DLayout*).

3. Select Create Component on the Assembly Panel Bar or toolbar to construct a part file (file name: *2DBase.ipt*).

Note: Construct this new part in the same folder (2DLayout) and use the metric *Standard.ipt* template.

4. In the part file, construct a rectangle and a circle in accordance with Figure 8.67. For easy reference, the size of the rectangle is 100 mm by 50 mm.

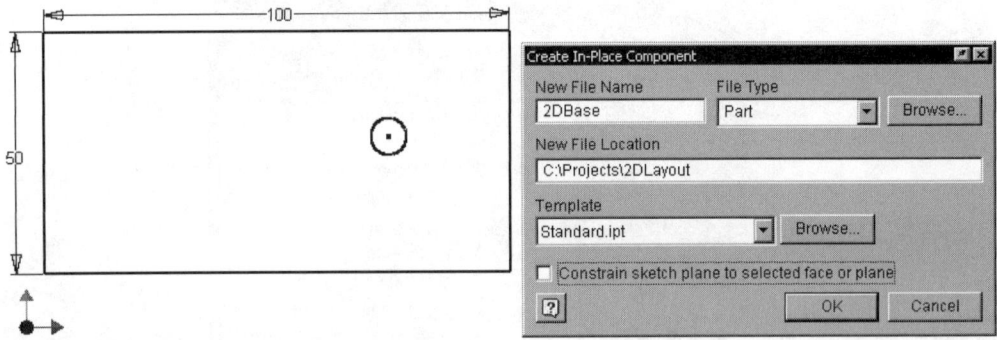

Figure 8.67 *2D layout of a solid part*

Now construct another 2D layout solid part. Again you will construct a simple sketch to depict the solid part.

5. Double-click the assembly on the Browser Bar to activate assembly mode.

6. Select Create Component on the Assembly toolbar Panel Bar or toolbar to construct another part file (file name: *2DCrank.ipt*).

7. In the part file, construct a line and circle in accordance with Figure 8.68.

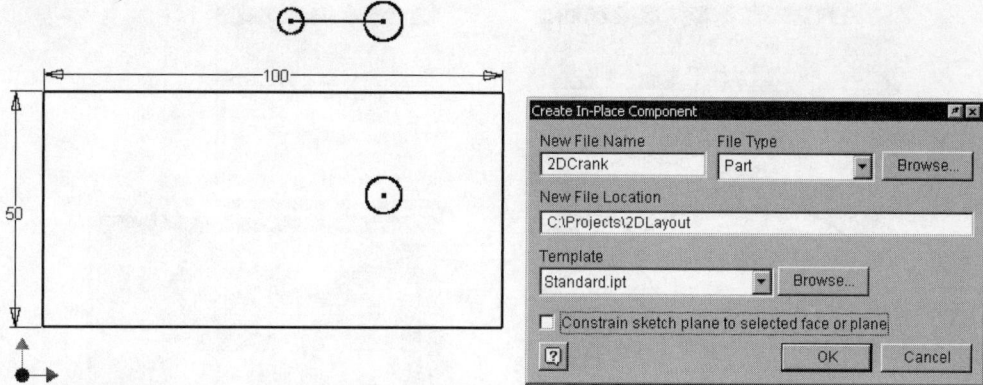

Figure 8.68 *Second 2D layout part completed*

Now construct the third 2D layout solid part file.

8. Activate assembly mode.

9. Create the third part file (file name: *2DLever.ipt*).

10. Construct two circles and a line. (See Figure 8.69.)

Three 2D layout designs are complete.

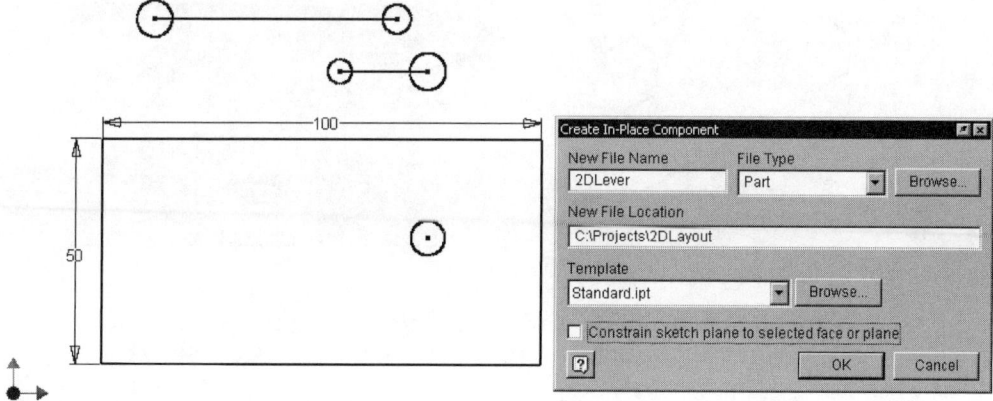

Figure 8.69 *Third 2D layout part completed*

Now assemble the parts together.

11. Activate assembly modeling mode and set the display to an isometric view.

12. Select Constraint on the Assembly Panel Bar or toolbar.

13. In the Place Constraint dialog box, select the Mate button.

14. Select circles A and B in Figure 8.70 and click the Apply button. The 2D crank is mated to the 2D base.

 Note: While you apply the assembly constraint, the axes of the circles are highlighted, denoting that they are selected.

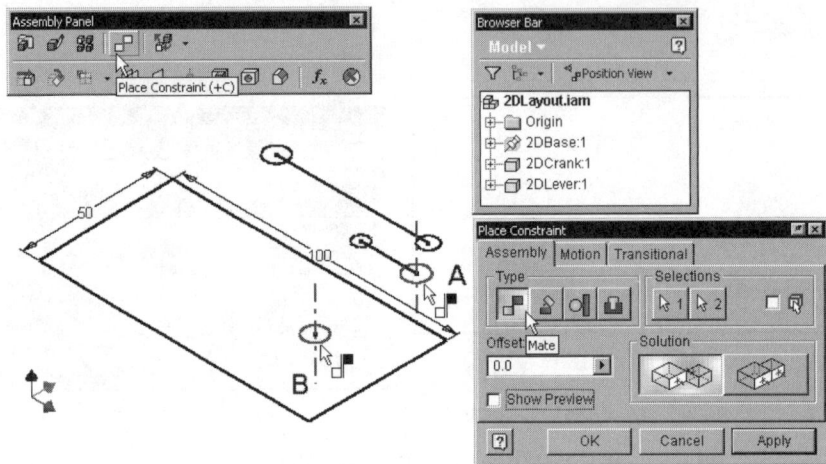

Figure 8.70 *2D crank being mated to the 2D base*

15. Select circles A and B in Figure 8.71 and click the Apply button to mate the 2D lever to the 2D crank.

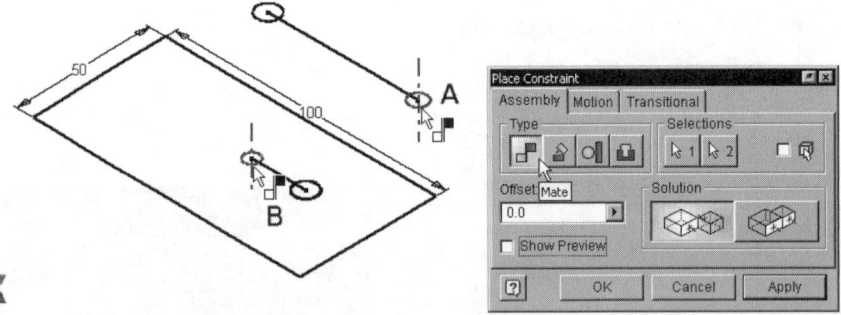

Figure 8.71 *2D lever being mated to the 2D crank*

16. Select the Tangent button in the Place Constraint dialog box. Then select the circle and the line at A and B in Figure 8.72. After that, click the Apply button to apply a tangent constraint.

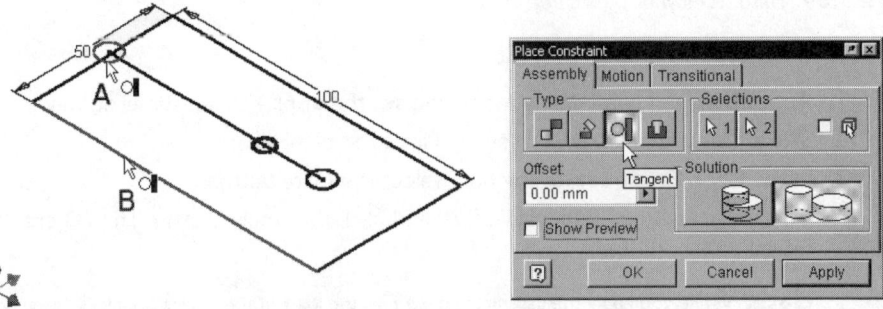

Figure 8.72 *Tangent constraint being applied*

To facilitate constraint driving to examine the motion of the mechanism, now apply an angle constraint to the assembly.

17. Select the Angle button in the Place Constraint dialog box. Then select lines A and B in Figure 8.73 and click the OK button. The assembly is complete.

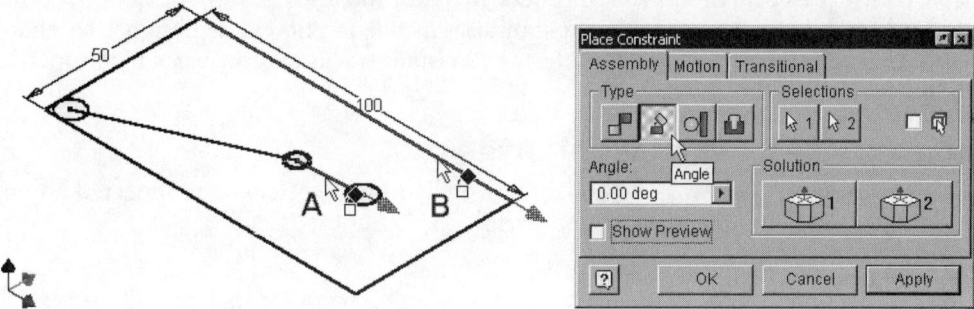

Figure 8.73 *Angle constraint being applied*

Now examine how the mechanism works by driving the angle constraint.

18. Select the Angle constraint on the Browser Bar, right-click, and select Drive Constraint.

19. In the Drive Constraint dialog box, set the end position to 360 degrees and select the Forward button. (See Figure 8.74.)

To further refine the mechanism, you can activate the part files one by one and modify the sketches. If you are satisfied with the mechanism, you can proceed to construct solid parts from the sketches.

Now save and close your files.

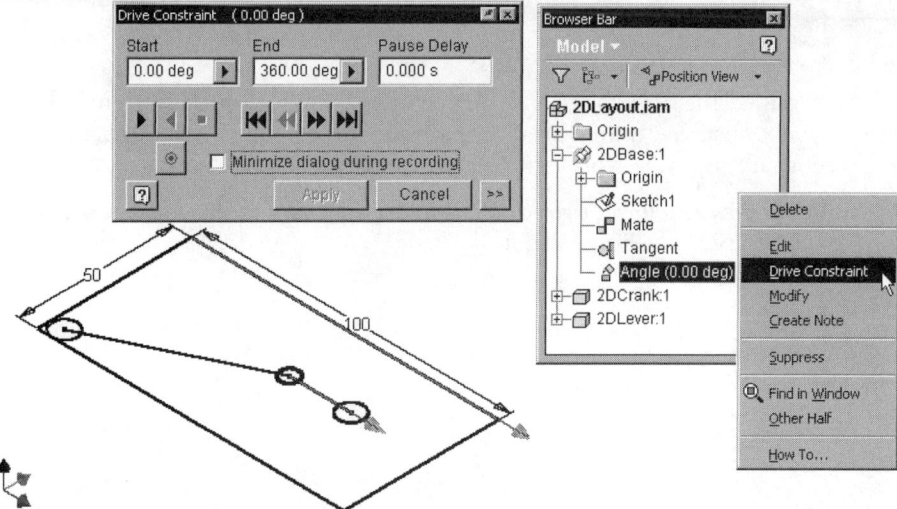

Figure 8.74 *Test driving the 2D mechanism*

DESIGN COLLABORATION

To enable a team of designers to work collaboratively on a project, you deploy a set of design collaboration tools. You use a project file system to manage local and network folders. You use Windows NetMeeting within the Autodesk Inventor environment to enable control of a peer designer's Autodesk Inventor file. In addition, the designer notebook and the Design Assistant also help in design collaboration. The designer notebook records design information, and the Design Assistant tracks and manages file properties and links among files.

PROJECT PATHS AND SHARE RIGHTS

A prior requirement for working collaboratively is to have a network connected among the designers, set up project paths, and permit shares.

Project Paths

As we already explained, you construct four sets of working directories that you will put in the project path file:

Workspace	A directory in each computer for use as each individual's workspace, the default location where each designer can construct new files on the local computer.
Local Search Paths	Search paths for locations assigned to each individual on their local computers, accessed only by the individual designer.
Workgroup	Search paths for all the designers in the workgroup in designated computers, which can be accessed by all the designers working on the same project.
Library Search Paths	Search paths for the storage and retrieval of standard library parts in designated computers.

Share Rights

Having determined the directories and file locations, you assign share rights. To assign share rights, you must have administrator or administrator-equivalent rights for the computer. To set share rights, perform the following steps:

1. In Windows Explorer, right-click the folder you want to share and select Sharing and Security. (See Figure 8.75.)

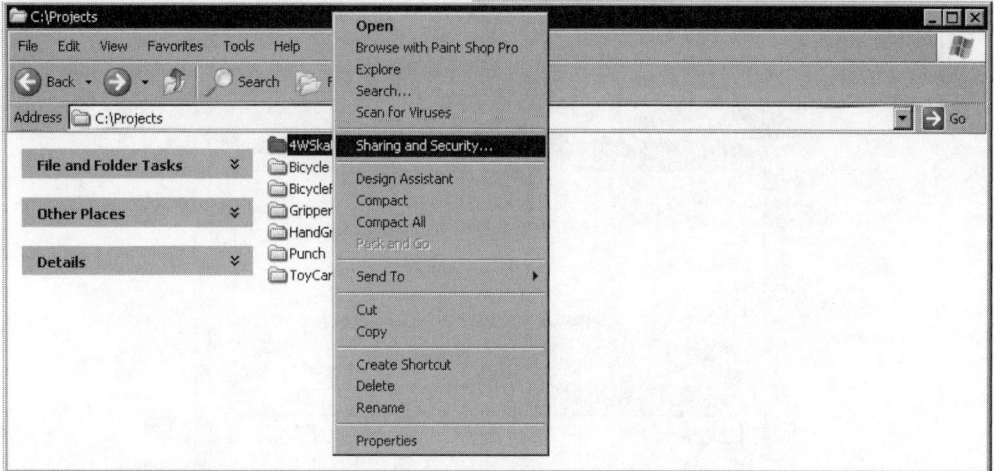

Figure 8.75 *Share rights being assigned*

2. In the Properties dialog box of the selected directory, select Share this folder, specify the share name, decide the number of users allowed, and click the Permissions button. (See Figures 8.76 and 8.77.)

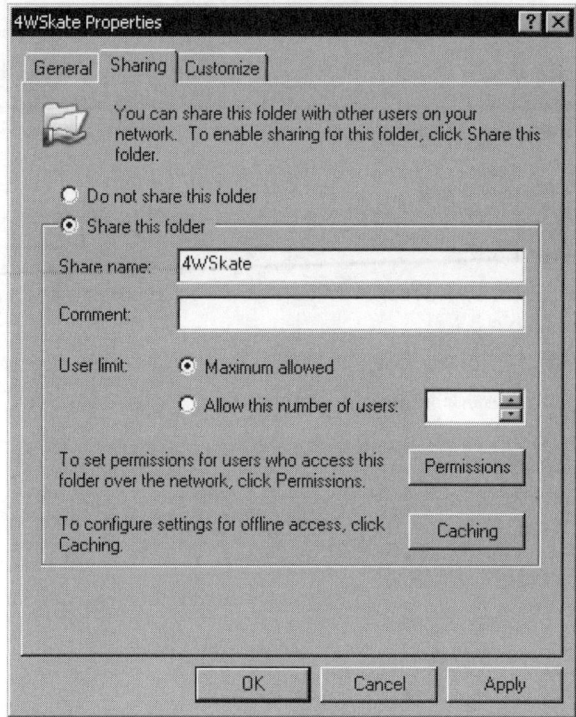

Figure 8.76 *Properties dialog box for the selected folder*

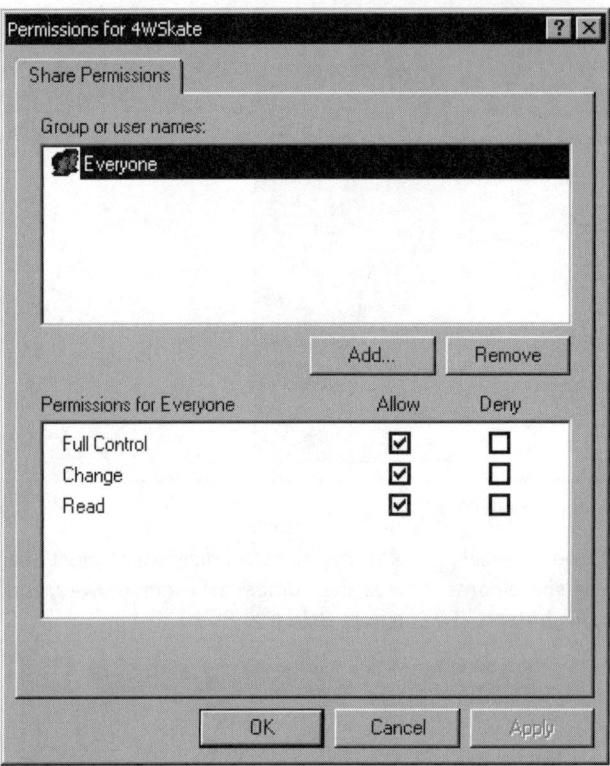

Figure 8.77 *Permissions dialog box for the selected folder*

3. By default, the share right is assigned to everyone with full control. Select Everyone in the Permissions dialog box and click the Remove button.

4. Click the Add button.

5. In the Add Users and Groups dialog box, select the users and assign appropriate rights. (See Table 8.1.)

6. Click the OK button to exit.

7. Click the OK buttons in each dialog box to complete the share right assignment.

Table 8.1 *Directory permissions*

Directories	Share Permissions and Suggested Type of Access
Workspace in each designer's local computer	No share
Local Search Paths	Designated individual designer (Full control)
Workgroup Search Paths	All designers (Full control)
Library Search Paths	All designers who use the library (Read);Designated designers who help to build the library (Full control)

WINDOWS NETMEETING

To communicate and share control rights in the process of designing, you use NetMeeting. You need to install NetMeeting on each designer's computer in order for them to participate. Figure 8.78 shows the NetMeeting dialog box.

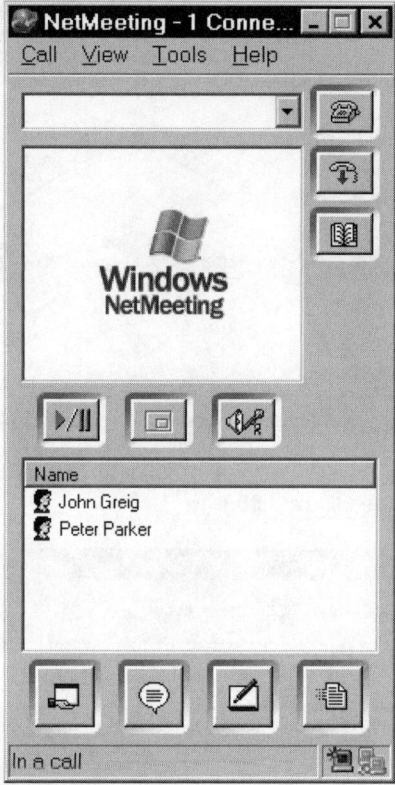

Figure 8.78 *NetMeeting dialog box*

Now suppose you, as John Greig, call Peter Parker to work collaboratively on an Autodesk Inventor file.

1. Open an Autodesk Inventor file.
2. Select View > Toolbar > Collaboration to display the Inventor Collaboration toolbar.
3. Select Add Participant on the Inventor Collaboration toolbar. (See Figure 8.79.) You can also select Tools > Online Collaboration > Meet Now.

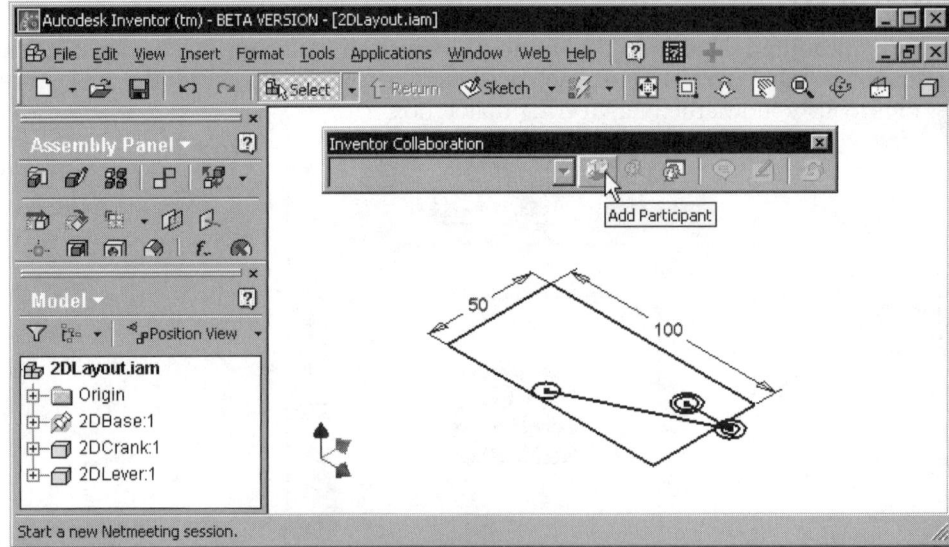

Figure 8.79 *A file opened and the Inventor Collaboration toolbar displayed*

> 4. In the Place A Call dialog box, type the IP address of Peter's computer and click the Call button. (See Figure 8.80.)

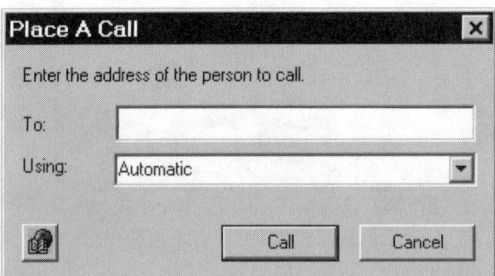

Figure 8.80 *Place A Call dialog box*

After you make a call, the NetMeeting dialog box displays a waiting message. (See Figure 8.81.) In the other party's computer, an incoming message is displayed. (See Figure 8.82.)

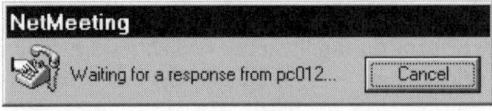

Figure 8.81 *Waiting message*

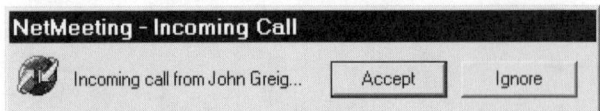

Figure 8.82 *Incoming message*

After the participant clicks the Accept button, the other participant's computer shows your screen display. If Peter wants to take control, he selects Request Control from the Control menu. (See Figure 8.83.) In your computer, a request message displays. If you agree, you click the Accept button. (See Figure 8.84.) Now Peter takes over the control and he continues the Inventor working session. In your computer, Peter's work is displayed. After Peter finishes the work, he selects Release Control from the Control menu. (See Figure 8.85.) You regain control over the working session.

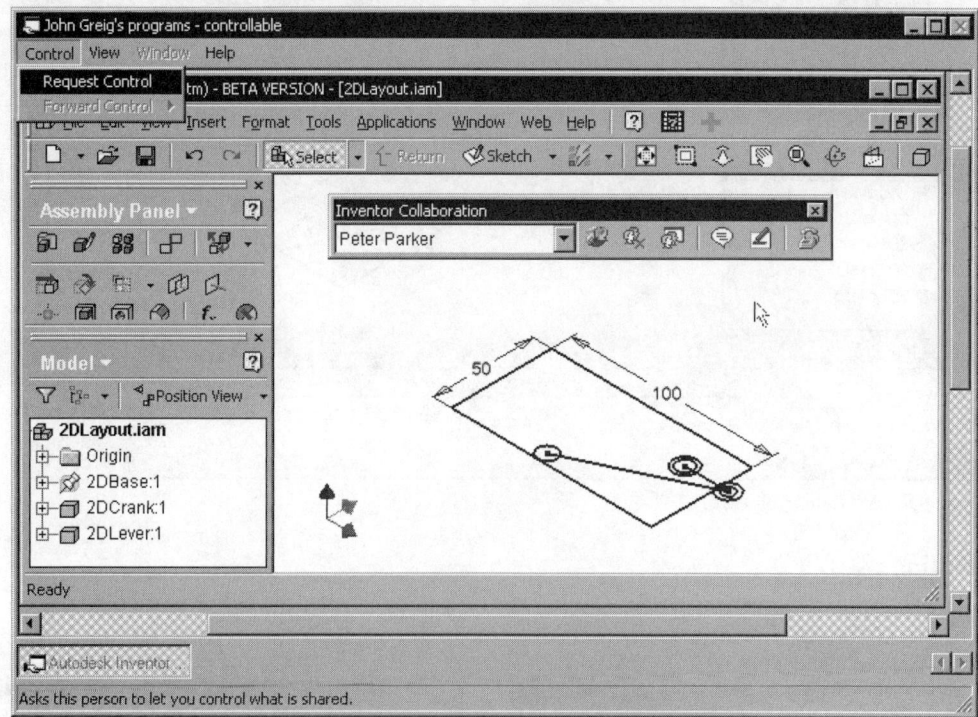

Figure 8.83 *John's Autodesk Inventor screen displayed in Peter's computer*

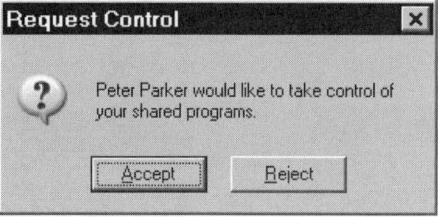

Figure 8.84 *Request Control dialog box*

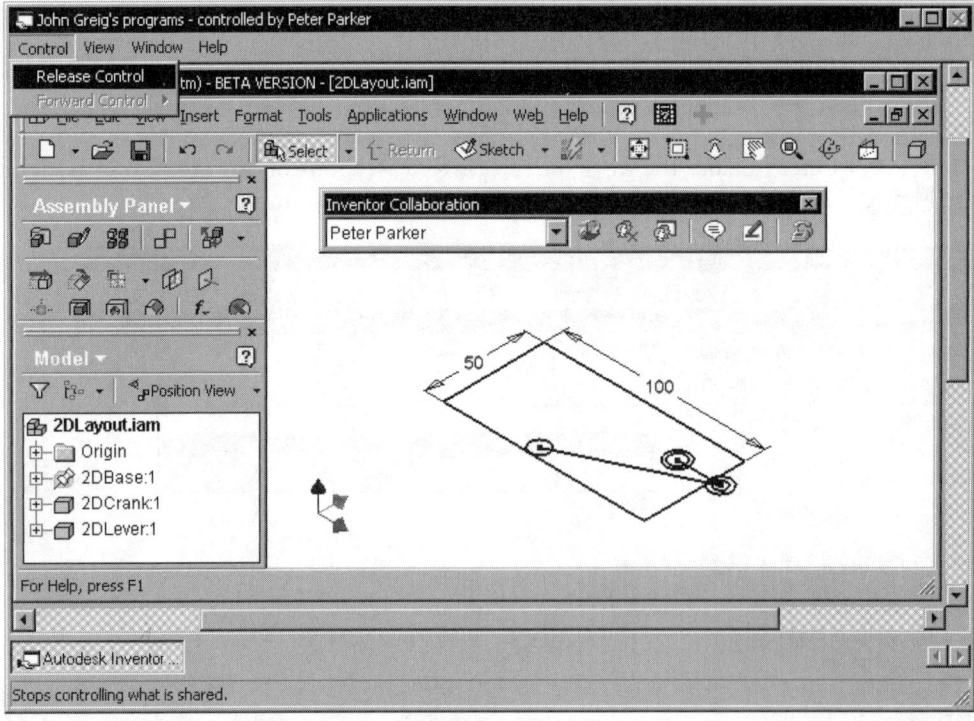

Figure 8.85 *Releasing control*

Chatting

Besides sharing an Inventor working session, you can chat with the other participant by selecting Chat on the Inventor Collaboration toolbar. Figure 8.86 shows the Chat dialog box.

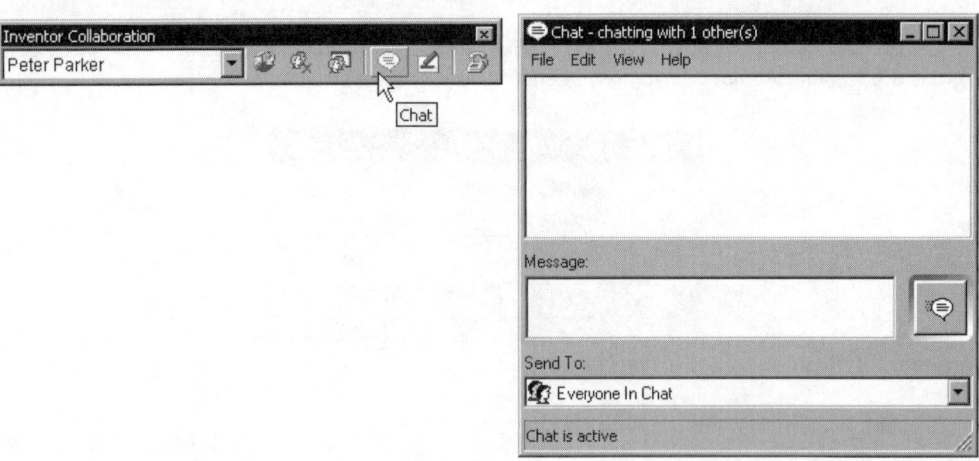

Figure 8.86 *Chatting*

Using A Whiteboard

To illustrate ideas, you use the Whiteboard. (See Figure 8.87.)

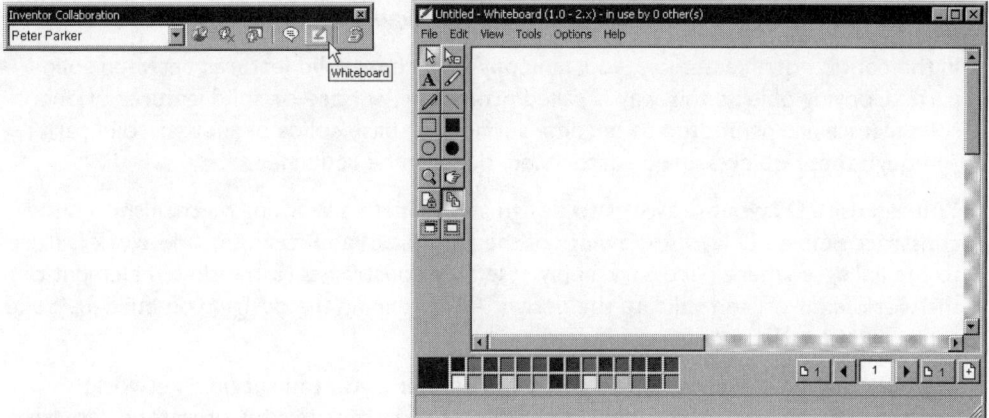

Figure 8.87 *Using Whiteboard*

Hanging Up

To end the NetMeeting session, select Hang Up on the Inventor Collaboration toolbar. (See Figure 8.88.)

Figure 8.88 *Hanging up*

SUMMARY

In the context of an assembly, it is very common that the size and/or shape of a feature of a component must match the size and/or shape of the feature of another component. To maintain this kind of relationship among components in an assembly, you use adaptive technology. After you apply an assembly constraint to an adaptive feature of a solid part for the solid part to be assembled with another component, the adaptive solid feature will change in size and/or shape to adapt to the mating solid feature. To construct an adaptive solid feature in the context of an assembly, you can construct adaptive work features and use the adaptive work features to construct solid features, or you can construct a sketch on the face of another component in the assembly while making a sketched solid feature. Besides setting parts in an assembly to be adaptive, you can also set the sub-assemblies to be adaptive.

You can derive a new part from an assembly as well as from an existing solid part. Deriving a solid part from an assembly enables you to construct a solid part of a very complex shape by first constructing a set of solid parts, putting them in an assembly, and then deriving a complex solid part from the assembly. While deriving, you can apply Boolean operations (union or subtract) on individual components from the assembly, but you cannot mirror or scale. The resulting derived object is a single solid part.

In the context of an assembly, you can copy surface or solid features between solid parts. Copying objects this way is called promoting. Surface or solid features of one solid feature are promoted to become surfaces or base solids of another solid part. Although these objects are un-associated, they can be redefined.

You can use 2D layout drawings to design and validate a working mechanism. You construct simple 2D layout drawings of the linkages of a mechanism, add work features to the linkages where necessary, apply assembly constraints to the sketch element or the work features, and validate the design. After refining the design, you build 3D solid parts from the 2D layouts.

To work collaboratively among a group of designers, you can set up a network, establish appropriate search paths, share files, set permissions for users, and reserve files. You need to install NetMeeting properly in all the participants' computers and to activate NetMeeting from within the Inventor environment.

REVIEW QUESTIONS

1. What is the major limitation to using a linear approach to construct a solid part?

2. Briefly explain the meaning of adaptive technology. State the advantages of using adaptive technology in design.

3. Briefly explain the two ways to construct an adaptive solid part in the context of an assembly.

4. Explain how to construct a complex solid part from a set of solid parts of simpler shapes.

5. Briefly describe the way to promote a surface or solid feature from one solid part to another solid part.

6. What are the key features that you need to construct on a 2D layout drawing for them to be assembled? Use an example to illustrate your answer.

7. How can you set up your computers to work collaboratively among a team of designers? Outline the steps for enabling a file to be shared in a NetMeeting.

CHAPTER 9

Engineering Presentation

OBJECTIVES

The aims of this chapter are to explain the concept of presentation and to delineate the procedures for constructing presentation views of an assembly. This chapter also explains how to animate a presentation view. After studying this chapter, you should be able to

- Explain the concept of assembly presentation
- Produce presentation views of assemblies
- Construct a presentation animation

OVERVIEW

An engineering presentation illustrates how components of an assembly are assembled by exploding or tweaking them to move them apart. To construct a presentation, you use a presentation file and link the file to an assembly file. In the presentation file, you explode or tweak the components. Optionally, you add trail lines between the exploded or tweaked components to indicate the relationships between the components. To better explain how the components are exploded or tweaked, you construct an animation of the explosion or tweaking motion.

PRESENTATION CONCEPTS

As explained earlier, an assembly file links to a set of component files. (The component file can be a part file if the component is a solid part or an assembly file if the component is a sub-assembly of the final assembly.) One of the prime functions of an assembly file is to indicate how the components in the assembly are put together. However, in many circumstances, some components placed far deep in the assembly might be totally or partly hidden by some other components in the assembly. As a result, you cannot clearly illustrate how these components relate to other components in the assembly. To explain how you put the components together, you use a presentation file to construct an exploded presentation of the assembly and generate an animation of the explosion.

In the presentation file, you select an assembly file that the presentation file will be linked to, select or specify a viewing direction, and explode or tweak the components to move them apart. To further illustrate the relationships between the components of an assembly, you construct an animation of the explosion or tweaking motion of the components and save it in AVI file format. The way you explode or tweak components in a presentation file is saved in the presentation file as a presentation view. In a presentation file, you can construct as many presentation views as you need. Figure 9.1 shows an assembly and a presentation view of the assembly.

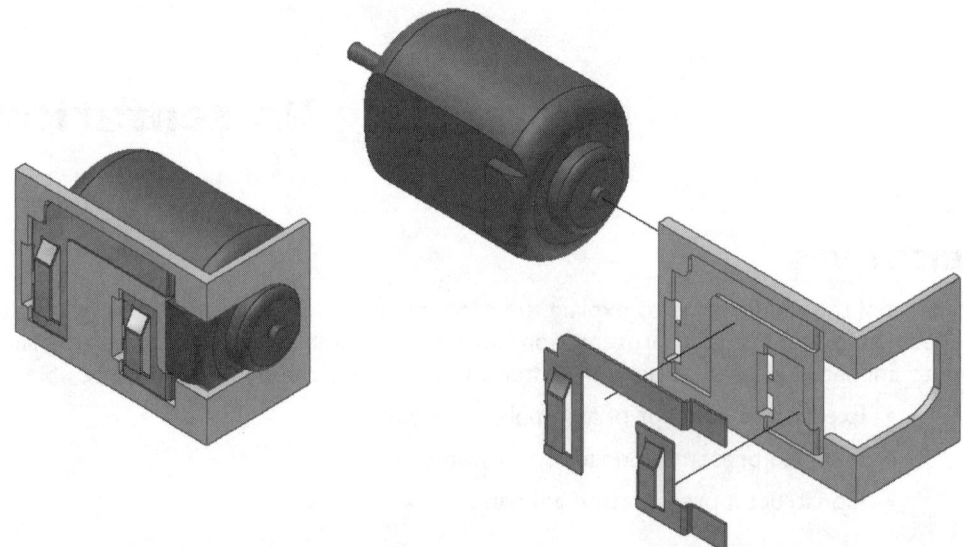

Figure 9.1 *An assembly (left) and an exploded presentation view of the assembly (right)*

PRESENTATION FILE

It must be stressed that the components exploded or tweaked in the presentation file will remain intact in the assembly file. In other words, the components properly assembled in the assembly file remain in position even if they are exploded or tweaked in the presentation file. Furthermore, changes in the components or the positions of the components in the assembly will be always reflected in the presentation file, because a presentation file is linked to an assembly file. Table 9.1 summarizes the information content in the part file, assembly file, and presentation file.

Table 9.1 *Information content*

File	Information
Part File	Definition and information about individual 3D solid part
Assembly File	Information about the location of the linked components and how the linked components are assembled together
Presentation file	Information about the location of the linked assembly file and how the components in the assembly file are exploded or tweaked

USER INTERFACE

Now you will start a presentation file to familiarize yourself with Autodesk Inventor's presentation user interface. (See Figure 9.2.)

1. Select New from the File menu or the New button on the Inventor Standard toolbar.

2. In the Open dialog box, select the Metric tab.

3. Select the *Standard.ipn* template and click the OK button.

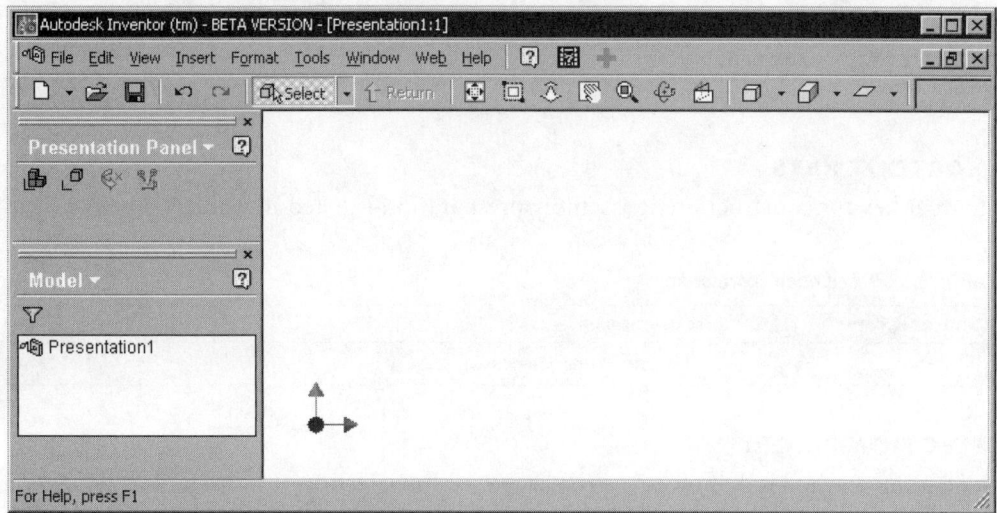

Figure 9.2 *New presentation file started*

PRESENTATION PANEL BAR AND TOOLBAR

The Presentation Panel Bar and toolbar display the icons for making a presentation. To display the accompanying Presentation Panel toolbar, select View > Toolbar > Presentation Panel. Figure 9.3 shows the Presentation Panel Bar and toolbar.

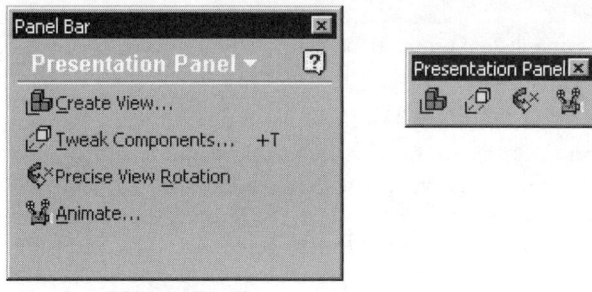

Figure 9.3 *Presentation Panel Bar and toolbar*

The Presentation Panel Bar and toolbar have four button areas. Table 9.2 describes the choices:

Table 9.2 *Presentation Panel Bar and toolbar options*

Button	Function
Create View	Constructs presentation views.
Tweak Components	Tweaks the components of an assembly apart.
Precise View Rotation	Precisely rotates the presentation view.
Animate	Animates the presentation.

SHORTCUT KEYS

Shortcut key for construction of assembly presentation is listed in Table 9.3 below.

Table 9.3 *Presentation shortcut key*

Shortcut Key	Function
T	Tweaks components apart.

SELECTION PRIORITY

Depending on the kind of objects are selected in the Selection Priority pull-down list on the Inventor Standard toolbar shown in Figure 9.4, placing the cursor on a presentation will select a component (sub-assembly) or a part of the assembly.

Figure 9.4 *Selection priority in a presentation file*

PRESENTATION BROWSER BAR

Objects displayed on the presentation Browser Bar include sets of explosion information depicting how you explode or tweak the components of an assembly. You can display these objects in three manners: Tweak View, Sequence View, and Assembly View. You select the Filter button on the Browser Bar to select a view. (See Figure 9.5.)

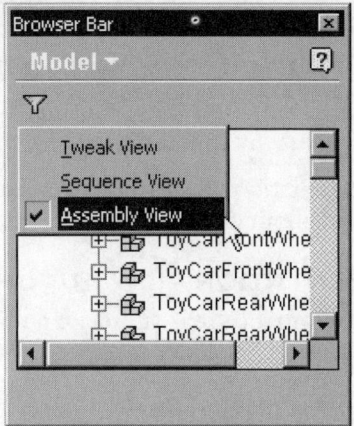

Figure 9.5 *Filter button on the presentation Browser Bar*

Tweak View displays the tweaks in the top hierarchy of the Browser Bar and the included components below each tweak. Sequence View displays the tasks in the top hierarchy of the Browser Bar and the tweak sequence below each task. Assembly View displays the components of the assembly in the top hierarchy of the Browser Bar and the amount of tweak below each component. (See Figure 9.6.)

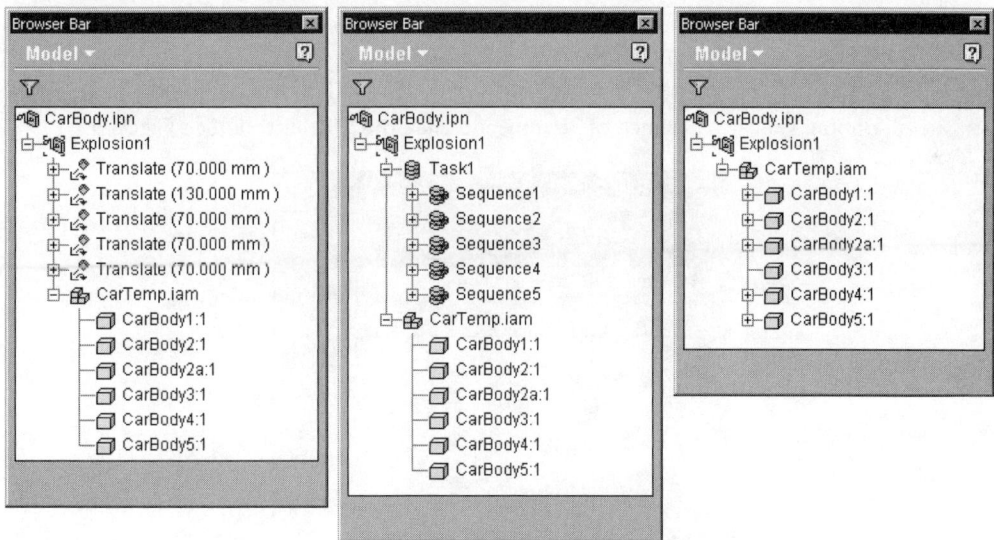

Figure 9.6 *Browser Bar showing Tweak View, Sequence View, and Assembly View (from left to right)*

CONSTRUCTING PRESENTATION VIEWS

You can construct a presentation view in two ways: automatically and manually.

In an automatically generated presentation view, you specify a distance, and all the components are exploded (moved apart) by that distance.

In a manually constructed presentation view, you tweak each component part individually along a linear direction and about an axis.

CONSTRUCTING A PRESENTATION VIEW AUTOMATICALLY

Making a presentation view automatically is simple. You select an assembly file on the Browser Bar, right-click, select Auto Explode, and specify the explosion distance for each component of the assembly.

TUTORIAL 9.1

In this tutorial, you will construct a presentation view for an assembly by specifying the explosion value.

1. Start a new presentation file, if you already close the presentation file that you started.

2. Select Create View on the Presentation Panel Bar or toolbar.

3. In the Select Assembly dialog box, select the assembly (*Pliers.iam*) that you constructed in Chapter 7, select the default design view, select the Automatic option, specify a distance of 30 mm, and click the OK button. (See Figure 9.7.)

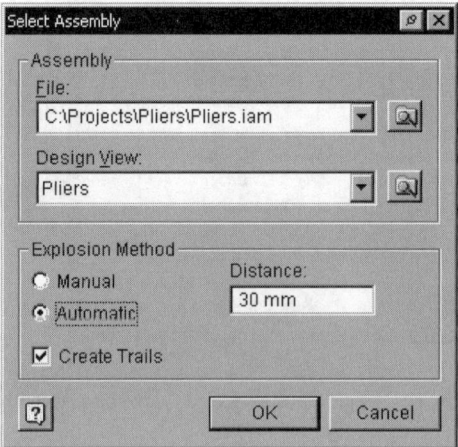

Figure 9.7 *Select Assembly dialog box*

An exploded view of the pliers assembly is constructed. In the exploded view, the components are exploded a distance of 30 mm from their constrained position. (See Figure 9.8.) Save your file (file name: *Pliers.ipn*).

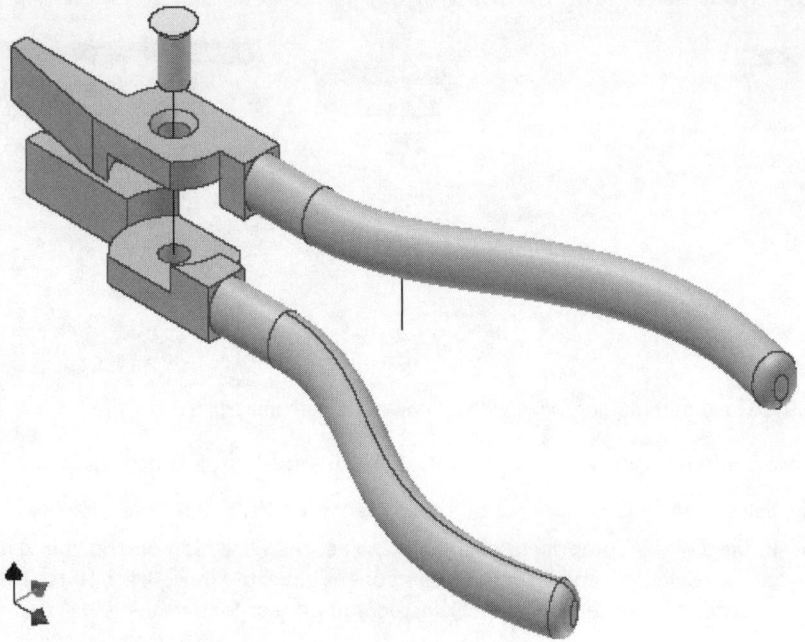

Figure 9.8 *Components exploded apart automatically*

CONSTRUCTING A PRESENTATION VIEW MANUALLY

To better control how each individual component in an assembly is exploded, you tweak each of them manually. You can tweak a component by moving it in a linear direction or rotating it about an axis.

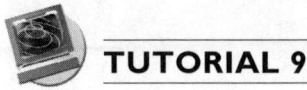 **TUTORIAL 9.2**

In this tutorial, you will construct a presentation view with the components tweaked apart manually.

1. Open the presentation file *Pliers.ipn*, if you already closed it.
2. Select Create View on the Presentation Panel Bar or toolbar.
3. In the Select Assembly dialog box, select the Manual option and click the OK button.

A presentation view without any explosion is constructed. (See Figure 9.9.) On the Browser Bar, you now see two presentation views. The presentation view with an explosion distance specified becomes inactive. If you wish to activate a presentation view, select it on the Browser Bar and double-click.

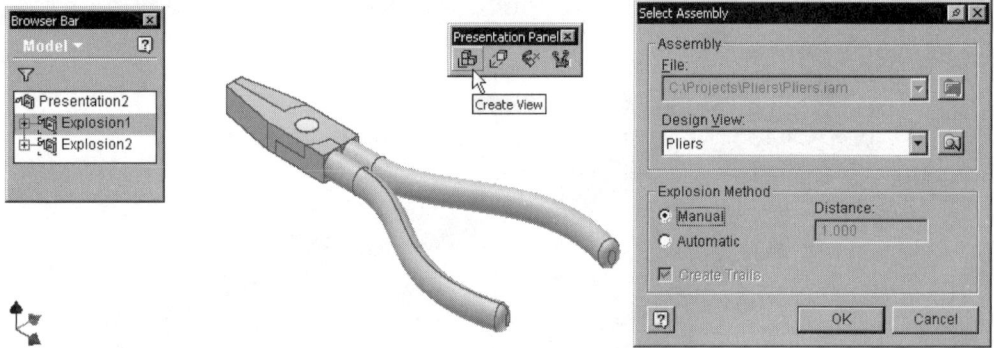

Figure 9.9 *Second presentation view (without any explosion) constructed*

Now manually tweak a component of the assembly in a linear direction.

4. Select Tweak Components on the Presentation Panel Bar or toolbar.

5. In the Tweak Component dialog box, select the Direction button, if it is not already selected, and select the component indicated in Figure 9.10 to use the coordinates of the face of the selected component for tweaking.

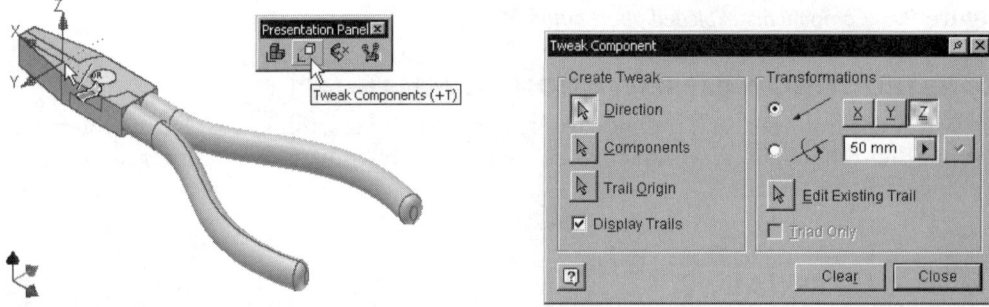

Figure 9.10 *Component selected for direction specification*

6. Select the Components button and select the component indicated in Figure 9.11.

7. In the Transformations area of the Tweak Component dialog box, select the linear translation option, select the Z button, specify a distance of 50 mm, and click the Apply and Clear buttons.

Tip: If you do not select the Clear button to clear the selection, the selected component will tweak together with the next selected component.

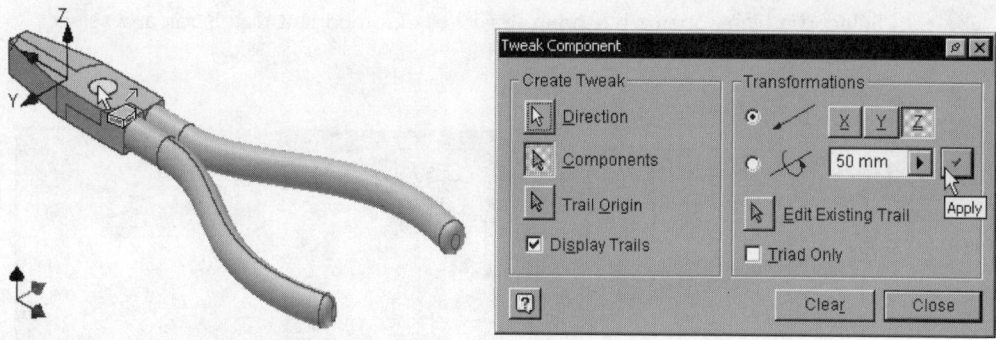

Figure 9.11 *Component being tweaked*

The selected component is tweaked. (See Figure 9.12.)

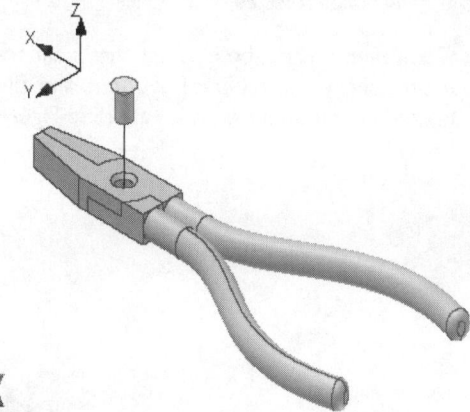

Figure 9.12 *Component tweaked*

Now tweak another component manually.

8. Select the Direction button in the Tweak Component dialog box and select the component indicated in Figure 9.13.

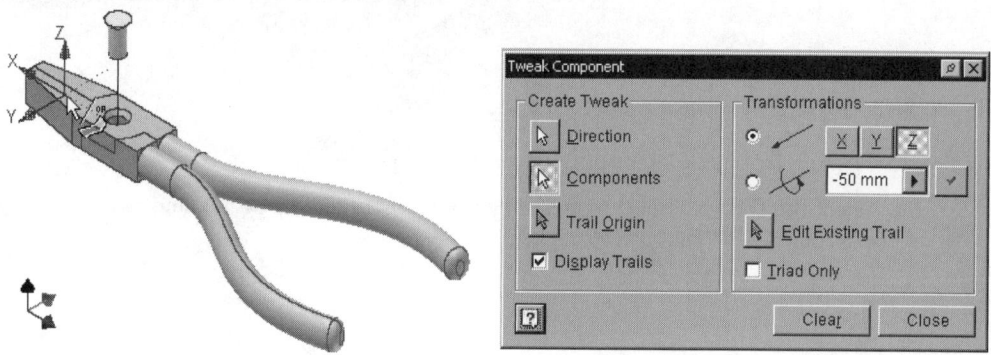

Figure 9.13 *Direction specified*

9. Select the Components button in the Tweak Component dialog box and select the component indicated in Figure 9.14.

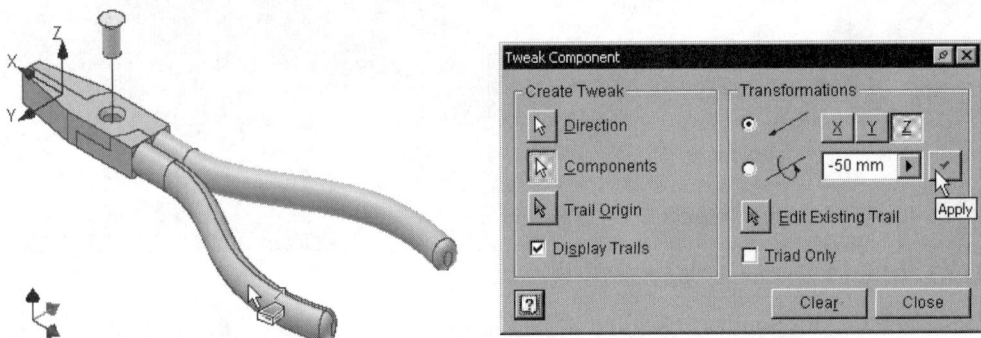

Figure 9.14 *Component to be tweaked selected*

10. In the Tweak Component dialog box, select the linear translation option and the Z direction button, specify a distance of −50 mm, and click the Apply and Clear buttons. The selected component is tweaked. (See Figure 9.15.)

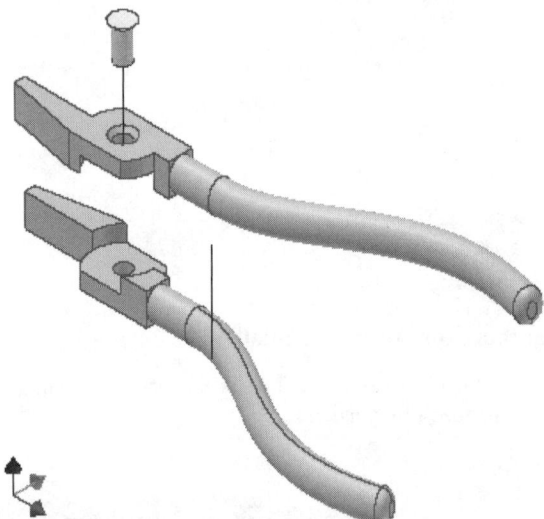

Figure 9.15 *Selected component tweaked in a linear direction*

Now manually tweak a component about an axis.

11. Select the Direction button in the Tweak Component dialog box and select the axis indicated in Figure 9.16.

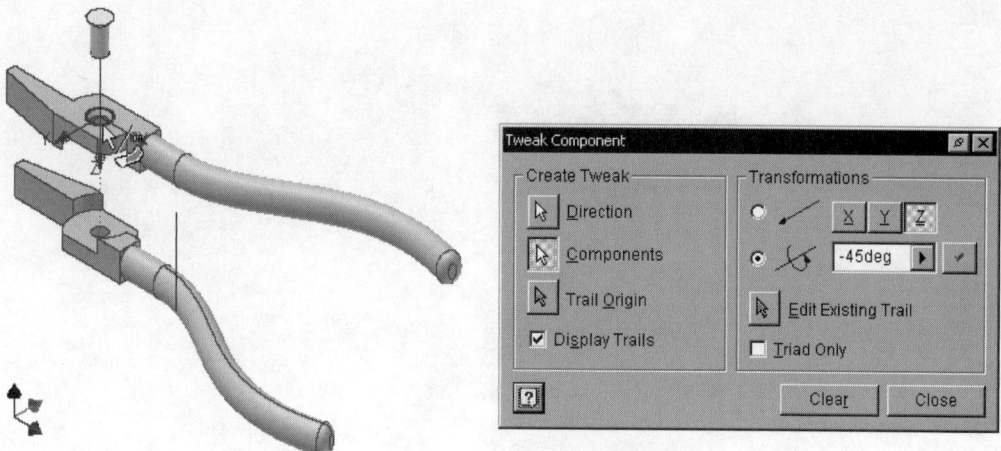

Figure 9.16 *Axis of a component selected to establish a direction*

12. Select the Components button in the Tweak Component dialog box and select the component indicated in Figure 9.17.

13. In the Tweak Component dialog box, select the rotation option, select the Z axis, specify a rotation angle of –45deg, and click the Apply and Close buttons.

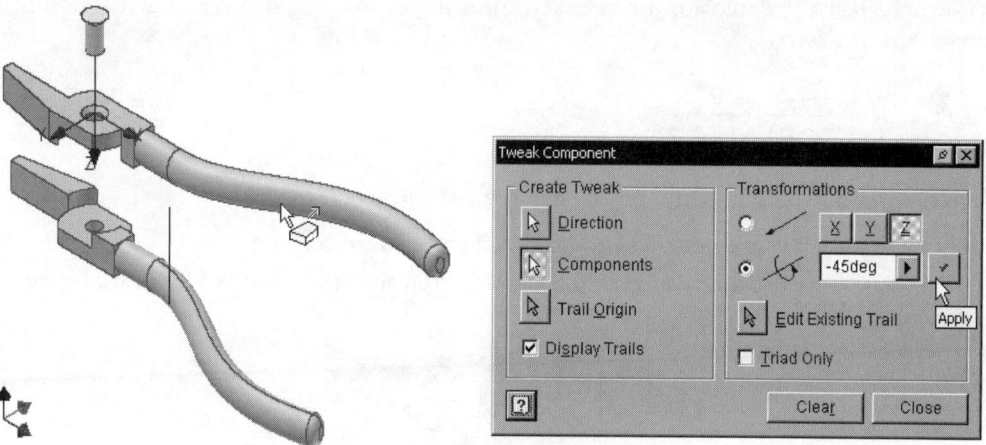

Figure 9.17 *Component to be tweaked selected*

The component is tweaked. (See Figure 9.18.) The second presentation view is complete. Save your file.

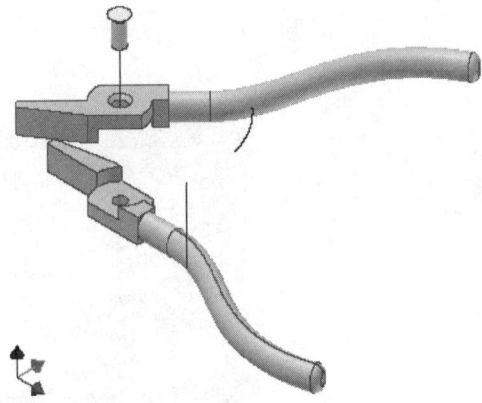

Figure 9.18 *Component tweaked about an axis*

EDITING PRESENTATION VIEWS

You can edit the presentation views by using the Browser Bar, the Tweak Component dialog box, or the trails.

USING THE BROWSER BAR

You can select the explosion or tweaked information on the Browser Bar to modify a presentation view.

TUTORIAL 9.3

In this tutorial, you will edit a presentation view.

1. Open the presentation file *Pliers.ipn*, if you already closed it.
2. On the Browser Bar, click the Filters button and select Tweak View. (See Figure 9.19.)

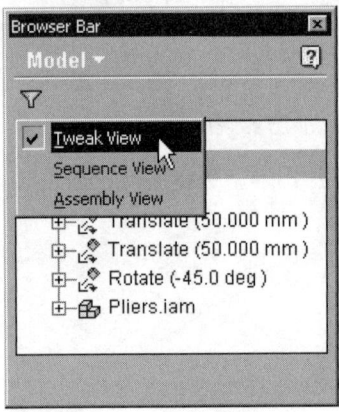

Figure 9.19 *Tweak View selected*

3. Select Explosion1 on the Browser Bar and double-click.
4. Expand the Browser Bar, select Auto Explode, and change the explosion distance to 50 mm. The explosion distance is changed. (See Figure 9.20.)

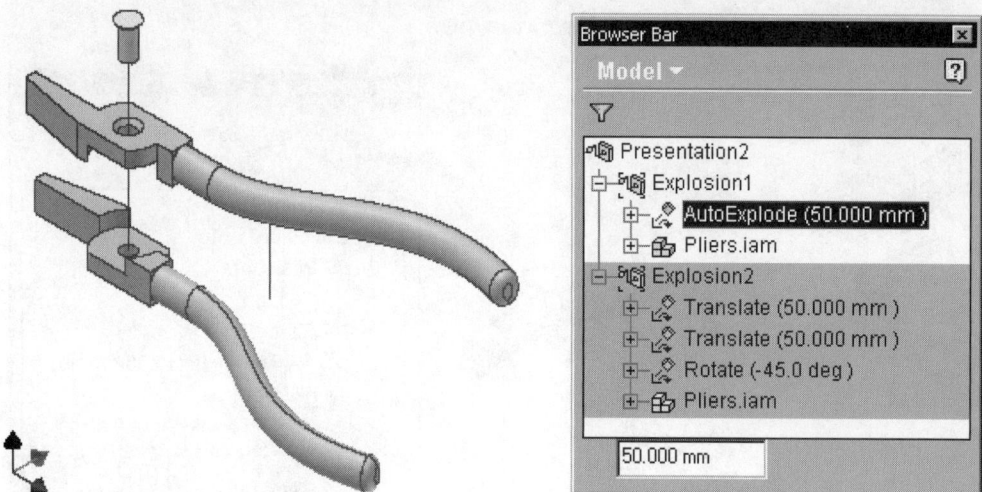

Figure 9.20 *Auto explosion distance changed*

Now display a sequence view and edit the presentation.

5. On the Browser Bar, click the Filters button, select Sequence View, and edit the explosion distances to 40 mm and 90 mm, in accordance with Figure 9.21.

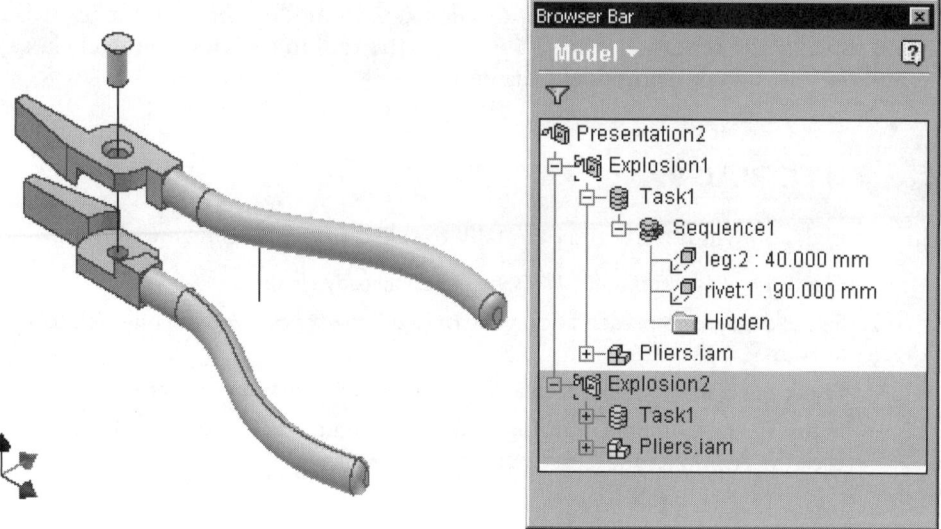

Figure 9.21 *Tweak amount modified individually*

Now display the assembly view and edit the presentation.

6. Click the Browser Filters button, select Assembly View, and edit the distance of leg 2 in accordance with Figure 9.22.

The presentation view is modified. Save your file.

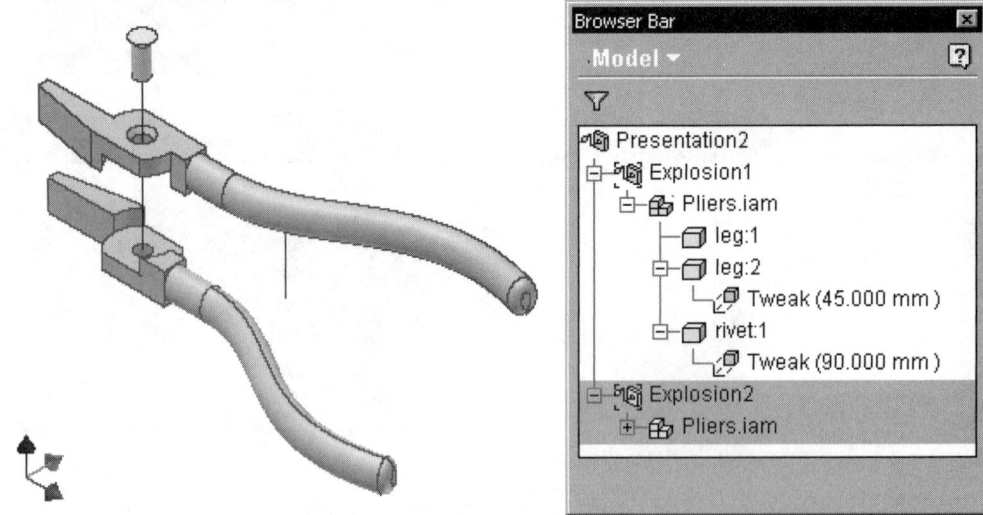

Figure 9.22 *Assembly View selected and presentation view modified*

USING THE TWEAK COMPONENT DIALOG BOX

You can activate the Tweak Component dialog box to modify the amount of tweak in manually tweaked presentation views. You select the trail in the view, right-click, select Edit, and use the Tweak Component dialog box.

 **TUTORIAL 9.4**

In this tutorial, you will modify the manually tweaked presentation view.

1. Open the presentation file *Pliers.ipn*, if you already closed it.

2. Select the second presentation view on the Browser Bar and double-click to activate it.

3. Select the trail line highlighted in Figure 9.23, right-click, and select Edit.

4. In the Tweak Component dialog box, modify the tweak amount to 60 mm and click the Apply and the Close buttons. Save your file.

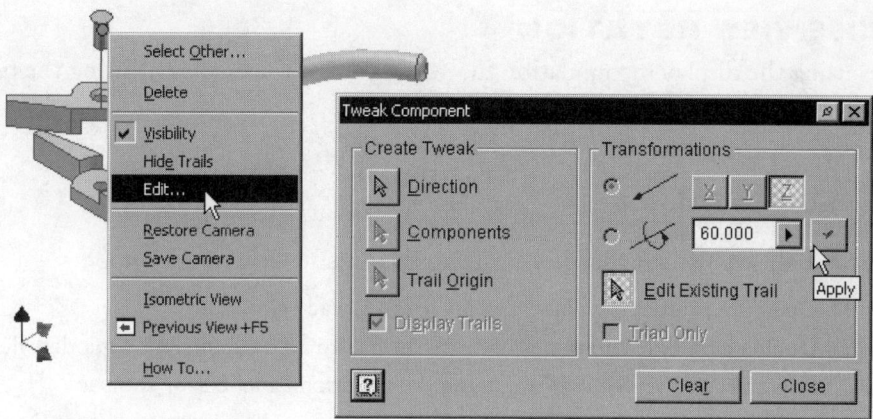

Figure 9.23 *Trail line selected and tweak amount being changed*

TRAIL EDITING

To modify the amount of tweak in a presentation, you select the trail in the presentation view and drag it to a new position.

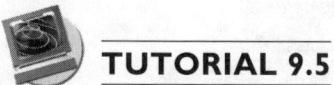 **TUTORIAL 9.5**

In this tutorial, you will edit a presentation view by selecting and dragging the trail lines.

1. Open the presentation file *Pliers.ipn*, if you already closed it.
2. Select the trail line indicated in Figure 9.24 and drag it to a new position. The amount of tweak is modified.

Save your file.

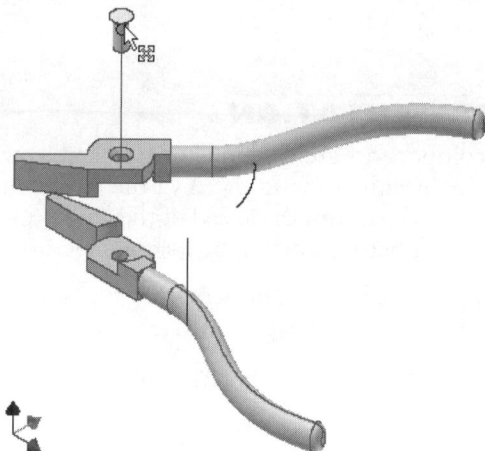

Figure 9.24 *Trail selected and being dragged*

PRECISE VIEW ROTATION

Besides using the display manipulation tools, you can set the display by using the precise view rotation tool.

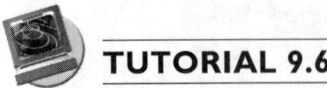

TUTORIAL 9.6

In this tutorial, you will set the view by using the precise view rotation tool.

1. Open the presentation file *Pliers.ipn*, if you already closed it.
2. Double-click Explosion1 on the Browser Bar to activate the first exploded view.
3. Select Precise View Rotation on the Presentation Panel Bar or toolbar.
4. In the Incremental View Rotate dialog box, there are six buttons: Rotate Down, Rotate Up, Rotate Left, Rotate Right, Roll Counter Clockwise, and Roll Clockwise. Select these buttons to set the display as shown in Figure 9.25.
5. Click the OK button when you finish.

Save your file.

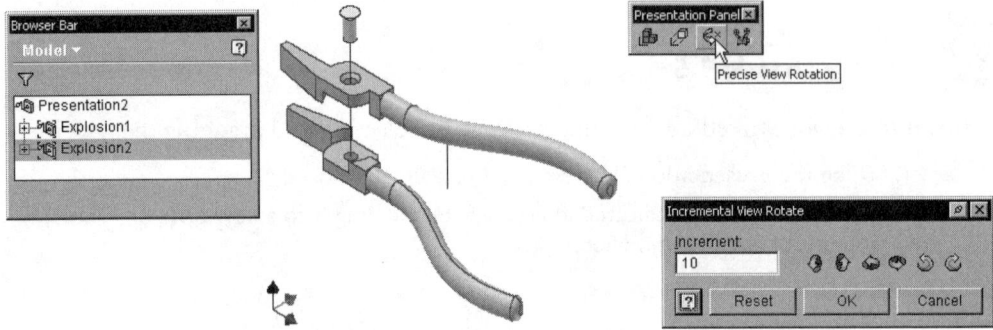

Figure 9.25 *Precise view setting*

ANIMATING A PRESENTATION

To appreciate how the components are exploded or tweaked, you animate the explosion or tweaking. You can save the animation in the AVI file format and play it back without running Autodesk Inventor. To control the speed of the animation, you can set the frame speed through the Document Settings dialog box, accessible from the Tools menu.

In the Document Settings dialog box of a presentation file, set the number of frames per second and the duration for each animated sequence.

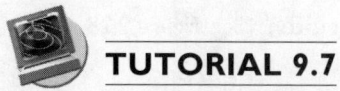

TUTORIAL 9.7

In this tutorial, you will construct an animation of a presentation.

1. Open the presentation file *Pliers.ipn*, if you already closed it.
2. Select Animate on the Presentation Panel Bar or toolbar. (See Figure 9.26.)
3. In the Animation dialog box, select the >> button to display the expanded dialog box.

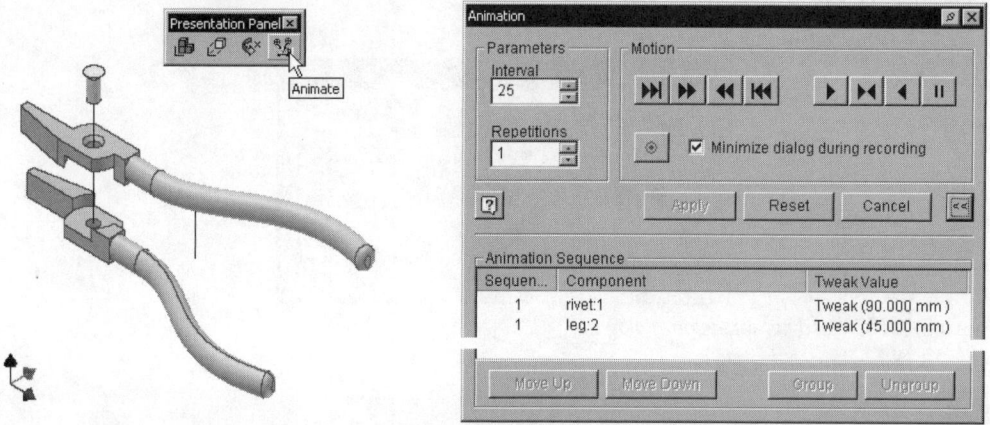

Figure 9.26 *Animation dialog box*

In the expanded dialog box, you can modify the animation sequence by selecting the tweaked component and selecting the Move Up or Move Down button.

4. Select the << button and then the Play button to see the animation.
5. Select the Record button and specify a file name (AVI file). (See Figure 9.27.)

Figure 9.27 *Record button*

6. In the Video Compression dialog box, click the OK button. (See Figure 9.28.)
7. Now select the Play forward button.

The animation is saved in an AVI file, and the presentation file is complete.

8. Click Cancel to close the Animation dialog box.

Save your file.

 Note: You can locate the new AVI file using Windows Explorer and double-click it to play it back in the Windows Media Player.

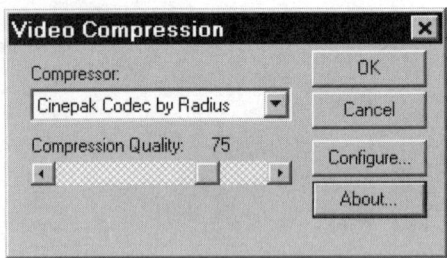

Figure 9.28 *Video compression dialog box*

 ## TUTORIAL 9.8

In this tutorial, you will reorder, group, and ungroup animation sequence.

1. Open the file *Pliers.ipn*, if you already closed it.
2. Double-click Explosion2 on the Browser Bar to activate it.
3. Select Animate on the Presentation Panel Bar or toolbar.
4. In the Animation dialog box, select the Play Forward button. (See Figure 9.29.)

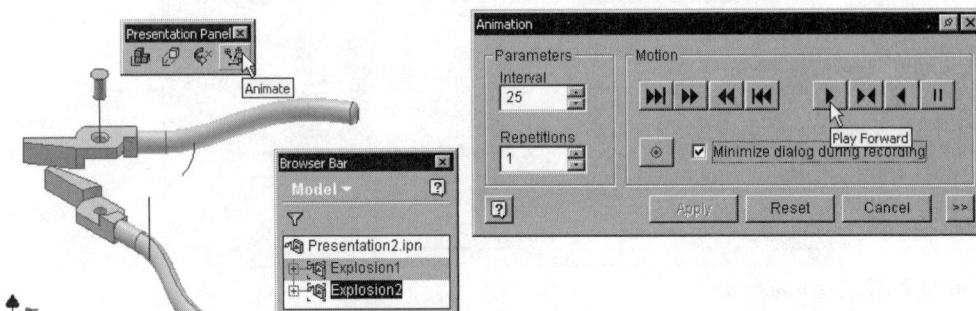

Figure 9.29 *Animating Explosion2*

As you might find from your screen display, the animation starts with one of the legs of the pliers rotating, then another leg moving up, and finally the rivet moving down. Now change this sequence.

5. Click Reset and then click >> to expand the dialog box.

6. Select leg:2 in the animation sequence list, select the Move Down button once, and click the Apply button. (See Figure 9.30.)

7. Select the Play Forward button.

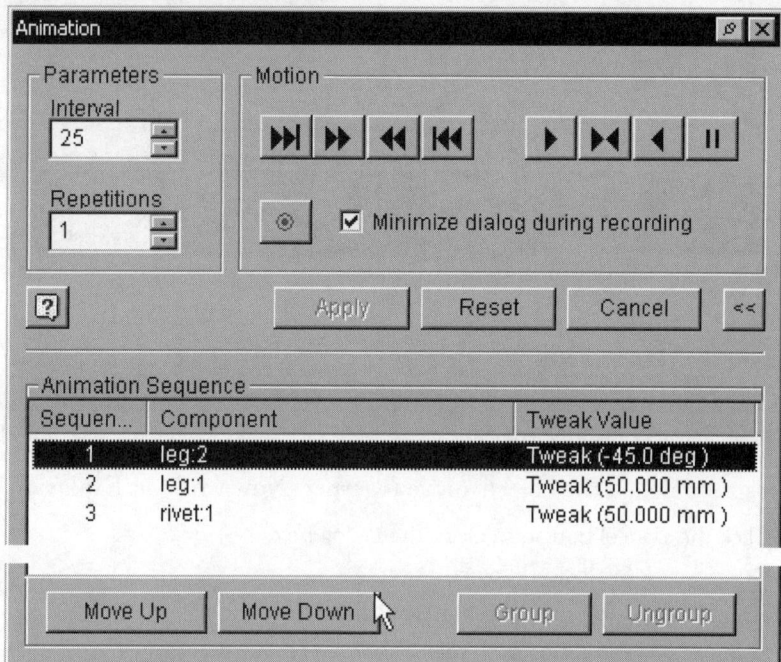

Figure 9.30 *Sequence being changed*

The sequence is changed. Now leg:1 moves upward before leg:3 rotates.

8. Click Reset, then hold down CTRL and select all three objects in the Animation Sequence list.

9. Click the Group and Apply buttons. (See Figure 9.31.)

10. Select the Play Forward button.

While you play the animation, the sequence that is in action is highlighted in the More tab of the Animation dialog box.

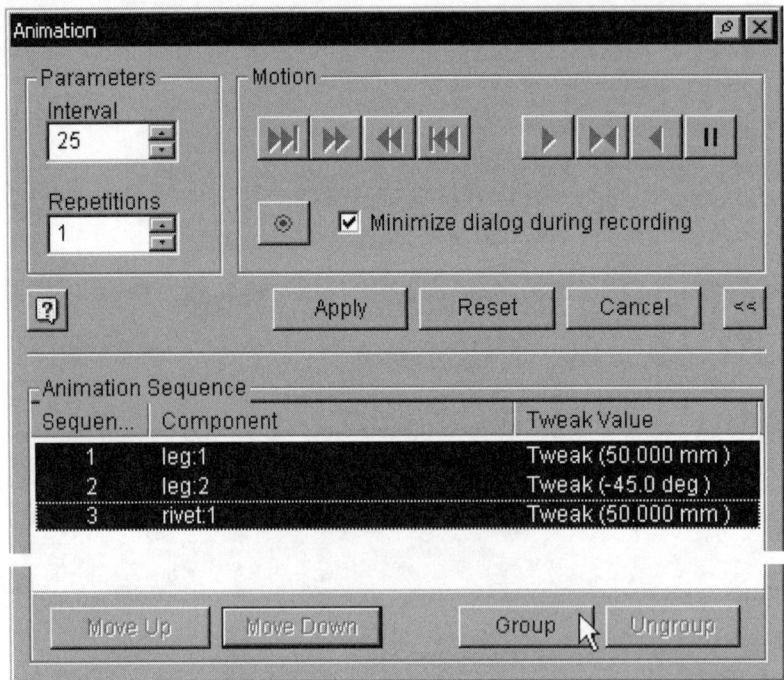

Figure 9.31 *Sequence being grouped*

After grouping, all the objects move together. Now work on Explosion1.

11. Click the Cancel button to close the dialog box.

12. Double-click Explosion1 to activate it.

13. Select Animate on the Presentation Panel Bar or toolbar.

14. In the Animation dialog box, select the Play Forward button.

All the objects move together because they are grouped. Now ungroup them.

15. Click Reset and then click >> to expand the dialog box, if it is not already expanded.

16. Select the objects in the animation sequence list.

17. Click the Ungroup and Apply buttons.

18. Select the Play Forward button.

Now they play in sequence rather than as a group together.

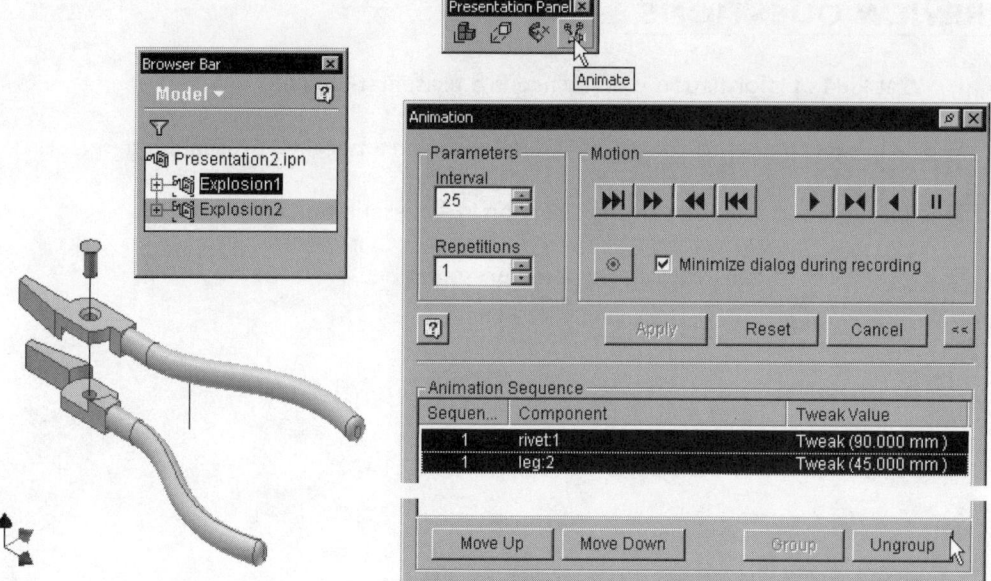

Figure 9.32 *Ungrouping sequence*

19. Click Cancel to close the Animation dialog box.

Save and close your file.

SUMMARY

A presentation file is linked to an assembly file that, in turn, is linked to a set of components, which can be part files or assembly files for the sub-assemblies.

To construct a presentation of an assembly, you link a presentation file to an assembly file, select a viewing direction, and explode or tweak the components to illustrate how the components are put together. In an exploded presentation, all the components are exploded by a specified distance. In a tweaked presentation, components are tweaked individually either in a linear direction or about an axis.

To explain how component parts are exploded or tweaked to move them apart, you add trail lines between the exploded or tweaked components and you construct an animation of the explosion or tweaking motion. The animation can be saved as an AVI file.

REVIEW QUESTIONS

1. What kind of information is contained in a presentation file?

2. Briefly explain how to construct an exploded and a tweaked presentation.

3. Explain how to select a viewing direction in a presentation file.

4. Explain how to construct an animated presentation.

CHAPTER 10

Engineering Drafting

OBJECTIVES

The aims of this chapter are to explain the concepts of engineering drafting and computer-generated engineering drafting, to illustrate how to prepare drawing sheets for constructing engineering drawings, and to explain how to construct a 2D draft drawing. It outlines the steps in constructing engineering drawing views from parts and assemblies, flat pattern views from sheet metal parts, and presentation views of assemblies, and the ways to add annotations to a drawing. After studying this chapter, you should be able to

- Describe the key concepts of engineering drafting and computer-generated engineering drafting
- Prepare drawing sheets for constructing engineering drawing views
- Construct 2D draft drawings without referencing any part, presentation, or assembly file
- Construct associative engineering drawing views of 3D parts, sheet metal parts, and assemblies
- Add annotations to a drawing

OVERVIEW

In a modern factory, you use the digital data about 3D parts and assemblies for downstream manufacturing operations. The need for and importance of 2D engineering drawings are diminishing. However, there are still many occasions when you need to produce 2D orthographic engineering drawings. Therefore, you will learn how to construct simple 2D drawings and to generate engineering drawings from the 3D solid parts and assemblies in this chapter.

With 3D parts and assemblies, the 2D engineering drawing is generated by the computer, and it is semi-automatic. You start a drawing file, select a 3D part file, assembly file, or presentation file and let the computer generate the orthographic views. Then you can add annotations to the drawing.

ENGINEERING DRAFTING CONCEPTS

Engineering drafting is an engineering communication tool in which a 3D object is represented on a 2D drawing sheet through a set of orthographic views. The traditional way to construct an engineering drawing is to think about how a 3D object will look when you project it orthogonally on a 2D plane, and to construct the orthographic views in accordance with your perception of the object's 2D appearance. You can construct the drawings manually or let the computer generate the drawing views for you if you have already constructed 3D computer models of the objects. After you construct the drawing views, you add annotations such as dimensions, text, geometric tolerances, surface finish symbols, welding symbols, and a parts list where appropriate.

ORTHOGRAPHIC PROJECTION

To depict a 3D object on a piece of 2D drawing paper, you use orthographic projection. The word "ortho" is a Greek word that means right or true. Orthographic projection is an engineering communication method to represent 3D objects on 2D drawing sheets by using multiple-view drawings. You project the 3D object perpendicularly onto a projection plane with parallel projectors. (See Figure 10.1.)

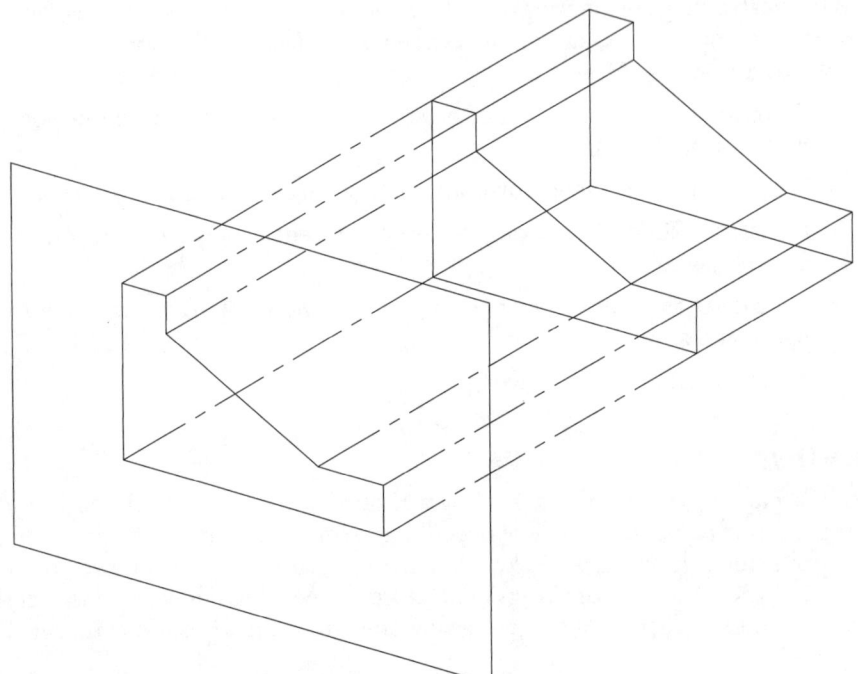

Figure 10.1 *A 3D object projected onto a plane*

Basically, you can use six projection planes that are mutually perpendicular to each other to construct six drawing views showing the front, right side, left side, rear side, top, and bottom of the 3D object. Now you can imagine the 3D object inside the box and project views orthogonally onto the six walls of the box. (See Figure 10.2.)

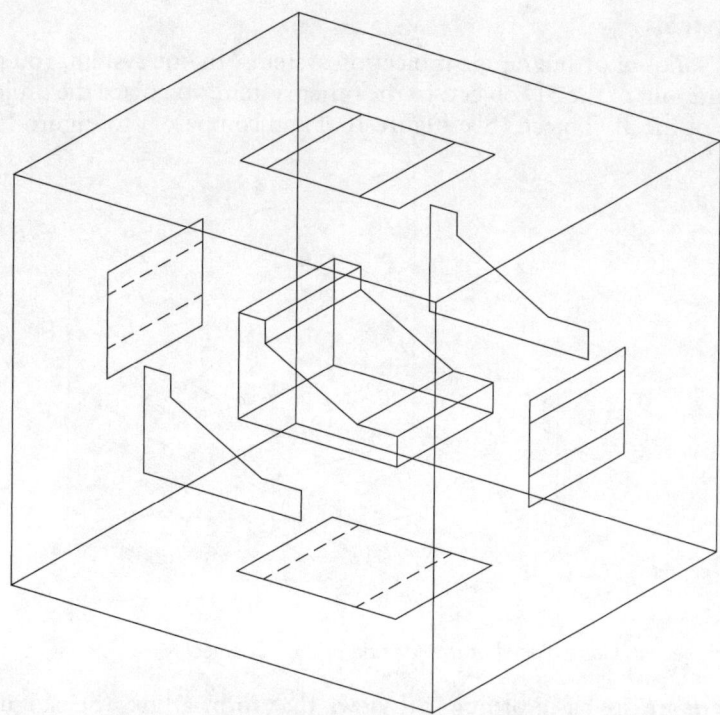

Figure 10.2 *Six projection planes forming a box with the 3D object placed inside*

Because it is inconvenient to carry the box around, you cut and spread the box onto a common plane to obtain a drawing showing the six basic views. (See Figure 10.3.)

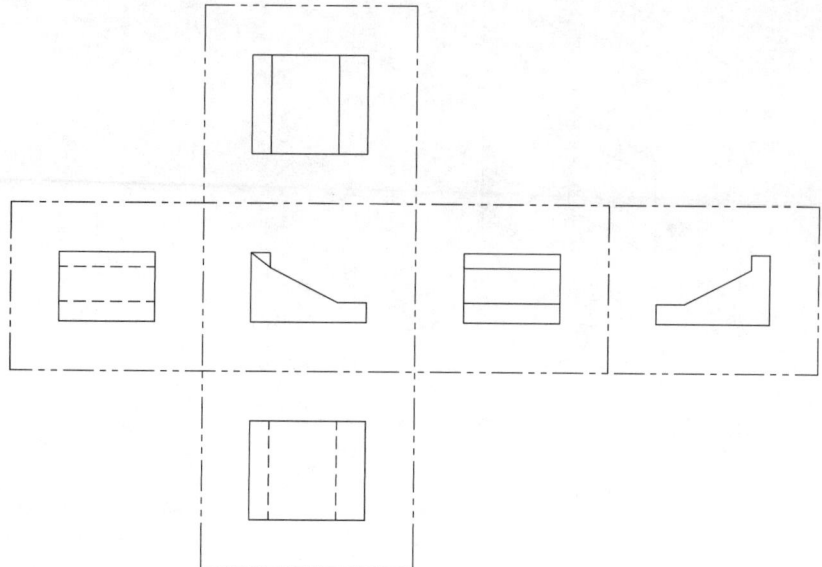

Figure 10.3 *Cutting and spreading the box*

Projection Systems

There are two kinds of orthographic projection systems. In one system, you put the projection plane in front of the 3D object. In the other system, you place the projection plane at the far side of the 3D object. (See Figure 10.4 and compare it to Figure 10.1.)

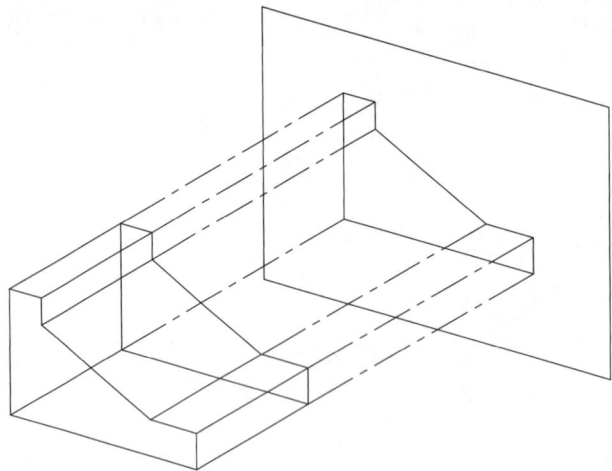

Figure 10.4 *Projection plane placed at the far side of the 3D object*

Similarly, there are six basic orthogonal views that form a box. (See Figure 10.5 and compare it to Figure 10.2.)

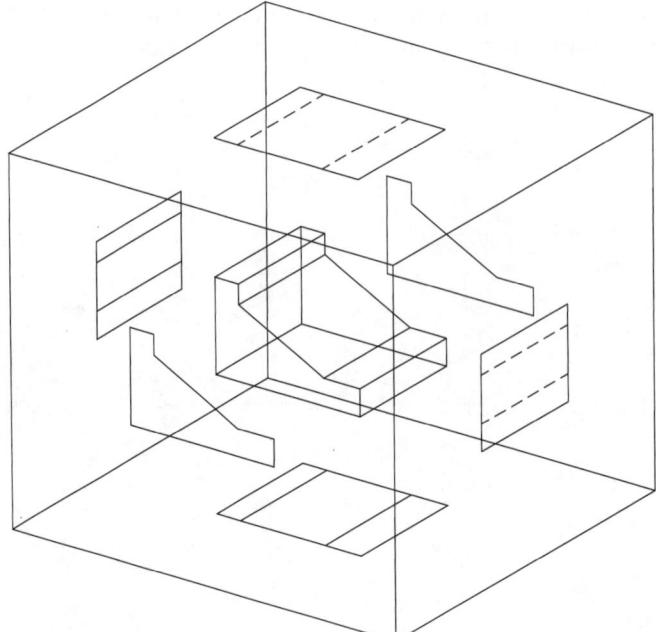

Figure 10.5 *Six projection planes*

Again, you will cut and spread the box onto a common plane to get six basic views on a drawing sheet. (See Figure 10.6 and compare it to Figure 10.3.)

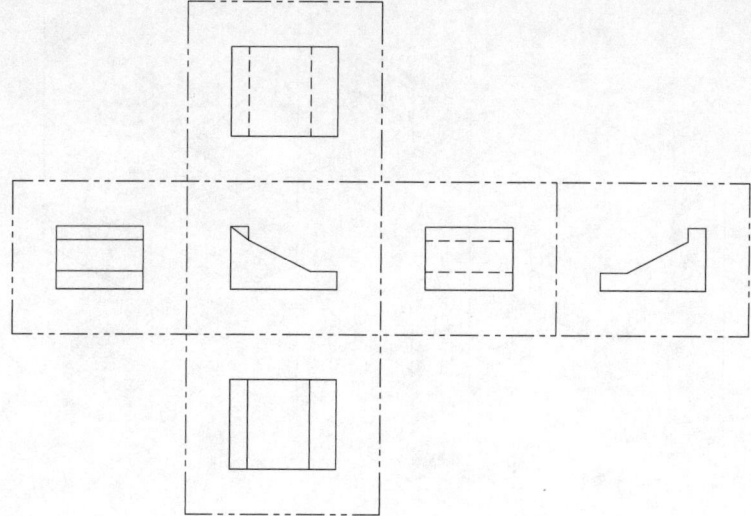

Figure 10.6 *Six basic views*

First and Third Angle Projection

In Figures 10.3 and 10.6, you can see that the front and rear views, the left and right side views, and the top and bottom views are quite similar. To describe this 3D object, three drawing views (front, side, and top) are sufficient. (See Figure 10.7.)

If you put the two projection systems together in 3D space, you will discover a very interesting picture. In Figure 10.8, one system falls neatly in the first quadrant and the other in the third quadrant of the 3D space. Because we have to give the two projection systems names to identify which system we are using, we call one system the first angle projection system and the other the third angle projection system.

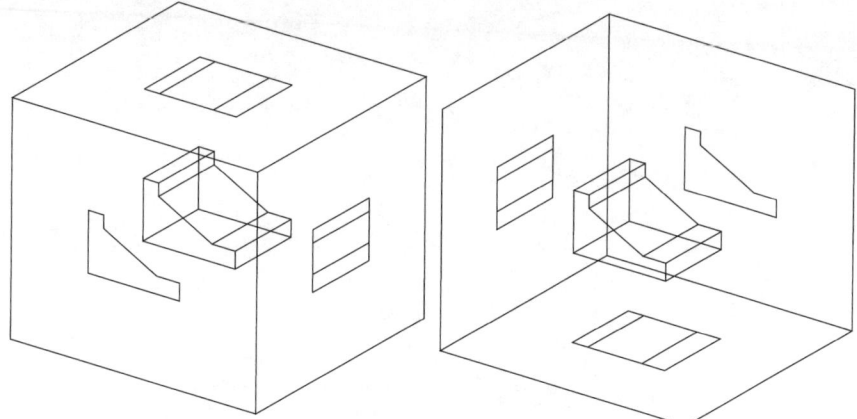

Figure 10.7 *Three views of the 3D object in two projection systems*

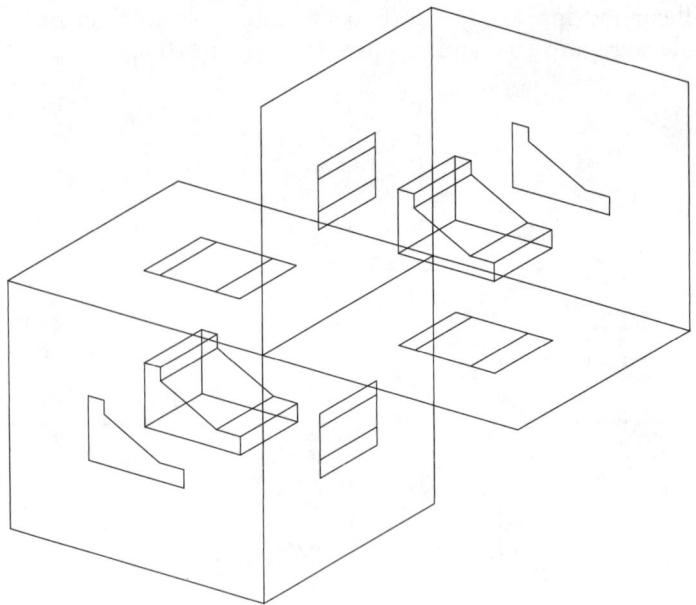

Figure 10.8 *Two systems put together*

Projection Symbols

To indicate the system of projection that you are using, you place a symbol on your draw‐ing sheet. The projection symbol is the engineering drawing of the front and side views of a conical object. (See Figure 10.9.)

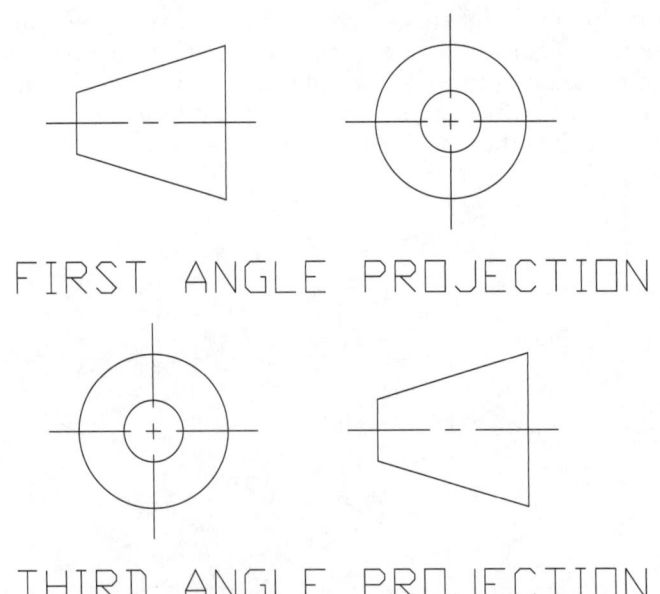

Figure 10.9 *Projection symbols*

COMPUTER-AIDED ENGINEERING DRAFTING

There are two major kinds of computer-aided engineering drafting. The first kind is purely 2D graphical elements and the second kind is associative engineering drafting.

2D Engineering Drafting

This is the oldest way of using the computer to construct engineering drawings. Here the computer is used purely as an electronic drafting board, and you use the sketch tools to construct an engineering drawing in accordance with your perception of how the 3D object will look when it is projected onto one of the orthographic projection planes. Figure 10.10 shows a manually constructed 2D engineering drawing in third angle projection. Autodesk Inventor calls this kind of drawing a sketched drawing. We should only use this kind of drawing in the absence of 3D objects.

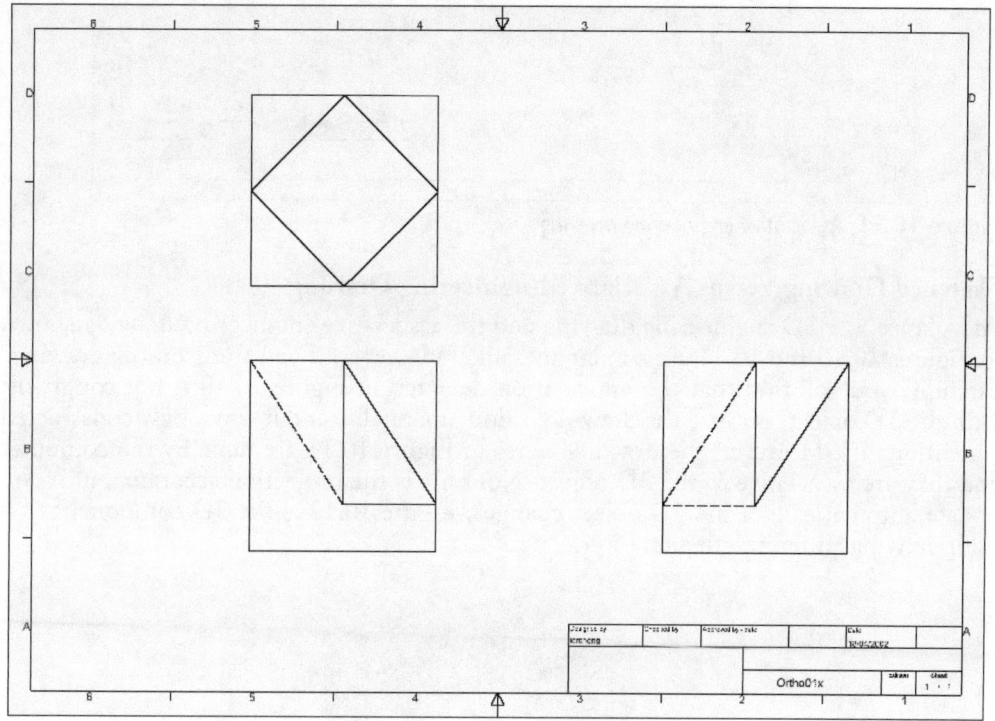

Figure 10.10 *2D engineering drawing*

Associative Engineering Drafting

If you have already constructed 3D solids and assemblies of 3D solids in the computer, the construction of an engineering drawing is very simple. You start a drawing file, construct a drawing sheet, select a solid part, a presentation, or an assembly, let the computer project orthographic views from the solid or assembly, and add annotations to the views. Figure 10.11 shows an associative 2D engineering drawing in third angle projection.

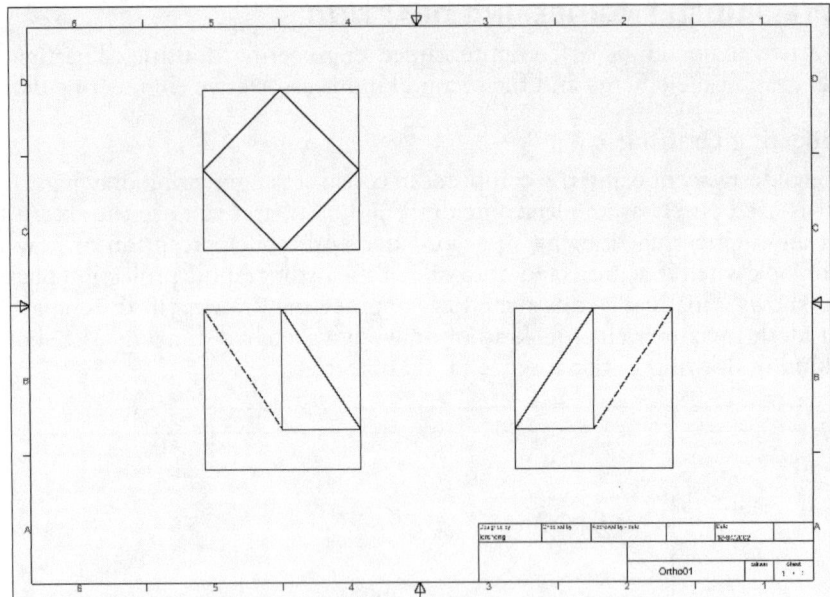

Figure 10.11 *Associative engineering drawing*

Sketched Drawing Versus Associative Engineering Drafting

At a glance, the 2D engineering drawing and the associative engineering drawing shown in Figures 10.10 and 10.11 are similar and alike. However, if you study on the drawings carefully, you will find that the information depicted in Figure 10.10 is not congruent with the 3D object, because the drawing is done manually, and it is wrongly constructed (intentionally). However, the drawing views in Figure 10.11 are done by the computer and they are associative to the 3D object. Not only is the projection accurate, but it will update automatically if the 3D object changes. Figure 10.12 is the 3D component that both drawings intend to depict.

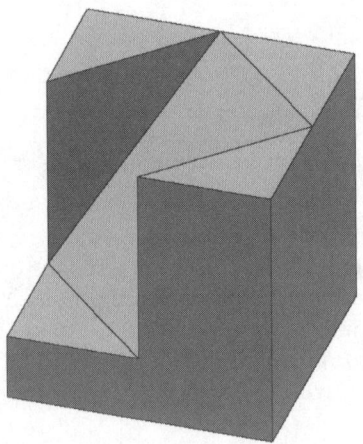

Figure 10.12 *3D component*

DRAWING FILE

To construct an engineering drawing, you will use a drawing file. Depending on which kind of drawing you construct, the drawing file contains either 2D draft sketches that are not related to any 3D object or associative drawing views that are related to a single 3D object or a set of 3D objects. Table 10.1 summarizes the information contained in the part file, assembly file, presentation file, and drawing file.

Table 10.1 *Information*

File	Information
Part File	Definition and information about individual 3D solid part
Assembly File	Information about the location of the linked components and how the linked components are assembled together
Presentation file	Information about the location of the linked assembly file and how the components in the assembly file are exploded or tweaked
Drawing file (Draft views)	Information about the 2D sketch drawing without any reference to 3D objects
Drawing file (Associative drawing views)	Information about the location of the linked part file, assembly file, and/or presentation file, engineering drawing sheet data, engineering drawing views of the 3D solid part, assembly, and/or assembly presentation, and dimensions and annotations

PROCEDURE FOR CONSTRUCTING AN ENGINEERING DRAWING

Constructing of engineering drawing through Autodesk Inventor involves three major tasks:

Task 1 Prepare a 2D drawing sheet in a drawing file.

Task 2 Construct sketched drawings or engineering drawing views on the drawing sheets.

Task 3 Add annotations to the drawing.

USER INTERFACE

Before you start constructing an engineering drawing, you should familiarize yourself with the Autodesk Inventor drawing file user interface. To construct an engineering drawing, you use the drawing template in the New dialog box:

1. Select File > New or the New button on the Inventor Standard toolbar.
2. Select the *Standard.idw* template, and click the OK button. (See Figure 10.13.)

To construct an engineering drawing, you use three sets of tools: drawing views, drawing annotation, and sketch. You use the drawing views tool to manage drawing views, the drawing annotation tool to construct annotations, and the sketch tool to construct the constituents of draft views.

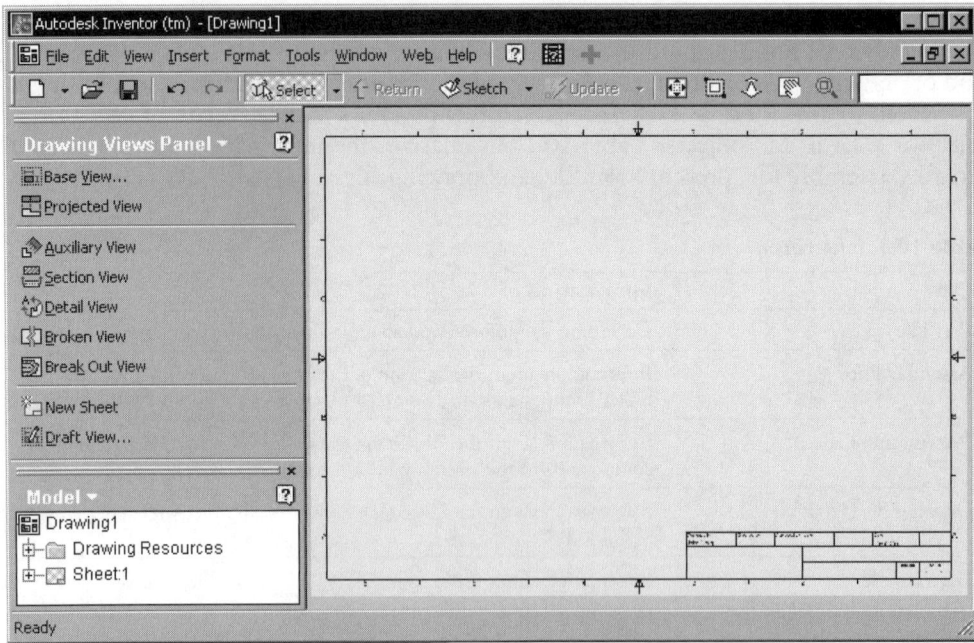

Figure 10.13 *A drawing file started*

DRAWING VIEWS PANEL BAR AND TOOLBAR

Initially, the panel bar displays the tools for making engineering drawing views. To display the accompanying Drawing Views Panel toolbar, select View > Toolbar > Drawing Views Panel. Using the Drawing Views Panel Bar and toolbar, you set up a new drawing sheet and construct engineering drawing views on the drawing sheet. (See Figure 10.14.)

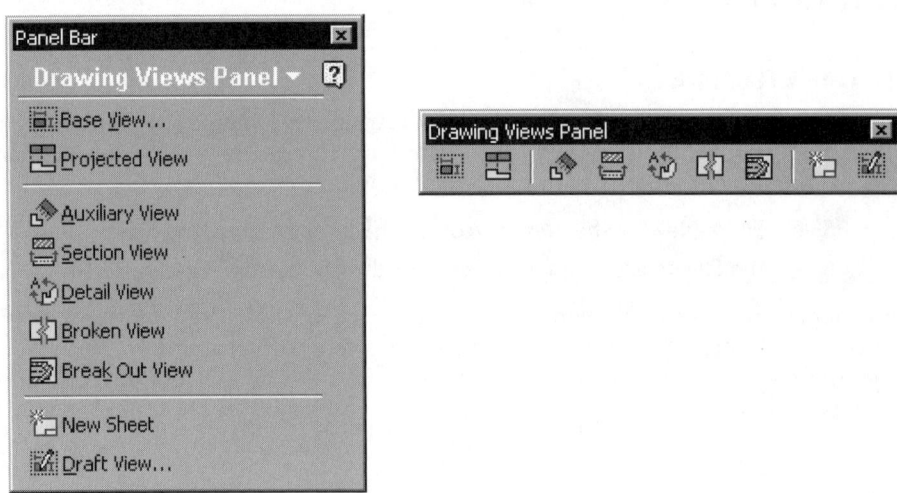

Figure 10.14 *Drawing Views Panel Bar and toolbar*

The Drawing Views Panel Bar and toolbar have nine buttons. Table 10.2 describes the choices:

Table 10.2 *Drawing Views Panel Bar and toolbar options*

Option	Description
Base View	Constructs a base view from a selected part, assembly, or presentation. The base view can be an orthogonal view or a perspective view.
Projected View	Projects drawing views from orthogonal drawing views. Projected views include isometric views.
Auxiliary View	Projects an auxiliary view from an orthogonal view.
Section View	Projects a section view from an orthogonal view.
Detail View	Constructs a detail view that is an enlargement of a portion of an orthogonal view.
Broken View	Converts an orthogonal view to a broken view.
Break Out View	Converts an orthogonal view to a break out view.
New Sheet	Places a drawing sheet on which you construct engineering drawing views.
Draft View	Sets the label and scale when you work on a draft view.

DRAWING ANNOTATION PANEL BAR AND TOOLBAR

To display the Drawing Annotation Panel Bar, move the cursor over the panel bar, right-click, and select Drawing Annotation Panel. To display the accompanying Drawing Annotation Panel toolbar, select View > Toolbar > Drawing Annotation Panel. (See Figure 10.15.)

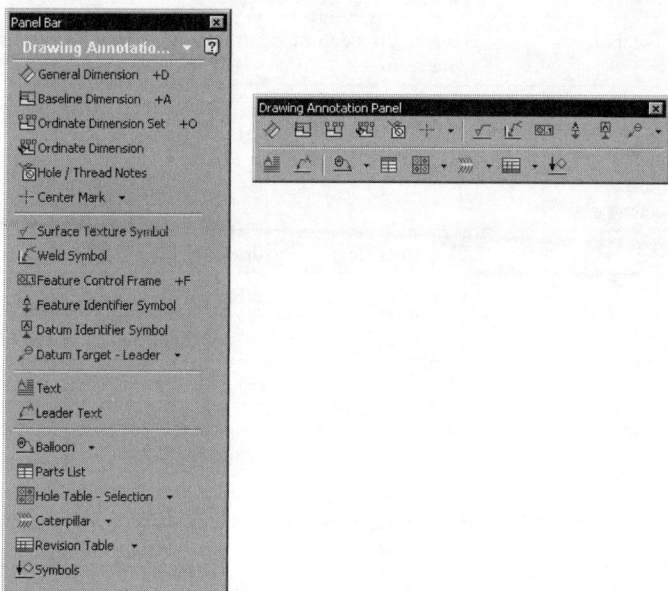

Figure 10.15 *Drawing Annotation Panel Bar and toolbar*

Using the Drawing Annotation Panel Bar and toolbar, you construct text objects, dimensions, geometric tolerances, welding symbols, surface finish symbols, and a parts list. The Drawing Annotation Panel Bar and toolbar have twenty button areas. Table 10.3 describes the choices.

Table 10.3 *Drawing Annotation Panel Bar and toolbar options*

Option	Description
General Dimension	Constructs general dimensions on the drawing sheet.
Baseline Dimension	Constructs baseline dimensions on the drawing sheet.
Ordinate Dimension Set	Constructs ordinate dimensions (with dimension lines) on the drawing sheet.
Ordinate Dimension	Constructs ordinate dimensions on the drawing sheet.
Hole/Thread Notes	Constructs a hole or thread note on the drawing sheet.
Center Mark/ Centerline/ Centerline Bisector/ Centered Pattern	Constructs various kinds of centerlines on the drawing sheet.
Surface Texture Symbol	Constructs a surface texture symbol on the drawing sheet to specify surface roughness requirement.
Weld Symbol	Constructs a weld symbol on the drawing sheet to specify welding requirement.
Feature Control Frame	Constructs a feature control frame to specify geometric tolerance requirement.
Feature Identifier Symbol	Constructs a feature identifier symbol to specify an identifier for a particular feature of the part.
Datum Identifier Symbol	Constructs a datum identifier symbol to specify geometric tolerance data reference.
Datum Target-Leader/ Datum Target-Circle/ Datum Target-Line/ Datum Target-Point/ Datum Target-Rectangle	Constructs various kinds of datum target symbols on the drawing sheet in order to specify geometric tolerance requirement.
Text	Constructs text on the drawing sheet.
Leader Text	Constructs a leader together with lines of text on the drawing sheet.
Balloon/ Balloon All	Constructs a balloon or a set of balloons in an assembly drawing sheet.
Parts List	Places a part list in an assembly drawing sheet.
Hole Table–Selection/ Hole Table–View/ Hole Table–Selected Type	Constructs a hole table depicting the size and location of the selected holes or all the holes in a drawing view.
Caterpillar/End Treatment	Constructs a non-associative caterpillar welding annotation in a drawing.
Revision Table/Revision Tag	Constructs a revision table or a revision tag.
Symbol	Adds a symbol from the drawing resource to the drawing.

DRAWING SKETCH PANEL BAR AND TOOLBAR

If you are going to construct a 2D drawing view by simply constructing sketch elements without referencing any part file, presentation file, or assembly file, you use the sketch tools on the Drawing Sketch Panel Bar or toolbar. (See Figure 10.16.) The Drawing Sketch Panel Bar and toolbar are similar to the set of tools you use for constructing sketches for making solid features.

Figure 10.16 *Drawing Sketch Panel Bar and toolbar*

SHORTCUT KEYS

Besides the appropriate panel bars and toolbars, Table 10.4 shows the shortcut keys available:

Table 10.4 *Engineering drawing shortcut keys*

Shortcut Key	Function
D	Adds a general dimension.
O	Adds an ordinate dimension.
B	Constructs a balloon.
F	Constructs a feature control frame.

SELECTION PRIORITY

When you move the cursor over a drawing, the kind of objects selected depends on selection priority and selection filter settings. Figure 10.17 shows the select pull-down list box of the Inventor Standard toolbar. Details regarding the selection priority and selection filter are listed in Table 10.5.

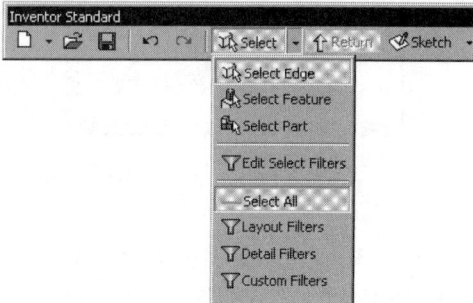

Figure 10.17 *Selection priority and selection filter*

Table 10.5 *Selection priority and selection filter*

Option	Description
Edge	Sets selection to edges of parts.
Feature	Sets selection to features of parts.
Part	Sets selection to parts.
Edit Select Filters	Enables you to specify the elements that can be selected when using the Select tool filter sets.
Select All	Applies a filter set to select all elements in a drawing.
Layout Filters	Sets selection to drawing views.
Detail Filters	Sets selection to details.
Custom Filters	Applies a custom filter set.

DRAWING BROWSER BAR

In the drawing file Browser Bar, there are two kinds of drawing objects: Drawing Resources and Sheet. Drawing resources consist of sheet formats, borders, title block, and sketched symbols. Sheet refers to the engineering drawing sheet where you construct 2D sketch drawings or associative engineering drawings. (See Figure 10.18.)

By expanding the Browser Bar, you will see four kinds of drawing resources (Sheet Formats, Borders, Title Blocks, and Sketched Symbols) and the drawing sheet with a default border and title block.

Figure 10.18 *Expanded Browser Bar of a drawing file*

Drawing Resources

The first set of objects found on the drawing Browser Bar is drawing resources. Drawing resources are a set of objects for you to configure a drawing sheet. There are four kinds of drawing resources: sheet formats, borders, title block, and sketched symbol.

Sheet Formats	A drawing sheet in the computer is analogous to a piece of drawing paper. To construct a sketch drawing or an associative drawing, you need a drawing sheet.
Borders	The four border lines around the edges of the drawing sheet and sub-divided zone lines along the border lines. You need to add a set of borders to each drawing sheet.
Title Blocks	In addition to the borders, you need to add a title block. In the title block, you include information regarding the name of the person who constructed the drawing, the date the drawing is constructed, part number, sheet number, and other information as may be required.
Sketched Symbols	2D sketched objects. You can also include special sketched symbols in your drawing sheet, for example, a company logo.

Sheet

The second set of objects listed in the drawing Browser Bar make up the drawing sheet. To construct sketched drawing or associative engineering drawing views of 3D objects, you need one or more drawing sheets. On a drawing sheet, you include a border, a title block, and appropriate symbols (sketched symbols) to comply with appropriate engineering drawing standards. When you start a drawing file, it already displays a default drawing sheet with default title block and borders. If you find these objects appropriate, you can start placing engineering drawing views on the drawing sheet.

DRAWING SHEET PREPARATION

Preparing a 2D drawing sheet concerns selecting a sheet format, inserting a set of borders around the edges, and inserting a title block. Sometimes, you can also insert sketched symbols in the drawing.

> To use a different kind of drawing sheet, select one from Sheet Formats in Drawing Resources or specify a new sheet by selecting Sheet from the Insert menu.

> To use a different kind of border, select one from Borders in Drawing Resources or define a new border by selecting Define New Border from the Format menu.

> To use a different kind of title block, select one from Title Blocks in Drawing Resources or define a new title block by selecting Define New Title Block from the Format menu.

> To include a symbol on the drawing sheet, define a new symbol by selecting Define New Symbol from the Format menu and insert it in the drawing sheet.

ENGINEERING DRAFTING STANDARDS

Because engineering drawing is an engineering communication language, it is important that the drawing comply with appropriate national and international standards. Therefore, you need to set drafting standards before making a drawing. To control the display of dimension and annotation text, you set dimension style and text style.

568

STANDARD PRACTICE

To comply with national or international standards, you manipulate the settings in the Drafting Standards dialog box.

1. Select Format > Standards. (See Figure 10.19.)

In the Drafting Standards dialog box, there are six kinds of drawing standards: ANSI, BSI, DIN, GB, ISO, and JIS. In addition to the six kinds of standards, you can create your own standard. In the Drafting Standards dialog box, select Click to add a new standard choice at the bottom of the standards list. Then specify a name for your new standard.

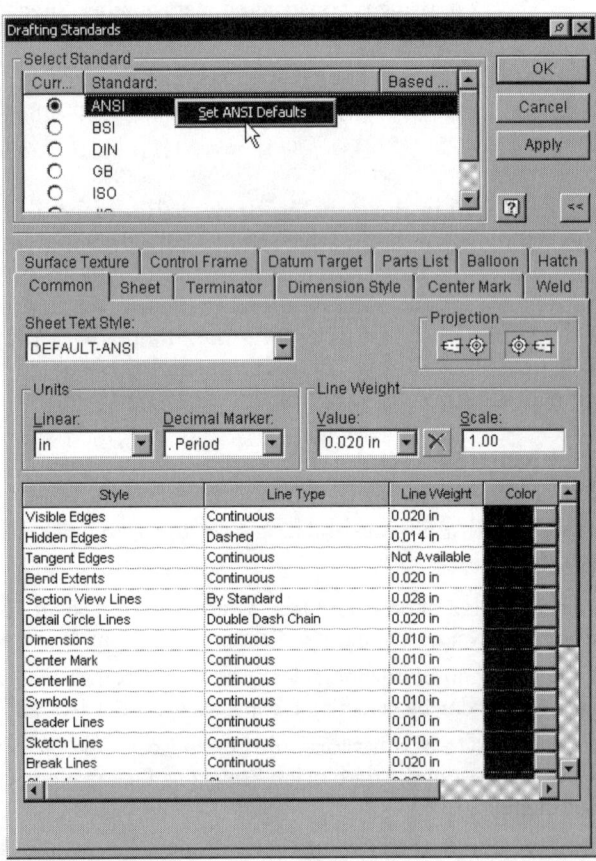

Figure 10.19 *Drafting Standards dialog box*

In the expanded dialog box, there are twelve tabs:

Common	Sets the text style, projection direction, units of measurement, and line style.
Sheet	Sets the sheet and view labels on the Browser Bar and the color scheme of the drawing sheet.
Terminator	Sets the style of arrows and datum references.

Dimension Style	Displays the default dimension style and characters that will be available to use in dimension text.
Center Mark	Sets the proportion of the dashes of a center mark.
Weld	Sets the weld symbols.
Surface Texture	Sets the surface texture symbols.
Control Frame	Sets the control frame of geometric tolerances.
Datum Target	Sets the datum target of geometric tolerances.
Parts list	Sets the styles of the parts list.
Balloon	Sets the styles of the balloons.
Hatch	Sets the styles of the hatch patterns.

Now you will select a standard.

2. Select ANSI in the list of drawing standards, right-click, and select the Set ANSI Defaults button to use default ANSI standard.

3. On the Common tab of the expanded dialog box, select the third angle of projection button.

4. On the Sheet tab, choose a color scheme to display the engineering drawing in your screen.

5. Click OK to close the dialog box.

DIMENSION STYLES

To set dimension styles in a drawing, you use the Dimension Styles dialog box.

6. Select Format > Dimension Styles. (See Figure 10.20.)

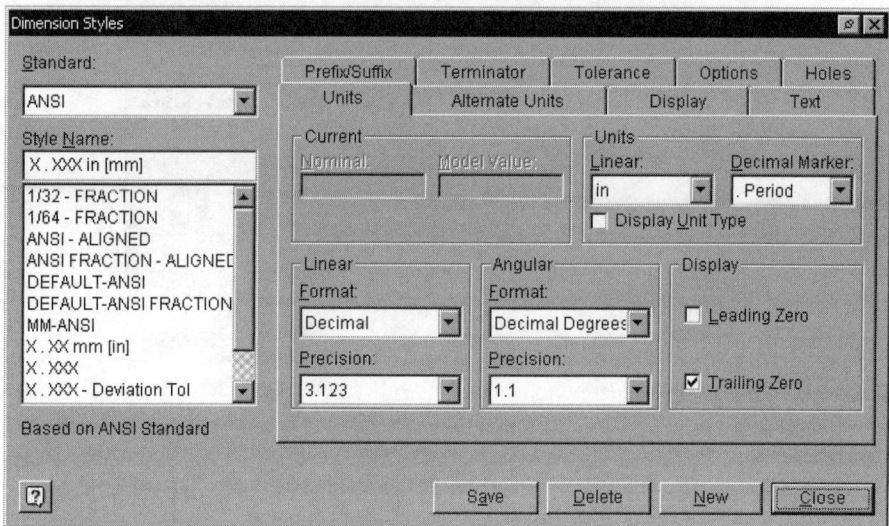

Figure 10.20 *Dimension Styles dialog box*

In the Dimension Styles dialog box, there are nine tabs. Their functions are listed below:

Units	Sets the units of measurement such as mm and inch.
Alternate Units	Sets and activates alternate units of measurement such as the display of English units in conjunction with metric units or metric units in conjunction with English units.
Display	Sets the display of dimensions, including line type, line weight, color, location of dimension value, and lengths of extension and offset of extension lines.
Text	Sets dimension text styles and orientation.
Prefix/Suffix	Adds a prefix or suffix to a dimension.
Terminator	Sets the arrowhead style.
Tolerance	Sets and activates dimension tolerance.
Options	Sets dimension methods and options for each method.
Holes	Sets hole note attributes.

7. Make appropriate changes and then click the Save and Close buttons.

TEXT STYLE

To set the style of text in a drawing, you use the Text Styles dialog box.

8. Select Format > Text Styles. (See Figure 10.21.)

9. Make appropriate changes and then click the Save and Close buttons.

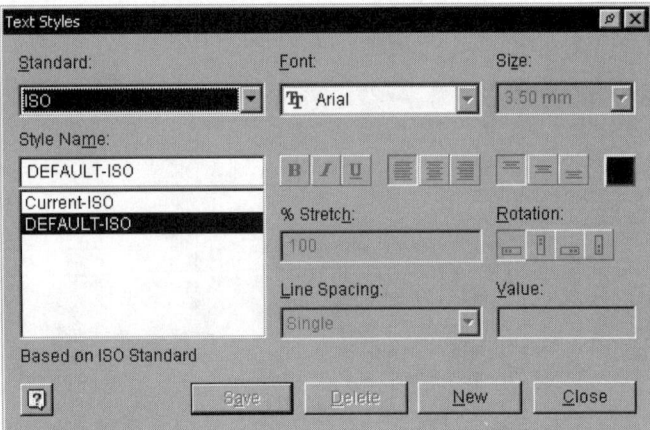

Figure 10.21 *Text Styles dialog box*

ORGANIZER

To save time in setting up dimension styles and text styles in every drawing sheet, you can save a template file and use the template for new drawings. To copy dimension styles and text styles from one drawing file to another, you use the Drawing Organizer dialog box.

10. Select Format > Organizer. (See Figure 10.22.)

11. Make appropriate changes and then click the Close button.

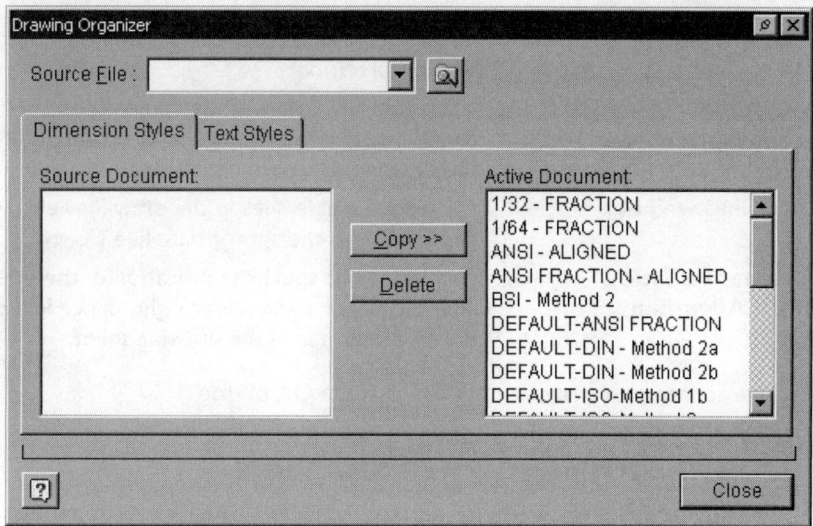

Figure 10.22 *Drawing Organizer dialog box*

SYSTEM SETTING

Before you use the drawing tools to construct 2D sketch drawings and associative engineering drawing, you should spend some time to study the related system settings and make changes if necessary.

1. Select Application Options from the Toots menu.

2. In the Options dialog box, select the Drawing tab. (See Figure 10.23.)

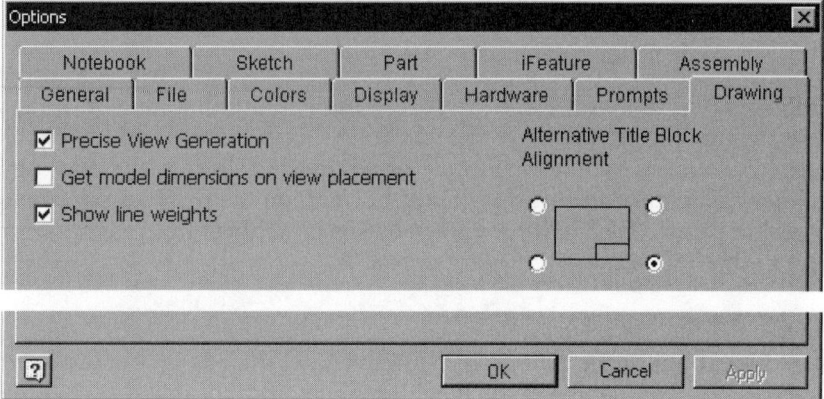

Figure 10.23 *Drawing tab of the Options dialog box*

The Drawing tab of the Options dialog box lets you set various settings regarding engineering drawing construction. The functions of the check boxes and options on this tab are explained below:

Precise View Generation	If selected, associative drawing views are constructed with full precision.
Get model dimensions on view placement	If selected, applicable model dimensions are added to the associative drawing views automatically when the drawing views are constructed.
Show line weights	If selected, visible lines in the drawing views will be displayed with the appropriate line weight.
Alternative Title Block Alignment	Enables you to specify the location of the title block in either upper right, lower right, upper left, or lower left corner of the drawing sheet.

3. Make appropriate changes and then click the OK button.

ENGINEERING DRAWING PREPARATION

To reiterate, a drawing file has two major kinds of objects: drawing resources and drawing sheet. Drawing resources are objects required by the drawing sheet. There are four kinds of resources: sheet formats, borders, title blocks, and sketched symbols. To construct engineering drawing views, you need a drawing sheet. When you start a new drawing file, a default sheet is given, with a set of default borders and a default title block. To comply with appropriate engineering standards, you need to select a proper sheet size, construct standardized borders and title blocks, and make sketched symbols on the title block.

If you are satisfied with the drawing resources provided by the drawing file template, you can proceed to the next section, "Draft View Construction." Otherwise, work on the tutorial below to set up your own drawing sheet, border, and title block, and construct sketched symbols, if necessary.

PREPARE DRAWING SHEET

A drawing sheet is analogous to a piece of drawing paper that you use to construct an engineering drawing. In a drawing file, you can insert more than one drawing sheet by adding new drawing sheets. You can delete drawing sheets, but you must have at least one drawing sheet. There are two ways to insert a new drawing sheet in the drawing file: you can select and double-click a sheet from Sheet Formats on the Browser Bar (to use one of the existing sheets) and you can insert a sheet by selecting Sheet from the Insert menu (to use a blank sheet).

INSERT BORDERS

A drawing sheet should have four border lines around the edges. To add borders to a blank drawing sheet, you can insert the default borders from the drawing resources of the Browser Bar or construct them by sketching.

INSERT TITLE BLOCKS

Now you will learn how to add title blocks to the drawing sheets in your drawing file. Again, there are two ways to use a title block: you insert a default title block or you construct a title block and insert it in your drawing sheet.

ADD SKETCHED SYMBOLS

In addition to the borders and title block, you can construct sketched symbols for insertion in the drawing sheet.

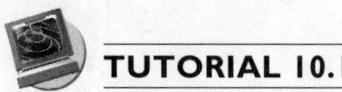

TUTORIAL 10.1

In this tutorial, you will learn how to construct a new sheet format and include custom-made borders, title blocks, and sketched symbols.

1. Start a new drawing file.
2. Select Insert > Sheet.
3. In the New Sheet dialog box shown in Figure 10.24, select A3 in the Size pull-down list box and click the OK button.

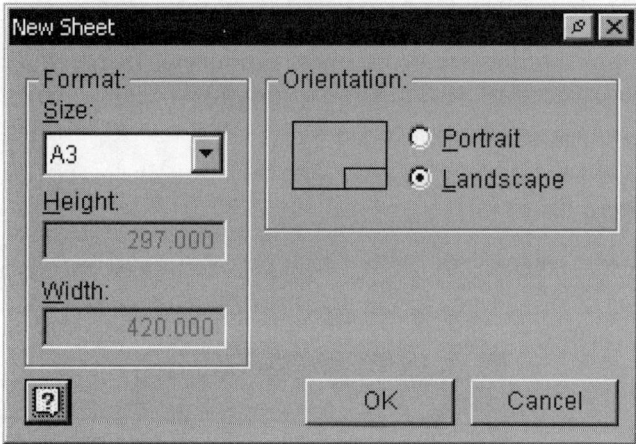

Figure 10.24 *New Sheet dialog box*

A new drawing sheet of A3 size is constructed. Now define a border.

4. Select Format > Define New Border.
5. Select Two Point Rectangle on the Drawing Sketch Panel Bar or toolbar and select two points on the sheet to construct a rectangle. (See Figure 10.25.)

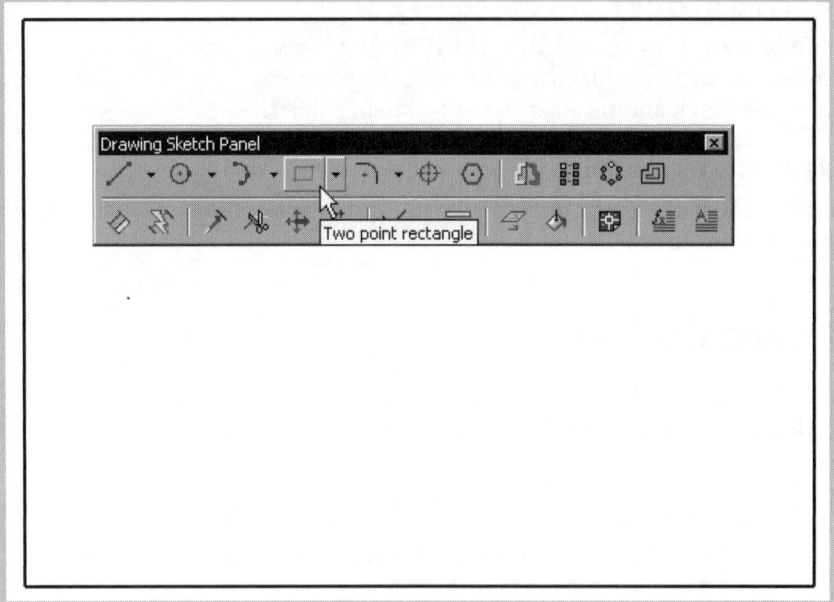

Figure 10.25 *Rectangle constructed*

A set of simple borders with four lines is complete. Before you can use this new border, you have to save it as a drawing resource.

6. Select Format > Save Border.
7. In the Border dialog box, specify a name (My Border) and click the Save button. (See Figure 10.26.)

 Note: After you save it, the border that you just drew will disappear.

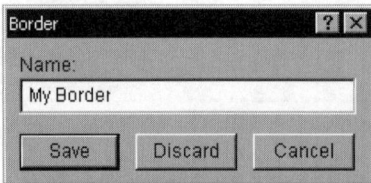

Figure 10.26 *Border dialog box*

The border is saved as one of the drawing resources in the file. Now use this set of borders in your drawing sheet.

8. Expand the objects on the Browser Bar.
9. Select the new border (My Border) from the drawing resources, right-click, and select Insert. (See Figure 10.27.)

The new border is inserted in the drawing sheet.

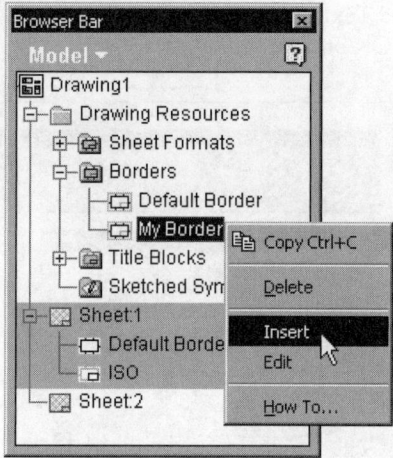

Figure 10.27 *Border saved as a drawing resource and being inserted*

Now define a title block.

10. Select Format > Define New Title Block.
11. With reference to Figure 10.28, construct a rectangle, a horizontal line, and a vertical line.

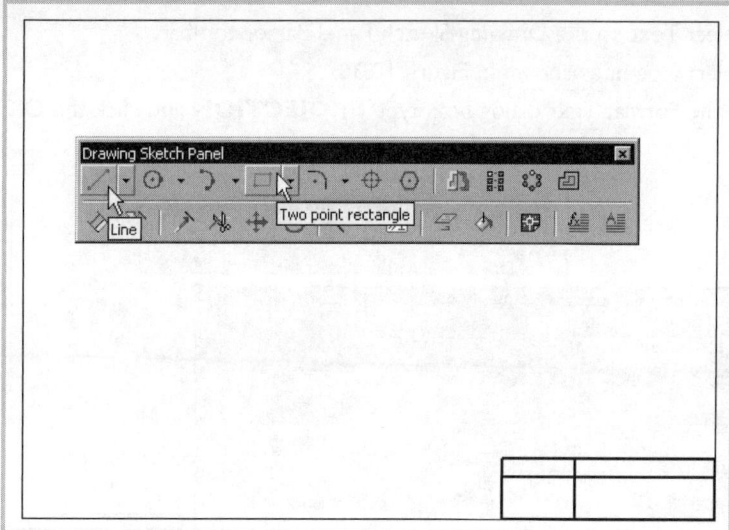

Figure 10.28 *Rectangle and horizontal lines constructed*

A title block should have textual information such as the name of the designer, and the title of the project. Now include these objects as property fields in the title block.

12. Select Property Field on the Drawing Sketch Panel Bar or toolbar.
13. Select a point as shown in Figure 10.29.

14. In the Format Field Text dialog box, select Title in the Properties pull-down list box and click the OK button.

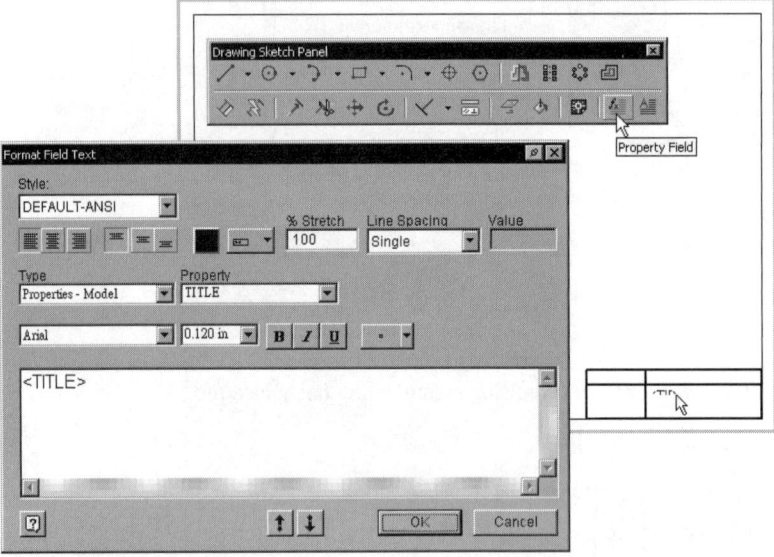

Figure 10.29 *Description field being added to the title block*

15. Select Text on the Drawing Sketch Panel Bar or toolbar.
16. Select a point as shown in Figure 10.30.
17. In the Format Text dialog box, type **PROJECTION** and click the OK button.

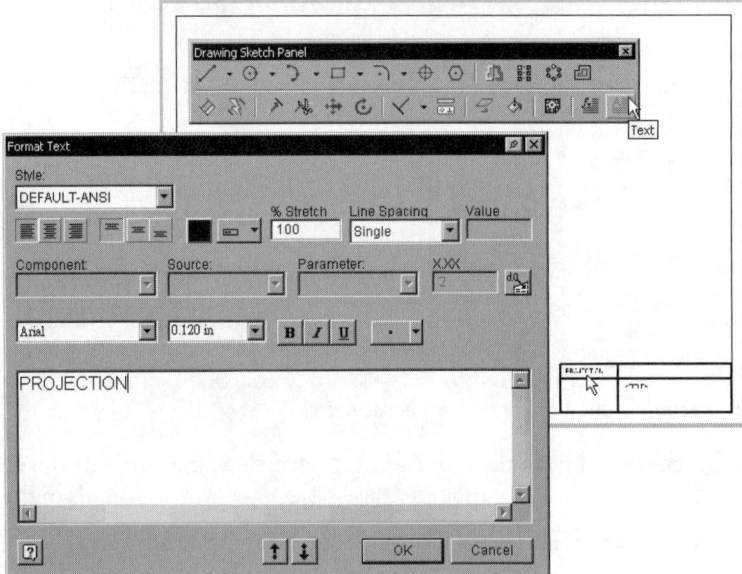

Figure 10.30 *Text being added to the title block*

A description field and a text string are added to the title block. In the Format Field Text dialog box, there are six sets of properties that you can add to the title block. Add other text and properties fields to your title block as may be required. Now save the title block as one of the drawing resources.

18. Select Format > Save Title Block.

19. In the Title Block dialog box, specify a name and click the Save button. (See Figure 10.31.)

 Note: After you save it, the title block you just drew will disappear.

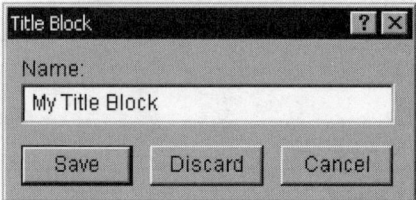

Figure 10.31 *Title block being saved*

The new title block is saved as one of the drawing resources. Now insert the new title block in your drawing sheet.

20. Select the new title block from the drawing resources of the Browser Bar, right-click, and select Insert. (See Figure 10.32.)

The title block is inserted.

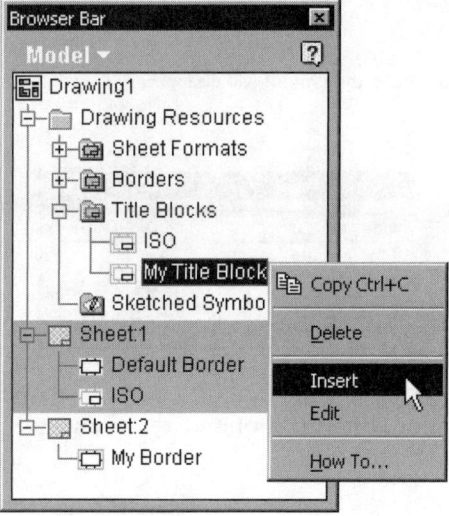

Figure 10.32 *New title block being inserted*

Now construct a sketched symbol.

21. Select Define New Symbol from the Format menu.

22. Use the circle, line, and other drawing tools on the Drawing Sketch Panel Bar or toolbar to construct the symbols shown in Figure 10.33.

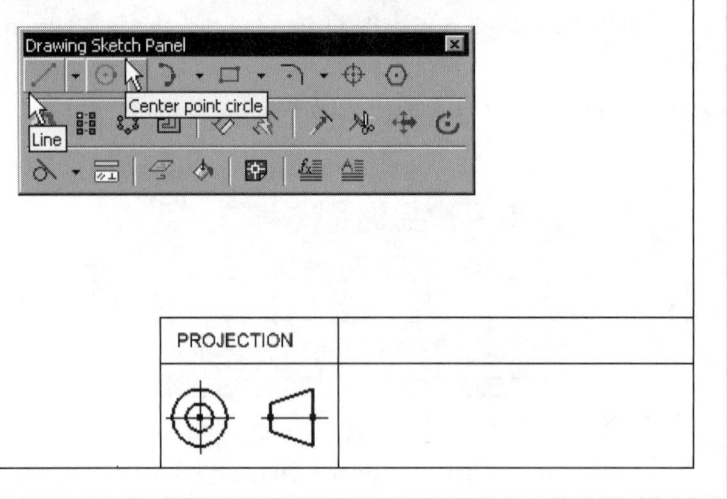

Figure 10.33 *Sketched symbol being constructed*

The symbol is complete. Now save the symbol as a drawing resource.

23. Select Format > Save Sketched Symbol.

24. In the Sketched Symbol dialog box, specify a name and click the Save button. (See Figure 10.34.)

 Note: After you save it, the symbol will disappear.

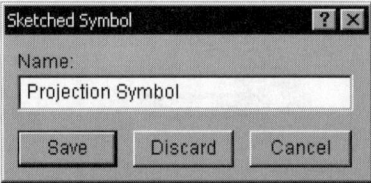

Figure 10.34 *Sketched symbol being saved*

The sketched symbol is saved as one of the drawing resources. Now insert the symbol in your drawing sheet.

25. Select the symbol in the Browser Bar, right-click, and select Insert. (See Figure 10.35.)

26. Select a point on your drawing sheet to indicate a location, right-click, and select Done.

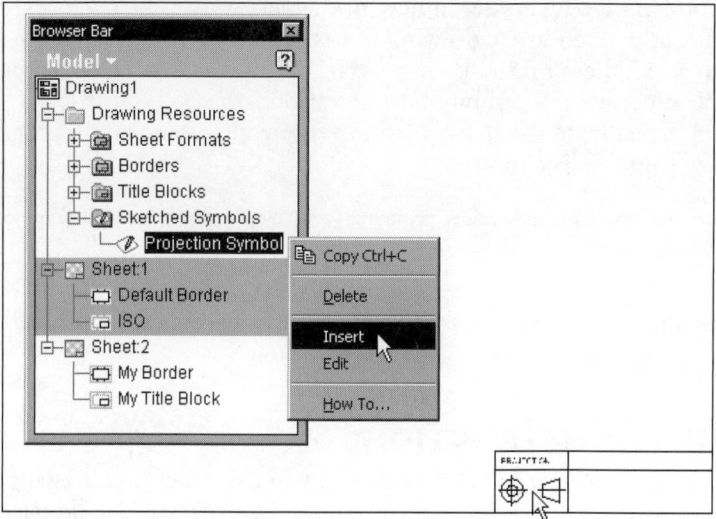

Figure 10.35 *Sketched symbol being inserted*

The sketched symbol is inserted. To use the new drawing sheet together with the inserted border, title block, and sketched symbol, you have to save the drawing sheet as a drawing resource.

27. Move the cursor over the drawing sheet, right-click, and select Create Sheet Format.

28. In the Create Sheet Format dialog box, specify a sheet name (My Sheet) and click the OK button. (See Figure 10.36.) A drawing sheet is saved as one of the drawing resources.

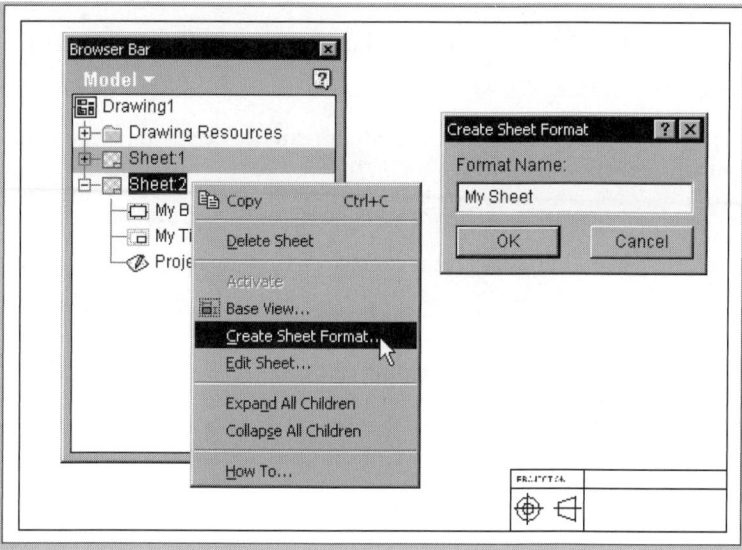

Figure 10.36 *Sheet format being created*

Now you have a new border, a new title block, a new sketched object, and a new drawing sheet saved in the drawing file as drawing resources. To use these objects, you can select them on the Browser Bar, right-click, and select Insert. To make a drawing sheet with a combination of various objects, you insert a new sheet by selecting Sheet from the Insert menu, and select these objects on the Browser Bar, right-click, and select Insert.

 Tip: To make these new drawing resources available to other drawing files, you can save the file as a drawing template.

29. Save the file in the Inventor Template directory (file name: *newsheet1.idw*).

30. Close your file.

DRAFT VIEW CONSTRUCTION

Draft view consists of simple 2D drawings that you construct in a drawing file. You use draft views to illustrate a sketch or a concept in the absence of 3D solid parts or assembly of solid parts.

To make a draft view, you select Draft View on the Drawing Views Panel Bar or toolbar to initiate a view, and use the sketching tools available on the Drawing Sketch Panel Bar or toolbar. You can refer to the sketching methods outlined in Chapter 3.

Draft views are parametric. In other words, you can change a draft view in exactly the same way as you modify the sketches of sketched solid features.

 TUTORIAL 10.2

In this tutorial, you will construct a sketched drawing.

1. Start a new drawing file. Use the template that you constructed in Tutorial 10.1.

In the drawing, there should be a drawing sheet activated. Now construct a draft view on this sheet.

2. Select Draft View on the Drawing Views Panel Bar or toolbar.

In the Draft View dialog box, you can specify a label of the drawing view that will be displayed in the Browser Bar and the drawing scale of the sketch objects in comparison to the title block.

3. Accept the default label and click the OK button. (See Figure 10.37.)

Figure 10.37 *Draft view selected*

A draft view is started. Now work on the content of the drawing view.

4. Use the tools in the Drawing Sketch Panel Bar or toolbar to construct a 2D sketch in accordance with Figure 10.38.

5. Select Return on the Inventor Standard toolbar.

A 2D parametric sketch is complete. Save and close your file (file name: *Draftview.idw*).

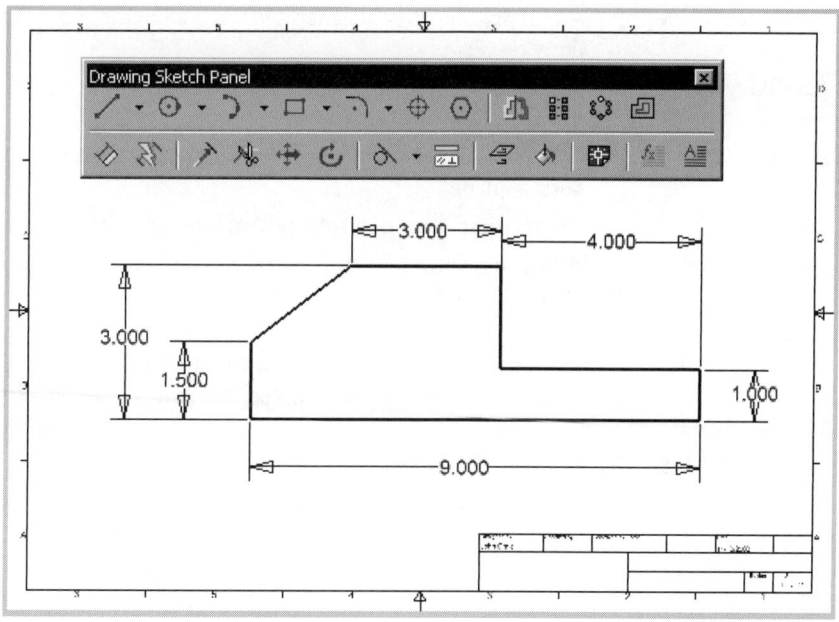

Figure 10.38 *2D sketch drawing constructed*

ASSOCIATIVE DRAWING VIEW CONSTRUCTION

The way to take true advantage of 3D computerized engineering design is to generate engineering drawings from 3D solid parts, assemblies of those parts, and presentations of those assemblies in a semi-automatic way.

To construct 2D engineering drawings from these objects, you do not have to think about how they will look when viewed from a certain direction and construct the drawing views in accordance with your perception of the objects. Instead, you need only to select the object (solid part, assembly, or presentation), specify the parameters of the engineering drawing (direction of viewing and scale of the engineering drawing in relation to the actual size of the objects, etc.), and select a location from the drawing sheet. The computer generates the drawing views accurately for you. All you need to do after generating the drawing is add annotations, which might be a tedious job.

DRAWING VIEWS

In a drawing sheet, you can generate eight kinds of drawing views. They are explained below.

Base view	The first drawing view of a set of orthographic projection views
Perspective view	A special kind of base view in which the drawing view is customized to display a perspective view.
Projected view	An orthographic view generated from an existing orthographic drawing view
Auxiliary view	A view projected at an angle from an existing orthographic drawing view
Section view	A kind of projected drawing view depicting a section across a cutting plane
Detail view	An enlarged view of a selected portion of a drawing view
Broken view	A drawing view with the central portion removed and the remaining portion put close together so that the drawing view occupies less space in the drawing sheet
Break Out view	A special kind of section view with only a portion of the drawing view sectioned to reveal the internal details of a view.

ASSOCIATIVITY AND DRAWING UPDATE

Engineering drawing views generated from a solid part, assembly of solid parts, and a presentation of assembly are associative to the source objects. Any change you make to the source objects will be reflected automatically. For example, if you construct a hole on a solid part after you make a drawing from it, the next time you open the drawing file, the hole will be exhibited in the drawing views automatically. You do not have to regenerate the drawing views.

However, if you prefer not to let the computer update the changes automatically, you can turn off automatic update selecting Defer Updates in the Drawing Document Settings dialog box (Figure 10.39) accessible from the Tools menu.

Turning off automatic update might be necessary when you are in the middle of modifying a set of solid parts, assemblies, and presentations and have not finalized the design yet. However, it must be noted that a drawing with Defer Updates selected will have the drawing view construction commands disabled. You need to clear the Defer Updates check box to re-activate automatic update before you can work on the drawing views again.

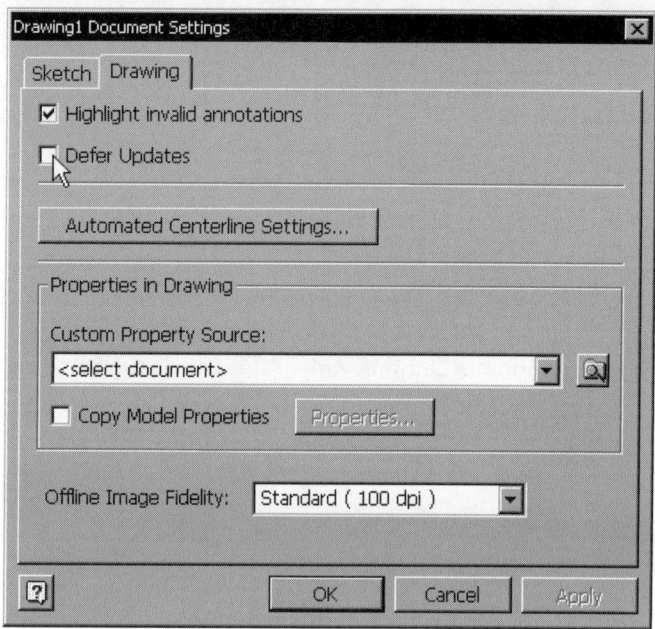

Figure 10.39 *Drawing tab of the Drawing Document Settings dialog box*

DRAWING VIEWS FOR A SOLID PART

To depict the details of a 3D component on a 2D engineering drawing, you construct an associative engineering drawing. You start a new drawing file and construct various kinds of drawing views, including base view, projected views, auxiliary views, section views, detail views, broken views, break out views, and perspective views.

Now you will learn how to construct these drawing views. Later in this chapter, you will construct annotations to detail the solid part.

Base View

We call the first drawing view of a set of orthographic views the base view. To make a base view, you select a file (part, assembly, or presentation) to generate a drawing view from the 3D objects in the file. After selecting a file, you select a viewing direction, display type, display scale, and location of the drawing view.

In a drawing, you can construct as many base views as you want and select different files in each view. However, you should refrain from incorporating base views of more than one 3D object in a single drawing file, because it is common engineering practice to have a drawing for each individual object. You should also note that individual base views in a drawing are unrelated to each other in terms of compliance with the orthographic projection system. If you want to construct a set of orthographic views, you should construct a number of projected views from a base view.

In a base view, you can project one of the standard drawing views or select your own viewing direction by selecting the Change view orientation button in the Drawing View dialog box.

TUTORIAL 10.3

In this tutorial, you will construct a base view from one of the eleven standard views.

1. Start a new drawing file. Use the metric *Standard.idw* (ISO) template.
2. Select Format > Standards.
3. Select the Third Angle Projection button on the Common tab of the Drafting Standards dialog box and click the OK button.
4. Select Base View on the Drawing Views Panel Bar or toolbar.
5. In the Drawing View dialog box, select the Explore Directories button. (See Figure 10.40.)
6. Select the part file *PunchPost.ipt* in the Open dialog box and click the OK button.

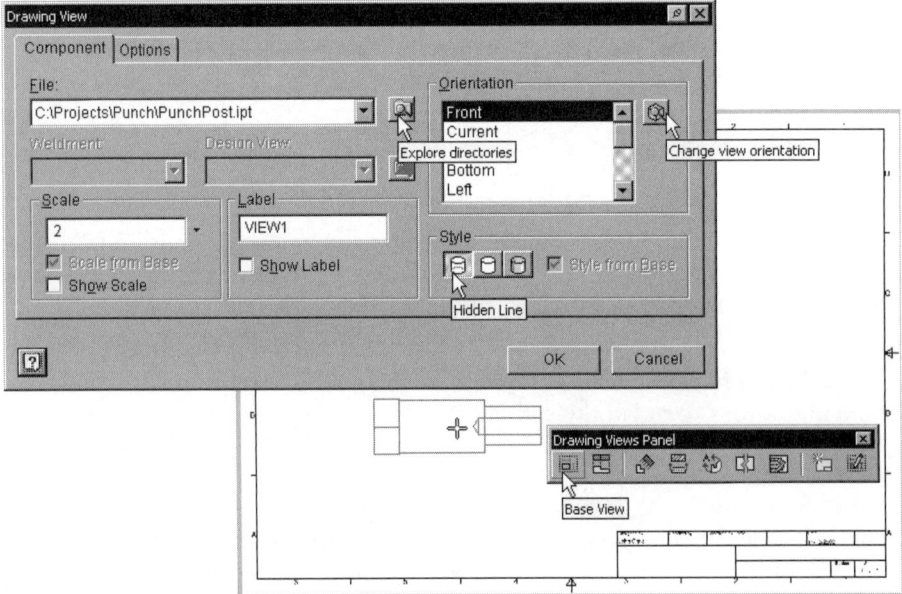

Figure 10.40 *Base view being constructed*

In constructing a drawing view, you can select one of the drawing views listed in the Create View dialog box or select the Change view orientation button and select a custom view. (See Figure 10.40.) Select Exit Custom View when finished.

7. In the Drawing View dialog box, select Front in the Orientation list. Under Style, select Hidden Line.

8. Under Scale, change the value to 2.

9. Click on the sheet to select the location for the base view. (See Figure 10.41)

 Note: You can change the location later if necessary.

A base view of the selected solid part is constructed. Save your file (file name: *PunchPost.idw*).

 Tip: You can use any custom view orientation to construct a base view, including isometric directions.

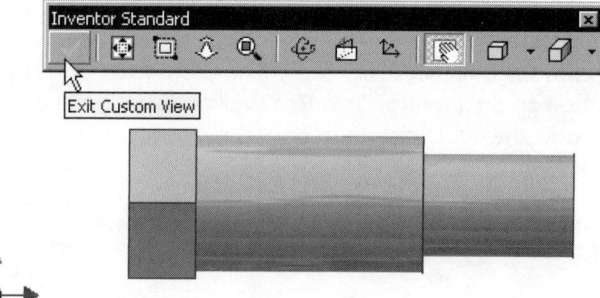

Figure 10.41 *Custom view constructed*

Perspective View

For the sake of illustrating the 3D object in a pictorial form, you project a perspective view in the drawing. To make a perspective view, you construct a base view and change the view orientation.

 TUTORIAL 10.4

In this tutorial, you will construct a perspective view.

1. Open the file *PunchPost.idw*, if you already closed it.

2. Select Base View on the Drawing Views Panel Bar or toolbar.

3. Select the Explore Directories button and select the file *PunchPost.idw*.

4. Set the drawing scale to 2 and select shaded style.

5. Select the Change view orientation button.

6. In the Custom View window, select Perspective Camera on the Inventor Standard toolbar.

7. In the Custom View window, rotate, zoom, and pan the view as necessary until it looks similar to Figure 10.42.

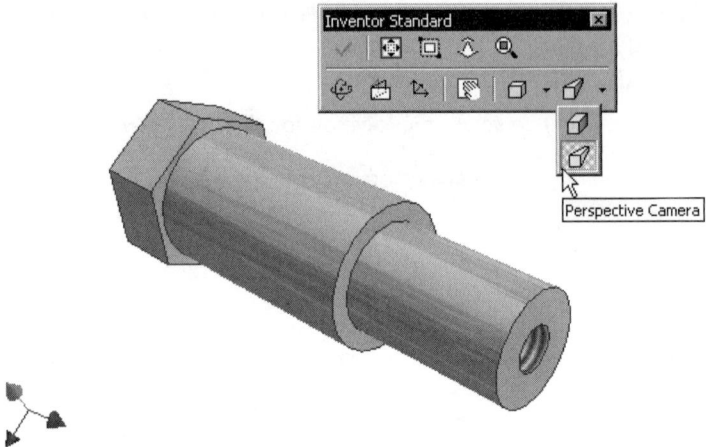

Figure 10.42 *Perspective view being customized*

8. From the Inventor menu, select File > Exit Custom, or click the Exit Custom View button on the Inventor Standard toolbar. Click in the drawing to position the view near the upper-right corner of the drawing sheet.

A perspective view is constructed. (See Figure 10.43.) Save your file.

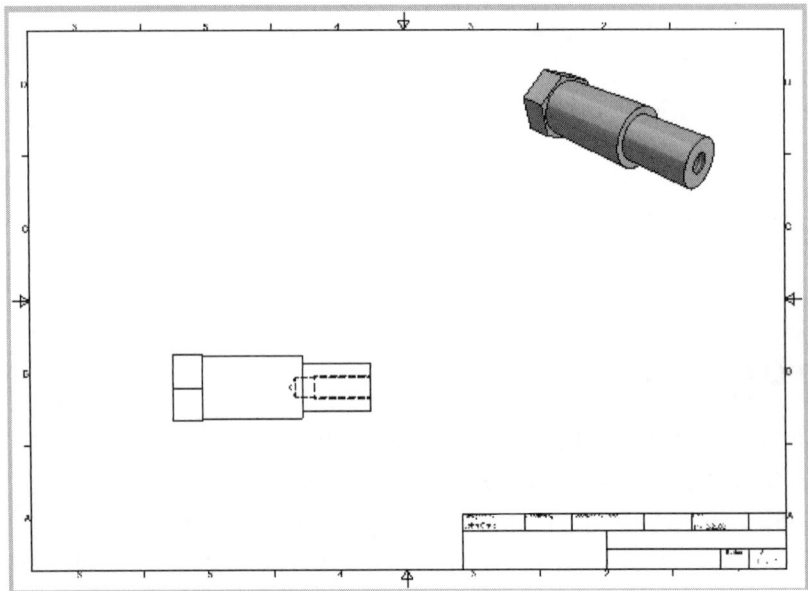

Figure 10.43 *Perspective view constructed*

Projected Views

To construct a set of orthographic drawing views, you construct projected views from any existing drawing views. You select a view to project and specify the locations. Projected views include the six basic orthographic views and isometric views.

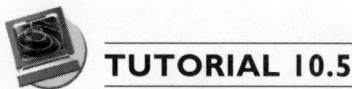

TUTORIAL 10.5

In this tutorial, you will construct a set of projected views from a base view. The projected views include a top view, a side view, and an isometric view.

1. Open the drawing file *PunchPost.idw*, if you already closed it.
2. Select Projected View on the Drawing Views Panel Bar or toolbar.
3. Select point A in Figure 10.44 to place the base view.
4. Select point B in Figure 10.44 to specify a location for the top view.
5. Select point C in Figure 10.44 to specify a location for the isometric view.
6. Select point D in Figure 10.44 to specify a location for the side view.
7. Right-click and select Create.

A set of projected views is constructed. Save your file.

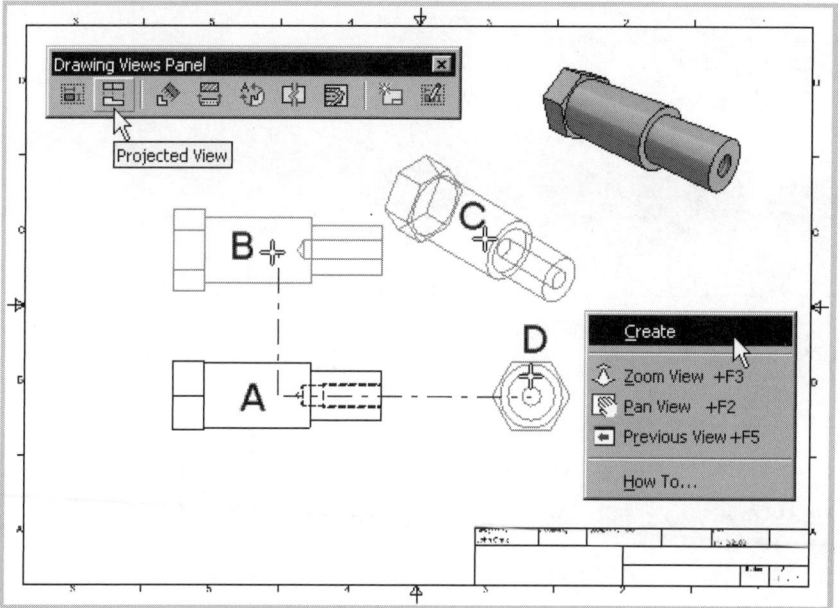

Figure 10.44 *Base view constructed*

 Tip: Now you have two base views and three projected views in the drawing sheet. If you save this file as a template and use the template in a new drawing file, you will have these views generated once you specify a 3D part or an assembly.

Auxiliary View

If you want to construct a drawing view that is projected in a direction other that orthogonal or isometric, you construct an auxiliary view and specify a custom direction.

TUTORIAL 10.6

In this tutorial, you will construct an engineering drawing with a base view and an auxiliary view.

1. Start a new drawing file. Use the metric *Standard.idw* (ISO) template.
2. Set the projection system to third angle projection.
3. Select Base View on the Drawing Views Panel Bar or toolbar.
4. In the Drawing View dialog box, select the part file *PunchHandle.ipt*, select Hidden Line style, set the scale to 0.25, and select location A in Figure 10.45.

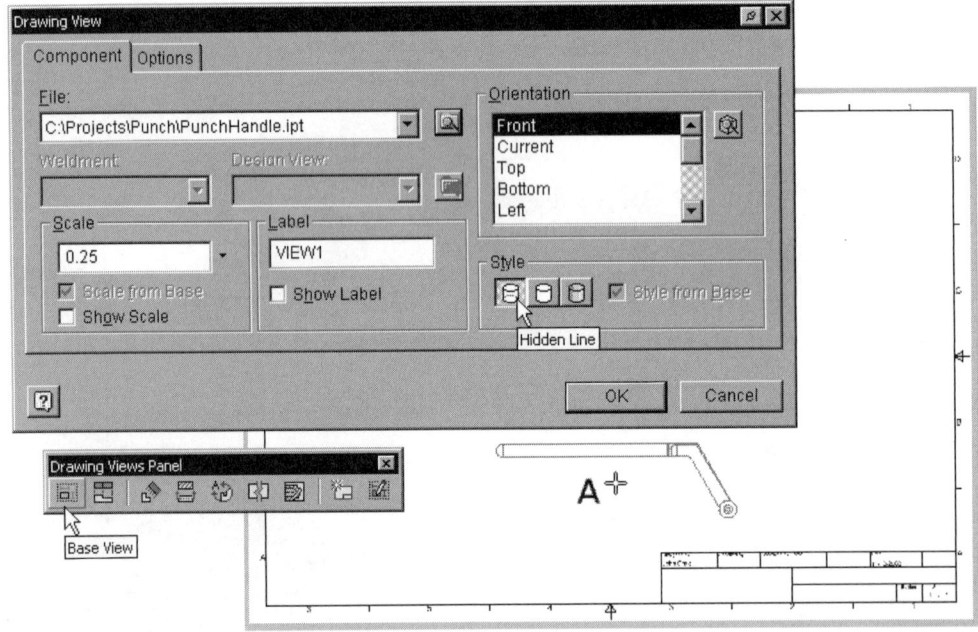

Figure 10.45 *Base view of the handle being constructed*

5. Select Auxiliary View on the Drawing Views Panel Bar or toolbar.
6. Select view A in Figure 10.46 to select a drawing view.
7. In the Auxiliary View dialog box, under Label select the Visible check box. Under Scale, clear the Visible check box.
8. Select edge B in Figure 10.46 to specify a direction.
9. Select location C in Figure 10.46 to specify the location of the auxiliary view.

A base view and an auxiliary view are constructed. Save your file (file name: *PunchHandle.ipt*).

Tip: You need to select an edge to depict the auxiliary view direction.

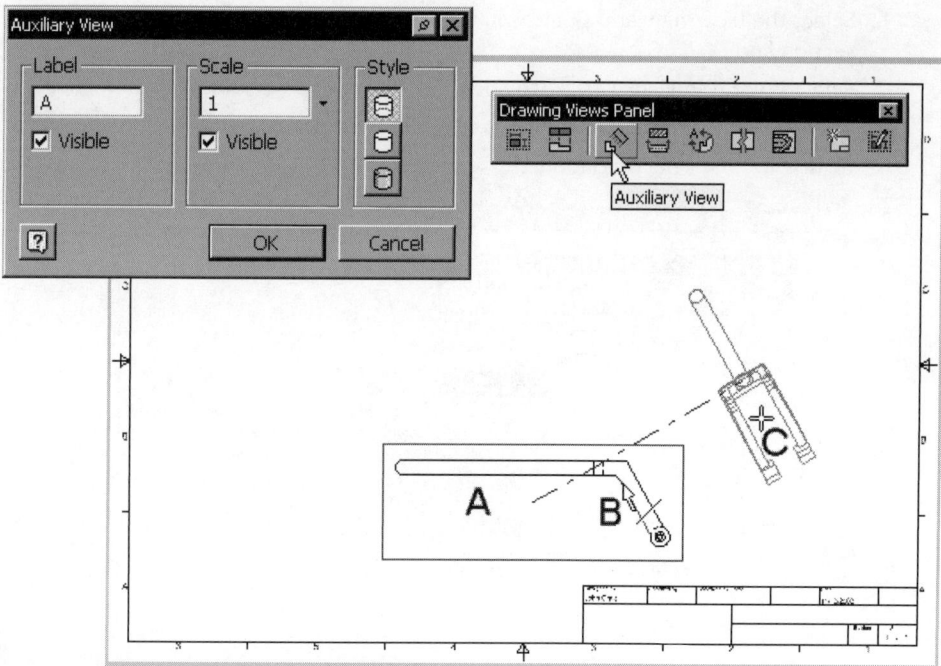

Figure 10.46 *Auxiliary view being constructed*

Section View

A section view depicts how a part would appear if cut by a cutting plane. You select a view to locate a cutting plane and specify a location of the section view. Because you will place hatching lines in the section view, you need to set the hatching style prior to making a section view.

Tip: You can set the hatching style by accessing the Hatch tab of the Drafting Standards dialog box.

TUTORIAL 10.7

In this tutorial, you will construct two section views in an engineering drawing.

1. Open the drawing file *PunchHandle.idw*, if you already closed it.
2. Select Standards from the Format menu. In the expanded Drafting Standards dialog box, select the Hatch tab.
3. Select a hatch pattern type and set the line weight, hatch pattern angle, and scale.
4. Click OK to have changes applied.
5. Select Section View on the Drawing Views Panel Bar or toolbar.

6. Select the base view and point A in Figure 10.47 to specify the start point of the section line.

7. Select point B in Figure 10.47 to specify the second point of the section line.

8. Select point C inFigure 10.47 to specify the third point of the section line.

9. Right-click and select Continue.

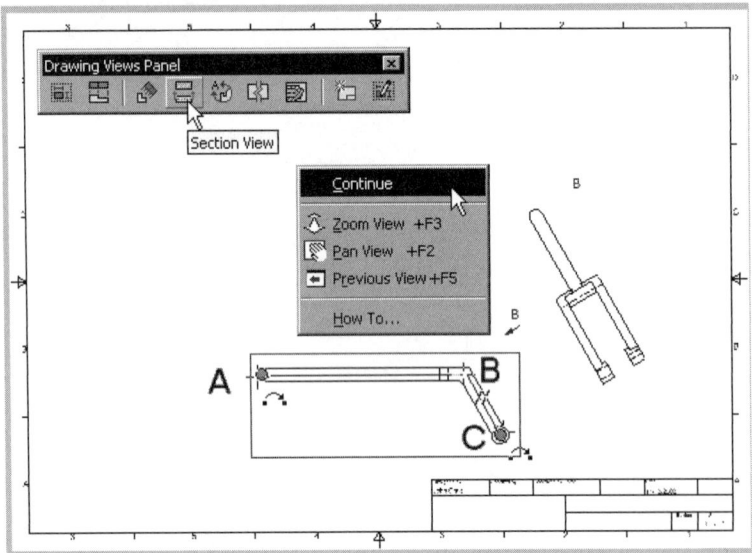

Figure 10.47 *Section line being defined*

10. Select point A in Figure 10.48 to specify the location of the section view.

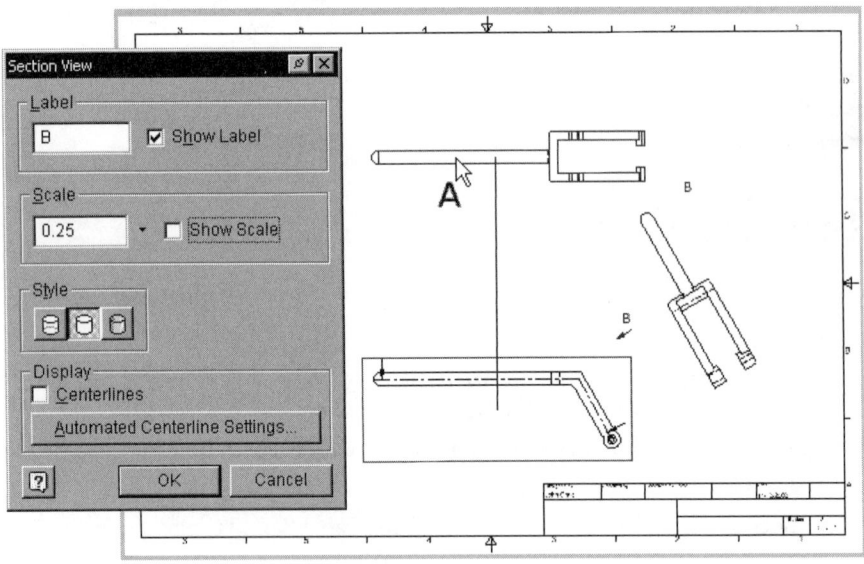

Figure 10.48 *Location of section view specified*

An aligned section is constructed.

Tip: By setting the location of the section plane, you can construct full, half, offset, and aligned sections.

Now construct a full section view.

11. Select Section View on the Drawing Views Panel Bar or toolbar.
12. Select the base view A in Figure 10.49.
13. Select locations B and C in Figure 10.49 to specify the section line.
14. Right-click and select Continue.
15. Select location D in Figure 10.49 to specify the location of the section view.

A full section view is constructed. Now save your file.

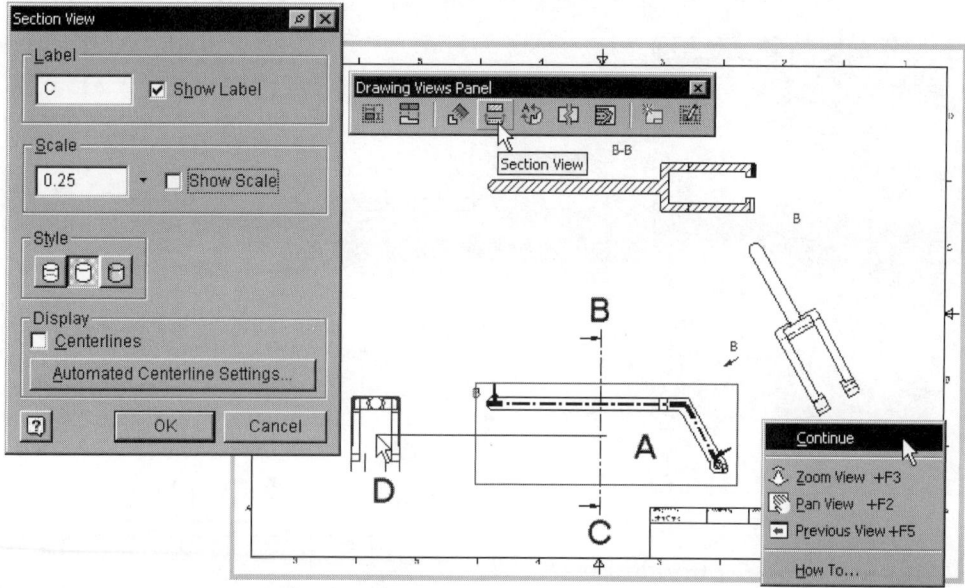

Figure 10.49 *Section view being constructed*

Detail View

A detail view shows an enlarged portion of a drawing view. By using a detail view, you maintain a drawing scale for all the views of the drawing and provide a closer look at a portion of the component. To make a detail view, you specify an enlargement scale, define a small portion of a view by describing a circle, and specify a location.

 **TUTORIAL 10.8**

In this tutorial, you will construct a detail view.

1. Open the drawing file *PunchHandle.idw*, if you already closed it.
2. Select Detail View on the Drawing Views Panel Bar or toolbar.
3. Select the base view W in Figure 10.50.
4. In the Detail View dialog box, specify the name of the label and set the scale of the detail view to be 0.5.
5. Select X in Figure 10.50 to specify the center of a circular zone of the detail view.
6. Select Y in Figure 10.50 to specify the radius of the circular zone.
7. Select Z in Figure 10.50 to specify the location of the detail view.

A detail view is constructed. Save your file.

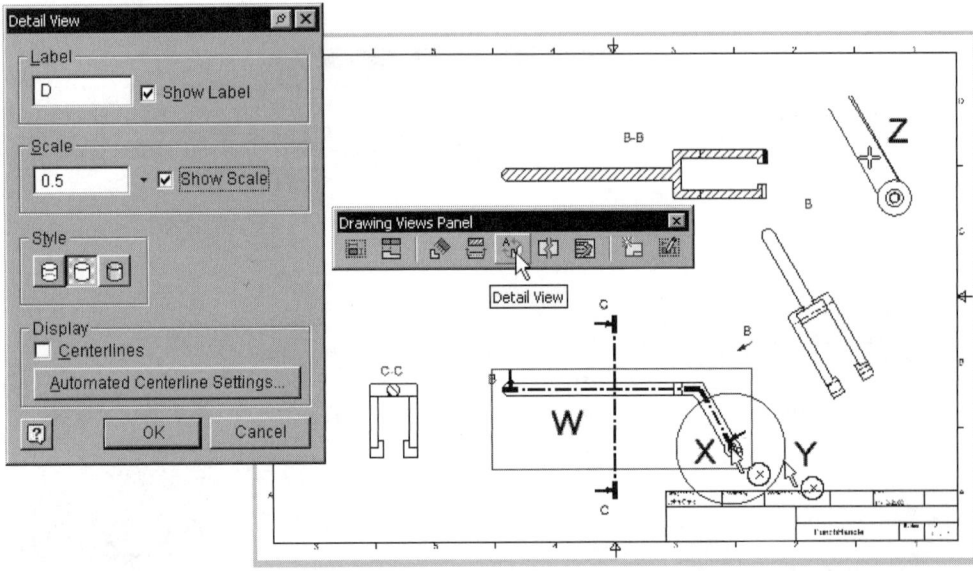

Figure 10.50 *Detail view being constructed*

Broken View

When we consider the scale of engineering drawing views, a dilemma exists if the component is a long and thin object. It is normal engineering practice to remove the central portion of the drawing view of a very long object and move the remaining portions of the drawing view toward each other to minimize the space requirement of the drawing view. This is called a broken view.

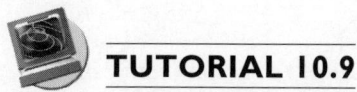

TUTORIAL 10.9

In this tutorial, you will construct a broken view.

1. Open the drawing file *PunchHandle.idw*, if you already closed it.
2. Select Broken View on the Drawing Views Panel Bar or toolbar.
3. Select the base view W in Figure 10.51.
4. Select X and Y in Figure 10.51 to specify the break lines.

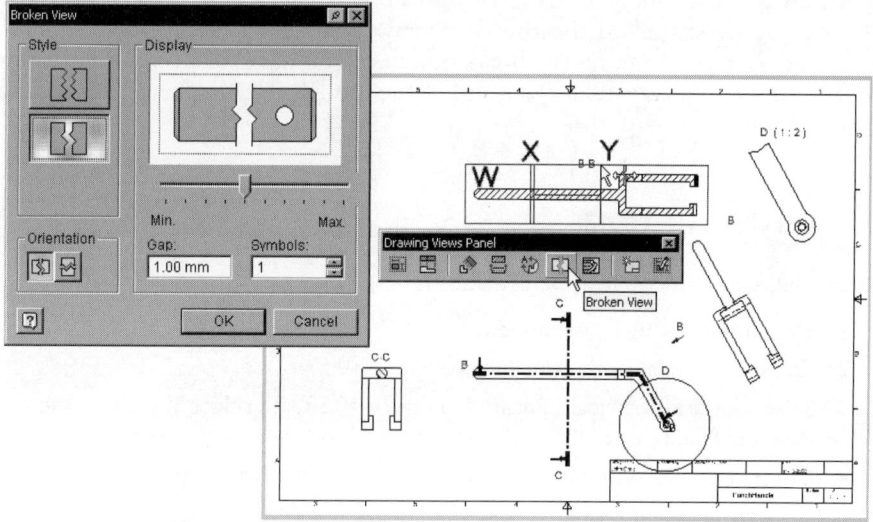

Figure 10.51 *Broken view being constructed*

A broken view is constructed. (See Figure 10.52.) Save and close your file.

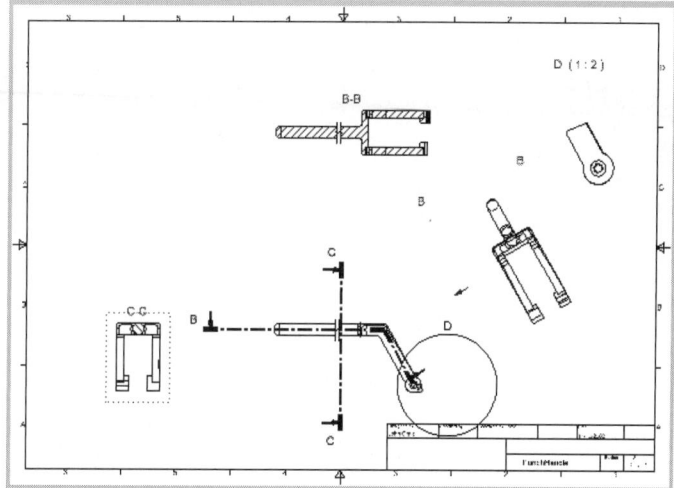

Figure 10.52 *Broken view constructed*

Break Out View

The disadvantage of a section view is that the external details of the component cannot be displayed in that particular viewing direction. As a result, you might need to add an additional drawing view. If the internal detail of the component that you want to display occupies only a small portion of the drawing view, you might consider using a break out view.

In essence, a break out view is an external view with a small portion sectioned to reveal the internal detail.

A break out view is constructed on an existing drawing view. In essence, constructing a break out view is converting a drawing view to a break out view. It involves the construction of a sketch associated to the drawing view. The sketch depicts the region to be broken. In the course of making the break out view, you have to specify the depth of the break out in one of four ways: select a point in the model, select a projected view, select a hole, or select a part.

 TUTORIAL 10.10

In this tutorial, you will construct a break out view.

1. Open the drawing *PunchPost.idw*.
2. Zoom the display with reference to Figure 10.53.
3. Select the drawing view indicated in Figure 10.53 and select Sketch on the Inventor Standard toolbar to initiate a sketch on the selected view.
4. Construct a closed-loop sketch in accordance with Figure 10.53.

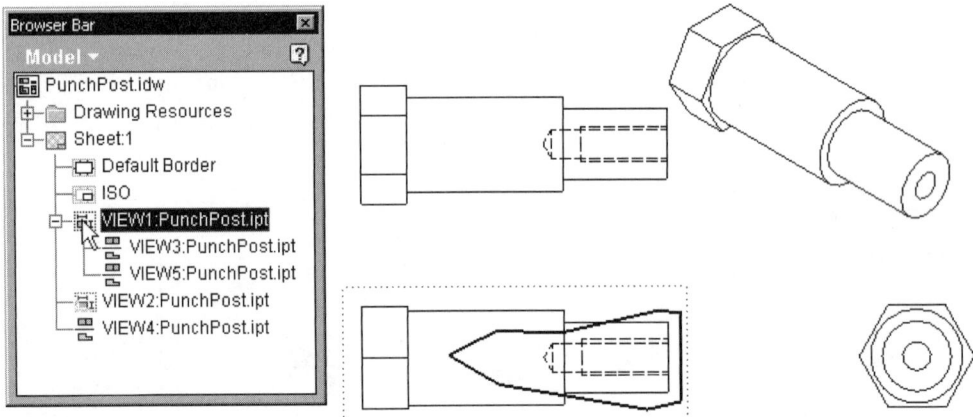

Figure 10.53 *Sketch associated to a drawing view being constructed*

5. Select Return on the Inventor Standard toolbar.

On the Browser Bar, your sketch should appear below the selected drawing view. (See Figure 10.54.) If not, your sketch is not associated with the drawing view. Delete it and repeat steps 3 and 4 again.

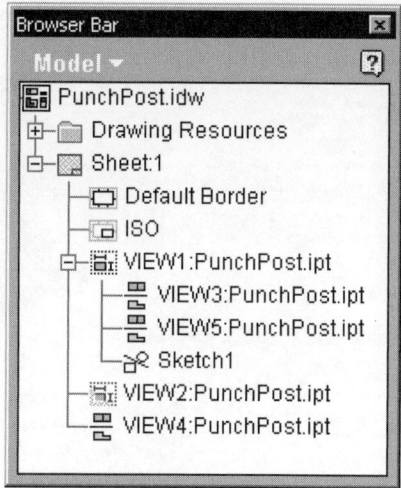

Figure 10.54 *Associative sketch constructed*

Now convert the view to a break out view.

6. Select Break Out View on the Drawing Views Panel Bar or toolbar.
7. Select drawing view A in Figure 10.55.
8. In the Break Out View dialog box, select To Hole.
9. Select hole feature B in Figure 10.55.
10. Click the OK button.

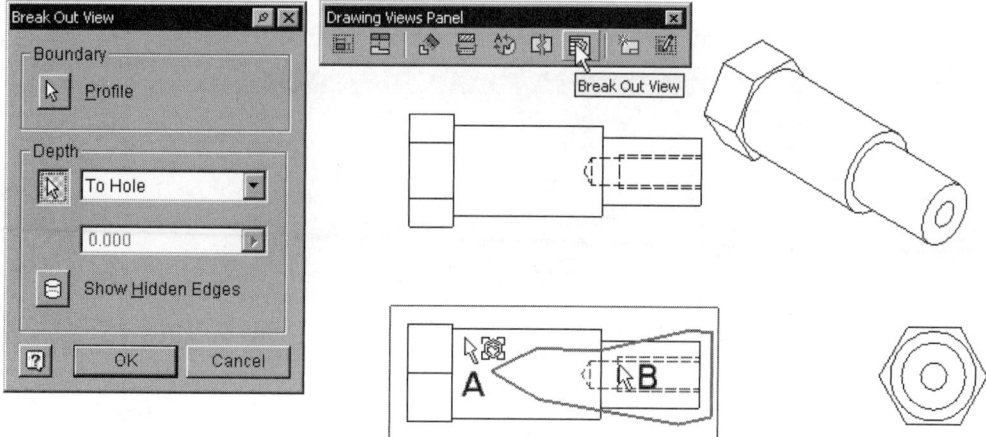

Figure 10.55 *Break out view being constructed*

A drawing view is converted to a break out view. (See Figure 10.56.)

Because a break out view is defined by a sketch associated to the drawing view, you have to edit the sketch if you want to modify the break out region. To edit the sketch, you select it, right-click, and select Edit. To reference the sketch to the drawing geometry, you can select Project on the Drawing Sketch Panel Bar or toolbar to project drawing view geometry as references in dimensioning the sketch. (See Figure 10.57.)

Save your file.

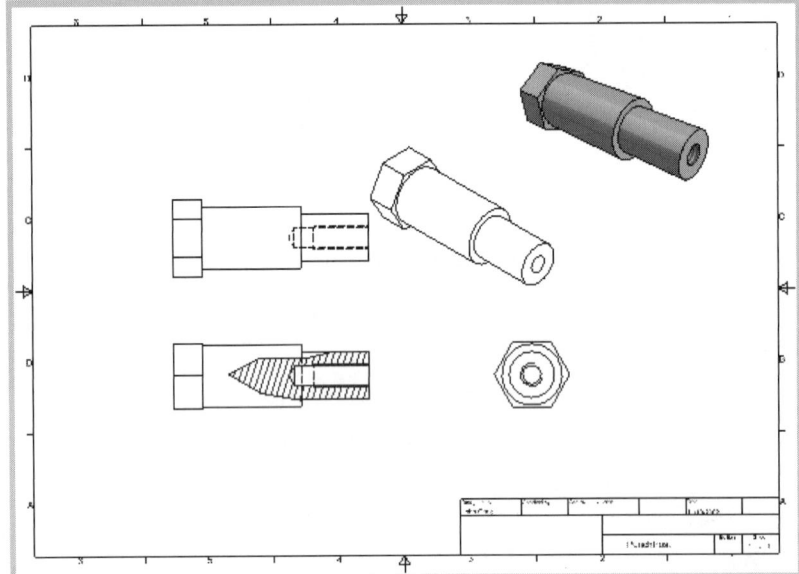

Figure 10.56 *Break out view constructed*

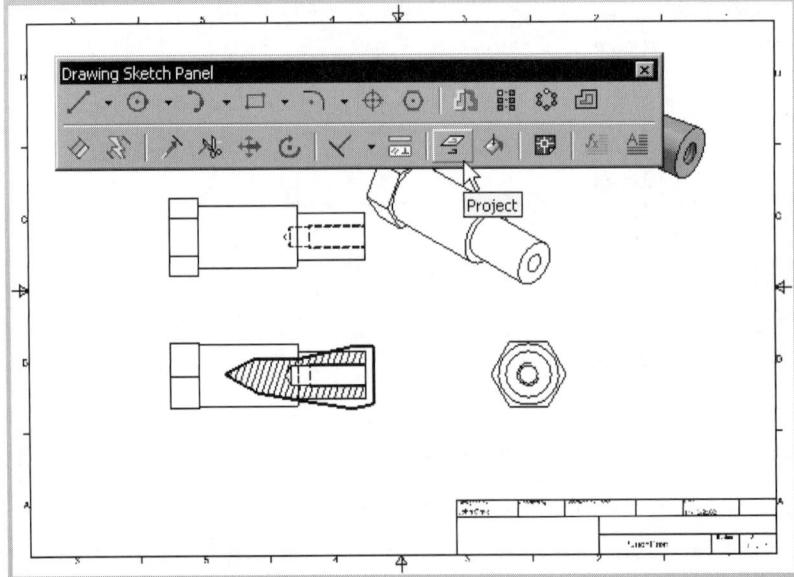

Figure 10.57 *Editing associative sketch*

Editing a Drawing View

You can edit the parameters of a drawing view by selecting it from the graphics area or the Browser Bar, right-clicking, and selecting Edit View. The dialog box for editing a drawing view is virtually the same as that for making a view. Consequently, you can modify any parameters that you entered when making the view in the first place.

TUTORIAL 10.11

In this tutorial, you will modify a drawing view.

1. Open the drawing *PunchPost.idw*, if you already closed it.
2. Select the base view, right-click, and select Edit View. (See Figure 10.58.)
3. In the Drawing View dialog box, select Hidden Line Removed button in the Style area.
4. Select the Options tab and clear the Tangent Edges button.
5. Click the OK button.

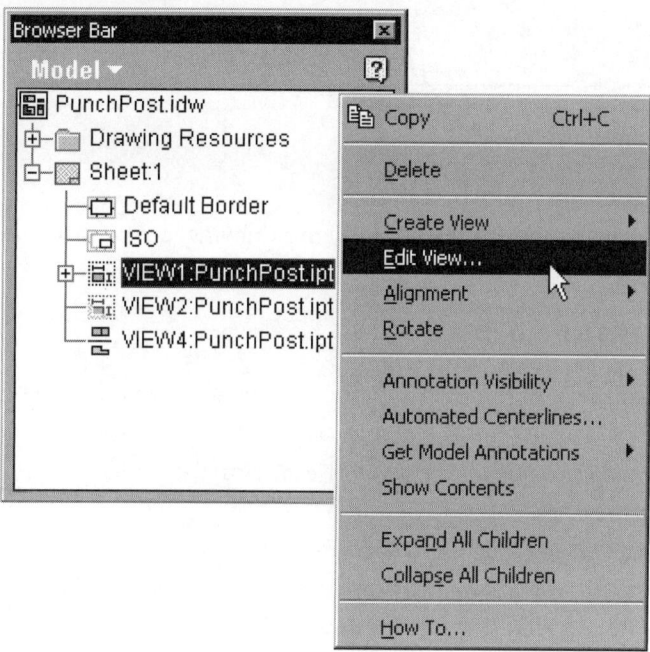

Figure 10.58 *Drawing view selected on the Browser Bar*

The selected drawing view is modified. (See Figure 10.59.) Save your file.

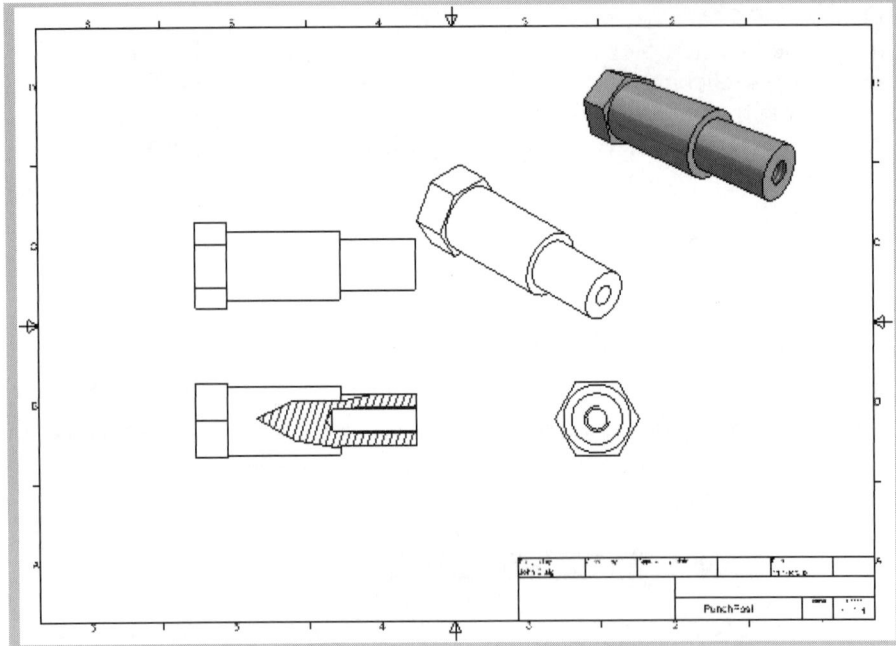

Figure 10.59 *Drawing view modified*

Drawing View Rotation and Alignment

You can rotate a drawing view, and you can align a drawing view with another drawing view.

 TUTORIAL 10.12

In this tutorial, you will learn how to rotate a drawing view and align one drawing view with another.

1. Open the file *PunchPost.idw*, if you already closed it.
2. Construct a new sheet.
3. With reference to Figure 10.60, construct a base view with a drawing scale of 2.
4. Select the drawing view, right-click, and select Rotate.
5. In the Rotate View dialog box, select Angle, set the angle to 45 degrees, and click the OK button.

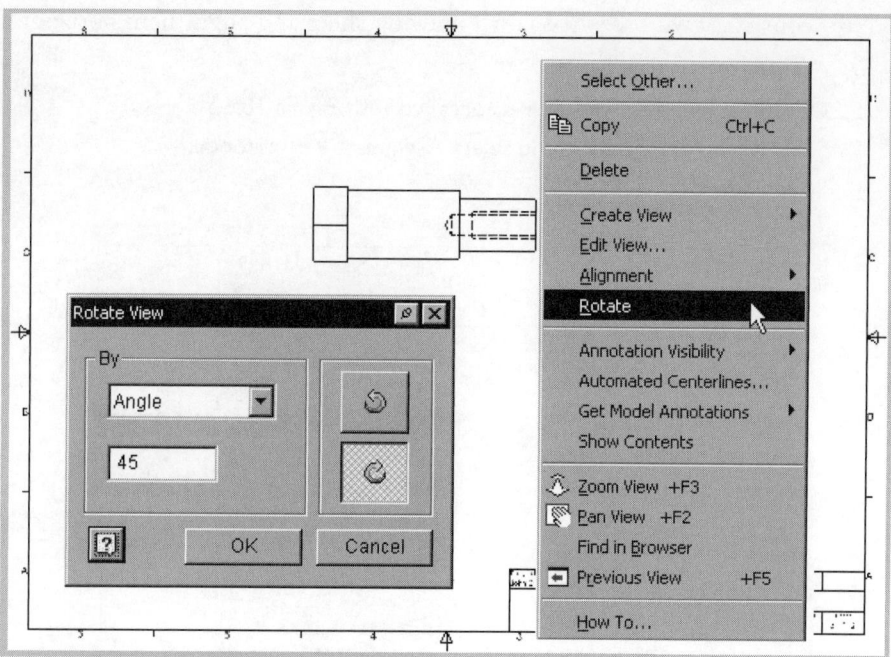

Figure 10.60 *Drawing view being rotated*

6. With reference to Figure 10.61, construct a projected view.

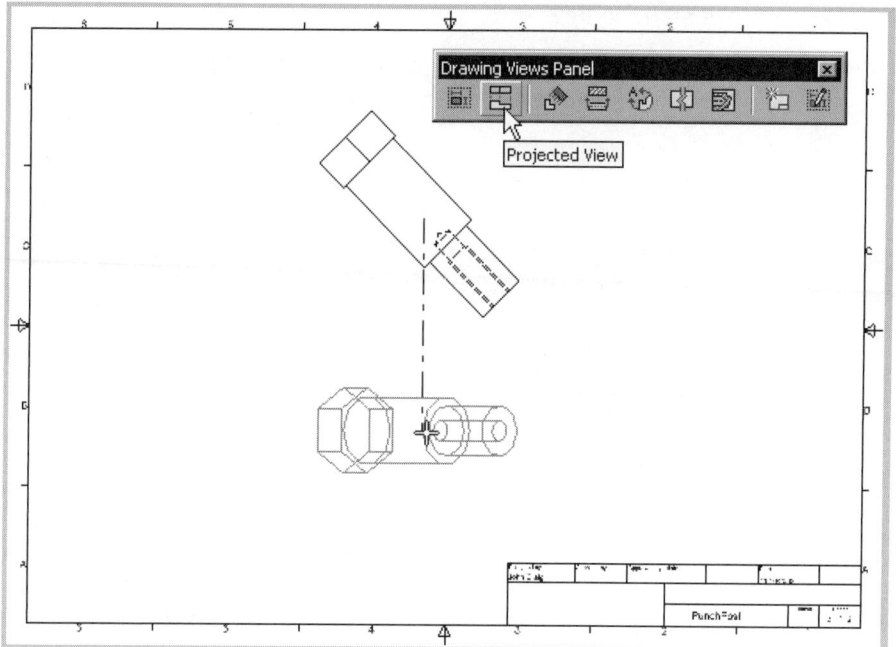

Figure 10.61 *Drawing view rotated and a projected view being constructed*

Now construct two base views on a drawing sheet and align them horizontally.

7. Construct a new sheet.
8. Construct two base views in accordance with Figure 10.62.
9. Select view A, right-click, and select Alignment > Horizontal.
10. Select view B.

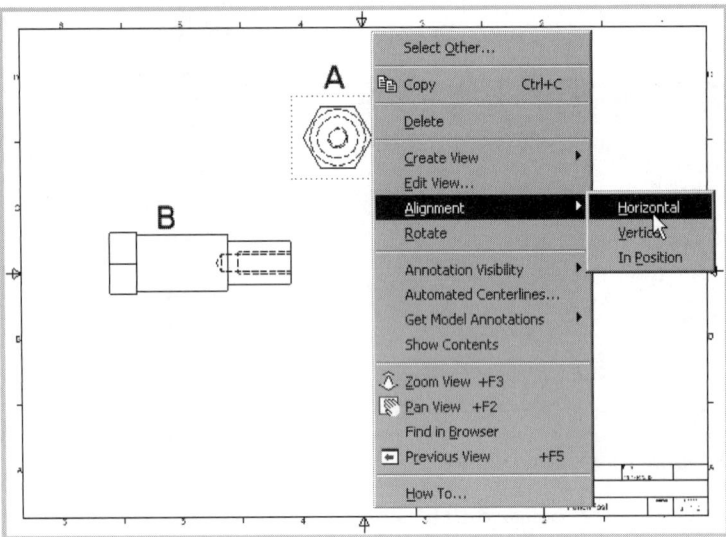

Figure 10.62 *Two base views constructed*

The views are aligned. (See Figure 10.63.) Save and close your file.

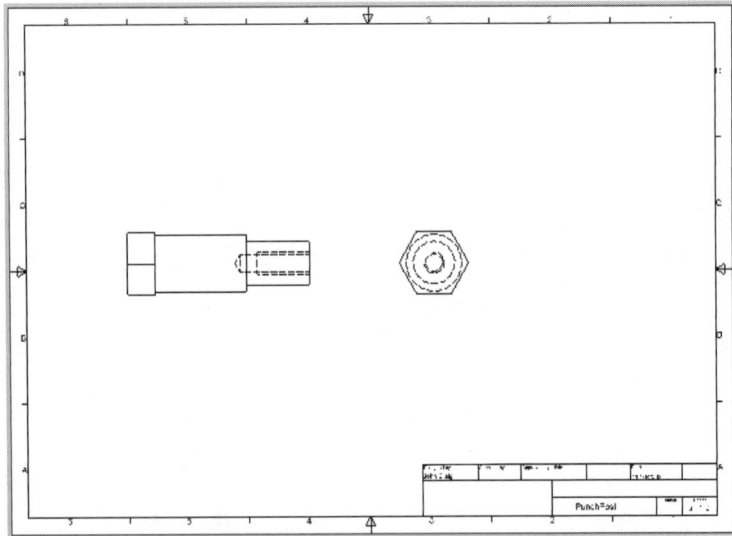

Figure 10.63 *Views aligned horizontally*

DRAWING VIEWS FOR A SHEET METAL COMPONENT

In an associative engineering drawing for a sheet metal component, you can, in addition to the drawing views that you construct for the folded sheet metal part, construct drawing views to show the flat pattern of the sheet metal part. In the drawing views depicting the flat patterns, you can suppress bend and zone lines.

TUTORIAL 10.13

In this tutorial, you will construct an engineering drawing for a sheet metal part.

1. Start a new drawing file. Use the metric *Standard.idw* (ISO) template.
2. Select Base View on the Drawing Views Panel Bar or toolbar.
3. Select the sheet metal file *SheetMetalA.ipt* that you constructed in Chapter 5.
4. In the Drawing View dialog box, select Flat Pattern in the Sheet Metal View pull-down list box. (See Figure 10.64.)
5. Select a location in the drawing sheet to specify a location of the drawing view.

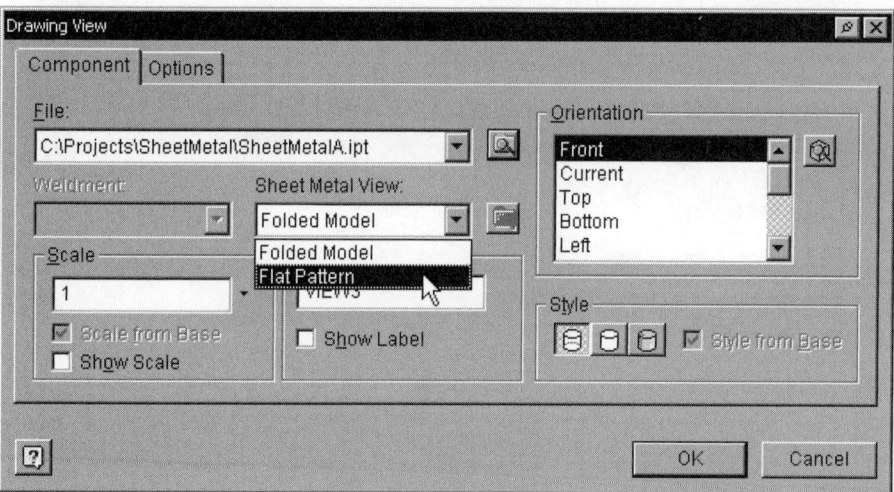

Figure 10.64 *Flat pattern drawing view being constructed*

A flat pattern drawing view is constructed. (See Figure 10.65.) Now display the bend extents.

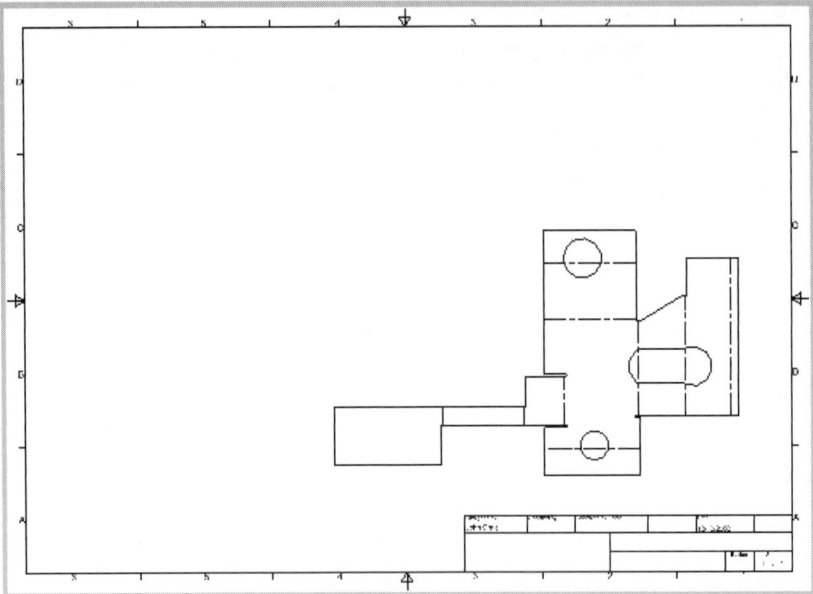

Figure 10.65 *Flat pattern drawing view constructed*

6. Select the flat pattern view, right-click, and select Edit View.

7. In the Drawing View dialog box, select the Options tab, under Display select the Bend Extents check box, and then click OK. The bend extents are displayed. (See Figure 10.66.)

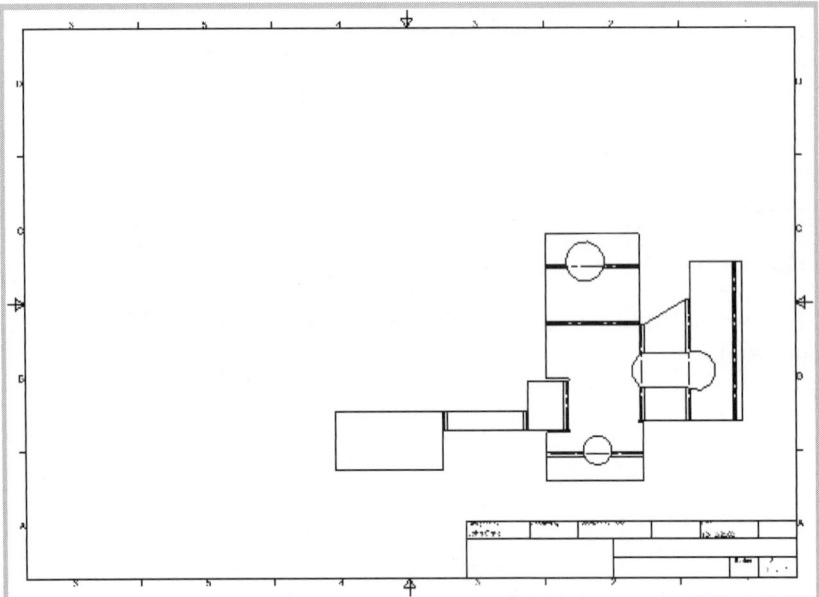

Figure 10.66 *Bend extents displayed in the flat pattern drawing view*

Now construct a base view and projected views of the folded model.

8. Select View on the Drawing Views Panel Bar or toolbar.

9. In the Drawing View dialog box, select the *SheetMetalA.ipt* file again. Select Folded Model in the Sheet Metal View pull-down list box. Select the Bottom orientation. Change the scale to 0.5. Select the Shaded style.

10. Click in the drawing to place the base view.

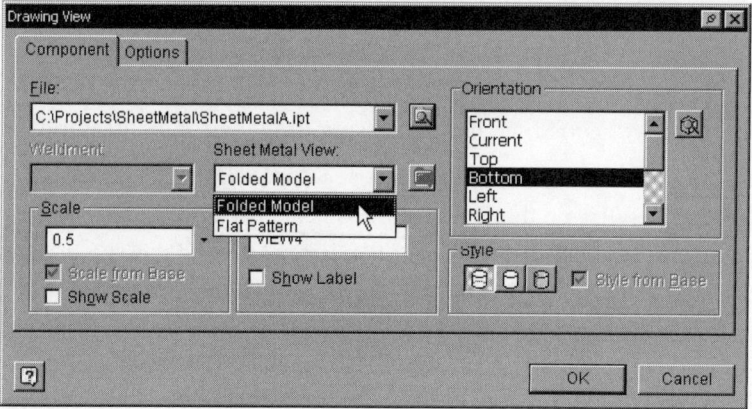

Figure 10.67 *Base view of the folded model being constructed*

11. With reference to Figure 10-68, construct three more projected views. Then right-click and choose Create.

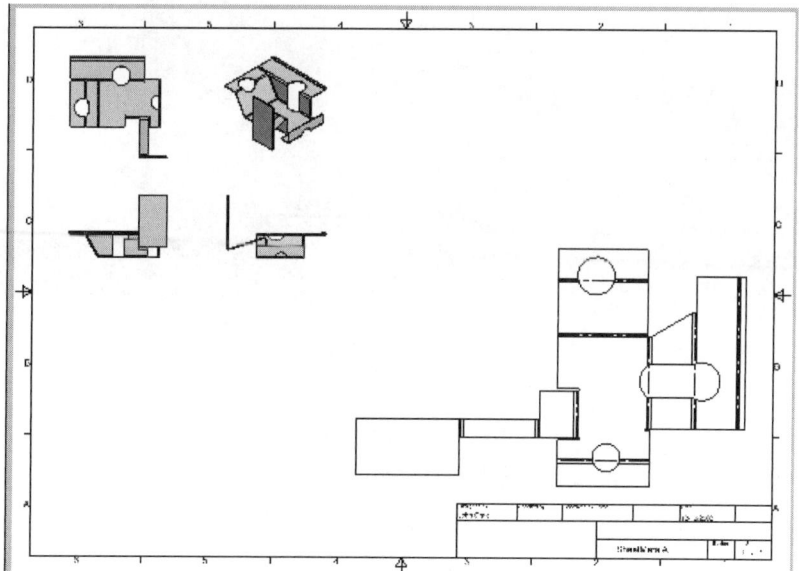

Figure 10.68 *Projected views of the folded model constructed*

The drawing is complete. Save and close your file (file name: *SheetMetalA.idw*).

DRAWING VIEWS OF AN ASSEMBLY

Constructing an engineering drawing for an assembly is similar to making a drawing for a solid part. You select or construct drawing sheets and construct engineering drawing views on the drawing sheets. You construct base, projected, auxiliary, section, detail, and broken views. In addition, you construct a bill of materials to tabulate the component parts in the assembly, a set of balloons to identify the component parts, and include presentation exploded views to illustrate the ways the components are put together.

TUTORIAL 10.14

In this tutorial, you will construct an engineering drawing for an assembly.

1. Start a new drawing file. Use the metric *Standard.idw* (ISO) template.
2. Select Base View on the Drawing Views Panel Bar or toolbar.
3. Select the assembly *Pliers.iam*. Specify a scale of 0.5 and the Top orientation.
4. Select A in Figure 10.69 to specify the location of the drawing view.

The base view of an assembly is constructed.

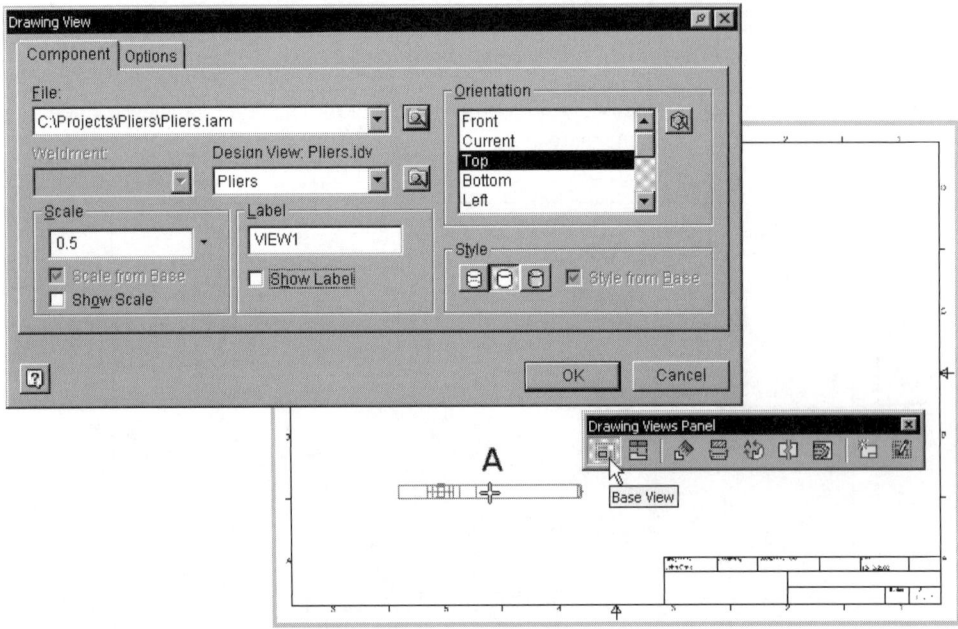

Figure 10.69 *Base view of an assembly being constructed*

Now construct a set of projected views.

5. Select Projected View on the Drawing Views Panel Bar or toolbar.
6. Select A in Figure 10.70 to select the base view.
7. Select B in Figure 10.70 to specify the location of the top view.

8. Select C in Figure 10.70 to specify the location of the isometric view.
9. Select D in Figure 10.70 to specify the location of the side view.
10. Right-click and select Create.

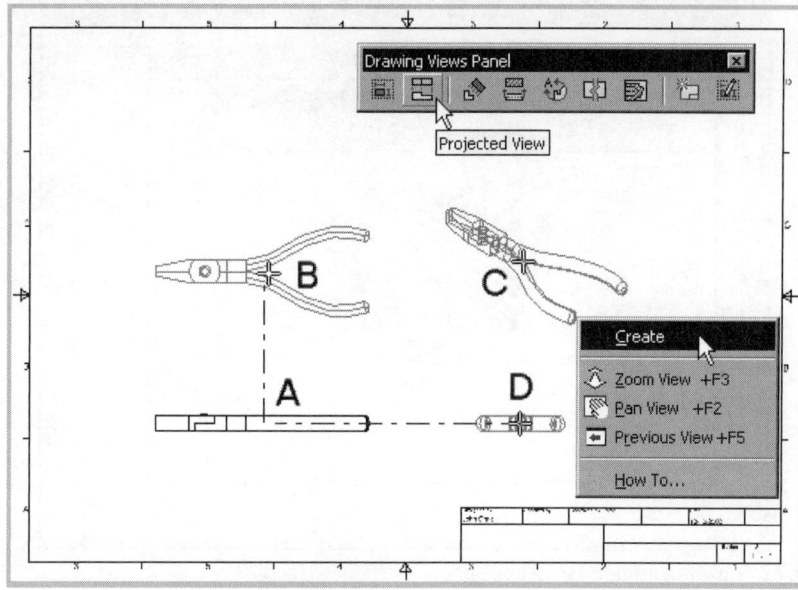

Figure 10.70 *Projected views being constructed*

A set of projected views, including a top view, a side view, and an isometric view, is constructed. Now construct an exploded view from the presentation file of the assembly.

11. Select Base View on the Drawing Views Panel Bar or toolbar.
12. Select the presentation file *Pliers.ipn.*
13. In the Drawing View dialog box, select Explosion 2 in the Presentation View pull-down list box and select the Change view orientation button. (See Figure 10.71.)

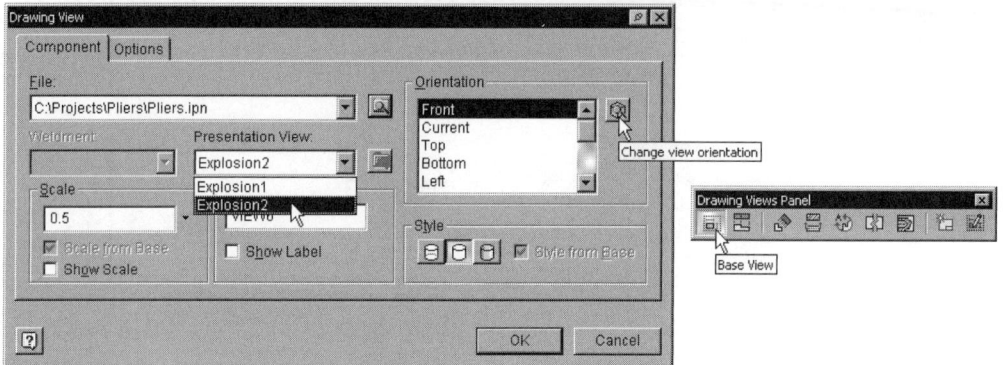

Figure 10.71 *Presentation file selected*

14. Rotate, pan, and zoom the custom view as necessary to orient the assembly so that it appears similar to Figure 10.72. Then click the Exit Custom View button on the Inventor Standard toolbar.

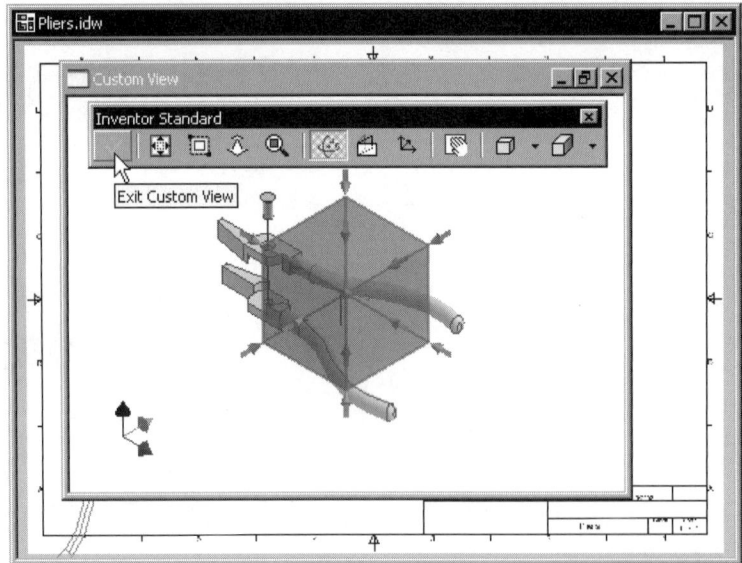

Figure 10.72 *View orientation being specified*

15. Select A in Figure 10.73 to specify the location of the exploded view.

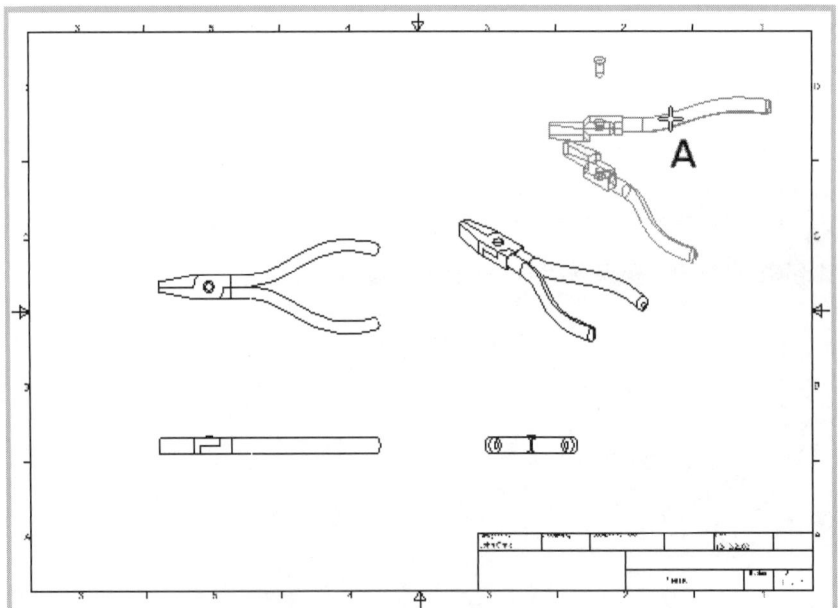

Figure 10.73 *Location of exploded view specified*

An exploded view is constructed. The assembly drawing is complete. Save your file (file name: *Pliers.idw*). Later in this chapter, you will construct a parts list together with a set of balloons.

SKETCHES AND WORK FEATURES IN DRAWING VIEWS

To help establish references in dimensioning and add annotations to a drawing, you can display work features and unconsumed sketches (sketches not yet used to create a feature) constructed in the model and use them as references.

Sketches that can be displayed in a drawing view have to be unconsumed. If you construct a sketch and extrude the sketch to a solid, the sketch is consumed. If you construct a sketch to depict the center location of hole and use the sketch in constructing holes, the sketch is, again, said to be consumed.

In your model, you use work features to help make other features. To help annotate the drawing views, you can display them in the drawing.

TUTORIAL 10.15

In this tutorial, you will display sketch and work features in a drawing view. Prior to constructing the drawing, you will construct work features and an unconsumed sketch in a solid part.

1. Start a new part file. Use the metric *Standard.idw* (ISO) template.
2. Construct a rectangle (50 mm by 30 mm) on the XY plane.
3. Set the display to an isometric view.
4. Extrude the sketch a distance of 20 mm.
5. Construct an offset work plane offsetting a distance of −10 mm from face A in Figure 10.74.

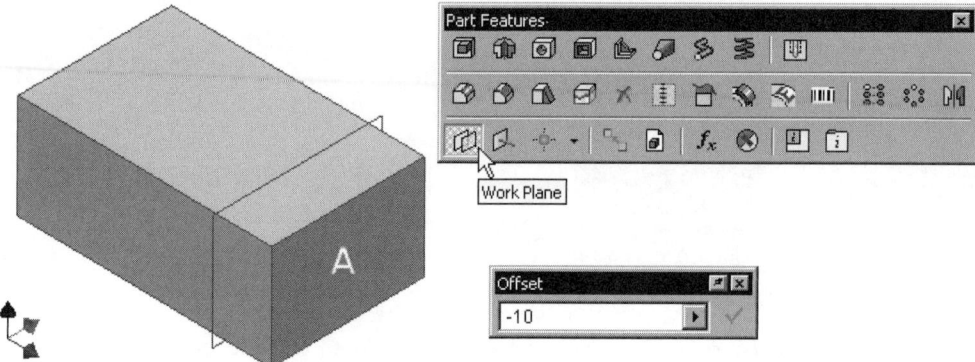

Figure 10.74 *Work plane constructed*

6. Construct a sketch on face A with reference to Figure 10.75.

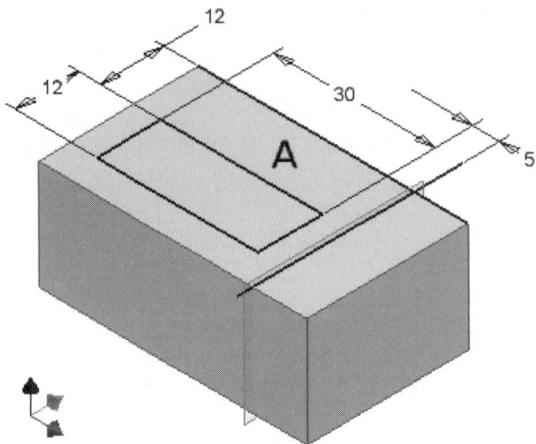

Figure 10.75 *Sketch constructed on a face*

A work plane and an unconsumed sketch are constructed. Save and close your file (file name: *DrawingSketch.ipt*). Now construct a drawing for the solid part.

7. Start a new drawing file. Use the metric *Standard.idw* (ISO) template.

8. With reference to Figure 10.76, place a base Top view and three projected views for the solid part *DrawingSketch.ipt* that you just constructed. Use a scale of 2.

9. Select view A in Figure 10.76, right-click, and select Show Contents.

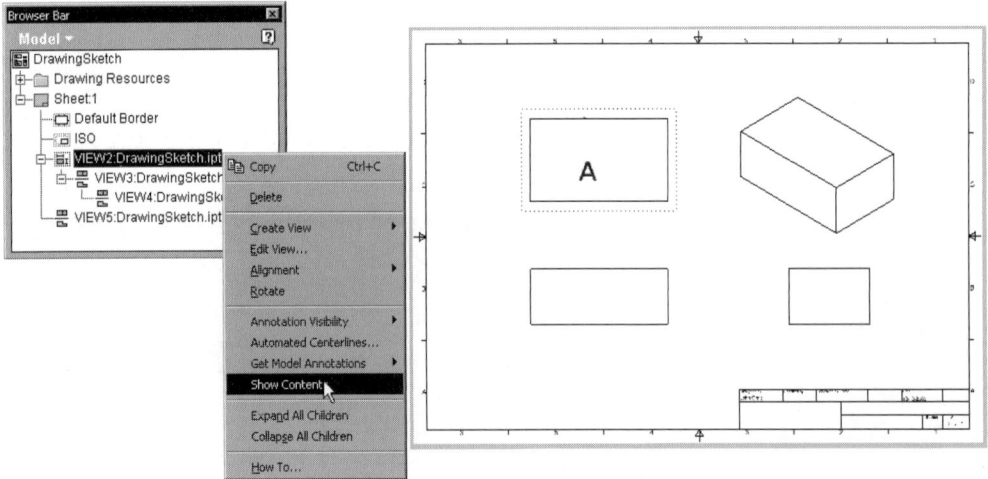

Figure 10.76 *Drawing views constructed*

10. Select the contents of the drawing view (the referenced solid part file *DrawingSketch.ipt*), right-click, and select Get Work Features.

11. Select the contents of the drawing view (the referenced solid part file *DrawingSketch.ipt*), right-click, and select Get Model Sketches.

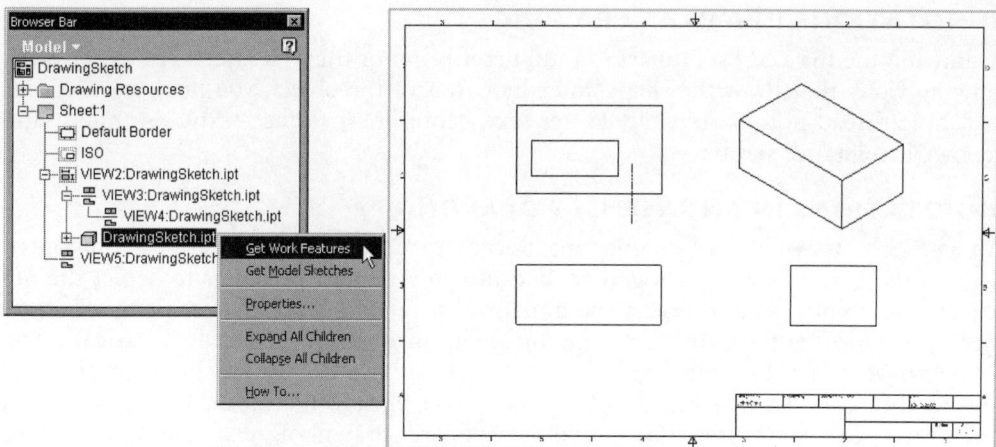

Figure 10.77 *Sketch and work features displayed in the drawing*

Sketch and work plane are displayed. Save your file (file name: *DrawingSketch.idw*).

ANNOTATIONS

Annotations serve as supplements to the information provided by the drawing views. They are dimensions, notes (including text, hole/thread note, hole table, and leader text), centerlines, surface texture symbols, weld symbols, geometric tolerance symbols, user-defined symbols, parts lists, and balloons.

Dimensions	Use to depict the size of the object in drawings (for part drawings) and to depict distances between objects in drawings (for assembly drawings).
Text	Use to provide a description in the drawing.
Hole/Thread Note	Use to delineate details about a hole or a thread.
Hole Table	Use to tabulate details of holes in the drawing.
Leader Text	Use leaders together with textual notes to illustrate and explain specific details of a drawing.
Centerlines	Use to illustrate axis and center locations.
Surface Texture Symbols	Use to mandate surface finish requirement.
Weld Symbols	Use to illustrate how you will weld the component parts together in the assembly.
Geometric Tolerance Symbols	Use to control the geometric shape of the object.
User-Defined Symbols	Define custom symbols and insert them in the drawing.
Parts List	Use to outline the particulars of the individual component parts.
Balloons	Use in conjunction with the parts list to illustrate the locations of the individual parts in the assembly.

ANNOTATIONS IN A PART DRAWING

A drawing file for a 3D solid part is a full description of the 3D object. Along with the drawing views that show the shape and silhouettes of the object, you place dimensions, text, hole/thread note, hole table, leader text, centerlines, surface texture symbols, and geometric tolerance symbol.

ANNOTATIONS IN AN ASSEMBLY DRAWING

An assembly drawing is an engineering document describing how various component parts of the assembly are put together. Because you use part drawings to depict the 3D object, you do not need to repeat the dimensions, surface finish requirement, and geometric tolerances of the individual component part in an assembly drawing. Basically, you need to have a parts list and a set of balloons included in the assembly drawing. In addition to the parts list and the set of balloons, you place dimensions, text, leader text, centerlines, surface texture symbols, geometric tolerance symbols, text, and weld symbols that are specific to the assembly.

DIMENSIONS

Although dimensional information is already an integral part of the 3D part and assembly database, the dimensions of the objects are not readily perceivable if you do not display them explicitly on the drawing. To depict the actual size of a 3D solid in order to eliminate any possible errors that might arise in measuring the drawing, you add dimensions to your drawing.

Components of a Dimension

A dimension has four components:

Dimension Value	Indicates the actual size of the described feature.
Dimension Line	Specifies the direction of the described feature.
Arrowheads	Indicates the extents of the dimension line.
Extension Lines	Projects from the feature to which the dimension refers.

In a dimension, there should be a small gap between the end of the extension line and the feature. The extension line should project a short distance away from the intersection of the dimension line. (See Figure 10.78.)

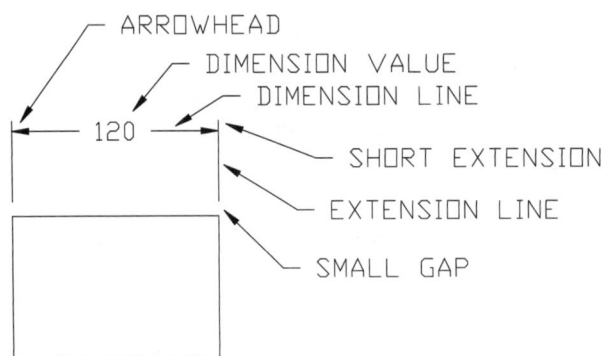

Figure 10.78 *Components of a dimension*

Dimensioning Principles

There are two basic principles to follow when you add dimensions to a drawing:

Appears Once	Each dimension required for the accurate definition of a feature should appear only once in the drawing. You should not assign more than one dimension to a feature.
No Calculation	As far as possible, you should not require the reader of your drawing to do calculation in order to obtain the dimension of a feature.

Because of the second principle, you might find it essential to add more than one dimension to a feature. In that case, you should put the additional dimension within parentheses to indicate an auxiliary reference dimension.

Dimension Styles

Before you add dimensions to your drawing, you should set the dimension styles by selecting Dimension Styles from the Format menu. (See Figure 10.20.) Using the Dimension Styles dialog box, you define units, alternate units, display, text, prefix/suffix, terminator, tolerance, options, and hole notes. A part of the dimension style is the dimension's terminator. (You can also set the terminator of the dimensions by accessing the Terminator tab of the Drafting Standards dialog box.)

In a drawing, you can maintain a number of dimension styles and apply different kinds of dimension styles to individual dimensions.

If you have several dimensions of different styles, you can select them from the drawing, right-click, and select New Styles to display the Dimension Styles dialog box. When you have finished editing the content of the dialog box, the saved dimension style is applied to the selected dimensions.

Dimension Types

There are two ways to add dimensions to your drawing. The first way is to select the drawing view, right-click, and select Get Model Dimensions. The second way is to add dimensions manually one by one. Dimensions that you add in an Autodesk Inventor drawing include general dimensions, base dimensions, and ordinate dimensions. To add dimensions to your drawing, select the General Dimension, Baseline Dimension, Ordinate Dimension Set, and Ordinate Dimension buttons on the Drawing Annotation Panel Bar or toolbar. (See Figure 10.79.)

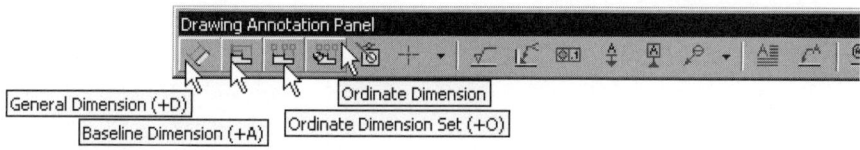

Figure 10.79 *Dimensioning buttons*

TUTORIAL 10.16

In this tutorial, you will add general dimensions to a drawing to the sketch displayed in the drawing view.

1. Open the drawing file *DrawingSketch.idw*, if you already closed it.
2. Select Format > Dimension Styles.
3. In the Dimension Styles dialog box, select the New button and specify a style name.
4. Make appropriate changes in the attribute tabs and click the Save and Close buttons.
5. Select General Dimension on the Drawing Annotation Panel Bar or toolbar.
6. Select A, B, and then C in Figure 10.80 to add a dimension.

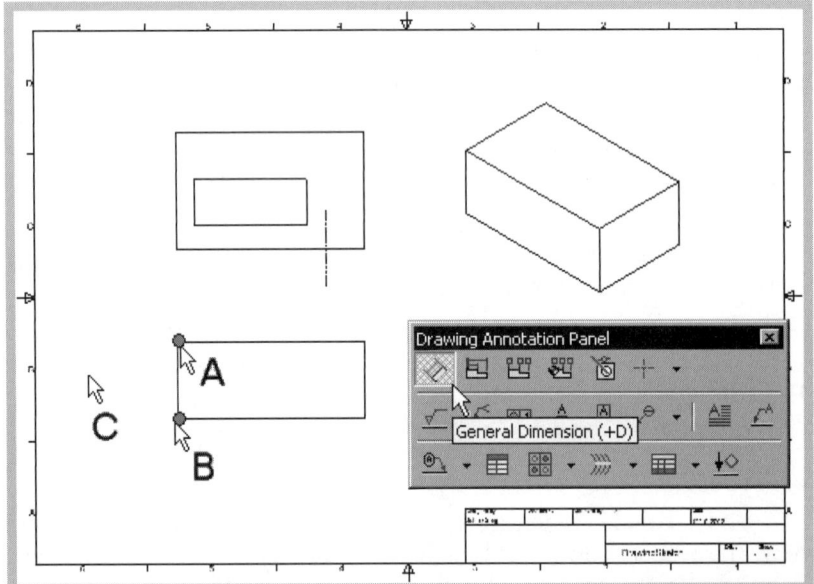

Figure 10.80 *General dimension being added*

7. With reference to Figure 10.81, add dimensions to complete the drawing.

The drawing is complete. Save and close the file.

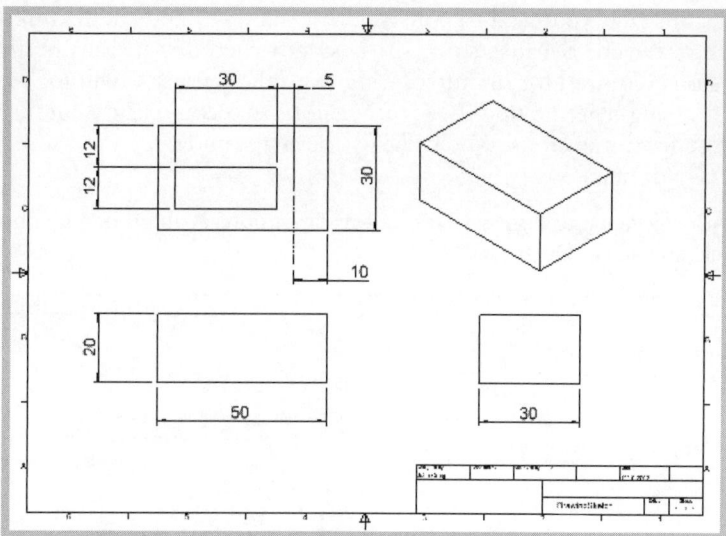

Figure 10.81 *Dimensions added*

 TUTORIAL 10.17

In this tutorial, you will get model dimensions and put them in the drawing.

1. Open the drawing file *PunchPost.idw*, if you already closed it.
2. Select the front view, right-click, and select Get Model Annotations > Get Model Dimensions. (See Figure 10.82.)

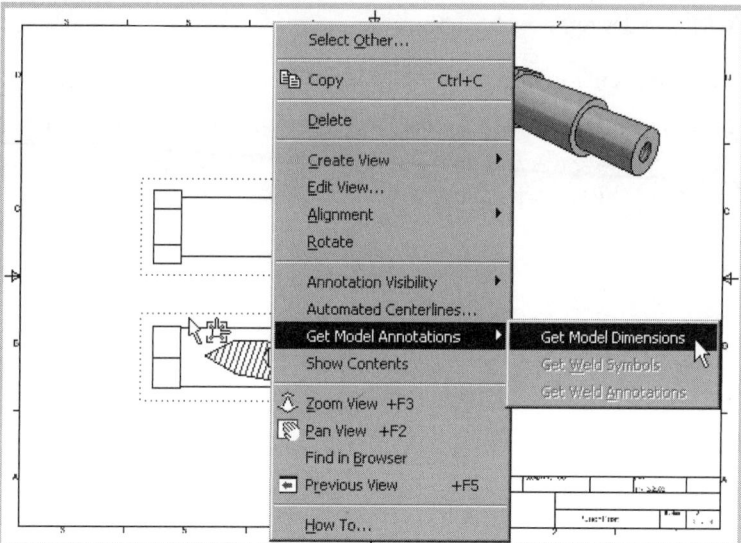

Figure 10.82 *Model dimensions being retrieved*

Dimensions that you used to construct the sketches and the features of the solid part are retrieved. Because some of these retrieved dimensions might not be the dimensions required for manufacturing the solid part, you will make adjustments to the drawing by removing those dimensions that could cause confusion and adding reference dimensions to help clarify the shape and size of the component. Now delete two dimensions.

3. With reference to Figure 10.83, select dimensions A and B one by one, right-click, and select Delete.

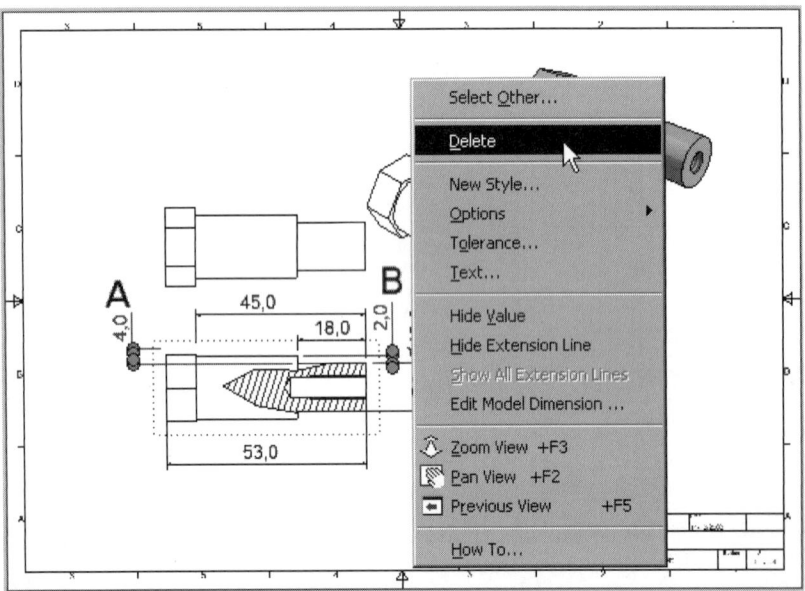

Figure 10.83 *Dimensions retrieved from the model*

Now add a dimension and then prefix it with the "Ø" symbol.

4. Select General Dimension on the Drawing Annotation Panel Bar or toolbar.

5. Select edges A and B and then select position C (in Figure 10.84) to place a dimension. When finished, right-click and select Done.

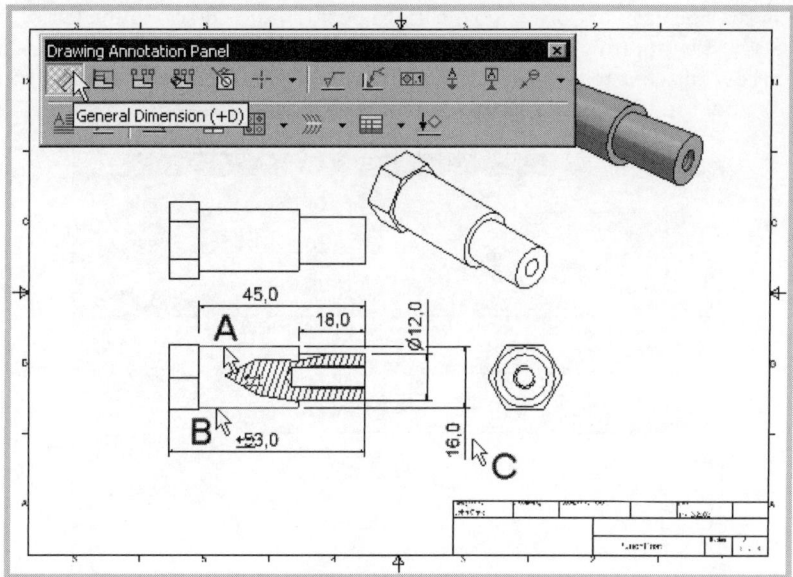

Figure 10.84 *A general dimension being placed*

6. Select the dimension, right-click, and select Text. (See Figure 10.85.)

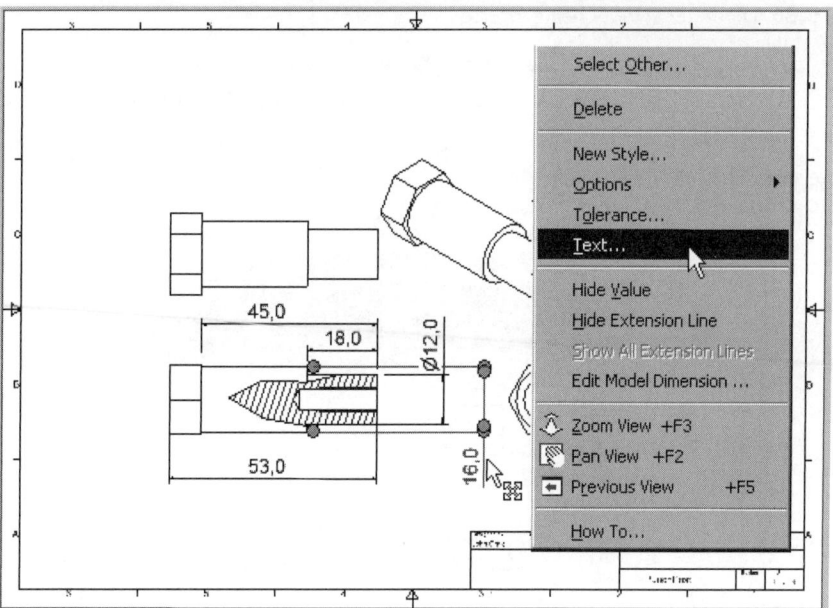

Figure 10.85 *Dimension selected*

7. In the Format Text dialog box, click to position the cursor to the left of the <<>> symbol (this symbol denotes the default value of the dimension), click the arrow adjacent to the Insert Symbol button, and select the Ø symbol from the symbol list. (See Figure 10.86.) Then click OK.

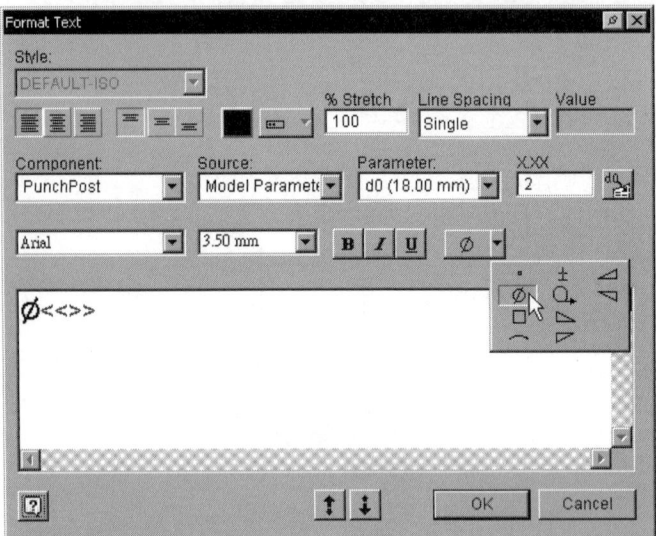

Figure 10.86 *Diameter symbol being inserted*

8. With reference to Figure 10.87, select General Dimension on the Drawing Annotation Panel Bar or toolbar, and select vertices A and B and location C to place a dimension.

Now save your drawing file.

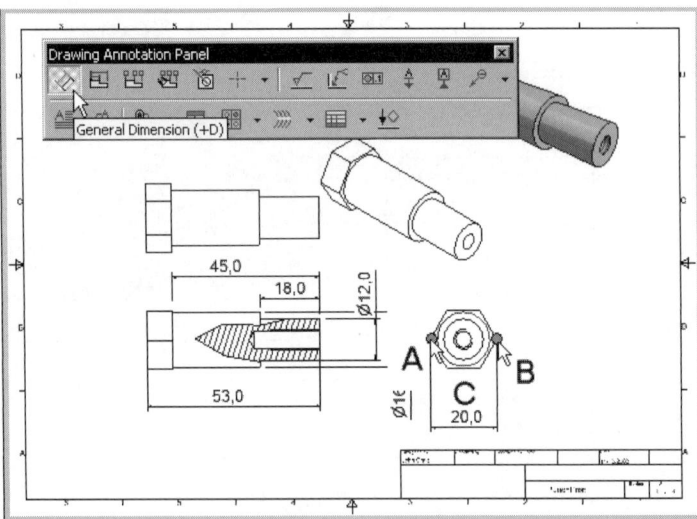

Figure 10.87 *Dimension placed*

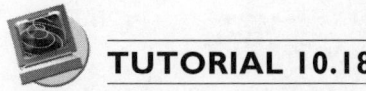

TUTORIAL 10.18

In this tutorial, you will add baseline dimensions to a drawing.

1. Start a new drawing file. Use the metric *Standard.idw* (ISO) template.
2. With reference to Figure 10.88, construct a set of drawing views for the part file *PunchHinge.ipt.* Set the scale to 1.
3. Select Baseline Dimension on the Drawing Annotation Panel Bar or toolbar.
4. Select edges A, B, C, and D in Figure 10.88.

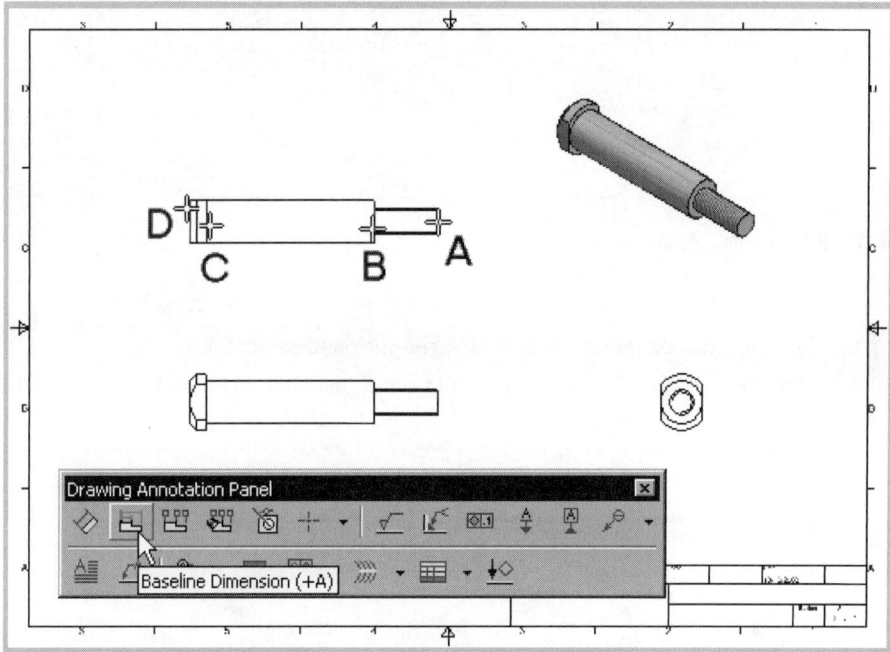

Figure 10.88 *Edges being selected*

5. Right-click, select Continue, and select location A in Figure 10.89.

A set of baseline dimensions is constructed.

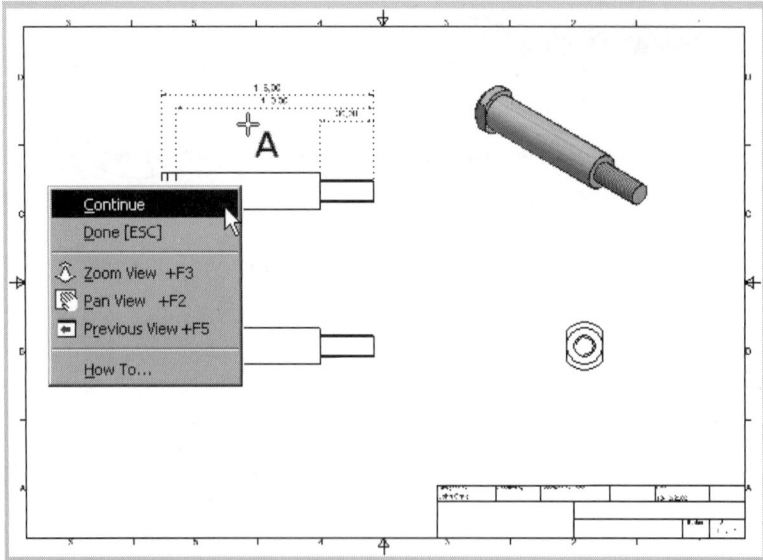

Figure 10.89 *Baseline dimension constructed*

Now add a dimension to your drawing and prefix it with a text string.

6. Add a general dimension, and then right-click and select Done.

7. With reference to Figure 10.90, select the dimension, right-click, and select Text.

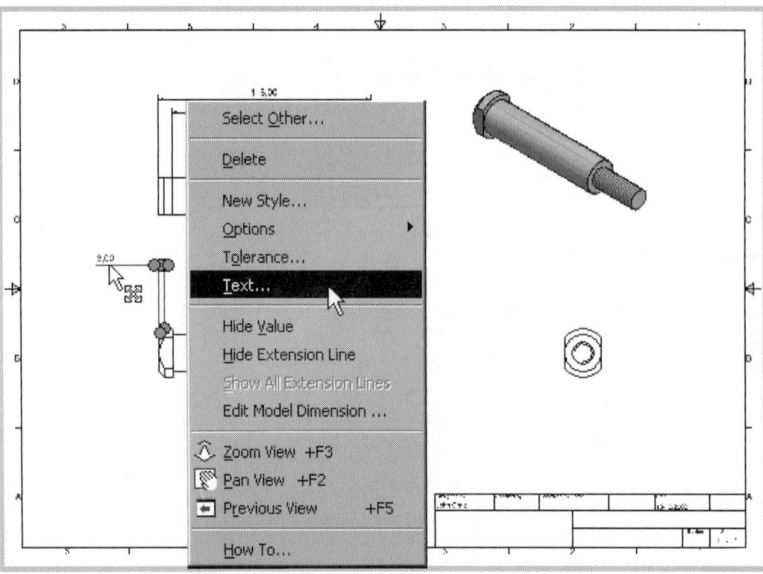

Figure 10.90 *Dimension added and selected*

8. In the Format Text dialog box, place a suffix for the default dimension value with the string x45 and the "degree" symbol. (See Figure 10.91.)

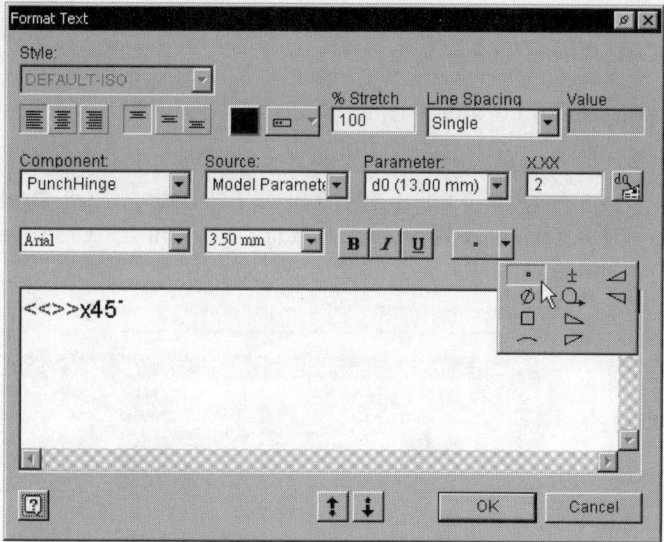

Figure 10.91 *Text string and symbol being suffixed*

9. With reference to Figure 10.92, add three dimensions, at A, B, and C. Right-click and select Done. Then select dimension A, right-click and select Text, and add the diameter symbol prefix. Repeat this for dimension B.

10. Hold down CTRL and select all the dimensions, right-click, and select New Style.

11. In the New Dimension Style dialog box, select the Text tab, change the Size to 7 mm, and then click OK.

Save your file (file name: *PunchHinge.idw*).

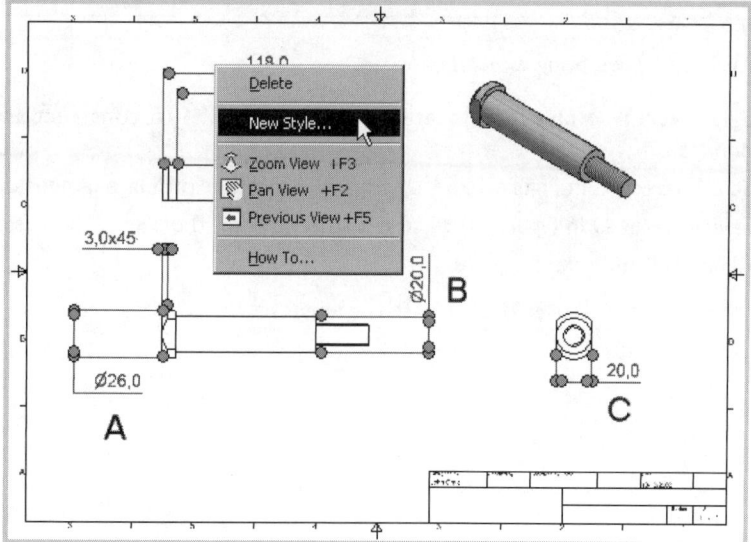

Figure 10.92 *Dimensions with prefix added*

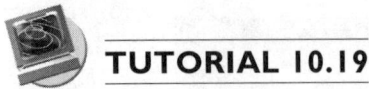

TUTORIAL 10.19

In this tutorial, you will add ordinate dimensions to a drawing.

1. Open the drawing file *SheetMetalA.idw*, if you already closed it.
2. Select Ordinate Dimension on the Drawing Annotation Panel Bar or toolbar.
3. Select flat pattern view A and select point B in Figure 10.93 to establish a datum point.

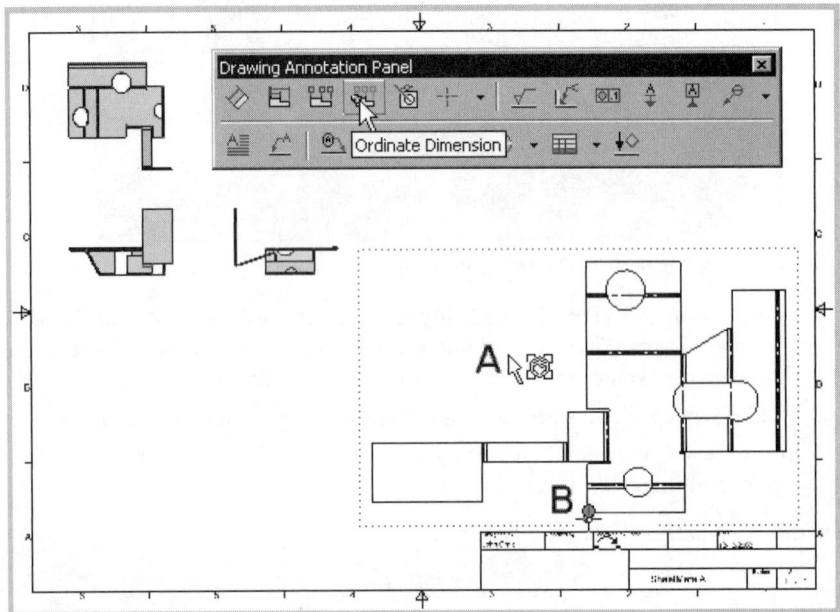

Figure 10.93 *Datum point being established*

4. Select vertex A and then location B in Figure 10.94 to construct an ordinate dimension.
5. Select vertex C in Figure 10.94 to construct another ordinate dimension.
6. Select vertex D in Figure 10.94 to construct the third ordinate dimension.
7. Right-click and select Create.

A datum point and a set of ordinate dimensions are constructed.

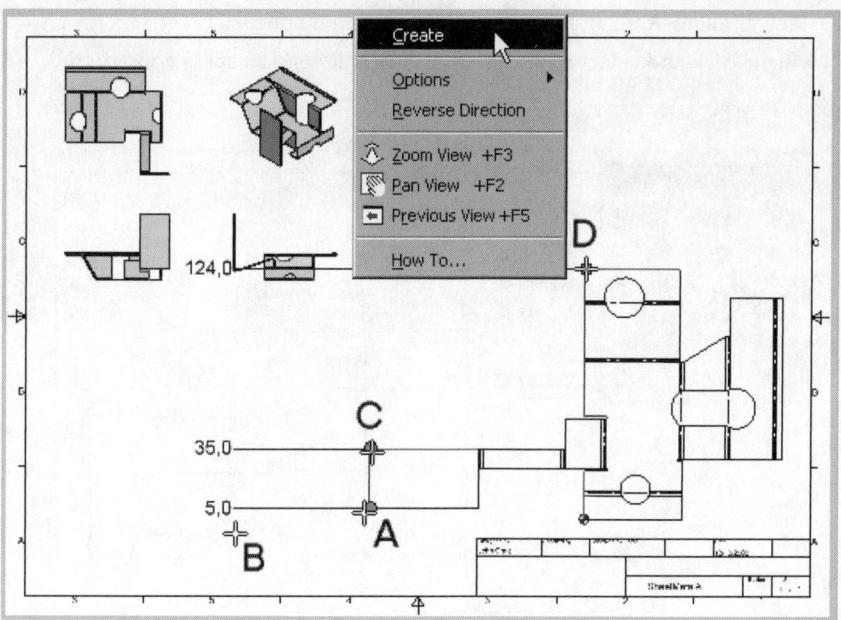

Figure 10.94 *Ordinate dimensions being constructed*

Now construct an ordinate dimension set.

8. Select Ordinate Dimension Set on the Drawing Annotation Panel Bar or toolbar.

9. Select edge A and location B in Figure 10.95 to construct a datum.

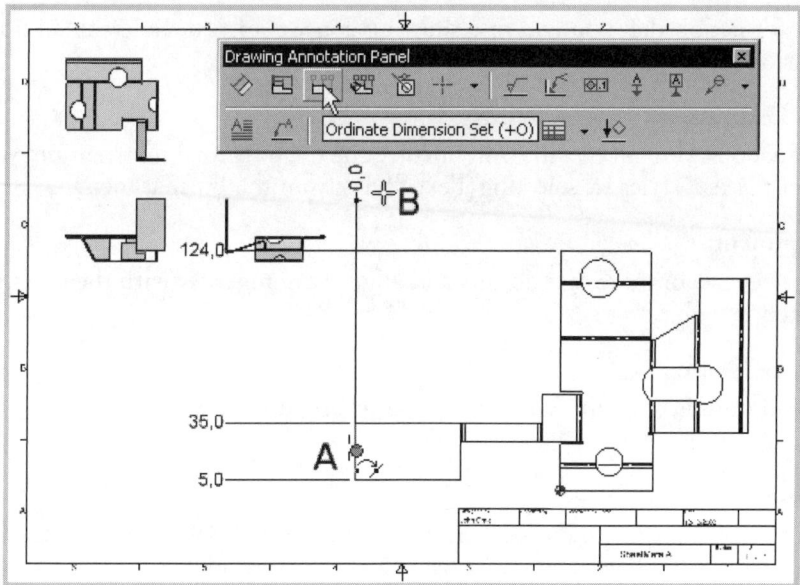

Figure 10.95 *Datum being constructed*

10. Select edges A, B, and C in Figure 10.96 one by one.

11. Right-click and select Create. An ordinate dimension set is constructed.

Save and close your file.

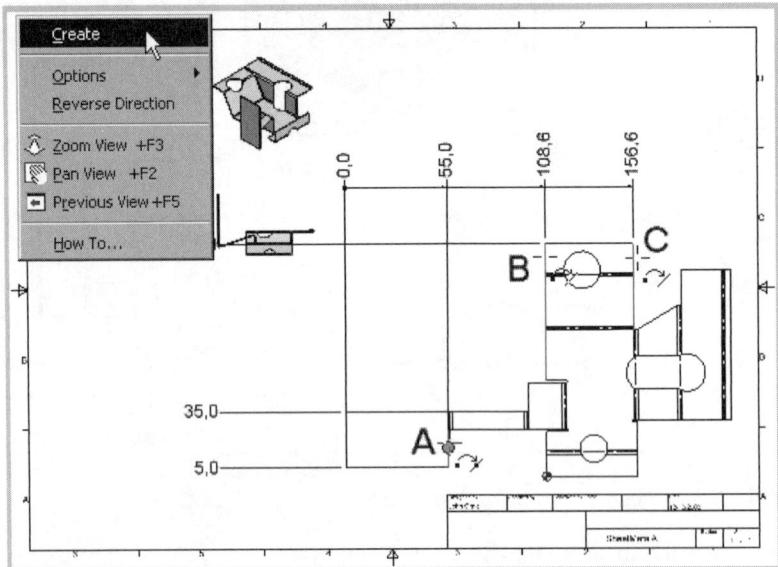

Figure 10.96 *Ordinate dimension set being constructed*

TEXTUAL INFORMATION

Textual information that you add in a drawing consists of text, hole/thread notes, hole table, revision block, and leader text.

Text and Text Styles

Because any notes you put in a drawing involve the use of textual information, you might want to set the text styles by selecting Text Styles from the Format menu.

Text Alignment

You can select two or more text objects and align them together with the position of the first selected text object.

Hole/Thread Note

Hole/thread note is a special kind of textual information depicting the size of a hole/thread.

Hole Table

A hole table depicts the size and coordinates of the hole features of the solid part in a drawing view with reference to a datum point. Using a hole table, you tabulate the size and location of holes instead of using hole/thread notes and dimensions. To construct a hole table, you select a datum point and select holes to include, or include all the holes in the table.

Revision Information

To delineate the revision number and revision information about a drawing, you insert a revision block together with a revision tag. The revision block outlines the changes in a table, and the revision tag highlights where the changes are made.

Leader Text

A leader text is a text string together with a leader.

 ## TUTORIAL 10.20

In this tutorial, you will add several text objects to your drawing and align them.

1. Open the drawing file *PunchPost.idw*, if you already closed it.
2. Select Text on the Drawing Annotation Panel Bar or toolbar.
3. Select A in Figure 10.97 to specify a location for the text string.
4. In the Format Text dialog box, set text height to 7 mm, type **PUNCH SET**, and click the OK button.

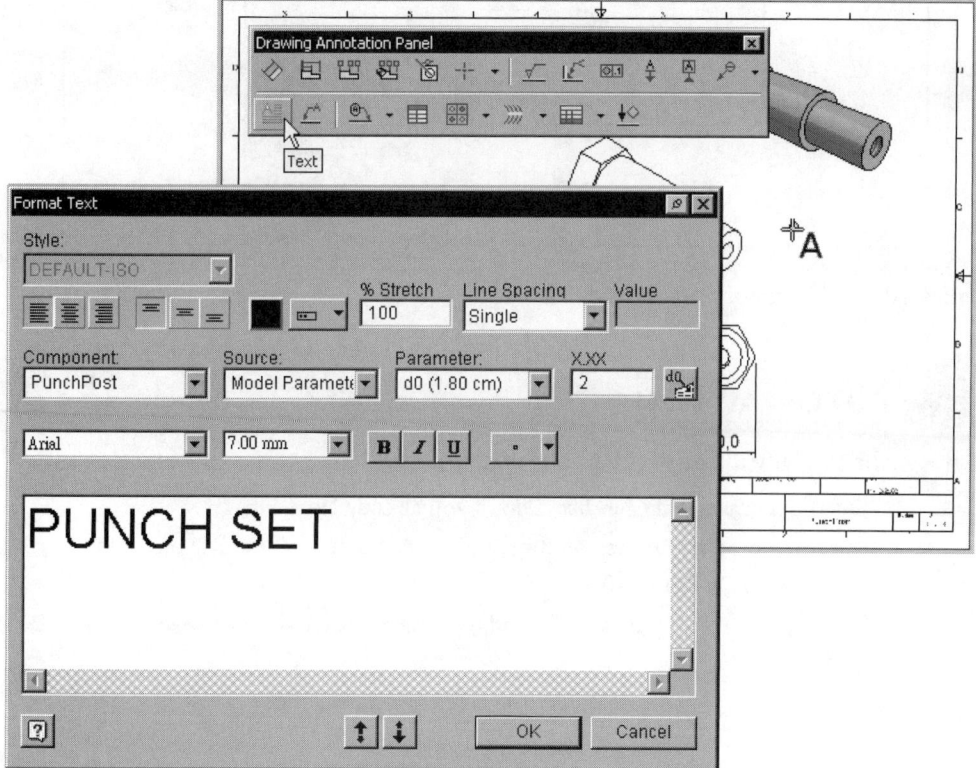

Figure 10.97 *Text string being constructed*

5. Construct another text string at location A in Figure 10.98. Then right-click and select Done.

6. Hold down SHIFT, select both text strings, right-click, and then select Align.

7. In the Align Text dialog box, select Align Left and click the Apply button.

Two text strings are constructed and aligned. Save your file.

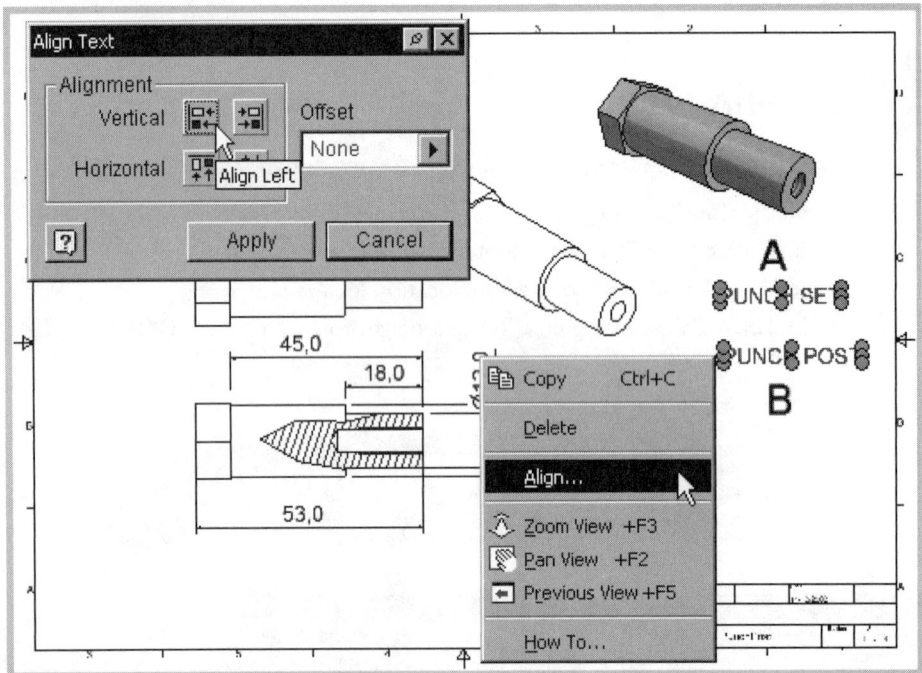

Figure 10.98 *Text strings being aligned*

 **TUTORIAL 10.21**

In this tutorial, you will construct a hole/thread note.

1. Open the drawing file *PunchPost.idw*, if you already closed it.

2. Select Hole/Thread Notes on the Drawing Annotation Panel Bar or toolbar.

3. Select hole A in Figure 10.99.

4. Select location B in Figure 10.99 and then right-click and select Done.

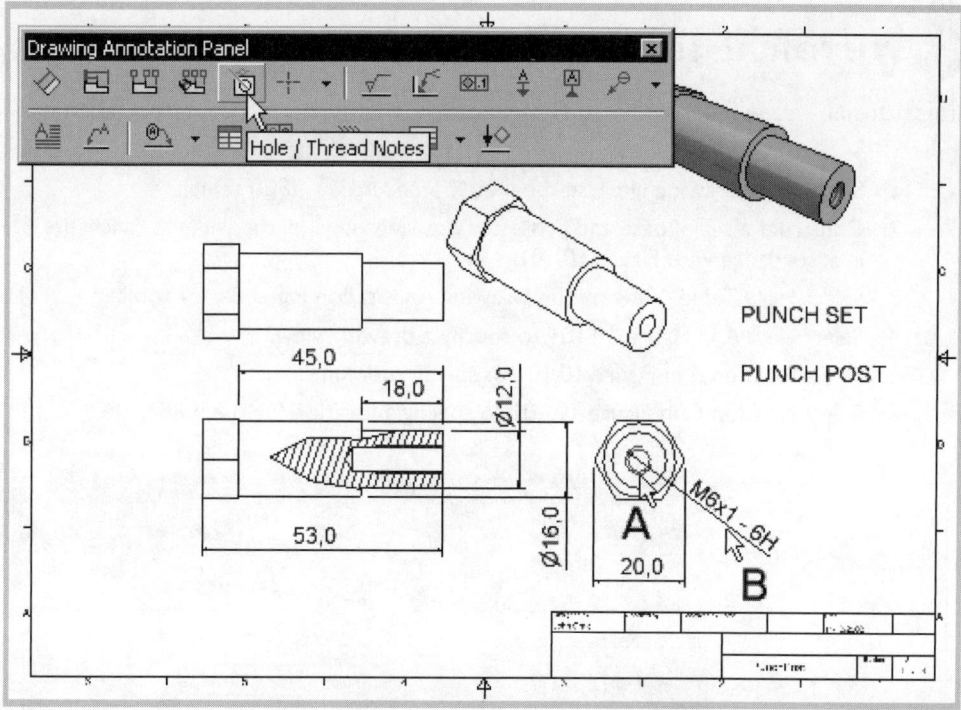

Figure 10.99 *Hole/Thread note being constructed*

A hole/thread note is constructed. Save your file. If you wish to edit the hole note, select it, right-click, and select Edit Hole Note. Figure 10.100 shows the Edit Hole Note dialog box.

Figure 10.100 *Edit Hole Note dialog box*

TUTORIAL 10.22

In this tutorial, you will construct an engineering drawing and incorporate a hole table in the drawing.

1. Start a new drawing file. Use the metric *Standard.idw* (ISO) template.
2. Construct a set of base and projected drawing views of the part file *PunchBase.ipt* in accordance with Figure 10.101.
3. Select Hole Table–View on the Drawing Annotation Panel Bar or toolbar.
4. Select view A in Figure 10.101 to specify a drawing view.
5. Select location B in Figure 10.101 to specify a datum.
6. Select location C in Figure 10.101 to specify a location for the hole table.

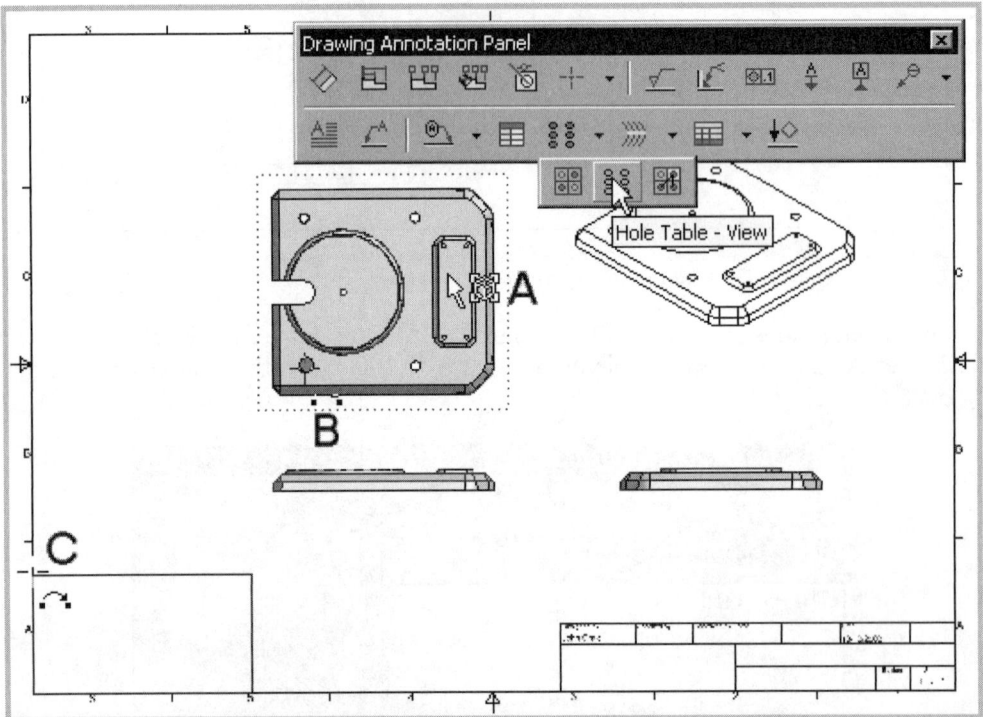

Figure 10.101 *Hole table being constructed*

A hole table is constructed. (See Figure 10.102.) Save your file (file name: *PunchBase.idw*).

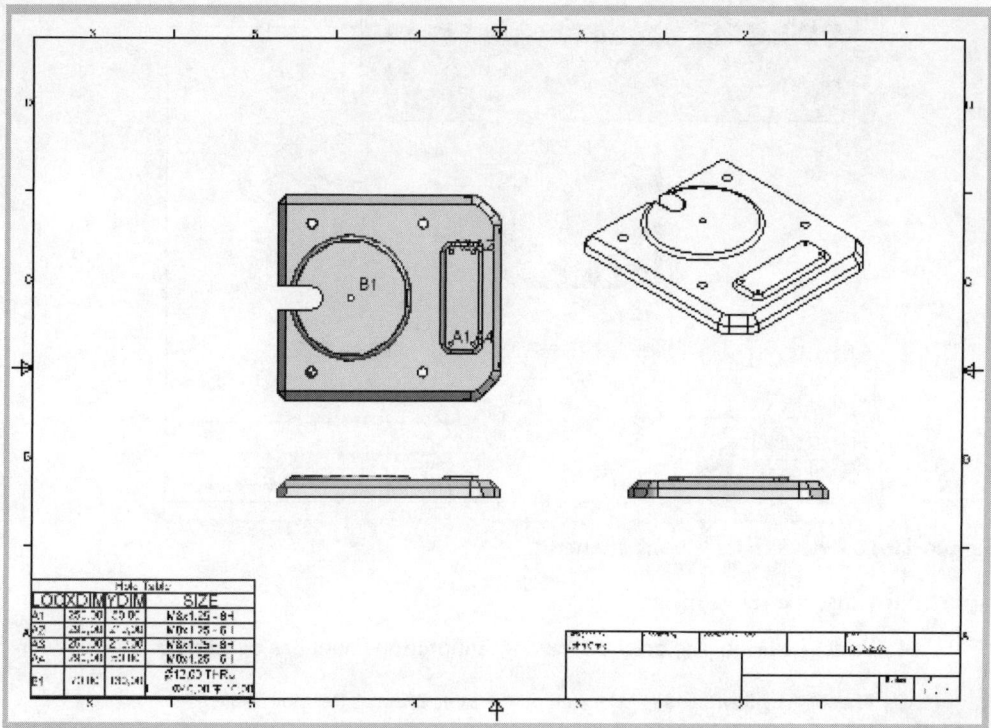

Figure 10.102 *Hole table constructed*

Tip: You can construct a hole table to include all the hole features by selecting Hole Table–View or selected hole features by selecting Hole Table–Selection.

Circular cut features from extruded circles or revolved rectangles are not hole features and are not therefore included in the hole table.

TUTORIAL 10.23

In this tutorial, you will construct a revision block in the drawing.

1. Open the drawing file *PunchBase.idw*, if you already closed it.
2. Select Revision Table on the Drawing Annotation Panel Bar or toolbar.
3. Select location A in Figure 10.103.

A revision table is inserted.

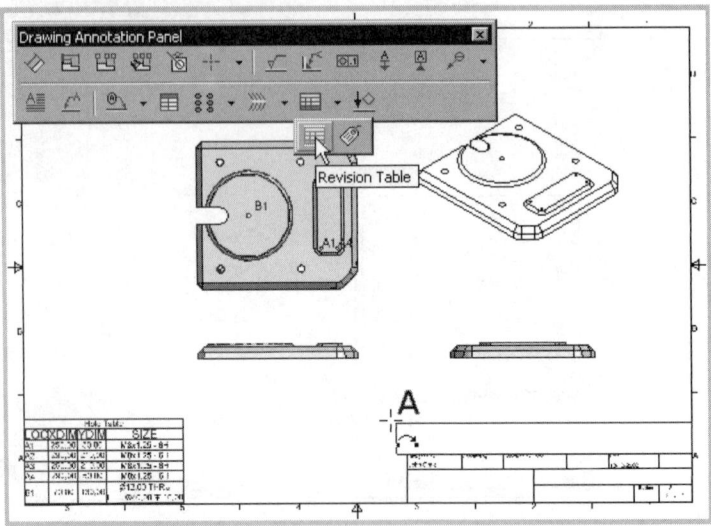

Figure 10.103 *Revision table being inserted*

Now insert a revision tag.

4. Select Revision Tag on the Drawing Annotation Panel Bar or toolbar.

Note: The Revision Tag is located on the same flyout as Revision Table.

5. With reference to Figure 10.104, select A to indicate the start point of the revision tag, select B to indicate the location of the tag, right-click, and select Continue.

The revision tag is constructed. Save your file.

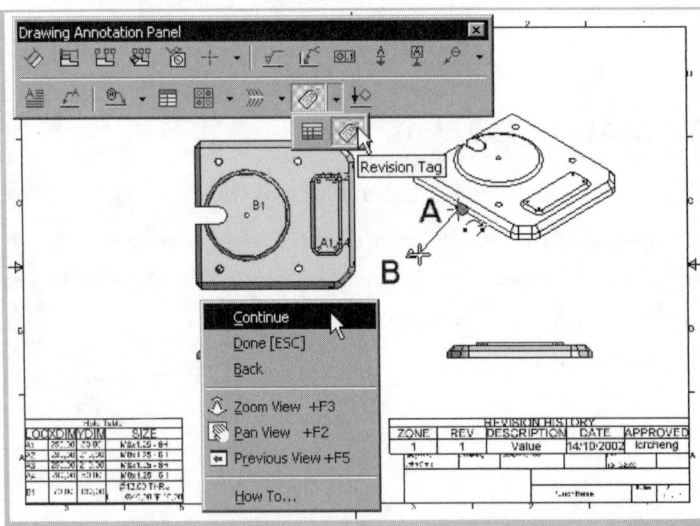

Figure 10.104 *Revision tag being placed*

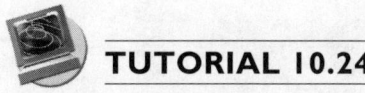

TUTORIAL 10.24

In this tutorial, you will add leader text to a drawing.

1. Open the drawing file *PunchBase.idw*, if you already closed it.
2. Select Leader Text on the Drawing Annotation Panel Bar or toolbar.
3. Select A and B in Figure 10.105 to specify the start point and a location.
4. Right-click and select Continue.

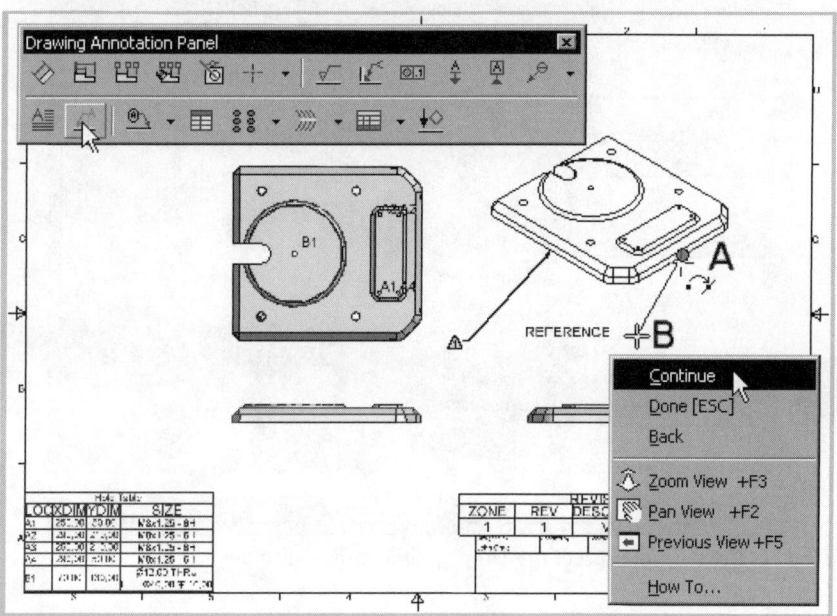

Figure 10.105 *Leader text being constructed*

5. Type a text string in the Format Text dialog box and click the OK button.

Leader text is constructed. Save and close your file.

CENTERLINES

You use centerlines to indicate a center point and an axis of a cylindrical object. There are four ways to construct centerlines. You select the Center Mark, Centerline Bisector, Centerline, or Centered Pattern buttons on the Drawing Annotation toolbar. Centerlines can be applied to the drawing manually by selecting an appropriate feature or automatically to the entire drawing view.

Set Centerline Style

Before you add centerlines to a drawing, you may need to set centerline styles by accessing the Center Mark tab of the Drafting Standards dialog box. (See Figure 10.106.) To open the Drafting Standards dialog box, select Format > Standards.

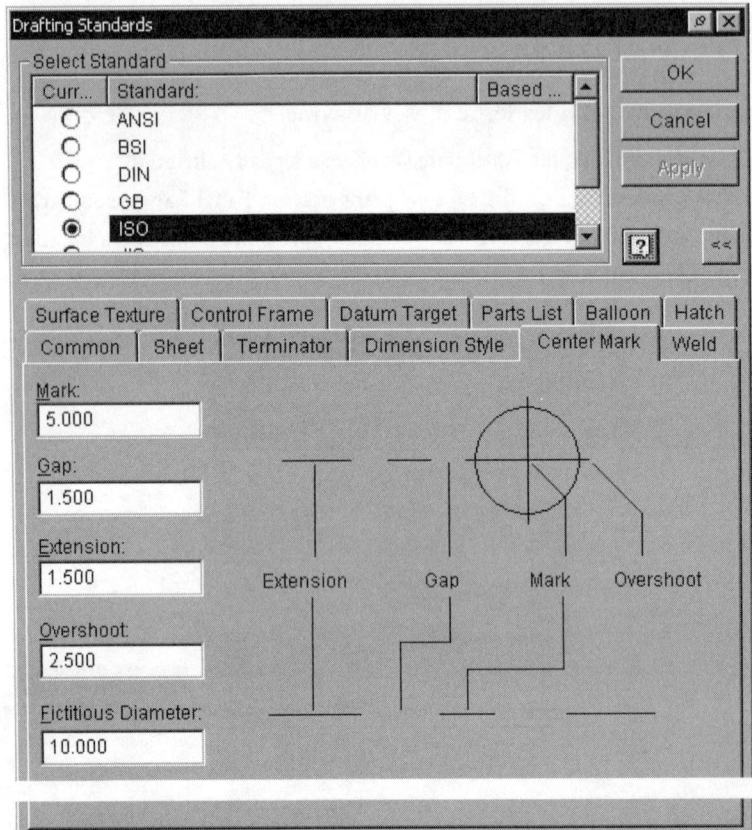

Figure 10.106 *Center Mark tab of the Drafting Standards dialog box*

 TUTORIAL 10.25

In this tutorial, you will add centerlines to a drawing.

1. Open the drawing file *PunchPost.idw*, if you already closed it.
2. Select Center Mark on the Drawing Annotation Panel Bar or toolbar.
3. Select the circular edge A in Figure 10.107.

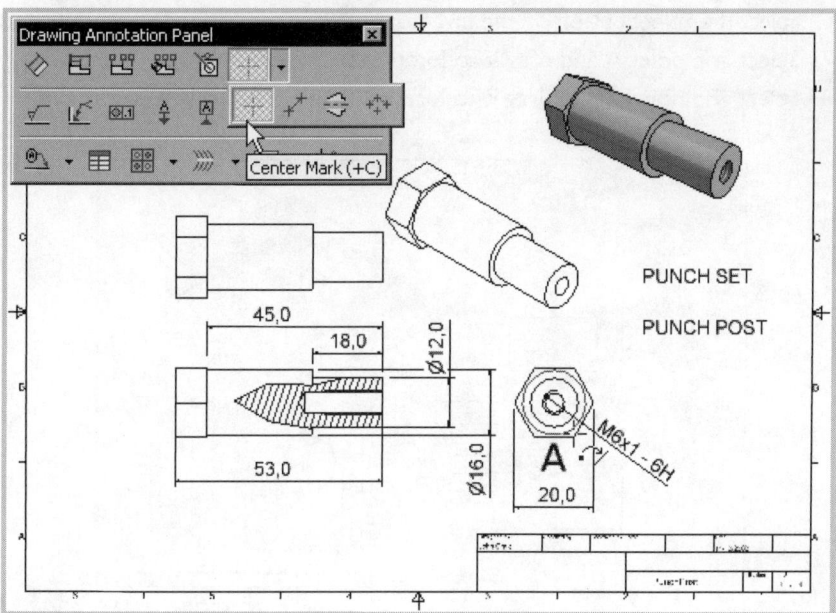

Figure 10.107 *Center mark being constructed*

A center mark is constructed. Now construct a centerline bisector and then lengthen it.

4. Select Centerline Bisector on the Drawing Annotation Panel Bar or toolbar.

5. Select edges A and B in Figure 10.108.

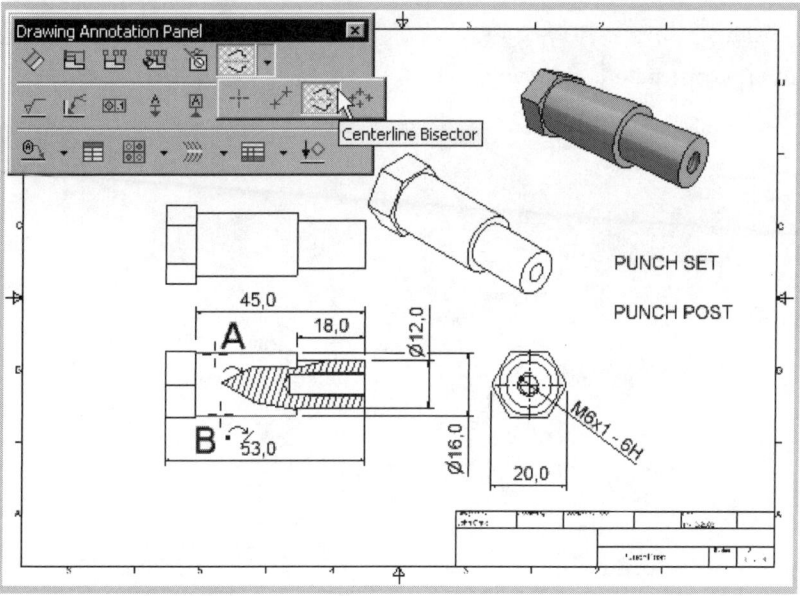

Figure 10.108 *Centerline bisector being constructed*

6. Select the Select button on the Inventor Standard toolbar and select the centerline.
7. Select grip point A and drag it to location B in Figure 10.109.
8. Select grip point C and drag it to location D in Figure 10.109.

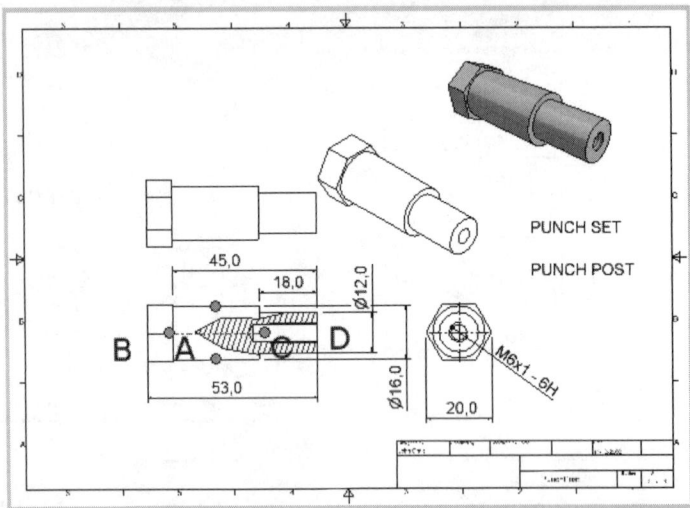

Figure 10.109 *Centerline being lengthened*

Now construct a centerline.

9. Select Centerline on the Drawing Annotation Panel Bar or toolbar.
10. Select edges A and B in Figure 10.110.
11. Right-click and select Create.

A centerline is constructed. Save your file.

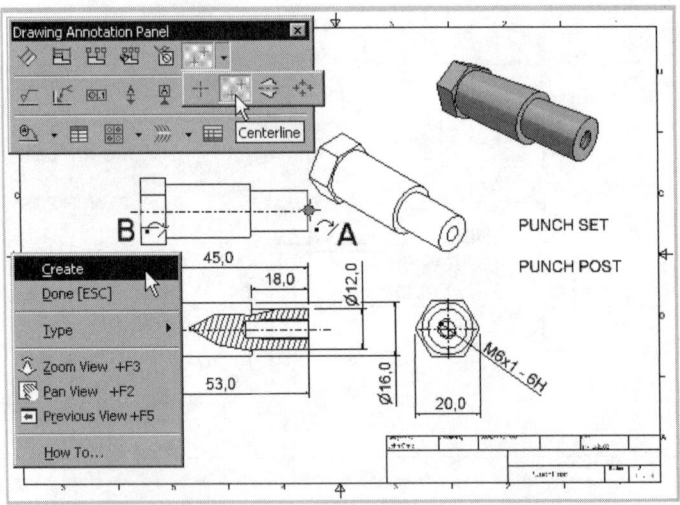

Figure 10.110 *Centerline being constructed*

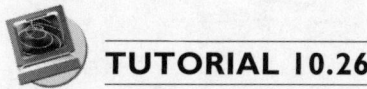

TUTORIAL 10.26

In this tutorial, you will add a patterned centerline to a drawing.

1. Start a new drawing file. Use the metric *Standard.idw* (ISO) template.
2. With reference to Figure 10.111, construct a set of engineering drawing views for the solid part *PunchTable.ipt*.
3. Select Centered pattern on the Drawing Annotation Panel Bar or toolbar.
4. Select circular edge A in Figure 10.111 to specify the center.

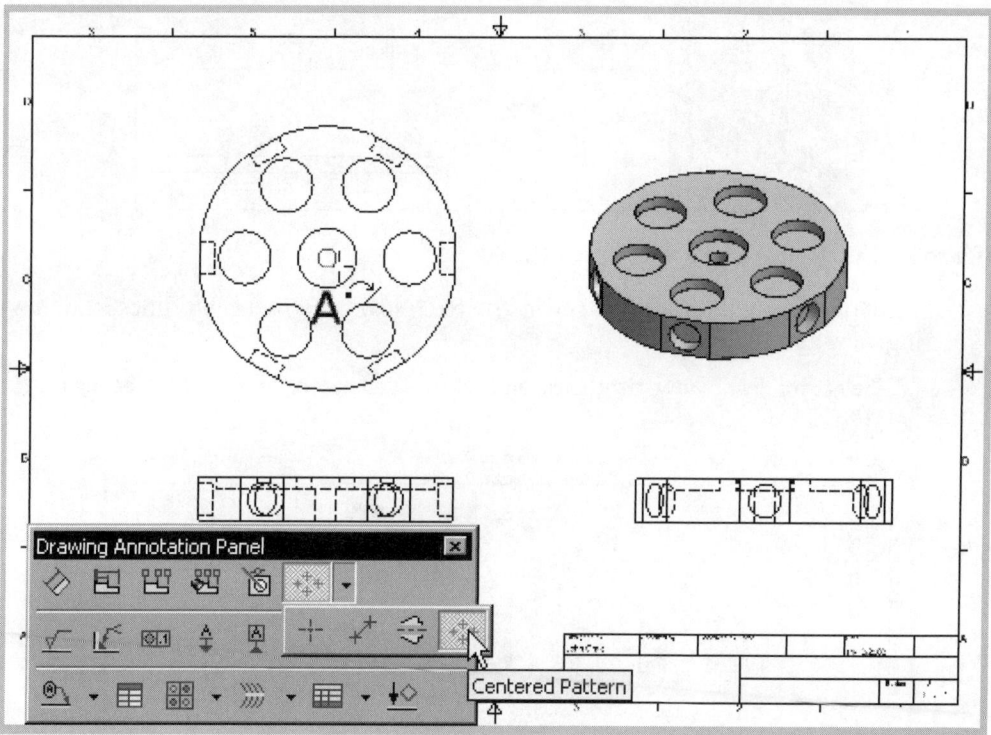

Figure 10.111 *Center of pattern being selected*

5. Select circular edges A, B, C, D, E, and F in Figure 10.112 one by one.
6. Right-click and select Create. Then right-click and select Done.

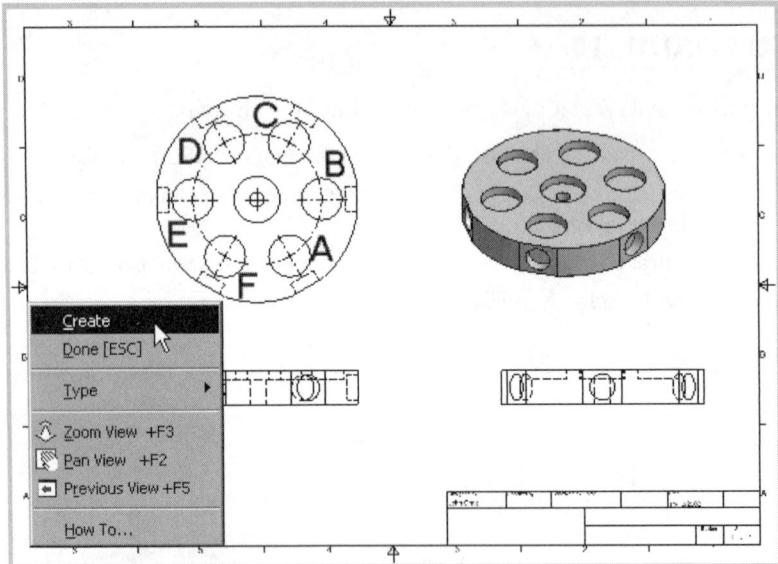

Figure 10.112 *Centered pattern being constructed*

A set of patterned centerlines is constructed. Now construct centerlines in a drawing view automatically.

7. Select the front view, right-click, and select Automated Centerlines. (See Figure 10.113.)

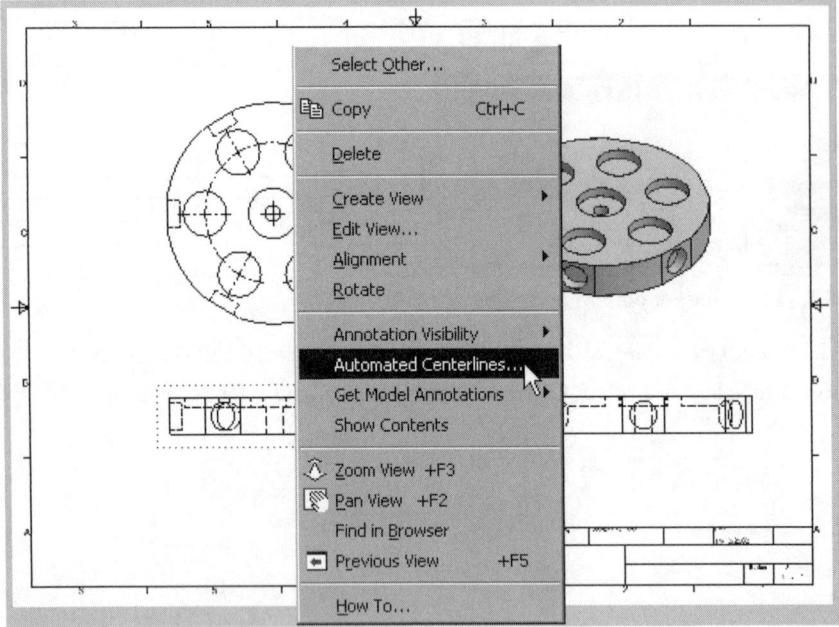

Figure 10.113 *Right-click menu*

8. In the Centerline Settings dialog box, under Apply To, select the box shown in Figure 10.114. Under Projection, also select box, as shown in the figure, and then click OK.

You can access the Centerlines Settings dialog box by selecting Document Settings from the Tools menu and selecting Automated Centerline Settings on the Drawing tab.

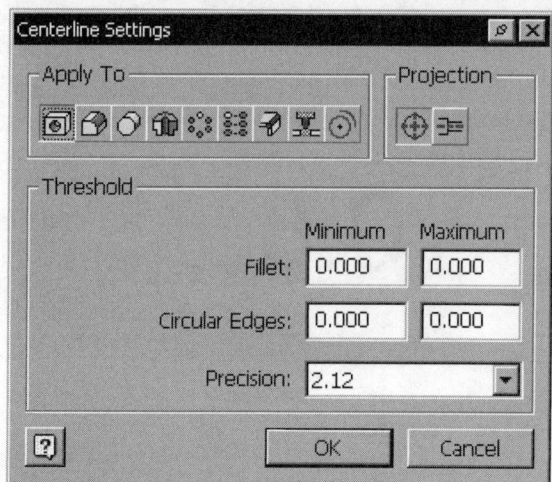

Figure 10.114 *Centerline Settings dialog box*

Centerlines are applied. Save your file (file name: *PunchTable.idw*).

SURFACE TEXTURE SYMBOL

To control the surface finish of a component, you specify surface finish symbols in your engineering drawing.

Basic Symbol

The basic form of a surface finish symbol is similar to the letter "v." In addition to the basic symbol, there are two derivatives: material removal required and material removal prohibited. (See Figure 10.115.)

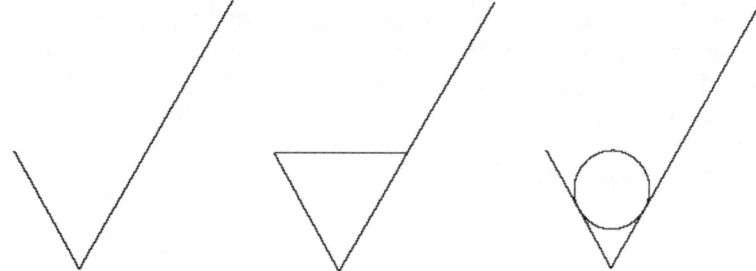

Figure 10.115 *Surface finish symbols (from left to right), basic, material removal required, and material removal prohibited*

With the basic symbol, machining of the component is optional, as long as the required roughness value is achieved. With the material removal required symbol, machining is mandatory. With the material removal prohibited symbol, machining is not allowed, but the required roughness value must be achieved.

Roughness Value

It is pointless to specify a surface finish symbol without stipulating the roughness value of the surface. You can simply state the maximum roughness or a range of roughness. Figure 10.116 shows a roughness value of 0.8 micrometers and Figure 10.117 shows a symbol mandating a roughness value between 0.8 and 0.4 micrometers.

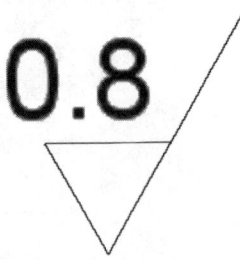

Figure 10.116 *Roughness value specified*

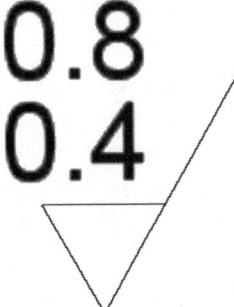

Figure 10.117 *Maximum and minimum roughness values specified*

Direction of Lay

Direction of lay refers to the machining mark or pattern left on the surface after it is machined. For example, if you use sandpaper to polish a surface and rub the sandpaper in a linear direction along the surface, the direction of lay on the surface is linear and parallel to the direction of your hand's movement. There are seven directions. (See Figure 10.118.) A symbol indicating a lay parallel to the plane of projection of the drawing view is shown in Figure 10.119. Specification of lay is optional.

═	Parallel to Plane of Projection
⊥	Perpendicular to Plane of Projection
X	Crossed in Two Slant Directions
M	Multidirectional
C	Circular Relative to Center
R	Radial Relative to Center
P	Particulate, Nondirectional

Figure 10.118 *Lay*

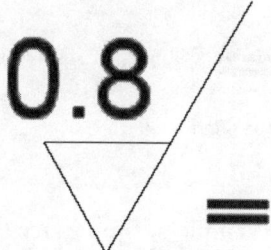

Figure 10.119 *Direction of lay parallel to the projection of the drawing view*

Machining Allowance

If a component has to be machined to achieve the required surface finish, you add an allowance on the component. To specify the machining allowance, you state it in the surface finish symbol. A surface finish symbol depicting a machining allowance of 2 mm is indicated in Figure 10.120.

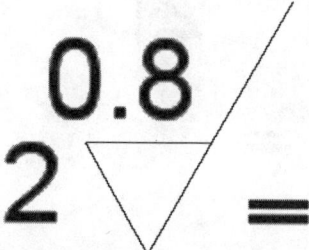

Figure 10.120 *Machining allowance specified*

Production Method

To stipulate the way to achieve the required surface finish, you specify a production method. Figure 10.121 shows a surface finish symbol with production method specified. Specification of the production method is optional.

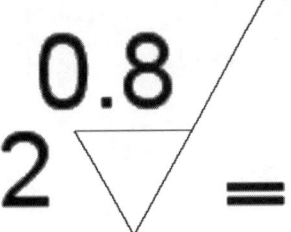

Figure 10.121 *Production method specified*

Sampling Length

To specify the length for taking a sample of the surface for roughness average value measurement, you specify the sample length. (See Figure 10.122.) As with the production method, specification of the sample length is optional.

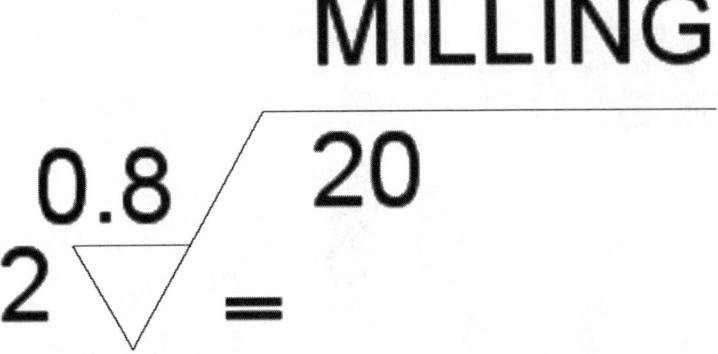

Figure 10.122 *Sample length specified*

Set Surface Texture Symbol Style

You use a surface texture symbol to specify the surface finish and the machining requirement for an object.

You can set the standard for constructing a surface texture symbol in your drawing by accessing the Surface Texture tab of the Drafting Standards dialog box.

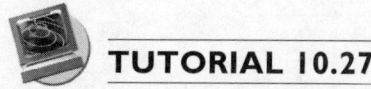

TUTORIAL 10.27

In this tutorial, you will add a surface texture symbol to a drawing.

1. Open the drawing file *PunchPost.idw*, if you already closed it.
2. Select Surface Texture Symbol on the Drawing Annotation Panel Bar or toolbar.
3. Select edge A in Figure 10.123.

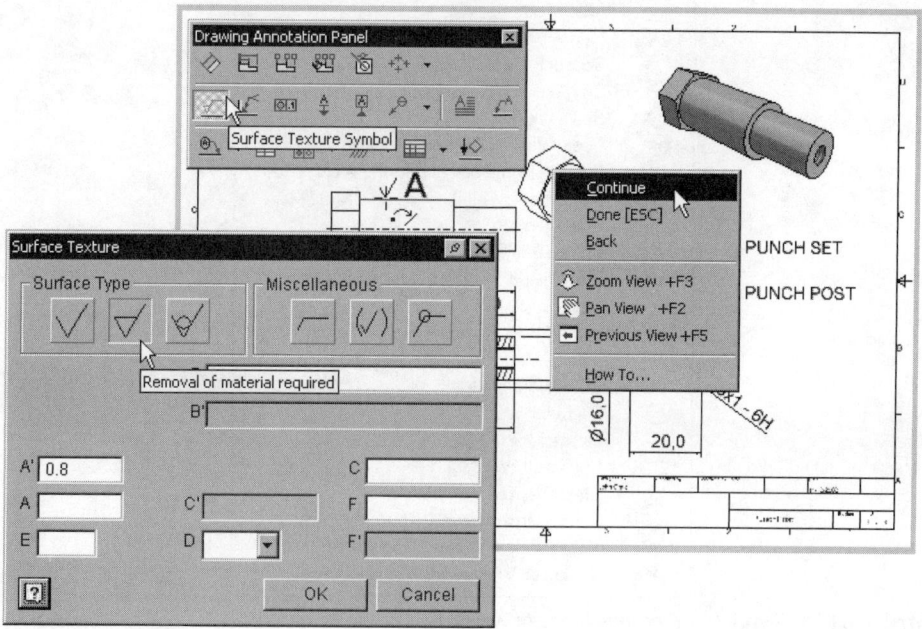

Figure 10.123 *Surface texture symbol being constructed*

4. Right-click and select Continue.
5. In the Surface Texture dialog box, set roughness value to 0.8, select the Removal of material required and the All-round buttons, and then right-click and select Done.

A surface texture symbol is constructed. Save your file.

WELD SYMBOL

To specify how two or more components are to be welded together, you add a welding symbol to your engineering drawing.

The basic form of a weld symbol consists of a horizontal line and a leader line. The leader indicates the location of the weld, and the horizontal line carries a symbol depicting the kind of weld. If the symbol is placed below the horizontal line, the weld is to be made on the near side of the leader. If the symbol is placed above the horizontal line, the weld is to be made on the other side of the leader. Figure 10.124 shows a fillet weld to be made at the near side of the leader. The symbols that you specify on the horizontal line depicting various kinds of weld are shown in Figure 10.125.

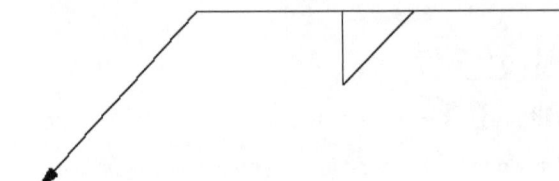

Figure 10.124 *Weld symbol denoting a fillet weld*

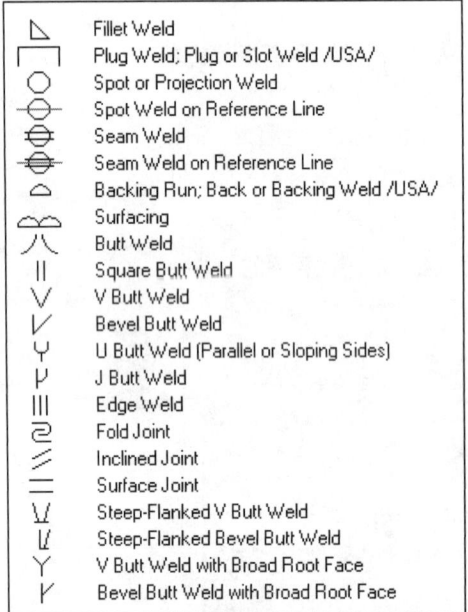

Figure 10.125 *Symbols for various kinds of welds*

Field Weld

If a welding joint has to be carried out on site, you specify a field weld indicator. The field weld symbol is optional. (See Figure 10.126.)

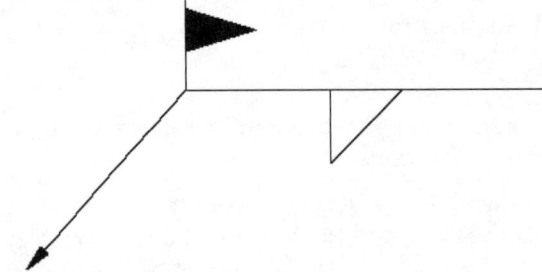

Figure 10.126 *Field weld indicator added to the weld symbol*

All Round Welding

If a joint is to be carried out all round an object, you specify an all round indicator instead of adding a set of symbols around the object. (See Figure 10.127.)

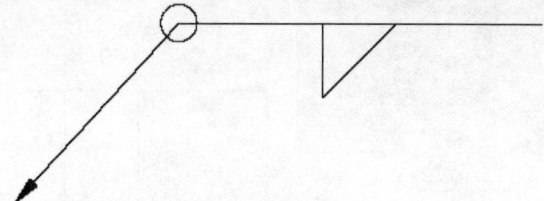

Figure 10.127 *All round indicator added to the weld symbol*

Notes

If additional information is required, you add notes to the weld symbol. (See Figure 10.128.)

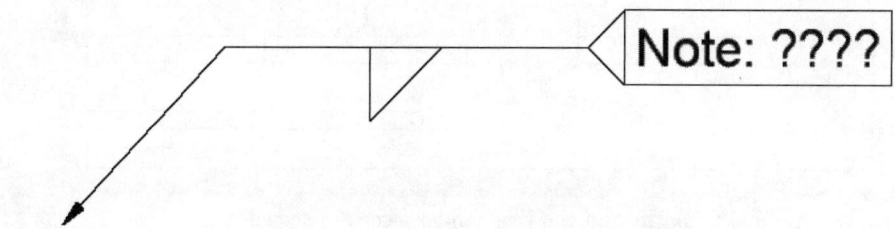

Figure 10.128 *Notes included in a weld symbol*

Adding a Weld Symbol

Putting welding symbol to a drawing can be done manually or automatically. You can select Weld Symbol on the Drawing Annotation Panel Bar or toolbar to place a weld symbol manually. If you already incorporated weldment information in an assembly of weldment, you can retrieve the weld information by selecting the drawing view, right-clicking, and selecting Get Weld Symbols and Get Weld Annotations.

TUTORIAL 10.28

In this tutorial, you will construct a draft view and add weld symbols.

1. Start a new drawing file. Use the metric *Standard.idw* (ISO) template.
2. Select Draft View on the Drawing Views Panel Bar or toolbar.
3. With reference to Figure 10.129, construct a draft view.
4. Select Return to exit sketch mode.

Tip: If you want to edit the sketch, select it, right-click, and select Edit View.

5. Select Caterpillar on the Drawing Annotation Panel Bar or toolbar.

6. Select edge A in Figure 10.129.

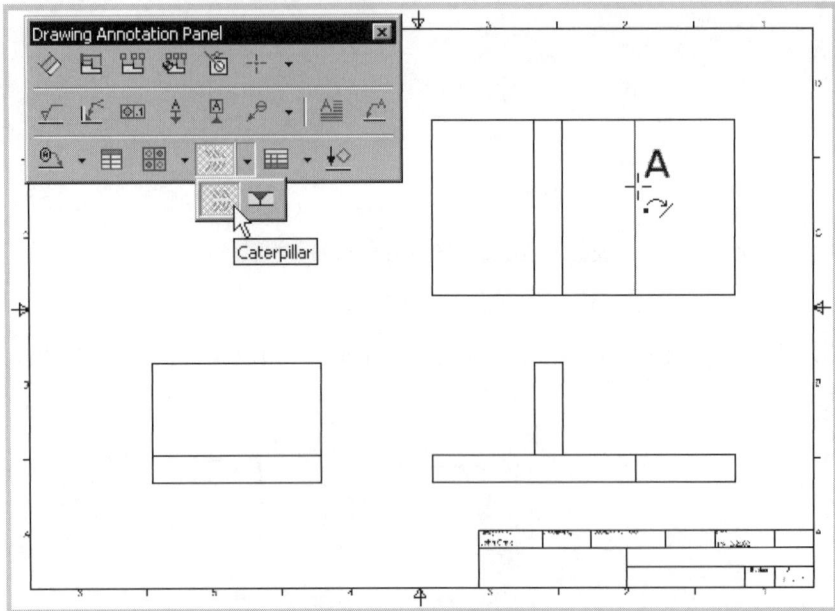

Figure 10.129 *Draft view constructed and weld symbol location selected*

7. In the Style tab of the Weld Caterpillars dialog box, select the Full button. (See Figure 10.130.)

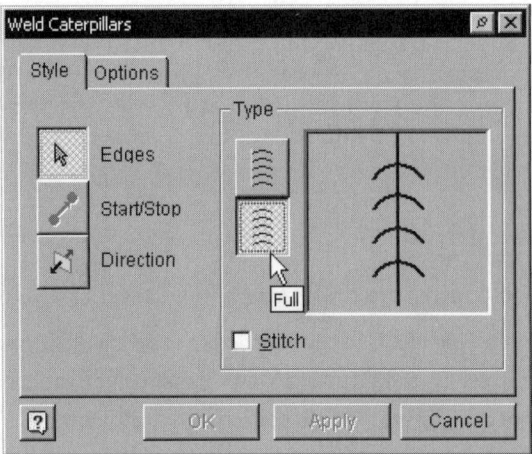

Figure 10.130 *Style tab of the Weld Caterpillars dialog box*

8. Select the Options tab and set legs width to 10 mm and legs spacing to 5 mm. (See Figure 10.131.)

9. Click the Apply button. A full caterpillar weld bead is constructed.

Figure 10.131 *Options tab of Weld Caterpillars dialog box*

10. Select the Style tab and select the Partial button.
11. Select edge A and place the cursor on A in Figure 10.132.
12. Select the Apply button. A partial caterpillar weld bead is constructed.
13. Select edge B and place the cursor on B in Figure 10.132
14. Click the OK button.

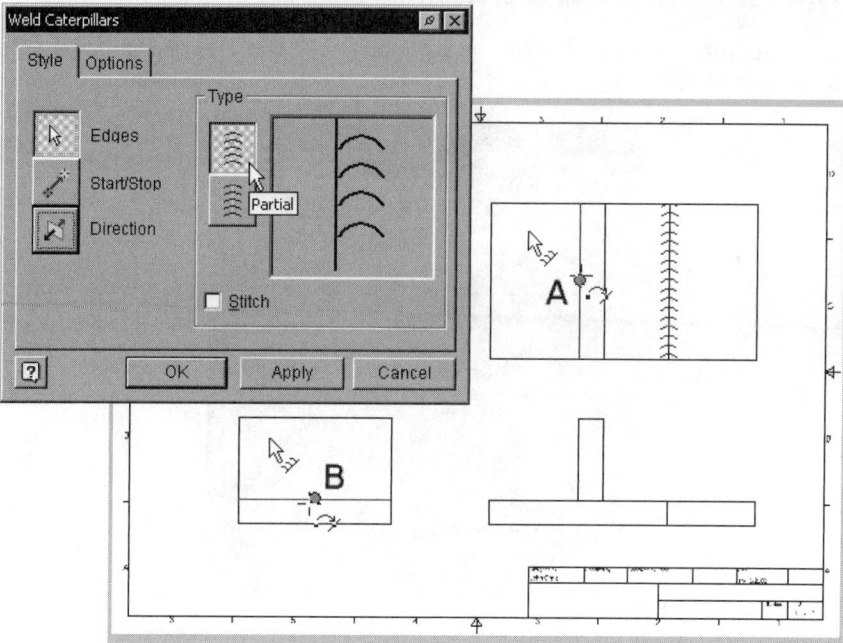

Figure 10.132 *Partial weld caterpillars being constructed*

15. Select End Treatment on the Drawing Annotation Panel Bar or toolbar. (See Figure 10.133.)

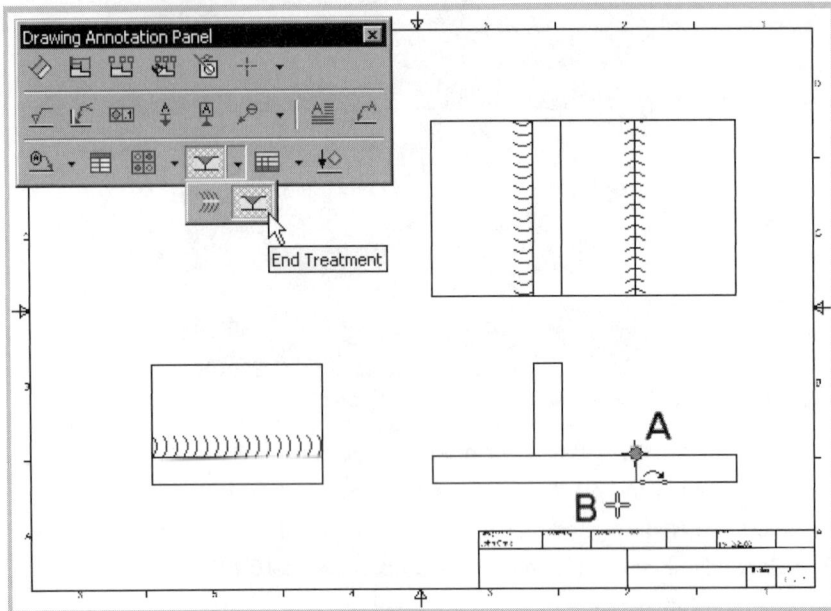

Figure 10.133 *End treatment being constructed*

16. Select the U-Type button on the Style tab of the End Treatments dialog box. (See Figure 10.134.)

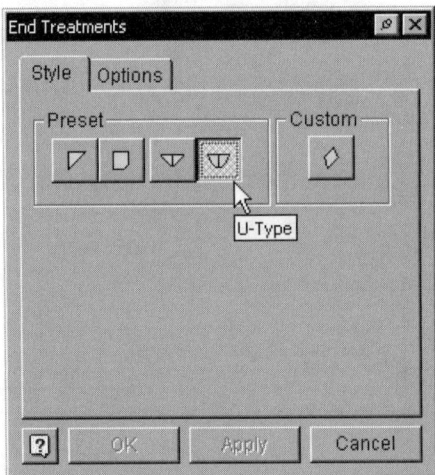

Figure 10.134 *Style tab of the End Treatments dialog box*

17. Select the Options tab and set the leg 1 and leg 2 to 10 mm. (See Figure 10.135.)
18. Select vertex A in Figure 10.133.

19. Move the cursor to B in Figure 10.133.
20. Click the Apply button.

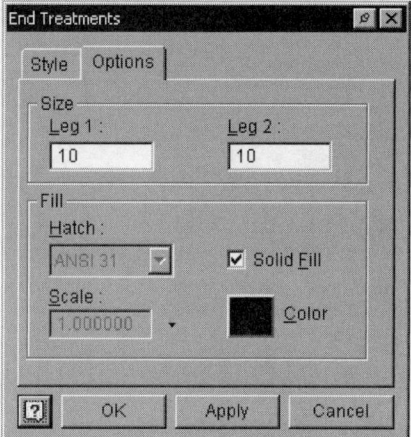

Figure 10.135 *Options tab of the End Treatments dialog box*

21. In the End Treatments dialog box, select the Style tab and select Bevel.
22. Select vertex A in Figure 10.136.
23. Move the cursor to B in Figure 10.136.
24. Click the OK button.

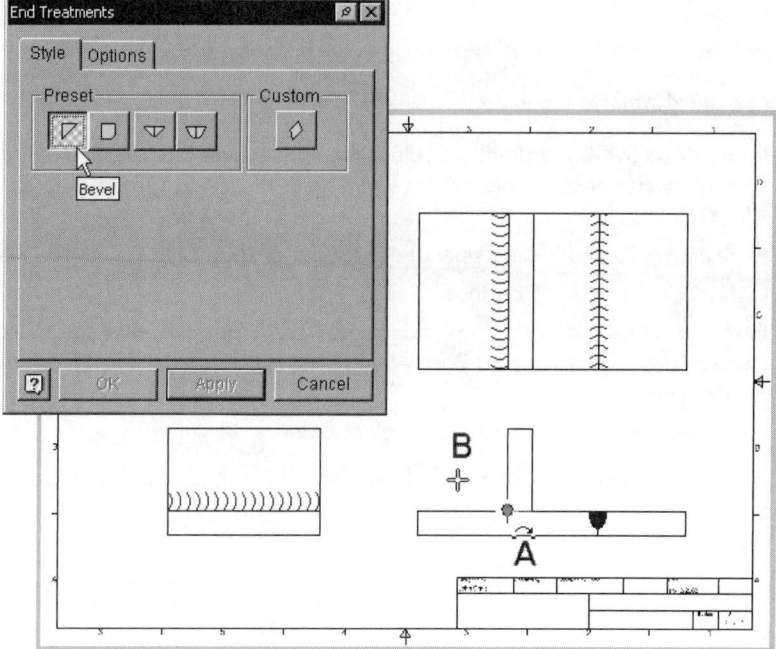

Figure 10.136 *Bevel end treatment being constructed*

25. Select Weld Symbol on the Drawing Annotation Panel Bar or toolbar.
26. Select edge A and location B in Figure 10.137.
27. Right-click and select Continue.

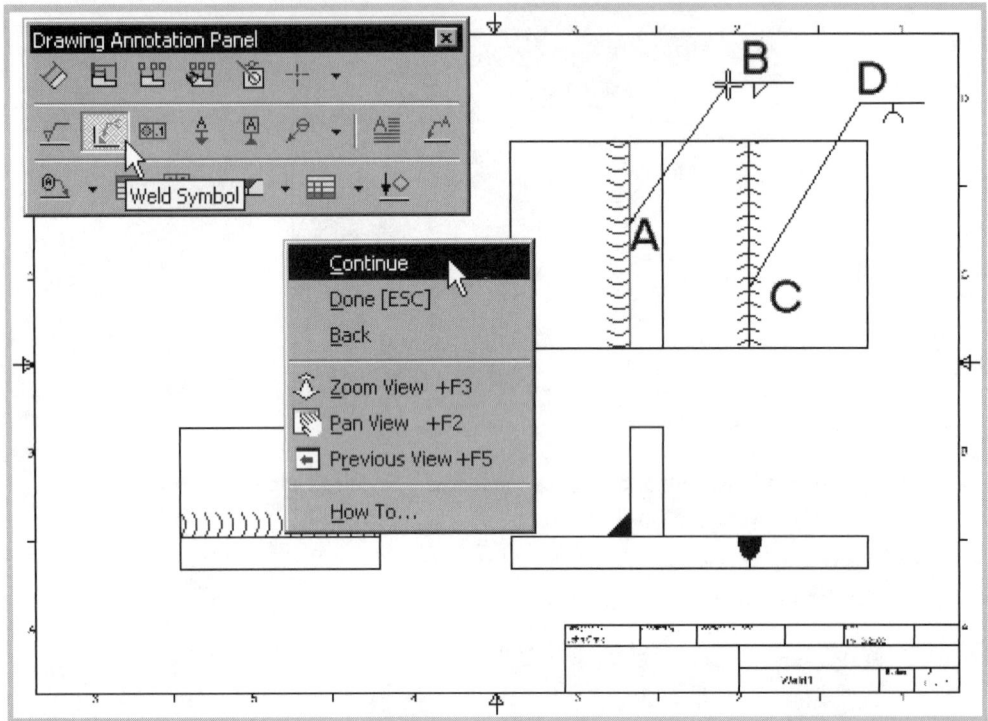

Figure 10.137 *Weld symbol being placed*

28. In the Weld Symbol dialog box, select the Arrow Side tab, select the Weld Symbol Palette button, select Fillet Weld from the Weld Symbol palette, and then click OK.
29. Repeat step 25 and 26 and select edge C and location D in Figure 10.137.
30. Right-click and select Continue.
31. In the Weld Symbol dialog box, select the Arrow Side tab, click the Weld Symbol Palette button, and select U Butt Weld (Parallel or Sloping Sides) from the Weld Symbol palette.
32. Click OK, and then right-click and select Done.

Weld symbols are constructed. Save and close your file (file name: *Weld.idw*).

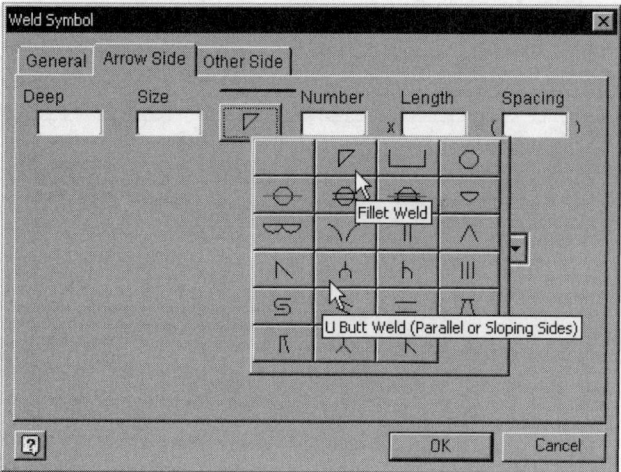

Figure 10.138 *Weld symbol dialog box*

 TUTORIAL 10.29

In this tutorial, you will retrieve weld symbol from the model.

1. Start a new drawing file. Use the metric *Standard.idw* (ISO) template.
2. With reference to Figure 10.139, construct a set of drawing views for the weldment assembly (*Weldment2.iam*) that you constructed in Chapter 8.

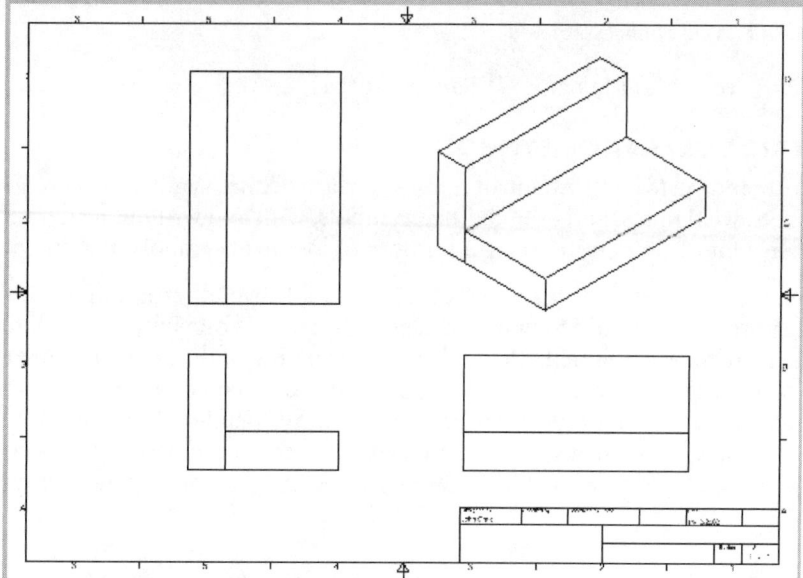

Figure 10.139 *Drawing views for a weldment model*

3. Select drawing view A in Figure 10.140, right-click, and Get Model Annotations > Get Weld Symbols.

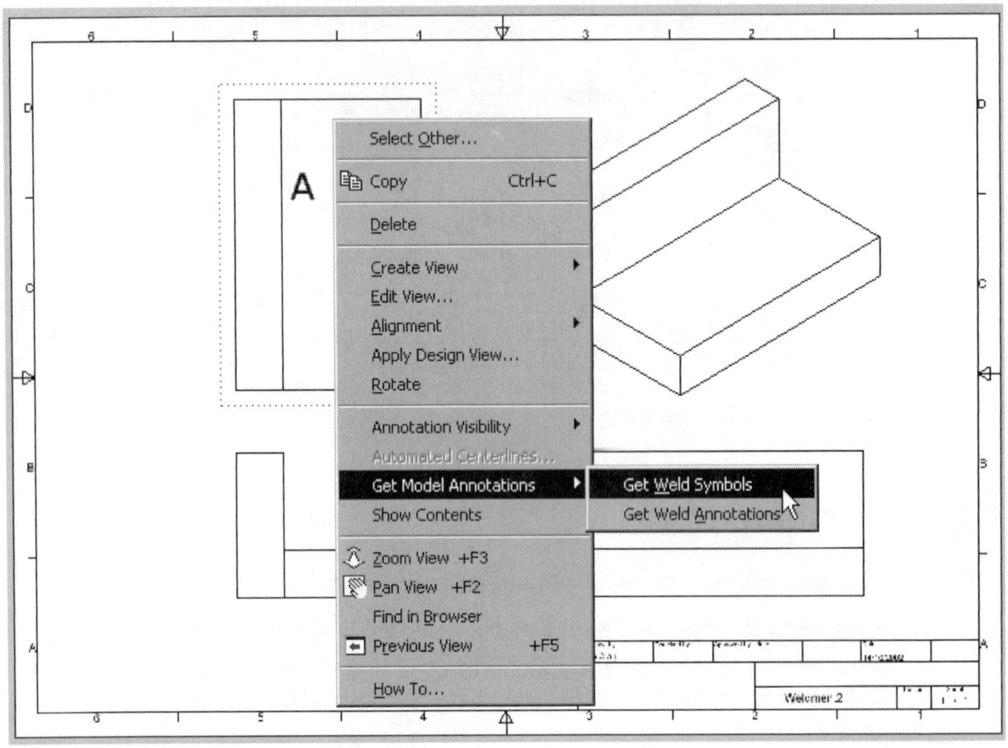

Figure 10.140 *Weld symbols retrieved*

Save and close your file (file name: *Weldment2.idw*).

GEOMETRIC TOLERANCE SYMBOL

You can copy and paste feature control frames, surface texture symbols, datum identifiers, datum targets, weld notes, and user-defined symbols, with and without leaders. This way, you can save a lot of time in inserting a number of identical symbols in a drawing.

Geometric tolerance concerns the maximum permissible overall variation of form or position of a feature. A geometric tolerance defines a tolerance zone within which features of a component are to be contained. Depending on the nature of the geometric tolerance, the tolerance zone can be the area within a circle, the area between two concentric circles, the area between two equidistant lines, the area between two parallel lines, the space within a cylinder, the space between two concentric cylinders, the space between two equidistant planes, the space between two parallel planes, or the space within a parallelepiped.

When to Apply Geometric Tolerance

There are four general conditions when geometric tolerances are required:

Form Control	You apply geometric tolerance when the application of dimension tolerances alone does not impose the desired control over the form and shape of a component. After you apply geometric tolerance to a feature, it will take precedence over the form control imposed by the size tolerance.
Machinery and Techniques	You apply geometric tolerance to mandate the use of appropriate machine tools and techniques to produce the product.
Remote Manufacture	You specify geometric tolerance in the engineering drawing to provide full information on the functional requirement of the product.
Features without Strict Size Control	You apply geometric tolerance to components without strict size control, such as to control the flatness of a surface table.

General Principles

There are five general principles, concerning scope, datum, tolerance value, feature shape, and form control.

Scope	You apply geometric tolerance to the whole length or surface of a feature, unless otherwise specified.
Datum	In choosing a reference datum for a geometric tolerance, you select a datum with adequate accuracy, and you consider the functionality of the part.
Tolerance Value	You specify geometric tolerance value relative to feature size, unless otherwise stated.
Feature Shape	The final feature shape can take any form within the tolerance zone if no further control is given.
Form Control	Some geometric tolerance can automatically control other kinds of form errors.

Symbol

You use geometric tolerance symbols to detail the tolerance applied to the geometric shape of the 3D object. The main body of a geometric tolerance is a feature control frame. In the feature control frame, there are two or more compartments: In the first compartment, you put a geometric tolerance symbol depicting the kind of control you are imposing on the geometry. In the second compartment, you put the geometric tolerance value that defines a geometric tolerance zone. The third and fourth compartments are optional;

you can add the datum reference name(s). When you specify datum reference(s) in the geometric tolerance symbol, you specify a datum by using a datum identifier or a datum target. A datum identifier specifies the entire face of the indicated feature as datum reference. A datum target specifies a zone of the face of a feature as datum reference.

Figure 10.141 shows a feature control frame controlling the position of a feature to fall within a circular tolerance zone of 0.05 mm diameter with reference to datum A and datum B. In the feature control frame, you can, as with dimensioning, include dual dimensions.

Figure 10.141 *Feature control frame with four compartments*

Figure 10.142 shows various kinds of geometric characteristics symbols that you can insert in the first compartment of the feature control frame.

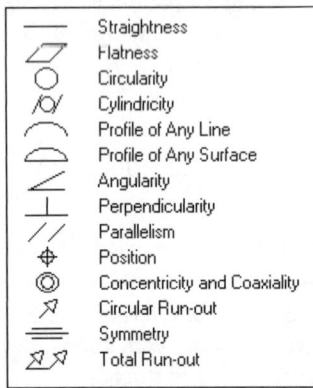

Figure 10.142 *Geometric characteristics symbols*

Set Geometric Tolerance Symbol Style

You can set the standard for constructing geometric tolerance symbols in your drawing by accessing the Control Frame and Datum Target tabs of the Drafting Standards dialog box.

TUTORIAL 10.30

In this tutorial, you will add a geometric tolerance symbol to a drawing. First, you will add a feature control frame to control the cylindricity of a feature.

1. Open the drawing file *PunchPost.idw*, if you already closed it.
2. Select Feature Control Frame on the Drawing Annotation Panel Bar or toolbar.
3. Select edge A and location B in Figure 10.143
4. Right-click and select Continue.

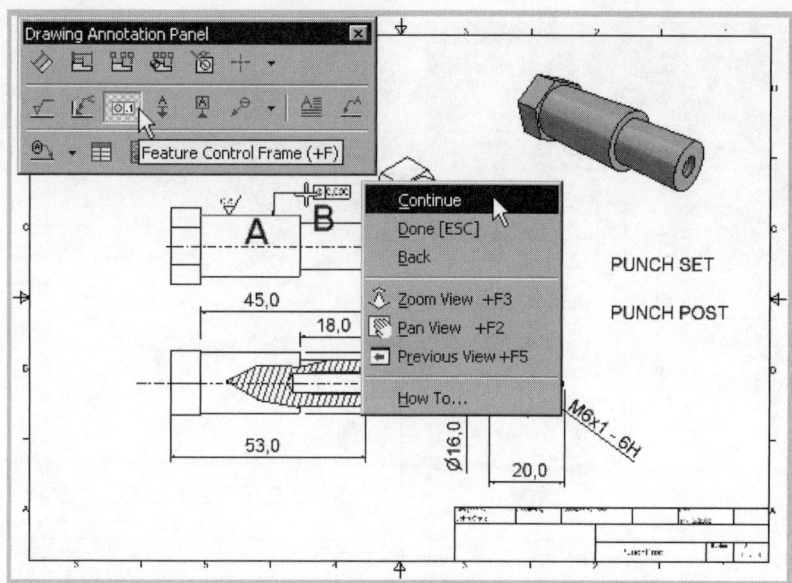

Figure 10.143 *Feature control frame being constructed*

5. In the Feature Control Frame dialog box shown in Figure 10.144, click the Geometric Characteristic Symbol button and select Cylindricity from the Geometric Characteristic Symbol palette. Under Tolerance 1, set the value to 0.2. Click OK, and then right-click and select Done.

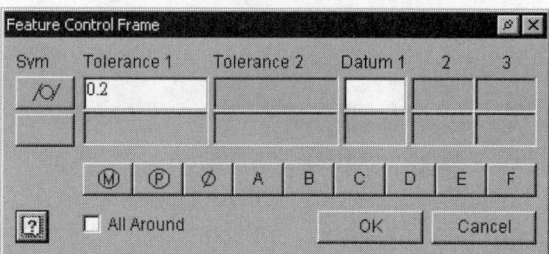

Figure 10.144 *Feature control frame dialog box*

Because control of cylindricity does not require any datum reference, the feature control frame has only two compartments, depicting the tolerance type and the tolerance zone.

Now add a concentricity control to the drawing. You will construct a feature identifier to specify a feature as datum reference and a feature control frame to control concentricity with reference to the datum.

6. Select Feature Identifier Symbol on the Drawing Annotation Panel Bar or toolbar.

7. Select A and B in Figure 10.145.

8. Right-click and select Continue.

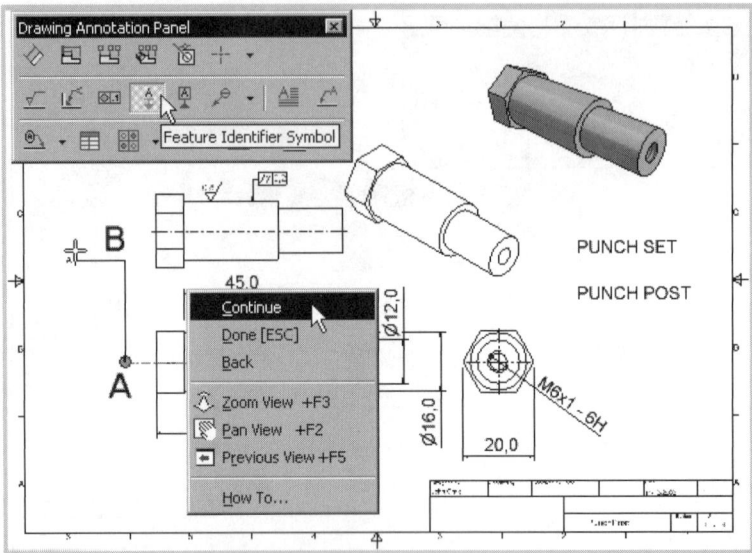

Figure 10.145 *Feature identifier being selected*

9. In the Format Text dialog box shown in Figure 10.146, type **A** (if it isn't already present), click OK, and then right-click and select Done.

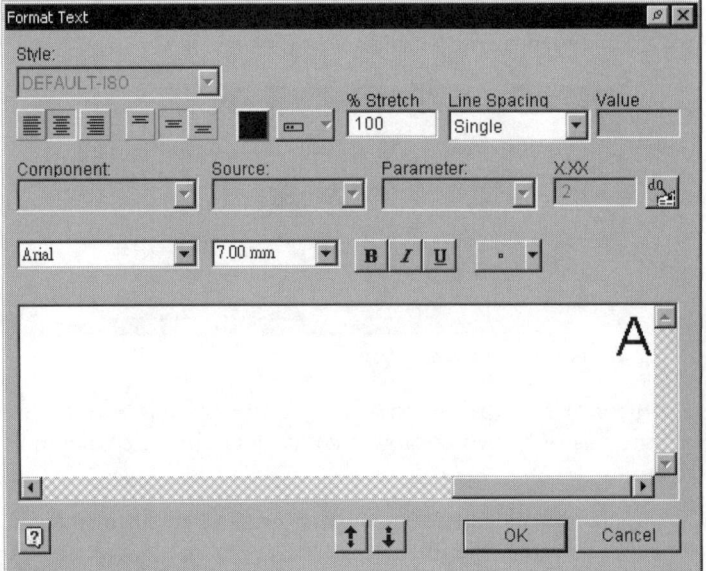

Figure 10.146 *Format Text dialog box*

10. Select Feature Control Frame on the Drawing Annotation Panel Bar or toolbar.

11. Select edge A and location B in Figure 10.147 to specify the start and location.

12. Right-click and select Continue.

13. In the Feature Control Frame dialog box, click the Geometric Characteristic Symbol button and then select Concentricity from the Geometric Characteristic Symbol palette. Under Tolerance 1, click the Ø button and then type **0.2** (so that the tolerance value appears as Ø0.2). Under Datum 1, type **A** (or click the A button) to specify a datum reference. Then click OK, and then right-click and select Done.

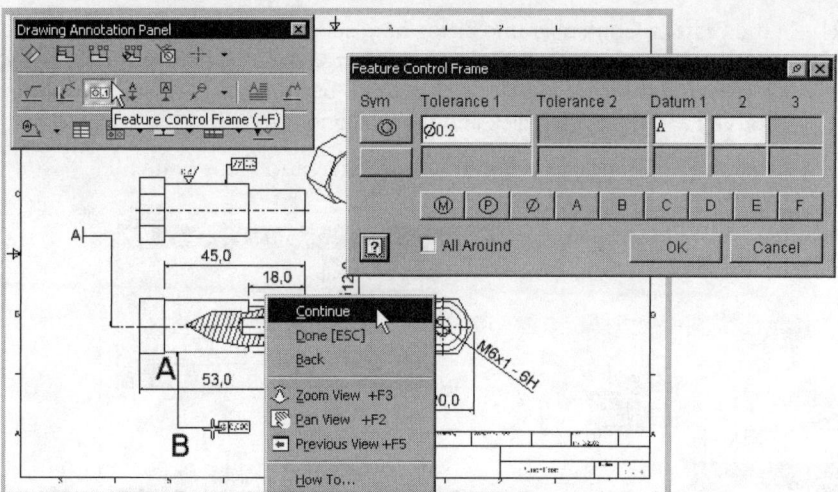

Figure 10.147 *Feature control frame being constructed*

Now apply a perpendicular control to the end face of the component by constructing a datum reference and a feature control frame.

14. Select Datum Identifier Symbol on the Drawing Annotation Panel Bar or toolbar.

15. Select A and B in Figure 10.148 to specify the start and location.

16. Right-click and select Continue.

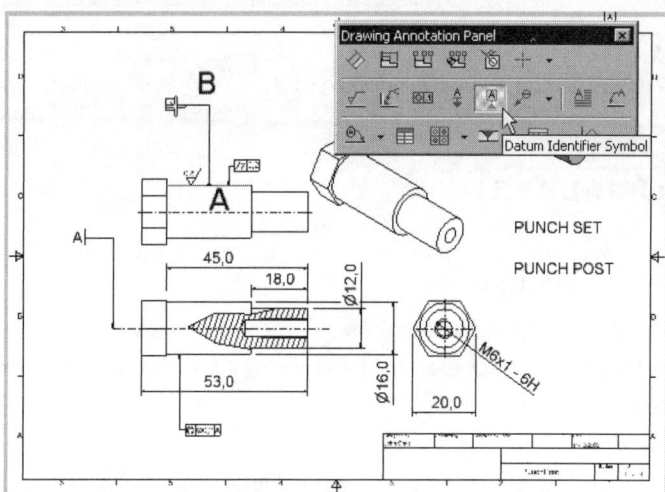

Figure 10.148 *Datum identifier symbol being constructed*

17. In the Format Text dialog box, type the letter **B** (if the letter A already appears in the text area, select it and then type **B**). Click OK and then right-click and select Done.

18. Select Feature Control Frame on the Drawing Annotation Panel Bar or toolbar.

19. Select A and B in Figure 10.149 to specify the start and location.

20. Right-click and select Continue.

21. In the Feature Control Frame dialog box, select the Perpendicularity symbol from the Geometric Characteristic Symbol palette. Under Tolerance 1, set the value to 0.2. Under Datum 1, type **B** (or click the B button) to specify a datum reference. Click OK and then right-click and select Done.

Geometric tolerances are added to the drawing. Save and close your file.

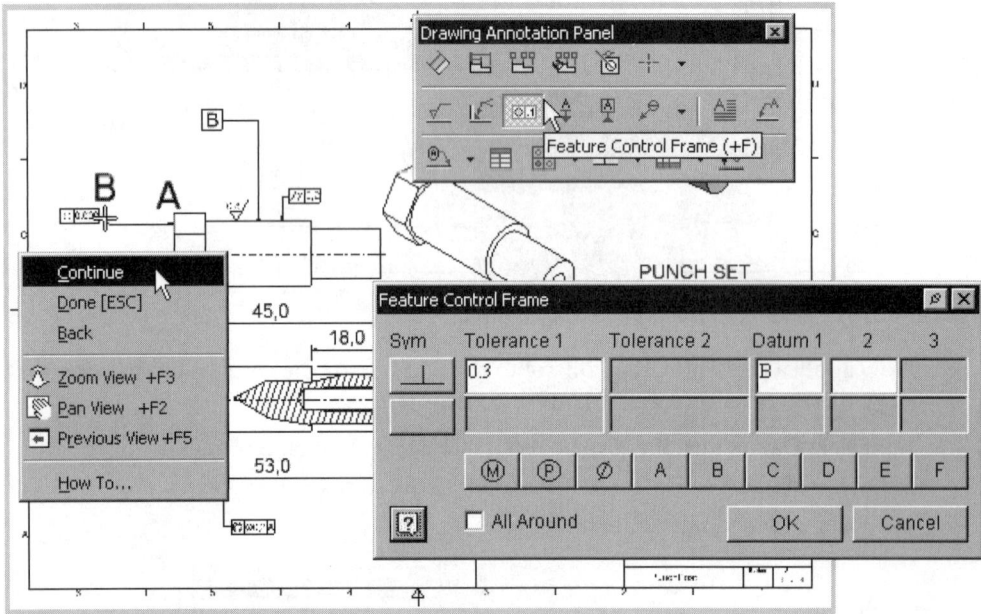

Figure 10.149 *Feature control frame being constructed*

 **TUTORIAL 10.31**

In this tutorial, you will construct a datum target to define a datum and a feature control frame to control the geometry of a component.

1. Open the drawing file *PunchBase.idw*, if you already closed it.

2. Select Datum Target Leader on the Drawing Annotation Panel Bar or toolbar.

3. Select A and B in Figure 10.150.

4. Right-click and select Continue.

5. In the Datum Target dialog box, set the dimension to 10. Click OK and then right-click and select Done.

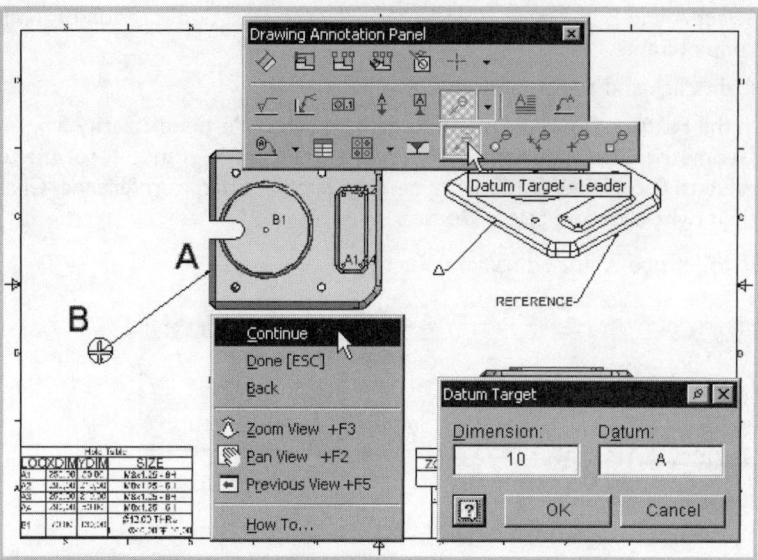

Figure 10.150 *Datum target being constructed*

6. Select the datum target, right-click, and select Copy.
7. Right-click and select Paste, then right-click and select Done.
8. Select the datum target at point A in Figure 10.151.
9. Select the other end of the pasted datum target and drag it to position B in Figure 10.151.

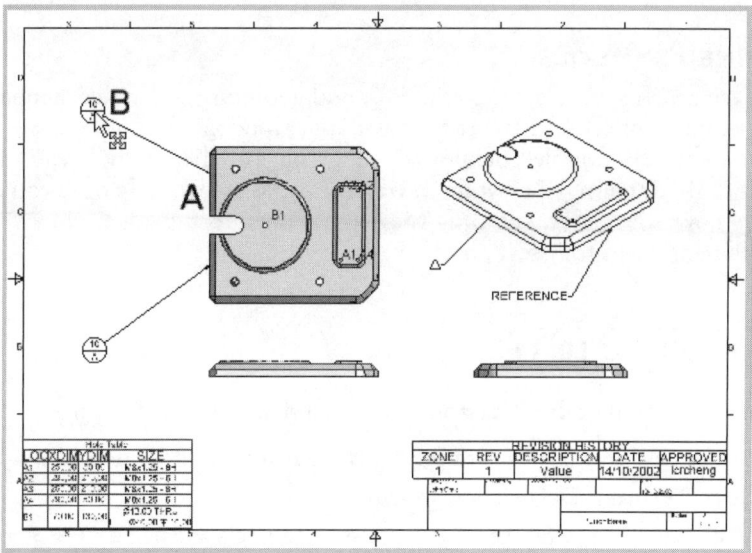

Figure 10.151 *Datum target copied*

10. Select Feature Control Frame on the Drawing Annotation Panel Bar or toolbar.

11. Select A and B in Figure 10.152 to specify the start and location of the feature control frame.

12. Right-click and select Continue.

13. In the Feature Control Frame dialog box, select Perpendicularity from the Geometric Characteristic Symbol palette. Under Tolerance 1, set the tolerance value to 0.2. Under Datum 1, type **A** to specify a datum reference. Click OK and then right-click and select Done.

A geometric tolerance is added. Save your file.

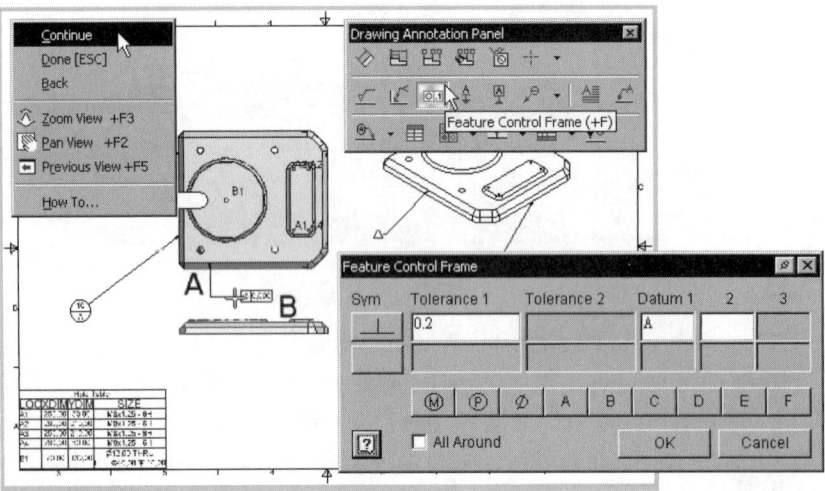

Figure 10.152 *Feature control frame constructed*

USER-DEFINED SYMBOLS

Besides surface finish symbols, weld symbols, and geometric tolerance symbols, you can construct custom symbols and insert them in your drawing. You can place a leaderless symbol on an edge by double-clicking. By dragging off the symbol, you construct an extension line. By dragging the symbol away from the associated edge, you construct a leader. To remove the leader, you drag the symbol back to the edge. Now you will perform the following steps to insert a symbol.

 TUTORIAL 10.32

In this tutorial, you will define a custom symbol and add it to a drawing.

1. Start a new drawing file. Use the metric *Standard.idw* (ISO) template.

2. Select Format > Define New Symbol.

3. With reference to Figure 10.153, construct a circle and a line.

4. Select Format > Save Sketched Symbol.

5. In the Sketched Symbol dialog box, type the name **CIR** and click the Save button.

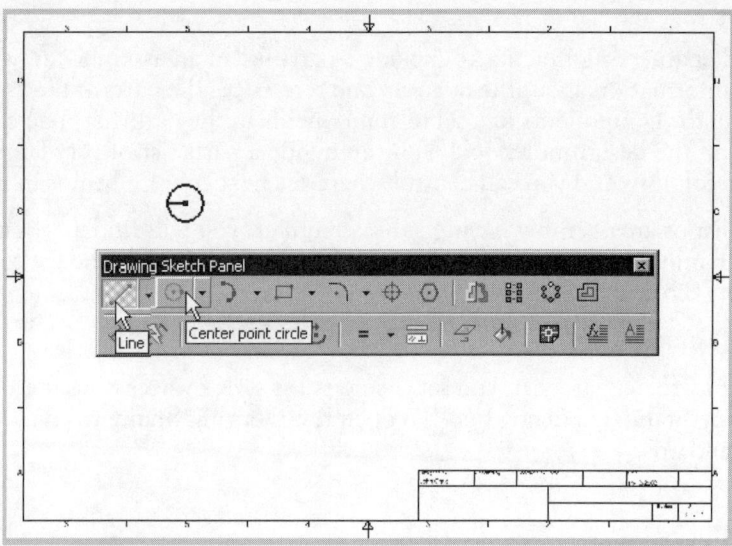

Figure 10.153 *Symbol being defined*

6. Select Symbols on the Drawing Annotation Panel Bar or toolbar.

7. In the Symbols dialog box, select the CIR symbol, click the Scale About Insertion Point or Center button and set the scale to 2. Click the Rotate About Insertion Point or Center button and set the rotation angle to 45. Then click OK.

8. Select location A in the drawing, right-click, and select Continue. (See Figure 10.154.)

9. Right-click and select Done.

Save and close your file (file name: *Symbol.idw*).

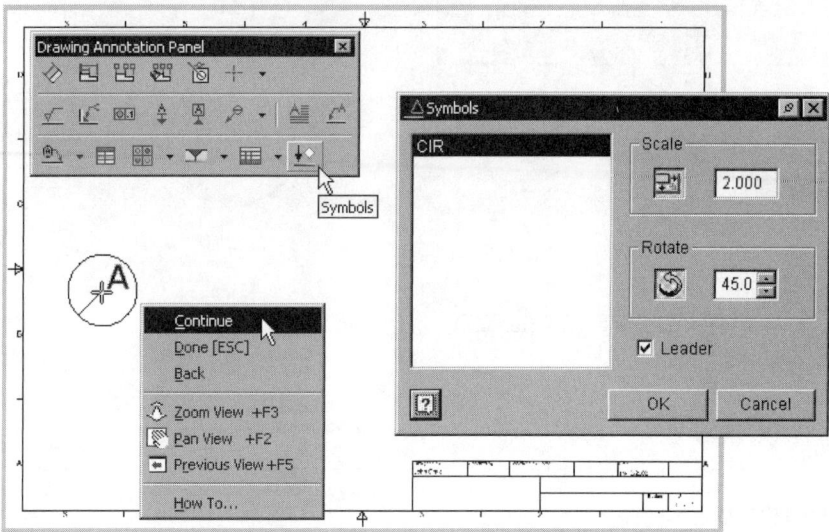

Figure 10.154 *Symbol being placed*

PARTS LIST

It is standard engineering practice to include a parts list in an assembly drawing. A parts list provides information about the quantity and references the parts of the assembly. You can include all the components or a set of components in the parts list. You can also place the parts list in the design notebook. You can divide a parts list of very large assemblies into multiple columns and you can control where the next column begins in the parts list.

In the parts list of an assembly having sub-assemblies, you will find a "+" sign prefixing the item containing sub-assemblies. Clicking the "+" sign expands the list to include the parts from the sub-assembly.

Set Parts List Style

Before you construct a parts list, you set the parts list style by accessing the Parts List tab of the Drafting Standards dialog box. To open the Drafting Standards dialog box, select Format > Standards.

TUTORIAL 10.33

In this tutorial, you will add a parts list to an assembly drawing.

1. Open the drawing file *Pliers.idw*, if you already closed it.
2. Select Parts List on the Drawing Annotation Panel Bar or toolbar.
3. Select view A in Figure 10.155.

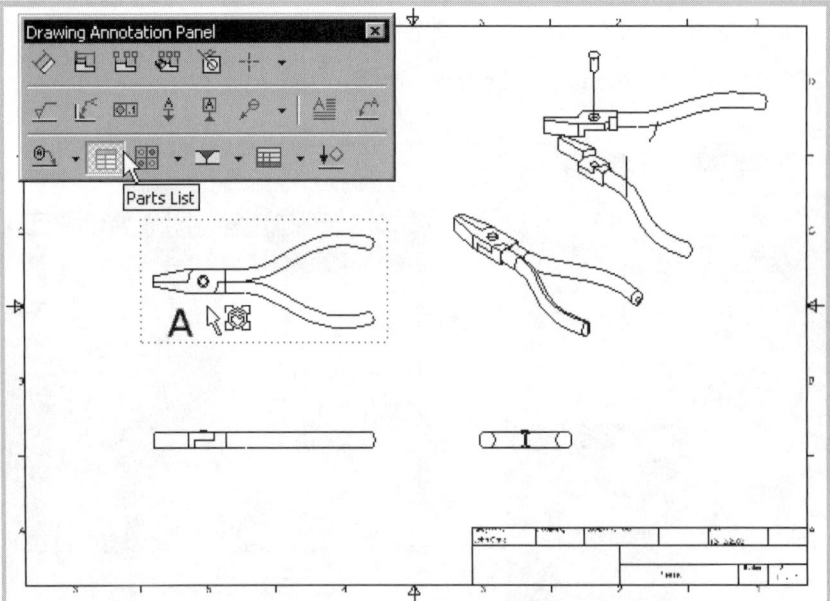

Figure 10.155 *Parts list being constructed*

In the Create Parts List dialog box shown in Figure 10.156, there are three areas: Level, Range, and Format. The Level area concerns the level in the hierarchy of an assembly. If you simply put all the parts together in an assembly, you have one level of parts. If you put some parts in sub-assemblies and put the sub-assemblies in the assembly, you have two levels of components. Because you can have an assembly that consists of sub-assemblies and parts, you have to choose whether you want the parts list to display only the parts or all objects in the first level. In the Level area, there are two options. If you select First-Level Components, the parts list will show only the top level of the components in the selected drawing view. If you specify Only Parts, you can choose to display all the parts or a range of parts.

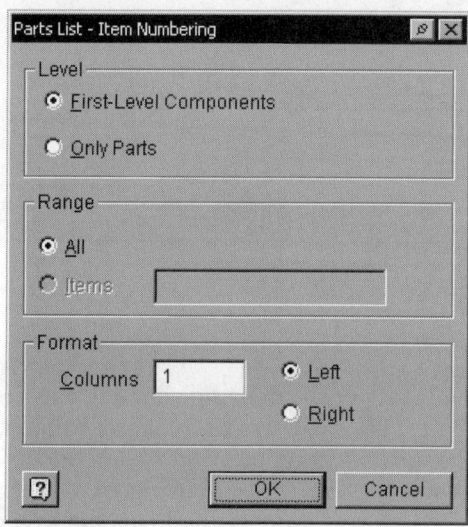

Figure 10.156 *Parts List dialog box*

4. Choose First-Level Components.

The Range area sets the range of parts to be included in the parts list. You can include all the parts or selected parts. Because you chose First-level Components in the Level box, there is only one option in the Range box: All.

The Format area specifies the number of columns into which to split the parts list. If you have a very large number of component parts, you might need to split the parts into several columns so as to reduce the number of rows in the parts list. The left and right buttons enable you to attach the cursor to the left or right of the parts list as you place it in the drawing.

5. Accept the default number of columns (1).
6. Select the Right button.
7. Click the OK button.
8. Select corner A in Figure 10.157.

A parts list is constructed. Save your file.

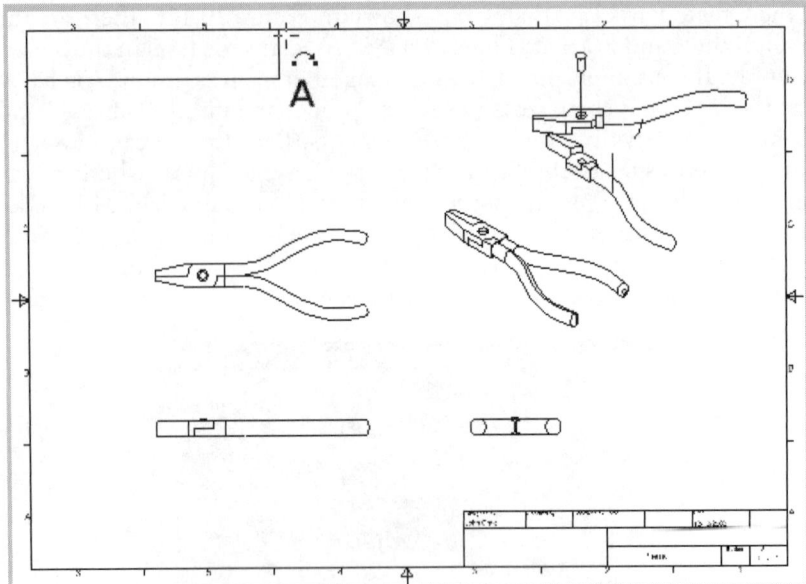

Figure 10.157 *Parts list location selected*

BALLOONS

To reference the individual parts in the drawing view and the parts list, you use a special kind of leader: balloon. There are two ways to construct balloons in a drawing: You can construct individual balloons by selecting individual components of the drawing, and you can construct a set of balloons by selecting a drawing view. To construct individual balloons, select Balloon on the Drawing Annotation Panel Bar or toolbar. To construct a set of balloons collectively, select Balloon All on the Drawing Annotation Panel Bar or toolbar.

Set Balloon Style

Before you construct a set of balloons, you set the balloon style by accessing the Balloon tab of the Drafting Standards dialog box. To open the Drafting Standards dialog box, select Format > Standards.

 TUTORIAL 10.34

In this tutorial, you will add a set of balloons to an assembly drawing.

1. Open the drawing file *Pliers.idw*, if you already closed it.

2. Select Balloon All on the Drawing Annotation Panel Bar or toolbar.

 Note: Balloon All is on the same flyout as Balloon.

3. Select view A in Figure 10.158.

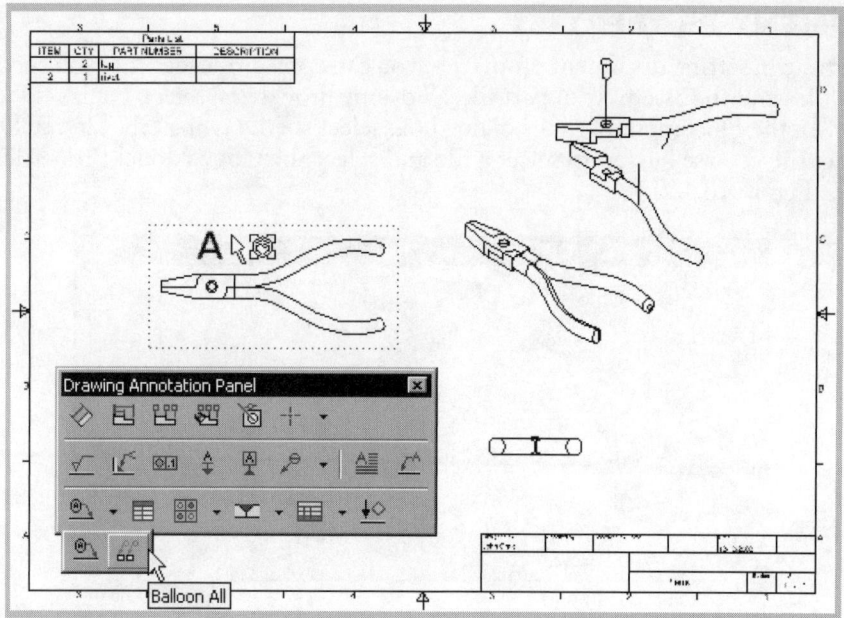

Figure 10.158 *Balloons for all the components in a drawing view being constructed*

A set of balloons for all the components in a drawing view is constructed. (See Figure 10.159.) The drawing is complete. Save and close your file.

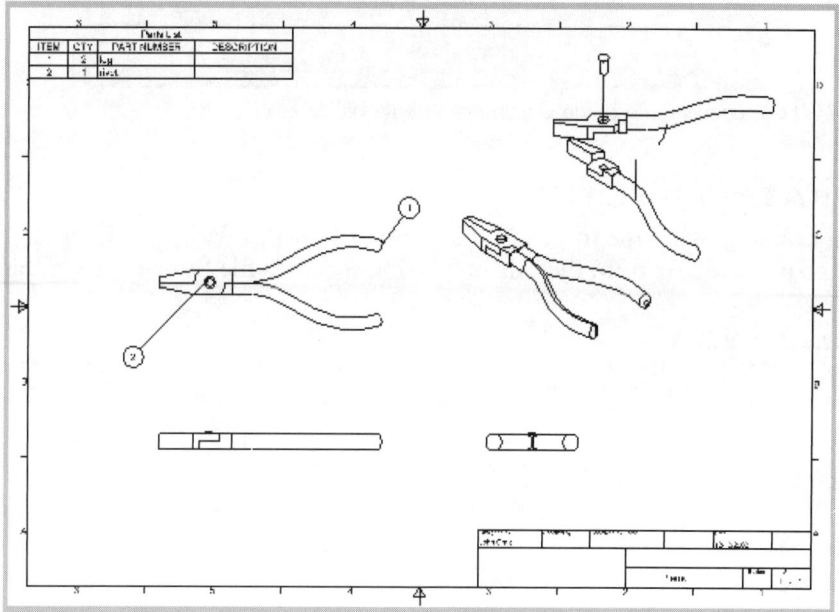

Figure 10.159 *Balloon constructed*

DOCUMENT PROPERTIES

To save time in setting document properties, you can copy properties that you specified in the part files and the assembly of part files. To copy properties, select Tools > Document Settings. In the Document Settings dialog box, select the Drawing tab. On the Drawing tab, select the browse button to select a file and select the Copy Model Properties check box. (See Figure 10.160.)

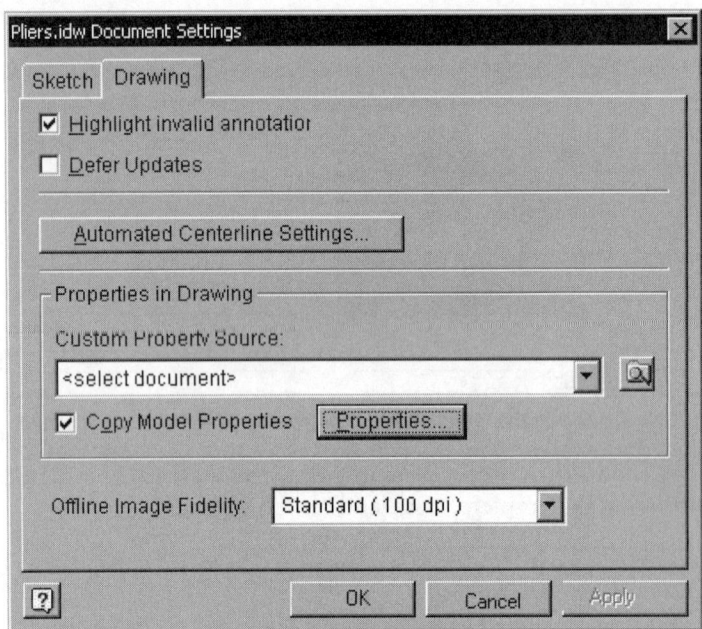

Figure 10.160 *Drawing tab of the Document Settings dialog box*

GENERATING DWG FILE

You can save a drawing file to various file formats: BMP, DWF, DWG, and DXF. By selecting Save Copy As from the File menu, you specify a file format. (See Table 10.6.)

Table 10.6 *Output file formats*

BMP	Windows Bitmap
DWF	Drawing Web Format
DWG	AutoCAD Drawing Format
DXF	Drawing Exchange Format

SUMMARY

An engineering drawing is a set of orthographic drawing views depicting a 3D object on a 2D engineering drawing sheet. To construct a drawing, you use a drawing file. (To reiterate, there are four kinds of Autodesk Inventor files: part, assembly, presentation, and drawing.)

Using a drawing file, you construct two kinds of engineering drawing: 2D sketched drawings and associative engineering drawing. You use 2D sketched drawing to depict concepts, ideas, or 2D drawing views of 3D objects in the absence of 3D objects. If you already constructed a solid part, an assembly, or a presentation of an assembly, you construct an associative engineering drawing by selecting one of the files, specifying a drawing view, and selecting a location of the drawing view.

A drawing file has two kinds of objects—drawing resources and drawing sheet. Drawing resources consist of standard drawing sheet templates, drawing borders, engineering title blocks, and sketched symbols. You specify a drawing sheet or select a drawing sheet from the resource. Then you add a border, a title block, and sketched objects. Besides the standard drawing resources provided, you can construct your own borders, title blocks, and sketched objects for insertion in the drawing sheet.

After setting up a drawing sheet with borders, title blocks, and sketched objects, you construct sketched drawings or select a 3D object (part, assembly, or presentation) and specify the drawing views.

In a drawing file, you can have more than one drawing sheet. Therefore, you can construct sketched drawings in one sheet and select different part, assembly, and presentation files for the other drawing sheets in the drawing file. In practice, you should use one part or assembly for each drawing file to avoid confusion.

There are six kinds of associative engineering drawing views—base view, projected view, auxiliary view, section view, broken view, break out view, and detail view. You construct these drawing views for part, sheet metal, and assembly drawings. In addition, you can include a flat pattern view for a sheet metal drawing and a presentation view for an assembly drawing.

To complete an engineering drawing, you add annotations: dimensions (including dimensional tolerances), notes (including text, hole/thread note, and hole table), centerlines, surface texture symbol, weld symbol, geometric tolerance symbol, leader, parts list, and balloons.

Before you add annotations to a drawing, you should select an appropriate engineering drawing standard (such as ANSI or ISO) and determine how they will appear in the drawing by setting their display styles.

REVIEW QUESTIONS

1. What are the four kinds of files? Which one will you use to construct an engineering drawing and how?

2. Distinguish between a sketched drawing view and an associative engineering drawing view.

3. What are the four kinds of drawing resources in a drawing file? Briefly describe the ways to construct these resources. How can you make these resources available to other projects?

4. How many kinds of drawing views can you construct in a drawing file? Use simple sketches to illustrate your answer.

5. What kind of annotations will you add to an engineering drawing depicting a solid part, a sheet metal part, and an assembly of parts? How would you set appropriate drawing standards?

Editing Imported Solids

You can import three kinds of solids by opening them: Mechanical Desktop solids (DWG), ACIS solids (SAT file), and STEP solids. If you open an ACIS or a STEP file, you get a base solid feature that is basically static and non-parametric. You can add sketched features, placed features, and work features to the base solid. The additional features are parametric and editable. (Mechanical Desktop files are the subject of Appendix B.)

 TUTORIAL A.1

In this tutorial, you will open and edit a SAT file. To learn how to manipulate a SAT or STEP solid, you will save an Inventor part file as a SAT solid file, open the SAT file, and edit the base solid feature.

1. Open the file *HandGripperBase.ipt* that you constructed in Chapter 2.
2. Select Save Copy As from the File menu.
3. In the Save Copy As dialog box, choose SAT Files (*.sat) from the Save As Type drop-down list, specify the file name (*HandGripperBase.sat*), and click Save.
4. Close the file.
5. Select Open from the File menu.
6. In the Open dialog box, select SAT Files (*.sat) from the Files of Type drop-down list, select the file *HandGripperBase.sat*, and click Open. The SAT file is opened.
7. Reorient the part so that it looks similar to Figure A.1.
8. On the Browser Bar, you will find an object Base1. This is the base solid feature. Select this feature, right-click, and select Edit Solid. (See Figure A.1.)

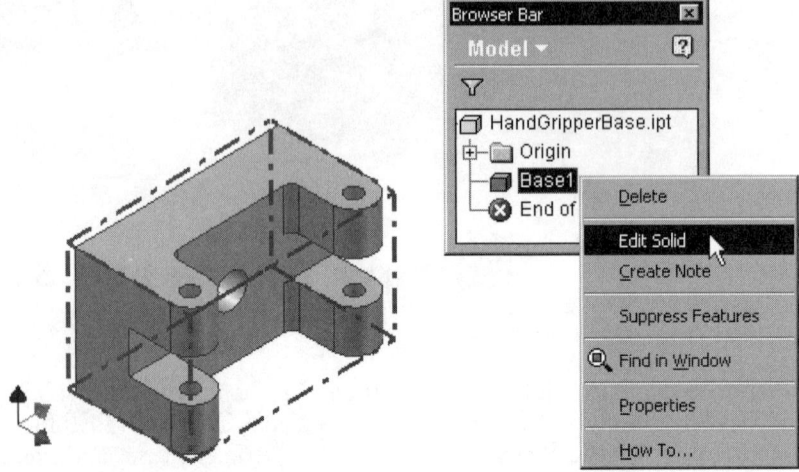

Figure A.1 *SAT file opened and base solid selected on the Browser Bar*

To modify the base solid, you use the Explicit panel bar or toolbar, which provides tools to manipulate imported non-parametric solid parts. (See Figure A.2.)

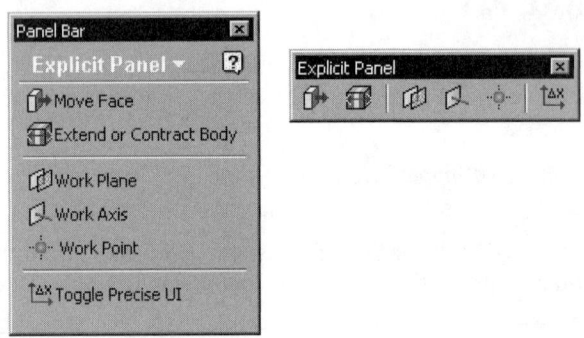

Figure A.2 *Explicit Panel Bar and toolbar*

The Explicit Panel Bar and toolbar have six button areas.

Move Face	Moves a selected face of the solid.
Extend or Contract Body	Extends or contracts the body of a solid.
Work Plane	Constructs a work plane.
Work Axis	Constructs a work axis.
Work Points	Constructs a work point.
Toggle Precise UI	Toggles the display of the Precise Input dialog box.

Now move a face.

9. Select Move Face on the Explicit Panel Bar or toolbar.
10. Select cylindrical face A in Figure A.3.
11. Select the Direction button, select edge B, and flip the direction in accordance with Figure A.3.
12. Then set the distance to 12 mm.
13. Click the OK button. The cylindrical face is moved.

Figure A.3 *Cylindrical face being moved*

Now extend the body of the solid.

14. Select Extend or Contract Body on the Explicit Panel Bar or toolbar.
15. Select face A in Figure A.4.
16. Set the distance to 10 mm and click the OK button.
17. The base solid is modified. Right-click and select Finish Solid Edit to exit edit mode.
18. The body is extended. (See Figure A.5.)

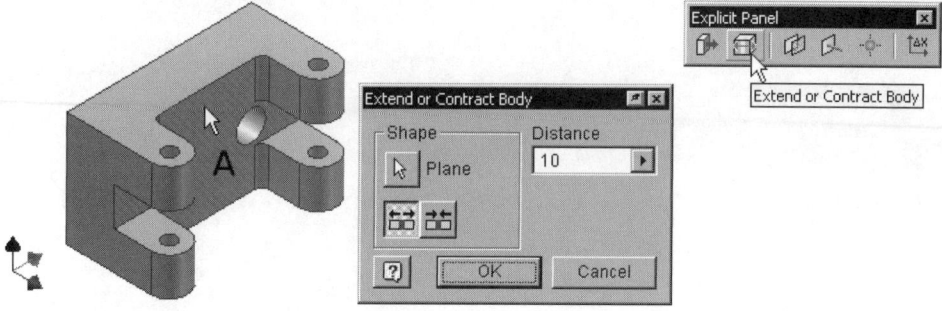

Figure A.4 *Face of a body selected*

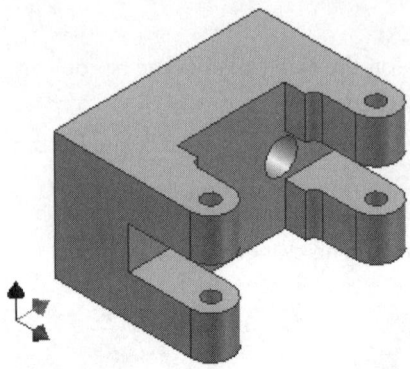

Figure A.5 *Body extended*

A SAT file is opened and modified. Now save the SAT file with a different name. *(HandGripperBase1.ipt)* . Close your file when finished.

APPENDIX

b

Importing Mechanical Desktop Files

If you install Mechanical Desktop and Autodesk Inventor on the same computer and start both applications, you can open a Mechanical Desktop solid part file in Inventor and retain all the parametric information of the original Mechanical Desktop file. When you open a DWG file, the wizard automatically determines whether the file is an AutoCAD file or a Mechanical Desktop file and provides appropriate options accordingly.

TUTORIAL B.1

To open a Mechanical Desktop file, perform the following steps:

1. Select Open from the File menu.
2. In the Open dialog box, select a Mechanical Desktop file and select the Options button.
3. In the DWG File Import Options dialog box, select the Mechanical Desktop File button and click the Next button. (See Figure B.1.)

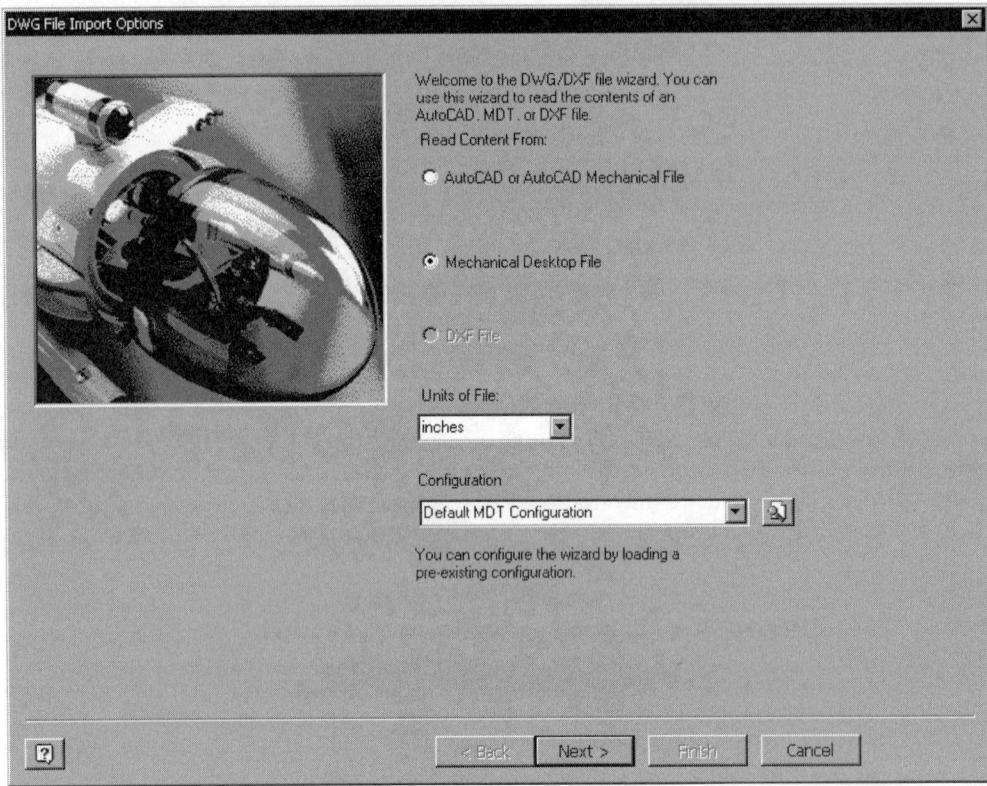

Figure B.1 *DWG File Import Options dialog box*

4. In the MDT Model/Layout Import options dialog box, check the Translate Parts And Assemblies and the Translate Drawing View check boxes, and click the Next button. (See Figure B.2.)

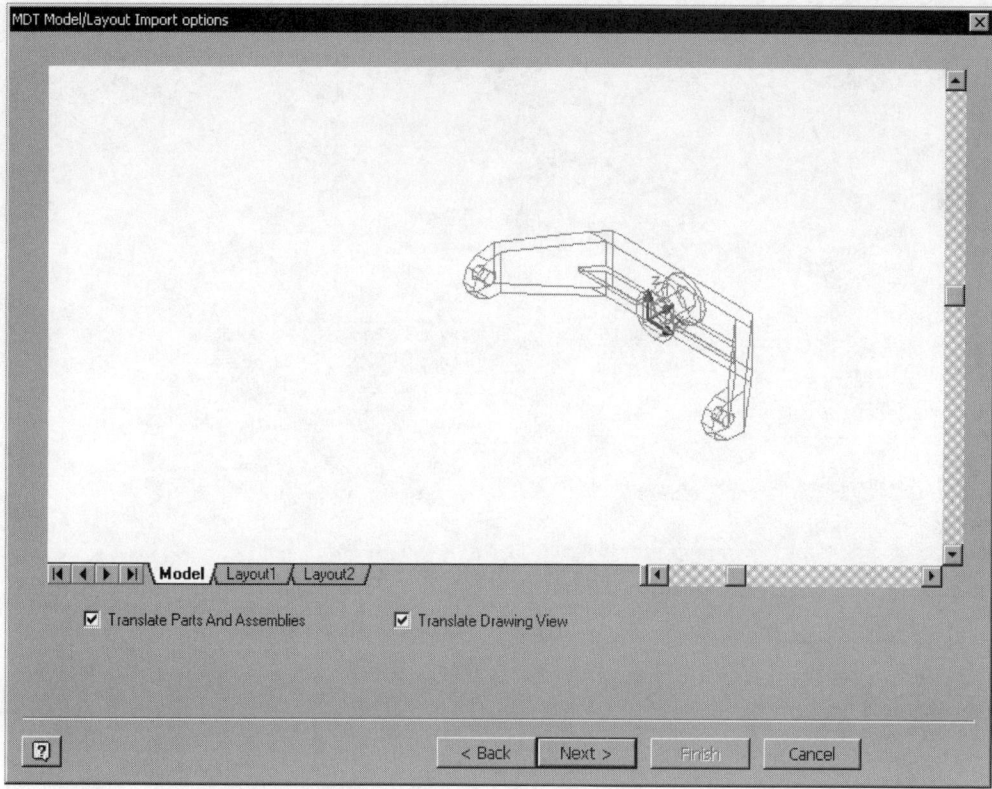

Figure B.2 *MDT Model/Layout Import options dialog box*

5. In the Import Destination Option dialog box, set options, select a destination directory, and click the Finish button. (See Figure B.3.)

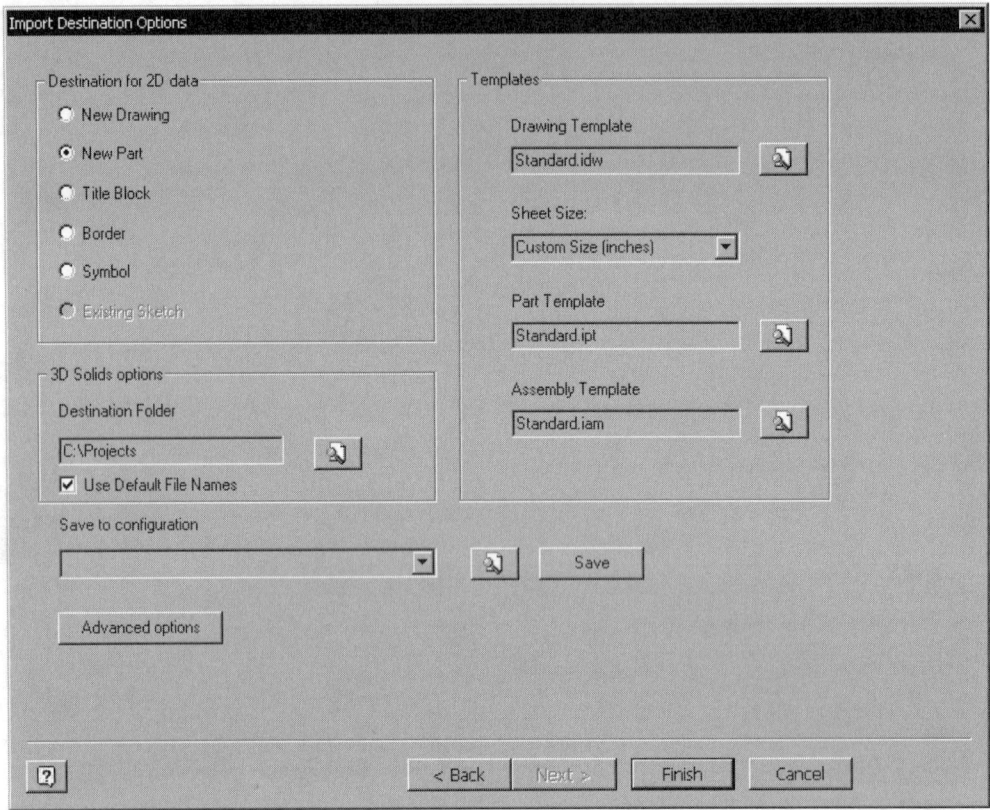

Figure B.3 *Import Destination Options dialog box*

The selected Mechanical Desktop drawing is opened and converted to an Autodesk Inventor file. (See Figure B.4.)

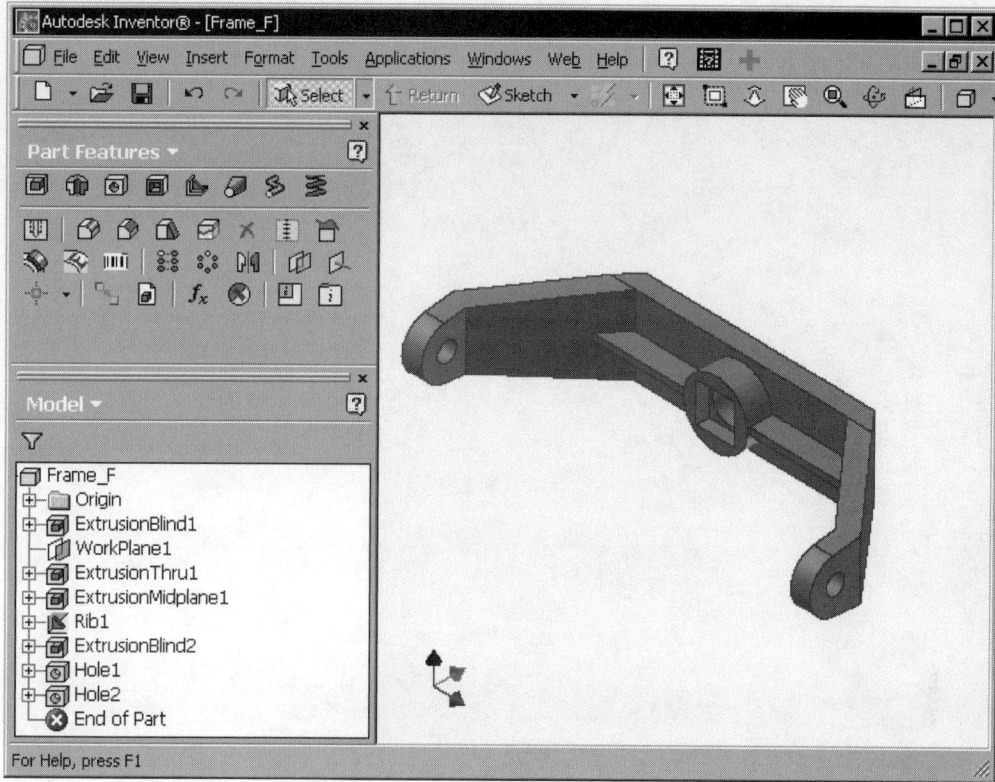

Figure B.4 *Mechanical Desktop file opened*

INDEX

Note: Page numbers in **bold** indicate tables.

LICENSE AGREEMENT FOR AUTODESK PRESS
A Thomson Learning Company

Educational Software/Data

You the customer, and Autodesk Press incur certain benefits, rights, and obligations to each other when you open this package and use the software/data it contains. BE SURE YOU READ THE LICENSE AGREE-MENT CAREFULLY, SINCE BY USING THE SOFTWARE/DATA YOU INDICATE YOU HAVE READ, UNDERSTOOD, AND ACCEPTED THE TERMS OF THIS AGREEMENT.

Your rights:

1. You enjoy a non-exclusive license to use the enclosed software/data on a single microcomputer that is not part of a network or multi-machine system in consideration for payment of the required license fee, (which may be included in the purchase price of an accompanying print component), or receipt of this software/data, and your acceptance of the terms and conditions of this agreement.

2. You own the media on which the software/data is recorded, but you acknowledge that you do not own the software/data recorded on them. You also acknowledge that the software/data is furnished "as is," and contains copyrighted and/or proprietary and confidential information of Autodesk Press or its licensors.

3. If you do not accept the terms of this license agreement you may return the media within 30 days. However, you may not use the software during this period.

There are limitations on your rights:

1. You may not copy or print the software/data for any reason whatsoever, except to install it on a hard drive on a single microcomputer and to make one archival copy, unless copying or printing is expressly permitted in writing or statements recorded on the diskette(s).

2. You may not revise, translate, convert, disassemble or otherwise reverse engineer the software/data except that you may add to or rearrange any data recorded on the media as part of the normal use of the software/data.

3. You may not sell, license, lease, rent, loan, or otherwise distribute or network the software/data except that you may give the software/data to a student or and instructor for use at school or, temporarily at home.

Should you fail to abide by the Copyright Law of the United States as it applies to this software/data your license to use it will become invalid. You agree to erase or otherwise destroy the software/data immediately after receiving note of Autodesk Press' termination of this agreement for violation of its provisions.

Autodesk Press gives you a LIMITED WARRANTY covering the enclosed software/data. The LIMITED WARRANTY can be found in this product and/or the instructor's manual that accompanies it.

This license is the entire agreement between you and Autodesk Press interpreted and enforced under New York law.

Limited Warranty

Autodesk Press warrants to the original licensee/ purchaser of this copy of microcomputer software/ data and the media on which it is recorded that the media will be free from defects in material and workmanship for ninety (90) days from the date of original purchase. All implied warranties are limited in duration to this ninety (90) day period. THEREAFTER, ANY IMPLIED WARRANTIES, INCLUDING IMPLIED WARRANTIES OF MERCHANTABILITY AND FITNESS FOR A PARTICULAR PURPOSE ARE EXCLUDED. THIS WARRANTY IS IN LIEU OF ALL OTHER WARRANTIES, WHETHER ORAL OR WRITTEN, EXPRESSED OR IMPLIED.

If you believe the media is defective, please return it during the ninety day period to the address shown below. A defective diskette will be replaced without charge provided that it has not been subjected to misuse or damage.

This warranty does not extend to the software or information recorded on the media. The software and information are provided "AS IS." Any statements made about the utility of the software or information are not to be considered as express or implied warranties. Delmar will not be liable for incidental or consequential damages of any kind incurred by you, the consumer, or any other user.

Some states do not allow the exclusion or limitation of incidental or consequential damages, or limitations on the duration of implied warranties, so the above limitation or exclusion may not apply to you. This warranty gives you specific legal rights, and you may also have other rights which vary from state to state. Address all correspondence to:

AutodeskPress
Executive Woods
5 Maxwell Drive
Clifton Park, New York 12065-2919